T0115799

CHRISTOPHER ISHERWOOD

LIBERATION

DIARIES, VOLUME THREE: 1970–1983

Books by Christopher Isherwood

NOVELS

All the Conspirators
The Memorial
Mr. Norris Changes Trains
Goodbye to Berlin
Prater Violet
The World in the Evening
Down There on a Visit
A Single Man
A Meeting by the River

AUTOBIOGRAPHY & DIARIES

Lions and Shadows
Kathleen and Frank
Christopher and His Kind
My Guru and His Disciple
October (*with Don Bachardy*)
Diaries, Volume One: 1939–1960
The Sixties, Diaries, Volume Two: 1960–1969
Lost Years: A Memoir, 1945–1951

BIOGRAPHY

Ramakrishna and His Disciples

PLAYS (*with W.H. Auden*)

The Dog Beneath the Skin
The Ascent of F6
On the Frontier

TRAVEL

Journey to a War (*with W.H. Auden*)
The Condor and the Cows

COLLECTIONS

Exhumations
Where Joy Resides

LIBERATION

DIARIES, VOLUME THREE:

1970–1983

CHRISTOPHER ISHERWOOD

Edited and Introduced by Katherine Bucknell
Preface by Edmund White

HARPER PERENNIAL

NEW YORK • LONDON • TORONTO • SYDNEY • NEW DELHI • AUCKLAND

HARPER PERENNIAL

First published in Great Britain in 2012 by Chatto & Windus.

First U.S. hardcover published in 2012 by HarperCollins Publishers.

FIRST HARPER PERENNIAL EDITION PUBLISHED 2013.

Library of Congress Cataloging-in-Publication Data has been applied for.

ISBN 978-0-06-208476-7 (pbk.)

13 14 15 16 17 OFF/RRD 10 9 8 7 6 5 4 3 2 1

Contents

Preface

Readers of novels often fall into the bad habit of being overly exacting about the characters' moral flaws. They apply to these fictional beings standards that no one they know in real life could possibly meet—nor could they themselves. They condemn a heroine, say, for not facing and condemning her lover's ethical cowardice on some fine point of inner struggle; both her failure and his would scarcely be perceptible in real life. Sometimes it almost feels as if readers, in discussing a book, are showing off, are eager to display a refinement that no one would bother with in the heat of actual experience. In real life everyone is too busy, too submerged in the murk of getting and spending, too greedy to survive and even to prevail to be able to make much out of moral niceties. In any event, friends are always too willing to forgive lapses that they scarcely notice and if they register are sure to share and eager to pardon. Real life is so rough-and-tumble, so clamorous, a bit like an over-amplified rock band on drugs; only in the shaded purlieus of fiction do we catch the tinkling strains of moral elegance coming from another room.

I mention all this because reading the several thousand pages of Isherwood's complete journals is an instructive corrective to the prissiness of reading fiction. Isherwood, whom most of us would consider to be nearly saintly if we knew him personally, had faults that we'd say were unforgivable in a novel (he was careful to distance himself from these very faults in his autobiographical fiction). He was seriously anti-Semitic and a year never goes by in his journals that he doesn't attribute an enemy's or acquaintance's bad behaviour to his Jewishness. I suppose some people would argue that the British gentry are or at least *were* like that, or that he grew up in another epoch and should not be held to the standards of today. I don't think that that defense quite works. After all, Isherwood lived in Berlin in the late 1920s and early 1930s and experienced first-hand the rise of Hitler to power and witnessed directly the appalling effect of the Nazis on the lives of his Jewish friends. Later he lived in Los Angeles for four decades and worked closely with many Jews; his milieu would never have accepted his anti-Semitism had it been declared. Moreover, he knew perfectly well that every word he wrote, even in these journals, would eventually be published, so one can't even argue that he was simply sounding off in notes not meant for anyone else's eyes.

Then he is a dreadful hypochondriac and in spite of all his much-vaunted spirituality terrified of the least ailment. He worries obsessively about his weight and berates himself when he's a pound or two up on the scale, though he seldom weighs more than 150. He worries constantly (with good reason) about his drinking ("I do hate it so," he admits).

And then he can be quite nasty about women friends and their writing. When he travels down to Essex to see Dodie Beesley he reads her novel and says, 'It is *exactly* what I feared: one of those patty-paws romances, a little kiss here, a little wistful regret there, one affair is broken off, another starts up. Magazine writing. What's wrong with it, actually? It's so pleased with itself, so fucking smug, so snugly cunty, the art of women who are delighted with themselves, who indulge themselves and who patronize their men. They *know* that there is nothing, there *can* be nothing outside of the furry rim of their cunts and their kitchens, their children and their clubs." Then, in a reversal typical of Isherwood, he writes, ". . . I am indulging in the luxury of being brutal about it because I know I will have to be polite about it to Dodie tomorrow—I also know that I shall *want* to be polite, because I do respect her and she is indeed so much wiser and subtler and better than this silly book."

Oh, yes, he's full of faults and yet I think any fair-minded reader who applies to Isherwood the very approximate demands of life and not the overly exacting standards of fiction will have to admit that he or she has seldom spent so much time with someone so generally admirable. To say so in no way mitigates the obnoxiousness of his real faults. But we should forgive him with the same liberality we apply to ourselves and our friends.

He loves his partner Don Bachardy with a constant devotion that is almost unparalleled in my experience. In the preceding volume, which covered the 1960s, Bachardy was endlessly quarrelsome and difficult. But in this volume, the last, which covers the final decade of Isherwood's life, Bachardy has achieved a measure of worldly success as an artist and has escaped the confines of domesticity enough to enjoy plenty of sexual adventures—enough to catch up with all the sex Isherwood himself had enjoyed in his youth in England, Germany and America. In total contrast to the anger and spite of the 1960s, in this volume Don is endlessly playful and affectionate and kind, and Isherwood (who was thirty years older) is deliriously happy. His main regret about dying is that he must leave Don, though as a Hindu he must have imagined he'd join Don in a future life.

After he has lunch with a friend called Bob Regester who is having problems with his lover, Isherwood writes: "So of course I handed out lots of admirable advice, which I would do well to follow, myself. Don't try to make the relationship exclusive. Try to make your part of it so special that nobody can interfere with it even if he has an affair with your lover. Remember that physical tenderness is actually more important than the sex act itself." We learn that Chris and Don no longer have sex but that they consider their relationship to be very physical; they sleep together and they are constantly touching each other. At a certain point Chris writes, "I'm glad people have had crushes on me, glad I used to be cute; it is a very sustaining feeling." I remember the ancient Virgil Thomson once telling me in Key West that he, too, had had a lot of sexual allure and success in his day.

We seldom count a happy marriage as a real accomplishment and yet it so clearly is—it is virtually an *aesthetic* achievement. It requires the same sense of proportion, creativity, empathy, patience, perseverance, equanimity and generosity of spirit as does the making of a novel or play. Isherwood's happy marriage with a tempestuous young man is, unlike the writing of a novel, a *collaborative* act (in that way it's more like preparing a play—and not incidentally Chris and Don were constantly working on film and theater scripts together). Anyone who has ever had a happy marriage knows that it is never stable, never finished; it changes every day and is always being created or at least celebrated anew. I suppose in that sense it is like cooking, something that requires a skill that can be acquired over time but that needs to be done every day from scratch. Isherwood understands the vagaries of love better than anyone and he feels (partly to Bachardy but largely to the gods) *gratitude*, the most appropriate of all the amorous emotions.

Another thing we admire about Isherwood is his seriousness and his curiosity. His reading lists reveal how far-flung his interests are and how deep they go. He is constantly reading demanding books that inform him about every aspect of the world past and present. At one moment, by no means atypical, he is reading Solzhenitsyn's *The First Circle*, Trollope's *The Eustace Diamonds* and *The Autobiography of Malcolm X*. His curiosity about the people around him is equally far-reaching. He wants to know what everyone is up to. His old friends—especially David Hockney, Hockney's erstwhile lover Peter Schlesinger, W.H. Auden, Tony Richardson, his neighbor Jo Masselink, and of course the whole Vedanta crew starting with Swami—make nearly monthly, sometimes weekly or even daily appearances. When Isherwood travels to New York he sees the composer Virgil Thomson and when he goes to England he sees E.M. Forster and the beautiful ballet dancer Wayne Sleep (portrayed in a wonderful canvas by Hockney) and travels north to visit his strange, alcoholic brother Richard.

For me this book sometimes felt like old-home week since I know or knew Virgil and Hockney and Howard Schuman and Gloria Vanderbilt and Edward Albee and Dennis Altman and Lauren Bacall and Allen Ginsberg and Gore Vidal and Brian Bedford and Lesley Blanch and Paul Bowles and William Burroughs and Truman Capote and Aaron Copland and on and on. I never met Jim Charlton but towards the end of his life he sent me dozens of letters that were nearly incoherent. I drop all these names because I suppose I feel that I can testify that Isherwood is accurate in his depictions and almost too generous in his assessments.

We learn how important Forster was to Isherwood; at one point he even considers writing about both Swami and Forster and calling it *My Two Gurus* (of course what he did was to write about Swami alone in *My Guru and His Disciple*, a book I praised in the *Sunday New York Times Book Review*). He seems delighted when Forster tells him that Vanessa Bell "was much easier to get along with than her sister, and how Virginia would suddenly turn on you and attack you." He admires Forster's equanimity, his relationship with his policeman friend, Bob Buckingham, and Buckingham's wife, May.

Isherwood was extremely important to me but I was just a blip on his screen as I learned reading his book. He gave me a blurb for my 1980 book *States of Desire*, though he told me he hadn't liked my earlier arty fiction. In those two novels he'd seen the bad influence of Nabokov, he claimed (*Lolita* he'd once dismissed as the best travel book anyone had written about America). I saw him and Don in New York and again in Los Angeles and I talked to him several times on the phone (he told me that he didn't have the patience to answer letters but that he was happy to receive telephone calls). In the years that followed I would mention in interviews that Isherwood and Nabokov were the two writers who'd had the most influence on me, just as a few years later Michel Foucault in Paris became my last mentor—in spite of himself, since he didn't believe in being or having a mentor. Perhaps it is my fate that Nabokov, Isherwood and Foucault, the three men who had the greatest intellectual impact on me, would have had to scratch their heads to remember anything about me or even my name.

We learn so much in this book precisely because it is so detailed and daily. We hear about the earthquakes, mostly small and soon over. We learn how much Chris hates to travel. We hear about his money fears (at a certain moment he is triumphant because he has $74,000 in savings). When he asks Don how he will respond to his death, Don assures him he'll give him a great send-off. Chris refers to himself several times as a "ham" who loves to show off in public and please crowds. We realize through a few hints that he, Chris, still has a sex life with various young and less young men.

Isherwood had spent most of his life in the closet, as anyone of his generation and social class would have, but in this volume he is relieved when an English journalist, almost in passing, refers to him directly and without hedging as a homosexual. In *Kathleen and Frank*, his memoir about his parents which he wrote during the period covered by this volume, he comes out in print for the first time. To be sure, he'd written frequently about homosexuals previously, notably in the groundbreaking novel *A Single Man*, but only now in the 1970s was he "out" in his own right, clearly and openly, without any screen of fiction between him and the reader. In the seventies he took an interest in gay politics, attended a few gay events and gave talks to gay groups. At one point he admits quite frankly that part of his original attraction to Vedanta lay in the fact that it accepted him as a homosexual.

Like any old man or woman he is surrounded by dying friends and family members. Isherwood is unusually calm and undramatic about these deaths (including that of his brother Richard), but he is never unfeeling. Perhaps because he thought so much about his own approaching death, he was able to take the death of his generation and of his elders in stride.

Even in old age Isherwood is still very much the working writer, sometimes collaborating with Bachardy (on a joint volume of texts and drawings called *October*, for instance), most often working alone. We learn that Bachardy had a true gift for naming things. Just as he'd thought up the title *A Single Man* in the sixties, now

My Guru and His Disciple and *Christopher and His Kind* were among the titles he suggested to Isherwood. Constantly Isherwood, like any writer, is lamenting his laziness and lack of progress, but somehow or other the old nag or "Dobbin" as he calls himself plods on toward the finish line. He is also hard at work on film scripts and theatrical adaptations of his various "properties," though he had nothing to do with *Cabaret*, the musical and movie that made him the most money and earned him the widest fame (nor did he much like *Cabaret*, though he was attracted to Michael York).

Isherwood had a personality that sparkled. When he entered a room everyone sat forward and smiled. He avoided all the accoutrements of the famous man. He asked questions and listened to answers. He refused to be complimented and if some earnest young admirer persisted, Isherwood broke into a whinny of laughter. His laugh could be deflating; once I called him from Key West to read him the end of Chateaubriand's *Mémoires d'outre-tombe* in which the author seizes his cross and walks bravely down into his grave. I was in tears but Chris thought it all so absurd that he laughed uncontrollably and I was puzzled then offended then (slightly) enlightened—I could just begin to see it was all pretty silly. When my ex, Keith McDermott, played in a theatrical adaptation of one of Chris's novels, I remember that all the Hindu monks in the play were endlessly laughing in ways a Christian divine would have considered beneath his dignity. Laughter for Chris could be deflating or just merry or impertinent—or divine, the very sound the planets make as they dance their eternal dance.

He was still startlingly handsome with his piercing eyes, shaggy eyebrows, straight nose and downturned mouth which was constantly turning up in a smile. He had a lot of charm but his charm did not stand in the way of his expressing strong opinions, especially about literature. How open writers are with one another is partly a question of nationality (the English are thick-skinned, the French thin-skinned, the Americans very thin-skinned) and generation (writers in the nineteenth century were much franker than twentieth-century writers). Isherwood came from a knockabout world of confident, upper-middle-class men and women from Oxford or Cambridge; they never minced words, any more than Martin Amis or Christopher Hitchens mince words now.

But of course he wasn't really one of his old London crowd. He'd been a pacifist and had moved to America on the brink of World War II, which had earned him the enmity of most of his countrymen—and of some Brits even to this day. He was openly gay; sophisticated heterosexuals treat gays as if they're a bad joke that's gone on too long. Tiresome. A bit silly. Tiresome and juvenile. And then, worst of all, he was a Hindu. That seemed a very "period" thing to be in Britain, something out of the dubious, not quite hygienic mysteries of Madame Blavatsky, a pure product of the 1890s. In America, Hinduism was more puzzling than anything ("Why didn't he go directly on to Zen?" most of us wondered; Hinduism seemed to Zen what Jung seemed to Freud: seedy, not very rigorous, slightly embarrassing).

He was a wonderful host, and that's how I choose to recall him. Carefully dressed, he'd climb out of an easy chair and greet a friend warmly, show him around his house. I remember seeing a photo of the very young Don with Marilyn Monroe and Chris with Joan Crawford ("our dates," Chris emphasized). In his study he showed me a school picture of Auden and himself, something he kept close by. He mentioned Tennessee Williams ("We had an affair in the 1940s when we were both still rather presentable"). There were Hockneys to be seen and works by other artist friends and of course Don's studio to be visited. If we drove to a nearby restaurant for dinner Chris lay down in the back seat ("I was driving Don mad with all my wincing, so if I lie down I don't see anything or complain"). It gratified me that even if I was a very marginal player in his drama, nevertheless he accorded me all his warmth and cleverness and kindness, if only for an evening.

Edmund White
New York City
February 25, 2011

Introduction

In his novels and autobiographies, Isherwood typically traces one thread at a time—a single character or relationship, at most a milieu in cross-section over a short period. In the pages of his diaries, he weaves together, entry by entry, week by week, the surprisingly diverse areas of his life, every thread touching upon, reinforcing, and contrasting with every other thread, so that the rich cloth of his own life also portrays the fleeting sensibility of his time. The pages teem with personalities, but even as Isherwood becomes an icon of the gay liberation movement and a sought-after participant in the celebrity culture which burgeoned in the 1970s, he continues to tell us as much about his housekeeper, his doctor, the boy trimming his hedge, or his weird and reclusive brother in the north of England as he does about David Bowie or John Travolta, Elton John or Jon Voight. Isherwood was fond of a great many people. He was a practiced, self-conscious charmer who worked hard to draw others to him. Some of his acquaintances and friends have been surprised and upset by what he wrote about them in his diaries, concluding that he withheld from them in life his true opinion recorded secretly. But what he wrote in the diaires is not what he *secretly* thought, it is what he *also* thought, on the particular day when he wrote it. It is part of a complete portrait that is perhaps never even completable. To him, a human individual was comprised of many traits; he found the so-called bad traits just as interesting and sometimes more attractive than the so-called good traits. Here is what he wrote about the woman doctor he selected in his old age to see him out of the world: "She's a nonstop talker, an egomaniac, a show-biz snob, and extremely sympathetic. Don's in favor of her, too."[1] And he criticized nobody more harshly in his

[1] Oct. 16, 1976.

diaries than he criticized himself and his companion, the American painter Don Bachardy, whose physical glamour and creative vitality transfigured Isherwood's last thirty-three years.

Throughout his writing life, Isherwood urged himself to work at his diary more often. In 1977, he wrote, "isolated diary entries are almost worthless ... the more I read the later diaries, the more I see how worthwhile diary keeping is."[1] He tracked his weight, his sleeping patterns, his trips to the gym, his illnesses and injuries; he recorded chance encounters, fragments of dialogue, and jokes, along with the more obviously important progress of his books and film scripts, his private life and his friendships, his teaching and public appearances. Thomas Mann once wrote that "only the exhaustive is truly interesting";[2] for Isherwood a more accurate phrase might be only the exhaustive is truly illuminating. He wanted to make a record of the whole human creature in context, in its natural habitat, so that he could consider and analyse its habits and commitments, its rituals and choices. He used his diaries as raw material for his novels and autobiographies, but also as a place to evaluate his life and decide whether to change his course. As a follower of Ramakrishna, he meditated almost every day for nearly fifty years, training himself to withdraw from his ego and study it from the outside; this complemented his diary writing, further developing his detachment and making his powers of observation the more acute.

But he was also looking for something more. He was a follower of Freud, too, and above all Jung, and he believed that he could edge the unconscious, the rich inner life, out into the open if he took note of everything it was delivering into the conscious arena. He kept a constant watch at the threshold between the inner and the outer worlds, impatient for new pieces of information, monitoring the revealing accretions of facial expression, posture, gesture, dress, casual gossip, dreams, all of which form the backdrop for premeditated speech and deliberate action. He jotted down coincidences, synchronicities, and numinous dates, trawling among them for clues to a hidden trajectory, an unrecognized mythology. Any stray detail could be the all-important detail that might unlock hidden meaning. His appetite for this hidden meaning increased as he grew older because he began to look upon the threshold separating the conscious and the unconscious mind as

[1] Aug. 7, 1977.
[2] Thomas Mann, Foreword to *The Magic Mountain* (1924), trans. H.T. Lowe-Porter (1927).

the very threshold which was separating him from death. Over his disciplines of observation and assessment hovered an ultimate goal: absolute knowledge might bring absolute liberation.

At the end of the 1960s, Isherwood and Bachardy began to work together writing plays and movie scripts. Collaborating brought them enormous pleasure, but in order to make money they had to keep several proposals going at one time. As this diary makes clear, they were well aware that whatever work they put in might eventually come to nothing because every project waited for the interest of a studio boss or a theatrical backer who could finance it, and then for a director and actors—preferably stars—who were equally committed and could make themselves available all at the same time. The diary charts how they and a number of their friends—writers like Ben Masselink, Jim Bridges, Ivan Moffat, Gavin Lambert, Gore Vidal, Tennessee Williams, directors like Tony Richardson, John Schlesinger, John Boorman—had to adapt their talents constantly according to changing tastes, changing social values, and new technology. Throughout the 1970s, formats also changed more quickly than ever before, sometimes for unpredictable commercial reasons, as story-telling possibilities diversified from feature films into television, including live T.V. drama, T.V. serials, made-for-T.V. films, and videotapes.

While Isherwood was working on his books and Bachardy painting or drawing every day, they remained alert to script writing opportunities. These came haphazardly, as individualized whims of the popular culture of the day, someone else's idea of what would sell. One of the few projects which they themselves originated, making a play out of Isherwood's novel *A Meeting by the River*, occupied them off and on for a number of years. It had some success in small staging, but made a resounding flop on Broadway in 1979. Actors, directors, and agents were mercilessly sketched by Isherwood in his diaries as he and Bachardy struggled to get the play put on. During the spring of 1970, Isherwood spent a month alone in England, waiting around for occasional meetings about a production that never happened; he endured this lonely episode of anticipation and disillusionment by socializing extravagantly and by making a vivid record of swinging London in his diary.

Such episodes amount to a kind of cautionary tale. In fact, Isherwood was almost indifferent to the ultimate fate of the stage and screen writing he did with Bachardy. A novel was entirely his to control, but plays and screenplays depended on the input of countless other people. As a longtime Hollywood writer, once he had sold his contribution, he tried to forget about it. In 1973, when he and Bachardy watched the television film of their "Frankenstein: The True Story," they were both horrified at what had been done with their work, yet he wrote: "the life we have together makes all such disasters unimportant, even funny...."[1] Nevertheless, their script went on to win Best Scenario at the International Festival of Fantastic and Scientific Films in 1976. In fact, the gothic fantasy was an ideal subject for them, and their version reflects their life together in a number of revealing ways. They made Frankenstein's creature beautiful, thus uncovering the Pygmalion myth latent in Mary Shelley's story and recasting it as a coded tale for their tribe. Their "monster" is presented as a suitable love object, and he stands for the "monstrous" homosexual—as George feels certain his neighbors see him in *A Single Man*—who in 1973 when "Frankenstein: The True Story" was broadcast in the U.S. would still have been a shocking subject for television. But their monster's beauty is betrayed by his makers. The creative process is reversed through a scientific error, and he begins to show on his face and body the moral decay of Dr. Frankenstein and his assistant Henry Clerval—as if he were the portrait and they the living originals in Oscar Wilde's *The Picture of Dorian Gray*.

Plenty have looked upon Bachardy as Isherwood's creature. Certainly he was a great beauty when Isherwood met him on the beach in Santa Monica in the early 1950s; he was preternaturally intelligent, strangely innocent for all his apparent sexual precocity, and still genuinely unformed. Isherwood once wrote that what he most adored about the very young Bachardy was that, "He is so desperately alive";[2] in a grimly funny sense, this is exactly the problem with Frankenstein's creature. But although Isherwood took Bachardy travelling, encouraged him to read, to converse, to go to college, to go to art school, to pursue his talent to the maximum, he could not make him happy. And so the creative process reversed, and moral ugliness began to show in the perfect boy. In their script, the struggles of Frankenstein's creature hilariously exaggerate Bachardy's predicament; and the sly, Edwardian charm of Dr. Polidori—the mad, malevolent scientist they added to the story and named after Byron's real life physician, John Polidori—mocks Isherwood's own. Indeed, Polidori and most of their characters seem to have walked out of the pre-Monty Python fantasy world which Isherwood invented with his boyhood friend

[1] Aug. 14, 1973.
[2] Feb. 14, 1960 in *Diaries, Volume One 1939–1960 (D.1)*.

Edward Upward in the 1920s and which they called Mortmere.

Isherwood was mildly contemptuous of the Universal executives and their enthusiasm for "Frankenstein." "When people say it is a 'classic,'" he wrote, "they really mean only that the makeup is a classic, as long as Boris Karloff wears it."[1] Nevertheless, he was flattered to be wanted by the striving world of commercial television, and it was easy for him to share in their nostalgia for the lost atmosphere of his own cinema-besotted childhood. "It pleases my vanity that I am still employable and that my wits are still quick enough to play these nursery games."[2] Outsiders suspected Bachardy's contribution, much as they had suspected the relationship since it began. In one studio meeting, an executive seemed to Isherwood to be "astonished to hear that Don had *any* opinion of his own. No doubt Don is being soundly bitched already as a boyfriend ... brought along for the ride."[3] Isherwood found the rudeness toward Bachardy close to unbearable, and neither he nor Bachardy had any illusions about the authenticity or indeed the difficulty of their collaboration. The diary reveals that they both felt impatient with Bachardy's inexperience as a scriptwriter: "Don is upset because he feels he is a drag on me. Actually he is and he isn't.... without him I wouldn't work on the fucking thing at all. And he does very often have good and even brilliant ideas."[4] As they thrashed about trying to bring their story to a close, Isherwood records proof of this:

> ... we have made one tremendous breakthrough, entirely due to Don. He has had the brilliant idea that the Creature shall carry Polidor[i] up the mast and that they shall both be struck by the same bolt of lightning—killing Polidor[i] and invigorating the Creature! This is a perfect example of cinematic symbolism. For, as Don at once pointed out, it was always Polidor[i] who hated electricity and Henry (now part of the Creature) who believed in it.[5]

Polidori is past his prime, and his hands have been eaten away to mere claws by an accident with his chemicals, symbolizing his moral deformity—his craze for power. But because of his hands,

[1] Jan. 11, 1971.
[2] Feb. 10, 1971.
[3] Apr. 24, 1971.
[4] June 3, 1971.
[5] Aug. 24, 1971. Isherwood typed "Polidor" as in drafts of the screenplay, but eventually they settled on "Polidori."

Polidori, the monomaniac, cannot work alone. As it happens, Isherwood was suffering from Dupuytren's Contracture, which was deforming and disabling his own left hand. He had been having trouble typing for some time, and so, experienced, bossy and power hungry though Isherwood was, Bachardy typed everything when they worked together. In September 1971, just after the "Frankenstein" script was finished, Isherwood had surgery to alleviate the condition.

The image of the aging scientist electrocuted by the very bolt which rejuvenates the Creature also makes fun of the changing dynamic in the sexual relationship between Isherwood and Bachardy, as Bachardy now had more partners and Isherwood fewer. And it resonates with the fact that Bachardy's career was taking off. About a year before writing "Frankenstein," in September 1970, he was the only portrait artist included in a group show organized by Billy Al Bengston and hung in Bengston's studio. The other artists were Ed Moses, Larry Bell, Joe Goode, Tony Berlant, Ed Ruscha, Ron Davis, Ken Price, and Peter Alexander. These were some of the most exciting and successful American painters, printmakers, and sculptors of Bachardy's generation, mostly Californians. A few of them became friends, and their names appear more and more in Isherwood's diary from 1970 onward as Bachardy introduced them into Isherwood's life. Bachardy recalls that they were a somewhat macho group in which being gay was barely acceptable, although it helped that Isherwood was an older, established writer.[1] Such a nuance demonstrates that Bachardy was admired by the best of his contemporaries for the quality of his work—in spite of being somebody's attractive young boyfriend rather than because of it.

Around this time, Bachardy was also taken on by a new dealer, Nicholas Wilder. During the 1970s, Wilder was to become the most influential contemporary art dealer in Los Angeles. He discovered and promoted a number of West Coast artists and showed Hockney, Helen Frankenthaler, Barnett Newman, and Cy Twombly. Wilder offered Bachardy an exhibition of his black and white portrait drawings, and Bachardy decided to show Wilder his recent color portraits, which nobody but Isherwood and Bachardy's previous dealer, Irving Blum, had seen: large-scale, head-and-shoulder images painted on paper in limpid water-based acrylics.[2] According to Isherwood:

[1] Conversation with me, Oct. 2006.

[2] The color portraits filled 25" × 20" sheets; the smaller black and white drawings were actually on larger paper, 29" × 23".

Don was very dubious about the paintings, afraid that Nick would be put off by them—which was only natural, after the negative reactions of Irving Blum. But I argued that a dealer is like a lawyer, you can't afford to have secrets from him if you want him to represent you, and Don agreed and we finally picked out about a dozen paintings—that's to say mostly the blotty watercolors.

> Well, to Don's amazement and to my much smaller amazement but huge joy and relief, Nick loved the watercolors and was altogether impressed by Don's versatility and said that he wants to give Don a show in which the whole front room of the gallery is full of watercolors with a few drawings in the back room. And, when we met Nick again, yesterday evening, at the opening of a show of Charles Hill's work, Nick told Don that he had nearly called him that afternoon, because "I can't get your paintings out of my mind."[1]

Only two of the paintings sold, yet Bachardy considers the exhibition was his successful public launch as a painter.[2]

Bachardy's growing self-confidence added to Isherwood's contentment. In December 1972, Isherwood describes their life together as "my idea of the 'earthly paradise.'"[3] He longed to write about the mysterious beauty of their relationship. In March 1971, he had begun a notebook about himself and Bachardy in their secret animal identities—Bachardy as the vulnerable and irresistible "Kitty" with unpredictable claws, Isherwood as the stubborn workhorse "Dobbin"—but he felt that any such project about their attachment was impossible:

> I shall never, as long as we are together, be able to fully feel or describe to myself all that our love means; it is much much too close to me. Don tells me from time to time that I should write about it, but how? Even my attempt to keep a diary of the Animals has failed. I can't see any of this objectively. Any more than I can really grasp what Swami means and has meant to me, in an entirely dfferent way.[4]

The following summer he observed, "the objection is, as always,

[1] Dec. 8, 1973.
[2] Conversation, Oct. 2006.
[3] Dec. 3, 1972.
[4] May 13, 1971.

that I feel it is a kind of sacrilege to write about the Animals at all, except privately."[1]

Only in his diaries was he able to record scraps of detail about himself and Bachardy. On Christmas Eve 1973, driving to a Palm Springs house party hosted by John Schlesinger, they talked at length of the form into which their relationship was settling and of Bachardy's present attitude toward various aspects of his life. Isherwood's account of their conversation implies they no longer had sex with one another and that, by mutual agreement, Bachardy looked for that with others:

> I asked how he feels about his meditation and he said that it is now definitely part of his life but that he doesn't at all share my reliance on Swami as a guru. "If anybody's my guru, you are." Well, that's okay, as long as he merely believes in my belief in Swami. Then I asked him about sex. He said that he doesn't mind our not having sex together any more; he agreed with me that our relationship is still very physical. The difficulty is that what he now wants is a sex object, not a big relationship, because he's got that with me. But no attractive boy wants to be a sex object; he wants to be a big relationship. I suppose I knew all this, kind of. But it was good to talk about it. Our long drives in the car are now almost our only opportunities to have real talks. As Don himself says, he is obsessed by time and always feels in a hurry, unless he is actually getting on with doing something. He says that there are now quite often moments, while he is drawing, when he feels that this is the one thing he really wants to do and experiences a great joy that he is actually doing it. But, even during the drawing, he says that he also feels harassed because he isn't drawing as quickly and economically as he could wish.[2]

They still slept together, and Bachardy recalls that this kept them physically close. Sex had dwindled only because it no longer seemed necessary. In fact, they did have sex on several occasions after this conversation, "as an instinctive means of reassuring ourselves that it was still plausible, that nothing had basically changed between us."[3] But the passionate sexual jealousy and conflicts of the 1960s were behind them, and other aspects of their relationship had become relatively more important. They identified more

[1] Aug. 26, 1972.
[2] Dec. 25, 1973.
[3] Conversation, 2007.

and more closely with one another until, as Isherwood wrote in 1975, "we are no longer entirely separate people."[1] Isherwood twice records in the diary that they could not tell their speaking voices apart, for instance, when they were revising their script of *A Meeting by the River*, "A weird discovery we have both made: since using the tape recorder to record our discussions of the play, we have both realized that we cannot be certain which of our voices is which!"[2]

But into the Animals' "golden age," as Bachardy once called it,[3] death was creeping. Isherwood was a year older than Bachardy's father. Fit and boyish as he was, his very body revealed the future bearing down on both of them; time together was short. Bachardy had the greater darkness to face, and he saw it clearly. Life with an old man, followed by the death of the old man. He says that he tried not to think about it.[4] Bachardy was more restless and more impatient than Isherwood by temperament, and whatever natural anxiety he possessed about the passage of time must have been exacerbated by living as he had done since youth, with a man thirty years his senior. If, as he told Isherwood during the drive to Palm Springs, the activity of drawing or painting lifted Bachardy out of time and freed him, at least a little, from this obsessive anxiety, nevertheless, his perception of what was to come is painfully evident in his work. He says that people praise his portraits generously as long as they are of somebody else. When they see their own faces emerge under his hand, they are often silent because they are shocked at how starkly the portraits reveal their advancing age.[5]

But of course, Isherwood also felt the passage of time, and in his diary he frequently mentions the poignancy this cast over his contentment: "the joy of waking with [Don] in the basket—the painful but joyful tenderness—painful only because I am always so aware that it can't last forever or even for very long, Kitty and Old Drub will have to say goodbye."[6] He knew that he was growing increasingly dependent on Bachardy, who drove him more and more often in the car and performed an ever greater share

[1] Oct. 5, 1975.
[2] Feb. 7, 1978. See also July 12, 1972.
[3] Jan. 7, 1965 in Christopher Isherwood, *The Sixties, Diaries, Volume Two, 1960–1969* (*D.2*).
[4] Conversation, Oct. 2006.
[5] Conversation, Oct. 2006.
[6] Dec. 8, 1972.

of domestic and social chores, and, as always, he recognized how difficult their situation was for Bachardy:

> Some of the inner rage he feels against me is because of the fact that I am going to leave him. He feels that this is a trick which I shall play on him—have, indeed, already played, by involving us so with each other. Any sign I show of illness, even of fatigue, makes him intensely nervous; he behaves as if it were a kind of bitchery on my part.[1]

And Isherwood's own friends were dying. Laughton in 1962 and Huxley in 1963 died before their time. Anyway, they were much older than Isherwood. So were Forster and Gerald Hamilton who died in 1970, and Stravinsky and Gerald Heard who died in 1971. But 1973 took contemporaries and friends of his youth—William Plomer; Jean Ross, who was the real-life original of Sally Bowles; and Auden, his closest English friend, whom he followed to Berlin in 1929, with whom he collaborated on three plays during the 1930s, and with whom he emigrated to America in January 1939:

> Wystan died yesterday—or anyhow sometime during the night of September 28–29.... This is still so uncanny. I believe it, I guess, but it seems utterly against nature. Not because I thought Wystan was so tough as all that. He seemed to have been ruining his health for years. And then, whatever he may have said, he was awfully lonely—isolated is what I mean—he made a wall around himself, for most people, by his behavior and his prejudices and demands. Perhaps he deeply wanted to go. His death seems uncanny to me because he was one of the guarantees that *I* won't die—at least not yet. I think most of us, if we live long enough, have such guarantee figures. On the other hand, the fact that he has gone first makes the prospect of death easier to face. He has shown me the way....
>
> An odd thing: That night he died—or rather, in the afternoon here, which might have been the exact time of his death in Vienna—I started a sore throat, the first I've had in a long long while, and it got so bad during *our* nighttime that I couldn't swallow. And today, despite doses of penicillin, I still have a fever and headache and feel lousy. This makes me glad. I like to believe that he sent me a message which got through to me.[2]

[1] Apr. 7, 1975.
[2] Sept. 30, 1973.

Did the message admonish Isherwood, like sore throats triggered by encounters with Auden in the 1920s, to be true to his inner nature, and to tell the truth in his writing? In his early autobiography *Lions and Shadows*, Isherwood writes of the Auden character, Hugh Weston, "I caught a bad cold every time we met: indeed the mere sight of a postcard announcing his arrival would be sufficient to send up my temperature and inflame my tonsils."[1] Isherwood was now struggling to get started with the book that was to become *Christopher and His Kind*. Auden's death not only warned him that he had better hurry, it also freed him to handle the material without anxiety that he might bruise Auden's feelings or invade his well-guarded privacy. In his lifetime, Auden never publicly acknowledged he was a homosexual, and he told friends he wanted no biography. Isherwood was among the few who could tell Auden's story—or his own story for that matter—and he knew this. As he wrote in his diary when *Christopher and His Kind* was nearly finished: "I am writing little bits about Wystan in my book.... I can't help feeling, wishes or no wishes, it is better if those who knew Wystan write now, instead of leaving it to those who didn't know him, a generation or two later."[2]

About a month after Auden's death, Isherwood saw that *Christopher and His Kind* must above all explain why he himself left England—to find somewhere that he could live as a homosexual. To his countrymen, to the press, Isherwood's departure in 1939 had long been seen as the turning point in his career and the decision on which both he and Auden had been publicly and harshly judged as war shirkers. But their emigration began years earlier, and it was a departure in their friendship just as much as it was a departure from their native land:

> ... I want to have this book start with our departure for America. But I have now realized that I can only put our departure in perspective if I begin with Germany—why I went there—"to find my sexual homeland"—and go on to tell about my wanderings with Heinz and his arrest and the complicated resentment which grew up out of it, against Kathleen and England, Kathleen *as* England.... I feel that it must start with my going to Berlin—not with my first trip out there to see Wystan, or with my visit to Wystan in the Harz Mountains that summer, but with my real emigration sometime later in the year....[3]

[1] Chptr. 5.
[2] Apr. 8, 1975.
[3] Oct. 29, 1973.

The book had to be a personal statement that he was a homosexual, and it had to show how this fact had shaped his life—how he had had to go abroad alone to Berlin to explore his sexuality freely, how once he had found a boy he loved, British immigration officials had denied the boy entry back into England and thereby forced Isherwood to go abroad to live with him until Hitler's rise made even their itinerant life impossible. Reluctant as he was to join any group, Isherwood accepted gay liberation as his own cause. But he was slow to engage with it because he feared to attract attention to himself as a member of Swami Prabhavananda's congregation. In the summer of 1970, he was invited to address the National Students Gay Liberation Conference, and he wrote in his diary:

> I feel quite strongly tempted to accept this invitation (as indeed I've often wanted in the past to accept others like it). I highly enjoy the role of "the rebels' only uncle" (not that I would be, this time—for there are scores of others—and Ginsberg their chief) and, all vanity aside, I do feel unreservedly *with* them, which is more than I can say for ninety percent of the movements I support. But something prevents me from accepting. Oddly enough, it all boils down to not embarrassing Swami by making a spectacle of myself which would shock his congregation and the women of Vedanta Place! I can admit this because I am perfectly certain there's no other motive. I am far too sly and worldly-wise to suppose that I'd be injuring my own "reputation" by doing this. Quite the reverse; this is probably the last opportunity I'll ever have of becoming, with very little effort, a "national celebrity." And I hope I'm not such a crawling hypocrite as to pretend I wouldn't quite enjoy that, even at my age![1]

For Swami, there was more at risk than awkward feelings. The Vedanta Society of Southern California was his life work; he was growing older, and it was unclear how the society would continue after he died. Nobody around him possessed his subtle understanding of how to make Vedanta accessible and plausible for Westerners. His separate insights about the personal character and temperament of each of his followers allowed him to recognize and love them as individuals; he applied no strict set of rules day to day, yet he never deviated from his own assured spiritual

[1] Jul. 30, 1970.

path. Many who tried to help him had far less patience and far less flexibility of mind. A few were against him. Some objected to the fact that he allowed the nuns to keep house alongside the monks in Hollywood. In India, women joined a separate order, the Sarada Math, but in southern California, two institutions were implausible because the few hundred devotees were not enough to populate them. Vedanta took root in southern California through the generosity of women, and there were generally more women than men in the congregation. Swami recognized that the order could not succeed in contemporary Californian culture unless it offered the same spiritual opportunities to women as to men. Only a few of his colleagues and superiors at Belur Math had been to the U.S., and most had not spent long there; they relied on his reports. But anything which might suggest to them that his style of leadership was giving rise to sexual impropriety threatened all that he had achieved. In 1970, Isherwood recorded:

> [Swami] says of Belur Math, "They are waiting for me to die"; in other words, they won't send him an assistant because, after he is dead, they can send one who'll do exactly what they want. And what do they want? Apparently to do away with nuns in the U.S. Swami takes all this quite calmly, seems to find it mildly amusing. But now he says he will seriously consider training some of the monks to give lectures. He remembers that Vivekananda said once that Vedanta societies should be run by Americans.[1]

Five years later, Swami made clear to Isherwood that he was consciously holding on to life until someone he trusted was sent to replace him:

> ... Swami had another spiritual experience....This time it was a vision during sleep. He was feeding Holy Mother and he began to weep. "I could have wept myself to death," he said. "When the doctor examined me, he said 'You have had a shock.' It was like a heart attack....When I wish to die, I can die. Whenever I wish. But I don't want to die yet—not until this place is saved."[2]

His love of Swami and his respect for the circumstances in which Swami was performing his life's mission had guided Isherwood when he decided not to write in *Ramakrishna and His Disciples*

[1] Sept. 8, 1970.
[2] Feb. 27, 1975.

about Ramakrishna's cross-dressing,[1] an episode which fascinated and inspired him and which was easily misunderstood. And even after Swami's death, when he used passages from his diary in *My Guru and His Disciple*, he altered details that might distress members of Swami's congregation or expose the Hollywood Vedanta Society to criticism or misunderstanding at the Math. For example, his description of the 1963 departure of a group from Hollywood for the *sannyas* ceremony at the Math of two American monks, Prema and Krishna, and for the simultaneous centenary celebrations for Vivekananda, reads like this in his diary:

> It is no annihilating condemnation of the devotees—about fifty of whom had come to the airport to see us off—to say that they would have felt somehow fulfilled if our plane had burst into flames on take-off, before their eyes. They had built up such an emotional pressure that no other kind of orgasm could have quite relieved it. The parting was like a funeral which is so boring and hammy and dragged out that you are glad to be one of the corpses. Anything rather than have to go home with the other mourners afterwards!
>
> Swami wouldn't leave until Franklin [Knight] arrived; he had to park the car which brought the boys from Trabuco. The fact that it was he who arrived last seemed to dramatize his role as The Guilty One, and his farewell from Swami was a sort of public act of forgiveness. He was terribly embarrassed, with all of us watching—especially all those [women] who knew what he did.
>
> So we got into the plane at last and it took off. Swami said, "To think that all this is Brahman and nobody realizes it!" I sat squeezed between him and Krishna; the Japan Air Line seats are as close together as ever. Despite my holy environment, I couldn't help dwelling on the delicious doings on the couch, yesterday afternoon. I didn't even feel ashamed that I was doing so. It was beautiful.[2]

In *My Guru and His Disciple*, Isherwood condensed the passage for good literary reasons, but in dropping sentences which might have caused offence—Franklin Knight reportedly behaved inappropriately toward a woman outside the congregation—he also muted its vigor and its hyperbolic wit, losing the potent atmosphere and the comedy of the original. He even changed the sensual and

[1] See Ramakrishna in Glossary.

[2] Dec. 1963, no day, p. 301, *D.2*.

sentimental convictions of the last three sentences into a relatively hollow pseudo-political statement:

> December 18–10. About fifty people came to the airport to see us off. The parting was like a funeral which is so boring and hammy that you are glad to be one of the corpses. Anything rather than have to go home with the other mourners afterwards!
>
> We got onto the plane at last and it took off. Swami said, "To think that all this is Brahman and nobody realizes it!" I sat squeezed between him and George; the Japan Air Lines seats are as tight-packed as ever. Despite my holy environment, I couldn't help dwelling on yesterday afternoon's delicious sex adventure. I even did so rather defiantly.[1]

He made no such concessions in writing *Christopher and His Kind*; even though he wrote it half a decade earlier, he didn't have to. He met Swami only in 1939, so he was free to be as candid as he liked about his life before that. And by the time he completed his final draft in May 1976, Swami Swahananda was already taking over from Prabhavananda at the Hollywood Vedanta Society. Prabhavananda died in July 1976, and *Christopher and His Kind* was published in the U.S. in November and the following March 1977 in the U.K. Swami was never to know that the book carried Isherwood into the heart of the gay political movement. The publicity was massive, both for and against it, the tours exhausting, and Isherwood's sense of fulfillment very great. Just before Christmas 1976, Isherwood wrote:

> San Francisco was drastic and New York even more so. Both were reassuring, because I found I could hold my own in the rat race. Indeed, I often surprised myself and Don because I was so quick on the uptake during interviews.... But I couldn't possibly have gotten through the New York trip without Don, who was sustaining me throughout. I have *never* known him to be more marvellous and angelic.
>
> Perhaps the most moving experience was going down to the Oscar Wilde Memorial Bookshop in the village and signing copies of my book, with a line of people, mostly quite young, stretching all the way down Christopher Street and around the corner. I had such a feeling that this is my tribe and I loved them.[2]

[1] *My Guru and His Disciple*, p. 260.
[2] Dec. 23, 1976.

Isherwood was a national celebrity after all, not as Herr Issyvoo, the narrative device distorted by Broadway and the movies into a popular bisexual mannequin, but in his own right, as the homosexual writer he had gradually brought into the open during his years in California. Even his rivalrous friend Gore Vidal acknowledged his fame, with the half-mocking line, "They're beginning to believe that Christopher Street was named after you."[1]

Isherwood's hard-won happiness with Bachardy was also celebrated, and it was to become a model of gay partnership among his ever expanding gay following—admired, envied, gossiped about, emulated. He went on reporting what he could about it in his diaries, and eventually Bachardy was to take over and continue the task, in his own diaries and above all in his paintings, producing perhaps his finest portraits ever during the last months of Isherwood's life. When Isherwood was too ill to do anything else, he could still sit for Bachardy, and he wanted to. So Bachardy painted him every day, on what proved to be his deathbed. He collected many of the portraits in a book, *Last Drawings of Christopher Isherwood* (1990), including nudes with their swags of given-out flesh and, over and over again, the haunted face of a creature stunned by the approach of the long-expected reality, the pain and darkness. The eyes question, even plead for mercy, but the spare, black acrylic lines—as if the brush itself were wearing mourning—are irrefutable. In late 1985, Bachardy began to notice that Isherwood seemed too ill to care about or comment on their shared project and he stopped showing him the daily results; whereupon Isherwood, with pictures drying on the floor around their bed on which he lay, said, "I like the ones of him dying."[2] Even after Isherwood died on January 4, 1986, Bachardy went on painting the beloved body, getting to grips with life and death, just as Isherwood had endeavoured to do through his relationship with Swami and in his diary through so many preceding years. Bachardy concludes in his introduction to the *Last Drawings*, "I was able to identify with him to such an extent that... [i]t began to seem as if dying was something which we were doing together."[3] Thus, the two longtime transgressors went together over the most awful threshold and made it into a work of art, illuminating their vigil, exposing every detail.

*

[1] Dec. 23, 1976.
[2] *Last Drawings of Christopher Isherwood*, p. xiv.
[3] *Last Drawings of Christopher Isherwood*, p. xviii.

Swami was an outpost of the Ramakrishna Order, working to fulfil the order's mission in a foreign culture, and relying on intermittent communication with colleagues in India whose day-to-day experiences moved them each moment along a separate trajectory, further and further from mutual understanding. Isherwood understood this because he was in the same situation with old friends and colleagues in England, with whom the finer filaments of intellectual and emotional harmony had long been severed. Even with Auden, once his closest friend, there were huge gaps. Brooding on their lives after Auden's death, he wrote:

> All yesterday and again this morning I have been looking through Wystan's letters and manuscripts—that tiny writing which I find I can, almost incredibly, decipher. He is so much in my thoughts. I seem to see the whole of his life, and it is so honest, so full of love and so dedicated, all of a piece. What surprises me is the unhesitating way he declared, to the BBC interviewers, that he came to the U.S. not intending to return to England. Unless my memory deceives me altogether, he was very doubtful what he should do when the war broke out. He loved me very much and I behaved rather badly to him, a lot of the time. Again and again, in the later letters, he begs me to come and spend some time alone with him. Why didn't I? Because I was involved with some lover or film job or whatnot. Maybe this is why he said—perhaps with more bitterness than I realized—that he couldn't understand my capacity for making friends with my inferiors![1]

Auden first decided to settle in New York when he and Isherwood visited there in July 1938, and he had told his brother as early as August 1938 that he intended to move to America permanently and to become a citizen.[2] The outbreak of World War II tempted him to change his mind, but he did not.[3] Isherwood judged both his and Auden's youthful behavior strictly, and also

[1] Nov. 6, 1973.
[2] See Humphrey Carpenter, *W.H. Auden: A Biography* (1982), p. 242.
[3] Auden travelled to Spain in 1936 and China in 1938 especially to observe the wars there, and he implied he might have returned to England for the same reason had he not fallen in love with and committed himself to Chester Kallman: "Trouble is attractive when one is not tied." See his questions and answers enclosed in a letter to his friend E.R. Dodds, *circa* March 11, 1940, in *W.H. Auden: "The Map of All My Youth"*, edited by Katherine Bucknell and Nicholas Jenkins, p. 113.

their subsequent rationalizations about it. In one of his diary entries, he records a sudden insight about himself: that once he had settled in southern California, he had deliberately tried to cut off any possibility of returning to England by sending home an offensive letter which he knew would be made public:

> ... the letter I wrote to Gerald Hamilton in 1939, attacking the war propaganda made by Erika and Klaus Mann and others, was really a device, to get myself regarded as an enemy in England and therefore make it impossible for me to "repent" and return. That was why I chose Gerald Hamilton to send my letter to—I knew he would broadcast it.[1]

Isherwood's observation is tidy with hindsight, but it is no self-reinvention. The letter had been rash and defiant, and he knew Hamilton was indiscreet. It was quoted in the *Daily Express* in November 1939, launching the worst of the public criticism of his and of Auden's decision to remain in America after the war started.

As Isherwood dug into the motives for his past actions and became increasingly candid about his generation, former friends in England seemed to understand him less and less. Some admired him for coming out as a homosexual, but they were bewildered and critical when he came out with other truths. In 1973, he asked John Lehmann not to publish their correspondence because when he reread his letters to Lehmann, Isherwood found that they said only what he had presumed Lehmann had wanted to hear at the time: "They are dull, mechanical, false. Don was horrified by their insincerity when he read them—he hadn't believed that even old Dobbin could be capable of such falseness."[2] He stood by the long-established friendship, but when Lehmann pressed for an explanation, Isherwood was evasive:

> Once started, I might have found myself cutting much deeper and telling John *why* my letters to him during the war were so false—namely because I knew he wasn't on my side, I knew he didn't believe I was serious about Vedanta or pacifism and I knew he would disapprove, on principle, of any book I wrote while I was living in America. I was false because I didn't want to admit how deeply I resented his fatherly tone of forgiveness of my betrayal of him and England—"England" being, in fact, his

[1] Jun. 27, 1981.
[2] Sept. 14, 1973.

> magazine....The stupid thing is that I'm fond of him in a way, and that I've often defended him, even. I think, as everybody in London thinks, that he's an ass and that he has almost no talent. But I am fond of him, which is more than most people are.[1]

In fact, it had been many years since Lehmann and others had understood who Isherwod was or what he believed. Although some of Isherwood's most beautiful writing in his diaries is about his childhood home in the north of England and about his visits there to his brother Richard, and although he chronicled the London social scene with energy whenever he visited, he never felt comfortable in England for any length of time. When he had returned to live there with Bachardy in 1961, he wrote:

> I realize now, on this trip, that my longing to be away from England had really nothing to do with a mother complex or any other facile psychoanalytical explanation. No, here is something that stifles and confines me. I wish I could define it. Maybe the island is just too damned small. I feel unfree, cramped. I long for California.[2]

He preferred California and its beaches even to Manhattan. The casualness and undress of Californian life freed him from any preconceptions about identity or social class—minimal clothes revealed little about status; names were pared away to a syllable or two and titles dropped; speech was relaxed towards a uniform drawl. The British theater critic Ken Tynan, who had known Isherwood since 1956 in London and who settled in Los Angeles for a time during the 1970s, remarked in his diary on "the classlessness that [Isherwood] shares with almost no other British writer of his generation. (I've seen him in cabmen's pull-ups and grand mansions, with no change of manner or accent.)"[3] Like Whitman or Kerouac, Isherwood had a promiscuous curiosity about his fellow men, and he knew he could find out more about them if he met them on an equal footing.

Perhaps the widest gulf between Isherwood and his English friends was religious. When he had nearly completed *My Guru and His Disciple*, he found that he could not summarize for publication the beliefs on which he meditated every day because he feared

[1] Sept. 25, 1973.
[2] May 15, 1961, *D.2*.
[3] *The Diaries of Kenneth Tynan*, ed. John Lahr, April 14, 1977, pp. 369–370.

derision back in England. In the past, Isherwood had been silenced and even made ill when he was unsure of what he wanted to say and whether he was justified in saying it. In the 1920s, it was "the liar's quinsy";[1] in the early 1960s, when his relationship with Bachardy profoundly wobbled, undermining his religious certainty and throwing him into conflict with Swami, he lost his voice for reading aloud in the Hollywood temple, and in India, where he was obliged to lecture to hundreds at Vivekananda's centenary celebrations, took to his bed with fever and stomach troubles. But by the late 1970s, his books had made clear who he was, and there was now no risk of hypocrisy in saying what he believed:

> What is holding me up? . . . surely I can make *some* statement?
>
> This block which I feel is actually challenging, fascinating. It must have a reason, it must be telling me something.
>
> Am I perhaps inhibited by a sense of the mocking agnostics all around me—ranging from asses like Lehmann to intelligent bigots like Edward? Yes, of course I am. In a sense, they are my most important audience. Everything I write is written with a consciousness of the opposition and in answer to its prejudices. . . . I must state my beliefs and be quite quite intransigent about them. I must also state my doubts, but without exaggerating them. . . .
>
> The doubts, the fears, the backslidings, the sense of alienation from Swami's presence, all these are easily—too easily—described. One mustn't overemphasize them. What's much more important is a sense of exhilaration, remembering, "I have seen it," "I have been there."[2]

In fact, Isherwood's initiation into Vedanta was no more mystical or irrational than Edward Upward's conversion to communism. Upward's autobiographical trilogy, *The Spiral Ascent*, shows that he followed his ideology inflexibly, sometimes at the cost of his writing and his personal happiness, and blinding himself to the crimes of Stalin. Auden also turned to religion, adopting again in the early 1940s the high Anglican faith of his upbringing. Why couldn't Upward or Auden take Isherwood's Hinduism seriously? In the historical drama of *their* 1930s, intellectual debate focused on Marxism, fascism, nationalism, socialism, even Freudianism and eventually Christianity. When Vedanta, Isherwood's deus ex

[1] *Lions and Shadows*, chptr. 5.
[2] Oct. 23, 1977.

machina, appeared from outside this cultural narrative, they simply rubbed their eyes in disbelief. They could not recognize it, and they made little effort to learn more about it.

Auden's personal myth proved to be the myth of the return, a circling back to the abandoned Christian beliefs of his childhood and even, at the very end of his life, a physical return to live in Christ Church, Oxford. Auden rediscovered his roots, and made much of their familiarity and exact fit. Isherwood kept progressing away from his roots, which is not a satisfying myth to those he left behind. If the hero of the quest never returns, how is anyone to know whether he really found the grail? Even Auden, convinced that Isherwood had a gift for befriending his inferiors, presumed that Swami was one of them. For one thing, Swami did not appear to Auden to be grim enough. The Anglican Church told Auden that as a homosexual he was a sinner; without repentance and without God's grace, he would go to hell. In Vedanta there is no hell, no place where one dwells eternally as a result of committing sins. Indeed, there are no sins, only errors. About his sexuality, Isherwood recalled in his diary, "right at the start of our relationship, I told Swami I had a boyfriend (and ... he replied, 'Try to think of him as Krishna') because my personal approach to Vedanta was, among other things, the approach of a homosexual looking for a religion which will accept him...."[1] To Swami, all sex was a worldly appetite which spiritual aspirants should try to overcome on the path to liberation, but it made no difference whether the sexual object was the same or the opposite sex. Moreover, as Isherwood realized when he was writing *My Guru and His Disciple*:

> ... this book differs in kind from books on other subjects because there is no failure in the spiritual life.... I can't say that my life has been a failure, as far as my attempts to follow Prabhavananda go, because every step is an absolute accomplishment.[2]

As these diaries affirm, Isherwood's vision is essentially a comic one, and it is based on love. His English friends were cynical about love. But love is what both Isherwood and Auden set out to find when they left England for America. Isherwood found not only love but also happiness. Auden proclaimed that if you had sex with your beloved, the vision of Eros (as he called it) would necessarily

[1] Nov. 26, 1970.
[2] Mar. 27, 1978.

fade.[1] His failure to find long-term reciprocal love provided him with plentiful material for austere lyrics about unrequited longing and love betrayed. In Isherwood's case, the vision stayed. In 1977, he wrote about Bachardy: "Our relationship is indescribable—though I guess I must try to describe it someday. It has moments which seem to blend the highest camp with the highest love with the highest fun and delight."[2] Happiness and religious joy are hard to write about; Dante is often excused for being boring in the *Paradiso* and most readers prefer the *Inferno*, just as they prefer Milton's *Paradise Lost* to *Paradise Regained*. In his diaries, Isherwood records plenty of hell with his heaven, but even so, happiness can also provoke envy and resentment and criticism. He knew that his life might seem schmaltzy to others, a Hollywood tale, too good to be true, and so he avoided exposing it or he cast it as camp. Just before publishing *My Guru and His Disciple*, he worried whether his readers would grasp what he was doing:

> I keep wondering how people will receive the book. I have almost no idea. Sometimes I think, after all, the material itself is so remarkable that some people must find it fascinating. But I think many others will be repelled by the weepy devotional tone of certain passages. Also by a very private kind of camp, which sometimes expresses itself in stock religious phrases which aren't meant to be taken literally—though how is the reader to know that?[3]

In 1962, the American philosopher and critic Susan Sontag wrote, "Camp is esoteric—something of a private code, a badge of identity even, among small urban cliques. Apart from a lazy two-page sketch in Christopher Isherwood's novel *The World in the Evening* (1954) it has hardly broken into print."[4] Since then, camp has broken not only into print, but into song, film, T.V., and merchandising for children. It is no longer an esoteric or private code, and yet Isherwood's use of camp to express religious feelings does remain private and unexplained. Sontag doesn't address it;

[1] For example, ". . . all the accounts agree that the Vision of Eros cannot long survive if the parties enter into an actual sexual relation." Introduction to Anne Fremantle, *The Protestant Mystics* (1964), rpt. in *Forewords and Afterwords* (1974), p. 64.
[2] Nov. 8, 1977.
[3] May 27, 1979.
[4] "Notes on 'Camp,'" *Partisan Review*, vol. 31, no. 4, 1964, pp. 515–530; rpt. *Against Interpretation* (1966), p. 275.

Isherwood himself in *The World in the Evening* had only touched on a connection between religion and camp when she sketched her "Notes on 'Camp.'" In *The World in the Evening*, Charles Kennedy—modelled on Lincoln Kirstein, a likely influence on Isherwood's own understanding of camp—contrasts camp with Quaker plainness, explaining to Stephen Monk that "a swishy little boy with peroxided hair, dressed in a picture hat and a feather boa, pretending to be Marlene Dietrich" is Low Camp. On the other hand:

> High Camp is the whole emotional basis of the Ballet, for example, and of course Baroque art.... true High Camp has an underlying seriousness. You can't camp about something you don't take seriously. You're not making fun of it; you're making fun out of it. You're expressing what's basically serious to you in terms of fun and artifice and elegance. Baroque art is largely camp about religion. The Ballet is camp about love...."[1]

Camp was a central resource in Isherwood's writing. And he balanced it with "tea-tabling," a kind of understatement which he and Edward Upward discovered in youth in the novels of E.M. Forster, and which Alan Chalmers, the Upward character, explains in *Lions and Shadows*:

> The whole of Forster's technique is based upon the tea-table: instead of trying to screw all his scenes up to the highest possible pitch, he tones them down until they sound like mother's meeting gossip ... In fact, there's actually less emphasis placed on the big scenes than on the unimportant ones ... It's a completely new kind of accentuation—like a person talking a different language.

Armed with these methods of highlighting and lowlighting, Isherwood was like an electrical technician with a complex set of dimmer switches. He could achieve fresh effects of great variety even with the most familiar materials.

His later style is far from baroque. In *My Guru and His Disciple* it is remarkably unshowy. It is intimate, frank, self-deprecating, and, in fact, its very plainness is like a disguise, camouflage even, for the remarkable thing it undertakes to reveal, the life of religious devotion. Its understated quality, achieved through the tea-table technique, requires us to lean in, to look more closely, to listen

[1] Prt. 1, chptr. 3.

more carefully. Baroque art was produced by institutionalized religion. The force and wealth of the Catholic Church, the state, the national culture was behind it. Isherwood's camp is a guerrilla effort. It is deployed on behalf of a minority, and it must do two things at once—edge into the light something marginal, unacceptable, beyond the pale of the culture, and protect that very thing from attack. It must be both expressive and guarded, articulate and secretive.

In his earlier work, Isherwood deploys camp just as Sontag observed, as the private code and badge of identity for the main urban clique she had in mind, homosexuals. Take *Mr. Norris Changes Trains*: how could anyone "in the know" avoid reading Mr. Norris as a homosexual like his real life original Gerald Hamilton? Consider his plump white hands, his wig, his over-large bottle of scent, his fretful anxiety—the exaggerated attributes of a woman of a certain age. Norris's young girlfriends with their whips offer an overt "perversion" which fulfils readers' expectation but also throws them off the track. Like a schoolboy trying out a big philosophical idea and following it with a dirty joke, Isherwood's narrative exaggerates to precisely the degree which permits it to claim, implicitly, it was all a joke. The author didn't really mean it; readers never really saw what they thought they saw. It is a small surprise attack from which he can pull back quickly.

Isherwood used camp to write about dangerous subjects, subjects which his audience would find morally, philosophically repugnant, subjects about which he could not risk talking straight. He used camp, with its lighthearted playfulness, to tempt the audience in, to suggest that what he was saying was unimportant, not serious, and that there was no risk in being exposed to it. He used it to write about what was nearest his heart. Once he was out as a homosexual, he used camp to write about the growing band of disciples who believed that Ramakrishna was an incarnation of God and could inspire them to spiritual liberation.

It was genuinely scary for Isherwood to come out as a Hindu, and his doubts heightened his vulnerability. His diaries reveal how difficult it was for him to argue on behalf of something he himself was continuously working to believe in. The moment of exhilarated certainty, the vision, is difficult to capture and easy to dismiss. Yet, in his diaries, Isherwood's religious life becomes convincing because acts of devotion are repeated again and again and again. Moments of vision and certainty regularly recur and increase in intensity as his love for Bachardy, Swami, and Ramakrishna are gradually manifested over time. This kind of repetition doesn't work in fiction, which calls for turning points, climaxes, denouements. And a simple statement, a summary of beliefs, cannot convince because it is intellectual and does not engage the emotions. Camp addresses the emotions, releases laughter and joy, and thereby disarms the reader.

Remarkably, Isherwood never mentions AIDS in these diaries. He was not aware of the disease while he was still well enough to write. Thus, his account of gay liberation has an almost innocent feeling of gradual and certain triumph, and it fulfils an expectation of social change with which he had been living ever since the 1930s: the Marxist dialectic still—finally—applies personally to him and to everyone like him. Homosexuals who were prepared to bring their private lives into the political arena, at the expense, for a time, of romance and even dignity, could at last participate in the revolution:

> A bit of paper has been lying on my desk for months. On it I have written: "For the homosexual, as long as he lives under the heterosexual dictatorship, the act of love must be, to some extent, an act of defiance, a political act. This, of course, makes him feel apologetic and slightly ridiculous. That can't be helped. The alternative is for him to feel that he is yielding to the compulsion of a vice, and that he is therefore dirty and low. That is how the dictatorship wants him to feel." I have copied this here so I can throw the paper away. It isn't clearly expressed but it means something important to me."[1]

The blossoming which Isherwood portrays of the London gay scene, in contrast to Los Angeles, is dandified and softened by an atmosphere of aristocratic aestheticism; Lord Byron in ruffled linen hovers in its background. Of a party in London he might tell us that the boys wore shiny eye shadow or silk trousers in appealing pastel hues and kissed on the lips at parting; in Santa Monica, he tells of meetings, parades, leaflets, and accommodating himself to people he must work with regardless of whether he finds it easy to be friends with them. Reading such passages now is like reading about the last summer parties before the Great War.

There are only a few hints at the beginning of the 1980s that the increasingly hectic sexual activity and drug-taking of some of Isherwood's circle might be starting to wear them out. Tony Richardson, for instance, seems to grow tired. And in September

[1] Aug. 8, 1976.

1981, Isherwood records that one young friend has a sore throat, "Poor angel, he really is a martyr to V.D." In the very next entry, another phones to warn that he has hepatitis; so Isherwood and Bachardy both get gammaglobulin shots because the boy had kissed Isherwood at a party.[1] In 1983, a boy who sits for Bachardy is suffering from cancer: "He carried on about the treatments he'd had for cancer and created a quite powerful death gloom, which he thickened by showing us two lesions on his legs."[2] Later the boy died after being ill for about ten years; he proved to be the first AIDS fatality Isherwood and Bachardy knew. Many of their friends were to die of AIDS thereafter, but Isherwood never wrote about their deaths or about the epidemic.

The disease of his generation, and the one he most feared, was cancer. In April 1981, he wrote: "Loss of hair. Loss of taste and consequent loss of weight—down to 147 and ½—no big deal, this."[3] But it was a big deal; the beautiful life he had forged with Bachardy could survive anything but "this." That autumn, Isherwood was diagnosed with prostate cancer. And he used the diary, as ever, to try to understand what he felt about it:

> Well, the moment has come when I must recognize and discuss the situation with myself, which means, as usual, writing it down and looking at it in black and white. I have got some sort of malignancy, a tumor, and that's what's behind all this pain. They will treat it, of course.... I shall get used to the idea, subject to fits of blind panic....
>
> What I have to face is dying....
>
> I pray and pray to Swami—to show himself to me, no matter how—as we've been promised that he will, before death....
>
> I get fits of being very very scared....
>
> Don says that I should work, and he is absolutely right. I remember how wonderful Aldous Huxley was, working right to the end. I have promised Don that I'll get on with my book.
>
> The love between me and Don has never been stronger, and it is heartbreakingly intimate. Every night he goes to sleep holding the old dying creature in his arms.[4]

His fears came and went, and he struggled humbly with them. He continued to be happy with Bachardy, and he experienced

[1] Sept. 15, 1980 and Sept. 25, 1980.
[2] Jan. 1, 1983.
[3] Apr. 19, 1981.
[4] Oct. 16, 1981.

powerful visions of Swami which he recorded in the diary and which suggest that what he had always hoped for from his religion was there for him at the end:

> Sometimes I feel the death fear bothering me again. I pray hard to Swami, asking him to make me feel his presence, "Now and in the hour of death." The response I get from this is surprisingly strong. I'm moved to tears of joy and love. I pray for Darling also, seeing the two of us kneeling together in his presence. Religion is about nothing but love—I know this more and more.[1]

In the 1980s, Isherwood made fewer and fewer diary entries. They have a tenderness and pathos unlike the smart and cynical writer of his youth. His psyche was so tuned and disciplined by a lifetime spent writing, that even as the grammar vanishes and the commas are no longer placed, he shapes his prose with a cadence of ending. As he fades physically and presses towards transcendence, he continues to report factually on his experience, finding words for the body's creaturely, unconscious existence on the threshold of an awesome physical transformation: "... I'm not in a good state. Death fears—that's to say, pangs of foreboding—recur often. They seem to be part of a quite natural physical condition; the pangs of a dying animal, thrilling with dread of the unknown."[2]

His final entry touches on the chief themes of his life, his love for Bachardy and his continuing pleasure in the literary rivalries of his youth which still seethe by letter backwards and forwards across the Atlantic. Spender, he reports, is too embarrassed to admit he has been knighted by the Queen because he mocked Auden in the leftist thirties for proudly collecting a Gold Medal for Poetry from the King, and he knows Isherwood will remember. Upward compares Spender's ambition to Macbeth's; he will stop at nothing to outdo Auden, the literary king of their generation: "(Edward Upward, in a letter, made us roar with laughter by quoting Banquo's line: 'Thou hast it now....')"[3] Isherwood made this last entry on Independence Day 1983, the anniversary of Swami's death in 1976 and of Vivekananda's in 1902; on this numinous date, he seemed already to be looking forward to his own liberation—and laughing.

[1] Apr. 11, 1982.
[2] Oct. 23, 1982.
[3] Jul. 4, 1983.

Textual Note

American style and spelling are used throughout this book because Isherwood himself gradually adopted them. English spellings mostly disappeared from his diaries by the end of his first decade in California, although he sometimes reverted to them, for instance when staying at length in England. I have altered anomalies in keeping with the general trend; however, I have retained idiosyncrasies of phrasing and spelling which have a phonetic impact in order that his characteristically Anglo-American voice might resound in the writing, and I have let stand some English spellings that are accepted in American since Isherwood had no reason to change these. I have generally retained his old-fashioned habit of liberal capitalization because this is often a key, like quotation marks but with less emphasis, to his private language of camp.

I have made some very minor alterations silently, such as standardizing passages which Isherwood quotes from his own published books, from other published authors, and from letters. I have standardized punctuation for most dialogue and quotations, for obvious typos (which are rare), and very occasionally to ease the reader's progress. I have usually retained Isherwood's characteristic use of the semi-colon followed by an incomplete clause. I have spelled out many abbreviations, including names, for which Isherwood sometimes used only initials, because I believe he himself would have spelled these out for publication, and I have corrected the spellings of many names because he typically checked and corrected names himself. Square brackets mark emendations of any substance or interest and these are often described in a footnote. Around his sixty-seventh birthday, Isherwood began to misspell names and occasionally transposed them or got them wrong in some other way; he had seldom made such errors before. For simple misspellings of this kind, I have generally not added footnotes. But for more significant and potentially confusing errors, I have done so, especially where Isherwood himself draws attention to them.

Square brackets also mark information I have added to the text for clarity, such as surnames or parts of titles shortened by Isherwood. And square brackets indicate where I have removed or altered material in order to protect the privacy of individuals still living.

This book includes a seven-month run of entries from Isherwood's 1976 pocket diaries; these were pre-printed appointment books, seven days to a page, in which he jotted down whom he met each day and, very briefly, main events. He called them his day-to-day diaries. The entries fill a gap in the first half of the year when he stopped writing in his diary while finishing *Christopher and His Kind*. The rationale for including the day-to-day diaries follows Isherwood's own example in handling a longer gap in his much earlier diaries, from 1945 until 1948. In 1955, he made an outline of the missing years based on his pocket diaries, and he drew on this outline later when he wrote the memoir published posthumously as *Lost Years*.

Readers will find supplemental information provided in several ways. Footnotes explain passing historical references, identify people who appear only once, offer translations of foreign passages, gloss slang, explain allusions to Isherwood's or other people's works in progress, give references to books of clear significance to Isherwood, sometimes provide information essential for making sense of jokes or witticisms, and so forth. For people, events, terms, organizations, and other things which appear more than once or which were of long-term importance to Isherwood and for explanations too long to fit conveniently into a footnote, I have provided a glossary at the end of the volume. The glossary gives general biographical information about many of Isherwood's friends and acquaintances and offers details of particular relevance to Isherwood and to what he recorded in his diaries. A few very famous people—Bette Davis, Liza Minnelli, Elton John, John Travolta—do not appear in the glossary because although Isherwood may have met them more than once, he knew them or at least wrote about them in their capacity as celebrities. Others who were intimate friends—Truman Capote, David Hockney, Igor Stravinsky, Tennessee Williams—are included even though their main achievements will be familiar to many readers. This kind of information is now easily available on the internet, but a reader of this diary should be able to find what he or she immediately wishes to know and to get a feel for what Isherwood himself or his contemporaries may have known, without putting the book down and turning to a computer. Isherwood has audiences of widely varied ages and cultural backgrounds, and I have aimed to make his diaries accessible to all of them. Where he himself fully introduces someone, I have avoided duplicating his work, and readers may need to use the index to refer back to figures introduced early in the text who sometimes reappear much later.

Hindu terminology is also explained in the glossary in accordance with the way the terms are used in Vedanta.

In any book of this size, there are many details which do not fit systematically into even the most flexible of structures, but I hope that my arrangement of the supplemental materials will be consistent enough that readers can find what help they want.

Acknowledgements

For me, this volume completes a task of more than fifteen years. My life has been transformed by the long and inspiring engagement with Christopher Isherwood, whom I never met, and Don Bachardy, whose friendship I cherish. I will always be grateful for their lives recorded in these diaries and for the opportunity entrusted to me by Don to prepare the diaries for publication. The passage of time has even brought with it a new life, my third child and youngest son, Jack Maguire; his father and I are proud that he is a godson to Don.

I am indebted again to the benevolence and generosity of Pravrajika Vrajaprana of the Sarada Convent in Santa Barbara. She has continued to guide my research in Vedanta, in small matters of terminology and large ones of ritual and doctrine. She has again read every page of this book in a foolhardy attempt to save me from error. I am extremely grateful to her, to Bob Adjemian, to the late Peter Schneider, and to the many other nuns and monks at the Vedanta Society of Southern California who have responded to my queries.

I have again had research assistance from Douglas Murray, Anne Totterdell, Gosia Lawik, and members of the London Library staff, and I thank each of them for their dogged professionalism. Tim Hilton methodically read for errors. Christopher Phipps contributed not only research assistance but also his sense of order and completeness in creating another fine index.

The list of those who have answered other questions is very long. Each person on it has made at least one trip to an attic or cellar to rummage through a box of old photographs or papers, even if only figuratively, and each has reported to me thoughtfully, on occasion hilariously, by phone, email, letter, or in person: Marie Mériaux Allemann, International Committee of the Red

Cross Archives; Julian Barnes; Keith Berwick; the late Thomas Braun; Jude Brimmer, Archivist, Britten-Pears Foundation; John Byrne; Lucy Bucknell; Leslie Caron; Patrick Cockburn; Camilla Chandon; Tchaik Chassay; Jim Clark; Mary Clow; Anna O'Reilly Cottle; Paul Cox, Assistant Curator (Archive and Library), National Portrait Gallery, London; Robert Craft; Linda Crawford; Stephen Crook, Librarian, Henry W. and Albert A. Berg Collection, New York Public Library; Evan Cruikshank; Santanu Das; Marie-Pierre de Lassus; Jane Dorrell; David Dougill; Chris Freeman; Robin French; Bella Freud; Jonathan Fryer; P.N. Furbank; Christopher Gibbs; Christian-Albrecht Gollub; the late John Gross; Joseph Hacker; Donald Hall; Nicky Haslam; Selina Hastings; Nancy Hereford, Press Director, Center Theater Group; Samuel Hynes; David Jenkins; Jim Kelly; the late Frank Kermode; Jane Klain, The Paley Center for Media; Mark Lancaster; John Lahr; George Lawson; Kristen Leipert, Assistant Archivist, Whitney Museum of American Art; Richard Le Page; Michael McDonagh; Lucy Maguire; Ian Massey; Edward Mendelson; Young Hoon Moon; Robin Muir; Melissa North; Annie Ochmanek, *Art Forum*; Richard W. Oram; Peter Parker; Michael Peppiatt; Philip Ramey; Andreas Reyneke; Rodrigo Rey Rosa; Giovanna Salini, Embassy of Peru, London; Peter Schlesinger; Howard Schuman; W.I. Scobie; Alexandra Shulman; Richard Simon; Wayne Sleep; Geoffrey Strachan; Leslie Kay Swigart, Librarian, California State University at Long Beach; Daniel Topolski; Jeremy Treglown; Gloria Vanderbilt; Hugo Vickers; Barney Wan; Edmund White; Deepti Zaremba.

For permission to publish excerpts from unpublished letters of Edward Upward, I would like to thank Kathy and Jeff Allinson, from an unpublished letter of John Lehmann, Georgina Glover at the David Higham Agency and for lyrics from his song "Honky Cat," Bernie Taupin.

My agent, Sarah Chalfant, has altogether changed my working life and made these diaries part of a new beginning for me. Thanks also to Luke Ingram, Kristina Moore, and Matthew McLean at the Wylie Agency. I remain forever grateful to Stephanie Cabot for her willing ear and good sense and to Caroline Dawnay for negotiating my role in this project. To the publishers of *Liberation*, Clara Farmer and Terry Karten, my warmest thanks for your conviction and your follow-through. In addition, I would like to thank Alison Samuel and the others at Chatto who have been such stalwart friends and hard workers in completing this long project, Sue Amaradivakara, Juliet Brooke, Anthony Hippisley, Alison Tulett and especially Rowena Skelton-Wallace.

Thank you once again, my dear friends, Richard Davenport-Hines, Isabel Fonseca, John Fuller, Bobby Maguire, and Robert McCrum for wading through and commenting on the parts of this book that Isherwood did not write.

To Jackie Edgar, Vilma Catbagan, Felisberta Rodrigues, Katrina Johnston, Elizabeth Jones, Susan Mellett, I have said thank you before, and I hope you know how much I have benefitted from your help. And to Bob, Bobby, Lucy, and Jack: No kidding, I really am done now. It's all over but the reading.

Liberation
January 1, 1970—July 4, 1983

January 1—March 2, 1970

January 1. We got up very late and have been fussing around, chiefly engaged in destroying old manuscripts; the early *Meeting by the River* material for instance. Now we have to go out and see people, including Margaret Leighton whom I like and Marti Stevens who bores me.

Bill van Petten called to wish me a happy New Year and to tell me that the new (sixteen-year-old) generation calls itself the Jam Generation. It relates to its parents, whereas the previous generation (twenty-year-old) didn't. It believes in pilgrimages; hence these huge gatherings as at Woodstock. Bill also says that the newsmen on the *Los Angeles Times* are very pessimistic about the seventies. They expect organized violence by the blacks.

Last night we saw the New Year in with Jack [Larson] and Jim [Bridges]. Several people connected with Jim's film[1] were there; he starts shooting on Monday. We both feel that the prospects of the film look dubious because Jim doesn't seem to have really thought through the material and found out what it's about. Surely it isn't *about* this girl at all, but about the weird married couple who decide that the husband must father the baby? The girl is just a human appliance, but it looks as if Jim is going to sentimentalize her. However, Jim feels he has got a very good cast. Sam Groom,[2] who plays the husband, was there last night. I liked him. We talked about Hemingway, on whom he'd written a thesis. He looks awfully young for the part, though, but Jim says he's a marvellous partner for mad Collin Horne.[3] One could see them as a wayout

[1] *The Babymaker*; see Glossary under Bridges.

[2] (B. 1939), a regular on American and Canadian soap operas from the mid-1960s, including "As the World Turns" and "Another World."

[3] Often known by her maiden name, Collin Wilcox; see Glossary under Wilcox.

Macbeth and Lady. The girl who plays the girl (Barbara Hershey)[1] and the boy who plays her boyfriend Tad (Scott somebody)[2] and a boy who is a friend of his and a girl who is Scott's girlfriend were also there, utterly ambushed in lank shoulder hair. I sure hope the Jam Generation will decree crew cuts. Don says that this is the most unflattering period for women in the whole of history. When Jim announced midnight and a toast in champagne, the young folk barely responded; it was too square for them.

Incidentally, Bill van Petten told me that, if you don't have any particular plans, this is now called being "unstructured"; in the sixties it was called "hanging loose." Bill is desperately in pursuit of all the very latest slang, the latest attitudes, the last "word."

January 3. I saw Swami yesterday. He doesn't seem sick, only tired. And of course there is a little bit of policy somewhere in it; for now Asaktananda has written a drastic appeal to Belur Math to send someone to be second assistant without delay!

Swami told me that when the palpitations, or whatever they were, came over him in Santa Barbara he felt quite detached, as if his body belonged to someone else. "I felt a flustering in my chest, and I was like an observer." He also told me he had a dream that he was swimming in the Ganges. He wasn't at all afraid of being drowned, he didn't know if he had clothes on or not, he couldn't see the banks of the river.

Yesterday we had lunch with Alan Searle. He is very red in the face (hinting at apoplexy) and very plump. He made big protestations of affection but the fact remains that he has been here eleven weeks and has only now made a move to see us, just before leaving! He says he's thinking of coming to live here. He drinks a lot. He says that Willie [Maugham] told him never to write his life, and that Kanin's book is made up of stories he told about Willie; that Kanin never really knew Willie as well as he makes out.[3]

1969 was actually a very happy year for me, mostly because of Don, and this despite the fact that it was a year of frustrations. We wrote the *Cabaret* treatment and got turned down by Tony

[1] (B. 1948), a T.V. regular in the mid-1960s and later a movie star in *The Right Stuff* (1983), *Hannah and Her Sisters* (1986), and *The Portrait of a Lady* (1996).

[2] Scott Glenn (b. 1941) later had roles in *Nashville* (1975), *Apocalpyse Now* (1979), *The Right Stuff* (1983), *Silverado* (1985), *The Silence of the Lambs* (1991), and others.

[3] Garson Kanin (1912–1999), American actor, writer, and director, described the friendship in *Remembering Mr. Maugham* (1966).

Harvey and the *Claudius* screenplay and got turned down by Tony Richardson. *Black Girl* got very disappointing notices. Ray Henderson's *Dogskin* production fizzled out after looking most promising, because Burgess Meredith deserted us. True, all these projects could easily be revived this year or later.[1]

I did quite a lot of work but not nearly enough. Only six chapters of *Kathleen and Frank*; that's disgraceful. The other chores were all for the Vedanta Society except for a foreword to Hockney's book of drawings, which I wrote with extraordinary difficulty, unwillingness and boredom, as a gesture of friendship, and for which I've never even been thanked, much less paid![2]

The best event was Irving Blum's decision to give Don this show. The most memorable days were the two on Tahiti and the day of our visit to Stevenson's grave—but altogether, that trip was really the best of my whole life, I think.[3] The chief disaster was the collapse of our hillside during the rains on February 25; this may well lead to much more serious slides nearer the house. The most boring thing that happened was John Lehmann's visit. The most interesting new person I met was Jim Gates (with Peter Schneider a runner-up); the most intriguing celebrities, Michael York and Jeanne Moreau. Relations with friends haven't been very intense; we have kept ourselves even more than usual to ourselves. When I ask myself who I've felt particularly drawn to, amongst people in this town, I'm rather surprised to find it's Leslie Caron—although I still don't dig her husband Michael [Laughlin]. I have been very regular in going to the gym but alas have gotten progressively heavier. I bulge with gas.

Never mind, it was a good year, with much beautiful quiet joy in it and loving snug closeness of Kitty and Dobbin.[4]

January 11. It's dripping rain and I have a cold lurking which I'm trying to keep at the back of my throat with Coricidin D. How I hate my huge belly! A lot of it is gas and I can dislodge it if I do

[1] Isherwood's adaptation of Bernard Shaw's 1932 story, "The Adventures of the Black Girl in Her Search for God," was staged at the Mark Taper Forum in Los Angeles, March 20–May 4. For more on this and the proposed musical of *The Dog Beneath the Skin*, see Glossary and Christopher Isherwood, *The Sixties: Diaries Volume Two (1960–1969)* (*D.2*.).

[2] David Hockney, *72 Drawings: Chosen by the Artist* (1971); Isherwood's introduction was not used.

[3] Isherwood and Bachardy travelled in the South Pacific and Australia, partly for the filming of *Ned Kelly*, July 20–August 13. See *D.2*.

[4] Isherwood and Bachardy in their private, mythological identities; see Glossary.

stationary trotting; haven't been out today except up to the trash cans.

It now seems almost definite that we'll go to London on February 2, after a stopover in New York, and that we'll seriously consider Clifford Williams as director[1] and, if okay, start casting and then rehearsals and then open, all being well, sometime in March or April. But Don's show is still set for the beginning of March, so he'll probably have to go back for it and then return for our opening. All this sounds as unreal to me at the moment as the plans to go with David Hockney and Peter [Schlesinger] on the trans-Siberian railway—except that Don and I have pretty well decided to back out of that.

We had supper with Jo [Lathwood] last night and yesterday afternoon I saw Elsa [Lanchester]. Much as Elsa would hate to admit it, she and Jo are now sisters in bereavement, because of Ray Henderson being about to get married.[2] Yesterday Elsa was giving a party too and her attitude was just like Jo's: all alone I'm doing it, poor little me, it seems so strange without him, etc. etc. Jo described to us how she had been given a surprise birthday celebration by "her" girls at the factory. Her slogan is, "They all love me." And Elsa described how she had gone to some celebrity dinner the other night and been applauded for five minutes; she said that this "made me feel worthy of my friends." Oh, the false pathos of these unhappy old girls! It doesn't move you in the way they intend, but it *is* genuinely, heartbreakingly squalid.

January 30. It's five fifteen a.m. and I'm up before dawn because today is the Vivekananda breakfast *puja.*[3] Shall be off there in a short while. Peter Schneider and Jim Gates will meet me there and afterwards we'll all drive to Claremont, where Jim will have the stitches taken out of his neck. He was operated on, messily, by a Dr. Joseph Griggs who is the brother of Phil who is now Buddha and a monk at the London center. The swelling is huge and looks nasty and Jim thinks it is maybe infected, but he seems serenely unworried. He told me that right after the operation, when it started to hemorrhage, and he really thought he was dying, he felt suddenly blissfully happy. Maybe he *is* Alyosha Karamazov—but

[1] For a proposed London production of Isherwood and Bachardy's stage adaptation of *A Meeting by the River.*

[2] Each had been deserted for a younger woman; see Glossary.

[3] See Glossary for Hindu terms.

oh wow man the pitfalls! Anyhow he is sort of dear to me and so is Peter. Peter is now coming up from behind, in racing parlance, I mean he had an interview with Swami, too, his second, yesterday and Swami told him to sit up straight and look after his body and Peter told Swami that his father had only told him to look after his mind and Swami had said but they are the same. (This from Swami himself, when I went to see him yesterday afternoon.) Anyhow, in a word, Peter isn't going to be left out of the *Atman* race. He is keenly competitive and quite jealous lest Jim should somehow get a bigger slice of me than he does; not that he wants me, or even maybe the Atman, but that's his character. And at his age it's lovable. He is therefore coming along with us to Claremont.

This evening I take off for New York, leaving Jim to house-sit here. He is rather thrilled at the idea of being alone in this (to him) vast palace. It is funny and interesting that there is very little involvement between the boys, seemingly; conscious involvement, that is. They are involved of course. And their life in that little shack beside the canal with the ducks and the kitty and the meditation hut at the back in the grassy yard is so idyllic, hippie in the best way and far more genuine and unpretentious than most hippies' lives are. I told them, they will look back and say they never had it so good, later on. People come to see them in droves and food is prepared and they eat, or else there is no one and they don't. They wander off to college or their jobs at these restaurants and it all seems so simple.

From New York, Don and I are to go to London on the 4th and then find out if the play is really going to happen. About all this I feel at present only the dislike of leaving the nest and the hate of flying and of the impending icy weather, etc. I called Don this morning, just to say hello, but I guess he is spending the night out; no answer.

Gavin [Lambert] just got back from Europe, last night. He says he's still ill and he looks it; he really seems quite fragile and his hands shake like an old lady's. He seriously considers buying a house in Tangier, in the kasbah, and spending several months there every year, chiefly amongst the Arabs. When he talks of this he *is* Lesley Blanch. He's also quite a wonderful person, so much style and courage. Style is a kind of courage, always, I suppose.

Am rambling to pass the time. Now I must call the boys and tell them the car will start (just tried it) so they needn't come around and pick me up.

8:07 p.m. A very quick goodbye. Charlie Locke called to say his wife is dying of terminal cancer and can I do something to arrange for them to spend her last weeks in Cornwall. But Jim Gates does *not* have anything serious the matter with him. So hurrah for that. Goodbye.

England, March 2—April 30, 1970

We arrived in England on February 5, 1970, from New York, where we had stayed since leaving Los Angeles on January 31.

All through the rest of February, Don was with me in London and we did a good deal of rewriting on our play, A Meeting by the River. *On March 2, after Don had gone back to Los Angeles because of his show at the Irving Blum gallery, I decided to start this diary and try to keep it every day until he returned or I left England to rejoin him in California.*[1]

March 2. I'll try and write this entirely at odd moments. Am now waiting for Bob Holness of the BBC radio program "Late Night Extra." (Long before that sentence was finished, he arrived, interviewed me—How is your play getting along? How do you feel about *Cabaret*?—and left within twenty minutes.)

Don left this morning for New York and Los Angeles. There was a nasty little snowstorm, then it cleared, then it snowed again. It's horribly cold. I rushed out and bought books—as people rush into pubs to get drunk for the sake of getting drunk: the script of Lindsay Anderson's *If*,[2] Frederick Brown's Cocteau biography,[3]

[1] Isherwood handwrote this diary on the rectos of a bound notebook with additions, usually marked with asterisks, on facing versos. The additions are printed here as a separate run of footnotes, above my numbered notes. On the first verso, facing the start of the diary, he wrote:

> From an article by Arthur Hopcraft on the Working Class in Britain (*Observer* supplement, February 22, 1970) which is headed "The Middle Class Get Psychotherapy and the Working Class Get Pills":
>
> > A woman social worker who takes part in Salford's psychiatric social service says: "You can't deal with marital problems because there's a housing problem; you can't deal with that because there's a wages problem; you can't deal with that because there's a health problem."

[2] Written with David Sherwin (1969).

[3] *An Impersonation of Angels* (1968).

Aldous Huxley's letters, *The Divine Comedy of Pavel Tchelitchew*,[1] Richard Neville's *Play Power.*

The heat is on and it's still cold. Clement Scott Gilbert and his secretary have taken away the new pages we did since the reading of the play on Wednesday last; they'll xerox them. Clement said that John Roberts and Clifford Williams have been "meaning" to ask me if I minded having all reference to the cubes removed from the version of the play which will be sent round to the actors when they start casting. I don't like their attitude, it seems sneaky. Are they so scared of me and, if so, why? What is Clifford up to? We have heard nothing from him since the reading and apparently he has gone to New York to see *Oh! Calcutta!* which he is to work on before he directs our play. I said, it's not that we're wild about having the cubes in, but how can you explain to an art director that he must devise a substitute when you haven't shown him what the substitute is a substitute for?

A BOAC plane with that terribly insecure-looking tail has just flown over. Like the one we flew here in. Thoughts of Don, flying in it today. Is he in New York already? He should be nearly, if not quite. Wish he would ring me from there.

An underground train rumbles below, shaking the house, as trains have been shaking it for exactly a hundred years.* This part of the inner circle is older than this street.

The girls on the street with their very long maxi coats which open to show madly indecent glimpses of miniskirts and endless leg beneath.

How relaxed the English are! As we were driving back from the Rodin exhibition yesterday, an anti-Vietnam-War procession came up Whitehall. So all traffic was halted by the police until it had gone by. The cars must have been backed up halfway down the Strand. Never mind. They just had to wait.

March 3. Sylvain [Mangeot] last night seemed quite middle-aged, slowed down, slow spoken, his deliberately told stories have almost no point because he goes into no details, "They had some incredible adventures," "There was an absolutely ludicrous scene." This

*No, says Norman Prouting, 102 years. This house was built in 1868. This whole area was built by some famous Australian cricketers. There is a pub round the corner called The Australian. Formerly, this land was marsh. [Isherwood was lodging with Prouting in Chelsea; see Glossary.]

[1] A biography by Parker Tyler (1969).

little household, the moustached Portuguese twenty-two-year-old student of theatrical design named Juan Melo and the Finnish girl, rather beautiful in her pale Arctic way, whose name I can't write (something like Kick-kicke-kee) help Sylvain look after André [Mangeot], who now can't use his legs or do anything for himself. He is downstairs and I didn't see him. Up to the age of eighty he was still playing tennis! Hilda [Hauser] died soon after Olive [Mangeot]. Sylvain thinks she just decided to, there was nothing for her to live for, so she stopped taking the pills the doctor had given her to keep down her blood pressure and died in the kitchen after getting home from a movie. Sylvain and Juan cooked. We had two kinds of wine, also sherry and brandy. (Sylvain told at great length how cognac is called cognac after the place called Cognac, because the original Hennessey had the idea, so all brandy became "cognac," although lots of other villages were making it.) A happy evening.

Terribly cold today. Last night I had to wear an undershirt and my bathrobe in bed. And to think a tiny kitty could keep Dub so much warmer!

It was cold at the Ramakrishna-Vedanta center too, where I've just lunched. Buddha looks starey-eyed, a little crazy, but doesn't seem so. Swami Bhavyananda is fat and laughing, a bit like Vishwananda. He is a *jnani*, so Buddha does all the pujas.[1] Four other men were there; they made an unusually good first impression—a Chinese, an Australian, an Englishman, a German; they all seemed genuinely friendly.*

March 4. Thick fluffy snow fell this morning. May it clear before I go north! A man from the BBC came to interview me, arriving more than half an hour early, just as I was in my bath. (The bath is one of the few appliances that really functions; you can have as much warm water as you like.)

Had supper with Robert Medley last night. He is planning to write a book about the Group Theatre and wanted my memories. We talked a lot and then found that his new tape recorder

*One of the boys at the Ramakrishna-Vedanta center, the English one, was told to take a photograph of Swami and me. He somehow got the camera jammed, being nervous. Buddha reproved him, addressing him as "brother," and one saw the grim bright-eyed smiling martinet. Buddha, speaking of his "lapse," said, "It's been a long way back." I think I pleased him by telling of the "lapses" of Franklin [Knight] and of [another monk he knew].

[1]In fact many jnanis do perform pujas; see Glossary under Bhavyananda.

wasn't recording.* Later, Gregory Brown came in and made it work. Robert was chiefly interested in the friction that developed between himself and Rupert on the one side and Wystan and me on the other. I said I really didn't feel hostility to Rupert (aside from the fact that he was one of the most infuriating prima donnas who ever lived) and even liked him. Wystan—as Robert agreed—really disliked Rupert because he was jealous of him.

Gregory is a silly billy, I think; a white goose with a strange look of a homely Hope Lange. Why am I being so nasty? Because he's married and yet comes around and smugly accepts the devotion of Robert—who told me he looked "ravishing." Is that my affair? No.

Robert is now painting hard-edge squares and other shapes. They looked a most awful lot better after some scotch—though it didn't take much, I admit, and I didn't get the least drunk. But liquor does help me to look at art, it always has. When sober, my eyes dart about so restlessly. That's why nowadays I find it a terrific effort to read *anything*.

Later. Still blasting this wretched snow. Have just come back from lunch with Gore [Vidal] at the Connaught. He looks thinner, hollow-cheeked but in a boyish way; terrifically attractive. He says he *feels* attractive, and he charmingly recalled that I was his present age when we first met and how attractive *I* was then. Indeed we were very pleased with each other. I love his consciously aggressive careerism. (Careerism is only loathsome when people are hypocritical about it, as they nearly always are.) Gore is now casting off the U.S.—as, he says, he threatened to if the Vietnam War continued six months longer after Nixon took office. But he also says the Nixon administration is out to get him, through income-tax audit. So he's planning to become a citizen of the Irish Free State, and leaving for Dublin tomorrow to buy a house. I urged him to go into politics there. "Yes," he said, "de Valera's about had it."[1]

The usual pronouncement that Truman Capote is a "birdbrain." Gore has finished a novel called *Two Sisters* in which he admits that he and Jack Kerouac went to bed together—or was that in an article? (Gore told me about so many articles he's written and talks

*How we found this out was that I said to Robert, "I think maybe you'd better switch that thing off for a minute; I want to say something very personal." (It was about Wystan [Auden]'s jealousy of Rupert [Doone]!)

[1] Eamon de Valera (1882–1975), Manhattan-born Irish revolutionary, then in his late eighties and still President of Ireland.

he has given that my memory spins.) Anyhow, Gore now regrets that he didn't describe the act itself; how they got very drunk and Kerouac said, "Why don't we take a shower?" and then tried to go down on him but did it very badly, and then they belly rubbed. Next day, Kerouac claimed he remembered nothing; but later, in a bar,* yelled out, "I've blown Gore Vidal!"

Talking about books, Gore recommended *The French Lieutenant's Woman* [by John Fowles] and said he likes everything by William Golding. He is to make a film called *Jim Now* which is to bring *The City and the Pillar* up to date and is set in Rome. Howard [Austen] will be the producer, and Gore says he's really doing it because he feels Howard needs something to occupy himself with.

Quite a flap over the play. Clement [Scott Gilbert] wants to get rid of Clifford Williams because he feels Clifford is trying to do two jobs simultaneously—this and *Oh! Calcutta!* John Roberts supports Clifford, but Clement believes he'll get a shock when he hears what a lot of money Clifford is demanding—the contract with him hasn't been signed yet.

Clement, Clifford and Roberts agree that, when our revised version of the play is typed up, all mention of the cubes shall be taken out of it. They say the *idea* of the cubes will scare off prospective actors! I say, if the cubes aren't mentioned, how can we explain what we want the art director to do about replacing them? (I've just realized that I have written all this already, which shows what a preoccupation it is with me at present!)

March 5. Last night, Norman Prouting and I saw *The Way of the World* at the Old Vic. It is depressing, how little Congreve's lines mean when they aren't spoken with style. Behind this shoddy clowning the old performance by Edith Evans and the others, the one I saw in 1927, kept appearing, their voices "came through"; I hadn't realized how well I remember their readings of many speeches.[1]

Norman Prouting is touching, kind, vulnerable—maybe he appears to be more vulnerable than he actually is. He says he has put his whole life into this house; that's conscious pathos, of course; but I don't see him as a tiresomely pathetic person. He fixed us a

*The San Remo?

[1]Nigel Playfair directed the 1927 production, in which Peggy Ashcroft also appeared with Godfrey Tearle, Scott Russell, and Dorothy Green; Michael Langham directed the later one, starring Geraldine McEwan.

little supper up in his flat after the theater. We talked about [J.M.] Barrie's plays.

March 6. A lady named Rosemary Ellerbeck[1] came to interview me. She was nice, and had written an article on lesbians. "We talked about why they were lesbians and why I am a heterosexual[.]" "And why are you?" "I found I had absolutely no idea." She says that the lesbians told her they hate male homosexuals and indeed all men. As she was leaving she was speared by one of the very sharp little horns on the back of the armchair in the living room. It made a hole in her dress, but she didn't complain.

I had lunch with Bob Regester. Neil [Hartley] is quite sick, he has a bloodclot in his leg. He returns from New York today. Tony [Richardson] has quarreled with Henry Geldzahler about Larry[2]—not for the obvious reason but because Tony invited Henry and Larry to come down to the Caribbean and then told Henry he wasn't wanted. Bob was dieting, so we ate beefsteak tartare, at a swanky club called dell'Aretusa.

With Peter Schlesinger to see *Hadrian VII*; almost incredibly poorly constructed.[3] Peter kept getting the shits. Patrick Woodcock thinks it's psychosomatic, and Peter certainly has homelife problems. The flat is always full of people, which he hates, and yet he can't go off alone much because David wants the two of them to go around town as a couple. Also different sex patterns—shall they do it evening or morning? So they end up hardly doing it at all. [...]

Peter says [Don's friend] asked him to find out if I'd see him. I said no—I wasn't a bit angry with him (this isn't *absolutely* true), I would be polite if we met socially, but I did *not* see any point in our getting together.

This morning Clement Scott Gilbert called to tell me that Clifford Williams wants the play rewritten; Penelope to come with Patrick to India, the Swami to be alive, Tom to be dropped and the mother too. John Roberts supports him in this; thinks the play as is isn't box office. Clement and Richard Schulman[4] declare their loyalty, however. I called Don, who is already in Los Angeles. He

[1] Author of *Inclination to Murder* (1965) and other thrillers, from Cape Town; she used the pseudonym Nicola Thorne.
[2] Stanton, a young American artist admired for his beauty; Geldzahler also admired his work. Stanton died of AIDS.
[3] By Peter Luke; directed by Peter Dews, who won a Tony for an earlier New York production.
[4] Scott Gilbert's producing partner.

was rather depressed, said that Clement and Richard are amateurs and that Clifford and John are the pros; but of course he agrees we can't consider reconstructing. (I see [Don's friend]'s shadow lying darkly over this. *He* didn't like the play either.) There is to be another meeting on Monday, and then the situation should be clearer at least. Nicholas Thompson maintains that Clement *is* experienced and professional and says he isn't at all disturbed.

Am just off to see Richard [Isherwood] at Disley. This gruesome cold! But no more snow, thank God.

March 7. Just before I got the taxi to Euston I rushed into a shop in the King's Road to buy some pajamas. Grabbed some rather sickening pink ones with stripes, paid for them and rushed out. Only later, in the taxi, I realized they had cost me ten guineas! Felt so disgusted I considered stuffing them into some trash can, lest Don should see them when he returns. But of course I shall tell him. I always do.

The electric train ride to Stockport was much smoother than the old steam ride used to be. Richard met me, looking a bit heavier but somehow much more distinguished and indeed, if one can use such an expression, more like other people—though he still twists his head about and blinks. Dan and Mrs. Dan [Bradley] looked the same, both blooming with health. Dan talks more than ever, plays the stereo for "background music" and has a little curly damp-looking Yorkshire terrier named George which barks and drives one slightly crazy. But the Bradleys are really lovable animal-people, cozy to be with.

An amazing father and son comedy act, "Steptoe and Son," on the telly, (Wilfr[i]d Brambell, Harry H. Corbett). They were so good it was even funny as intended, but moving, like Chekhov. The story anyhow wasn't in the best of taste; an old horse dies and is made into cat food.*

This morning I woke, and there had been a quite heavy snowfall during the night. I leaned out of the window to breathe in the beautiful pure Brontë air, and saw dark drops on the snow along the windowsill—BLOOD! I was having a profuse nosebleed, my first in I don't know how long. And what a *Wuthering Heights* thing to do! Cathy bleeding into the snow!

*The Bradleys' T.V. set has something wrong with it, the picture is distorted vertically into a rectangle. This makes the dobbins particularly adorable, with their very short legs and immensely long bodies.

Am reading *The First Circle* by Alexander Solzhenitsyn and Trollope's *The Eustace Diamonds*. Also brought *The Autobiography of Malcolm X* here with me. Trollope goes down easily; Solzhenitsyn as tough as his name.

March 8. Woke with such a feeling of the hills all around in the heavy snow. They are so powerfully present, so aloof and yet so suburban, and really so small; but I have never experienced any hills like them. And there comes a sense of how Kathleen [Isherwood] saw them, from the Wyberslegh windows, and how her wish was granted, to end her long life amongst them and die amongst them, grumbling but finally contented.

What I actually see from *my* window here at the Bradleys' are seven minigarages, some brick council houses, and a shed, a little tree with large very black rooks (from the churchyard rookery) in it, and the hillside of small gardens under heavy snow, black sticks sticking out of it. The air of the hills smells strongly of cow.

Richard really does seem much less bizarre; perhaps I am comparing him with a fright-Richard of my imagination, but I don't think so. We drove over to tomb-chilly Wyberslegh and came back with Kathleen's news cuttings album, her diary for 1924 (for details of Granny Emmy's death) and the Marple-Wyberslegh book,[1] which Richard will let me take to London and get xeroxed. Also he has told me quite a number of valuable extra details—such as that Emily, at the end, would only eat food that was yellow.*

Later: 10:40 p.m. Don is running around somewhere in Los Angeles, getting ready for his show later in the afternoon; it'll be about 2 a.m. tomorrow, English time. I mustn't pray for his success, only Swami can do that, but I can give thanks that I know him and love him. How amazing he is!

Have accomplished a good deal while here, but today was pretty

*Richard is obviously drinking a little, offstage, during my visit, but very little. I have smelt beer on his breath a couple of times; but he even refused a glass of sherry to drink to Dan's birthday. Ordinarily, he drinks mostly at Wyberslegh, during the afternoons.

Dan is worried because he feels that Thomas [Isherwood] is using his influence on the lawyer to save up money to pay death duties; which means Richard won't be so likely to build a house for himself and the Bradleys near Wyberslegh, when Dan retires.

[1] Kathleen's history, added to over many years, of Marple, Wyberslegh, and the Bradshaw and Isherwood families, including family trees, floor plans, newspapers clippings, and her own watercolors and pencil and ink illustrations; Isherwood describes it in *Kathleen and Frank*, chptr. 13.

trying. It was Dan Bradley's birthday, he was sixty. He is wonderfully vigorous and so truly honest and courageous, and Richard certainly should give thanks for *him* and Mrs. Dan, who is adorably good-natured and looks absolutely marvellous for her age, unless she's far younger than I think she is.* *But* today the children were here, in two shifts most of the time, and this produced maximum schizophrenia—the attention torn seven ways at once by grown-ups, kiddies, the little dog, Dan's memories of being buried by a bomb at Plymouth during the war, Mrs. Dan bringing in food, the telly or radio played nearly full strength and Richard coughing just when you were trying to listen to anyone or anything else. Finally, glutted and dazed and slightly asphyxiated by the gas fire, I went out for a walk, just up the hill as far as the turn off of the road to Macclesfield. Snow was falling lightly but it was thawing. The air was so pure and full of strength. The snow hills with the dark crests of copses and the blackish-green stone walls and the black telegraph wires and barbed wire and fence posts. The farm where I stayed when the Monkhouses were at Meadow Bank and I had a crush on Johnny and Rachel had a crush on me. I'm glad people have had crushes on me, glad I used to be cute; it is a very sustaining feeling. Of course I often behaved like a little faggot bitch, but no tears need be shed over that. I too have been bitched—more than once.

The most interesting members of Dan's family are the Danish boy, Bent Nilson (Nilsson? Neilson?) and his wife, Dan's daughter Elizabeth. She met him in Australia, fell in love with him and had a child. She told him he didn't have to marry her; she didn't want him unless he wanted her. But he did marry her and they had another child after returning to England—both sons, both called by Danish names, Nils and Bjorn (this maybe proves something). Elizabeth adores him. [. . .] He is really quite powerfully sexy, with his smooth sulky face still boyish, his pretty carefully arranged fringe of wavy brown hair, his broad shoulders, thick wrists, tall lithe body, straight athletic legs with big knees and sturdy thighs. [. . . T]hey are soon leaving for a visit to his family in Denmark by ferry and car through Belgium and Holland and Germany.

Heard Richard laughing in the kitchen with Mrs. Dan. He sounded just as he used to, long ago, when he was laughing with Kathleen.

*Richard says no, she's fifty-eight.

March 9. Back at Moore Street after a slow train trip during which I read myself a little more deeply into *The First Circle* and continued to enjoy *The Eustace Diamonds*.

Clement Scott Gilbert came by: Clifford Williams is definitely out. Now he's eager to get Alec Guinness (and I suddenly remember how Guinness wrote me that fan letter about *Meeting*). Am to talk to Don in the morning. He sent a handsome announcement of his show, says he has several commissions already.

March 10. Have been talking to Clement, Don and Nicholas Thompson (in that order) about the play etc. John Roberts will stay with us *if* we can cast it. (I remarked to Clement that this was like the Los Angeles City Council, which took a pledge of allegiance to the flag on V-J Day[1]). Now I'm to dictate the necessary changes to a typist, who will then type up the copies to be sent around. Clifford Williams has sent a note of apology to us; so that's that.

Don's show seems to have been a success, whatever that means. Don was as cagey as usual about it. Mrs. Blum told him Irving thought this was the best opening he'd ever had. (How utterly unimportant this play is to me, compared with the prospect of Don's achieving something like this, all on his own!) So we've decided to stay each of us where we are, for the present, and await developments.

Don says Evelyn Hooker has emerged into the light of social intercourse. She called, after having been all this time in the funny ward at Mount Sinai! She had been suffering for two years from depression because she couldn't write her book; now she has resolved not to write it at all and go into private practice. Don says her face has quite lost its deathly "terminal" look. Is she a demonstration of Homer Lane's statement that you can cure yourself of cancer by going mad?

Don has seen Jim Bridges' film, still dislikes the story but thinks it really remarkable and Jim has got a great performance out of Collin Wilcox. The question now arises, should Jim direct the play after all, if he's willing? I said well, you ask him and, if he's free and wants to, let him get in touch with Nicholas Thompson at once, but meanwhile we'll be looking elsewhere.

Talking of performances, Don says my T.V. talk about

[1] I.e., once victory over Japan in W.W.II had already been declared.

homosexuality[1] was thought very highly of by Evelyn, Gavin and—George Cukor!

P.'s confidences at supper last night about poor D.* make D. seem terribly like an image of me and my past (I hope) behavior. P. says that whenever he tells D. he'd like to get away on his own and have more freedom, sexual and otherwise, D. always reacts by saying, "You hate me." P. wants to have his own place and go on visiting D., without all this social involvement. Another point of similarity is the money question; D. is completely generous and wildly extravagant, but P. is nevertheless shy of behaving as if D.'s money really is his, as well. P. has only a small allowance from his family, doesn't earn anything. [. . .]

I think it very likely that P. will end up by marrying.

March 11. Last night—oh Kitty forgive me!—I got very drunk with a young man named John Byrne. He wrote to me several months ago because he's working with Alan Clodd on a bibliography of me.[2] He has a job with an antiquarian and first edition bookseller called Bertram Rota in Savile Row (I went there today and we had lunch and I was given a copy of Romer Wilson's *The Death of Society* because I said I wanted it, largely for nostalgic reasons; I'm sure it isn't good). I'm not particularly thrilled by him and certainly not attracted—the poor boy has bad scars on his face, due to an accident with a heater, I think—but he's good company (or should I say a good audience?). We stayed up till the small hours and today I've had a terrible hangover, which made me visit a Turkish bath, the one in Jermyn Street, and have my hair cut. It's been raining on and off, which is depressing. Today Clement came by and went through the script of the play, now that the cubes have been removed. It will be typed up again as soon as possible. Only one alteration; I have added the two final lines spoken by Patrick and Oliver. I think they help a lot.

A note from [Don's friend] asking if we can meet. Now I must write and tell him no. I only hope he won't be at Tony Richardson's tonight!

*P. = Peter Schlesinger, D. = David Hockney.

[1] An interview with correspondent Piers Anderton about sexuality, filmed by NBC at Anderton's house on November 19, 1969. Others involved from NBC were Lew Rothbart and Mike Gavin.

[2] Clodd, Irish librarian, bookdealer, publisher (1918–2002), began collecting Isherwood's work in the 1950s and built a modern literature collection of 20,000 volumes. He also founded the Enitharmon Press in 1967. The bibliography was never completed.

March 12. He wasn't. And now I've written him saying that I don't want to see him. He should get my letter tomorrow; so that's that.

It was quite fun at Tony Richardson's, because David was so sweet, as usual, and adorable wriggly little Wayne Sleep was there, and [Rudolf] Nureyev, with his current friend* and some other guy. I think Nureyev had a sort of suspicion of me or thought I was some obsolete old Establishment fart. Anyhow, after dinner, while we were talking, he suddenly twisted my wrist with really cruel violence and in order not to let him hurt me I had to swing around with the result that I lost my balance and fell across the cocktail table, providentially not breaking anything. Actually I did skin my shin and drew blood but I made like it was nothing and apologized for my clumsiness. The others were aghast for a moment. Immediately, Nureyev's manner changed, he became mock-affectionate, hugged me to his cold breast, covered my face with vampire kisses. He really is a macabre absurd nineteenth-century vampire, but at least he has great style and he dresses most elegantly. I felt quite warmly towards him, but it was much nicer to cuddle with Wayne, who has an admirer from New York who owns nineteen ships and has given him a gold watch. Peter defied Tony, when Tony said that Andy Warhol is no good at all; but Tony didn't seem to mind a bit, and we are all to visit him for Easter in the South of France. The house is full of rather terrifying masks he has brought back from New Guinea. Also he has a construction of colored lights which flash on and off in turns and varying combinations, made by an artist named [Vassilikas] Takis.

This afternoon I got Kathleen's manuscript book xeroxed.

March 13. I feel a sudden black depression. This weather is so wretched and the play is dragging its feet (not even ready to be typed till Tuesday) and I have just talked to Don and he hasn't sold any pictures from his show and has hardly any commissions. He also says that Jim Gates has got his call-up and now will soon have to leave, unless he can get approved C.O. work near home. Also, to be frank, I minded because Don seemed quite casual about my returning or not returning—yet I know so well how easily one can give that impression without meaning. ... Well, fuck all that. Courage! Now I must go out in the drizzle and visit Hermione Baddeley. And tomorrow Cambridge and Morgan [Forster].

Edward [Upward] came up to see me today, which *ought* to have left me cheerful because his visit was really all that our meetings

*Bob Coh[a]n.

are at their best. He seems fatter and speaks with half-closed eyes, sometimes sleepily, sometimes excitedly and inaudibly. One always has a tremendous sense of his vocation as a writer. *Nothing* else matters to him. (This actually isn't true; he is devoted to Hilda and the children.) But this intellectual passion is immensely stimulating and we rattled away, hardly noticing the hours pass.*

Last night I had supper with Phillip Foster and his Finnish wife Eija, a tiny blonde with hair hanging down.[1] I couldn't help feeling a tiny falseness in her—or is it merely a hardness? Yes, she is hard, has probably had to be. Phillip is sleak, well fed, quite fat with a jowly face but still fat-sexy. Everything is Finland—he is learning the language and the flat is full of Finnish fabrics and artifacts. A very snug little couple; there seemed almost no difference in their sex.

Gerald Hamilton came on on T.V. and was quite marvellously himself (now eighty-two) so polished and gross and charming and hideous. He rolls up his eyes until you see nothing but the whites; it's almost as terrific as the picture of Dorian Gray.

March 14. Another little nosebleed this morning and it's raining and cold weather is forecast and I'm off to Cambridge, but my mood is good—partly because, after a very short visit to a bash given by Hermione Baddeley, at which I knew no one except Victor Spinetti,[2] I went to supper with Patrick Woodcock and met a really beautiful boy named [K]arl Bowen who kissed me, in the

*Edward's third volume of the trilogy seems to be a flashback which starts from childhood and brings him up to the point at which volume one begins. He says he is having just as great difficulty with this volume as with the others, but of course there *is* a difference; the block has been at least partially removed, he now knows he can actually get a book finished and published. To me he seems to have fears of rejection by publishers on political grounds and fears of prosecution on grounds of libel which verge on paranoia. Perhaps this is the result of the kind of life he has led—always feeling himself to be an illegal underground worker. But, without the life, Edward wouldn't now have his own personal myth; and, lacking that, he'd write quite differently or not at all.

[1] Eija Vehka-Aho, a model, mostly for Helmut Newton, used only her first name professionally. She was married briefly to Foster who also worked in fashion.

[2] Welsh-born half-Italian actor, writer, director (b. 1933); he won a Tony Award in *Oh! What a Lovely War* when it transferred to Broadway in 1964 and appeared with the Beatles in *A Hard Day's Night* (1964), *Help* (1965), and *Magical Mystery Tour* (1967).

nowadays style, in front of the taxi driver as we said goodnight. I do wish I could stop drinking though. I do hate it so.

March 15. When I got to Cambridge I saw Mark Lancaster; we met in the middle of the grass of the main court, joking about whether or not Mark was a senior member of the college and thus entitled to walk on it. Mark has Lowes Dickinson's[1] old rooms; they are above the archway through which you see a view of the Backs[2] and therefore look straight across at the college gate, commanding a view of everyone who goes through—also, at present, of two monster cranes in the background. Cambridge is being rebuilt but not nearly as fast as most places.

Morgan looked almost exactly the same in the face, the clear blue eyes, the long nose, the pink complexion, the mussed-up hair (except that it's white) but he is fatter and more stooped—he looks almost as if he had a hump—and much shakier. He moves insecurely with a stick; but he does move and the sight, at least of one eye, is actually better. In Coventry, when I last saw him,[3] he was being read to; now he reads to himself. In Coventry—probably partly because of more drink, more comfort and it being a less chilly time of year, he seemed drowsier, lazier and less mentally alert than now. Today, most of the time, he was obviously able to follow all that was said and join in the conversation whenever he wanted to.

His affection, as always, was touching, childlike. He loves being hugged and kissed. "How extraordinary!" he kept saying; and I took this to mean that because he sees me so seldom and therefore keeps me as a creature of his memory, the fact that I actually exist in the flesh seems extraordinary to him. I asked him if I had changed a lot. "A bit thicker, that's all." He made me turn around to look at me. While we were embracing I felt a sort of fake, because I was consciously going through the motions, wishing only to do what would please him and also very much aware of Mark looking on. But actually I'm not faking, on such occasions, it's only that I take so long to come to a boil. I only felt the emotion of this meeting when I was in the train going back to London.

[1] Goldsworthy Lowes Dickinson (1862–1932), political historian, philosopher, pacifist; author of *The Greek View of Life* (1896) and, based on his work toward founding the League of Nations, *The International Anarchy* (1926). At King's, he was Forster's teacher, and Forster became his executor and biographer.

[2] I.e., the back sides of the seven Cambridge colleges situated along the river Cam.

[3] At Bob and May Buckingham's in May 1967; see *D.2*.

Of course it was hard work. I reached sweatily for scraps of news, anecdotes, questions about mutual friends. And he reached too. Do we embarrass each other a bit? Yes. Have we always? Somewhat, maybe. But oddly enough that has little to do with affection. And isn't the same thing true, to a much smaller degree, of me and Edward?

Talking about his health, Morgan said, "I have been a little displeased with myself lately." But he made it clear he was only referring to his physical health. (I wondered if he has a skin cancer problem; there seemed to be something growing in his cheek.) I reminded him of how he had said to me long ago: "I hope I shan't get depressed—no, I don't think I shall." And he told me that, on the whole, he hadn't. I also reminded him of a letter he had written in which he said that he was staying with [Leonard and Virginia Woolf] and must therefore be careful to seal it up at once, not leave it lying around open. "I'm *glad* I wrote that," Morgan said, and this was one of the few times he showed any resentment. Speaking of Vanessa Bell he said she was much easier to get along with than her sister, and how Virginia would suddenly turn on you and attack you.

Morgan said he had liked getting the Order of Merit. I said, "It's the only decoration really worth getting," and he said, "I've come to feel that—now I've got one." He said he had been rereading some of his early writing (I should have asked him what, exactly, but like an idiot I didn't) and he had liked it very much. He added, "The creative power has gone now, but I don't mind." He has a marvellous flair for saying such things without the least pathos—merely with mild surprise. He then said that he hoped he'd "pop off quickly" when the time came.

We had tea with Mark in his rooms. A gaunt, long-haired rather attractive [...] geneticist from Caius named Richard Le Page was there. Morgan remarked three times to me how pale he looked, seeming quite concerned; "He must be ill."

Our meal in hall in the evening was bad; the meat was tough. Morgan got quite enraged. "How awful that I should have brought you here to eat this *filth*." There one saw a characteristic flare-up of senile, childlike rage, and he sulked a bit during the rest of the meal. There we did seem out of the picture. Term is just over and the only remaining dons were young—they seemed to be mostly mathematicians with Ban-the-Bomb attitudes. (The cutest of them, Denis Mollison, volunteered to drive me back to the station and was charmingly friendly and genuinely concerned about Morgan's health.) The two estates of the college are now divided in a different way. The undergraduates don't have to stand when the dons come in. They eat between certain hours as they like, with self service. The high table is down on their level, moved to the other end of the hall. In a year or two there will be coeducation. And the lady dons will move the high table back and reintroduce all the protocol and ceremony, no doubt.

This visit to Morgan was really very moving. Yes, he has survived, he is past ninety and he functions. But at what a price! How slow and how alone! It is his speed that isolates him, for he is surrounded by people. He has fallen out of the running. I think he would really like to stay with the Buckinghams all the time and be made comfortable. But perhaps not. He looks after himself with amazing doggedness, taking ages to switch off the light, pick up clothes from the floor, shut the door of his rooms. He is under sentence of death, just as visibly as if he were lying on his deathbed. And yet he enjoys conversation, affection, food, sherry. He told me he had put his homosexual stories out of his mind (I think he only meant, had ceased to think of them) but when I talked about *Maurice* he showed pleasure and he told me he was glad to think of all this again and wished he could write another such story.

March 16. A girl named Catherine Cook came to see me about a thesis she is writing on my work. She wore a maxi coat or robe or whatever, which looked like a tacky black velvet you'd get from the Goodwill.[1] She was pretty, quite intelligent, but overly gushy. I was just too informal for words—served her coffee in the kitchen and we sat at the kitchen table. Her father and mother (a Belgian, and even gushier on the phone than Catherine) were waiting in a car several blocks away, up towards Sloane Square; maybe they had instructions to call the police if she didn't return before nightfall. Anyhow, it made her seem helpless and coddled—which she didn't at all have the air of being. I talked volumes, chiefly about other people, and she never tried to bring me back to the subject. Perhaps she merely wanted to meet me.

Then I went to see Nick Furbank. I have met him before but I didn't in the least remember what he looked like. He is pale and he stammers. Am not sure what I feel about him. His face isn't altogether a face one trusts. I told him right away that I don't want Morgan's letters to me to go into the book he is writing. I said they were too personal and really much more about me

[1] I.e., second hand; see Glossary.

than Morgan; but of course what I mean is that I would like to turn them into a book, myself. Nick seemed rather surprised when I said I felt we couldn't publish *Maurice* without Bob [Buckingham]'s permission. And perhaps I am being overconsiderate about this; because Bob's objections, if any, would really be May [Buckingham]'s. (Joe Ackerley told me, during that meeting in Coventry in 1967, that May was now declaring that Bob had never known what homosexuality was until quite recently, when Morgan had told him—and Nick says she takes the same attitude now.) But Nick also says that Bob has so far raised no objections to his book, in which Nick intends to be absolutely frank about sex.

He lent me two of Morgan's later stories, the one which takes place on board ship and one I haven't read before, called "Dr. Woolacott."[1]

Then I had supper with Bob Regester and Neil Hartley at a Greek restaurant just below the Post Office Tower. So Bob and I went up the tower, mildly drunk and daring each other, but it really isn't very giddy making—especially at night—unless you stand right up against the railings at the open section. Bob is fun to be with on such occasions but I feel I'm apt to overdo the aged schoolboy role. Also my conscience is pricking me because I haven't been working at anything but just pleasuring myself, as Don calls it. I keep thinking about him and hoping he's all right. Sometimes I wonder, is he unwilling to come back here because of [his friend]?

Last night I dreamt of being in an earthquake. Wasn't really scared.

March 17. 6:45 p.m., have just finishing talking to Don on the phone. This time it was much more cheerful and it seems he got a very good notice and also Irving Blum wants to take the show to San Francisco and New York. But Don wants to come back here soon, so maybe he'll join us all on this trip to the South of France for Easter.

Yesterday David and Peter and I had lunch at Marguerite [Lamkin]'s. I'm really fond of her but this lunching is ghastly. She *cannot* resist inviting lots of other people and stuffing us with food and creating an anti-oasis in the midst of the day.

Then we went to the Richard Hamilton show at the Tate

[1] "The Other Boat" and "Dr. Woolacott" both published posthumously in *The Life to Come and Other Stories* (1972).

and Peter and I went to *Zabriskie Point** and had supper together at Odin's, and Patrick Procktor came in, fresh from India, and covered me with effusion. I didn't snub him but it was embarrassing, after the vile way he behaved to Don.

Today Clement brought the newly typed scripts around. Very few typos and I think it reads well. He *still* wants it lengthened, but I said not until we are going into rehearsal. The play is now being sent to Alec Guinness and to Donald McWhinnie and Peter Gill as possible directors.

March 18. Peter Gill is out of it, I just heard from Clement. He'll be working in Canada until much later.

Have been lunching with Bob Regester. Vanessa and Corin Redgrave† and their daughters were there. Also Rory Cameron, whom I met at Marguerite's lunch. He's rather a nice man, writes travel books, lives on Cap Ferrat. Vanessa seemed quite ridiculously large and thick thighed and hoydenish and she sort of mother-handled the little girls, like a great big tactless nanny. We went ice skating, which I enjoyed greatly, though I thought old Drub's shaky ankles would never survive. Have been so hugely fat, lately, what with enforced drinking, bread eating and rich desserts.

David Plante really is a darling. I'd forgotten how dark-eyed and vulnerably American he is. I spent a lot of yesterday skimming at top speed through his novel, which I'd tactfully bought. And then

*I suspect I'll like *Zabriskie Point* better the second time, as often with Antonioni. Of course the facile oh-it's-all-ghastly attitude to California is a bit irritating, but I liked the student riot sequence better than the Death Valley part which I find, oddly enough, visually disappointing. And that celebrated orgy seems arty and unsexy in the worst way. And the boy [Mark Frechette] is a bore. His act is "Son of James Dean." As for Richard Hamilton, I want to see that show again too, with Don. He slows me down and makes me look. At present I like best some of the beach scenes, the "Swingeing London" series, some of the automobile pictures and the "Cosmetic Studies." Also that afternoon I discovered Atkinson Grimshaw's "Liverpool Quay by Moonlight, 1887." [For "Swingeing London," see Glossary under Richard Hamilton.]

†I was watching Corin Redgrave carefully, wondering how he'd do for Oliver. He looks very young, too young, but is the right type more or less. He is now acting in a T.V. version of [James Elroy] Flecker's [play] *Hassan*. So is Gielgud. Corin remarked that he felt Gielgud was doing the part of the Caliph all wrong (not making him into a perverse and exhausted old man who can only get kicks out of torture) but that of course he couldn't tell Gielgud this. He and Vanessa are writing a play which brings in Garibaldi.

he brought me a signed copy.* We had supper together. Nikos [Stangos] didn't come, because an aunt of his has died. David said that the aunt was a wonderful woman but her death (cancer) had been long expected and he couldn't *quite* understand the tremendous upset it caused Nikos. We agreed that Mediterranean people use grief as a ritual to somehow propitiate the spirits of the dead—that there's this always in addition to what the mourner naturally feels. David is eager that we shall all get together, but I have a hunch I won't like Nikos so much, this time around. He sounds so domineering. David says of himself that he is very jealous—so much so that he never wants to do anything to make Nikos jealous of him, for fear of reprisals! Stephen [Spender], that old monster, has been urging David to marry—telling him that if you don't you miss a great experience. The boys see through him but adore him.

March 19. It turns out that Rory Cameron was the driver of the car which wrecked and so severely injured Norman Prouting, years ago, on the Riviera. Like the other passengers, all rich people, he apparently neglected Norman completely while he was going through the subsequent operations—Norman still limps slightly—and never offered to help him with money. At least that's what I infer, because of Norman's reaction when I mentioned Rory's name to him last night; he froze up solid.

We went together to see [Shaw's] *The Apple Cart*. John Neville was really excellent as King Magnus. He *could* play Patrick, but his face is wrong, there's a dryness in it, hard to imagine him being physically vain.

This morning Clement told me that two other German theaters are interested in our play. The script is now with Alec Guinness and Donald McWhinnie.

I've just had lunch with Moore Crosthwaite, at his house near Clapham Common. He has much of the style of a former British Ambassador to Stockholm and Beirut but this merely modifies what might otherwise be a too screaming queenishness; it strikes a balance and makes him human. He's got an American friend living in the house and old Herbert List staying there for a few days.

*David Plante's novel *The Ghost of Henry James* certainly has something; it is about a family and its interrelations, written impressionistically, in very short chapters, skipping from one character to another. It has a lot of skill in it and is sophisticated—in a rather old-fashioned way—but not the very least bit pissy assed or closet queeny. Bits of description are often good. It reminds me much more of Virginia Woolf than of James. I wouldn't be a bit surprised if he turns out to have considerable talent.

Herbert, increasingly pouchy and full bellied, is keener than ever on collecting prints. Much was spoken against Warhol's films and in praise of Hockney's pictures.

Speaking of Hockney reminds me that he told me today on the phone that he had just called Don to try to persuade him to come to France with us. I don't quite see how Don can do this unless he takes off tomorrow, since he wants to use an excursion ticket and that is no good for weekends. I'll probably hear from Don himself in the morning. He had told David that he would first have to ask Irving Blum if he was needed in Los Angeles, but maybe this was an excuse. David had told Don to come and draw French food, and then show his drawings in the States to attract American gourmets to France. This is the sort of superficially silly sounding remark which actually reveals the shrewdness of David's character, because one can quite imagine him literally doing that and making money out of it. (Indeed, the other night, when I was having supper with Peter at Odin's, the proprietor Peter Langan asked me to contribute to a pornographic cookbook he is preparing, and said David has promised to illustrate it!)

March 20. Talked to Don this morning. I don't think he really wants to come back here, and not particularly to come with us to France, which seems set for Monday. We had one of our best kind of conversations; everything we said, even the details about calling the Maltins[1] to see about the second payment on the property tax, was full of love. I can truly say, with Patrick in our play, "How lucky I am!"

Last night I took Nancy [West], Joe Ackerley's sister, out to supper at The Hungry Horse, along with Nick Furbank and a quite nice young artist who knows Mark Lancaster, named Richard Shone. Nancy was very lively and seemed quite contented with her life, though she talks about Joe continually. She really is an amazingly handsome woman, for her age. Shone, who's about twenty, is a great talker and all went merrily. We even got the desirable table in an alcove. But Nancy is a demon pourer of drinks—I had the wits to resist them, but Nick Furbank turned clay pale in the restaurant, said he had to get some air and then fainted. However, this morning on the phone, he sounded all right again.

March 21. Shortly before eleven this morning, I decided to dial

[1] Arnold and Mrs. Maltin (not their real names), Isherwood and Bachardy's accountant and his wife.

Tony Richardson in London and ask him what his number is in the South of France, so I could give it to Nicholas Thompson and Clement, in case of emergencies. When I called the number there seemed to be a lot of confusion, a foreign voice answered, I asked was this 536 6933 and the voice said yes. I asked for Tony and the voice said, "Have you seen the paper this morning? I suggest you look on the front page of the *Daily Mirror*—there's a picture of him." I tried to ask more questions and he repeated, "I suggest you look in the paper."

Well then of course I thought, Tony's been killed in a car wreck—or at best he's been involved in some gruesome scandal, and I decided it must have been the voice of Jan [Niem], Tony's Polish chauffeur, telling me this. But then I rang Peter Schlesinger and he got a *Mirror* and there was no picture of Tony and nothing about him, and *he* rang Tony's number and the housekeeper told him Tony is expected back today, so the whole thing seems to have been a false alarm. Maybe I got a wrong number and the man who answered was a malicious practical joker, or drunk or high on something.

Now we have air tickets for Monday, with a stop in Paris. We're supposed to arrive in Nice late, around ten or eleven, and find a hired car waiting for us in which we'll drive off somewhere without delay. Obviously these arrangements may well break down and so I'm somewhat dreading the trip but at the same time looking forward to travelling with David and Peter.

Yesterday Bob Regester and I had lunch on the Post Office Tower. The outer part of the dining room, on which the tables are, revolves; the inner part doesn't, which is somewhat sick making. Also, there is one phase of the revolution which is unpleasantly bumpy. The food is terrifically expensive; our meal cost six guineas. The overall prospect of London has been largely spoilt by all these towers. St. Paul's, Tower Bridge, Westminster Cathedral, the Abbey and Parliament are now completely dwarfed. Afterwards we joined an attractive young Australian, John Hopkins, who helped me buy a Burberry. Hopkins worked on the crew of *Ned Kelly* and we met him in Australia, though I didn't remember him exactly.[1]

A very happy supper with David Plante, Mark Lancaster and Peter Schlesinger at the Carrosse. We had a table downstairs, half a long Victorian desk-table in fact. It was divided in two by artificial plants and a rampart of leather-bound books, including [Willard]

[1] He was an actor, and played O'Connor in the film.

Motley and Montaigne, so that you could hardly see the people dining at the other end. This kind of device, so amusing and original, is absolutely unthinkable in the States. It would terrify them if you suggested it. It's an utterly alien style of camp.

I must keep reminding myself how disgustingly fat I am, about 156 pounds. All very well to say don't eat, don't drink; all these meals and drinks are practically forced on me, socially. And tonight there's this big party being given "for" me by Robert Medley. The only thing is moderation, but that's hardest of all.

God is very far. Don is near. I think of him increasingly. The day before yesterday, I think it was, I saw [his friend] on the street, between the Royal Court and the Underground entrance. I was going into the Underground and had to pass quite close. Probably he saw me. He looked very bald.

God, how wretchedly cold it still is! The heaters don't help much and anyhow I begrudge the money. This is what spoils London, despite all its other charms. Never underestimate the power of this mild dampish cold. It creates puritanism, primness, disbelief in Love and God and a taste for mini-art.

March 22. A little sun, this morning. I've started writing this in the hopes that it'll make Patrick Woodcock arrive; he is coming to take me to lunch. Tomorrow morning, David Hockney, Peter and I are to meet at the West London Air Terminal at 10:30, to start our trip. It's really very exciting and I am looking forward to it, despite my constitutional dislike of travelling.

Have talked to Amiya [Sandwich] in the country (Chard, Somerset) and to Dodie Smith; shall have to see both of them when I get back.

Lamont Johnson has been here. He has some prospects of making a film out of *Black Girl*.

Later. Have had lunch with Patrick. He would like me to take his friend David Mann out and find out what the problem is between them. I asked him, "Do you like him?" He said, "Just for a snap judgment, I should have to answer no." He said of Don (admiringly) that he was "beady," and that he knew Don hadn't liked David. Don had given him some good advice: "Your tempos are different."

Bob Regester says he probably isn't coming to the Nid de Duc because Tony doesn't want him. He thinks this is because Tony doesn't like John Hopkins, saying, in effect, that he's a cockteaser.

Last night I had a drink (soda water) with John Cullen of

Methuen. John Cullen doesn't like Henry Heckford's book,[1] says it simply isn't intelligent enough and that he doesn't understand how to write a critical essay. He wanted me to read it. But I don't really want to, because then I will have to condemn it, which might even mean my rewriting it. I told him about Alan Wilde's book[2]—then felt rather a traitor because I'd done so. Cullen is wild about Brecht.

A big party last night, given for me by Robert Medley. He had worked for hours and produced an astonishingly impressive amount of edible food and delicious boys. But oh, the terrible schizophrenia of parties! One is pulled in all directions by the chatter. My haven was the Hockney gang, including Wayne Sleep, Mark Lancaster and his twins,* Karl Bowen and a black-haired boy† who shares a flat with Wayne. False Patrick Procktor danced memorably with closed eyes, like a very tall zombie; Christopher Gibbs[3] in strangely embroidered boots, streaming with sweat, brought a greeting from Mick Jagger; a boy in silver pants contorted like a serpent; a Congo prince wore a sweater which showed a bare midriff; Keith Milow was one of the most attractive people present, in the manner of Mick Jagger; a bearded bore named Philip Matthews[4] wanted to know what had become of Bill Coldstream's painting of me.‡ (This morning, in *The Observer*'s colored supplement, is an article on David called "No Dumb Blond," with his painting of Don and me included.) Throughout the four and a half hours I stayed at Robert's party, I didn't drink anything but a little grapefruit juice! (But today at lunch, wine, alas.)

*David (Mark's friend) and Tony Meyer.

†Jolyon Jackl[ey]. [An aspiring actor; son of music hall star and comic film and T.V. actor Nat Jackley (1909–1988).]

‡Bob Regester, in a crimson pirate blouse and a jewelled belt; John Hopkins with a gold band in his hair, Gregory [Brown] looking like a plump white-skinned lesbian in a mannish velvet jacket; a black boy in a shimmering white coat, diamonds on snow; Patrick Procktor's friend Ole [Glaesner] with heavy silver and blue eyelid-paint.

[1] About Isherwood's work.

[2] *Christopher Isherwood* (1971); Wilde had already published *Art and Order: A Study of E.M. Forster* (1964).

[3] London antique dealer (b. 1938), intimate of Jagger and other members of the Rolling Stones.

[4] British painter (1916–1984); he studied at the Euston Road School and privately with William Coldstream and taught at the Camberwell School of Art, 1946–1981, serving as head of Fine Art from 1964.

March 23. 7:30 a.m. and pouring down rain, the heaviest since I've been in London! Thanks a million. So now I'm washed, shaved, dressed, with nothing to pack but a little bag and nothing to do but go. I wish I could hear from Don first but somehow I feel unwilling to call, it'd be just spending more phone money. I long to tell him how much I love him.

Supper last night with Robert Moody and his wife Louise. Like so many of my contemporaries, Robert looked like a little old man when I first met him, the pretty blond hair turned white and gone from the crown. But then the boy reappeared from behind the wrinkles. Robert's mannerisms, I now realize, always had something in them of the old man, he was like a young actor playing old age and poking fun at it—screwing up his face and puttering around. He has had heart attacks but is all right now. He is a Jungian, has actually known Jung and has been president of the international society of Jungian analysts, whatever that is called. He told how he had sat next to Jung at a big dinner and Jung had said to him, "You needn't feel you have to talk to me—when one gets old one is grateful if one isn't talked to." Jung had confided to him that, "When one's old it is over very quick," referring to sex acts.

Our evening was evidently a great success, as far as Robert was concerned. I found it hard to read Louise. She sat there, motherly, rather plump. Robert kept referring, rather awkwardly, to his two other wives. All his stories seemed to include them. It was impossible to say if Louise minded this, but Robert was certainly apologetic about it.

8:30—Don has just called! He seems to be seriously considering coming over, via Nice, in a day or two. I only hope, if he does, that there won't be some ghastly mix-up. Everything you do in France is just twice as difficult.

March 24. Prophetic words! We did get to Nice yesterday and indeed all the way to Tony Richardson's house, but both our planes were hours late due to an airport strike at Orly and we didn't reach Tony's till nearly three this morning. In Paris we only had time to wander about for a while in the Latin Quarter, see the newly cleaned Notre Dame and visit two friends of David and Peter's—Jean Léger and Alexis[1] who were getting ready to attend a supper party in drag. The guests, they explained, were all "serious" business people who had never been in drag before. They had got

[1]Jean Léger worked at Helena Rubinstein, the skin care and cosmetics company, and Alexis Vidal, his lover, at the couture house of Philippe Venet.

themselves lent very handsome Chanel clothes for the occasion, and some shoes which had been made for black actresses and were extra large.

It rained heavily in the evening and the wait at Orly seemed interminable, almost.* To my surprise, when we did at last get to Nice, the rented car was still available, and David, with magnificent determination, drove us all the way to Le Nid de Duc, through fog and up into the mountains. Today, following his "whiz tour" schedule, he drove us to Carcassonne. I'm writing this at the Hotel Donjon in the Old City. No more for now, because I'm exhausted.

March 25. Yesterday, at the Nid de Duc,† I was woken by a rooster and several peacocks; this morning (around six) by a water lock in the next-door bathroom pipes—I suspect David of getting up to scoop the sunrise in his eager beaver way.

I must say, after yesterday, I feel as fond of him as ever or rather more so, and this despite all the frictions of travelling. He actually drives very well but very fast and the constant overtaking of cars on the narrow three-lane roads was a great strain on my nerves. What *is* the sense of rushing like this, just for the sake of making it to Carcassonne? What a joyless kind of pleasure, tearing past the umbrella pines, the vines, the hilltop towns, the red Cézanne landscape, the narrow streets and the plane trees. Yet David himself remains lovable, so good-humored even when he is scolding Peter, so full of fun and enthusiasm.

And Peter is lovable; I really like him better than before. He doesn't sulk, he isn't merely vain, he wants to know things, his squabbles with David also have an underlying good humor, he nags nicely. His nagging is chiefly about David's untidiness‡ and David's needless anxiety that we'll miss the way, and David's overeating. The question of food affects both David and me. David has been so looking forward to the great French meals, and yesterday we

*David said later that he didn't really mind waiting at Orly because, after all, we *were* being delayed by a strike and he was on the side of the strikers.

†Yesterday morning, as I lay in bed after waking, I had a strong but not frightening feeling that the Nid de Duc bedroom was haunted—that is, I felt a crowded sense of *presences* there. I stupidly told this to David who was simply bewildered. He couldn't believe I meant it. To him, such things are unthinkable.

‡Also, Peter has style, he keeps up an appearance, washes his hair, makes an effort—in this he is a disciple of Don. His seemingly indestructible snakeskin jacket (which even resists rain) always looks good on him.

actually ate lunch and supper at two restaurants with one Michelin guide star*—the Jules César at Arles and the Logis de Trencavel on the outskirts of Carcassonne. We had cassoulet at the latter, which David loves; at lunch I had rabbit. Was careful not to eat much of anything but one gets stuffed nevertheless; at supper we all consumed a fat-making but disappointing bombe.

At Arles it rained, but we managed to see the Roman amphitheater. By far the most attractive place was Sète, a harbor town between Arles and Béziers with real busy untidy life in it. Most of the places we passed through seemed strangely shut up and dead.

David keeps running his fingers through Peter's hair, no matter how many people are around. This embarrasses Peter but he doesn't get really annoyed. He still loves David very much, I feel.

As for Carcassonne, we shall see more of it this morning.† But its great walls, though awe inspiring and even magnificent from some angles, depress you. How awful to have to be shut up inside this gloomy place, amidst this beautiful landscape, for fear of the neighbors! And nowadays we are shut up too, inside ramparts of rockets.

March 26. Have just spent the night at Douglas Cooper's improbable home, which is a château and also a museum of Picassos, etc. I really don't know that I even want to describe it, yet. Like all such places it is basically uncomfortable and intimidating.‡

*We decided that one of the marks of a "great" restaurant is that the glasses are not merely clean but flash in the light like diamonds.

†8 a.m. (March 25) David and Peter not up yet. Thought I would go and pray for us all in the church. But the church wasn't open, only the curiosity shop around the corner. Everything else, the restaurants, the other hotels, are locked up. The place is quite dead. But, in the morning, death seems calming and even refreshing, you breathe the cold pure breath of the dead stone.

‡At supper at Douglas Cooper's there was a French senator, Madame Crémieux [Suzanne Crémieux-Schreiber (1895–1976), Radical Senator for the Gard region, 1948–1971] and her friend Madame Bernier. It wasn't Cooper's fault they were there, they had been invited long before, and he hadn't been expecting us definitely. The two women spoke English and yet, like most Frogs, they wouldn't, despite the fact that David and Peter have hardly a word of French. So I got bitchy and would only answer them in English, until Madame Bernier asked me about my books and what they were called in their French translations. That started me off and I betrayed too much knowledge of the language, so that Madame Bernier said bitchily, "You speak beautifully—you speak as much as you wish to." David didn't condemn me however.

And indeed this trip has made me love them both even more than before.

Douglas Cooper talks French abundantly, with an American (Texas)-Australian accent. He bitches nearly everybody. When I said how kind

On the way here yesterday, we had a succession of memorable experiences—lunch at a dear little port in the Camargue called Palavas, the mad-looking freak lopsided new workers' apartment houses at La Grande Motte, a glimpse of the wild white horses of the Camargue at some place in between (most of them were black or brown) and marvellous Aigues-Mortes. (Aigues means "waters" in Provençal.) Then we took to the Roman ruins of Nîmes, which I find a drag, then we reached Douglas Cooper's.

(Actually this passage was interrupted by the charming entrance of Peter through the window of my room, soon followed by David; they had been walking around the battlements.)

March 27. So then we went off to Les Baux and ate at the notorious three-star restaurant,* the Baumanière, and then visited Natasha Spender† at their nearby house and then drove back to Nid de Duc, where we found a huge cast of guests, including Edward Albee, his friend Roger Stock, Bob Regester, Tony's friends Will Chandlee and Giancarlo Cesaro, Anna O'Reilly the ever faithful secretary, Diana Dare, a girl who is said by Bob R. to be stuck on Tony, and Jan the chauffeur, and Jean-Pierre the housekeeper[1]—oh yes of course and the two little daughters of Tony and Vanessa who I went skating with on March 18.

Much to be written about all this, but for the moment suffice it to say that I decided last night I couldn't bear to stay in my room, because I could hear every word that was being said in the living room below where they were playing cards, so now I've moved to the bottom of one of the other buildings where at least I am quite private.

Today I went into Nice with Bob Regester, in the hope that Don might come in on the plane from New York via Paris (he didn't) and we met Neil Hartley and Nureyev's friend Bob Cohan. Now there is a mistral. I realize I am being incoherent and there is

Maugham had been to me, he said, "It's curious how nasty people do kind things."

His house is *stiff* with Picassos, not to mention all the other treasures. The wall with the sandblasted Picasso drawings is wonderful and unique and somehow hateful.

*I fed the two fat three-star pussies bits of my Coquille St. Jacques, to the disgust of the waiters.

†Natasha seemed calmer than she usually does. The only somewhat hysterical thing she did was to make a fire by burning an old chair in the fireplace! David promptly photographed it. His camera clicks continually.

[1] And estate manager.

so much more to write, but I'll leave it for now. Am rather drunk because of Blanc de Blanc[s] at lunch with *moules*.

March 28. The mistral has raged all night. Now it's a quarter past seven. I feel very snug in the tiny bedroom I've moved down to; after that other room I was in, which had three doors (one of which wouldn't stay shut) and the living room below and David walking in and out, I can appreciate the pleasure of privacy. This is actually quite a big building I am in, but it can be entered at different levels and I might as well be alone.

Le Nid de Duc is a former village of about eight houses, old stone buildings on the slopes of a steep valley, amidst cork oaks, with a stream tumbling in occasional waterfalls. Low wooded hills rise into a gap of light toward the sea. Ninety people used to live here, once.

In this building are (I think) Edward Albee, his friend Roger Stock, Bob Regester and Neil Hartley. Edward is a bit of a square, likeable, professionally aggressive, "regular"—he plays bridge, for example, as though on principle, and he gets politically angry on principle. Last night at supper there was a loud mock quarrel (well, fairly mock) between him and David. David was supporting the Soviet Union and China and calling the USA fascist, Albee was saying he wouldn't have the U.S. attacked, it was the only place where true socialism was being practised, at least to some extent. They were both absurd and much shouting followed, in which Tony joined. Tony began taking that masochistic attitude toward art, saying that it isn't important, only feeding and breeding are important. David said the arts would become far greater when everybody in the world is free. Nobody talked about religion. It disgusts Tony and David scorns it. Albee might make more sense. I'll bring up the subject if I get a chance. After supper "the men" (Tony, Albee, Neil, Jan) played bridge. I proposed to David that he should write a poem* while I drew a picture. I drew two—so bad they weren't even funny—of Diana Dare talking to Giancarlo Cesaro and of David himself. Then we tried to play guessing games and failed. David got drunk and flashed a flashlight at people saying, "We need some light on this subject."

*David said his "poem" was actually the beginning of a novel—it's a description of me drawing. Here it is: "He held the red pencil like a piece of heavy chalk, only with the black pencil did I notice any delicacy. Peter thought this was as bad as his drawing. Now I can see that it is. I am impressed though by the straightness of his back and the fact that he looks for five seconds at his subject before translating his marks from his mind. What a mistake I am making. I have no need to look when I am writing."

Bob, while driving me down to Nice yesterday, said that Diana Dare is out to get Tony. She would like to be a bitch but can't "get it together." She thinks of herself as a beauty and expects homage. Giancarlo *is* a beauty of sorts—he is a very good-natured hustler with Italian good looks and a huge splendid body (guessed at, admittedly, beneath a thick sweater). Will Chandlee is very good-natured also. Bob Cohan a bit pretentious. He is very much the official friend of Nureyev (for the past few months) very protective and possessive. He told how Nureyev is constantly afraid of being kidnapped by the Russians. They tried it when a plane he was on made a forced landing in Egypt. He locked himself in the john. Bob says that Roger Stock is very shy because he feels out of his depth. He is nearly pretty* and yesterday, when he came out of the bathroom in his jockey shorts, I saw that he has very well-made smooth strong legs and sexy buttocks. (The boys, of course, all dress in shimmering cocktail pants, big jeweled belts, blouses, scarves—which they change throughout the day.)

Bob says he had a great showdown with Tony, who begged him to come to Le Nid de Duc and promised to be pleasant. Bob thinks Tony is jealous of him because of his relations with Neil, Vanessa, Jeanne Moreau, Tony's daughters—and even with me! Tony vents his spleen by calling Bob a kept boy.

Yesterday, as we approached Nice, the Alps were magnificent, finer than I have ever seen them, in thick snow.

Where are you, Don? I had such a strong feeling yesterday, almost a certainty, that you were coming. In a way, your being here would make things more difficult—that's to say, as soon as you were here I'd want to go away with you and leave all these people. And yet I'd like them to see you in all your wonderfulness, your magic and style and beauty.

March 29. A bad hangover. Last night I got stupidly drunk on wine, chiefly because I'm so disappointed Don isn't here. Now I want to get as quickly as possible to London, where I can at least talk to him on the phone. Drunkenly careless, I let my pillow fall against the electric fire while I was out of the room brushing my teeth. So it was burnt.

*The defect in his face is his ugly nostrils and badly placed nose but he does have charming blue eyes. His voice is Cockney and his manner ungracious, trying hard to play it cool. Tony is said to dislike him. He used to be an English disk jockey in the States. Now he sells British men's clothes to America. Albee says that, when he goes away for the weekend, he brings ten times as many clothes as he can possibly wear.

David and Peter and I drove into Cap Ferrat yesterday and had lunch at Rory Cameron's beautiful little house.* Marguerite [Lamkin and her companion] were staying with him. [Her companion] nervously active, walking all around the countryside and playing tennis; Marguerite insisting on having her hair done and going to parties she pretends to hate.

It was sadly nostalgic to pass the faded sign on the gateway to the Villa Mauresque and to look down from the garden of Rory's former house[1] on those very rocks where Don and Willie sat together and were filmed, during a walk.

This morning we went up to the village, La Garde-Freinet.† It was icy cold, with a strong mistral blowing. But then the sun got hot and we all lay in it, and some swam in the heated pool—but not Roger Stock, unfortunately. What with him and Peter and Giancarlo (the splendor of whose body is no longer unguessed at, he stripped for sunbathing) and fattish but muscular Will Chandlee, this is quite a sexy gathering, and there's nobody here I don't like, in fact I definitely like most of them. Linda Austin, the children's nurse, is very sweet. Her face and nice candid eyes are quite adorable. Peter and David and I are thicker than ever.

Today, for the first time, the peacocks opened their tails—at least one of them definitely did. He only likes the white guinea hen and shudders all his feathers at her, constantly inclining his tail to present her with the maximum hypnotic spectacle. But Tony says they can't mate.

While we were sunbathing, Tony decided he wanted to draw on Peter's bare back, so colored pencils were brought and he did. This was so typical of his milder sort of sadism. It apparently hurt a bit but Peter rather enjoyed it; it excited him sensually to have this done to him in front of so many people. Tony then began to suggest that maybe it wouldn't come off. He had drawn a rhinoceros, a flag and a flower.

*Rory has rather horrid treasures. I mean there is something distasteful about a grown man collecting such very expensive toys. This is unjust but I felt it. David's socialistic puritanism was also aroused by Rory. And on the way home he said, every time a car with a Monaco sign passed, "There goes another tax dodger!" But his indignation is always good-humored and self-mocking.

†While there we saw the Easter Parade, the girls stepping like drum majorettes all in white, many quite pretty, the men of the band ugly, gnomelike. Tony says that this was originally a Saracen village and its inhabitants are many of them obviously Arab.

[1] La Fiorentina; see Glossary under Cameron.

An Easter breakfast this morning with many chocolate eggs for the little girls. Bob Regester had bought a scarf for each of us, we had to choose which one we wanted. Tony chose the same one as I had, without hesitation; but he never knew this because I hastily picked another.

March 30. The little white cell with black-grey damp stains on the rough plaster, the small terra-cotta tiles, the tiny narrow bed, the warped reflection of the gilt-framed mirror of an old thing in his spectacles with a ropey neck who writes these words compulsively, to fill this book, to do his daily task, because that is his nature.

This morning is glorious. No mistral. The sunshine coming up hotly. The peacocks crying, the Chinese fighting cocks crowing.

I now feel real goodwill from Albee and some guarded friendliness from Roger Stock. They leave for England at midday. Albee has never kept a diary, says he can never make his characters use bits of prerecorded dialogue. Roger described a game they all played last night—someone was blindfolded and the others touched him and made noises and he had to guess who it was. This, it seems, was terribly difficult.

I had gone to bed because I had a stomach ache and felt sick. Today I'm pretty much all right—this seems to have been a warning against drinking and general gluttony.

Giancarlo Cesaro gave a truly marvellous performance, yesterday evening after supper, in bits of *Carmen* and Italian popular songs. He is a classic androgyne, with his Michaelangelo body and feminine soul. It wasn't merely inspired camping, though much of it was very funny it was also startlingly beautiful.

Later. This afternoon we (David, Peter and Bob Cohan) drove down to St. Tropez and wandered amongst the crowds. The old port has been restored after World War II damage and it looks just like I remember it when I came here with Olive [Mangeot] and the boys in the twenties. So does the square where we stayed. But of course it is now crammed and very expensive.

Bob Cohan wanted to talk about Krishnamurti, in whose teachings he has been interested for years. I like him much better now.

The builder came to give David an estimate for fixing up one of the buildings for him as a studio. It amounted to about six thousand pounds, exclusive of the bath fixtures and the installation of floor heating. I can't help feeling that this project is another of Tony's snares. He would like to have David around as a tame painter, but is it worth all that money to David? Won't it end up badly, like my garden house at the Hookers'? And, talking about money, I find it just a trifle unsplendid that we have to pay for our air tickets both ways—sixty pounds in my case.

March 31. Last night Tony said that he hadn't made any money on a film for seven years and that he lost all the remaining money that would have come to him from *Tom Jones* (a million) because he had to put it into *Laughter in the Dark* when the backing was withdrawn after he'd gotten rid of Richard Burton.[1]

He says he doesn't like Albee personally; "I don't like anything about him."

He very firmly made it clear to David that it wouldn't be convenient for him to hear the work on David's studio going on in the early summer, because he, Tony, will be here and it would disturb him. David took this quite calmly, but then he said that he will have to consult his brother about the money, and I began to hope he will wriggle out of the deal, in his canny Yorkshire way. What would happen if Tony sold the property at any time?

Well, today we're off, and I'm glad to be getting back to London, although this trip—particularly the tour with David and Peter—has been memorable and pleasant. Tony is in a strange withdrawn mood, jealous perhaps, resentful perhaps—but of whom or what? He has never mentioned *Claudius* or excused his arbitrary and irrational behavior to us about it. Yet I'm still more than willing to be friends with him, any time he wants.

I quit drinking two days ago.

April 1. Feeling a bit depressed. The weather is so loathsome, wretched snow and cold and rain. And although David's show*

*Drove to the show with Marguerite in her Bentley and a girl called Sandy Campbell and Peter and his friend Jean Léger (the one we met in Paris). Marguerite suggested that they should bring the silver goblets and shaker with their Bloody Marys in the car. This made two black hippies in a neighboring car clamor for a drink and then the Bloody Marys were upset over Marguerite and Peter.

Nothing to say about David's pictures, because I hardly saw them for the crowds. Must go again. The show disturbed me for another reason: David's overwhelming success must make anyone who loves him (and I do) afraid. Surely the world will make him pay for it, cruelly. For David is not only a "golden boy," as the press calls him, but a crusader for his way of life, for our minority. (He kissed me on the mouth, without the least affectation, when I came into the gallery.) Many people must be gunning for him. [Hockney's retrospective "Paintings, Prints, Drawings 1960–1970" was at the Whitechapel until May 3.]

[1] Nicol Williamson replaced Burton several weeks into filming Richardson's 1969 adaptation of Nabokov's novel.

was really impressive when I saw it this morning, the total effect of meeting the people at it was depressing—my contemporaries looked so beat-up and the few beautiful young were unattainable and I felt excluded from both worlds. [Don's friend] was there. I made rather a thing of shaking hands with him, which I now regret. Why not have just turned my back on him, the silly faggot? (When I got to the flat, yesterday, I found a note from him saying he is "very unhappy and distressed about everything; whatever my failings have been I've always wanted the friendship between you and Don and me to be there." Ha ha. She ends, "I'm simply writing now hoping your feelings will change and that we can make it up.")*

My only world is Don and he isn't here and here is no place for me to be. But I must sit it out until something is settled about the play. We talked on the phone this morning. Jim Gates is now working for the Goodwill as an accepted form of alternative service for conscientious objectors. He has to rise at dawn, go clear downtown, work till 5:30, for which he gets only $1.50 an hour.

It poured yesterday morning on our way to the airport but cleared for the flight. We only had a short stopover in Paris and got to London about 6:00 p.m. Anna O'Reilly† and Bob Cohan came back with David and Peter and me. Anna is a very curious girl. [...]

April 2. Last night I went with Peter to *Tiny Alice*. Neil was there too and Bob Cohan and Albee and Roger Stock. I liked the play far more in every way than when I saw it in New York. David Warner[1] is perfect, perfect casting and an excellent performance. I've never liked him before. Irene Worth is as good as ever, though now she looks rather too old. (I went backstage with Neil and was surprised to find that she knew me quite intimately and

*Another unwelcome item in the mail when I got back was Dodie's new novel *A Tale of Two Families*. It makes my heart sink.

†As we were landing in Paris through an unusually thick cloud, I told Anna, who seemed slightly nervous, that there are two apparently crazy sayings I keep remembering in appropriate moments of stress—a pilot once told me, "You're really safer when you're inside a cloud than anywhere else," and Patrick Woodcock's dictum, "Cancer is really the easiest way to die." (He later modified this, referring specifically to lung cancer, see April 16.)

[1] British actor (b. 1941), trained at RADA, a member of the Royal Shakespeare Company; his films include Richardson's *Tom Jones*.

remembered Don; with actresses you never know.)[1] Also I thought the set was marvellous, especially when it was representing the library, with long dreamlike vistas of reflected bookcases.

Albee's dialogue impressed me. It is tricky and seems to belong to the school of Christopher Fry, the word mincer. But it has great style and atmosphere. I now feel I *almost* know what the play is about but I doubt if Albee does. We all had a pleasant dinner together afterwards. Roger was drunk and friendlier and now I feel I really like Albee and also Bob Cohan. And I always like Neil. But, oh dear, why do we have to go around in these tribes? I would so much rather be alone with any single one of them and *communicate*.

Dreams about Don last night. I was talking to him and he said, "I'm a terribly jealous person." I felt very happy because he had admitted this and so I could talk of *my* jealousy without its being one sided. Then I was surrounded by eels which kept changing into kittens. The eels repelled me until they changed, but when they did I loved them as kittens, although they still had a white transparent look, sort of eely, and indeed kept changing back again into eels. I commented on this to Don. Was it a dream about the doubleness of love-hate?

A charming note from Dirk Bogarde this morning, saying that he can't/won't do our play, which he claims to have "enjoyed enormously" ("I rather feel that I have done my bit for this particular theme—if you know what I mean"[2]) and going on to say how he wishes he had played Chris in *Camera*—"He is my Hamlet."

April 3. Still this hellish cold. Tomorrow I have to go down to Essex and see Dodie and Alec [Beesley], where it'll be snowing for sure. I have just finished her novel. It is *exactly* what I feared; one of those patty-paws romances, a little kiss here, a little wistful regret there, one affair is broken off, another starts up. Magazine writing. What's wrong with it, actually? It's so pleased with itself, so fucking smug, so snugly cunty, the art of women who are delighted with themselves, who indulge themselves and who patronize their men. They *know* that there is nothing, there *can* be nothing outside of the furry rim of their cunts and their kitchens, their children and

[1] Worth (1916–2002), American stage star, played opposite Gielgud in the original New York production half a decade before, December 1964–May 1965. Isherwood and Bachardy first met her in London in 1961, and Bachardy drew her twice that summer.

[2] Bogarde (1921–1999) played a lawyer pursuing blackmailers responsible for the death of a boy with whom he was chastely in love in *Victim* (1961), and he played Gustav von Aschenbach in Visconti's *Death in Venice* (1971).

their clubs. The book is rotten with Christian Science and I am indulging in the luxury of being brutal about it because I know I will have to be polite about it to Dodie tomorrow—I also know that I shall *want* to be polite, because I do respect her and she is indeed so much wiser and subtler and better than this silly book.*

Talked to Nicholas Thompson this morning. He feels that if we don't get some action by the end of next week on the play we should consider postponing it. And I really want this. I want to go home. I need Don.

Had supper last night with Richard [Le Page], the nice (and attractive) geneticist[1] I met in Mark Lancaster's rooms at Cambridge. He told me about the research which is going on to discover how the genetic information is encoded in the cells—in other words, how do the cells know the particular way in which they're to combine to promote a certain kind of growth. He's very tall and has a long nose and reminds me just a little bit of Jim Charlton. I took him after supper (at Provans) to meet David and Peter who were having a birthday (Peter's) supper at Odin's. It was a success, though perhaps not entirely with Peter; I felt he hadn't had much of a birthday, under the shadow of David's show. That's the blind side of David. He can't quite realize how Peter feels.

Richard [Le Page] has one very British mannerism which could become annoying. When you're telling him something he is apt to keep breaking in with "Yes ... yes ... yes ..." which sounds as though he had heard it all before and must therefore be bored.

April 4. Supper with Albee last night and Roger Stock and a friend of his named Brian. Not being alone with Edward, I couldn't ask him about *Tiny Alice* but I enjoyed the evening. We ate at the Hungry Horse. Roger was very friendly, almost flirty; he suggested we should have a drink together after Edward leaves on Monday. Today he is going to show his whippet at a dog show.

Heard last night that Ian [McKellan][2] has refused to play Oliver, says he isn't right for it. Meanwhile he praises the play. Incidentally, Albee was telling me how much trouble he had, getting *Tiny Alice* put on. His producers were all afraid of it.

*The most basic criticism of the book is, as so often, contained in the question: "Yes, but *why* have you told me all this?"

[1] Here and below Isherwood mistakenly wrote "La Plante"; also, Le Page is a microbiologist, not a geneticist. See Glossary.

[2] British actor (b. 1939), especially of Shakespeare, and, later, film star; educated at Cambridge. Isherwood wrote "McClellan."

April 5. On his way back from seeing me on the 2nd, Richard [Le Page] skidded on a patch of ice, knocked down a telephone pole and wrecked his car. He writes from Cambridge, "Rarely in one evening has my elation been so rapidly quenched!"

Yesterday I saw Dodie and Alec. Alec much fatter, quite bulging in both face and figure, but still tanned brick red and very healthy looking. Dodie has aged greatly; she's now almost a freak with her little white waving arms and her head sunk into her tiny trunk. But our meeting had a marvellous sense of continuity. Nothing had to be rediscovered or reassessed—we simply picked up the threads. The weather was foul, so we couldn't go out for a walk and get relief between the big meals. They have now adopted a second dalmatian, one of those problem dogs. It was alledgedly thrashed by a wicked U.S. sergeant at the air base, which made it for a long time impossible to put on a leash. Then various trial ownerships ended because people didn't like the way it broke things. So someone conned the Beesleys into taking it on. It jumped up on me and put its muddy paws on my precious Swami's scarf, so I was not charmed. (Whenever I see people like the Beesleys, whose "dog love" is really a blind spot of utter insensitiveness toward the feelings of other human beings, I start by feeling very affectionate toward their dogs and then begin to hate them; very soon I was flinging this one away from me with actual brutality. Dodie registered this and said to herself, for the one thousandth time, "Chris doesn't like dogs." And when, later, I really drooled over their four adorable little donkeys, I could see that Dodie found this a bit hysterical and an indirect insult to the dogs!)

She is wonderful, though, in her beelike energy. Now she has written two more plays! And her novel has been getting good notices, which excused me from praising it. Indeed she excused me herself, saying at once that she knew it wasn't the sort of book I'd like. I have given her our play, which excited her interest greatly in advance, but I can hardly believe she'll like *that.*

She almost never comes up to London, she says—with a certain implication of pathos. Don't know why. We spent a lot of time trying to remember the name of the author of *The Lord of the Flies.* (I still can't!)* Talk about the theater, John van Druten, death.

*William Golding—this name was censored for at least a week, until I finally read it on the back of some other book. Reasons for the censorship? I can only think of two possibilities: When *A Meeting by the River* appeared, it was reviewed in both *The Sunday Times* and *The Observer* (I believe) second to a novel of Golding's—I remember reading the reviews as I travelled down to see the Beesleys! Also, perhaps, because of a general boosting of Golding far a

Dodie is obstinately pessimistic; she wants to survive *as herself* and she is sure she won't. She memorizes Eliot's "Four Quartets" every few months. She says they are "the nearest I can get to religion."

Lunch today with David and Peter; Peter wearing a beautiful white suit. The Sunday papers are raving about David's show; he is England's pet, this is truly the springtime of his fame. Showed me the photos they took on our trip. The ones of me are grotesque, and so fat. There's one of me kissing Peter on top of the tower at Aigues-Mortes which looks like an assault by the Hunchback of Notre Dame.

April 6. Yesterday I went to see Paul Taylor, who I visited in hospital years ago and thought to be dying of T.B. He now evidently isn't; he is a very gracious furniture-queen and an obsessive talker. But I spent a quite agreeable two hours with him and his friend Ian Grant, who is an architect, plump and merry.[1] They were inclined to be shocked by the goings-on of David and thought that queers should keep it to themselves.

Then I saw Larry Madigan, the New Zealand seaman who has been writing me for a long while. He is a big dark-haired man of thirty, not good-looking but I suppose very attractive to some people. He is Catholic, has tried being a monk, writes stories and poetry, drinks a lot and gets into fights. He is obviously well educated though he mispronounces some words. He has been at sea since he was sixteen and says that is the only life for him. He is queer—at least I presume so from the way he talks—and describes it as being "bent." While we were in the restaurant, Provans, David and Peter came in with Patrick Procktor and his friend. Peter, bless his heart, is now wearing silver makeup on his eyelids. Their appearance shocked Madigan a bit, but he is dutifully going to the Hockney show today.

This morning the rain has been pouring down and I suddenly feel I've got to make definite plans for leaving—if only as a challenge to Fate to come up with some good news about our play. So I've written [my brother] Richard asking if I can come up next Monday 13th and leave Thursday 14th and tomorrow I'm going to get air reservations for April 19, back home.

above his deserts, in my opinion; Morgan praised *Lord of the Flies* very highly.

[1] Isherwood visited Taylor in the hospital in 1956; see *D.1*. He saw him again in 1961, with Grant (1925–1998), an interior designer, furniture collector, preservationist, co-founder of the Victorian Society and the editor of *Great Interiors* (1967); see *D.2*.

An actor named Basil Hoskins has read the script and is eager to play Patrick. Clement says he's tall, good-looking and quite talented.[1]

April 7. Terry Hands[2] has just refused the play, Ronald Eyre (*Three Months Gone*[3]) has also refused it, sayings there's no conflict and it's too literary and he doesn't understand all the light changes! It has been offered to Robert Chetwyn[,] and Richard Chamberlain is still reading it. Through all this I must praise Clement for his courage; he still declares the play is good and is still determined to do it. But I am all the more decided to book my plane ticket today.* Must rush out now, to see Sam Shepard's *Red Cross*.

Later. Red Cross was performed in a walkdown, a Chinese restaurant-club at the bottom of a steep flight of stairs at 6A New Compton Street, called the Soho Theatre. There are three characters, played by Briony Hathaway, Katherine Tracy and James Gary, the young man who already wrote to Clement and to me, asking us to come and see him because he wants to play "The American Boy" in our play. All three of them are Americans, and the two girls were quite good; Gary wasn't, he is a hit-or-miss performer, and anyhow out of the question as Tom because he's a very oriental looking Jewboy with a huge nose and slitty eyes. *Red Cross* isn't a play in the ordinary sense of "and then what?" It is "about" behavior simply and solely and its rule is that, if one of the characters starts something, the other goes along with it and joins in the game, as children do. The first girl involves the man in her fantasy of skiing and being broken to bits. The second girl is involved by the man in an imaginary swim in a lake, which ends with her getting drowned. Aside from this, the first girl is living with the man in a hut with two beds in a forest; it may or may not be part of a mental hospital. The second girl comes in to put fresh sheets on the beds. The man has crab lice and the first girl gets them from him.

The only way to enjoy this kind of thing (and I quite did) is to take it just as it comes and be content to accept it from moment to moment, because there is no apparent development and no

*I did book it.

[1] Hoskins (1927–2005) trained at RADA, acted in Shakespeare and other classics, and worked regularly on T.V.

[2] British theater director (b. 1941), Artistic Director of the Royal Shakespeare Company, 1978–1991.

[3] By Donald Howarth; Eyre directed it at the Royal Court that year.

"ending"—except that the man suddenly shows the first girl what seems to be a bloodstain on his forehead. But when I said this later (politely) to James Gary he obviously didn't agree. Obviously he felt the whole play was interwoven with meaningful symbolism—and it didn't matter anyhow what symbols you found in it, because today's symbolism is strictly do-it-yourself. When I said (testing) that I thought the bloodstain looked like a Hindu caste mark, he agreed, graciously, that yes, some others had thought that too.

He says the play got very good notices in the *Financial Times* and that, last week, they had had "a crowd." Today there was no "outside" audience, I suspect, except ourselves. The rest of the maybe dozen people looked very much part of the Sam Shepard world and maybe were members of the company. As I said to Clement, *we* could only be taken for managers from the West End—or the police.

This morning Don sent me a copy of the first issue of a magazine called *In Unity* which represents the homosexual Community Church, with Don's portrait of the Reverend Troy D. Perry on the cover. It is a revealing drawing, which shows perhaps a little bit more about Perry than he'd like, but he does look very determined and quite attractive. Those silly faggots have credited it as "Sketch by . . . Don Bachady![1]"

There was also a letter from Jim Gates, in which he says that he's already working for the Goodwill and that he likes Jim Bridges so much. "I've gotten to know Don a little more over the past weeks but it's still not often that I really feel as if I see *him*. We're sort of 'polite,' do you know what I mean? He's really kind of shy seeming to me."

Yesterday, Peter, Wayne Sleep and I saw *The Damned* again. A lot of it is very slow and somewhat simpleminded but I still find two sequences—the orgy-massacre and the wedding-suicide—really tremendous. The boys singing the "Horst Wessel Lied"[2] in drag, so mournfully, Germanically drunk, while the launch with their S.S. killers comes stealing in across the lake. And Ingrid Thulin's grotesque finery and blank white death-face, surrounded by the wedding party of whores and hustlers fucking all over the room.

After the movies (we also saw *Entertaining Mr. Sloane*, about

[1]Bachardy's name was misspelled by the magazine, but recording the error in his diary, Isherwood mistakenly typed the correct spelling, Bachardy. He and Bachardy also objected to the word "sketch" for a drawing.

[2]The Nazi anthem, launched at the funeral of Horst Wessel (1907–1930), a Storm Trooper, who wrote the verses not long before he was shot by communists in Berlin; banned in Germany after 1945.

which nothing need be said) we had supper at Odin's with David. He was feeling sick in his stomach and made such a tragedy about it that we were all under its shadow. The kind of fuss which is only made by someone who has basically an iron constitution—the kind of fuss I used to make.

April 8. Today it actually snowed, hard! Last night I had supper with Patrick Woodcock and David Mann at Odin's. The evening was a tremendous success, despite the tiresome attentions of the owner-chef, Peter Langan, because I somehow quite unintentionally became the go-between who brought Patrick and David more closely together than ever before—this was what Patrick told me in a note, this morning: "The old magician hasn't lost his power to cast spells." Actually, I doubt if this particular spell will be very binding, but who knows. David is a rather sadistic tease, but maybe he's far more involved with Patrick than he cares to admit.

Patrick says that Edward [Albee] took Roger Stock away from Binkie (?) Beaumont![1]

This morning Dodie phoned to say how much she likes the play. Alas, I don't like hers (have forgotten the title and now it's in its envelope ready sealed up to go back to her, but anyhow it's about *Measure for Measure* and the headmistress of a girls' school) one little bit.[2] It has a severe case of the cutes. Dodie and Ale[c] both think Oliver should explain more clearly what he values about monastic life, what the point of it is. I think this is a valuable suggestion.

April 9. Still this wretched cold, though no snow. Yesterday afternoon, Catherine Cook came round again, masochistically drenched (she hadn't brought a coat) and maddeningly apologetic. I rattled off a lot of glib answers to her thesis questions and caught sight of myself in the mirror, doing it. It's high time I got out of this place and stopped prancing and settled down to work while I still have my health.

Lunched yesterday with Nick Furbank. He brought me xerox copies of two more of Morgan's stories. When he arrived he seemed drunk, maybe he wasn't. I get along well with him and feel at ease. I find it very hard to believe that his book on Morgan will be good, but he is one of these quiet, perhaps deadly, mice and you can *never* tell what they will or won't produce.

[1] Hugh "Binkie" Beaumont (1908–1973), the most powerful theatrical impresario in London during the 1940s and 1950s.

[2] *Between You and Me*, never produced.

With Cecil Beaton to see *Conduct Unbecoming*, a melodrama about the British Army in India in the 1890s. It's silly and far-fetched and Paul Jones kept slipping right out of period during the trial scenes and bullying the witnesses exactly like a modern district attorney in an American film.[1] But, just the same, some kind of magic was made. Later Cecil took me back to his house and we had supper there with Patrick Procktor, who seemed a bit crazy and was so full of blatantly insincere compliments to us both that I didn't know where to look. Was he always like this? Maybe so.

A letter from Don arrived yesterday, with many enclosures—six pages of a translation Swami has made of Shankara's "A Garland of Questions and Answers"[2] (which I'm to fix up; Don calls it a "Dubtask"); a letter from the Society of Authors' Representatives, urging me to write to a senator about the Copyright Revision Bill;[3] a drooling letter from David Smith to Don about his show; a letter from George Hayim (whom we met with James Fox and Andee Cohen) saying he has written a book;[4] an application for amateur rights to perform *Black Girl* at Harvard and a brochure from *One*, announcing a preview of a film called *That's the Way It Goes* with an actor named Dale Stephens; the brochure has two shots from the film; in one of them, an Hawaiian-looking youth in a flowered shirt is starting to make sex with Stephens, who is naked to the waist, with his pants open and his shorts pushed down by the youth's hand, baring his belly right down to the bush and groping to pull out his cock. Stephens stands languidly passive, with his arm around the youth's shoulder. The story of the film is described as being, "The search for love by a young boy being used—as only grown men can use the young—getting tossed aside like a used and tired condom after each encounter, but the search continues—only to momentarily stop—then it continues, continues."

Don writes that Jim Gates deeply offended Jack Larson the other night during an argument about astrology—Jim Bridges and Nellie [Carroll] were also there. "I, too, though I thought Jim surprisingly

[1] Jones (b. 1942), once lead singer in Manfred Mann, was not a trained actor. Later he founded The Blues Band, worked as a radio and T.V. presenter, and appeared in West End musicals.

[2] Included in paperback editions of Prabhavananda and Isherwood's translation of Shankara's *The Crest Jewel of Discrimination*.

[3] To extend author's copyright from twenty-eight years, renewable for the same, to fifty years after the author's death; the bill was introduced in 1964, bogged down over cable T.V., computer data storage, and other new technologies, and finally passed in 1976.

[4] Hayim was a middle-aged mentor to Fox before Fox met Cohen; the book was *Obsession* (1970).

articulate and sensible, thought him the tiniest bit glib and impertinent. I am still (or again) undecided about him. Every once in a while I hear the sound of Goody Two-shoes being punctuated by a cheeky kind of arrogance, a faint smug intellectual pride. . . ."

Later. Clement called to say that a director named Robert Chetwyn is "enthusiastic" about the play but "doesn't understand" the lighting. We are to have a talk the day after tomorrow.

When I spoke to Dodie Smith on the phone she said that she wanted me to stay in England and Don to come here. "I have a strange feeling that you will. I'll try and do something about it." I begged her not to attempt any of her Christian Science magic. "Remember 'The Monkey's Paw,'"[1] I said, "suppose they fulfil your wish *literally*. All they have to do is wreck the 2:00 p.m. train on Monday for Manchester. Then Don'd come over for my funeral." Does Robert Chetwyn's interest mean that Dodie's spell has begun to work already?

Have just seen Christopher Isherwood off. He's a disc jockey from Bristol who wrote me several times. He has a Beatle haircut and spectacles, is very thin, energetic, has very bad teeth, is thirty-five years old, married, with a daughter. He has just got a job on a Mediterranean cruise boat for the whole summer. He says of himself that he loves being in front of an audience. He has also managed a building company, been a professional photographer and in the air force. He has a sort of egocentricity which seems quite attractive and wholesome, not at all repulsive. Only, it's hard to imagine that he could be a good entertainer; one doesn't feel he has enough temperament.*

The car he was driving in broke down, so he arrived nearly three hours late, and fucked up my day. A cute youth, a would-be disc jockey, had been driving him. Christopher Isherwood said of him, "I'm his idol."

April 10. I'm ashamed to say I got drunk last night, having dinner with Richard Simon. That was stupid and unnecessary, because I really do like him. He seems genuinely good-natured. The fact

*He calls himself professionally "Chris Connery," Isherwood being too long for the entertainment business.

He looks inescapably like a *clerk*, and so respectable. His hairdo is grotesquely unsuitable, it makes him look like an earnest games mistress in the twenties.

[1] W.W. Jacobs's short story in which the monkey's paw grants three wishes, leading to tragedy.

that I found him nicer than I remembered may be partly due to the fact that he is very happy with a boyfriend.

Before we left for the restaurant (Provans, where the beautiful tall girl-boys with their flowing locks now know me well and greet me with discreetly understanding smiles) I was talking about *Kathleen and Frank* and said something that seemed so good to me that I wrote it down, namely that we sail out into life with, as it were, sealed genetic orders. When we finally break the seals (as I did, so to speak, by reading through Frank's letters), we find that the orders have often already been obeyed. For example, Frank "wanted" me to go and visit Vailima for him, and Japan too.[1]

April 11. Still horribly cold but at least a little pale sunshine today without rain. Walked up to the bag shop to pick up my bag; it mysteriously got a slit in it during our French trip.

Yesterday I had lunch with Amiya; she came up to London specially. (She remarked that she'd sold a piece of furniture in order to do so but I think she is talking a poor mouth partly in order to convince herself that she must not give money to [her sister] Sally and her other greedy relatives and hangers-on. Anyhow, I paid for our lunch, which we ate in a gaming club called The Nightingale in Berkeley Square; a peculiarly dreary place, as I suppose they all are, reeking of polite gangsterism: they addressed Amiya as "My Lady.")

Amiya seemed much fatter, quite piggy, and drunk of course, but with her skin still white and smooth. She rambled on without stopping, about herself, her tears and tribulations and spiritual insights and good deeds and general belovedness; I barely got to talk to her. The funny thing, though, is that she *is* lovable, her egotism somehow doesn't matter, and though she only talks about herself she actually makes you feel that she cares for you.

Later in the afternoon I went to the notorious Strand Sauna, to pass the time and to get in out of the cold and rain. It was very disappointing. Two dreary-looking young men in the sauna handled their own cocks as a signal to each other and then retired to the toilet to make out, where they were watched by a third, but without much apparent interest.

Then to the ballet, with David, Peter, their friends Mo [McDermott] and Celia [Clark] and Patrick Procktor's friend Ole [Glaesner]. Wayne Sleep danced Puck in *The Dream*. As before,

[1] Isherwood visited Vailima, Robert Louis Stevenson's house in Samoa, in July 1969 with Bachardy (see *D.2*); he visited Japan in June 1938 with Auden.

I cried most of the way through—though not so much at the end.* Wayne was absolutely enchanting. He was so erotic with Oberon that you couldn't imagine why Oberon wanted the little changeling. He was also ridiculously girly as a Scotsman in *Façade*. David Ashmole (a very attractive boy), danced in the *Lament of the Waves*, replacing poor Carl Myers who was hurt in an auto wreck.

Later. Had lunch with Marguerite and [her companion] who told me that Kate Moffat has been having an affair with a young man named Peter [Townend] (who has written a book which has been accepted by the publishers),[1] because Peter makes her feel attractive and because Ivan doesn't love her and she doesn't love him. Ivan has gone off to Spain to think things over. Marguerite predicts that he will return to Kate but will punish her for humiliating him in public and that later they'll split up. [Marguerite's companion] predicts that they will split up now, because Ivan is getting to enjoy his freedom. It seems that Kate complains that Ivan has become so stuffy and conventional. Now [Marguerite's companion] expects him to turn into a bohemian. [He] feels rather smug about all of this, I could see. He feels smug that his [relationship] with Marguerite is one of the very few successful ones. "Only three or four of the [women] we know don't have lovers," said Marguerite. "Present company excepted," said [her companion].

April 12 [Sunday]. Yesterday afternoon, Robert Chetwyn and Clement came here. Chetwyn talked quite intelligently about the play, asked searching questions and *seemed* really interested, though I feel he has reservations still. He seems a nice man, middle thirties perhaps, but with a face too old for his long hair. He is to read the novel today.

*For me, the magic quality of *The Dream* is that it seems to express pure happiness—not the bliss of *ananda* but nevertheless unalloyed (it's like when, in the Katha Upanishad, they say "Death is not there" ["In the realm of heaven ... You, Death, are not there ..." (I.12).]). The world of Fairyland is full of tiffs and feuds but they aren't serious, they are high camp, and high camp is happiness. Puck's relations with Oberon are so beautiful and animal, his vanity is beautiful; and what's so terribly moving, when you watch, is that Wayne *isn't* immortal like Puck; *his* marvellous grace and fun and sexiness will only last for such a short time, then they'll be lost or rather inherited by another dancer. Also terribly moving, for a reason I can't quite analyse, are the first modest steps which Alexander Grant [as Bottom] essays on his hoofs; at that point the tears pour down my cheeks. On the other hand, the relationship between Oberon and Titania seems merely ceremonious and frigid and uninteresting.

[1] *Out of Focus* (1971); see Glossary under Smith, Katharine.

Then I phoned Don. He is eager to come back and ready to leave on Thursday, if I can assure him before then that Chetwyn definitely means business. It'll be awkward to do this, as I'll be up with Richard and will perhaps have to call him from Wyberslegh—cables sent from the country are so unreliable. Well, we shall see.

Saw *Widowers' Houses* last night, alone.[1] It is surprisingly shocking. You feel the real naked evil of being a slum landlord, and the cruelty and vicious anger of Blanche, the spoilt daughter, was shocking too. (Nicola Pagett played her very well.) This play made me respect Shaw from a slightly different angle—respect him as a man. He isn't kidding. This is no shit. He really *is* indignant.

It's raining mournfully and God bless dear David who is coming with Peter to take me out to lunch with the artist who wants to illustrate my "Gems of Belgian Architecture"; it's for the same publisher who did David's illustrated book of Grimm's Fairy Tales. The artist's name is Howard Hodgkin.[2]

April 13. We reached Howard Hodgkin's house after a long drive;* it's away off in Wiltshire, on the road to Bristol and then down winding lanes—an eighteenth-century building which used to be a mill, by a stream. The place was full of men, women and children. Kasmin and his wife were there. I liked Hodgkin, he has greying hair and a ruddy face and paints near-abstract interiors, landscapes and portraits, in very bright pleasing colors. (Keith Vaughan, whom I saw at Patrick Woodcock's later, doesn't think much of them.) I can't imagine how he will illustrate my story, but he certainly wants to. We all went for a real (rather self-consciously) country walk, crossing the stream, splashing through wet meadows and

*We stopped for lunch at a ridiculously depressing bungalow hotel, which I tried to make interesting to David and Peter by telling them it was exactly like a building in Australia. The diners inside were pure *Separate Tables* characters, and we three had so much fun giggling about them that the meal was quite a success. Also, in that fascinatingly unpredictable English way, the food (roast beef) was good. [*Separate Tables* (1954) was two short plays by Terence Rattigan, and later a film set in a shabby genteel English hotel.]

[1] A revival of George Bernard Shaw's 1892 play, directed by Michael Blakemore, at The Royal Court.

[2] Hodgkin (b. 1932), educated at Eton, Bryanston and Camberwell School of Art, became widely known only in the mid-1970s. Hockney's *Illustrations for Six Fairy Tales from the Brothers Grimm* was published by the Petersburg Press in 1970; the publisher, Paul Cornwall-Jones, approached Isherwood and Auden unsuccessfully in 1969 to make new translations for it. He does not recall the plan, evidently abandoned, for "Gems of Belgian Architecture."

slogging across ploughed fields, and climbing up through a wood. There was also much lifting of, and crawling under, barbed wire.

A rather tiresome "expert" on country matters talked to me about the habits of crows, which only nest in elms or beeches, never in oaks or conifers. He also told me, more interestingly, that the kind of "battlement" on the gables of Wyberslegh is called "crowsteping" (I wrote this with two "p"s in my notebook and he instantly corrected me.)[1]

The rain had stopped by then and we had a beautiful drive home, the country suddenly looked marvellous. A tall obelisk and a white horse cut in the downs.

Then supper with Patrick Woodcock, his friend David Mann and Keith Vaughan. The evening degenerated into pot and was rather a bore,* later. Patrick was charming as always but Keith was rather prissy. He had met Dodie Smith and Anthony Page at a lunch in the country. Anthony had apparently spoken quite favorably of our play.

April 14. Yesterday afternoon I came up to Disley to stay with Richard and the Bradleys. We are both in the tiny hot sitting room, writing our diaries. The weather is suddenly mild, with pale sunshine, and I feel reprieved; now that it is warmer I realize how passionately I have been loathing the climate here. Could I ever bear to settle on this island again?

This afternoon I am to phone Don from Wyberslegh—a terrific production. Richard thinks it may take hours or perhaps a whole day and night before we get through. But I have to lay all the pros and cons before him. I talked to Chetwyn again yesterday before leaving London and I'm sure he's scared by the vagueness of the situation; that we have no cast as yet. Also he has a job in New York in August, directing a play by Tom Stoppard.

April 15 [Wednesday]. I talked to Don yesterday and got the strong impression that he will come right away, be here when I talk to Robert Chetwyn on Friday, in fact—the day after tomorrow. Got through to Los Angeles in about forty-five minutes.

Yesterday was quite mild, but the tomb chill inside Wyberslegh was as deadly as ever and I had to get it out of my bones by walking briskly back to the Bradleys'. Yet Richard spends hours in the

*I mean actually that *I* was a bore. I tried to hypnotize a rather charming boy who had the grass giggles, named Eddie MacAndrew.

[1] But Isherwood's double "p" is correct.

house, sort of communing with it. Mrs. Bradley says he doesn't realize, or rather simply isn't aware, of what terrible shape it is in. He invited some Americans (alleged distant cousins) to see over it and Mrs. Bradley was dismayed because she felt it had been a shock to them. She says that Richard is really unwilling to have it repaired in any way.

This time I feel the spirit of place very powerfully here. And it *is* a spirit or it's nothing. Physically, Disley is just a rather smug little suburb. When I looked at postcards, down in the village, to which I also managed a quick walk yesterday morning, the hills seemed flattened and utterly undistinguished, the Ram just another little pub, Lyme Cage a tiny dump and Lyme itself quite lacking in grandeur.[1] And yet, despite the cheerless ugliness of the stucco and brick villas which are steadily crowding in, the spirit of place is powerful indeed; the rooks caw fatefully around the church with its big gold ball below the weathercock, the little sweet and cigarette shops seem stoically North Country, the Ring o' Bells[2] is so sturdily ancient, Lyme Cage is sinister and numinous, and the air of the hills is still poignantly refreshing and stirs longings—even if they are longings for escape!

April 16. The *Apollo 13* astronauts are wobbling back toward earth. For the past two days we have been watching their troubles on the telly—or rather, we have watched various types of grim ham reporters telling us about them. If this produces a really violent reaction against these money-wasting public circuses I suppose it will have been almost worthwhile. But what a nightmare way to die![3]

Yesterday afternoon I took a walk right up to the crest of the Old Buxton Road. There was a strong wind and the grey clouds streaming across England, the Cage against a yellow steak of light, the chimneys of Manchester like huge mysterious monuments in the blue beyond. I told the shaggy horses in the field that Kitty is coming to be with us again.

Later. 6:30 p.m. Back at Moore Street, to find a cable from Don saying he'll come immediately the director says yes. Am disappointed of course, because I'd felt sure he would come anyway,

[1] Lyme Cage is a sixteenth-century tower built as a hunting lodge in the medieval deer park surrounding Lyme Hall; Lyme Hall, a Tudor house, was lavishly rebuilt in the early eighteenth century as an Italianate palace.

[2] The Quaker Meeting House.

[3] An explosion on April 13 crippled their command module, and the astronauts were in jeopardy for four days; see Glossary under *Apollo 13*.

but it's the only sensible decision unless he feels he wants another visit here. Because it's by no means certain that Chetwyn *will* say yes, and if he says no I may well be on that plane home on Sunday. The great thing now is to get a definite answer from Chetwyn, so I'm pressuring Clement to make Chetwyn's agent say how much Chetwyn will cost, if he *does* say yes; there must not be any last-moment backing down because he's too expensive.

Meanwhile Nicholas Thompson, who has done absolutely nothing for us, is leaving for New York tomorrow. Fuck all agents[.]

I'm having supper with Richard Le Page tonight, chiefly because he made such a thing out of it and is coming up to London on the train to see me (well, partly at least).

Spent much of yesterday and the day before reading Kathleen's diaries for 1936 and 1937—all that sorry tale of my efforts to get Heinz a new nationality, and of how Kathleen was induced by me to pay the thousand pounds to Salinger to give to the Mexicans.[1] I must say, it now seems to me more probable than ever that the money ended up in the pockets of Salinger and Gerald [Hamilton]. Oddly enough, Richard rather opposed this idea and kept raising points to prove that it couldn't be true. His points were all based on ignorance of the facts; it was his attitude that interested me. Goodness knows, I don't feel particularly vindictive toward Gerald—and as for Salinger he was punished by other circumstances, far more than I could ever have wished.[2] I was a silly little ass, anyhow, thinking I could fool around with these sharks and not get nipped, and it was I who made Kathleen go through with it.

Kathleen disliked Heinz, because he was socially unsuitable and a nuisance. She really hated Olive Mangeot, as a domestic subversive influence. She had a life-long feud with Nanny. She latterly referred to Amiya as "that bitch"; she also disliked Richard's "adopted aunt," Miss Colvin. She was very impatient with me when I got so ill in 1937 and writes that I behaved like a pasha or sultan, I forget which. She deplored Wystan's untidiness. She mistrusted Edward. Perhaps Britten is the only one who gets her top marks as a guest.

In a way I'm rather sorry I read these two diaries. They disturb my image of the younger Kathleen and also that of the old woman

[1] Salinger, introduced by Gerald Hamilton, was the English lawyer in Brussels who promised to obtain, through an acquaintance at the Mexican embassy there, a Mexican passport for Heinz. See Glossary under Neddermeyer.

[2] Salinger was a Jew; this may refer to his fate during the war. Possibly he went to prison for other reasons.

with whom I became affectionate again from 1947 onwards. They remind me of the mother figure I often hated so much throughout the twenties and thirties. And I want to forget her. I have no use for her, whatsoever; she doesn't come into the scope of my book.

Before I left this morning, Dan Bradley told me that he expected Richard would have a big booze-up tonight or tomorrow as a release of tension after staying dry during my visit! The joke is that I too was staying dry during my visit to Richard, not to impress him but just because it was a convenient excuse for taking a rest.

This morning saying goodbye I was struck once more by the sweetness and innocence and real goodness in Mrs. Dan's face.

Patrick Woodcock has just (7:50) been by to see me, to hear about the latest developments connected with our play. He doesn't recommend Gielgud as a director (this was an idea of Clement's) because he's losing the power to concentrate. Noël Coward is far worse; when he starts out alone to visit someone he often arrives not knowing who he's visiting. He can't remember who he had lunch with, that same day. The experts, Patrick says, recommend giving the ageing brain cells a workout by learning something altogether new, like a foreign language.

Patrick quite approves of another idea of Clement's—that we should try getting the play put on by a theatrical group—like the Hampstead Theatre Club. They have done good things and they often go on to the West End.

Patrick says he is anxious to reassure people about dying. He hates it when he hears that someone "died in agony"—"that's nonsense." He thinks lung cancer is the best, today he told a very intelligent patient (sixty-two) that she had it. There is no pain and no breathlessness.

I feel a great warmth from Patrick, nowadays. He says of himself that all he cares about is intimacy.

April 17. Last night I had supper with Richard Le Page at Odin's. David, Peter, Kasmin, Ossie and [Celia][1] Clark and a lot of others were there. David was drawing them and giving away the drawings; he gave me one of Peter. Then Wayne Sleep arrived. The other night he was at Odin's with just a few other people—I suppose it was very late —and he got drunk and stripped all his clothes off and danced stark naked on the tables. Too many strangers were present for him to do it yesterday evening, however.

Later we all went back to David and Peter's flat, where they

[1] Isherwood wrote "Sylvia."

presented me with an album full of pictures taken during our tour in France and stay at Tony's—a marvellous thing to have. I believe I can give them a suitable counter-present; a copy of the big illustrated book on Klimt.[1] Kasmin was very drunk and desperately eager to belong to the gang; he wanted to strip and dance with Wayne!* [Don's friend] was also there, but we avoided each other except for polite smiles at parting.

It was so late that I urged Richard Le Page to stay the night, which he willingly did. There was some necking later, also quite willing on his part, though he obviously didn't want anything beyond that. We parted very friendly after a sausage breakfast.

Am now reading *The First Circle* with much more pleasure. It seems to me to pick up at chapter 34 with Nerzhin's interview with his wife Nadya; and all the stuff about the daughters of Makarygin, the State Prosecuting Attorney, is interesting so far.

April 18. An old lady, hobbling briskly along, said, as we passed each other on the street this morning, "Isn't it awful, getting old? I'm absolutely worn out. How old are you?" "Sixty-five." "Oh, you're a baby. I'm eighty-five." I said: "Do you think they ought to finish us off?" "Oh, I shouldn't like that at all! Well—at least we can laugh about it, can't we?"

Later. Have just had lunch with Robert Medley. He told me that Eliot sent Rupert Doone an earlier draft of *Murder in the Cathedral* which was in prose and Rupert told him it was a bore and why didn't he write it in verse, since he was such a good poet! Also, that Eliot once told Rupert that he was a genius.

There is a bitter layer in Robert. He talks bitterly of Keith

*Another of Kasmin's gestures was to give away some valuable books he'd just bought from a man who works at Bertram Rota's bookshop. (Maybe this *was* Rota;** anyhow he was with Kasmin and falling-down drunk.) Kasmin gave me a first edition of Sidney Keyes' poems, inscribed, "For Christopher I. from Kasmin with wanton fondness."

**I found out later that it *wasn't* Ro[t]a but a George Lawson—this from a note which John Byrne delivered to me in person after I got home from supper with Neil and Bob on the 17th. (He said he'd seen my light was on.) I was too tired to ask him in. It was quite a nice note, to say goodbye, because Lawson had told him I was leaving on Sunday, but it irritated me, because Byrne admitted that he'd forgotten all my answers to the bibliographical questions he'd asked me!

[1] *Gustav Klimt with a Catalogue Raisonné of His Paintings*, Fritz Novotny and Johannes Dobai, translated from German by Karen Olga Philippson (1968); Isherwood took another copy back to Santa Monica.

Vaughan, how prim and cagey he is. Robert is lonely of course, but he says he finds his relationship with Gregory [Brown] and his wife satisfactory and permanent.

I have cancelled my flight back to Los Angeles for tomorrow and indeed I really believe we have got Robert Chetwyn; the only remaining obstacles are financial. After talking to him at some length yesterday, I'm now at least ninety-five percent convinced that he's the right man for us, intelligent and cooperative and not a trickster. Now the question comes up, how soon can I risk telling Don that it's okay for him to take off?

Supper with Neil Hartley and Bob last night—at Odin's again, I seem to live there. I didn't get drunk but I did drink too much, which I do so loathe doing; it results in my not remembering anything that was said. All I have left from the evening is an impression of Neil; very weary, really worn out by his life, gentle and rather beautiful in his (perhaps terminal) exhaustion. He alluded to Tony's cantankerousness—Tony was in one of his "anti-Anna" (O'Reilly) moods. Neil said he didn't want to see Tony while he was like this, but of course he has to be with him constantly. On the other hand, I felt great warmth toward me from Neil—perhaps a kind of fraternity in decrepitude, but anyhow pleasing because I am really fond of him.

April 19. It's an unsatisfactory day. Phases of weak sunshine followed by sunless cold. Have had to switch on the heating. But the tree outside the window is in leaf.

Have talked to both Clement and Bob Chetwyn (he wants to be called Bob) this morning, trying to get them to commit themselves sufficiently so I can tell Don how soon to come, when I call him this evening. I now believe that Bob *does* mean to go ahead; Clement refuses to commit himself until he has talked the deal over with Richard Schulman tomorrow morning.

Richard Le Page was here again last night. David had invited him to come to the ballet and when he showed up he had his overnight bag with him and made it clear he expected to stay with me, not with the friend he has in Clapham, where he usually stays. However, after the ballet, and supper at Odin's, he decided to go off to a party—one of the dancers in *La Boutique Fantasque* (Anthony Rudenko) who came to supper with us, had invited us all but I didn't feel like going. So I didn't see Richard again until this morning, when I fixed him scrambled eggs. I like him, and could easily get to like him a lot more, but I rather doubt if I ever shall. Circumstances are against it. We may see each other in California, however, when he comes out to lecture at UCLA, later this year.

The ballet seemed insipid. Not one of the dancers appealed to me especially. This is the touring company; the regulars (including Wayne) are off in New York. *From Waking Sleep* features the Buddha; *Lazarus*, Jesus.[1] They are both deathly respectful and serious but, at the same time, their style is essentially campy. That's what is wrong with them, I guess.*

I gave David and Peter a copy of the Klimt book. Peter was annoyed with David during supper, and no wonder. David keeps retelling, somewhat braggingly, the story of how they met, to semi-strangers at table, and then fondles Peter with thoroughly self-satisfied possessiveness. He also announced that, when Peter goes to California, he'll go too—although Peter doesn't want him to come, as he freely admits. Once, when David put his arm around Peter, Peter jerked angrily away. Sooner or later, David will have to face the fact that their relationship is in quite considerable danger. I have seldom seen such a clear demonstration of the very mistakes I have made—and still sometimes make—with Don. Ought I to speak to David about them? Probably not—unless Peter asks me to.

April 20 [Monday]. Peter did sort of ask me to, but not definitely enough, in the middle of a supper party at Patrick Woodcock's, last night. He repeated that he wanted a separate place to live and he again referred to the unsatisfactory sex situation.

Actually the party was quite a success, although Patrick had invited old Gladys Calthrop and two Frogs from the French embassy. Poor Gladys got a terrible headache and had to leave, but the Frogs were nice, even charming, Jean-Pierre and Odile Angremy; she was both attractive and friendly and he was blond, plump but sexy, and quite intelligent—by which I mean, among other things, that he admired my work. He is a writer himself, and has published what he claims is a pornographic novel.[2] He greatly recommends

*There was also *La Boutique Fantasque*. This performance of it was perfectly all right, but it didn't charm me, either.

[1] New ballets, the first choreographed by David Drew, the second by Geoffrey Cauley.

[2] *Et Gulliver mouret de sommeil* (1962) under his pen name, Pierre-Jean Rémy (1937–2010), followed by other prize-winning novels, as well as T.V. and film scripts, magazine pieces and works on opera. He was then first secretary at the embassy; later, he was ambassador to UNESCO, director of the French

a modern writer named Bataille. He spent several years in China, and is preparing an anthology of writings about it. He hadn't heard of *Journey to a War.*

John Gielgud and Martin Hensler were also there; John as amusing as usual, and so genuinely benevolent. Martin was in an exceptionally gracious mood and made himself very agreeable. David was David. Tonight we are going out to dinner together again, with the novelist Julian Mitchell.[1] I sometimes begin to wonder if David proposes all these threesomes because Peter and he quarrel when they are alone!

When I called Don yesterday I told him the situation between Clement and Bob Chetwyn and we've agreed that he shall book a passage on a plane leaving Wednesday, the day after tomorrow, and arriving early Thursday, the 23rd. Meanwhile I have to send him a cable as soon as possible, either confirming this or telling him to wait. At present the situation doesn't look good. Richard Schulman, Clement's partner, feels that Bob Chetwyn is far too expensive. They are obviously going to bargain. Bob's [. . .] agent Elspeth Cochrane has lost his former contract and another copy of it has to be picked up and copied for Schulman and Clement. So nothing can be settled till tomorrow. Of course I *can* still reach Don in time but I hate to make such a cliff-hanger out of the proceedings, when the delay is all due to the inefficiency of this [agency]. Clement said to me on the phone: "It's a war of nutrition."

An attractive dark lank-haired young man from Hull, with a high forehead, crooked teeth and a tape recorder interviewed me for two hours this morning to get material for a thesis he is writing on four novelists of the thirties and their attitude to political questions—Edward and me, Anthony Powell and Henry Green. His name is David Lamborne. Then a big fat blond pleasant but unattractive young German-American named Otto Friedrich followed him immediately, with questions for a book on pre-1939 Berlin.[2] Mrs. Gee[3] contrived to interrupt us several times; she comes in very quietly unannounced, often startling me violently. I want to

Academy in Rome, president of the Bibliothèque Nationale, and elected to the Académie Française.

[1] Award-winning playwright, screen writer, and T.V. writer (b. 1935); educated at Winchester and Oxford, later known for his 1981 play *Another Country*, filmed in 1984.

[2] *Before the Deluge: A Portrait of Berlin in the 1920s* (1972).

[3] The cleaner, whom Isherwood was obliged to hire as a condition of renting the flat.

kick the shit out of her—there's something masochistic about her attitude—and yet, whenever I settle down to talk to her, she's nice and very sensible.

Don told me on the phone that Swami is already beginning to fuss about the *Garland of Questions and Answers* by Shankara, which he sent me to rewrite—although the deadline isn't till June 1! Started work on them yesterday.

April 21. Everything still hangs—it's 12:20 p.m.—and Clement still hasn't had his conference with Miss Cochrane about the terms.

Last night I went with David and Peter to have supper with Julian Mitchell. Peter told me that [Don's friend] complained to him about Don—how Don had "stood over him in the middle of the night and watched him while he read the play" and how [the friend] hadn't been able to make up his mind under such pressure and so he'd shown the play to someone else at the Royal Court, which had made Don furious. So I said to Peter that I dislike [the friend] and could never under any circumstances work with him—which I hope he passes on [. . .].

Meanwhile, in Shankara, I come upon this question and answer: "What is sin?" "Jealousy." This baffles me, it seems so sweeping and suggests that "jealousy" is being used with some much broader, deeper significance. Is Shankara saying that *all* sin (and that word itself is a bit vague) is jealousy in some form? Then, obviously, "jealousy" isn't merely jealousy in the ordinary sense. Would "envy" be better—meaning that desire for what belongs to other people is the root of all attachment? It isn't, though. What about attachment to the things which *do* belong to you? Must write to Swami about this.

I am jealous of [Don's friend], because Don still cares for him. I hate the thought that when Don comes to London he'll be wishing he could see [him]—that is, if he does wish it. And yet I know that, the more I can make Don feel perfectly free, the more he will love me. And now Shankara tells me the jealousy *is* sin! Ha.[1]

Also at the party last night was that curiously attractive rather battered-looking artist, Keith Milow. He always plays it very cool when we meet, but this time he invited me to some get-together next Sunday, and he kissed me when we said goodbye.

Another of the guests, James somebody (the friend of John

[1] Isherwood and Prabhavananda agreed to translate as: "What are the evils most difficult to rid oneself of? / Jealousy and envy."

Golding who's some kind of art historian)[1] told me that he'd been waiting to meet me for thirty-five years!

The party itself was quite unmemorable, as far as I was concerned. Mitchell is pleasant enough, but such a professional entertainer and host; there's a slickness in his geniality which, you feel, must surely infect his books as well.

Another day of miserable cold.

April 22 [Wednesday]. The situation has, as they say, deteriorated. Yesterday I was told by Clement that Chetwyn's agent Miss Cochrane told him Chetwyn has another (previous) engagement which may be coming up—it's a two-character play with Lynn Redgrave in one of the parts; he worked on it last year and now suddenly it's a possibility or even a probability. It would take Chetwyn away from us from the end of May until about July 20. However, the Tom Stoppard play has been postponed until September 1, so there would be more time for our play at the other end.

Hearing this, I blew my top, exclaiming that this was Clifford Williams's betrayal being repeated. When I talked to Miss Cochrane on the phone (she was unexpectedly sensible and nice—they always are) she assured me that all this was very uncertain and that anyhow Chetwyn would never leave our play if it could be cast before the Redgrave play came up; only in the event of a total failure to cast it would he leave. However, when I repeated her words to Chetwyn he stammered a good bit, then muttered that this wasn't quite the case and that he must talk to Miss Cochrane again. Now it seems clear that the Redgrave play *is* part of his program, if he can get it. He wants to fit both of us in.

So this morning I had a very painful conversation with Don, who is terribly disappointed but says there's no question of his coming just for the trip—unless we are going ahead right away, he won't. And I can't advise him to do otherwise. Obviously he should stay where he is and I should stay here a little longer—but only a little.

So I'm endeavoring to scare Clement into action by threatening to leave on Friday and by talking constantly about the huge costs

[1] James Joll (1918–1994), British historian, educated at Winchester, the University of Bordeaux and Oxford, then a professor at the London School of Economics; he published major works on modern European and British history and politics. Golding (b. 1929) is a British art historian, curator, and painter, educated in Canada and at the Courtauld, where he taught. He also taught at the Royal College of Art and at Cambridge.

of my staying here, including all these phone calls to Santa Monica (there have been ten of them, to date). And I'm going to talk things over very frankly with Clement and Bob Chetwyn today. Richard Schulman will be there too.

The latest candidate for Oliver is Tom Courtenay. (Have I already commented on the extraordinary run of Cs in this affair? Clifford Williams, Clement, Chetwyn, Miss Cochrane, me—not to mention other people we offered the play to: Richard Chamberlain, Christopher Plummer. Corin Redgrave we also considered. And then, of course, Clifford Williams left us to do *Oh! Calcutta!*!)

A severe shock on the scales this morning; I weighed nearly 160!

Last night a quiet rather sad little supper with Neil Hartley and Bob Regester. I felt that Neil is an increasingly unwilling slave of Tony. He really resents Tony's caprices, or is worn out by them. He feels that the Nijinsky film will run on the rocks because of them; he expects temperament clashes and overspending. Tony merely says, "It'll be fun."[1] Neil said to me how nice it would be if Don and I and Bob and he could get together on a film, "You're such creative people." But he quickly added, "That is, if you don't have to have a huge amount of money"!

Both he and Bob liked the extreme "pudding-basin" haircut I got yesterday from an Italian barber far down the King's Road. "It wouldn't look good on anybody but you."

5:30 p.m. Feeling blue, a cold starting. I'm increasingly miserable because I had to tell Don not to come, this morning. And yet I'm sure I did right. Even if the whole situation changes, it was right, at present. And if we take it as possible that I may leave now and we may both have to come back later—when the play has been cast and rehearsals are about to begin—then Don will have saved an entire round trip.

The utter foulness of this miserable spring! I long to get off the island. Am now planning to leave next Wednesday the 29th.

Had lunch with Bob Regester at a club he has just joined, Burke's. He says it's very smart, but the general effect is more Mafia. We talked about his difficulties with Neil, who is possessive but unemotional and undemonstrative. Bob finds the relationship confining. And then, when I went round to Woodfall to get the Tom Cuthbertson story about Peter Schlesinger xeroxed,[2] Neil started, in an oblique way, to ask my advice about Bob, without

[1] Nureyev was to play Nijinsky and Paul Scofield, Diaghilev.

[2] Possibly "Las Cruces Springs." Cuthbertson was a California friend of Schlesinger's; see Glossary.

ever mentioning his name. So of course I handed out lots of admirable advice, which I would do well to follow myself. Don't try to make the relationship exclusive. Try to make your part of it so special that nobody can interfere with it even if he has an affair with your lover. Remember that physical tenderness is actually more important than the sex act itself. Etc., etc., etc.

Roger Stock had been in to see Neil and had told him that he and Edward [Albee] can't make it together. What slums of deprivation other people's private lives seem to be—when you're in the sort of mood I'm in today!

April 23 [Thursday]. I had rather a nice evening, yesterday, with Bob Chetwyn and the friend who is living with him, Howard Schumann(?).[1] Howard is American, Jewish New Yorker, with possibly some Negro in him. He writes for T.V., films and off-Broadway revues. He comes on super-friendly and cozy, clasped my hand with both of his when we said goodnight, and seemed to be telegraphing repeatedly that Bob and he are a domestic couple. Bob did and said nothing to confirm this, however. But they have evidently been together for quite some time and Bob told me, while he was driving me home, that Howard wants to settle permanently in this country. Howard has a darkish skin, very bright dark eyes, close curling perhaps Negroid hair and a thick moustache. I suppose he's in his late twenties. Yesterday, viewing Bob's face in the new frame of reference, as a guy with a houseguest, I thought he looked like Alec Guinness.

Earlier, we had our conference; Chetwyn, Clement and Richard Schulman. Richard, who is very dapper, white haired and queenish, led off with a real loyalty pledge—never never would he cease to believe in our beautiful play and strive to put it on. So I commended his loyalty, and Clement's; and then we got down to discussing who we'd like for a director if we ditch Bob Chetwyn! Chetwyn arrived a bit later, which was convenient. But he too pledged loyalty. He now talks as if it'll be no sweat to fit in the Lynn Redgrave play and then our play before doing the Stoppard play in New York. He agreed that he and I could list the alterations he would like to see made and then I could take them back to Don in Santa Monica and work on them there. In other words, no efforts were made to restrain me from leaving. When I rubbed in once again all my expenses they merely clucked sympathetically.

Came home feeling bluer than ever. The fact that I had told

[1] Schuman; see Glossary.

Don not to come seemed extraordinarily painful. I don't really understand why, myself—for, after all, it now also seemed to have been the obviously right decision.

Then this morning I got a cable from Don, "Arriving Saturday morning unless you call." So I called and we talked it all over. He was so sweet—sweet like nobody else on earth can be, assuring me that it didn't matter *where* we were together as long as we were together; and sly old Dub said (with doubts still flickering about [Don's friend]) why *don't* you come to London—there are so many people who'd love to see you—and Kitty purred, there's lots of people everywhere, but he only wants his old Dub. So it was settled—barring last moment changes—that I'm to leave next Wednesday the 29th.

What does concern me is that Don hasn't sold any more pictures as the result of the show and that Blum doesn't seem to be able to set up any shows in New York or San Francisco for him. Also, he says he isn't working at all. When he told me this he didn't betray any concern, but of course he *is* concerned, he must be, being Don. So I'm worried. I only hope I can help him, somehow, by being with him.

This business of work is such a bewildering circle, it takes you round and round. Work is important, yet only as much as you wish it to be—because it's symbolic, never an absolute. Then again, work is desirable (or so we say) for an artist because it makes him happy to be working. But then, is happiness important? When I say that I want Don to be happy, he always retorts, but why, what's so special about happiness, why should happiness be important, why isn't it just as important to be unhappy? But then one says, the unhappiness of not working is due to guilt. If you aren't working, you feel guilty. But is guilt wrong? No, not necessarily.

The Gita says, "Never give way to laziness,"[1] but the Gita is speaking of the constant effort one should make to realize the meaning of life, the Eternal within. This effort can go on throughout periods of inaction. Inaction is not the same thing as laziness. And certainly the effort can go on throughout periods of unhappiness. Indeed, unhappiness is possibly less stultifying than happiness, provided it isn't too intense.

Such a burst of philosophy, all of a sudden! It is partly because I'd like to fill up the pages of this notebook before I leave. But that too has a motive. I have kept this diary up doggedly, day by

[1] 2.47; in the Prabhavananda-Isherwood translation, see "Yoga of Knowledge," p. 46.

day, because I believe that a *continuous* record, no matter how full of trivialities, will always gradually reveal something of the subconscious mind behind it. I've never regretted keeping a diary yet. There are always a few nuggets of literary value under all that sand.

The most difficult thing to make oneself record is the day-to-day climate of the mind, the contents of its reverie. Today I kept remembering that I'm still grossly overweight, about 158, and that I ought not to eat starches or drink alcoholic drinks. I have a vague desire to go to a steam bath which is partly sexual, partly a wish to get in out of the rain and cold.

April 24. I did go to a steam bath, the one on Jermyn Street, where amidst many of my fellow toads I ran into an autumn crocus, Horst Buchholz! What a cheerful bright-eyed shameless rogue he is, and how attractive still. He was much interested to hear about the play and eager for a part. I told him maybe he could play Patrick in German, and he replied, "First in German and then in English!" He and his old lover Wenzel Lüdecke are still together—in business. They recently started to shoot a film in Egypt. The Egyptians assigned them a place to shoot exteriors, without mentioning that it was part of an army camp. And then the Israeli planes flew over and bombed it! This shattered Horst's nerves so completely that he quit. And now they are being sued by their backers for not delivering the film and are suing the Egyptians for exposing them to this danger.

He made me feel very cheerful. I asked Clement to send him a script.

Then I went with David and Peter to the National Film Theatre to see some quite unbelievably boring German "underground" films. We—and enough other people to cram the theater—went (presumably) because of all the advance publicity put out by the maker of *Sisters of Revolution* and *Rosy Worker on a Golden Street*—the maker is called Rosa von Praunheim but is a young pale arrogant-looking man with shoulder-length hair, who is married.[1] *Films and Filming* carried an article by him, attacking nearly all other German filmmakers. It was illustrated by three stills of scenes allegedly contained in *Sisters of Revolution*—and not one of them appeared on the screen! Since two of the stills showed naked or

[1] Holger Mischwitzky (b. 1942), gay, Latvian-born film director and activist, took this "artistic" name in 1964. Rosa refers to the "rosa Winkel" (pink triangle) which Hitler made homosexuals wear; Praunheim is a town near Frankfurt. His fifty-some films eventually included two about Magnus Hirschfeld, *Gay Courage* (1998) and *The Einstein of Sex* (1999).

nearly naked men about to make love or rape each other, the audience might well have demanded their money back, but they bore it all with British resignation. There were only a few protests.

Then we went to eat at Odin's and were joined by Patrick Procktor, who was so irritating that I spoke to him quite angrily, which delighted him. He is a really malicious bitch. This morning I ran into him at Ron Kitaj's show.* Patrick had brought his dog with him into the gallery, and Ron came out, quite annoyed, and said, "I gave orders that no dogs were to be let in," to which Patrick replied, "I felt sure you couldn't be serious." Later, I'm almost certain I heard him running down Ron's painting to a young man, on the street, near the gallery entrance, as I was coming out.

Then I went hunting for a gold chain for Don, along Bond Street and in the jewellers' shops of the Burlington Arcade. Nothing even remotely suitable. The jewellers here say that the chain Truman Capote had from Bulgari's in Rome must have been copied from an antique one.

Then I had lunch with Jean Cockburn[1] and her daughter Sarah and three of their friends at a little restaurant in Chancery Lane. Jean looks old but still rather beautiful and she is very lively and active and mentally on the spot—and as political as ever. Sarah is a barrister and, according to Jean, hasn't cared to marry because "since she took to the law, she has seen so much of what marriage lets you in for." Sarah is rather plump but quite nice looking.

Seeing Jean made me happy;† I think if I lived here I'd see a

*The Kitaj painting I liked best is called "Outlying London Districts (in Camberwell)." Its power is of a kind one simply can't describe in words; the large figure is, in every sense, larger than life; it is powerfully numinous, its mere presence is like a statement about the world it inhabits. No, I shouldn't say "is like," it *is* a statement. (When I told Ron how much I admired this picture he was greatly pleased, and largely, I felt, because it is a recent one, painted in 1969. I suspect he is going through a period of unproductivity and doubt about his later work.) Ron certainly gives me a sense of being menaced by our present world, his art is profoundly disturbing. But often the pictures seem cluttered up with merely intellectual symbols, and the titles are so literary, "Apotheosis of Groundlessness," "Trout for Factitious Bait," "His Cult of the Fragment," "Primer of Motives [(Intuitions of Irregularity)]"—a truly American portentousness. One title I like, though: "Where the Railroad Leaves the Sea"; I like the painting too.

†I was particularly pleased that Jean had liked *A Single Man*; she even quoted from it the remark made by the Californian professor about the Oxford don's book on Quarles, that you need so much background to write a book like that. I suppose what pleased her was my taking the side of the underdog against the upperclass don!

[1] I.e., Jean Ross; see Glossary.

lot of her—that is, if I could do so without being involved in her communism. Her devotion to Sarah doesn't seem repulsive, either. Peggy [Ross] is an invalid most of the time, because of her arthritis, and their mother has had a stroke. So Jean is kept very busy.

This afternoon I worked with Bob Chetwyn on the play. Again I liked him very much and found him really intelligent in his comments and suggestions. It worries me a bit, though, that he now says that he doesn't think he *will* do the other play—they are making too many conditions—and so he is now determined to push on with the casting of ours. Also, he added that he wished Don could be here when he hires an art director. So, suddenly, I begin to wonder if I *shall* leave on Wednesday, after all! Suppose Tom Courtenay unexpectedly says yes? And, even if I do go, how long will it be before Don and I have to come back here again?

April 25. More chilly rainstorms, and just now some thunder.

Supper with Tony Richardson last night was not a success. Tony was glum, and we courtiers—David, Peter, Neil, Bob and me—were too familiar and perhaps too attentive for his taste. I should never have let David persuade me to push myself in, especially as I am to go there tonight as well. I wouldn't drink (which displeased Tony but has already got me back down to 154 pounds), and I felt dull. We went to an Indian restaurant (the Tandoori) and ate Tandoori chicken. Tony quite casually admitted that he has heard nothing from Albee yet and that Edward Bond's Nijinsky script is useless.[1] I couldn't help hinting, very very discreetly, that I'd be happy to take a shot at it.

Afterwards we went to a queer pub, the Coleherne Arms, and obediently waited around in the huge pre-closing-time crowd, until our Leader, who was wearing a fringed leather jacket, decided that he would stick around for a while and perhaps collect another recruit. I was somewhat favorably impressed by the clients—nobody smashing, but lots of possibles. David was unwell; "I have a *rotten* headache," he kept repeating, in dismay. Peter would have liked to go on the town.

Today I had lunch with [Marguerite and her companion] and with "Spider" Quennell[2] (whom I quite like) at a little restaurant called Parkes on Beauchamp Place which Marguerite says has five stars and is one of the best in London—a place run by (I guess)

[1] Albee eventually wrote the final script, but finances for the project fell apart.
[2] Sonia Leon Quennell (b. 192[8]), Soho beauty; from 1956 until about 1967, she was the fourth of five wives of Peter Quennell (1905–1994), Oxford-educated poet, literary biographer, magazine editor.

queens. The food was certainly okay—but we merely had steak sandwiches. Marguerite says Ivan is back, and is off with Kate at some country house party; but she doesn't believe there is a real reconciliation. Ivan looks very well and enjoyed himself in Spain and is inclined to think marriage is merely a habit and that maybe he has just broken himself of it. There was some discussion of his financial state; can he *afford* to leave Kate and just play around?

[Marguerite's companion] seemed very pleased to see me, sorry I'm leaving England and altogether affectionate. He told a very curious story—how he heard two middle-aged Jewish business men ("They were *Jewish*," [he] said, looking me in the eye and slightly underlining the word) in Fortnum and Mason's, discussing the best way of travelling from Calais to Switzerland. This seemed ordinary enough, although they did it in great detail, showing off a lot of information about roads, hotels, etc. But *then* they discussed how best to travel from London to Eastbourne—disagreeing with each other in both cases. And *then*—one of them asked the other, "How would you get from here" (meaning their table in the tea room) "to your car?" and they discussed *that*!

April 26. More greyness and rain. I *know* that I shall have to leave before they have any spring here—so I'll be doing the British a real service on Wednesday next.

Yesterday afternoon I went round to see Alexis Rassine. I'd resisted calling him simply because that infuriating old John [Lehmann] told me to, but really he is very pleasant to be with and I'm fond of him. A bald star of yesteryear in gracious and dignified retirement, with his old man's face and his youthful well-made body, he is still sexy and by no means on the shelf—indeed he is planning to open a dance school of his own and already has backers to form a company which will take care of it as a business operation. He sipped bourbon and repeatedly tried to get me to join him. I stuffily refused, because of my efforts to lose those sinful pounds. (Was *just* below eleven stone this morning!) Alexis likes to talk about actresses, what was Garbo like when I knew her, etc. He had seen some beautiful legs from behind in a shop—and the lady turned and *it was Marlene* [*Dietrich*]! He had met Ava Gardner too, she looked "glamorous" but she drank a lot and her language was terrible. (He pronounced the name as "Arva" and, like nearly all the English, called Houston "Hooston.") A big dog, referred to as "the Monster," kept trotting in and out, wanting to be played with and fetching articles of clothing. Alexis brushed me up the right way by saying he thought *A Meeting* was one of the best of my books. I answered that John doesn't think so; he much prefers the prewar ones, he once told me. Alexis protested. He also told me, "You never change—I always tell John you're one of the indestructibles." When we said goodbye, he kissed me warmly and I even think that, if I'd had that drink, a dissolve of ten minutes might have discovered us rolling naked on John's double bed. It would have been sort of suitable and maybe a barrel of fun.

I went on to Tony Richardson's, where I found Patrick Woodcock. We had had a little conspiracy about this, Patrick was to ask if he could come by for a drink—because there seemed no other chance of our meeting—and hope he would be asked to stay to supper. But he had never had the chance of talking to Tony, who was now upstairs resting.

Patrick was in a rather naughty-boy mood, rarin' to get drunk because he wasn't on call that evening. We immediately got onto the subject of dieting because *I* wouldn't drink, and exchanged confessions of our gluttony. Patrick doesn't eat *any* breakfast, but then he's apt to break down and stuff himself with cakes and candy. Tony came in and found us giggling and, I think, suspected some conspiracy against his dictatorship, but he became gracious and held forth to Patrick about a theory he had formed last night in the Indian restaurant—after looking through David Hockney's glasses he decided that they filter out color, and that this is why David's colors are so "cool." Tony also talked of his terror of blindness. And of his untidiness; he simply *refuses* to put his clothes away, just throws them on the floor, because "life is too full of things to do."

Then the two other guests came in, [...] Billy McCarty [and a friend]. They both seemed to know me or Don or both, I couldn't quite be sure. [The friend] is a very good-looking oldish youngish man with a slim strong figure, conservatively dressed. Billy McCarty is a decorator, pretty but not very, extremely tall, with a messy physique, skinny with a belly on him, hips too wide, wretched arms, no torso—dressed in a white sweater with a big fancy belt. However he melted me by praising Don's show—he had been out there when it opened, visiting the Duquettes. And he was jetting over to New York tomorrow to take on some extremely chic job; he's an American who lives in London but earns his money in the States.[1]

Tony talked amusingly of his awful experiences, years ago, when they took *Look Back in Anger* to Moscow and were at first treated like mud, given no transportation and almost nothing to eat—with

[1] He was later adopted by Douglas Cooper; see Glossary.

the food served hours after it had been ordered—and then, much too late, when the bureaucratic machine had at last registered the fact they were VIPs, given the red carpet and the grandest suite in the biggest hotel and "the best table" in the restaurant—which was right under a blaring band!

[Billy's friend], who seemed to be an authority on such matters, told us that, for a tourist, Stockholm is the most expensive city in the world.

Billy then began bragging about his girlfriends, including a man who'd had a sex-change operation. "Has she a proper cunt?" Patrick asked. "Oh yes," said Billy, "I *know*, because I've fucked her." "Has she a clitoris?" I asked. "Well no—no, not exactly," Billy admitted. "Thank you," I said, "the case for the Crown rests." Patrick, now quite fairly drunk, drove me home. He denounced [Billy and his friend]; they were ruthless spongers. Billy would latch on to anyone of either sex, if he or she was rich. They had both been living in Tony's house and had without doubt got money out of him. "Aren't I a shit?" said Patrick, repenting. "Tell Don I'm still a shit." "He doesn't think you're a shit." "Oh," said Patrick, as we hugged each other goodbye, "Don knows me very well indeed."

A talk with Jean Cockburn on the phone, this morning. She says it's quite untrue that the journalists got her to go to a performance of *Cabaret* and meet Judi Dench; she refused to do so. She thinks it was Claud Cockburn who gave her away. But she says that Claud's son (by his present wife) who is now in college is often asked if Sally Bowles is his mother![1] She told me that Sarah had liked me and said I was "very self-contained and quick on the uptake" and that she'd asked Jean, "Was he always like that?"

This morning, some facts of life. Norman Prouting came down to talk about the amount he calculates I owe for telephone, heating, gas. We decided on $200, to be adjusted later, if too much or too little. This seems to me conservative, considering all those phone calls. What I do begrudge are the $173 I shall have paid out for the largely unneeded services of Mrs. Gee. This means that the (not quite) eleven weeks I shall have spent in this apartment will have cost me about $34 a week. Well—and suppose I'd lived in a hotel? Don't be such a miser, Dub.

Also talked on the phone to Dodie. She told me that (as I already knew) she'd met [Don's friend] and had told him I'd told her he

[1] Cockburn and his wife, Patricia, had three sons, later leftist journalists: Alexander had already left Oxford, Andrew and Patrick were still there.

didn't like our play. He had denied this strongly—and had repeated this story about having been shown it by Don "in the middle of the night" (who but a born mischief-maker would say that?) and how he'd had no chance to consider it properly. But that didn't stop him writing me that idiot letter, suggesting that it should be given as a sort of dramatic reading. Dodie thinks that he was a bit shaken because she had praised it so much in his presence.

Nearly midnight. Have spent the latter part of the afternoon talking to Bob Chetwyn about the staging of the play. Then I took him and Howard Schuman out to supper at The Hungry Horse, where we found ourselves next to Tom Courtenay, who recognized me at once; I didn't recognize him, he had a beard. So, at the end of the meal, Courtenay told me he'd heard his agents had a play of mine for him to read—this after we'd been told he'd been reading it for the past week! He vowed to get to it at once.

The relationship between Chetwyn and Howard is now clear. Howard giggled when offered treacle tart, saying Bob had told him to take off ten pounds. And, snooping along their bookshelves, I opened an edition of Donne and found it inscribed to Bob, "The reason for this is explained by the great Donne on page nine, H." The poem on page 9 was, "For God's sake hold your tongue and let me love."[1]

April 27. This from Dicky Buckle's *Sunday Times* article on the Polish Mime theater's ballet program at Sadler's Wells: "There are four outstanding performers ... Stefan Niedzialkowski, the blond shepherd boy, whose little nose might cause a Trojan war, and whom I should confess to being a little in love with if I were not afraid to embarrass my grandchildren." I'm going to see them tomorrow night, if possible, with Bob Chetwyn and Howard.

A sad letter this morning from Harry Heckford. He really hasn't a chance of getting his book published, I fear—must call Cullen a bit later to find out about this—and now he needs money to take a degree at the University of Liverpool. Do I know of an American scholarship which could help him?

April 28. Today it's a "beautiful" morning, but I don't trust it. This foul climate will probably assert itself as we drive to Cambridge. Bob [Regester] is coming to pick me up in about an hour, and we're to see Morgan and I hope Bob Buckingham, who wrote me he was coming down today.

[1] "The Canonization," 1.1.

David (who, like all of us, has the defects of his virtues and simply can't understand why we friends shouldn't want to get together in any conceivable combination at any time) has wished Peter on to us as a passenger; he isn't able to come himself. I don't feel this matters as far as Morgan is concerned; he likes bright eyes and bushy tails around him, as I do. But Bob Regester may well resent it, and by the worst of luck his phone is out of order this morning so I can't reach him and explain. Bob may well want to be alone with me to talk about his domestic situation. While he was at Le Nid de Duc, after most of us had left, he went down one evening into St. Tropez alone and met a French-American boy named Barry and this was a very big thing. Bob says he's never had such absolutely mutually perfect sex in his life, and of course Barry is already getting a bit possessive and starting to influence him. Barry doesn't like hippie clothes and long hair, so Bob is going to cut his hair and shave off his moustache and dress more conservatively. (He took me to a wholesale warehouse where they sell the latest in way-out shirts and pants, and it was amusing that he bought exactly the same more or less sober things I'd picked out for Don!) Neil objects to all this, strenuously, though he is trying hard to be reasonable. (My remarks to him the other day seem to have made an impression.) However, the other night, he announced that he was bringing a boy back to the house. (N.B.: It is the injured party who *always* behaves worse than the other!) So Bob, hurt by this proposed violation of the sanctity of the home, said he'd spend the night with a (non-sex) friend. And then of course Neil wouldn't go through with it, which merely made Bob angry.

During our time together, we passed Cranley Mansion (It *is* still Mansion, not Mansions) and got out to look at it. It's a very well-preserved tall brick building with nicely designed ironwork on its balconies (flowers in, maybe, the manner of William Morris). I noticed that the balcony rails were rather low and not very massive—so no wonder Frank felt vertigo while standing by them.[1]

Had lunch with Nick Furbank, who brought me a (very poor) xerox copy of Forster's story about the lovers on the ship.[2] He is a strange pallid sly little thing and I imagine he's expert at winkling information for his Forster book out of his informants. From me he got a memory of the Spanish Civil War period—how, when

[1] As Isherwood tells in *Kathleen and Frank*, his mother lived with her parents in Cranley Mansion on Gloucester Road, South Kensington, from the early 1890s and throughout her difficult courtship with Isherwood's father, who suffered from vertigo all his life.

[2] "The Other Boat."

everyone was planning to go out to Spain and showing off a bit, Morgan was asked, "Why don't you come?" and he answered, quite simply, "Afraid to."

Nick wanted to know how intimate Wystan had been with Morgan. I said not very, and added that Wystan has always found it difficult to be intimate—he's shy in that way. Even this little confidence I somehow regretted as soon as I'd made it. But perhaps I'm being unfair to Nick.

Dicky Buckle told me that David Hockney likes Don's paintings far more than his drawings—says they're so much freer and more original. I went out with him last night to the ballet; it was *Giselle*—with its *They Shoot Horses, Don't They?*[1] scene at the end of it.*

Dicky had an art student with him named Ian Lewis, a brown-eyed tall boy with heavy cheeks and shoulder-length hair (nearly) who was quite pleasant but rather stupid-looking, like a dull girl. (Actually I don't believe he is dull at all. Probably Dicky reduces him to silence by his ill-advised showing off, he parades his literary and artistic knowledge but never really tries to entertain Ian on Ian's own terms as he should—for he claims to be in love with him.)

Later we had a quite horrid snack of scrambled eggs and tepid mini hot dogs at the flat of Dicky's secretary, David [Dougill][2] and his friend Richard Davi[e]s.[3] I didn't mind the scantiness of the snack, however, for it was all in aid of my dieting—am now almost down to 152 again. And the boys are both bright and friendly. (Don thought David Dougill would be a nice playmate for me while he was away; and David is indeed the kind of playmate that someone's lover would pick for him under such circumstances! He's adequately nice-looking but just quite hopelessly unattractive.)

The boys are late; it's ten to ten. Have just talked to Neil on the

*And I remembered Bill Caskey's intense indignation when the Wilis finished off Hilarion [by dancing him to death]. Right in the midst of a soft passage in the music he exclaimed furiously, "The *cunts*!" and several people around him laughed.

[1] The 1969 film about a dance marathon.

[2] British dance writer and broadcaster (b. 1944); he was Buckle's research assistant, 1969–1975, and then succeeded him as ballet critic at *The Sunday Times*. Isherwood wrote "Dugal."

[3] Teacher (d. 1991), of Religious Studies at a North London school; in the 1970s, he also became known as a ballet writer for *Classical Music*. Their flat was in Hampstead High Street.

phone. He is a bit mournful. Said the older he gets the more he hates to be alone.*

April 29. Again a "beautiful" morning, like yesterday which ended in rain however. But we did have a fine drive. Bob didn't mind having Peter along and Peter was as sweet as usual. He admitted he had sulked during their weekend with Cecil Beaton. Cecil, he said, always treats him as the boyfriend of the great painter, and when Cecil was photographing David he made Peter stand a few paces into the background. Dicky Buckle and Ian Lewis had been there for a meal. Cecil had described Ian as having "a Brontë face," which I find a quite brilliant, if flattering, description.

At Cambridge were Mark Lancaster, Richard Le Page, Richard Shone (whom I met with Nancy Ackerley on March 19) and a tall dark boy with beautiful blue eyes named Paul Wheeler. He is a singer, who composes his own songs. Richard Le Page is maybe stuck on him. Anyhow they are planning to come out to Los Angeles together, when Richard lectures at UCLA later this year. Paul hopes to earn money singing.

Of course the boys wanted to see Morgan, but they obviously couldn't sit round him for hours. So I arranged it that they came in to get me, just before 4:00, when we had to drive back to London and Morgan had to go out to tea with an Indian.

Nick Furbank had told me he was sad, and Morgan told me so too; he admitted to being sad and lonely and said he hoped he'd "go" soon. I asked him why he was sad and he said he felt "so empty." He kept repeating, "There's nothing new in this room," and, "I've got nothing to show you." But he also repeated, "I've had such a nice sleep, I feel so comfortable," and he showed me how he napped on his lopsided broken-down sofa (quite a feat) with one foot on the floor, a sort of sidesaddle position which prevents him from rolling off. I sat down on the floor beside him and tried to reassure him that I didn't have to be entertained; but the problem wasn't so easily solved. We are simply in different predicaments. "We perish, each alone,"[1] is too melodramatic for this case, but, let's say, we are both awfully busy being ninety-one and sixty-five respectively.

*I told Bob about this conversation, later. He replied that Neil loves being miserable and will find excuses for it on all occasions.

[1] Cf. William Cowper, "The Castaway," "We perish'd, each alone," the line which Mr. Ramsey repeats to himself in Virginia Woolf's *To the Lighthouse* (1927).

Morgan was pleased when I told him how much I like his story about the lovers on the ship. And he was perfectly aware that *Howards End* has just been performed on T.V., though he hadn't seen it. Indeed, as when I saw him last, he seemed well aware of everything, and the boys, when they came in, found him much more alert and generally in better shape than they'd expected. Bob said later that Morgan laughed at me, when I was switching off the electric fire for him, "As if you were the village idiot." I do clown for him a lot—that's largely my nervousness.

While I was seeing Morgan, after lunch, Peter went into town with Bob and bought a boater for himself, a real old-fashioned hard straw hat, now back in style. Bob bought me a tie and a scarf with the Corpus Christi colors, they are quite a pretty red.

Before this, we all went for a walk over the bridge and into the Fellows' Garden.* Paul Wheeler had read in *Lions and Shadows* how I left Cambridge and he told me he felt just the same; he longed to get out. When indirectly flattered like this by a beautiful boy (well, he doesn't even have to be beautiful) I feel overweeningly pleased with myself and hastily pray to be released from vanity. But, really, it *is* satisfactory to walk these paths again amidst the prison-past and feel that I escaped in time and that Edward and I were *right*!

We got back in time for me to rendezvous with Robert Chetwyn and Howard Schuman and drive with them to the Polish Mime Ballet. In the intermission we met [an English boy I used to see in Los Angeles in the late 1950s. He was] at the bar, drunk. He is coarser looking, thick necked, getting plump; his nose is now definitely much too short. But I mustn't forget all the fun we had together. (I remember asking, "Does it hurt?" and him saying, "I *want* it to hurt.") Howard Schuman knew him and Norman Prouting says I actually came to visit [the boy] in this house, when he was staying here in 1961! I dimly connect this with another memory; [the boy] wearing nothing but a leather jacket. But last night he was just rattling on tiresomely and name dropping and being loud.

The Mime Ballet was unforgettable—at least *Gilgamesh* was; the other one, *Baggage*, wasn't quite so exciting and showed more of the negative or minor qualities of the Polish approach; rigid stances, capricious body-jerks, poses with open-fingered hands,

*Peter turned eyes and heads and maybe some hearts in his beautiful white suit. I said to him, "It's sad, the time in your life when you can really wear clothes is also the time when you oughtn't to be wearing any." This dazzling epigram didn't have the success it deserved, when I repeated it later at Odin's.

eccentricity, distortion, campy-macabre satire. *Gilgamesh* really does convey an epic archaic quality and at the same time its psychology is altogether modern. The hero-brothers wrestling naked, falling in love, sleeping with hands clasped, becoming involved in and escaping from the snares of women, dying with desperate spasms, wandering into terrible underworlds or dreams where a huge bird flutters its wings with a most intense menacing vibration. I have no idea what this ballet was about, except that it was about all of us. Pawel Rouba and Stefan Niedzialkowski (the one Dicky Buckle likes) are savagely beautiful, Pawel dark and Stefan blond. Pawel's body is dark brown, hard and faultless, Stefan's is white and voluptuous, all the more so because he has just a tiny bulge of flesh above the flat, sometimes concave belly with its deep navel—he'd better watch it.

I went on to Odin's for a farewell supper with David, Peter, Ron Kitaj, Melissa(?) and Chaik(?).[1] Peter Langan tried hard to make me drink. David ordered champagne. But I meanly wouldn't—I even slipped my champagne to Peter while pretending to sip it.

Kitaj drove me home. He is furious with Nigel Gosling[2] for saying in *The Observer* that, "Kitaj lacks the gift (often granted to lesser artists) of catching a likeness." Not, Ron explained to me, that he *wants* to get a likeness, but he always could, any real artist could. So he rang *The Observer* and spoke to Gosling and told him that he had no right to write this because he'd never seen the originals of the portraits. Gosling is to be brought down to the gallery and made to eat his words—"like you rub a puppy's nose in its shit."

Becoming less upset, Ron said that David is the greatest artist of our time, or his generation; I forget which. When we said goodbye, Ron said, "It's been a good time."

This morning, Clement told me that Tom Courtenay has turned the play down.

3:00 p.m. Here I am, out at the airport. Packing was difficult, because the bag I bought in exchange for the one that got damaged coming over here is smaller than the old one. I talked to Bob Chetwyn before leaving the house and tried to encourage him to take a personal interest in casting—that is, see prospective actors and talk them into taking the part before their wives or agents can talk them out of it. Clement came round to see me and really I

[1] Melissa North (b. 1944) was a rock and roll booker's assistant; later, she ran a furniture shop, designed clothes and, eventually, interiors. Tchaik Chassay, an architect and a founder of the Groucho Club, was remodelling Hockney's flat in Powis Terrace. They married in 1975.

[2] British art historian and art and ballet critic (1909–1982).

must admire the way he bounces back from all these disappointments. But I think he is now beginning to think in terms of a production which won't be in the West End. So much the better, probably. Richard Schulman also called. So did Richard Shone. And I called Patrick Woodcock, Marguerite and David Hockney and Peter to say goodbye. "You cheered me up," Peter said.

(A lady here at the airport just asked me if I was Christopher Isherwood and then explained that her father, a New Zealander, is an Isherwood. Did we come from Lancashire, or where? She then asked, "What's David Hockney like, is he as mad as a hatter?" I said, "He's super.")

Have been having lunch at the Vedanta Center. Swami Bhavyananda was wearing a sort of smoking jacket over his *gerua*—it made him look like a very good-humored gangster. Buddha is skinnier than ever. He has now fixed up the shrine on a proper pedestal and was very pleased when I commented on this. I ate some deadly fattening little cakes at lunch because they were *prasad*; Buddha then told me, "You didn't have to." The swami had been giving a talk in Liverpool, at a college. They had asked him if it wasn't selfish to retire from the world and become a monk who no longer does anything useful to help others. The swami retorted that pure scientists were also engaged in research which didn't help others, yet nobody criticized *them*. I must say, I found this argument somewhat specious.

Buddha and another monk, the big English boy, drove me to the Pan American terminal. I was grateful to Buddha when he said he couldn't stay and chat; he had work to do. Buddha says that when he's a swami he hopes to be allowed to stay here at the London center. He has a horror of India, says the centers are all too hot or too cold, and he's always getting sick there.

April 30. Safe back here in the beautiful casa, with the sun getting slowly ready to set after a perfect day. It is chilly for California but so warm and heavenly after England, which now seems every inch of its six thousand miles away. As for this time yesterday, it's like something which happened a month ago.

The flight was deadly dull and the plane crammed. This is one of the most uncomfortable periods of travel we have had, probably, during the past seventy years. The seats are squeezed together so tightly that you can barely get in and out of them, and your briefcase or bag takes up half of your legroom. No doubt, in a few years, these old-fashioned jets will have become quite spacious again. I read in a magazine article that they will redesign the interiors and take out a lot of seats in order to offer a counter-attraction to the jumbo jets.

To make matters worse, Jean-Louis Barrault[1] and his entire company were with us, flying out to San Francisco to give their *Rabelais* show there. The Frogs were aggressively noisy and jokey and they walked up and down the aisle and stood over us, gossiping. At one point they got *so* noisy that some member of the crew switched on the "Fasten Seat Belts" sign to make them return to their places and sit down—at least, I can't think of any other reason for switching it on; the air at that time was totally calm.

At last, at last, after about eleven and a half hours, a movie (*The Molly Maguires*) a terrible meal and a worse snack, and around fifty pages of *The First Circle*, we saw the shining sea and the vast constellation of the city and bumped to earth (the Frogs clapped) and as I came out of the plane I looked at the big window and there was Don, standing just where I've so often stood waiting, his face dark golden and his hair parted in the middle and more silvery than ever, smiling that marvellous smile that every human being must wish to see on the face of someone waiting for *him*—what happiness that was! Even the passport examiner was pleasant this time, he knew who I was and said it was a pleasure to have met me, and the customs inspector hardly poked into my bags at all and I got out quite quickly and there was Don himself, no longer behind glass, and I was home again and the luckiest old dobbin of them all.

(The funny thing is, I wanted to fill this notebook exactly and I have; I dislike having half-filled notebooks lying around. Yet I didn't make any effort to work out the number of pages to be written per day. I couldn't have, until very near the end, because I didn't know exactly when I should be leaving—even at the last moment the acceptance of a couple of actors could have changed our plans. It just happened like this.)

[1] French actor, mime, director, writer (1910–1994), best known for his film role in *Les Enfants du Paradis* (1945).

May 27, 1970—August 26, 1972

May 27. This is to launch another volume of diary. It's exactly four weeks since I got home from England. During my time alone there, after Don had come back here for his show, I kept a diary every day; oddly enough it was only a short while ago that it occurred to me I'd kept it in handwriting—which means that my much-complained-of arthiritic thumb must either be much better or so much a part of my normal experience that I don't notice it any longer! I do still prefer handwriting for a diary, this typing *is* an obstacle and makes you self-conscious, and I seriously considered going out and getting a notebook and writing in it this time too. But somehow that seems infantile here. When I'm away and haven't a typewriter it's necessary. Here, it's just a caprice. Handwriting gives me a sense of privacy, but who's kidding who[? N]othing I write will be permanently private unless I burn it; and I can just as easily burn a typescript.

Am now really plugging away at *Kathleen and Frank*. My objective is to finish the three chapters covering the Wyberslegh period before my birthday, three months from now. Well, one chapter is roughed out already. But now there's this prospect of maybe working on this film for Dean Stockwell.[1] More about that after I've talked to him tomorrow.

No more news from London. Clement Scott Gilbert writes saying why don't we send the script to Richard Burton, who's actually in town here at present. Robin French discourages me from doing this, perhaps because he has other plans for Burton. Anyhow neither Don nor I feel Burton would be any good, even

[1] Child actor and, later, leading man (b. 1936); his films include *The Secret Garden* (1949), *Kim* (1950), *Sons and Lovers* (1960), *Long Day's Journey Into Night* (1962), and *Married to the Mob* (1988).

if he wanted to do it. Clement has nobody else in prospect—except Albert Finney! Chetwyn hasn't written at all to say what he thinks of the rewrites we've done since I've been home. Clement thinks it's wrong for Tom to speak directly to the audience at the end and wrong for him to dance with Penelope and Mother. I'm *sure* we're in the right about this. In fact, I think the idea of their dancing together is brilliant!

May 28. I went into my workroom after breakfast this morning and, after a few minutes, I heard Don talking to someone and he called to me and I came out and there was David Hockney. I could hardly believe my eyes. David had just flown over nonstop from London. He looked younger and sprucer, although his jacket was stained, and he smiled in a self-contained manner. He seems to have left England quite suddenly. Don feels there is something wrong between him and Peter. David merely said he felt he had to get away and draw. He is staying at the Miramar.

Dean Stockwell called to say he can't come and see me today.

We had Evelyn Hooker and Jo Lathwood to supper last night. While we were fixing the food, Evelyn and Jo had a big heart-to-heart. That was the point of the evening; Evelyn had said, rather rashly, that she'd like to give Jo some advice as a psychologist. So of course Jo needed no prompting to pour forth her woes. I heard Evelyn saying, "But why do you *torture* yourself with such thoughts?" At the end of the evening she had to concede that she didn't think she could help Jo at all. At least not as long as Jo maintains her present attitude. When Evelyn reminded Jo that she too had lost someone, Jo implied that that wasn't at all the same thing and not nearly so bad—in other words, better Ben [Masselink] dead like Edward [Hooker] than married.

I have made a resolve, to read the rest of Dante before my birthday. Paradise or bust.

May 29. Last night we went to a lecture by Shirley, Irving Blum's wife and my former colleague at U.C. Riverside, on modern American art. It was very disappointing, deeply infected by that dreary heresy which regards artworks as automobiles which keep rendering earlier models "obselete." Thus, Monet's waterlilies were made obsolete by Jackson Pollock and his haystacks by Larry Bell's glass boxes—that was the implication. Shirley is a really very sweet girl, though. We had supper with her and Irving and David and Brooke Hopper (who had gatecrashed the party) afterwards. The evening was originally designed as a gesture toward the Blums, a sort of thank-you for Don's show. Don is mad at Brooke because she told him she'd buy the drawing he did of her if it was in the show and then she backed out and pretended that it wasn't flattering enough! She is terribly stingy.

Meanwhile, David has already started working. Asked what he would draw at the Miramar, he said everything in the room, the furniture, his clothes, the view from the window. How simple he makes life seem! Still no hint that there is anything wrong between him and Peter.

A nice businesslike letter from Bob Chetwyn this morning. He is quite pleased with the rewrites and he mentions again the possibility of trying the play out in a small theater without any stars. We favor that too, of course.

May 30. Am depressed, because it's a grey morning and a public holiday, Memorial Day, and because Don is depressed about his work. In moods like this I always feel acutely the nervousness and instability of the life everybody is leading here. I don't mean just the war[1] and the recession[2] and all the other political tensions—I mean the jitters of nowadays, the strain of living now and here. Partly, of course, this rattles me because I'm getting old; I feel I can't keep up with it all. Why do things have to change so fast? It no longer seems exhilarating that they do. For instance, I mind enormously that they finally are going to put up this monster apartment building at the end of the street, two twenty-floor towers. And yet, why not? Why shouldn't we have to move? We've been here ten years, already.

Every day I say in my prayers: help me to know that you are my *only* resource.

After all this long time, Don decided to put a piece of plywood under his half of the bed; I've had one under my half since the beginning. It is a great success; he says he sleeps much better and it helps his back.

David, Jack [Larson] and Jim [Bridges] to supper last night. I barbecued swordfish steaks. Don and I definitely do not have the art of casual cookery; we always make such a production out of it and I'm sure this gets on our guests' nerves. Jim is deep in rehearsals of a lot of short plays, including two of Jack's; he only brought part of himself to the house. And David wasn't entirely present either; at least, I felt he wasn't, having become accustomed to seeing him

[1] In Vietnam and, since April 29, Cambodia.

[2] The 1969–1970 recession was a mild one; see Glossary under recession.

always with Peter. Jack rattled on about Nixon and the students.[1] He feels a better age is at hand, when all the old conservative farts will have died off.

May 31. Soon after I finished writing the above, Swami called on the phone and read me a passage from a letter Swamiji wrote to the Hale sisters on July 31, 1894. Swami said, "I was reading it this morning and it made me cry":

> Say day and night, "Thou art my father, my mother, my husband, my love, my lord, my God—I want nothing but Thee, nothing but Thee, nothing but Thee. Thou in me, I in Thee, I am Thee. Thou art me." Wealth goes, beauty vanishes, life flies, powers fly—but the Lord abideth for ever, love abideth for ever. If here is glory in keeping the machine in good trim, it is more glorious to withhold the soul from suffering with the body—that is the only demonstration of your being "not matter," by letting the matter alone.
>
> Stick to God! Who cares what comes to the body or to anything else! Through the terrors of evil say—my God, my love! Through the pangs of death, say—my God, my love! Through all the evils under the sun, say—my God, my love! Thou art here, I see Thee. Thou art with me, I feel Thee. I am Thine, take me. I am not of the world's but Thine, leave not then me. Do not go for glass beads leaving the mine of diamonds! This life is a great chance. What, seekest thou the pleasures of the world?—He is the fountain of all bliss. Seek for the highest, aim at that highest and you *shall* reach the highest.[2]

While Swami was reading this, I kept saying to myself: He is telling me this because he knows I need spiritual instruction. I am being instructed by a saint. Even if I can't feel much, I do believe that things like this are being stored up inside me and that they are valuable, in a way I can't yet imagine.

Don says he is more depressed when he's with me because I'm so optimistic. When he's alone he knows he can't afford to let himself get depressed. But you can't help that, he says. It's just role playing.

In the afternoon we went with David to Griffith Park, where there was a Gay-in. Only it wasn't very gay or very well attended. The police had been by, earlier, harassing them because they were

[1] Antiwar protesters; see Glossary under Nixon and the students.
[2] *The Complete Works of Swami Vivekananda*, vol. 6, pp. 259–262.

distributing leaflets without a permit. Nobody got arrested but it scared a lot of people off. Lee Heflin was there, and a friend of his stamped our hands with the sign of a hand in purple ink, denoting some gay-liberation front group; they took a lot of scrubbing to get off. Lee introduced me to an elderly man named Morris Kight(?)[1] who was wearing a silk dressing gown and a funny hat and who appeared to be directing the proceedings. He married two pairs of girls, explaining that this wasn't a marriage but a "mateship." We had to join hands and chant something about love. Kight also introduced me publicly and called on me to speak, so I said, in my aw-shucks voice, "I just came here because I'm with you and wanted to show it." There were several journalists with cameras and quite possibly Don and I will appear in the *Free Press* or *The Advocate* or elsewhere. Well, at least it was a political gesture of sorts. David was taking photographs too, as usual, and a black boy came up and protested that he didn't like his picture to be taken without his permission. I thought maybe there was going to be trouble but there wasn't. The black boys were almost the only attractive ones there. There were lots of dykes, black and white.

Then we went to see a film about Vietnam, [*In*] *the Year of the Pig*. The most damning parts of it were simply clips of various politicians, making speeches. It was being shown at the Bay Cinema in the heart of Republican Pacific Palisades, but the audience was strongly in favor of the film, quite big and mostly young.

Have finished a rough draft of chapter 12 today—the second chapter I have finished in rough since I got back here and started working again. Now I'm wondering if maybe I won't try to get right through to the end of the book in rough and then go back over the whole thing later.

June 8. Yesterday morning, a woman called me from New York—I think she was representing United Press—and told me Morgan had died that day. Of a stroke or a heart attack or maybe both, I forget. She said he died at Coventry, which means with the Buckinghams. I was talking about it to Don and I said, "He really had a very happy life, he was very lucky." As I said this I began to cry a little, and Don kissed me and said, "So is Kitty."

But I can't feel sad about this. When I last saw him, on April 28, he told me he wanted to go. And he didn't die alone in that big chilly bedroom in college; he was snug and warm and tucked up and looked after by May and Bob.

[1] See Glossary.

He is absolutely alive inside my head as I write this. I have been living so long with him in my head that I know he won't fade out until I do.

This morning we went down to see Michael [Barrie] and Gerald [Heard]. Michael told us that Gerald became quite lucid yesterday for a short while and told him that, living as he does between the two worlds (or however he put it), he sees quite clearly that this world is held together by "the demiurge." So Michael asked him, "Do you mean Ishvara?" and he nodded.

David Hockney went back home yesterday, planning to make a stopover in New York. We both of us feel very fond of him, and admiring, too. Don said that he has an "easy grip" on things.

On Saturday night, when we were giving him a goodbye supper, Nick Wilder (whom David had asked us to invite) arrived wearing a kind of formfitting dark female gown, and leather boots. He also forced us to have his friend Jason in after supper. Jason talked a lot of shit about art which greatly irritated Billy Al Bengston. Billy Al said, "I'm an autocrat," and added that nobody was going to talk to him about motorcycles who couldn't get out there and ride one—"faster than I can" was implied.

June 12. Today I finished chapter 13 of *Kathleen and Frank* in rough. That means that I've covered more or less all of the family history material, Marple Hall as symbol of the Past, the ghosts, etc. Now my feeling is that I'd like to push on hard and try to get as much roughed out as I can, before some new interruption takes place. Playing projections is really a meaningless game, *but*—if one projects two weeks per chapter, then I ought to be able to get five more chapters done in rough before my birthday. That would mean I'd have eighteen out of twenty chapters done, assuming that I can do the rest of the book in seven chapters, which I think is really quite possible. I don't want to dwell nearly so much on Strensall, Frimley and Limerick.[1]

Yesterday, Peter Schneider came over with Jim Gates, and Don did three of his very best pen and ink drawings of Peter. I honestly can't see how anyone wouldn't have to admit that they put David's work to shame; but that's a silly remark, because of course David's not being able to draw too well is the whole point. Don caught so beautifully the grossness and the cuteness of Peter, his thick gross lips and his big dirty feet. Peter actually looks adorable now, with

[1] Towns in Yorkshire, Surrey, and Ireland, where Isherwood lived in childhood while following his father's regiment before W.W.I.

his beautiful golden skin; while Jim seems to be getting homelier by the week. Peter has decided to go and live in the back country someplace and grow nuts and meditate, and Jim seems to be thinking of coming with him. Peter is very grand and superior about the "typical American adolescent" he has staying with him now (and paying him eleven dollars a week rent). This youth smokes, and he says fuck and shit, and it's so awful to watch him getting up in the morning that Peter has taken to sleeping out in the yard in a blanket. Nevertheless, Peter feels that the Ramakrishna atmosphere is starting to work; the youth is slowly improving. He is the same age as Peter, nineteen.

June 15. Gloom today, despite the glorious weather, because a writers' strike is almost certain to be called for tomorrow[1] and somehow or other we have to find a way of wriggling out of picket duty. Don has a much better case than I do, because this really isn't his profession, only a sideline, and he can maintain with a great show of reasonableness that he has to keep making appointments to draw people and really cannot be expected to wear himself out trudging back and forth in front of a studio. They have already asked us both to picket and I have said I will, tomorrow, just as a token of willingness. I'm only supposed to do two hours. They have asked Don to do three, and to show up at six in the morning, but that's the day after, so maybe he can get out of it.

Swami had a mild attack of flu, so I had to give a reading at the temple yesterday. I read Vivekananda, Brahmananda and the Gita. Swami much better already. I think this is one of his demonstrations to nudge the Belur Math about his assistant.

We had supper with Gore, who showed up in town as unexpectedly as usual and is leaving today. He said, of Morgan's death, "Well, we've all moved up one rung higher." That's such a typical remark; he sees everything in terms of competition. Now he's not so anxious to withdraw from the States to Ireland; he wants to start a third party with Dr. Spock.[2]

May Buckingham writes: "Dear old Morgan was taken ill in King's, in Hall actually, a fortnight ago. His legs gave out but his mind was perfectly clear. We brought him here at his urgent request

[1] Writers Guild of America contracts with producers expired June 15; writers and animators wanted a bigger share in arcane but valuable areas like possessing credit, runaway production and residuals.

[2] Benjamin Spock (1903–1998), pediatrician, psychoanalyst, and author of the bestseller *Baby and Child Care* (1946), campaigned against nuclear weapons and the Vietnam War and ran for president in 1972.

and he gradually got worse. This morning he died. We were able to look after him to the end, which was what he so wanted."

June 16. Last night Don and I spent a good deal of time discussing whether or not we ought to go to the Writers Guild meeting—we neither of us wanted to and we were so snug lying on the bed watching T.V. and eating corn on the cob and fried eggplant. I argued that we actually should not vote against the strike because our motives were personal, we'd be simply voting against it as a personal inconvenience, and we were planning to get out of picketing anyhow, if we possibly could. This was maybe disingenuous but it convinced both of us. Later Jim Bridges called to say that he had arrived at the meeting to find that it had been dissolved without a vote, because someone had raised the alarm that there was a bomb planted in the room!

Today, Jim tells me that there'll be another meeting on Thursday night and that a strike vote seems nearly certain. As for the bomb[,] it must have been a particularly false false alarm, because the fire department wasn't sent for and no one was cleared out of the rest of the hotel (the Beverly Wilshire). Jim says he heard that the meeting last night was very bitter on both sides. One can't help wondering if the bomb scare wasn't a trick to avoid the possible antistrike vote, relying on the fact that something of this kind can be blamed on the producers (vaguely) and so used to stir up hawkish sentiment. Also, when a vote is deferred, it is probable that a lot of people won't show and thus leave the field clear for the determined minority which runs the guild most of the time without opposition. This all sounds fantastic, almost paranoid, and it's characteristic of the present atmosphere in this country that I should be thinking such things even half seriously.

June 19. Jim Bridges has just called to tell me that the strike is off, at least temporarily; he doesn't know any details. Of course I am delighted. No excuses to be made, no sulks to be sulked. I can get on with the book, which is all I want at present; and it begins to look as if it may be shorter than I expected. I may even get through Frimley in one and a half chapters and Limerick about the same, which would mean six chapters in all, including the three about Wyberslegh and Marple. Then, surely, there would be only two more to finish the book.

Dr. Allen rather tiresomely discovered that my cholesterol count is too high and said I must go on a diet. And Dr. Ashworth says my contracture is slightly worse and must be closely watched.

Yesterday a man from the BBC in London named John Drummond called me about a program they are doing on Morgan (he wants me to do a bit of it) and told me that Gerald Hamilton died, on the 17th. Now I feel sorry I didn't get in touch with him before I left London. I don't give a damn whether or not he swindled us—though I would love to know, simply out of curiosity. But I do remember all the fun we had together. He had a very cozy personality and an animal innocence, you felt he was only acting according to his nature.

The day before yesterday, Don had a conversation with Irving Blum. Don has been so miserable lately, feeling that his life is all wrong, not at all what he would have wished for himself, that he isn't free, that he's getting old and ugly and, above all, that he is coming to the end of the kind of drawing he has been doing and doesn't know what to do next. All these statements are unanswerable, because if he believes them they are true, so you cannot argue with him. So I begged him to talk to Irving, since, after all, Irving has believed in him and has arranged these shows. Of course, I don't know exactly what Don said to him—Don was very much afraid of insulting him by suggesting that the "failure" of the Los Angeles show was somehow Irvings's fault[1]—but, anyhow, Don did ask for advice and Irving rather wonderfully made a definite suggestion; Don should draw a team of women roller skaters. This seems to have turned Don on at least a bit, and he is investigating their doings, where they race, etc. The idea is to get away from portraits as such (that is, pictures of people who demand that they shall be likenesses). Personally, I feel that Don achieves this every time with his new pen and ink technique; these last drawings have been most of them marvellous. It's largely, I think, because they are full figure and often surrounded by an environment, while his pencil drawings tend to be heads with very little else. When the pencil drawings succeed, it's because of a psychological penetration; therefore he has to draw people who are worth penetrating. The ink drawings make a more generalized poetic statement, so they can be of almost anybody.

June 21. Thick white fog. Don has gone off to see the women roller skaters this afternoon, then have supper with his parents. But meanwhile something has happened which I feel, hope, is most

[1] Only one picture sold—of Salvador Dali—to a friend of Bachardy's rather than to a client of Blum's. And Blum failed to press Brooke Hopper, his own friend, to buy the portrait of herself which she asked Bachardy to do and to hang in the show. This portrait was praised in the *Los Angeles Times*.

awfully important; Don talked to Billy Al Bengston on the phone this morning (we were there last night for supper) and told him all about Irving's suggestion and asked his advice. And Billy Al (who doesn't like Irving anyhow) said, "Only do what you *want* to do," and went on to tell Don that all the artists he knows who have seen Don's work think very highly of it and that the opinion of artists is all that matters in the long run, Don just has to wait until their opinion is accepted which will probably take at least two years or more. Furthermore, Billy said that he'd like to show Don's work in his own studio, which consists of several very large rooms and which he now uses instead of a commercial gallery.

So of course Don is immensely encouraged, as well he may be, for indeed what Billy says about his work is far more impressive than any of the wretched little press notices he has so far had. Don says he now feels he can go ahead. Also, after the talk with Billy, he admitted that he does feel he has made a great breakthrough by starting these ink drawings and particularly by doing them on smaller paper, the paper he was working on, he says, was too big for him. Don wouldn't be Don if he didn't add a note of doubt, so he points out that he is now using exactly the same paper and the same pen as David Hockney uses—implying that he is therefore just a copycat! But he *is* pleased. Billy Al has given him courage. So now I love Billy Al, though he embarrasses me terribly to be with.

Yesterday was Father's Day (as far as the Vedanta Society was concerned) and I went there for the lunch. Swami pretty much recovered, but Pavitrananda, who is staying with them, looks ghastly and Swami says he is in serious trouble with his prostate gland, because he insists on being treated by a homeopathic doctor in New York. Swami will take him to his doctor tomorrow.

There was singing, of course, and Jimmy Barnett surpassed himself. In the midst of it, I got a sudden comic vision of a sort of last-act-of Faust scene. Jimmy is being condemned for his sins, when God's voice exclaims, "No—he is saved! Because he is shameless. He who is utterly shameless is pure." This amused me so much that my eyes filled with tears, I was really moved, and the boys, who were singing *to* us, *at* us, right *into* our eyes and hearts like crooners, were very pleased, they felt they were truly hitting the button.

June 24. Have just finished chapter 14 of *Kathleen and Frank* in the rough draft. Tomorrow we are probably going up to San Francisco for a couple of days, chiefly because there is a postmidnight showing of *The Shanghai Gesture* at a theater there! But we'll also see

Ben Underhill[1] and maybe I can show Don the [Frank Lloyd] Wright Marin County Building. Would like to see it again.

On Monday I did a short speech for a BBC T.V. film about Morgan. A man named Eric Timmerman shot it and he had the really absurd idea of doing so in the midst of the Japanese garden at the top of Orange Drive above Hollywood Boulevard. So I squatted on a Japanese toy bridge over a miniature stream with a waterfall, which had been turned off so as not to interfere with the sound reception, and spoke about Morgan to Timmerman and his cameraman and a girl with nearly no skirt—none of whom had even read him. (Timmerman confessed that he mixed Morgan up with [C.S.] Forester and—Conrad!) It was very hot and I had to shout because a Mexican was working with a vacuum cleaner in the background and in the middle of it all a plane flew over.

Yesterday, Phil Carlson came and interviewed Don for *Esquire*—at least that was what Don had been told in advance. But it was just on speculation and had been suggested by Elsa. Don was on his guard and how rightly, because Carlson tried to get him on the subject of our friendship, how did we meet, etc. Also, the day before, he'd brought a cameraman to get Don's picture with a drawing of me; he took no interest in any other work by Don! Don refused to sign a release on the photos until he'd seen them, which rather shook Carlson. He is probably just an innocent. He is certainly an amateur. I'm still waiting to decide just how furious to be with that bitch Elsa.

Don went to take a look at the Roller Derby, downtown, on Sunday afternoon. He decided that he really didn't want to have anything to do with it because it is so faked, the skaters playact fights and feuds and the audience *pretends* to believe that they're real. However, that was just his first reaction. It's entirely possible—as so often when an idea is suggested to him—that, after turning it down, he'll keep on thinking it over and finally come up with another idea which is an adaptation of it.

June 29. Two girls named Mercedes are cleaning house this morning! One was supposed to show the other the ropes, but neither of them knew how to work the washing machine.

Clement Scott Gilbert and Bob Chetwyn now want to open the play at the Phoenix Theatre at Leicester on August 19, without stars, just to see how it works. Rehearsals would begin about July 27. So back we shall go, I suppose, at our own expense. It is a

[1] Not his real name.

nuisance but a good thing in the long run, in fact what we've always wanted.

We went up to San Francisco on the 25th and came back here early on the 27th. It was chiefly to see *The Shanghai Gesture*, which turned out to be terrible. We also saw Ben Underhill who, as always, produced some awful friends. But San Francisco is still beautiful, despite all the new towers, and I feel stimulated by the visit.

Rushing on with the book. Don has offered to help but now he has such a bad cold. Even if I can get as far as the outbreak of war before we go to England it will be a great advance.

Last night we had supper with Julie Harris and Jim Murdoc[k]. Jim is almost babylike in his vanity and general behavior but I can imagine getting fond of him and even not being embarrassed by Julie's embarrassment. We talked a lot about his nose, which has just been fixed (though Don says, with his expert eye, that this obviously wasn't for the first time) and about diets. Am still hanging on to mine but with no loss of weight to speak to.

July 6. The dreadful holiday is over and now I must work harder than ever to get this book finished. There are still some very big humps of raw material to be worked over. I feel the ending will be the most inspiring part to write.

Woke up with more than usually acute worries about Don and his future. From a "sensible" point of view, I really ought to die now, while the going's good. It would be a bad jolt for him but better in the long run. But it's no good talking like that, even to myself. We love each other in spite of all our conflict of wills. All I probably *can* do for him is to turn more and more toward God. ("More and more" sounds outrageously complacent; in actual fact I don't seem to be turned one millionth of an inch toward him right now.)

We heard from Clement a few days ago that the whole Leicester project is off; not a big enough audience in the summer. So that will no doubt mean that we won't get Bob Chetwyn after all, because soon he'll be going to New York. And that will probably lead to other jobs.

We saw Harry Brown last night, with a moustache. He is getting sick of Mexico. The only thing he cares about is his young son Jared, he says. He drinks again but not much. Why is he such a deeply depressing character? Because of a built-in and almost subconscious self-pity. For the first time, last night, I realized the power of his mother over him. She is dying now. The other day she told him she could never forgive him because of the terrible language in his new novel, part of which was published in *Playboy*.

July 12. Finished chapter 15 today. Maybe only two more, or, if there is a third, it'll be quite short. So it still seems possible to get through before my birthday.

Am happy working, as always. And happier than ever with Don. He is seeing a lot of Mike Van Horn, drawing with him and going dancing and to the club, all of which is good; Mike is a sort of ideal friend to have. So ideal in fact that we both find him a tiny bit mysterious.

A disturbing letter from Bob Chetwyn, telling us that it wasn't the fault of the Leicester theater at all, that the deal fell through. They were ready to come up with an adequate sum of money, Bob was quite satisfied with his share, but they did ask for two hundred pounds from the producers, and this Clement and Richard [Schulman] absolutely refused to pay! My God, we'd probably have agreed to raise the money ourselves if we'd been given the chance. But we weren't. Now we start to wonder, should we refuse to renew Clement's option when it comes up in a month. But how are we to find anybody else, from six thousand miles away? Meanwhile, Camilla Clay is very sweetly agitating to get the play read by Bill Ball of San Francisco.[1]

July 19. It turned out that Bill Ball is away, so Camilla said we should show the play to Ellis Rabb. (We have always supposed that Ellis read the play a long time ago and didn't like it, but Ellis told Camilla no, he had never even heard it existed.) So I made a great fuss with Chartwell to get them to xerox some copies in time to give one to Ellis last Monday, the day before he returned to San Francisco, and I talked to Ellis on the phone and he declared he was longing to see the play and—since then, not one word! Even Robin French hasn't condescended to call and tell us how *he* likes the revised version. However, Ellis returns here this coming week, so at least we'll hear something from Camilla, who'll be seeing him.

We never replied to Clement's letter, to punish him a bit. Since then he has written another, obviously feeling a bit guilty, to say that he has sent the play to Paul Scofield, now that Scofield won't presumably be playing Diaghilev in the Nijinsky film (since Tony

[1] William Ball (1931–1991), American stage director, founded the American Conservatory Theater in 1965 in Pittsburgh, moved it to San Francisco in 1967, and continued as artistic director for two decades.

Richardson is no longer directing it). (Which reminds me, Neil Hartley wrote and asked if I could "use my influence" to get Woodfall [Productions] the film rights to *Passage to India*![1])

Kathleen and Frank drags on. My unconscious resistance to this long chapter of copying letters and diary extracts is terrific. "I" make mistake after mistake, get the order of the extracts mixed up and generally do "my" best to sabotage the whole project. "The Devil" really is much more *tamas* than *rajas*.[2] However, it is going ahead. I am halfway through chapter 16, which will cover Frank's last nine and a half months. After that, there's the chapter about Kathleen's life immediately after his death until the end of 1915. And then one or two more chapters, concluding. I still don't see why I shouldn't finish all this before my birthday, in a rough draft.

On the 15th, a stranger called me from New Jersey. (I couldn't quite get his name but I think it was Scott Dancer.[3]) He said he lived in a farm near Sussex and that he had a friend named Eric who he'd lived with for many years and that he'd had a vision in which he saw what death is all about and what the *answer* to it is, and he knew he had to get in touch either with R.D. Laing[4] or with Auden or with me. "And now," he said, "what are we going to do about it—are you coming to me or shall I come to you?" I tried to wriggle out of this, suggesting that he should communicate with me telepathically rather than physically. He said, "I can tell from your voice that you're tense, you're holding back." I said, "Of course I'm tense, I'm trying to get on with my writing." He then suggested he should call me when I was more relaxed—why not that evening? "This evening," I told him, "I'm going to see my guru." I knew this would shake him up and it did, having a guru is a kind of checkmate move in dealing with this type of communication, because of course Mr. Dancer was in fact offering to be my guru, himself. However he wasn't discouraged. He told me, in a "hypnotic" tone, repeating each sentence three times, that I mustn't resist, I must open myself to what he had to tell me, I must give up writing, writing was no good for me, I had come to the end of writing, and now I must *act*. "Tell your guru everything I've told you." I promised I would. "And, before you

[1] Isherwood was named in Forster's will; see Glossary

[2] More laziness than passion; see Glossary under *guna*.

[3] Not his real name.

[4] Scottish psychiatrist and psychoanalyst (1927–1989) whose radical approach to psychosis emphasized social interaction and the patient's existential choice to be mad or to be cured.

go to see him, smoke a joint." I told him severely that my guru strongly disapproves of grass. However, we parted quite friendly, after I'd assured him that I would let him know instantly when I got a "sign" from him. He had told me he would send me one.

I really *was* going to see Swami that evening—he's staying at the house on the Old Malibu Road where he was last year—and I did tell him about Mr. Dancer. He laughed, but he was much more amused by a letter he had had from that exhibitionist clown, Peter Schneider. Peter said he was worried because he found that he sometimes repeated "his" *mantram* (he hasn't got one, since he isn't initiated yet) "automatically"! His letter ended, "I love you." (A couple of weeks ago, Peter wrote Swami asking if he could come to Trabuco but stipulated that it must be sometime when Cliff Johnson wasn't there. This didn't displease Swami because he thinks Cliff is such an egomaniac and alienates everybody—which may well be true; but just the same I rebuked Peter for writing it.)

Swami told me how he was lying in bed, not long ago, in the middle of the night, and his little finger began to twitch, and suddenly the thought came to him: I have no control over this body, it is the Lord who controls it. And this made him ecstatically happy and he was awake for a long time, "having a wonderful time." He also told me that often while he is meditating he imagines that he is in the Ramakrishna *loka*: "They are all there and I am their servant."

July 23. Such a strong disinclination to write up this diary and yet I want to, really; there's so much to record. I'll try doing a very little, but something, each day. Yesterday I finished chapter 16, which is almost entirely copying. The war material speaks for itself. So will chapter 17, which is Kathleen's account of her efforts to find out what had happened to Frank. This nearly makes me cry whenever I read it. It mustn't be commented on, or I shall spoil it. I think one more chapter after that will finish the book. I see that I can't write very much about Christopher at the end; it would be beginning another story.

Today is Gavin's birthday, so we're taking him out tonight to have dinner at La Grange. Not a word about our play, either from Ellis Rabb or from England, but now we have the top copy and a xerox from Chartwell. From Robin French we hear that the Chris in the filmscript of the *Cabaret* film is homosexual but makes it with Sally! Also that Gertrude Macy's lawyer has started dickering to get back part of the money that's being withheld from her, rather than accept arbitration. Robin says this proves she knows that her case is weak.

July 24. What's the use of getting up at half past six to turn off the alarm? I stagger back to bed, and then we snooze (sensually speaking, this is the most beautiful part of the twenty-four hours) and then we do get up at maybe half past seven or more likely a quarter to eight, and then there's a bleared "meditation" (anyhow as far as I'm concerned) chiefly concerned with thoughts about my book, and then breakfast on the deck (beautiful, too) and then around nine we're ready for action.... Well, now it's just after eleven and what have I accomplished? Called Cukor to apologize for not having shown up last night with Gavin for a birthday drink (we didn't get out of the restaurant till ten forty-five), arranged with Swami for Jim Gates to visit him on Sunday next with his adored Gib[1] who is down for the weekend from up north (this visit will probably be regarded by the powers of Vedanta Place as a flagrant breach of security regulations and I shall be blamed by Ananda[2]). Then I've been to the mail and found a letter from Bob Chetwyn more or less offering to find us another producer if we ditch Clement. And now I'm writing this diary before starting work on chapter 17.

Cukor took advantage of our talking on the phone to ask me [to] come for a drink to meet some clergyman he knows from Vancouver, which sounds too tiresome. When I politely said of course I'd like to come, in the first place because I want to see you again, Cukor obviously meant to return the compliment, but what he said was, "I want to see me again too!" Now and then he seems quite gaga. He also contrived to wish on me the chore of refusing to organize a T.V. show about Aldous [Huxley], which Laura [Huxley] had suggested *he* should do!

The birthday evening with Gavin was quite a success. We opened the bottle of Dom Pérignon which Jennifer [Selznick] gave Don for his birthday and this pleased him as a symbolic act, though he didn't finish his glass. Dinner was good too, at La Grange, which is in many ways the best restaurant we know around here. Gavin was probably feeling a bit depressed because he has just had a fuss with Christopher Wedow [...].[3] Christopher descended on him a short while ago and I think Gavin already feels stuck with him and anxious to get him out of the house.

Jim Gates, talking to me on the phone the other day, said,

[1] Gilbert Peters.
[2] Swami was vacationing at a devotee's house in Malibu; Anandaprana forbade visitors so he could rest.
[3] An aspiring critic who was writing an essay about him.

referring to his job at the Goodwill: "I'm really a good employee because I'm so likeable." This is a perfect specimen of his sincerer-than-thou dialogue, the tone of voice which so disgusts Don. Me rather, too—but I can never quite make up my mind about Jim, and I am fond of him in spite of it.

Poor Jim Bridges is threatened with appendicitis and may have to have an operation; he's being examined today. Jack is also suffering, psychologically, from being bitten by his beloved Paco[1] (whom we hate). Paco was apparently about to hurt herself on a saw and Jack grabbed her from behind and she bit him in the chin. It pains Jack that Paco should bite him under *any* circumstances but he excuses her by saying she was hysterical because of all the workmen in the house. The workmen were, and still are, there because dry rot has been discovered in some of the woodwork. That miserable house of theirs seems to be chronically damage prone but they go on pouring money into its repairs.

A third Jim, Charlton, has been much in my thoughts lately. It often astonishes me how much I still love him, or rather, how romantically I still feel toward him. Camp-romantically almost, since I also see him preeminently as a comic figure. Nevertheless, there's something unique in this feeling. It's not at all that I want to see him often. I most definitely do *not* regret that we never lived together. My romanticism about him is concerned with his essential aloneness. When I think of him alone, I love him; when I think of him married or otherwise involved with either sex, I laugh at him. But some memories of being with him—driving up to Palomar or down to Mexico in the old days, or spending nights at his apartment on the beach, or visiting that waterfall last year on Oahu,[2] even—are still astonishingly vivid and beautiful.

July 25. Have stayed in all day. I seem to be starting a cold, have taken two Coricidin tablets. Have got as far as Kathleen's entry for June 24, when she receives the notification from the Red Cross that they have found Frank's identity disk. Must pause here to give Hugh Gray[3] time to find out what "siche" means; "siche 5" is engraved on the disk.[4] Hugh seems to love finding out things like

[1] A puli (Hungarian sheepdog) with a shaggy corded coat.

[2] August 11–12, 1969, on the return from Australia; see *D.2*.

[3] Oxford-educated British screenwriter and university professor (1900–1981); he taught film, theater, aesthetics, and humanities at UCLA and Loyola Marymount University. He appears in *D.1*.

[4] Isherwood copied Kathleen's 1915 diary entry into his own diary, October 24, 1966 (see *D.2*), and it is unlikely this was engraved on the disk. The

this; the only trouble is, once he starts explaining he can't stop—a complaint made also by Don about a certain old Horse.

Don and Mike Van Horn made an agreement, each to do a group of paintings; each group was to have its own general theme, or manner. Don did some of his movie heads; these were of Barbara Stanwyck and two of them at least were quite remarkable, one of these, Don says, shows him an approach to a whole new way of working. Mike took some pieces of canvas, had them stitched together in various arrangements (two were like outlines of a very fat man's trousers) and then treated them with gesso and then dyed them. I am very happy that Mike and Don are turning each other on like this. But alas part of the arrangement is that each can pick one of the things the other did, so Mike (who has very good taste) is pretty sure to take away Don's best painting and we shall have to find a place to hang up one of Mike's rather large canvasses!

July 30. My cold came on quite bad, I even missed two days of work on the 27th and 28th and spent a morning in bed. (Now I'm on Vibramycin[1] capsules and feel much better though still shaky.) However, yesterday I finished chapter 17, a short chapter consisting almost entirely of Kathleen's diary entries down to the end of 1915 and her final acceptance of the fact that Frank is dead. And it's clear that I have only one more chapter to finish the book. The problem is, just how shall I finish it? At present I think the logical ending is a description of Christopher's attitudes towards Kathleen and Frank. But it will be hard to cover Kathleen's later life briefly and yet not make the reader feel cheated. I must give some significant glimpses of her in later life, but where can I find them? Perhaps I can get more from Richard, if I can ask him the right kind of questions. And I'll dig into my diaries to see if there's anything there.

Don has gone down to Laguna Beach today to see Jack Fontan and Ray Unger, and Mike Van Horn has gone with him. I am so happy they are seeing so much of each other; tomorrow Mike is

phrase appears to be a mistranscription from Frank's German-language death certificate held in the archives of the International Committee of the Red Cross in Geneva. The certificate gives Frank's details on a pre-printed list. Item 6 on the list, "*Inhalt der Erkennungsmarke*" (contents of identity disk), is filled in as "siehe 5 C of E" (see 5 Church of England). The instruction to *see item 5* was simply miscopied, *siehe* becoming *siche*. Item 5 on the list "*Truppenteil*" (unit) is filled in as "Y. & L. Rgt." (York and Lancaster Regiment).

[1] Antibiotic in the Tetracycline family.

going with him to Santa Barbara to draw with Bill [Brown] and Paul [Wonner] and maybe stay the night. I am as happy with Don right now as I have ever been—sometimes when I can draw back for an instant and look at us both I am absolutely awed at the miracle of having him with me. We had a wonderful talk the other day and seemed to achieve a real advance in frankness with each other, but I don't want to write about that, not yet. I want to let more time pass and see what develops[.]

A letter from Charles Thorp, head of the National Students Gay Liberation Conference, asking me to go and talk to them in San Francisco on August 23. "Give a short talk and then just rap with us. What I'd like is to have you come 'hold-our-souls,' hold our hands." Instead of "my best regards" or whatever, he writes "my gay-love."

I feel quite strongly tempted to accept this invitation (as indeed I've often wanted in the past to accept others like it). I highly enjoy the role of "the rebels' only uncle" (not that I would be, this time—for there are scores of others—and Ginsberg their chief) and, all vanity aside, I do feel unreservedly *with* them, which is more than I can say for ninety percent of the movements I support. But something prevents me from accepting. Oddly enough, it all boils down to not embarrassing Swami by making a spectacle of myself which would shock his congregation and the women of Vedanta Place! I can admit this because I am perfectly certain there's no other motive. I am far too sly and worldly-wise to suppose that I'd be injuring my own "reputation" by doing this. Quite the reverse; this is probably the last opportunity I'll ever have of becoming, with very little effort, a "national celebrity." And I hope I'm not such a crawling hypocrite as to pretend I wouldn't quite enjoy that, even at my age!

July 31. Suddenly there's so much to record, or so it seems.

Yesterday I had a classic day of failing to get started with the last chapter of *Kathleen and Frank*. I made every possible excuse: I couldn't begin without consulting my diaries, until I'd got a letter from Richard answering my questions, until I'd read through the whole manuscript first. But this last alternative just depressed me so much that I ended by rereading the script of *The Monsters*[1] and "Afterwards"[2] and doing nothing. (The two chief characters of *The Monsters* seem utterly unconvincing but just the same I wouldn't

[1] A play he and Bachardy wrote, beginning in October 1958; see *D.1*.
[2] A "queer" story he wrote in the summer of 1959; see *D.1*.

be surprised if the thing didn't play well and even have a success with a West End matinée audience; the twists of the plot are still fairly surprising. "Afterwards" in its present form won't do at all. Wystan was right, the chief character is so unpleasant. Anyhow, I've used much of it far more successfully in *A Single Man.*)

In the evening I went down and had supper with poor old Jo. And by God she has two new subjects for *moaning*, and one of them is truly serious. First, less seriously, Ben's father in Florida has had a stroke and won't be able to live alone any longer, and Jo would so gladly have rushed to him ("he really cared for me far more than he did for Ben, and I just adored *him*") but no, she couldn't, because Ben and Dee are on their way there. Jo says that of course they ought really to bring Dad back with them, but even if they did they'd be hopeless at looking after him. So now, the old wound is open again and Jo, who was beginning to get used to the situation, she says, hates Dee more than ever because, just because she exists, Dad won't get properly taken care of. Ben had called Jo when he got the news—probably half hoping that somehow Jo could be conned into coming along with them to Florida and taking charge—and when he found she wasn't about to, all he could say was, "I'm so *sorry* about everything" ... But what's ten million times worse, what *is* really ghastly, is that her old enemy Louis [Gold], the owner of the Tumble-Inn and the motel opposite, has decided to build a three-floor block of apartments on the parking lot right smack outside the windows of Jo's apartment! She won't be able to see anything, except the channel at one end and the street at the other,[1] her view of the sea and about one third of the daylight will be cut off completely and people will be staring in on her. Jo was in tears and no one could blame her. Of course she's got her little house, she can turn the tenants out of that and move in, but she doesn't want to, says it is too big for her alone and besides she'd be "away from everything." Curiously enough, she really loves all the noise and traffic and hippie music on Channel Road! So poor old Jo, I'm glad I went down to see her.

This morning I did finally manage to get the chapter started and then I went to the gym and then I took Forster's short stories to Gavin, who's been sick. He was being visited by a young man who dances at the Honey Bucket and the young man came on very strong in the Sincere Young Nature Boy style, asking me what I thought of him when I came out into the garden and saw him and

[1] Her apartment, on Channel Road, was between the street and the eponymous channel of water, concreted since 1938 to contain flooding.

why it was that I didn't look at him while I talked to him, so I said, you mean why didn't I flirt with you? So then the conversation got quite relaxed. The young man (wish I could remember his name) is actually far from being a simple little enormous stud nude dancing boy; he's an actor with quite a lot of experience in the theater up in San Francisco and he's been in films and whatnot, and I kind of think Gavin is considering him for the lead in that adaptation from Colette, *The Cat*, which he originally did for Clint [Kimbrough] (Christopher Wedow is *out*![1])

Yesterday Don went down to see Jack Fontan and Ray Unger, as I said. What happened down there is what I mostly want to write about—but that must wait until tomorrow, because it's so late, tomorrow in fact. Today, I mean yesterday, Don and Mike drove up to Montecito together to draw with Bill and Paul and they're spending the night and coming back after breakfast.

Oh, one other piece of news, we did hear from Ellis Rabb, expressing great interest in our play; he excuses himself for his silence by saying he's been trying to get it put on up there but he wants to see if he can do anything for it elsewhere, etc. We neither of us think, from reading his letter, that he likes it.

August 2. This morning I finished a first draft of what will most likely be the end of the whole book, an analysis of Frank and Kathleen as influences on Christopher and as symbolic figures in his myth. What remains to be done is an outline of Kathleen's life from the end of 1915 to her death. This is either a small chapter all by itself or it fits into chapter 17 immediately after the diary entries of 1915.

Don and Mike got back around noon yesterday; they had been to see John Ireland[2] in the morning to deliver a portrait Don drew of him several years ago, which he had paid for at the time but never collected! Don said he'd had a nice evening with Bill and Paul, but he didn't come back happy and stimulated as he did after being with Jack and Ray. Talking to Jack Fontan about his horoscope was an important experience for him.

Don has made four pages of notes of the things Jack told him—they talked for several hours. I'm not going to try to write down even as much as I remember of what Don told me, and it's quite possible there are some very important things he *didn't* tell me,

[1] I.e., out of the house.

[2] Canadian-born stage and screen actor (1914–1992), nominated for an Academy Award for his supporting role in *All the King's Men* (1949).

for one reason or another. Jack said to Don, for example, "You don't have to worry about losing him," explaining that I would live to be very old and then die very suddenly. He also seems to have said that Don will die or be in danger of death a few years from now but that he will probably survive the crisis and will then completely change his way of life. (Don may well have done some censoring there.) Jack said Don would fulfil his ambition, to be taken seriously as an artist by those whose opinion he values. Jack thinks that Don is very ambitious in this way and very determined. He said that Don is an "evolved soul" (maybe that's not the phrase he used) and that even when we first met[,] Don and I were equals (this seems to me to be absolutely true, in a paradoxical way).

After hearing all this from Don, it struck me that analyzing a horoscope, when it's done by someone with Jack Fontan's perceptiveness and empathy, is really quite as good or better than a session with a psychologist. Don strongly agreed, saying that he had got far more out of talking to Jack than he ever got from Oderberg.[1] I suggested that this is because the psychologist is fundamentally dealing with your hang-ups, inhibitions, phobias and other weaknesses, while the astrologer is helping you to create your life myth, to see your life in terms of poetic significance and creative potentiality, so that even your weaknesses are exciting, like obstacles and hazards on a knight's quest, and even impending dangers only stimulate you to make a greater effort to struggle through them. "Why," I said, "Jack makes our getting together sound like the meeting of Tristan and Isolde!" "Well, and so it was," said Don.

August 12. Poor old Jo called me this morning to report another disaster, she has lost her job! The firm is closing down her whole department because it isn't making money and every business is economy minded these days. Jo doesn't know what she will do next. She hates the thought of working at home because it's so lonesome—and besides, her home's being taken away from her, too!

I have been going ahead revising the early chapters of my book, one a day, daubing the pages with "liquid paper" and typing in corrections on top of it. I want to make a version which can then be xeroxed, rather than copied by a typist. Today I finished chapter 8. Two more to go, and then I have to rewrite the rest of the book

[1] Dr. Phillip Oderberg, a Santa Monica psychotherapist Bachardy began seeing in 1963; see *D.2*.

and write the passage about Kathleen's later life, for which Richard has just sent me a lot of details I asked him for.

Don is very active, painting. Also he writes down his dreams for Ray Unger and he is enthusiastic about Adelle Davis's theories in her book *Let's Eat Right to Keep Fit*. He is putting us on a protein diet, with some homemade breakfast food, etc. He is being absolutely adorable.

A cliché of nowadays: "Have a nice day." Even kids say it to you, in filling stations and markets.

How-dumb-can-you-get department: A girl came into the gym the other day. About a dozen of us were working out. She looked at us, pulled out her card from the file and was actually heading toward the locker rooms when someone stopped her and pointed out that this was a men's day. She was shocked and utterly amazed.

August 17. On Wednesday I saw Swami. He is up at Vedanta Place again, because Pavitrananda has been having his prostate operation and is still in hospital. He asked me about my meditation and I told him it was as bad as usual and he told me to try to make my mind a blank before meditating, and then to try to feel that I am in the presence of Ramakrishna, Holy Mother, Maharaj and Swamiji—only I think he said to think of Maharaj first, and then the others. I asked, is it all right if I think of you there, too, and he said yes. He also told me, as he has told me before, to say to myself that I am the Atman and that it doesn't really make any difference whether I am aware of this or not, I am the Atman and that's that. (This I do find helpful. As for the rest of his instructions they just don't *seem* to help, at all. But then again, what do I mean by that? The point is, I am trying to meditate and that in itself is a great grace—a grace arising out of the infinitely greater grace of having met Swami and become his disciple in the first place.)

Two things distract me chiefly at present; lust and my book. Right now, I am thinking of the extreme shortness of Michael York's shorts which, despite the extreme longness and badness of the film we saw him in last night (*Something for Everyone*) kept me sufficiently stimulated throughout. And I am always thinking about *Kathleen and Frank*. I have now finished revising the first ten chapters and am about to start writing the section about Kathleen's later life—how I wish I could start it today, instead of toiling down to Laguna and wasting time while Gavin gets his horoscope read and Don tell[s] his dreams to Ray. But Don for some reason wants me to go, so there it is.

A fearful row is building up with Phil Carlson and probably with

Elsa as well, over the article he wrote about Don. Phil asked for the interview letting Don suppose it was for an article on California artists, and now it's revealed that what the magazine (*Esquire*) really wants is an article on "the friends of the great" and how their talents are being overshadowed by the friends' reputations.[1] In other words, it's just another anti-fag operation. Phil knew this all the time and it does seem more than likely that Elsa sicked him on to Don; her enmity knows no limits and never tires, we are all to be punished for her sufferings during her marriage with Charles. Oh, cunts, cunts, cunts!

And talking of cunts, I can definitely say, having now reached canto 23 of the *Paradiso*, that Beatrice is the character I most hate in the whole of fiction.

August 19. Don is most terribly upset about the Carlson article. It has shaken his security to the very foundations; he feels that if it is published and Billy Al and Irving Blum read it he will be humiliated in a way which he can't even face. When I try to get him to face it, he accuses me of treating him as if he were being a bit insane. Actually, he has talked to the editor (who suggested this himself, th[r]ough Phil) and been assured that the article won't include him; but he feels he should discuss the whole thing with Arnold Weissberger and get his advice. This was Jack Larson's advice and I think it's a good idea.

Talking of getting upset, I had an almost crazy outburst yesterday when one of these representatives of the oil company came around—"to improve public relations." I slammed the door in his face, instead of being wonderful and subversive and fatherly and sending him away with a worm of guilt in his mind.

I am full of nervous energy, these days, but so deeply weary underneath it all. We get very little sleep because we go to bed so late and Don insists on having the alarm go off at six-thirty, but the real trouble is, I don't sleep soundly anyhow. And I keep bleeding from the rectum. I *think* it's only piles.

The visit to Laguna Beach was a success. Gavin was very pleased with Jack Fontan's reading, thought him much the best of the astrologers he had met. And I got through a lot of nuisance-letter-answering. Jack and Ray's new octagonal house is really attractive. But oh how they do talk, in their carefully quiet sweet calming voices!

[1] The other lovers and assistants of famous men included Chester Kallman (Auden), Robert Craft (Stravinsky), Howard Austen (Vidal), and Frank Merlo (Tennessee Williams).

August 23. Today I finished in rough the extra part (Kathleen's later life) which I'm adding on to chapter 17; so now I have covered the whole book, more or less. What remains to be done is rewrite from chapter 11 on. This means actually that chapters 11, 12, 13, 14, 15, 18 and the second half of 17 will have to be pretty thoroughly rewritten, but that 16 and the first half of 17 don't need to be touched, except for a couple of inserts, because they are over ninety-nine percent direct quotation from diaries and letters. So I can still hope to finish this fall, barring serious interruptions.

On the 19th, I went to see Dorothy Miller. She looked far better than I expected and although the doctor has given her a crutch she wasn't using it. As usual we talked a lot about Elsa. Dorothy said, "If you talk to her a lot she *bores* you." She is very happy about a Unity church in the neighborhood which she has just joined. The pastor is white and so are most of the congregation and it seemed that what Dorothy really appreciates is that she feels completely accepted by them.

When I told Swami that my meditation isn't any better he said he would write me out some instructions. I'm to get them when I see him next.

The day before yesterday, a goodbye dinner for Peter Schlesinger, who returns to England today. We are both fond of him and rather concerned about his future, maybe without cause. He looks like "a flower of the field" but is probably a very tough little evergreen. Gavin's friend, the one who dances (or maybe *has* danced, for now it's banned) naked at the Honey Bucket, came with Gavin. His name is Mark Andrews.[1] He is an overwhelming talker and flirt—he flirted with *everybody*, Jack, Jim, Camilla, Linda [Crawford], Peter, Truman Brewster,[2] Ralph Williams, Don, me; but there's something amusing and quite delightful about him; he made our party go.

August 27. A happy busy birthday yesterday. Did some quite good work on chapter 11, I have to be careful and slow about this, and went on the beach with Don and in the ocean and then on to the gym (where damn it my weight has gone up again—I was down just a shade below 150, now I'm just a shade above) then we went up to have supper with Jennifer. This was quite delightful though sad because it'll be for the last time at the Tower Grove Drive house; she's selling it in two weeks. Such memories of that place,

[1] Not his real name.

[2] A longtime friend of Jack Larson.

it really is the only "Great House" I have ever known throughout my life in Hollywood. And last night the living room was as beautiful as ever—all the more so for being empty—lit by a beige-gold radiance from the lamps. Don's two great drawings, the one of David with the blank lens on one side of his glasses and the tip of his little finger appearing through the other, and the one of me, so cruelly and exquisitely exact. The dining porch with all its reflections of trembling candle flames amongst the dark bushes outside. The "hashish room" which Tony Duquette designed. And the chairs and table out on the porch all set with lighted candles and quite empty, as if it were a shrine. One had a sense that Jennifer is living here alone amidst her ghosts. Dinner was served by a maid, but you felt the house was *vacant*.

Jennifer has a new deadline (October 15) on which her lover[1] has to decide between her and his wife—which of course is the very last thing he wants to do. It's curious how absolutely convinced she is that, *if* he does finally come to her, they will be happy for the rest of their lives! But she really is adorable and so delightful to be with. We stayed on until late.

Don has now heard that the *Esquire* office is mailing his release on his photo back to him. (I still don't trust them one inch.)[2]

Also, which really *is* good news, Billy Al Bengston is holding a group show at his studio of the work of his friends and he is exhibiting three of Don's drawings—this after only asking for one to begin with. So Don's morale is considerably boosted.

I want to discuss with myself the problem which immediately faces me in chapter 11; exactly what was the private mythology I created out of *Wuthering Heights*?

Well, first of all, *Wuthering Heights* showed me how to see the Disley landscape dramatically; that much is obvious. It did this by showing me an approximately similar landscape which was related to a drama, a love drama. At the time when I first read *Wuthering Heights* I was (it seems to me) in love with Johnny Monkhouse, "Mr Honeypot," with his blond hair and his grin and his beautiful long legs and his hockey stick.

Johnny was a "hopeless" love; actually I didn't even seriously want to make him and fear absolutely prohibited me from ever trying to. I enjoyed suffering and mooning around and watching the house to see if he'd come out. Yes, and talking about my feelings, of course, to Paddy [Monkhouse].

[1] A psychiatrist who had been treating her.
[2] The piece never ran because of Bachardy's objections.

Heathcliff doesn't (apparently) make Cathy. Not because he's shy but because she won't cooperate, she's married, she's fatally sick and, finally, she's his sister, so to speak. Their relationship is presented as being infinitely deeper, more violent, more binding, more everything than the mere sexual relationships which both he and Cathy enter into, quite irresponsibly, almost casually. Of course there are suggestions here of Emily's relationship to her own brother; not to mention Byron's relationship to his half sister. So there's the incest thing as a barrier.

When I took on the fantasy role of Heathcliff, the "hopeless" love wasn't incest but homosexuality; which was all very well while I was very young and inhibited. Later, when the love turned out to be not in the very least hopeless, I had to drop that part of the Heathcliff role.

But Heathcliff has another aspect. Like Byron, he is a mysterious traveller; he has been away somewhere, "in foreign lands," but he won't say where. And then he *returns*. That part of the role was what really appealed to me—the returning traveller from romantic journeys and that part I still play whenever I go up to Disley on visits.

Heathcliff wasn't visiting, however. He came back to stay. And this stay was tragic and ended in death. This suggests that what I have latterly made out of the role is a Heathcliff who refuses to stick around and get involved in the tragedy. After enjoying the emotions of the returned native son, he leaves again while he still can, and returns again and leaves again, over and over.

August 31. On the 27th, Swami gave me new meditation instructions; he had written them out for me. They are (abridged):

> Cover the whole universe with the presence of Brahman as Light, and repeat mentally: "I meditate on the glories of that Being who has produced this universe. In Him we live and move and have our being. May he enlighten my mind."
>
> Feel the presence of Ramakrishna with you, and talk to him: "Oh Lord may I serve Thee in every way, whatever I do, whatever I give, may I do and give as an offering unto Thee. May I be truly an instrument in Thy hands. Thou art my only refuge. Enlighten me for I am Thine."
>
> Feel that you are free through His grace and rest for a while thinking that you have His peace and bliss.
>
> Send love and goodwill to every being.
>
> Think of your guru, feel his presence. Then of his guru,

Maharaj. Now again feel the protection and guidance of Ramakrishna, Holy Mother, Swamiji and Maharaj. "May they inspire me with love, truth and purity. May I feel their grace." Then think of the mind as pure and perfect and that no evil can enter your mind.

Now hold the mind in the lotus of the heart and think of your Chosen Ideal for a few moments. Then let the mind run as it wills for about ten minutes and you stand as a witness and watch. Do not seek to control it even if the thoughts are bad.

Then take a firm hold of the mind, concentrate on your Chosen Ideal, seem to see him as bright and luminous and hear his voice. Repeat the mantram as you meditate. Perform mental worship, offer flowers at his feet, place a garland around his neck, wave the light before him, burn incense, wash his feet and wipe them.

(I meant to write a lot about my problems in connection with this, but must wait till tomorrow because I have been interrupted so much by people coming in.)

September 1. I find the above desperately frustrating, almost meaningless. All my old resistance to things Indian comes back upon me and I keep remembering Aldous's objections to formal meditation (which actually made me rather angry when he used to speak of them) and even someone at La Verne (it was Harold Stone Hull, I've just looked this up in my journal) saying sarcastically, "When do we stop the motion pictures and start meditating?"[1]

The only part which is relatively easy and seems helpful is mentally doing the acts of ritual worship. The hardest part is thinking of the universe covered by Brahman as Light. And when I try to assemble the two Ramakrishnas, Maharaj, and Swamiji in the room I get a feeling, *where are they all going to sit?*

Well, I really do want to watch my reactions to this problem so I'll write down something about it every day. I know that will be helpful.

Hunt Stromberg called up out of the blue at the end of last week and asked would I be interested in doing a T.V. series, "the definitive 'Frankenstein'"?! So he's coming round this afternoon.

Don much upset yesterday, said I am never frank with him, jolly

[1] In La Verne, California, at the July 1941 seminar on the active and contemplative life; attended by Gerald Heard, some of his followers, and members of the Quaker American Friends Service Committee. Stone Hull was a Christian minister and a pacifist. See *D.1*.

him along, why don't I tell the truth—that his paintings are faggy. I said, even if they are, why the hell shouldn't they be if that's your myth? He said, I'm ashamed of my myth.... I write down these lines and they sound funny, but there is a terrific deep problem here which I only partly grasp. ("Because of a basic inattention"? Yes, there's that, and there's my desire to get on with my work undisturbed and not have the boat rocked. True. But it is also true that I'm not nearly so perceptive about Don's problems as Don thinks I ought to be and secretly am and won't let on to being, lest I have to get involved with them. Of course you can say my stupidity is self-protective and so we go round in a circle.) All is love again now for the moment. But I must pursue this problem. I must pursue it particularly at the times when Don is *not* disturbed about it—though usually when I try to do this he isn't inclined to talk about it.

September 2. Saw Swami last night. I complained about my difficulty with the meditation on Brahman and he said, then just say Om three times and feel that it resounds throughout the universe and that Brahman is Om. I complained about the "crowding" when I try to see Ramakrishna and the others all together. He said, I should try to see them one after the other in succession. (This certainly is easier.) He also said that, when mentally doing the ritual worship, I should sometimes do the offering up of the physical body, the subtle body etc. etc., until Brahman alone remains.

Hunt Stromberg came by yesterday with a boy from Florida [...]. Hunt talked about the "Frankenstein" T.V. special (it isn't a series) and made it sound as if it was a *fairly* firm offer. He has broken up with Dick Shasta,[1] whom we liked; said Dick was impossible to live with and talked vaguely about a car being blown up—he didn't actually say Dick did this. As a result, Hunt hired a bodyguard, a seemingly very square ex-marine. Now, he claims, the bodyguard is so fascinated by the "fag scene" that he is showing signs of "coming out" himself. Hunt described how [the boy from Florida] was sunbathing by the pool and the bodyguard came out and didn't see Hunt and started lustfully looking [the boy] over. Then he saw Hunt and immediately began shooting all over the place with a BB gun to cover his embarrassment!

This morning in the mail came the announcement of the group show at Billy Al's studio: Moses, Bengston, Bachardy, Bell, Goode, Berlant, Davis, Ruscha, Price, Alexander.[2] (In that order.) This is

[1] Not his real name.

[2] Ed Moses, Larry Bell, Joe Goode, Tony Berlant, Ron Davis, Ed Ruscha,

really a very distinguished gathering and Don is pleased. Irving Blum was very nice about it when Don phoned him in New York to ask if he objected to Don's exhibiting at another gallery; and Irving thought it was good for Don's prestige too.

September 6. Such an effort to write anything in this book! Because whenever I'm at the desk I have this blind urge to get on and on and on with *Kathleen and Frank.* And rightly. There's still a tremendous lot of work to be done on it. Am slowly coming to the end of chapter 11.

On the 2nd, we had supper with Gavin, Mark Andrews and Rita Hayworth at Trader Vic's (which I hate, it's so depressing). Mark made rather an ass of himself; we had been expecting him to be marvellous with Rita and charm her. But he really is very much more naive than we'd supposed. He made corny advances and asked tactless questions, such as, "Where are you from?" This got under Rita's skin—I hadn't realized how sensitive she still is about being Mexican—and she told him loud and clear, "I'm from *here*, I'm North American!"

Felt sort of protective towards her but not sorry for her; she can so obviously look after herself. Anyhow we are far too easily sentimental about the beautiful who have ceased to be beautiful—and Rita still looks very good, in a different way.

On the 4th I went with Swami and Krishna to have supper with Len, Peter Schneider's father. His girlfriend, a nice middle-aged woman, cooked, and Peter and his little brother Danny and Jim Gates (who is now working at the West Hollywood Goodwill and back living in the shack with Peter) were also there. Peter had just bought a *chadar.* He camped around in it, calling as much attention to himself as possible, and then Danny put it on. Swami took the whole thing in his stride as usual and I think made a good impression on Dr. Schneider, for whose benefit our get-together was really being staged; Peter is determined to "convert" him. Jim has just cut his hand severely, on the shower faucet. I can't help seeing him more and more through Don's eyes, that is, seeing his prim mischief-making side. Later, and this was the second objective of our meeting, the boys got Swami to come back for a moment to visit their shack, and thus bless it and the meditation shed behind.

September 8. Today I finished chapter 11 and sent a copy of it off to Richard.

Ken Price, Peter Alexander; see Glossary for more about these artists whose works are widely exhibited.

The day before yesterday was the preview of the show at Billy Al's. Was much disappointed; there seemed to be so little to see, in those two big rooms. Billy Al is really the only one of the group whose work I like. Don only had two of his drawings up on the wall. They looked fine, but why not three? Also I wish he'd had a couple of paintings. He has just done a marvellous one. Crowds of art squares and squaws, with kids, were lounging around and drinking wine or sitting stoned on the floor and not really looking at anything. Leslie Caron came, which was a gesture, but then she missed going to Don's show in March.

We had supper with her last night. She is a very remarkable woman; just the same type as Madame Chiang [Kai-shek], and maybe Lady Macbeth, but benevolent—well, fairly. But she's capable of imposing her French way of life with overwhelming power. Her beauty and her trimness and her energy are so awe-inspiring. She is sticking burlap all round the dining room with her own hands. Michael seems such a slob beside her. He was shot at by a Texan gunman sent out to San Diego by the father of a girl he was going with. The gunman just barely grazed him with one shot after three misses. Since then, if a car comes up close behind them when they're driving, Michael gets scared. Jack Larson says Michael sees the United States as a place of extreme danger—as of course it is. He's now working on several films; one about juvenile dope takers.[1]

Then tonight, I had supper with Swami. (Don has gone to Laguna for the day to see Jack Fontan and Ray Unger.) He says of Belur Math, "They are waiting for me to die"; in other words, they won't send him an assistant because, after he is dead, they can send one who'll do exactly what they want. And what do they want? Apparently to do away with nuns in the U.S. Swami takes all this quite calmly, seems to find it mildly amusing. But he now says he will seriously consider training some of the monks to give lectures. He remembers that Vivekananda said once that Vedanta societies should be run by Americans.

A day of energy. I am really in pretty good shape now, though Leslie's supper put a couple of pounds back on me. I went to the gym, ran round two blocks, shopped, polished the bumpers of my car with chrome polish. We keep getting different maids. Today was Josephina, for the second time; Leslie says she's "sly" and always getting pregnant.

September 11. Swami said of himself the other night that he could

[1] *Dusty and Sweets McGee* (1971).

never have run this center without Maharaj: "I had no education." This surprised me. I have often heard Swami express spiritual humility, and one expects no less of him; but I've never heard him say this about his education, before. I've always thought of him as having had rather a good one; those English-run Calcutta colleges had very high standards in his day. I suddenly saw that he must have felt his lack of education strongly in the presence of Aldous and Gerald—but then, anyone who didn't would have been a fool. No, what I mean is, I'm surprised to find that Swami values education for itself; I suppose I expected him to find it slightly ridiculous, as Ramakrishna did.

Talking about humility, he also rather stunned me by saying, "But Chris, you're the most humble man I ever met—all that fame, and you're so humble." He also said (referring to his need for an assistant) that he wished so much I had stayed with them, so I'd now be able to give lectures! We were talking about this at breakfast this morning; Don asked me if I regretted having left the center. I said I find it impossible to say I regret *anything* I've done—that is, anything not definitely "bad." Yes, I turned my back on the boot camp which would have probably drilled me into swami material, and perhaps I would have had a marvellous life and even been thrown out of the order for giving my blessing to homosexuals—either that or just died young of misery and gone straight to the loka or committed suicide by living in India. But, on the other hand, haven't I, by doing what I did do, fulfilled at least a part of my *dharma*? How can I regret the books I've written since then? And as for the life I've led! And I'm not now speaking of mere "happiness." My life with Don has been happiness–unhappiness raised to an intensity I'd never dreamed of before; much more and better than that, it's been a discipline in which I've often failed but from which I've learnt a great deal of what I now know. Well anyhow—

Tennessee Williams and Oliver Evans[1] had supper with us the night before last; now they have left for Japan on a boat. Tennessee got a bit drunk (which apparently is nowadays unusual for him) and talked about how his brother got him put away in a looney bin. He is obviously unreliable in his account of this, yet the horrors don't seem like paranoia. There is a sanity and an intention behind them, they are a sort of sketch for a future artwork, a play or a

[1] Poet, biographer, literary critic (b. 1915), from New Orleans, a longtime friend of Williams. He wrote about Carson McCullers, Anaïs Nin, and E.M. Forster, and worked as an English professor at several American universities.

story. I do brighten up and expand in his presence; I love him. He is so essentially joyful.

September 13. Finished chapter 12 today. It only took me five days, which sounds good but actually there wasn't that much to be done to it.

Yesterday we got up extra early. By seven thirty I'd nearly finished my meditation when there was an earthquake, the second strongest one I've ever felt, I think; only the 1952 quake seemed stronger. The bookcases creaked and swayed. I sat there absolutely suspended, as it seemed, beyond feeling, thought or action. Didn't even have the presence of mind to worship the earthquake as Shiva! Elsa, whom I talked to today, says she said to herself: how good that this is natural, not human—it might have been a man coming in with a knife!

Supper with Sandy Gordon and his friend Bob Griffith. A good comedy situation, namely that Bob and to some extent Sandy are star worshippers but they expressed it by telling *me* everything they'd read in books about the habits and life of Garbo, Monroe, Garland etc.; they realized I'd met these women but that didn't particularly interest them. They preferred to know their stars thirdhand, so to speak, rather than second. Sandy has just recovered from hepatitis; he had that rose-bloom look one only gets for a few days after an illness.

September 18. On the 14th I got a traffic ticket for making a sharp right turn at the bottom of the off ramp from the freeway onto Santa Monica Boulevard, just as the light changed to red. It astonishes me how comparatively much I mind a thing like this; *any* contact with the police makes me feel slightly nauseated.

Then on the 15th I went to a meeting of the Society of David. It turned out to be quite a production, with sophisticated sound recording equipment to pick up everything we said. (So sophisticated, it turns out, that a lot of offstage whispering of boys in the kitchen is also on the tape!) As far as I was concerned, this was a kind of trap; my being there was turned into a confrontation between an old liberal square celebrity and the young activists of the Gay Liberation Front. A big swarthy baldish guy named Don Kilhafter (I think)[1] put me down, without absolutely directly attacking me personally. Old Kight aided him without seeming to. Most of the others were genuinely friendly and pleased that I'd come. And the

[1] Kilh*efn*er; see Glossary.

boy who has organized this society, Gary Hundertmark, is only twenty and quite a *jolie laide* doll; I liked *him*. But all in all I regret having gone, rather. (Don warned me not to.) It will all be used in some indiscreet way, for one thing. And for another I feel I fell stupidly into their trap and began defending myself, as I so often do, with fake humility under which there is a cold determination to relieve my wounded vanity by hitting back hard at my trappers much later when they have forgotten all about it and are not on their guard. I really did dislike this Don [Kilhefner]; he had the more-engaged-than-thou, dogmatical rudeness of a thirties left-winger. No doubt he does a lot of good work, helping queers out of jams, but oh why can't he be nicer about it!

September 20. Yesterday evening we went with Billy Al Bengston, his girlfriend Penny Little, Joe Goode and his wife, to see motorcycle races at Ascot Park in Gardena. We did it, as Don said, for "business"—that is to say, Don feels he should respond to any overtures Billy makes to include him in his circle. It was bitterly cold, and it went on endlessly—not the racing but the waiting for the racing to start. But I enjoyed seeing it greatly; the only one drawback was that we two couldn't leave when we'd had enough. The boys who take part are, many of them, curiously Early American types; gaunt thin faces, with a quiet determined look of preparedness, like young soldiers in the Civil War or boy outlaws like Billy the Kid. The race itself seems almost superhuman. The incredible daring and speed of it. The riders seem to be straining to hurl themselves to destruction. When you stand down by the wire fence you feel quite sick (though you soon get used to it) seeing them hurtle past you, crowding each other so tightly; and then, when they've passed, the shattering noise—that adds to the feeling of animal terror. And when they move ahead into the straight they accelerate until you feel *no, it's physically impossible!*

Billy Al and some of his other friends who were there played it very big as aficionados, they were really just like bullfight fans. And as always with aficionados you wondered, justly or unjustly, can they *possibly* care that much?

A word about my meditation. It's almost absurdly bad. Yet I am going through the motions and isn't that something? It comes to me often how hard it is to meditate unless you are leading a disciplined life related to meditation, i.e. in a monastery. Still, I know that that's neither here nor there. I am not in a monastery. I am not leading a hellishly wicked life, either, however, and I am living with someone who also practises meditation, which is a great blessing and help. So let me make up my mind to go on and on trying.

Don arranged for Jo [Lathwood] and Alice Gowland to have their horoscopes done by Jack Fontan. This was a huge success. Jack told Jo she was a "crape-hanger" and that she had got to stop it, which impressed her and filled her with good resolutions. In order to get the horoscope drawn up, Jo had to give Jack her real age, of course. She wrote it in a sealed note which she gave to Don to pass on to Jack. While Don was down there the other day he read it. She's sixty-eight. He had guessed she was older.

October 2. This is just to announce that I bought a new typewriter this morning, another Smith-Corona Electric. For some reason, because the keys are enclosed perhaps, I don't make nearly so many mistakes as on the old one, particularly those slips off *a* with my contractured little finger which kept landing me on *z*. But why can't I underline properly, I wonder? No time to find out now, because Bill Inge, Paul Dehn and Leonard Stanley[1] are coming to supper....

October 3. It's so typical of me in my old age, this fussing about details. In this case, the underlining key on my new typewriter. I simply could not meditate this morning because I was worrying about it and impatient to call the typewriter shop and ask about it. Now I have done so and find that there will be no one in the repair department till Monday. So relax, buster.

Our dinner was quite a success, because Paul Dehn talked a blue streak and Leonard Stanley knows how to be a pleasant guest when older queers are present. Poor old bulky Bill looked, at moments, as if he had just had a stroke—that staring expression. Don thinks he won't last long. You feel his terrible melancholy.

The smog was so bad yesterday that Bruce Connors at the gym said one really shouldn't go out jogging in it; making yourself breathe heavily and inhale all that stuff does you much more harm than the exercise does you good.

Don is doing such interesting work. I keep urging him to show it to Billy Al Bengston. He does so need some strong professional encouragement—but he hesitates; he's afraid Billy won't like it.

October 23. This morning I finished chapter 15 of *Kathleen and*

[1] Interior designer and photographer, a close friend of Tony Duquette, who introduced him to Isherwood, and of Irving Blum.

Frank, though of course Don may suggest some revisions when he reads it. Now there are the two big chapters of direct quotation to proofread; very little can be done to them.

Still nothing definite about "Frankenstein," but Hunt Stromberg did call and say the studio wants to go ahead. Am very glad I've had this break, as it's enabled me to get ahead with the book.

Swami has been ill again, or weak, anyhow. But this is partly a means of exerting pressure on Belur Math. Now his doctor has written them a letter and we have sent cables, begging and blackmailing them for an assistant.

Had a dream about Swami a few nights ago. He was making a speech, in which he seemed to be attacking Ralph Hodgson (the poet). When he did this, people hissed. I knew he didn't mean it, that this was a misunderstanding. I tried to clear things up. And woke. Well, if not spiritual, this was at least a "loyal" dream!

Whenever I see Swami nowadays, I take the dust of his feet on arriving and leaving. I do this because I need to think of him definitely as The Guru, so I can meditate on him, according to my new instructions. When you think that I really do feel about him in this way and believe in him and know how privileged I am to keep having these meetings alone with him, it's a *sheer miracle* how unspiritual I still am.

October 24. Yesterday at the gym, my weight was down to just a bit over 147. I have hardly drunk any wine and no hard liquor for several weeks, and it does seem to be steadily dropping, though with pauses. I manage to jog three or four days a week. Am fairly careful about what I eat, though not extremely. I think my weight loss is also due to not getting much sleep. Our alarm rings at 6:30 every morning and we're usually out of bed by 7:35. None of this seems to affect Don, however. He is up high, for him: 145.

Was over at Elsa's today. Al[le]n Drury[1] was there. Elsa wanted me to come and encourage Al[le]n to take on the Charles Laughton biography; he's interested. So I encouraged him—because after all why not? He's big enough to look out for himself and cope with Elsa, or he should be. Elsa says he's an extreme right-winger, but I still like him rather.

Mark Andrews really seems to have moved in with Gavin, solid; he's brought a wonderful wine-red couch which belongs to him and now it's in the living room, with Mark draped all over it. I

[1]American political journalist (1918–1998); the screen adaptation of his Pulitzer Prize-winning novel about Washington politics, *Advise and Consent* (1959), was Laughton's last movie.

continue to like him, quite definitely, and to think what a good Frankenstein's monster he'd make. But we won't tell him about that until we know definitely whether or not we have the job.

October 27. Watching T.V. with Don tonight—an idiotic ghost story called "The House That Would Not Die," with Barbara Stanwyck, after a supper of swordfish steaks cooked under the broiler, and salad. Felt a pang of such painful fear and dismay because I soon have to die and leave him and such perfect moments of happiness behind me.

The day before yesterday, I finished revising chapters 16 and 17; there was very little to do to them. Am now well started on chapter 18.

Beautiful warm weather. Last night, Jack Larson made our flesh creep with tales of the coming economic ruin of the country. Oughtn't we to put money in Swiss banks while it's still allowed? But we don't know how to. And Robin French, whom I asked about it today, isn't nearly so pessimistic.

November 1. 6:30 p.m. Have just finished my second draft of the two last chapters, 18 and 19. The next time around they *should* be pretty much as I want them—though you never know; I have a feeling that I might get some last-minute revolutionary notion.

Have been talking to Hunt Stromberg, who still thinks they will want to start on "Frankenstein" very soon. Am glad at least that the book is as far along as this.

The day before yesterday, we drove down to San Diego and saw John Lehmann who is teaching at the State College and his friend from Austin, Texas[,] Chick Fry.[1] Quite a nice homely little boy. But John is a menace. He wants to come and stay here at Thanksgiving. Don thinks he has no feelings at all and is only interested in symbolic events, like meetings with well-known "old friends."

Heard on the car radio yesterday: The "Zodiac Killer" in San Francisco has sent a Halloween card to a newspaperman there, saying, "Peekaboo, you are doomed!" The newspaperman (for whom I felt very sorry) came on the air and said, quite seriously and as though this were surprising, "The police interpret this as a threat against my life"!

A notice from One Incorporated to say that they are appealing

[1] California-born Morris Fry catalogued the *New Writing* papers which Lehmann sold to the Harry Ransom Humanities Research Center in Austin; they met in 1969 and began an on-off affair.

to all us queers to buy lots of Hamm's beer during this Halloween weekend because Hamm's have bought advertising in *The Advocate* and we ought to demonstrate Gay Purchasing Power. So we bought a six pack.

Don is eagerly awaiting the end of my book so we can start the house painting. Meanwhile we are going to get the hallway properly lit so one can see the pictures. Also the living room.

Swami is taking off at the beginning of this week for a holiday in Arizona, Oak Creek Cañyon. Nothing is settled yet about an assistant. They offer to send one but Swami suspects he won't be suitable. So now it's we who are stalling, not they! When I took the dust of his feet he said, "You don't have to do that every time!" so I explained that I am trying to fix the image of him in my mind as The Guru. This seemed to amuse him.

I jog nearly every day. My weight hovers around 147, though I seem to be awfully hungry and eat an awful lot. Hardly drink, though. It is weeks since I had hard liquor or more than a couple of glasses of wine at most.

November 23. On November 17, I finished revising the last chapter of *Kathleen and Frank.* (It was chapter 19, but Don advised me to change its title to "Afterword," pointing out quite rightly that the Frank–Kathleen narrative ends with Kathleen's death at the end of chapter 18.) On the 19th, I got the manuscript xeroxed (at the Pacific Palisades stationer's, Sanders', which does quick and fairly efficient work and has a very sexy assistant, the manager's son). On the 20th, I mailed the copies off to New York and London. There is lots more to do, but I had to get it off.

This is the history of the book. The first draft of it, called *Hero-Father*[,] was begun on January 2, 1967 and never finished. On June 15, 1967, I began copying selections from Kathleen's diaries and Frank's letters for the book and finished doing this on August 5, 1968. I began the book itself on September 19, 1968. There were big interruptions, of course, throughout all of this work. Indeed, I think it went rather quickly, considering.

What remains, aside from corrections, is to get permission for all my quotes and check a couple of research points. I still don't know what "siche" meant on Frank's disk. And I don't know about the false Judge Bradshaw's activities in Maryland.[1]

[1] A John Bradshaw emigrated to the U.S. in the seventeenth century and pretended to be Judge John Bradshaw (1602–1659), Isherwood's ancestor born at Wyberslegh, who sentenced Charles I to death. As he tells in chapter 13 of *Kathleen and Frank*, Isherwood was unable to find out more.

John Lehmann came to stay with us for the night on the 21st and left yesterday. Throughout all of his visit, my nerves were in a constant state of irritation and tension. He is fantastically thick-skinned and so selfish. You have to wait on him without ceasing and he never offers to pay for anything. There is more to say about him than this—I'm merely letting off steam—but I'll do that tomorrow.

November 24. One thing John told us is nearly incredible but obviously true. While teaching at San Diego he met a woman professor who has spent three or four years already comparing published versions of Virginia Woolf's work with actual manuscripts in some college collection. And this professor told him she has found that Leonard actually made alterations in Virginia's diary and in her fiction and essays published after her death! John doesn't find this impossible to believe, because Leonard was so envious of Virginia and so arrogant and convinced of his own genius.

John himself certainly isn't arrogant in that way. He is very complacent but I don't think he really believes all that much in the value of his own work. He believes in his position in the literary world and his "services to literature" (which no one denies) and his CBE.[1] Don thinks he is really quite desperate, beneath all this.

Perhaps John's best quality is his capacity for feeling literary enthusiasm and for doing something about it. I showed him the story Morgan wrote in 1957, about the two boys who fell in love with each other on the ship going to India.[2] John immediately resolved to publish it, if he can get the permission and the money to do it. He has just succeeded in getting his publishing firm back under his own name, after years of legal tie-ups.

Which reminds me, I have had several calls already from publishers in this country, about *Maurice*; also two applications from students who are writing Forster theses and want to read *Maurice* before publication and include comments on it.

I think it must have been because John had got us both so rattled that we made an idiotic mistake, the day before yesterday, when we drove him up to Santa Barbara to have lunch with Bill Brown and Paul Wonner. We set off over the Sepulveda pass, meaning to take the inland freeway route to Ventura, and drove and drove up into the hills. Soon we began remarking how amazingly everything had been altered; they seemed to be building new freeways

[1] I.e., the formal recognition of these services by the Queen in making him a Commander of the Order of the British Empire.
[2] I.e., "The Other Boat," mentioned above.

everywhere and tearing up the landscape, so you couldn't even recognize it. And then at last we were confronted by a sign which said Bakersfield and another which marked the turnoff to Ventura via Santa Paula—so we had made a huge detour and arrived more than half an hour late. Although we know the Sepulveda pass like the inside of our hats, we had stayed on the San Diego North freeway instead of turning off onto the Ventura freeway!

Don is leaving for San Francisco for his show[1] at the weekend. I still don't know if I shall go up with him. Then a fearful period of upset impends—the repainting of the inside of much of the house and the refinishing of the hardwood floors; this will mean taking down *all* the books in my room!

We've heard nothing more about "Frankenstein" from Hunt Stromberg.

Last night, going into a movie, we met Renate [Druks] coming out and she told us that she'd heard Ronnie Knox has been put in Camarillo by his parents. The last time she saw him he was rather crazy, she says. Am going to call the head psychiatrist there, Dr. Benjamin Siegel, and find out for sure. Evelyn Hooker, who has now settled in Santa Monica and is in practice as what she describes as a homosexual-marriage counselor, gave me his name; and it now turns out that I have talked to him before this, about Michael Leopold.

November 25. Dr. Siegel was absolutely charming, remembered my calling about Michael and didn't even act too busy to talk (what a lesson to cross old Dobbin!). He did however tell me that there is a new law which forbids him to give out information about his patients. This morning he called back and in an indirect way indicated that Ronnie *had* been there at the hospital but only for one day; also that he had either insisted on leaving or had actually escaped—but Dr. Siegel had to be very vague about this part of it and I'm not quite sure what he meant.

Today I took the three folders of my *Kathleen and Frank* manuscript over to Gavin to read. Have had a cable from Richard that he has received the last two chapters. Nothing from New York, but that probably means that Curtis Brown *has* got the manuscript.

In the paper this morning there is this hideous news about Mishima's death, as a demonstration against Japan's no-war constitution:

[1] A joint exhibition with Ed Ruscha, opening December 1 at the Hansen-Fuller Gallery.

... Mishima drew out a samurai sword, bared his stomach and pulled it across his middle, drawing blood. A youthful disciple standing beside him then chopped off Mishima's head in the approved samurai hara-kiri tradition.

Then [Masakatu] Morita, twenty-five, a leader in Mishima's private army, plunged a short sword into his own abdomen. One of the other three severed Morita's head ...

Mishima had been closely tied to the Japanese army in recent years ... his eighty-man Shield Society spent some time training with the self-defence forces. Mishima had outfitted the tiny force in uniforms of his own design and written a song for it.[1]

I suppose he had become completely crazy. I just cannot relate this to the Mishima we met.

November 26. Thanksgiving this year is particularly for having finished the book. Now that it is finished I am terribly at a loose end, however. What shall I write next? The idea of a novel about Swami and me no longer seems so appealing. Surely it would be better from every point of view to do this as a factual book? Well of course there is the difficulty of being frank without being indiscreet; but that difficulty always arises in one form or another. For example, it is absolutely necessary that I should say how, right at the start of our relationship, I told Swami I had a boyfriend (and that he replied, "Try to think of him as Krishna,") because my personal approach to Vedanta was, among other things, the approach of a homosexual looking for a religion which will accept him. Another difficulty, far more serious, is that the book couldn't be truly complete until after Swami's death. (But that's no reason why I shouldn't make a rough draft of it now. Then, if I die first, there will at least be something left behind which may be of interest to others.)

I have also had the idea that my memoir of Swami might be published with a memoir of Morgan, based on his letters to me. So that the book would be a *Tale of Two Gurus*, as it were.

Then there is a fairly big chunk of diary fill-in which I might do, covering the scantily covered period between January 1, 1945 and February 1955—or maybe February 1953, when I met Don, because that's the beginning of a new era. This would be quite largely a sexual record and so indiscreet as to be unpublishable.[2] It

[1] *Los Angeles Times*, November 25, 1970, A4.

[2] This reconstructed diary appeared posthumously as *Lost Years: A Memoir 1945–1951* (2000).

might keep me amused, like knitting, but I should be getting on with something else as well.

Have I given up all idea of writing another novel, then? No, not necessarily. The problem is really as follows: The main thing I have to offer as a writer are my reactions to experience (these *are* my fiction or my poetry, or whatever you want to call it). Now, these reactions are more positive when I am reacting to actual experiences, than when I am reacting to imagined experiences. Yet, the actuality of the experiences does bother me, the brute facts keep tripping me up, I keep wanting to rearrange and alter the facts so as to relate them more dramatically to my reactions. Facts are never simple, they come in awkward bunches. You find yourself reacting to several different facts at one and the same time, and this is messy and unclear and undramatic. I have had this difficulty many times while writing *Kathleen and Frank*. For instance, Christopher's reactions to Kathleen are deplorably complex and therefore self-contradictory, and therefore bad drama.

Well, I just wanted to think aloud for a while, even though it hasn't gotten me anywhere.

Last night we left in the middle of the new *Julius Caesar* film because it was so bad.[1] Driving home, I happened to notice that *Ned Kelly* is on at last. The script is awful; you have no idea what is happening, and there is no clear progression. The photography is sometimes marvellous but mostly foggy. I did get a terrific feeling of Australia and even of period Australia—though that's partly because we could recognize some of the film locations we had actually visited last year. I still think Mick Jagger is very interesting as Ned, though he is obviously quite miscast. He isn't nearly big and strong enough and he doesn't have the right sort of Irish warmth (or much warmth of any sort for that matter). But he does seem aloof and dangerous. There is one scene, when they try to put handcuffs on him, in which he really is like some wild animal, perhaps a bobcat, when it is fighting with pure instinctive savagery against being trapped. Don felt that Tony Richardson shows a kind of inattention, a lack of real interest, throughout. He says he really wonders if Tony will ever make another film. Still we have to remember that his *Hamlet* film is quite recent and certainly one of his best; Don thinks it the best Shakespeare film ever made by anyone and I guess I agree. Perhaps the truth is, Tony should never be allowed to mess around with the script; in Shakespeare's

[1] Adapted from Shakespeare's play, with John Gielgud as Caesar, Charlton Heston as Mark Antony and Jason Robards as Brutus.

case he doesn't of course. But then, a director *ought* to mess around with the script. Otherwise he's merely a putter-on-scene, the film isn't fully his work.

We had supper with Chris Wood, still wondering if he shall stay here or go back to England when Gerald dies. For some reason he seems unwilling to accept the obvious common-sense suggestion that he should go over to England on a trial visit and see what he thinks of it. His nature demands an either-or.

An absurd predicament I got into, last night, after I'd parked my car to go to the theater where *Julius Caesar* is playing. I absent-mindedly locked both doors (as you can without a key by pressing down the buttons on the windows and then slamming them) and then found I'd left my key in the ignition and the engine still running! I knew I had an emergency set of keys hidden in a magnet box somewhere by the engine, but I couldn't remember where. I groped frantically around the running engine for what seemed an age, before I found them.

November 28. It's raining and I'm unoccupied and therefore blue, but not very. Have just been reading through the entries in my diary between April 1948 and February 1953, the Caskey period. Such desperation! Surely in those days I was suffering in a way I can hardly even understand, now? And what got me through? Grace. And work.

Camilla Clay is ready to produce our play, she says. On the understanding that we begin with a rehearsed reading of it, to see how it goes in front of an audience. She doesn't really believe in it, yet. The reading is Jim's idea; he suddenly came up with it yesterday. Now the problem is to find our two leads.

Don leaves for San Francisco the day after tomorrow. If he gets any commissions after the show opens, he will stay on there; but this does seem awfully doubtful. I'm not going. I should have nothing to do and only be in the way.

Every morning when I wake, I dread the disturbance of the upcoming painting and floor sanding. It will be utter hell and probably last at least three weeks.

My next chore—however much I loathe the prospect: to produce a "definitive" version of *Black Girl* which can be shown to the Shaw Estate, passed, published and then, we hope, performed all over the world in all possible languages.

November 29. We had a very heavy rainstorm all night and this morning. Now it looks as if it might lighten but you can't tell.

Don's father came and they took up all the matting in the front and back bedrooms and hallway, a terrific job, five hours' worth, which would have cost sixty dollars if the floor finishers had done it. Now that we see the nice hardwood floor again we wonder why we ever had it covered!

During the rain, two leaks developed, one in my workroom, one in the living room. This was one of those providential disasters, because it warns us that the roof must be fixed before those ceilings are painted. Irving Blum just called Don to tell him he can't come up to San Francisco with him tomorrow because the storeroom at the gallery here has been flooded, damaging a large number of paintings and prints which aren't sufficiently insured. Don says nothing will go right for him until the middle of January because Saturn is fucking things up; Jack Fontan told him this. Negative astrological predictions make you philosophical. You expect no good, but at the same time if good does come you have the added satisfaction that the astrologer was wrong.

December 1. Yesterday started by raining, then cleared, but it's still damp and more showers are expected. This weather is exactly what I fear for our house painting. Everything will be moved out onto the deck and then it'll pour.

That event still impends like major surgery. A very nice man named Norm Berg came by to arrange about fixing the roof. He said not to worry about the cracks in our pavement outside the house; he thought the land on this side of the Canyon is pretty firm, on the Palisades side it is unsafe. He says he has inside information that the building at the end of Adelaide will begin very soon, and that the two towers will be only seventeen stories high. Only! But that's a reduction of seventeen stories on the original estimate!

Meanwhile we are preparing to get rid of all kinds of junk. The St. Vincent de Paul people[1] are seemingly ready to take away the carpets, and we're trying to wish all sorts of other items on to them, including the broken wicker chair from the kitchen, my globe of the world and the statuette I won in a Save State Beach lottery. Then there are artbooks from the bookcase in my room which Don has now moved into his back room. The grimmest deed (which I performed myself) was twisting up Henry Guerriero's horrible mobile, ramming it into a carton from the market and leaving it out for the trash collectors. I still feel that it is a Crime Against Art to destroy any kind of artwork.

[1] The Society of St. Vincent de Paul, a Christian charity.

The removal of the second bookcase alongside my couch (Don was forever objecting to it) really has improved the room, making it much lighter and more spacious. I think I wanted to cut the room in half because of my memories of Emily's sitting room in Buckingham Street; I thought it would be snugger that way but actually I only made part of it unusable.

By yesterday's mail I got a letter from Richard Simon of Curtis Brown in London saying he likes the book very much but feels the first (pre-marriage) part is too long. Then Perry Knowlton[1] phoned from New York to say Peter Schwed has read it and wants to publish it, but feels it will only do well as a hardcover and be difficult or impossible to sell as a paperback. So some cutting seems indicated. I have told both Simon and Knowlton that I'm ready to consider this but want to have the publishers make their own suggestions about cuts before I decide.

Don put off going to San Francisco yesterday and is flying up today with Irving Blum, just for the opening of the show, and coming back by the midnight plane tonight.

December 2. Don had a horrible experience coming back; the plane took off in a rainstorm and lurched about and groaned and struggled until everybody was scared and Irving got terrified and kept saying he was going to be sick. This was all the more sinister because the cunty stewardess had just caused a drunk passenger to be removed from the plane right before takeoff, which made it seem like one of those Ancient Mariner stories (as Don said). He would have spent the rest of his life bragging about how his crime caused him to escape death and destroyed everybody else—by making them late and thus get caught in the storm!

Don doesn't think the show did him any good at all. Irving apologized to him for the failure of the Los Angeles show but said nothing about doing anything for him in New York.

Meanwhile I puttered around, took books to Needham[2] to be sold, got our driver's licenses for 1971, etc. etc. My only worthwhile exploit was to jog from the gym all the way down to Olympic [Boulevard] and back, a record. But this was maybe unwise, because since then I've had several severe twinges in the upper buttock!

Gavin called to say he'd finished the book. He seems to like it but doesn't rave. Admits parts were too long. However he does like the end, which is important.

[1] Isherwood's New York agent; see Glossary under Curtis Brown.
[2] A secondhand book dealer.

Poor old Jo has now definitely heard that Louis [Gold] is going to build on his parking lot very soon, thus driving her out of her apartment. She wept and told me she couldn't move into her own house because she was afraid to live on the ground floor and because she wouldn't get a view of the sea, which she *has* to have because she has always had one.

December 4. St. Vincent de Paul's people, a young Negro and a rather terrifying old man with one eye, who looked like Jean Genet, refused to take away our carpets and then refused the rest of the stuff saying they had a rule that they either take everything or nothing! So today we are trying the Salvation Army.

John Bleasdale can't come and paint until December 18, because he has to redo some work for Lon McCallister;[1] the rains ruined the first job. So the floorers, a Mr. Cvar is one of them, are coming in first, on the 14th. We are still waiting to hear when Mr. Berg can get our roof leaks fixed.

The weather is cold but absolutely classical in its beauty; that rose-golden desert light on the palms and white buildings and on the mountains. [Mount] Baldy is under a heavy snowfall.

(At this point Glenway [Wescott] called about *Maurice*. He is longing to write a huge piece on Forster to be published at the same time as the book. He said of himself that he has a terrible habit of jumping onto bandwagons. I am encouraging him to write a foreword to the book because I know that if I write one myself I shall offend the touchy and devious Furbank; yet Furbank's foreword, alone, simply isn't right for America, it is too inner-circle. We shall have to publish it with something else by an American writer.)

The day before yesterday I saw Swami, for the first time since his holiday in Arizona. He seemed tired and is still engaged in this tug-of-war with Belur Math about the demanded assistant. It will have to stop before so very long, however, because he doesn't want to be on bad terms with them when Len [Worton] and Mark[2] and Paul [Hamilton] go over there in February to take *sannyas*. He gives me the impression that he is play acting his indignation, a bit; and he has written a very warm note to Gambhirananda, whom he regards as his arch-opponent, appealing to him to cooperate. I asked him if he had had any spiritual experiences while he was

[1] Teenage film star (1923–2005) in *Stage Door Canteen* (1943), *The Story of Sea Biscuit* (1949), and others; he retired at thirty and began investing in real estate.

[2] I.e. John Markovich.

away. He said no. While I was sitting there I tried to meditate on the fact that I was in the presence of The Guru. I find that I can now meditate on the memory of having done this, just as I meditate on the memory of having sat in front of the shrine.

Peter Schneider has now applied for reclassification as a C.O. This last-minute move is really shameless and I can't believe it will work. But he is so cute. Jim Gates has strangely obliterated himself by shaving his head and wearing glasses. He told me after the reading that he wants to talk to me. I wish he would. I'm curious about him. I want to know what he's up to.

December 8. Swami is sick, up at Santa Barbara, but Ananda doesn't think it's serious.

Have just finished the second folder of *Kathleen and Frank* (through chapter 13) which I'm rereading for possible places to cut. The most obviously cuttable part is chapter 9, which consists largely of wearisome details about Frederick's financial demands on the Isherwoods; these are undramatic because Frederick is really only bluffing anyway, as I finally have to admit. I think a lot can be taken out here, but even so it is very little in relation to the size of the book.

No news from either Methuen or Simon and Schuster.

On Saturday next, the 12th, Mr. Lemke the carpenter is scheduled to come and take the big bookcases and my desk apart, so they can be removed from the room before the floorers come on Monday. So that's the beginning of chaos—for *at least* three weeks (during which time, a heavy rainstorm could fuck us up, but good)! All we can do to prepare for it is to collect cardboard boxes, lots and lots and lots of them, from markets etc., to put the books in.

Seth Finkelstein gave me my dolphin clock back yesterday. He only wanted ten dollars for all that work; I insisted on his taking fifteen. It now runs beautifully again, but it's slow![1]

Don has heard nothing whatever from San Francisco about his show. It might as well never have happened. He's disgusted, naturally, and says what's the use of going on with this noncareer as a portrait artist.

Yesterday I called Hunt Stromberg for news of "Frankenstein," having heard nothing for weeks. To my surprise, Dick Shasta

[1] The clock was given to Isherwood by Fräulein Thürau and is described in *Goodbye to Berlin*, p. 15. Finkelstein, a neighbor, cleaned and repaired it on at least one other occasion, in 1968.

answered the phone, so apparently they've made it up. Hunt told me that now someone else is trying to finance the "Frankenstein" project and that he's to hear news this week.

December 11. This may be my last entry for a while, because tomorrow this desk will probably be moved and the top taken away to be refinished and I shall lose touch with most of my other possessions.

Got a contract to sign from Simon and Schuster; am to be given an advance of seven thousand five hundred dollars, which is good. But no word so far from Peter Schwed, or Michael Korda.

Saw Evelyn Hooker yesterday. She wants me to work with her on a "popular" book on homosexuality. She seemed very emotional still; once she actually shed a few tears while describing the goodness of her sister to her during her breakdown. I am doubtful about the project. It seems that I shall have to read through sixty case histories and then write about them—which really means retell them, and what the hell is the use of that? Nonwriters never understand what writers can and cannot do. They think they can tell you exactly what to say and that you will then somehow magically resay it so it's marvellous. However, I didn't want to refuse straight away. I'll read some of the stuff first and try to find out more exactly what it is that Evelyn expects. She is a very good woman and her intentions are of the noblest and I would like to help her, if I can do so without becoming her secretary.

After threats of rainstorms, the weather is brilliant; you can see the whole arm of the bay in clearest detail. I feel we are "wasting" this weather; it ought to be saved until we really need it, during the repairs to the house. The roofers are supposed to come in and do the job tomorrow.

Gavin and Mark Andrews are going to Tahiti for Christmas. Gavin saw an ad for a round-trip on the French airline; you get a week on Tahiti and a week on Moorea with hotels and food, plus the fare, for six hundred dollars—cheaper than staying at home! We wonder if they won't be terribly bored, however. (This reminds me that Jim Bridges had an outburst, quite violent for him, about the relationship between Brian Bedford, Gavin and Mark; he said he found it disgusting that they call each other by women's names—Gavin's is "Dora." "It offends my dignity as a homosexual," Jim said.)

Jim is busy trying to organize this reading of our play for after the holidays. Yesterday we saw a young British actor named Ter[r]ence Scammel[l]; but he isn't right. Scammell told us he has read the script of the *Cabaret* film (because he's up for the part of Chris) and that "Chris" (now called Brian) is queer, that's to say he can't make love to Sally at first and then later he can and then Sally does it with a mature but very attractive baron and Chris is jealous and makes a scene about it with Sally, and Sally exclaims, "Oh, fuck the Baron!" (meaning that he's unimportant) and Chris replies coyly, "I do." That's the kind of thing which offends *my* dignity as a homosexual. The queer is just an impotent heterosexual; that's what these Jews keep saying, over and over again.

December 24. Just to record that I am sitting down at my "re-finished" desk today for the first time since this upheaval began. My workroom is now painted. So is Don's little back bedroom and so is the hallway. That is all. John Bleasdale works very slowly but he is pretty thorough it seems and easy to get along with. The girl who did the desk isn't much good and she still hasn't delivered Don's desk and his chair. Her name is Carol Palermo. Cliff Lemke, the carpenter recommended by Lon McCallister, is no good at all. The work he did on my bookcases had to be redone by another carpenter, Bob Main, whom we like. The floors look beautiful but they will mark easily and occasion much sweeping and polishing. My room seems staringly white at present but no doubt I'll get used to it.

Today, I got the advance money from Methuen and the money from Simon and Schuster is promised before the New Year. Both contracts are signed.

So far, this has been an unexpectedly happy period, despite the discomfort. It was snug sleeping out in the studio. And, up to today, we have had our mattress on the floor in the front bedroom instead of on the bedframe.

There is much more to say, but time is slipping by and I must get on with tidying the desk and stowing everything away again in drawers and cupboards and on shelves.

December 28. From December 14 until the 21st, while the floors were being done, I didn't get to meditate at all, and meditation is still badly disturbed by this sense of *house*, this preoccupation with endless domestic details—the sanding, the painting, the light fixtures, getting rid of unwanted furniture, fussing over this and that. It's more alienating than lust or pride or vanity or anything else. I have got to evolve a better way of making acts of recollection. In the mornings there is always the problem of getting the show on the road; if Don has finished his meditation and gone into the kitchen, then I ought to hurry up and get in there too, to help him.

Yet this has been such a happy time and my health has been good, thanks to jogging and perhaps also the endless vitamins we take, and my weight has been down sometimes as low as 146, which would have seemed miraculous at the beginning of this year. Don and I seem to come closer and closer together, in between spats. Sometimes the sense of our togetherness is terribly painful because it gives me such a sense of human impermanence.

Don is bothered by the sense of *house* quite as much as I am. While we were grappling with the mattress of our jumbo bed, trying to erect it again after the painting, he said, "There must be something desperately wrong with one's life when one has huge pieces of furniture like this." We have discussed getting a proper Japanese bedroll which we can spread on the floor and stow away in a chest.

We had Christmas dinner with Jennifer Selznick and her children, including Bobby Walker[1] who has decided to settle with his wife and children on one of the islands of the Seychelles group. They are going to get back to nature. He is humorless and a bit too holy but admirable and he said one thing which impressed me a lot, that in a situation like that you have to learn to accept boredom and make it part of your life.

Swami's birthday party was rather a success, less embarrassing than usual. He is exactly the same age as Mao Tse-tung! They are still waiting for news about the promised swami.

At Jennifer's we got a lot of presents, all unwanted except for a gaudy but pleasing waistcoat with great silver buttons and tassels on it, made of leather dyed in various colors. The rest of the stuff, plates, red glasses and a huge useless thing to hang on the kitchen wall in order that cups, corkscrews, knives, etc. may hang on *it*, we hope to trade in at Van Keppel Green's[2] today.

Yesterday a Michael Silverstein came to see me—he had interviewed me a few years ago for a student paper at UCLA. He is active in Gay Lib. I liked him. He is very intelligent, in a belligerent Jewish way. He dislikes the idea of presenting "normal" queers to the world (as Evelyn Hooker wants to do in her book) because he says how can any of us be normal as long as we are subjected to this persecution. He believes that a revolution is inevitable. He

[1] Robert Walker Jr. (b. 1940) trained at Actors' Studio in the early 1960s and appeared on stage, in a few films, and often on T.V.

[2] Hendrik Van Keppel (b. 1914) and Taylor Green (b. 1914) ran a shop on Santa Monica Boulevard in Beverly Hills selling furniture of their own design along with housewares, lighting and jewelry by others.

says that psychoanalysis is ultimately political because it judges the patients according to the standards of the establishment; those who are for it are "normal," those who are against it are "neurotic."

December 30. One effect on me of Silverstein's visit is that I have rewritten a couple of sentences on page 279 of *Kathleen and Frank*, making them much more aggressive. Instead of:

> Without even trying to decide between the relative disadvantages of alimony and police persecution, he is now quite certain that heterosexuality wouldn't have suited him. And he has always felt content and well-adjusted, being as he is. . . .

I have written:

> Despite the humiliations of living under a heterosexual dictatorship and the fury he has often felt against it, Christopher has never regretted being as he is. He is now quite certain that heterosexuality wouldn't have suited him; it would have fatally cramped his style.[1]

Incidentally, Silverstein's personal "discovery" seems to have been that you could go to bed with your friends, not just with one-night stands from bars. He was quite astonished when I told him I had done this all along!

Don is in a fury against Carol Palermo, who, he now finds, left his desk out in the rain and ruined it and then didn't dare tell him so and stayed away from home on the day when he picked it up. He left a note on her door, telling her what he thinks of her.

Yesterday evening I had supper with Swami. Telling me the familiar stories about Maharaj, he shed tears. How amazing he is! In one respect I find him much more wonderful than those first disciples, who lived together and helped each other and who had all known Ramakrishna. For Swami has lived all these years in an alien land, amidst the most alienating people and surroundings, and now he is an old man and all alone—for however much we love him we can none of us really understand him—and behold, "The inner life has paid,"[2] his faith is its own absolute reward. What else need any of us do but meditate on him and his achievement? He told me that if you think about Ramakrishna you are thinking

[1] Chptr. 15.
[2] Cf., "The inner life had paid," Forster, *Howards End*, chptr. 37.

about God, even if you don't regard Ramakrishna as God. When you think about other people, you have to think of them as God, consciously.

1971

January 1. First lines of New Year dialogue on waking:

"Beautiful Kitty."
"Dobbin always notices Kitty's beauty when he hasn't got his glasses on."
"From here, his profile is so beautiful."
"Dobbin likes half a cat much better than a whole one."

We didn't get home till nearly four this morning, from a nice party of young things, mostly stoned Buddhist boys, to which Jack [Larson] and Jim [Bridges] took us to see the New Year in. The only jarring note was Jack's enthusiasm. Dear Jack! He would make heaven absolutely intolerable by raving about it.

The reading of our play is now definitely off, because a Patrick can't be found. However, Camilla told Don over the phone, from Mt. [K]isco, that she has "a new idea."

Yesterday Hunt Stromberg called to make an appointment for us to talk to the new boss of Universal about the "Frankenstein" project.

January 11. The "Frankenstein" project now seems to be sanctioned by the head of Universal, Sid Sheinberg, who is not too bright but quite pleasant—he has a mad idea that the story should be written in four [one-]hour sequences! Personally I am still doubtful if the story can be written at all. Have been reading Jim Bridges' treatment which only shows up its weaknesses. When people say it is a "classic," they really mean only that the makeup is a classic, as long as Boris Karloff wears it. But now Hunt has another project which appeals to me far more, to do a film about *The Arabian Nights*.

Yesterday I went to see poor old Charlie Locke, who is in a convalescent home in Santa Monica. He had a heart attack after his wife died and also I suspect a nervous breakdown, but now he is fairly all right except for lapses of memory and a state of anxiety which reminds me of Granny Isherwood;[1] he kept asking me if

[1] Elisabeth Luce Bradshaw Isherwood (1836–1921), his father's mother.

there wasn't something I was looking for, had I ordered "something from the kitchen" or was it a telephone call? The room where we talked was crowded with old men and women listening to some ghastly-colored golf tournament on T.V., mostly lilac. The old women were like greedy seagulls; one of them swooped down and carried off the ashtray from right under Charlie's nose. Poor Charlie talked cancer solidly for an hour and a half, and how it is increasing everywhere. At the same time, he seems to understand quite well how psychologically dangerous it is to dwell on it. The surgeon had told him, "It tries to scare you to death."

On the 14th we shall have been involved in this house decoration for a whole month, and it seems unlikely that John Bleasdale will finish the painting by the end of the week, even. He is still not through with the service porch and has much of the outside trim of the windows to do. He is slow and tactless and infuriates Don, but his work seems to be fairly all right. Don says he looks like an old koala bear.

Last week we were haunted by a sweet little cat, which marched into the house and was so loving and eager to be petted. I really suffered terribly because I had to be the one who said absolutely that we mustn't adopt it; Don didn't want to really, but he liked to play with the idea a little. And then John Bleasdale fed it, which made me really furious, because everybody says that is the one thing you must not do if you want a cat to go away. And then the next day, quite inexplicably, it did go away! It had a flea collar on and was quite sleek, so maybe it belonged to someone.

Gavin and Mark are back from the South Pacific trip. They say it was marvellous, but we suspect otherwise. They left Moorea without even spending a night there and they never went to Bora Bora, and when they got to Honolulu it rained and they didn't even see the Pali![1]

January 13 [Wednesday]. It's pouring down rain. The weatherman promised water-spouts but none have appeared so far, and the rain is a terrible nuisance at this point because John Bleasdale was just about finished inside the house (only the service porch remains unpainted) and we were hoping he would be doing the outside windowsills today and that we'd be rid of him by the weekend. Yesterday he didn't come at all because he broke a tooth and today (11:40) he hasn't shown up so far.

The day before yesterday, it was announced that Derek Bok

[1] The cliff on Oahu, over a thousand feet high.

has been chosen to be president of Harvard. He is forty—and I remember him so vividly as a little boy jumping up and down and rushing around the garden at Peggy [Kiskadden]'s Alto Cedro house. It makes me curiously delighted—partly because I feel he has become admirably sly, a rogue in the right cause. The paper (*Los Angeles Times*, January 12) tells how, during trouble on campus in 1969, "law students held a study-in at that school's library to protest grading. Bok" (who was then Dean of the Law School) "was summoned [at 12:30 a.m.] to handle the crisis. He calmly ordered coffee and doughnuts, climbed atop a library table and announced, 'I want to thank you all for coming here to show your concern about the law school.'" Derek also travelled to Washington last year to join Harvard groups protesting the sending of troops into Cambodia. That to me is the typical action of a very shrewd person—seemingly a dangerous, perhaps career-ruining move but in fact one which was sure to please and impress the governing board; in these days of terror of the Young, anybody who can manage them is in, even if he shares their dreadful revolutionary opinions! So I've written Derek a note of congratulation. I wonder how Peggy will react to that, if she hears about it.

Nothing about "Frankenstein" yet, but Hunt did call this morning to ask if Universal had got in touch with Robin French to arrange a deal. They hadn't. However this is explained by the fact that Frank O'Connor[1] has gone to New York, and he was supposed to handle it.

A fan in Derbyshire has sent me the interview Brendan Lehane did with me for the *Telegraph* when we were in London last spring. It is well-informed and fairly well written and it gets Don's name right and mentions that his Auden drawing was bought by the National Portrait Gallery. (The way it does this is amusingly ambiguous, because it calls the drawing "his *Auden*," as though the picture were an unrecognizable abstraction which Don might as well have named *Solitude* or *Fish*!) The article also says I am homosexual, quite flatly, without further explanation. I think it is the first time anyone has said this right out, in print.[2] I'm glad he did. It sort of prepares readers for my remarks in *Kathleen and Frank*.

Elsa is lonely during her weekends, down at 147 [Adelaide], because Ray [Henderson] is now mostly with his wife, and Jack[3] has paired off (so Elsa thinks) with another of her male attendants

[1] A Universal executive.
[2] *The Daily Telegraph*, August 7, 1970.
[3] Grinnich, an actor who occasionally spent weekends with her.

and is therefore ashamed (so Elsa thinks) to appear with him in her company. She is left to make do with David Graham[1] but she can't bear the way he smells and he has a trick of chewing bits of food and finding them tough and spitting them out again and arranging them around his plate! In spite of all this, however, Elsa declares that, "I am the happiest person I know"!

My New Year's resolution is to read all the stories by Chekhov which we have in the house—that's to say all the volumes of Constance Garnett's thirteen-volume translation but one. It's really the only way to enjoy him properly, because you realize that these stories are all part of a world and indeed you hardly notice when one stops and another begins. He is tremendous. I feel as if I could go on indefinitely—maybe read the entire oeuvre at one sitting. He never bores me for one instant.

January 28. After being miserably cold it is heavy and hot and still—ninety degrees today downtown! The work on the house was finished the day before yesterday; Bob Main finished putting new asbestos wrappings on the heating pipes downstairs. John Bleasdale finished painting on the 23rd.

Swami is in Mount Sinai Hospital, having some tests; he seems to be fairly all right, however. I'm going up to the vespers of the Brahmananda puja and may see him later.

At last, the "Frankenstein" project has been set in motion. We are to have our first story conference with Hunt Stromberg tomorrow.

On the bad side: I have to pay over nine thousand dollars state and federal income tax tomorrow morning, plus whatever the Maltins have the gall to ask for "managing" our affairs. Arnold is useless, either sick or out of town; Mrs. Maltin does all the work and I do rather like her. Her twin boys, no longer cute, alas, have become the most fantastic junior athletes. The whole mantelpiece is already covered by their trophies and they are only eleven!

Also on the bad side: They have at last started, as of last Monday, to get ready to put up the high rise, or two high rises at the end of our street. This really sickens me. I feel deeply *invaded*. I keep imagining that I can hear them banging away up there. One can, sometimes, but it's very mild compared to the noise made by helicopters, motorbikes and other nuisances.

Yesterday I went with Ray Henderson to play his tape of the *Dogskin* musical to Ed Parone down at the Music Center. I'm sure

[1] Her agent; he represented actors, writers, and directors in Hollywood and New York and later became a casting director for T.V. and films.

Ed didn't like it and will do nothing about it—but I was surprised how good it sounded and indeed the half dozen assistants he had brought in to listen to it kept breaking out into obviously quite spontaneous laughter.

Hunt Stromberg has called a couple of times since making the date for tomorrow, because he keeps getting ideas for "Frankenstein." He wants it to be a musical, with [Leonard] Bernstein composing—no, he doesn't—but he wants Elizabeth Taylor to play the bride of the monster—and Burton is to play Pretorius (the "mad scientist" who appears in the *Bride of Frankenstein* film). And then there's Rex Harrison. Or maybe Albert Finney.... I just keep my mouth shut. But I do hope he isn't going to continue bugging us like this.

January 29. 9:00 p.m. Gore just called. He's in town with Howard; they just got back from India which he says he loved. He has decided to write a novel about the last incarnation of Sri Krishna, as a white soldier, a G.I.! As always I am pleased at the prospect of seeing him.

The puja (vespers) last night was spoilt for me by Swami's not being there, and, as usual, by the gabbling whispers of the women waiting to get their turn at being touched by the relics. Afterwards I went to Mount Sinai Hospital and saw Swami—so tiny and bird-legged and so beautiful with his silver hair. He was attended by two Asian nurses, one from Thailand, the other Japanese, I think. He said, "They are my daughters"; and already they had somehow become devotees, without quite knowing of what. Swami didn't seem as weak as I'd expected. He took my arm and walked right down the passage and back to his room. But they are not going to let him return till Monday. They have put a catheter in him.

I forgot to mention that I've had a pain in my side since about the 16th. I went to Dr. Allen about it and he seemed to think it is diverticulitis(?)[1] and prescribed a diet—which, characteristically, I haven't been following. Sometimes I feel something hard in there, then it seems to disperse, as though it were a blockage in the intestine.

We had our talk with Hunt over lunch, and now, inescapably, the moment has arrived for us to have some truly brilliant ideas. Am hoping for them tomorrow morning.

Reading Chekhov, with joy. Also dipping into Henry James's letters, just bought.

[1] Inflammation of a small sac or sacs formed on the outside of the colon.

February 1. We started working on "Frankenstein" the day before yesterday and, as of now, I really do think we are doing well; that's to say, we have given the Creature an extra dimension. Frankenstein has a rather older college friend named Rudolf, who is the sinister influence. It is Rudolf who gets Frankenstein involved in monster making. Then Rudolf dies, of blood poisoning caused by cutting himself while dissecting one of the bits of bodies they have stolen from a churchyard. And Frankenstein then takes the brain out of Rudolf's corpse and puts it into the Creature. So the Creature is partly Rudolf, but only partly. (He points to the Bible, opening it at the text: My name is legion, for we are many.[1]) Rudolf is imprisoned within the Creature, struggling to communicate—because of an injury to its vocal cords, it speaks with difficulty. This solves quite a lot of problems, right away.

We saw Gavin and Mark the night before last. It seems that their film project is snagged. He can't get the money from Jim Davison,[2] whose status as a dear friend is thereby endangered. Gavin takes this setback very casually, as he always does. Mark is far less noisy and demonstrative, nowadays, and nicer. But he does smell so bad; like some creature in a cage.

Supper last night with Joe Goode and Mary Agnes Donoghue. Billy Al and Penny were there. Gradually we are relaxing towards them all; indeed, I feel almost at ease with them. But for some reason their heterosexuality keeps showing, like a slip, and one looks tactfully in the other direction. In Billy's case, this is because he is deliberately showing it. What a masked, playacting character he is!

Still feeling the lumpiness within the left side of my guts. It worries me, of course.

Have finished the first third of Hemingway's *Islands in the Stream* with far more pleasure than I'd expected. Writing about the father–son relationship he seems far honester and more frankly sentimental than he does when he writes about women. But his killing off of two of the boys at the end of this episode is an awful tear-jerking trick, surely. However, I'll wait and see if he makes anything out of it, later.

Got a slightly gracious, slightly distant brief reply to my note of congratulation from Derek Bok. In spirit, it was like the gracious note Peggy sent me when I tried to reestablish relations with

[1] Mark 5.9, when Jesus casts out the demons.

[2] Tall, good-looking, and wealthy; once a companion of Speed Lamkin. He is mentioned in *D.2*.

her—the note which made me feel we had better not meet again.

February 2. Gore and Howard had supper with us last night. Gore had the shits, after getting all the way through India without them. Howard didn't look nearly as much older as I expected. Gore says that Anaïs is furious with him for his description of her in his last novel.[1] Don says it was cruel of me not to tell him anything nice about it; but, really, there are limits. It was nice seeing them—all the more so because they now seem more than ever a devoted couple, despite Gore's outward coolness. You might say that the fact we are both old couples is almost the one important thing we have in common. But I felt embarrassed, because I couldn't ask Howard, "What are you doing now?" for fear he would have to say, "Nothing." They have just lost their beloved old dog. That was what made them take this trip around the world, they told us.

We are plugging along with the "Frankenstein" treatment; this is our fourth day on it.

February 3. A beautiful morning but cold. Am writing this during the reading period which Don and I have been observing, from immediately after breakfast until about ten, when the mail comes and we begin work. Don is just starting *The Brothers Karamazov*, after finishing *The Autobiography of Alice B. Toklas*. He says he has read more, these last months, than at any other time in his life. I have just finished Chekhov's marvellous "[The] Black Monk."

A call from poor tiresome Charlie Locke. He talked to me in a state of mysterious agitation yesterday, saying that his daughter has been behaving "like a daughter" and was about to move him to another nursing home. Now I discover that this home is in Burbank, so there's nothing for it but to go there today, before going to Vedanta Place, which is a nuisance. He says he must talk to me.

I still have my little lump. Today I'm worried about it. Sometimes I'm not. Have been doing my yoga stomach twists energetically, trying to dislodge it. Otherwise my health seems excellent and my weight is down nearly to 146.

There's Isabel, our Mexican maid, so must stop.

February 4. I found Charlie Locke in a much less grand place, made of spit and chandeliers, right by the Golden [State] Freeway, which reeked of urine but had unusually friendly nurses and at least one

[1] Marietta Donegal in *Two Sisters* is modelled on Nin.

groovy young orderly. Charlie, poor old dear, was terribly anxious to be reassured that he hasn't become senile—he has, a bit, but not too seriously. Then we got onto Beatrix Potter and talked about her works for a solid hour. And then his daughter Mary Schmidt arrived and was really much nicer than I remembered. Obviously Charlie is a terrible handful, and she is a teacher and has three children to look after. She greeted me warmly, probably with the mad hope that I'd somehow make myself responsible for some of Charlie's problems. But I can't.

Then on to Vedanta Place, where Swami seemed wonderfully better. (Yet he too—which is unlike him—wanted to be reassured, after the reading and his answers to questions, that he had been up to the mark!) He told me the depressing fact that this year at the Belur Math there will be only eleven monks taking sannyas—and this includes the four from the West, our three and Buddha from London! And only twelve are taking *brahmacharya*. Unless these numbers are going to double or treble, the order will gradually shrink away to nothing.

We keep on with "Frankenstein." Today they at last brought the shades for the bedroom and Don's studio. The ones in the bedroom look terrible, they don't fit together properly. Don is disgusted.

Have just been talking to Mary Schmidt on the phone. She says that Charlie sometimes gets violent and that therefore they can't get him a room in a private house. She told me that, on Charlie's first night at this new place (The Burbank Palms) he was terribly indignant because one of the two male patients who were in the room with him "molested himself, while the other watched and enjoyed it!"

February 5. A day of worry and depression. The lump in my side is still there and seems bigger, though I have the illusion that I can somehow disperse it by manipulation after I have been in the steam room at the gym. (Weighed 148 today.) And then Don came back after seeing Dr. Kafka, the foot specialist, who told him that the growths in his foot are papillomas and sometimes "don't yield to treatment," whatever that means!

We *may* go up to San Francisco tomorrow, chiefly in order to see Chaplin's *Monsieur Verdoux*, which is on a midnight show. Am in a flap about "Frankenstein." We haven't finished it, and we ought to have, by now.

Have finished *Islands in the Stream* and although the main character is fakey in parts I still think it is one of his best books. Largely that's because it's about the Cuban area; I like his Cuban writing far better than his Spanish descriptions—he is so awfully dutifully holy about Spain and its bulls.

Eddie Anhalt says he'll go to visit Charlie Locke. By great good luck he happens to be working at Warner's, so it's only a short drive.

February 6. We decided not to go to San Francisco, because Jack and Jim didn't want to go. This morning we finished the first section of our rough treatment of the Frankenstein story and this afternoon Don took it up to Hunt's house. So we'll hear from him tomorrow what he thinks.

They've been fooling around on the moon again.[1] And there have been two quite big earthquakes, in Italy and in the Aleutians.

We had a delicious supper of hamburgers at home and watched *The Night of the Iguana* on T.V. Gore and Howard were to have called us before leaving tomorrow, but they haven't. A dull but quite happy day, with weather to match.

February 7. Don, who is worried about his foot and keeps expressing his worry in jokes about having his leg amputated, saw a very bad omen today. A little grey cat *with three legs* ran across the road in front of his car!

Gore called from the airport to say goodbye. Says he's still undecided whether to become Irish, Swiss or Italian.

Hunt Stromberg seems genuinely pleased with what we have done on "Frankenstein." As usual he made a couple of mad suggestions: that the kid brother William's arm should somehow be used—it keeps grasping at things, fighting for life. And that, when Victor and Henry (as he is now called instead of Rudolf) swear blood brotherhood and cut their arms, the blood runs down and somehow brings the butterfly to life!

February 8. We had supper with Jo last night. She has had a reprieve; Louis isn't going to build outside her windows for another year, maybe two. And now she and Alice Gowland have decided to give cooking classes in her apartment! We had cracked crab, which was quite good, but unfortunately Anne Baxter and her lover Dick Linkroum[2] were there and Anne was in a super-manic state, she

[1] *Apollo 14* made the third moon landing on February 5.

[2] T.V. producer and director, for instance of "The Jack Benny Show" in the 1950s.

couldn't stop talking about herself for a moment—which meant chiefly talking about *Light Up the Sky*, this dreary play she was in, and reciting great chunks of it. Dick, when he got a word in, kept quoting lyrics from Gilbert and Sullivan!

This morning we drew up our lists of nominations for the Academy Awards—best pictures, *Satyricon*, *Women in Love*, *Zabriskie Point*, [*The*] *Baby Maker*, *I Never Sang for My Father*; best screenplay from another medium, *Satyricon*, *Women in Love*, *Something for Everyone*, *Colossus*, *I Never Sang*; best original story and screenplay, *Baby Maker*, *In Search of Gregory*, *Zabriskie Point*, *Adam at Six A.M.*, *The Swimming Pool*.

February 9. Don says he woke up just a minute or so before the quake began at 5:59. He said, "It's an earthquake!" and then I was awake. It seemed to go on endlessly, in fact it was just sixty seconds. There were several quite loud crashes, and I took it for granted that the mirrors had broken; actually the glass shade of the light on the service porch had fallen and a coffee jar and some pictures, nothing serious. A very bad omen for Dobbin was that my two photos in Don's room both fell down and also the horse on my desk![1]

As before, I was too dazed to be frightened, but I was about to get really scared as the quake continued, and the later small jolts made me progressively more and more jumpy. Don says he thought, "This is it," and that he now no longer feels he wants to be in an earthquake, he *has* been in one. It was 6.5 on the scale, not bad at all, and certainly without question the biggest I have ever felt; but tonight commentators are becoming rather superior about it and saying it wasn't among the real great quakes.

February 10. Last night there was an eclipse of the moon. Some people believe that eclipses trigger earthquakes. An earthquake expert who was interviewed on T.V. wouldn't absolutely deny this but wouldn't confirm it. The experts we heard yesterday evening (whoops, there was an aftershock, quite a strong one, as I wrote that!) were all a bit reproachful and demurely bitchy, as much as to say, Oh yes, you are very interested in seismology all of a sudden, but you never let us have enough money for research, you built towns all along the faults, and by next weekend you'll be all steamed up about something else, children that you are.

[1] A Japanese papier-mâché horse which Bachardy gave Isherwood for his birthday August 26, 1962 (see *D.2*). It was about six inches high, seven inches from head to tail, and had one foreleg daintily raised.

Meanwhile, life has to go on, as they say, and we are already trying to figure out how to reconcile some of the things Hunt Stromberg wants us to do with "Frankenstein" (we saw him the day before yesterday) with our own story line. At the moment, I feel discouraged and think to myself, why, at the age of nearly sixty-seven, do I have to be involved in projects like this? The answer is, I don't. Actually it pleases my vanity that I am still employable and that my wits are still quick enough to play these nursery games. And, although we have about $74,000 in the savings accounts, plus the Hilldale property half paid for,[1] I always feel we ought to have more, or at least not start living on capital.

A Richter Scale joke: If Fellini made a film about the San Francisco earthquake he would call it *8½*.

February 11. Had tea with Larry Holt, whom the quake had made so jittery that he kept feeling new jolts every other minute. He was most indignant because his friend Tom had told him that Swami, at the end of his last Sunday lecture, had said, "I'd better stop," and someone in the congregation had said, "Please go on!" Larry thought this cruelly inconsiderate, since Swami's health is so poor. But the truth is, Swami would never have said "I'd better stop" unless he wanted to be asked to go on—it seems obvious to me that he was enjoying himself and very characteristically flirting with his audience. And Larry's indignation is all part of his deep resentment at being excluded from the Vedanta circle.

Went on to Vedanta Place for supper and the reading. Swami told us that during the quake he got out of bed, put his robe on and went to stand in the doorway to his bathroom. As he did so, he thought of a text in Sanskrit: "If Krishna saves, who can kill; if Krishna kills, who can save?"[2] One member of the congregation had his house jolted right off its foundations. Larry had a friend who lost his liquor store; everything was smashed. Thousands of people are still evacuated from their homes, lest the dam on the other side of the San Fernando Valley should break, before it's been emptied. On Hollywood Boulevard quite a lot of shop windows were smashed. But, alas, almost no damage is reported from any of the high-rise buildings; so the work at the end of this street is presumably going ahead.

Toward the end of the day I noticed that the earthquake atmosphere was getting me down. Guts very gassy and an overall feeling

[1] Some apartments they had bought as an investment, on Hilldale Avenue in Santa Monica, financed partly with a loan.

[2] A Hindu truism.

of angst. One doesn't mind a bit of worry about one's possessions at first, but then the act of worrying becomes in itself humiliating; it is so unworthy of the Atman to fuss over all this fucking real estate.

February 12. A quite sharp earthquake jolt this morning, while I was trying to meditate, "Trying" is nearly always the operative word. This meditation period isn't satisfactory, because I usually want to shit soon after getting up, and because there is a lot of traffic on the hill below our house at that hour, and because there is an underlying psychological pressure to get the show on the road, have breakfast, start working and worrying. It is very good to say, "Whatever I do, whatever I give, may I do and give it as an offering to You," but how often do I remember to try for this attitude? Almost never.

Am taking Maalox to reduce the gas, which it does; but I still have this little sensitive knotty lump on my left side, it seems to be in the muscle rather than in the gut, but how can I tell? I keep looking in my old diaries to find if I have had this symptom before. It seems to me that I have, but I haven't yet turned up any reference to it.

Supper with Gavin and Mark Andrews. Mark talked throughout the meal, about himself and his reactions to his drama coach. (He has only lately started taking these lessons.) He was lively but boring and we might almost as well not have seen Gavin, we hardly got to talk to him. It was like a meal with Frankenstein and the Creature: out of politeness to Frankenstein, you had to let the Creature ramble on and on, lest you should seem to suggest that Frankenstein had made a mess of him. Gavin still has no definite offer for his screenplay, but now he has an offer, or offers, for a film of *The Goodbye People.* This would also probably include a part for Mark and possibly a chance for Gavin to direct. Gavin is such an intensely private person anyway that you don't really communicate with him unless you see him alone, and I haven't done that in months.

Am still reading Chekhov. And Richard Brautigan's *Trout Fishing in America.* Am just about to start a book of short stories by Dostoevsky. Don is fascinated by *The Brothers Karamazov.*

John Lehmann called from Texas yesterday morning to know how we had come through the earthquake. Of course his experiences during the bombing of London were brought in. But still, John has changed. He no longer looks down his nose at us who live in America. He has fallen in love with it, or rather, with American campus life. So our relations are very cordial.

I am thankful to say that Dr. Kafka thinks the growths on Don's foot are responding to treatment.

We are taking a fantastic amount of vitamins, prescribed by Ray Unger. Can't say I feel any better for them, as of now.

February 13. Yesterday afternoon at four, the people who had been ordered to leave their homes were allowed to return to them. At the gym, Bill van Petten told Don that "the very rich" are leaving Los Angeles, because they are expecting another, much worse quake.

A sudden fit of worry about the little lump in my left side. So I went to Dr. Allen again. He said, "I can't feel any mass there." He suggested a barium enema but I said let's wait a while. Don got terribly worried about this. He is being angelically sweet. Last night, after I got back from having supper with Chris Wood, he showed me some new paintings, two or three of them marvellous, all of them interesting.

Chris Wood says Peggy wrote a note to Michael, begging to be allowed to see Gerald. (Chris is allowed to see him every week.) Peggy's note was very tactful, according to Chris, but this I don't believe; she is incapable of tact and she is too hostile to Michael to be able to conceal it. Michael is now thought to be very angry with her.

As usual, Chris talked about going to England after Gerald dies, but he also said he is so lazy. And then there's the question of Paul [Sorel]. Chris says Paul is still guilty because he lives on Chris's money. Chris says he is so accustomed to living alone that he couldn't possibly live with anyone else now. He says that having lost interest in sex makes a great gap in your life. "It use to take up so much of one's time." I have given him the first two folders of *Kathleen and Frank* to read.

February 14. Our eighteenth anniversary. Go back eighteen years from the day we met, and Don was only just born and I was—where was I? Somewhere in Europe with Heinz, making myself miserable and worrying about a lot of things which never happened and a lot which did and weren't so ghastly, after all. I wish I could find out exactly what I was doing on the day Don was born. It seems to me that the rising of that tiny but powerful star on my horizon must have begun to influence my life immediately.

Anniversary dialogue: "Merci, Kitty." "Mercy, Dobbin."

Yesterday I at last persuaded Don to take some of his paintings to show to Billy Al Bengston. Billy seems to have been quite impressed, told Don he ought to show them. So now Don feels encouraged to start work again. Billy and Penny are leaving for Mexico tomorrow, eager to get away before the bigger quake which is promised us takes place.

Last night we went to Royce Hall and saw Emlyn Williams in his Dickens recital. It is real old-fashioned exactly contrived acting, absolutely brilliant in its own way. Emlyn older, fatter, self-indulgently weepy, yet sort of delighted with his grief or rather with all the sympathy and attention it has brought him.[1] I am not being brutal. This is actually a very good way of recovering from a loss, and why shouldn't one recover? One owes it to one's fellow creatures to recover, to recover *at all costs*. The alternative is to be like old Jo. We took them later to expensive inferior Stefanino's, where we spent fifty dollars on a mere snack for Emlyn, his son Brook and Marti Stevens. Emlyn and Brook sort of flirted throughout the meal. Indeed, Emlyn joked that the restaurant people, seeing them kissing each other, would think Brook was his pickup and Brook said well, why not? He is kind of attractive in an ugly funny-faced way. Is here working on a film which Peter Ustinov is about to direct in Mexico.

February 15. Don gave me various anniversary presents, such pretty light and dark blue polo shirts to wear as undershirts, and a choice of beautifully patterned Indian rugs. I had nothing to give him, but he reminds me that he gave me nothing last year and I gave him a gold toothpick! We had a happy day of work on "Frankenstein" and then Don painted and we ate at home, our special hamburgers, with parsnips and salad, and watched *Fantastic Voyage* on T.V.

Horrors, at the gym this morning I weighed over 149, a long-time high. Nowadays I go for quite long jogs, always two blocks and sometimes more. Saw Dr. Kafka who has now fixed up my toe, at least temporarily; he told me to come back only if I have trouble with it. He also told me that Don's growths on his foot are tumors but that they are never fatal and he thinks he can get rid of them altogether. As for my gut lump, I can't say it has disappeared but I have lost interest in it for the time being.

Two aftershocks of some size, during last night. Meanwhile bitter complaints of the sightseers who crowd into the disaster area. The announcer on the radio said, "They went to church in the morning and then they said let's take the kids and go and look where all those folks were killed." So the police are setting up roadblocks.

[1] His wife had just died.

It is so sad to see the builders at the end of Adelaide tearing down dear old Moon Manor and excavating its site,[1] ready for these stinking high-rise towers.

February 16. 1:45 and it has just started to rain, dripping down out of gloomy fog. This morning Hunt called to say he is going away for about two weeks, the day after tomorrow; so we are off the hook, as far as the "Frankenstein" treatment is concerned—except for the fact that being off the hook creates a temptation to finish off the whole thing in record time and get it out of the house.

Last night Don went to supper with his family to celebrate Glade's birthday. Glade sent me a sweetly intentioned ethnically outrageous Valentine card of a little Redskin girl saying, "Valentine, you nice and how! You have-um heap fine time today, because you nice in every way." I had supper with Marti Stevens, Emlyn Williams and Brook. Emlyn was in great spirits. He told us how Princess Margaret had said to him about Elizabeth Taylor, "She's a common little thing, isn't she?" I must say, this struck me as a real new low for the monarchy—but then, the princess is herself a new low; Emlyn described her as a ridiculous tiny creature in pink with an enormous cigarette holder. Brook treats his father with a kind of nineteenth-century respect, jumping up from his chair as soon as he enters the room. Emlyn is a bit of a "Public Baby" and I am sure he will be coddled and cherished more than ever now that he's a widower; women will fight each other to look after him. He is so pretty and naughty and feminine. Poor Marti was in pain from her leg; she went to the San Fernando Valley to help a friend who was an earthquake refugee and he hadn't any water, so she asked a man to give her some; he had a big container of newly boiled water and he upset it over her, scalding her thigh badly. I greatly amused Emlyn by telling the story about the man on the plane who jacks off in front of a *Playboy* pinup picture, then puts the picture carefully away in his briefcase and says politely to the girl sitting next to him, "Do you mind if I smoke?"

February 17. At the gym today I weighed just under 148. Saw Evelyn Hooker, who gave me two of the files from her archives of case histories. After I've read them, we're to meet and talk. I feel more doubtful than ever about doing the book with her and I tried

[1] Four two-story deco buildings with duplex apartments, mostly hidden by shrubs, on the corner formed by Adelaide Drive, Ocean Avenue and San Vicente Boulevard.

to prepare her for this, this morning. I'm determined not to take on another job in which I'm just a ghost writer—unless, of course, it's for the Vedanta Society!

Rather miraculously this morning, despite the British postal strike, the missing Chekhov volume, *The Schoolmistress*, from the Constance Garnett edition, arrived, sent by Heywood Hill.[1]

Don has hung two of his paintings, which I call the "white" ones, in my workroom. One of them is certainly among the best things he has ever done—the white woman with the pink-face man looking over her shoulder within a blue frame which seems to be partly made of spider's webs.[2]

February 18. Yesterday Don went over to the house on the opposite hill from which we hear the terrible dog barking—the house he once visited in the middle of the night, without being able to rouse anyone. This time a man was there; Don describes him as a middle-aged hippie. He wasn't at all apologetic, but Don feels some impression was made on him.

This morning I ran a whole class of painters with their easels off the top of our ramp and the top of the steps down to Don's studio. I feel ashamed of my ill humor and yet, why the fuck do they think they can encamp themselves there without even coming down to the house and asking permission?

We are pushing ahead with "Frankenstein," rather grimly. Probably we'll enjoy it more later.

John Lehmann called yesterday. He wants to stay here, early next month, when he gives a lecture. Don is furious.

February 19. At the gym I weighed only a fraction over 147. (These gym weights are always with my gym clothes on but without shoes; the terribly high weights of this time last year in London were naked on the bathroom scale.) I believe my piles bled, because I found some blood on the bench where I was sitting after being in the hot room and under the shower; a bit worried about this.

A gruesome party last night given by Rachel Roberts and her lover Darren Ramirez. Darren is a very sweet-natured boy, slender and well built and pretty in a Mexican El Greco way. He designs clothes and often wears something striking; last night it was a marvellous shirt with a pattern like the markings on a tropical

[1] London bookseller (1906–1986); *The Schoolmistress and Other Stories* was the volume Isherwood didn't have when he resolved for the New Year to read his Constance Garnett translation.

[2] From a non-portrait series inspired by still pictures of movie stars.

butterfly. The shirt was open nearly to the waist, showing his long slim body with curly black hair on it, which was quite sexy but embarrassingly faggy. I'm sure everybody in Rachel's circle dismisses him as a queer who is simply after her money—but life isn't that simple. Marti Stevens is sure that he is devoted to her, and has helped her stop drinking. Darrin and his mother fixed a first-rate Mexican meal. I wish they had fixed the guest list too. Some time ago, Rachel tried to invite me to meet the British Consul, Andrew Franklin, and I refused. Well, he was there and he is a stunning bore of the old school—he reminded me a bit of Sylvain [Mangeot] and even of Basil Fry. We talked about Berlin and China and the Manson case,[1] and he knew a great deal—about the first two subjects at any rate—and made his knowledge absolutely intolerable. Tony Shaffer, the one who wrote *Sleuth*,[2] was there too and was also a bore, though much more sympathetic.

Have just talked to Rachel on the phone. This morning, her divorce proceedings went through, at the court in Santa Monica. Told her how much I like Darrin, which pleased her.

(Marti surprised me yesterday by telling me she had a short affair with Brook Williams; that was how she got to know Emlyn and Molly. She was at the party too, looking ghastly. I feel sure the scald she got at the time of the earthquake is festering and starting to poison her.)

When telling me about the court proceedings, Rachel kept saying how ridiculous and obsolete and meaningless marriage and divorce are, nowadays; and yet, a moment later, she remarked that one item in the divorce settlement hadn't been clear to her and she had asked the judge to explain it—namely, how was the money for her support going to be paid in the event of Rex's death! The answer, it seems, was satisfactory.

February 20. Terrific wind began as soon as it got dark and grew stronger and stronger; a man on T.V. said it was blowing in gusts of seventy-five miles an hour. The door to the cellar won't shut properly and kept slamming. The wind makes both of us nervous. But this time I noticed that I didn't mind it quite as much as usual,

[1] Charles Manson and his gang were convicted on January 25 of murdering seven adults and Sharon Tate's unborn child; in early February, the court reconvened for a jury trial to decide punishment.

[2] Anthony Shaffer (1926–2001), twin brother of playwright Peter Shaffer, studied law at Cambridge and worked in advertising and T.V. production; he adapted his Tony-Award-winning play *Sleuth* for the screen. Other screenplays include Hitchcock's *Frenzy* (1972).

although it was shaking the house, because it seemed an anticlimax after the earthquake. This morning there is still a lot of wind and it is clear and brilliant; the light is almost Australian.

Gore was on "The Merv Griffin Show," almost unbelievably fatter than when we saw him, less than three weeks ago. (Of course I don't know when this show was taped.) Griffin tried to get him to talk about Merle Miller, the writer who has just declared himself to be homosexual;[1] but Gore was cagey, perhaps because he is now halfway into politics again, getting together this third party with Dr. Spock[,] and Ralph Nader for their presidential candidate. Last night, Don decided that *he* was overweight, 143; so we had nothing but soup for supper—only, the way he fixes it, it is a meal, anyhow.

Jess and Glade Bachardy came down early and Jess proceeded to fix the hum on our T.V. within a few minutes; we've had it for months.

February 21. I went to see Swami yesterday afternoon, having been put off from doing this because Swami was feeling tired and his pulse was missing beats and he had slight asthma (which the doctor said was due to his heart condition). I arrived early at Vedanta Place, so I asked Ananda for the key to the shrine gate and went in and sat right up close in front of the shrine. I don't know when I did this last; it is quite different, and not nearly so satisfactory from my point of view, to sit at the side as I do at the Vivekananda breakfast pujas. And then of course it makes all the difference, being alone in there. When I am alone I get the sense of confrontation, the "setting face to face" which is so wonderful and which I try to recapture when I'm meditating here at home. Well, it began working almost at once and almost without effort; I just reminded myself that it was before this shrine that Swami had his visions and that Sister used to see "the light" and that Krishna has chanted day after day. I felt the Presence and exposed myself to it, making no demand whatever, except to include Don in the exposure; it was like a radiation treatment and I knew that it was getting to me, all I had to do was stay there. And then, just when I was really open to it, in came someone and sat just behind me to the side and began whispering "Chris" and I turned and it was Ananda to say apologetically that Swami was ready to see me.

[1] Miller (1919–1986), journalist, screen and T.V. writer, novelist, and biographer of Presidents Truman, Eisenhower, and Johnson, published "What It Means to Be a Homosexual" in *The New York Times Magazine* on January 17, 1971.

So I reminded myself that Swami is a human shrine and therefore really much more worth visiting, and that he contains his relics too, his memories of Maharaj and the others. I found him looking not only beautiful but surprisingly well. He described his treatment as he always does, in great detail, and then told me the doctor had asked him, "Are you depressed?" and that he'd answered, "Oh *no,* I'm *never* depressed." (The smile with which Swami told me this was so marvellous, not in the very least superior but gently amused, as much as to say how ridiculous to ask someone if he is depressed, when that person has seen Maharaj!)

Last night we had Camilla Clay and Linda Crawford and Gavin and Mark to supper, and barbecued a butterflied leg of lamb. I had feared that the high wind we had earlier would make this dangerous but it dropped to a calm which was so dead that I had to use the bellows to get the charcoal going. Indoors, we had a cheerful fire fed by bits of my old bookcases. Camilla is on the wagon but Linda got fairly drunk. Both Gavin and Mark are dieting. Mark wants to lose twenty pounds, Gavin just a few. I asked Gavin what weight he wanted to get back down to and he surprised me by saying 158. Gavin and Mark had been to Las Vegas to see Elvis and had both liked him, but Gavin had been disgusted by Las Vegas itself. During the evening there were two earthquake jolts which I, who was in the kitchen and standing up, didn't even feel; but the others felt them and this morning we hear that the bigger of the two was 4.3. Everyone is still quake conscious and inclined to be jittery and there is much talk of the "big one" that is coming.

February 22. I have now finished another volume of Chekhov, the fourth, and two Richard Brautigan books, *Trout Fishing* and a volume of poems called *The Pill Versus the Springhill Mine Disaster.* I still love reading Chekhov as much as ever and am starting on the next volume right away. Brautigan amuses me much of the time, when I'm not being repelled by his whimsey. This kind of writing is sweaty, though; I am so aware of the author knocking himself out keeping me amused.

This morning I also finished the second of the two files I borrowed from Evelyn Hooker. What a plodding old donkey psychology is! Evelyn's questions are full of phrases like, "His own processes of sexual arousal are on the ascending incline," "I don't have a very clear picture of how much mutual stimulation is going on," "The primary stimulation is on the head of the penis. Would that be true?" "While I have asked you many questions about sexual preferences and gratifications, I have not really asked you questions couched in his terms of the basic mechanics of sex." I really can't imagine myself working with Evelyn on this sort of thing; it would be like having to write a book in a foreign language. But I mustn't prejudge the issue. I must wait until we have had a talk and I have found out just exactly what it is she wants me to contribute.

Just a shade over 148 pounds at the gym.

February 23. Yesterday afternoon, Hunt Stromberg's secretary called, asking if we had any work for her to send on to him in Texas—this is the first jog he has given us. So we have been hurrying to get a faircopy of our opening finished, as far as the attempted suicide of the Creature and Frankenstein's return to Elizabeth. That ought to be ready tomorrow.

A series of tornadoes in Mississippi and Louisiana has killed more people and done more damage than our earthquake. Hope this doesn't make the San Andreas Fault feel competitive!

Chris Wood came by this morning, with the first two folders of my book, to collect the third. He has had flu and looks terribly stricken and skinny, though sunburnt. His back is getting so rounded that he almost has a hump. He told us proudly that he hadn't gone to bed or seen a doctor and that he had thought he was dying. He was on his way to see Gerald but thought perhaps he ought not to go in, in case he was still infectious. It would certainly be ironic, beautiful almost, if Chris were to be the one who gave poor Gerald his release. We joked about this—you say such things to Chris without embarrassment—and yet I suppose if it really happened he'd feel guilty.

An item in the London *Sunday Times* says David Hockney is going to Japan later this year, so maybe we'll get a visit from him here.

February 24. Weight at gym, 148. Today we finished and sent off a new draft of the first section of "Frankenstein" to Hunt in Texas.

Last night we saw two rather dreary films with Gavin and Mark, *Babes on Broadway* (with [Mickey] Rooney at his sweatiest) and *The Harvey Girls*, which really is a deeply shocking story about the triumph of respectable girls over whores; the only sympathetic character was Angela Lansbury and she didn't have a proper part. And then the terrible extravagance of eating out, at Frascati's on the Strip—seventeen dollars for the two of us. It is wrong not to be stingy in these cases. Mark Andrews wore a scarlet sweatsuit; it's curious to think that, ten years ago or less, he would never have been allowed into any restaurant in the area, dressed like that! Again, he hogged the conversation, which didn't matter except that it was so late and we were exhausted.

Gavin was shown the paintings by Don which are now in my room. He liked one of the "white" ones (the one I like less) and the big head. We weren't sure how much—but he repeated that he liked them as we were saying goodnight; which, as Don said, was tactful at least.

February 28. I had made one of my compulsionistic resolves to keep this diary every single day throughout February. Well, I haven't.

There's not really much to record. I got a copy of a pamphlet written about me by an associate professor of English at Columbia University named Carolyn G. Heilbrun. It's only forty-six pages but at least it's *a book*, and it's in a quite distinguished-looking series called Columbia Essays on Modern Writers. I can't judge how "good" it is. Shall be interested to know what other people think of it. No news of Alan Wilde's book on me yet.

On the 25th we had Leslie and Michael Laughlin and Jack Larson to supper. It was blowing a gale and the whole house shook. I had to barbecue outside the front door and even with that much shelter it was quite dangerous. Jack made the evening by announcing, as soon as he arrived, that Jean Dickson (I think it was)[1] had prophesied there was to be a ghastly earthquake complete with tidal wave right here between 8:00 and 10:00 p.m. this very evening! Jack said he had been told this by phone by Jim who was in New York and said the whole city was talking about it—and gloating no doubt. Anyhow, both Leslie and Michael got a bit nervous and the gale added to our state of tension. It was a relief when 10 o'clock passed. Both Leslie and Michael, Michael particularly, praised the new paintings which Don has hung in my workroom. Michael liked the same "white" one that Gavin liked and thought it much the best of all Don's "movie star" paintings.

Today's *Los Angeles Times* has more reports of the psychological damage done to people by the Sylmar quake. One lady showed extreme calm for four days after the quake (during which she had screamed). On the fifth day her hands "froze" to the steering wheel of her car; it took three men to pry them loose. Then she became sexually cold toward her husband. Eleven days after the quake she told her husband she wanted a divorce. They had been

[1] Jean*e* Dixon (1918–1997), astrologer and Washington socialite, famous for her prediction that Kennedy would die in office.

married for eighteen years. The husband reported, "She says when she was lying on that kitchen floor, she realized she'd 'never really had a chance to live.'"

Have just reread Gerald's *The Gospel According to Gamaliel.* The idiom in which the dialogue is written is entirely unconvincing, the characters talk like no human beings on land or sea or in any historical period, except Gerald himself. And yet, partly because the whole performance is so utterly Gerald, it has its own power, and it does make its points very clearly, much more clearly than Gerald usually does. How amazing he is! (I make myself write "is," though I am thinking "was.") I read it very quickly, with ease and enjoyment.

Went to the vespers of Ramakrishna puja yesterday. The wait to get touched by the relics was again spoilt by whisperings of women. Asaktananda called me up into the shrine right after the monastics; he did the same thing at the Brahmananda puja. Rightly or wrongly I feel that this is part of a protocol of politeness which we are evolving in our relations with each other, in preparation for the day when he will become head of the center. (He may well not be aware that he is doing this, but I am that I am; I realize that, as a senior member, I must build his position in advance by showing him extreme respect. I always rise—although it's an effort because I am sitting on the floor—when he comes into Swami's room.)

But I did once again get a strong sense of contact by just sitting outside the shrine gate for a few minutes, the day before yesterday.

Not one word from Hunt about the "Frankenstein" material we sent him, to Texas. And meanwhile we seem stuck. I am rather worried about this, can't see any ending to the story yet. And I know we ought to have something ready for Hunt when he returns, otherwise he will bug us with more mad suggestions.

March 2. Hunt hasn't called but he has sent us a message through Ann, his secretary at Universal. He likes what we have written very much but makes two suggestions: the butterfly Elizabeth sees should turn into a horrible bug of some kind and terrify her—the severed arm, when Henry finds that it is becoming monstrous, should attack him and beat him, damaging his brain, so that it is no longer a normal brain when it is put into the skull of the Creature. The first of these suggestions seems merely inartistic, the second misses the whole point of the story as we wish to develop it. Still, Hunt can't be expected to know how the story will develop because we don't yet really know that ourselves.

I see a little more light on this than I did two days ago, but we are still fumbling about with a lot of characters at cross-purposes. What is Pretorius (now provisionally called Dr. Thorndyke) up to? Does the Creature want a bride, or does Pretorius create one to please himself? What part does Elizabeth play? Does the Creature rape and or murder her? Does the character of Henry emerge more and more within the Creature as the story comes to an end? And how is this demonstrated?

Yesterday morning I saw Evelyn Hooker and told her that I can't write her book with her. I think I explained why I can't quite lucidly and I think I convinced her. The analogy of *Kathleen and Frank* was very useful, in doing this, because Kathleen's diary can be likened to Evelyn's files of case histories. The diaries, like the case histories, can be commented on, they can be elucidated and conclusions can be drawn from them; but they can't be rewritten because nothing can be as good as the source material itself. What is embarrassing—and what I think sticks as a reproach against me in Evelyn's mind—is that I told her, in the Saltair Avenue days,[1] I was prepared to write a "popular" book about homosexuality with her. Of course I was always saying things like this, quite irresponsibly, subconsciously relying on the probability that I wouldn't ever be taken up on them. To Evelyn yesterday I said, "Well, you know, in those days I was nearly always drunk"; which, the more I think of it, was a silly tactless altogether second-rate remark. (But why should Drub expect that his remarks should always be first-rate? Vain old Ninny-Nag.)

March 4. John Lehman arrives this morning; more on that later.

Last night, at the end of the question period after the reading, Swami said, "Let me speak out, you have no one but the Lord, no other refuge, your own nearest and dearest, they will desert you, only the Lord will never desert you." The "Venice Gang"—Jim Gates, Peter Schneider, Gib [Peters], Beth [High], Doug [Rauch] and all their friends—now form a majority almost of the group and indeed the older people seem only half there by comparison; these young people, with all their faults and phoniness, are nevertheless the involved ones, the ones to whom Swami literally means contact with the living truth, or so it seems to me. I am not being sentimental about them, indeed I probably view them with more realism than most older members of the group do. It makes me glad that Swami has them now at the end of his life; it is a little reward

[1] I.e., August 1952 to September 1953, when he lived in her garden house.

for his long years with those frumpish self-indulgent women; like Ramakrishna at last getting his young disciples.

Yesterday I went to the place recommended by Penny Little, Billy Al's girlfriend, to buy organic nonpollutant soap for the Laundromat and for dishwashing, Shaklee's Basic-H and Basic-L. The sales are made in a garden house behind a quite grand home on Devon, just off Wilshire in Westwood, and the ladies who run the place have the air of amateurs, they might equally well be engaged in local politics. Well, this *is* politics.

After leaving Vedanta Place I picked up Don at his parents'. Both Jess and Glade had made a point of having me invited in, to sit with them and watch T.V. for a few minutes; this was a symbol of reconciliation and no doubt one of these days we shall actually have a meal together. The funny thing is, I really like them and even feel quite at ease with them. Don had been down to San Pedro where they exhibited three of his drawings at the art museum, along with work by Billy Al, Joe Goode, Peter Alexander.

March 6. Tonight the sun set just behind the headland for the first time this year. Shortly before it did so there was what seemed to me a quite sharp earthquake jolt. But Don, who was out in the studio, said he didn't feel it: "I must have been dancing."

We have now put up a fence across the steps; it was finished yesterday. This morning Don had to yell at the first boys who were opening its gate. Since then, the gate has been padlocked. This is all part of our futile but passionate defence against the invading Others.

John Lehmann's visit wasn't so bad. He was away all evening on the 4th, after reading a lecture on the Woolfs and the Hogarth Press at which I had to introduce him, and he left yesterday afternoon. Our conversation, on his side, was mostly a leering inquisition. Grotesque questions like, when did you and Wystan last have sex? No, such questions aren't at all grotesque in themselves, only the way he asks them, with the air of a dirty old bishop. Still—"Friendship never ends."

Still wrestling with the plot of "Frankenstein." It's funny how the fact that one's dealing with monsters and murders (and early nineteenth-century ones at that) seems to make no difference to the laws of probability. One keeps dismissing ideas because they are "impossible," "too farfetched," "unconvincing." But perhaps that's our mistake. Perhaps we ought to be wilder and sillier.

March 8. Am worried about the hard, seemingly swollen area in my intestine. I shall go and see Allen again tomorrow and then probably have X-rays, unless I can talk him into telling me it's nothing.

David Sachs called yesterday to tell me that Richard Montague, the mathematical philosopher at UCLA, has been strangled, seemingly by a hustler. Have heard no further news about this.

Evelyn also called yesterday to say that she had been to the opening of the Reverend Troy Perry's new church. Perry introduced her to the congregation and she had a terrific reception. But when she stood up to speak she was so moved that all she could think of to say was, "I want to correct one misstatement which the Reverend Perry made—my research was not sponsored, as he said, by President Johnson but by the National Institute of Mental Health"! So she has to write a letter to Perry thanking him and them all, to be read out in church next Sunday. I told her this was the kind of thing you find yourself saying in a speech you make in a dream.

Last night Paul Wonner had a show at the Landau Gallery, the last show before it closes. Some of his paintings I didn't like at all; they seemed cluttered with gimmicks. But a few seemed better than anything he has ever done; they had a sort of prophetic authority, I felt they were genuine visions. I have a feeling that he will soon find himself able to paint simple landscapes without any gimmicks at all which have this same quality. (What I mean by gimmicks are, for example, a self-portrait face in a cloud, butterflies, flowers, a helicopter—all these superimposed upon landscapes. What he does convey is the *awe* of the wilderness; he makes you aware of his strong religious feeling when he is out in lonely places.)

Later we all went to a party at the home of a Mr. and Mrs. Lusk in the rich ghetto off Wilshire called Fremont Place. The Lusks, who have just split up, were both there as hosts and seemed on the best of terms.

Finished Romer Wilson's *The Death of Society*. Why on earth did I ever like it? It's one of the most bogus books I ever opened. The love scenes are so idiotic that I wouldn't like them even if they were about two boys. Maybe the book appealed to me because I thought the Norwegian background glamorous. Only it isn't. And it got the Hawthornden Prize! Here are a few lines, taken at random:

> "To be mad, Rosa!—it's life itself. Rosa, I am mad—crazy. I am clean crazy," he cried. "Aren't you afraid of a mad man? Hi, Rosa, what do you think!" and continued in a worse strain.

After an impudent peroration he went on gaily: "I am your lover, Rosa—as mad as a bee—my delight! my fairy! my shepherdess! my princess!"[1]

March 12. Cassius Clay's defeat last Monday was sort of saddening—the more he bragged, the more one wanted him to be unbeaten. Now, even if he wins the title back, it won't be the same.[2]

On the 9th I finally went to see Dr. Allen, determined to talk him out of giving me another barium enema X-ray examination. So he agreed to something else which is really a whole lot more trouble; I have to lay off eating meat for six days and, after the first three days (that's tomorrow), I have to collect three daily specimens of my shit. If there is no blood in it, then I won't have to be x-rayed. Dr. Allen says that all meat contains blood and that even the tiniest trace of blood can be detected in the shit; two statements which I find hard to believe.

Swami seemed tired, the day before yesterday; perhaps he will go away to the desert for a few days. We hear from India that Len, Mark and Paul have become Swamis Bhadrananda, Tadatmananda and Amohananda respectively. They have been kept inside the grounds of the Math, except for a boat trip to Dakshineswar, because of the outbreaks of violence in Calcutta. But now that Mrs. Gandhi has won the election things may get better, Swami thinks.[3] (I forgot to mention Buddha; he's Swami Yogeshananda now.)

The experts have decided that the Sylmar earthquake was much bigger than was supposed. Although it was only 6.5 at the epicenter, it set up a sort of diagonal thrust under the hills which hit Sylmar with a force of nearly 8. It must be *some* consolation, when your house has been wrecked, to know at least that you have been in a major earthquake. Also, it makes the major earthquake we are promised seem slightly less alarming, since we evidently had a worse shaking than we thought we were having!

The British postal strike is over at last, but no mail from England yet and so I hesitate to send the illustrations and other material for my book to Methuen.

[1] Chptr. 20.

[2] Clay (b. 1942), known as Muhammad Ali since joining the Nation of Islam in 1964, lost the World Heavyweight Championship to Joe Frazier on March 8.

[3] The Bengali independence movement was reaching crisis point in Pakistan; see Glossary under Bangladesh.

March 19. The day before yesterday, I think it was, someone rang me up in the morning and said he had a friend who wanted to illustrate a porno poem by Auden[1] and he understood that Auden (he pronounced it Ow-den) and I were intimate friends. I took it for granted that this was Gore (whom we'd met on the 15th at the [Paul] Newmans') and that he was giving one of his imitations, so I started playing the Dirty Old Man and asked this guy to meet me at the men's room of the Catholic Church on 7th Street, etc. etc. It was only right at the end of the conversation that I came back out of my playacting and (assuming still that it was Gore) asked him to have supper with us tonight. He accepted. Then I had doubts and called the hotel and got hold of Gore, who assured me it hadn't been him on the phone. (I'm still not sure he wasn't lying.) Anyhow, we are having dinner with Gore tonight and we may also expect this guy to show up!

I delivered the hemoccult slides (this process is called an "occult blood test") to Dr. Allen, dully daubed with shit. The instructions tell you to "dispose of the applicator" after use—as though you'd be apt to treasure it in your hope chest! Yesterday Dr. Allen told me that the test was okay. So I'll leave my gut to produce some further symptoms before I bother with it again.

On the 17th, I started a sort of notebook on Kitty and Dobbin—I'll try to write it rather like a study in natural history; their behavior, methods of communication, feeding habits, etc. I had a very strong feeling that I ought not to record all this, that it was an invasion of privacy. But where else have I ever found anything of value? The privacy of the unconscious is the only treasure house. And as a matter of fact, Don is always urging me to write about us. I have no idea, yet, what I shall "do" with this material after I've collected it. I'll just keep jotting things down, day by day, and see what comes of it.

March 25. We saw Gore, he took us out to supper at Chasen's. If the non-Gore guy showed up, expecting a meal, it must have been after we had left the house. But there was no note from him when we returned and there has been no call since.

Gore has lost a lot of weight and isn't drinking at all; he was full of energy and ideas. He repeated one marvellous line which he says has been attributed to him (he couldn't bring himself to state quite definitely that it *wasn't* his): "In this country we now have socialism for the rich and private enterprise for the poor."

[1] "The Platonic Blow," written in 1948 and circulated in typescript; it was first printed, without Auden's permission, in 1965.

As usual, he kept repeating that the novel is dead; and as usual one got the impression that what he really meant was that one couldn't nowadays become a nationally known personality just by writing novels! Gore, like Mailer, has gotten around this difficulty by becoming a journalist and a T.V. interviewee.

All this while Hunt has remained in Texas. We have come along slowly but steadily with the treatment and are now practically at the end of it. I do hope he stays away another five or six days.

David Hockney arrived yesterday and is coming to supper tonight. He has come here to renew his California driver's license, because it is much easier to do this than to get a British one!

Am keeping up my Kitty and Dobbin notebook.

My gut is about the same.

Am reading Ray Bradbury's *I Sing the Body Electric!*, Graham Green[e]'s *Travels with My Aunt*, Aaron Latham's *Crazy Sundays: F. Scott Fitzgerald in Hollywood*, Chekhov's *The Witch* [*and Other Stories*], and Dostoevsky's *A Gentle Creature* [*and Other Stories*].

Last Monday there was a review of the San Pedro Art Gallery show by William Wilson in the *Los Angeles Times*. He praised Don highly, making it obvious that it was Don's work that he really *liked*—although he had to drag out that tired old comparison to Ingres! Also Don's drawing (of Peter Alexander) was the only drawing reproduced in the paper.

Dr. Kafka says that the growth on Don's foot is returning—but now he thinks it may be only a corn. He doesn't seem unduly alarmed, told him to come back in a month.

I now have five of Don's paintings in my workroom, plus two drawings, one of them a really magnificent nude. Paul Wonner has seen them and seemed truly impressed. He told Don he ought to show them to Blum and try to have an exhibition.

April 5. We did show the paintings to Blum—that's to say, he came to supper with David Hockney and Ron Kitaj and went into my workroom during the evening with David and Ron, so he certainly saw them. But he made no comment, and neither did David. Ron at least had the politeness to tell Don he liked them. This was a severe setback though not too surprising, since David almost never praises (he did, however, tell Don he liked his drawing of Andy Warhol, which has been lent to Jack [Larson] and Jim [Bridges]) and since Blum would hardly praise anything in David's presence unless David praised it first. Don is very upset again, as well he may be, and I am upset because there is absolutely nothing I can do about it.

Otherwise, David was as endearing as ever. We did think we detected a slightly increased sense of importance, but in a likeable Yorkshire way. He told us a funny story about de Gaulle and Madame, at the time of de Gaulle's retirement. A journalist asked Madame, "What do you look forward to, now that you are out of public life?" Madame replied, "A penis." Everyone was surprised and de Gaulle hastily corrected her, "*No*, my dear, it is pronounced 'appenis...."

David left last night. He is planning to go to Japan with Peter in the fall.

Hunt Stromberg likes our treatment very much, he says. There are various small additions and changes to be made to it this week, then it will be sent in to the front office at Universal.

Swami really does seem very weak and tired. This time, I don't see any policy in his tiredness and am therefore getting anxious about it. Yesterday I had to appear for him at the temple and read one of his old lectures. Afterwards, a girl who belongs to what we now describe as The Venice Group—one of Peter and Jim's friends—put her baby into my arms and asked me to take it into Swami's room and have him bless it. So I did. The baby (which has been named Rakhal) was very plump and heavy and absolutely quiet. I felt sure it would begin to cry when I carried it away from its mother, but it didn't. ("We're going to Heaven," I told it, "but we're coming right back, don't worry.") Swami blessed it and I brought it out of his room to find all the Venetians waiting, and Peter led the applause. (He and his friends would be quite capable of taking over the Vedanta Society and running it, in case of an emergency!) And later, when the baby was safe back with its parents, I did hear it crying.

My gut is just the same, or worse. I keep saying to myself I must do something about it and keep putting it off.

Graham's *Travels with My Aunt* is quite entertaining but only that. It doesn't mean anything except its own high spirits and they aren't high enough. I don't think an elderly writer can ever get away with this kind of sportiveness; the bones in his skeleton can be heard clicking, now and then. I don't know what to say about the Dostoevsky. It is wonderful now and then, I feel the greatness which created *Karamazov*, but his meandering gets on my nerves, in short stories it seems tedious, in a huge novel it doesn't. When I'd finished, I felt I'd merely read it out of compulsiveness.

April 6. Igor [Stravinsky] died early this morning in New York, of a heart attack. Don thinks we ought to go to his funeral, and of course he is quite right. And of course I don't want to. So I've been gloomy. And Don says I am behaving as though he had done it all to me. Which I probably am.

We are hurrying to get the "Frankenstein" treatment fixed up. This adds to my feeling of having the screws put on me.

That's all. Just wanted to write this down.

April 8. Dobbin's sulks were all for nothing, as so often. Because yesterday, when we talked to Lilian Li[b]man,[1] she assured us that it was quite unnecessary for us to come and that, if we did come, we wouldn't be able to see Vera; the doctor has forbidden her to have any visitors until she returns from Venice, where Igor is to be buried next week. (Maybe Vera is remaining in seclusion partly because she doesn't want to be involved in any ugly scenes with Igor's children.) So we have decided to go to New York later, at the end of this month, and see Vera then.

Michael Barrie called yesterday to tell us that Gerald had another stroke, his twenty-first, a couple of hours before Igor died. Gerald is now more or less in continuous coma. When Chris Wood last came to see him, Gerald didn't know him.

Swami, on the other hand, seems much better. But Asaktananda is trying to hurry up the arrival of Chetanananda, the new swami. Peter Schneider is staying at the Hollywood monastery for his Easter vacation from college. He is playing it very cool, coolly observant, but at the same time he makes himself thoroughly popular by doing lots of manual work. We had quite a long chat, yesterday, but I still don't know if he intends to try being a monk, or not. I think he has a solid layer of his father's skepticism, deep down.

The hateful building continues at the end of the street. As soon as you go out of doors as far as Don's studio, or up the ramp from the carport, you can hear the clatter of the cranes. The latest news is that the towers will be *twenty-six* stories.

My weight is going up again. Can't get it below 149.

I feel I "ought" to write something about Igor, but I won't till I feel something. The truth is, he had been dying—that's to say, separating himself from us—for such a long time. The cuddly adorable affectionate bitchy brilliant heavy-drinking Igor I used to know died at least five years ago, though there were glimpses

[1] Once a publicist for the Santa Fe Opera; she was retrained by the Stravinskys in 1959 to manage some concerts in New York and became an unofficial press liaison. Isherwood typed Lilian Litman.

of him after that. (I *never* felt Morgan was dead before he actually died.) No doubt when we see Vera and Bob [Craft] again, in these new circumstances, a lot of the real Igor will come back.

April 13. Doug Walsh came around this morning to try to clear out the drain outside the front door with a snake or maybe it's called a rooter. He brought a man named Matthewson with him to help, someone he had known a long time who was out of work, a hard drinker, just over fifty. I went out and said hello to him, and he seemed quite sober, polite, small, quiet. About half an hour later, Doug called me because Matthewson had collapsed. I phoned the fire department for an ambulance. He'd had a heart attack. Doug was efficient, worked to keep his lungs going by massage, put a wooden spoon in his mouth to prevent him from choking and a cushion under his neck to make his head hang back. He isn't dead yet, but when I called the emergency ward at Santa Monica Hospital they said his condition was very serious.

The day before yesterday we finished a revised treatment of our Dr. Frankenstein story and now it's being typed ready to be shown to the Universal bosses. Neither of us really quite believes that it will be accepted and that we shall have to do the teleplay. It's not that the story is bad—though it can be greatly improved—but that the Universal people will surely realize that Frankenstein is now hopelessly old hat.

On the 8th, Don finally had his long beautiful mane of hair cut shorter; it is now a more usual collar-length "shag cut." He still has a close-trimmed black beard. I never liked the mane, it seemed to me distastefully feminine. The shag cut and beard I can live with, but I do find that they somehow come between Don and me; it's like being with an actor who's always in makeup for a role.

April 14. When I called the hospital later in the day, they told me that Mr. Matthewson had been dead on arrival, so I suppose they hadn't decided to "release" his death when I called earlier.

Hunt Stromberg seems slightly insane on the subject of "Frankenstein"; casting turns him on like dope. Latest suggestions—Chamberlain for Frankenstein, Finney for the Monster, Burton for Polidor,[1] Liz Taylor for Agatha and Prima! I suggested Lieutenant Calley for the Monster, saying that I felt sure Nixon

[1] An invented character based on Byron's real-life doctor, John Polidori, who appears in the film's historical prologue and epilogue. In later versions, both the invented and the real characters are called Polidori.

would release him to play it, under armed guard.[1]

To come back to Mr. Matthewson, my feelings about his death puzzle me. I feel a sort of exhilaration. It was "exciting." That's on the surface. Beneath this is perhaps a sense of reassurance; how easy it was! Doug Walsh's behavior showed such genuine goodness; he was so big and clumsy, kneeling over the little man (who already looked flattened, deflated by death, except for his potbelly) massaging the heart, slapping the bluish cheeks, muttering, "Come on! Come on, Matt!" When Doug asked for a wooden spoon, I brought him the salad spoon. Afterwards I boiled it. My first reaction was to throw it away—I can't imagine why, for I am simply not squeamish like that. As I boiled it, I said to myself "the death spoon!" and "dead man's spoon!" Doug knew "Matt" fairly well and was quite upset; he drank two snorts of bourbon afterwards. The men from the fire department were admirable too—their absolutely matter-of-fact efficiency in giving the oxygen made you feel the unsentimental beauty of a helpful act.

April 19. This morning, Jim Gates has been initiated. Swami (or Ananda, I suppose) suddenly decided that there would be two or three initiations this week, each one on a separate day. I feel slightly proud in a paternal way, since I was at least partly responsible for bringing the boys to Vedanta Place. (I'm astonished to find that Peter doesn't even have his name on the list, yet, but no doubt he'll manage to push his way into the queue before long.) I called Jim yesterday to ask if he would house-sit for us in the event that we go to New York to see Vera. He says he will.

Still waiting to hear what Universal thinks of our Dr. Frankenstein.

I gave a reading to the Society of David the day before yesterday at the house where I met them before, Hank Barley's. But there was no Gay Lib opposition present, this time, and the reading was a huge success, I think. Read them bits from *Kathleen and Frank*—also various snippets from my commonplace book.

Am going through some conscience searchings about taking part in the Christopher Street West Parade which is scheduled for June 27.[2] Ought I to take part in it? If it were being held in some other city where I happened to be visiting, I don't think I should

[1] U.S. Lieutenant William Calley received a life sentence on March 31 for murdering hundreds of Vietnamese civilians in March 1968 at the village of My Lai in South Vietnam.

[2] Second annual Gay Pride parade commemorating the June 29, 1969 riots outside the Stonewall bar on Christopher Street in New York City.

hesitate at all. But here, in my home town, I can't help thinking that my participation might embarrass Swami—for some busybody would surely find out and report it to him; indeed, my name and photograph would probably be in the papers. Then, again, my wish to take part in the parade is largely vanity; I'd enjoy posing as one of the Grand Old Men of the movement. By the same token, my refusal to do so will hurt my "image" and hence my vanity.

A board has been put up on the construction site at the end of the street with a picture of the proposed twin towers. It says they are to be *sixteen* stories!

April 24. The day before yesterday, we went to Universal and were told about all the alterations Hunt wants in the Frankenstein treatment. They don't really amount to much. Taking out the whorehouse, because you can't have that on T.V., and some of the gore, for the same reason. And the Creature mustn't smell bad! Charlie Engel,[1] who is assistant to someone and a special friend of Hunt, sat in on the talk. He is supposed to be "bright" [. . .]. He has a weak messy moustache—how profoundly this era of face hair repels me, physically *and* psychologically, and I probably shan't even have the satisfaction of outliving it! Engel uses all the latest slang, of course; my unfavorite expression is "right on!"

I felt very proud of Don, who spoke up, very clearly and to the point, every so often. Engel was offensive about this, too, suggesting that he was astonished to hear that Don had *any* opinion of his own. No doubt Don is being soundly bitched already as a boyfriend who is being brought along for the ride. Hunt realized that Engel was being rude and tried to make things better by saying that, when he used to work with Dorothy Parker and Alan Campbell, it was always Alan who had the bright ideas, "I never heard Dorothy say *anything* clever." However we just sat and smiled and took it.

We called New York a few days ago and were told by Lilian that Vera won't be back until the middle of next week, so those plans are still up in the air.

A boy who had come down to sit for Don brought his dog with him. The dog began by putting its paws on a wet painting. Then the boy tied it to the gate just outside the front door, and it pulled the gate down!

Elsa's house next door has been robbed. The thief forced the back door but left no mess. He simply took the color T.V. set and

[1] Longtime studio executive in T.V. production at Universal.

two Polish statuettes. The police think it was an inside job, and Elsa suspects one of [her] gardener's stepsons, who has a record as a juvenile delinquent.

Mike Laughlin (on Gavin's advice) offered Don a job on his latest film, as a sort of art director. Don is asked to see that the sets have "a forties atmosphere." Don feels that Mike doesn't know his own mind, and the script is simply awful, and poor Leslie is to be in it. Don's instinct is to refuse the job—yet it just *might* be a lead-on to similar but much more promising assignments.

We watched a very good English T.V. film about George Eliot.[1] I found it painfully touching when she spoke of her happiness with George Henry Lewes and how she treasured it because she knew it must be short. That's exactly how I feel now about Don; I grudge every one of these beautiful snug evenings together at home, watching T.V. and eating eggplant or hamburgers on the bed. We have never been happier together.

A very sad confused rambling letter from Lee Prosser, who set forth from Missouri to come back west and then abandoned the trip to San Francisco, and has now returned home—chiefly, as far as I can judge, because, while there, he got the news that Mary his ex-wife has been killed in a head-on auto collision. From the way Lee writes, he seems to be on the verge of a breakdown.

May 1. Vera is back in New York and we have spoken to her and she says she would like to see us. Probably we won't go until after the 10th, however, as we have to be available to make whatever rewrites the front office at Universal demands.

Michael Barrie wants to know the exact time of Igor's death. Because between 9:30 and 10:00 p.m. on the night of April 5, Gerald became excited and worried in a quite uncharacteristic way and began to talk a lot, though unintelligibly. Michael believes that Igor, realizing he was about to die, sent out telepathic distress signals and that Gerald picked them up and was therefore trying to tell him not to be afraid. The three hours' difference between Los Angeles and New York time would fit in with the fact that Igor died early on the 6th.

On the 29th we went to a supper at the Bel Air Hotel to be talked to by James Fox about his conversion to Jesus. It happened in Blackpool—James had come there with a touring company in an elsewhere successful comedy, and it flopped. He met a man

[1] "The Confessions of Marian Evans" (1969), directed by Don Taylor for the BBC arts program "Omnibus."

who said that *he'd* come to Blackpool "to spend a day with Jesus." The man opened his Bible and read texts and showed Jimmy that the wages of sin is death and that he, Jimmy, was dead but could be alive in Christ. And Jimmy realized that this must be true and ever since then he has "lived with Jesus," and he has been in Australia and New Zealand lecturing about this.

It is important not to sneer, not to be superior. Indeed, when I first heard of Jimmy's conversion, I was pleased and excited and I never doubted that it was genuine. I don't doubt that it is genuine now, up to a point. But certain things Don and I saw that evening were very disturbing.

James looked marvellously young, at first glance, and very slender. But when he began to speak, he seemed frantic and his face became tortured. Don said he was like a confirmed stammerer. His mouth kept pulling sideways. The first part of his talk was all about his career; I suppose he intended to reassure us that he was a normal sane successful actor, but he merely seemed to be bragging in an embarrassing way. And then he is such an out-and-out fundamentalist; everything is backed up by Bible quoting. He has absolutely no use for our Vedanta, or any other religion. It all *has* to be through Christ—no other way is possible. And that was sinister, especially in view of the people who had arranged the supper, his new friends. Pasadena people, with money. They call themselves The Nagivators, because the movement began in the Navy. They are smiling people, with self-assured little in jokes, but underneath they are fanatical [...]. And we felt James was captured by them. After supper he pleaded with us tearfully to live with Jesus. It was almost as if they'd told him to go get us. No use my pointing out that we had a religion already that suited us, or that Swami had written a book about the Sermon on the Mount. Finally Don said, "If you'll read the Gita, I'll read the Bible."

Richard Chamberlain was at the party. There was no opportunity to ask him why. But he looked embarrassed and I don't believe that he shares James's views.

May 2. My lump in the gut is back again after quite a long remission. Don's foot produced some more growths, new ones. Dr. Kafka still doesn't seem worried but he put on much stronger medication and poor darling Kitty has been limping around. He's to see Kafka again tomorrow.

We spent the evening with Gavin yesterday, without Mark, which was nice for a change. Gavin says that Natalie Wood no longer sees him. Gavin thinks it's because of Mark, and because her husband doesn't approve. That's so typical of The Others—they just love bachelor queers but find it distasteful when they find a steady and bring him around; as long as the bachelor is playing the field that's fine, and even cute and amusing. Also, as long as the queer is a bachelor, they can be sorry for him and reflect on the tragedy of his life.

Have just finished reading *Dracula.* It really tells you little or nothing about how it feels to be a vampire. What turns the author on is the predicament of the two beautiful *bitten* women. Of course they both remain perfect ladies, except at moments when they are "not themselves." Poor Lucy! "Her eyes blazed with unholy light, and the face became wreathed with a voluptuous smile.... There was something diabolically sweet in her tones—something of the tingling of glass when struck."[1] Mina never sinks *that* low, but you feel she rather loves having the vampire scar on her brow, and being oh so brave, and having these good-looking guys running around protecting her: "Oh, it did me good to see the way that these brave men worked. How can women help loving men when they are so earnest, and so true, and so brave! And, too, it made me think of the wonderful power of money! What can it not do when it is properly applied; and what might it do when basely used. I felt so thankful that Lord Godalming is rich...."[2] (This last remark refers to the fact that Lord Godalming could afford a steam launch to pursue Dracula's coffin up the river. *What might it do* is exquisite. Was Bram Stoker making fun of Mina? Maybe so. He was Irish. I would like to read his life, if there is one. Also his *Dracula's Guest* and his memories of Henry Irving.[3]

Ackerley's *We Think the World of You*, just reread, seems as excellent as ever. This is the very best kind of realistic writing, so bluntly unsentimental and funny and understanding and unapologetic. Then, quite suddenly, comes that strange final paragraph which begins with a humorous description of the dictatorship of the dog and the author's masochistic enslavement to her, and ends on the brink of an abyss:

> Not that I am complaining, oh no; yet sometimes as we sit and my mind wanders back to the past, to my youthful ambitions and the freedom and independence I used to enjoy, I wonder what in the world has happened to me and how it all came about....

[1] Chptr. 16.
[2] Chptr. 26.
[3] *Personal Reminiscences of Henry Irving* (1906).

But that leads me into deep waters, too deep for fathoming; it leads me into the darkness of my own mind.

May 11. Don left for New York last night. I'm to join him there on the 14th, unless Vera has a relapse and is too ill to see us. Now I'm in a mad flap, getting things done.

All I want to record now are some things about Swami. When I saw him on the 5th, he told me he had just initiated a woman who he thought was slightly crazy. He did it because her husband begged him to. But he didn't like doing it. He said she smelled of garlic all over her body! (Could she have decided he was a vampire, I wondered.) During the initiation the figure of Krishna fell off the shrine and broke its pedestal, and Swami was so upset he couldn't think of a mantram to give her; at least, not for a long while! And she didn't know what chosen ideal to have! He told me a story I don't remember having heard before, about Maharaj. Maharaj, when he was president of the Math, had to sign some important document. He kept putting it off, until the deadline day arrived. His secretary urged him to sign and gave him the document and a pen. But Maharaj hesitated. Then he said, "I've forgotten my name," and the secretary knew that this was because his mind was absorbed in God.

At the end of the question period after the reading, Haridas, who is always trying to provoke Swami, in his complicated masochistic-hostile Jewish way, asked: "Does the Guru ever withdraw his love from his disciples?" And Swami answered: "I don't know yet." The lazy good-humored get-you-Mary tone of voice in which he said this was absolutely adorable. It was like Jesus kidding with Satan in the wilderness. It came out of his mouth so spontaneously, with such perfect timing, that everybody laughed and a lot of us clapped. But what *exactly* did he mean? Did he mean that he wouldn't be able to answer the question until after he had died?

May 13. So I'm off to New York tonight at 11 p.m. Don is already in Maurice Grosser's flat, so we'll be quite on our own. I dread the trip, as always, but maybe it'll give me a needed shaking-up and I will try to make my darling Kitty's birthday fun for him and me both.

I shall never, as long as we are together, be able to fully feel or describe to myself all that our love means; it is much much too close to me. Don tells me from time to time that I should write about it, but how? Even my attempt to keep a diary of the Animals has failed. I can't see any of this objectively. Any more than I can really grasp what Swami means and has meant to me, in an entirely different way. If prayers are answered—and I believe mine have been, in almost every respect, throughout my long indecently lucky grumbling life—that doesn't mean that you can always be fully aware that they have been answered. Just at moments you're aware that they have been, but the enormousness of the answer is beyond your grasp.

The day before yesterday, I finished correcting the proofs of *Kathleen and Frank* and mailed them back to Simon and Schuster. What a huge undertaking that was! When I think of all the reading and copying and cutting I had to do, before I even started writing anything of my own. I ought to give thanks that it got done as quickly as it did. And yet I realize that I merely crawled along, when compared with the working habits of an Allen Drury.

Jim Gates is going to house-sit for us. I saw him and Peter last night for a short while after the reading at Vedanta Place. They both remarked on the fact that we have seen so little of each other lately. "You're elusive," they said. Oddly enough, I feel a stronger affection from Peter than from Jim, though Peter is the cool one. Poor Jim, I had to read a letter from Selective Service which arrived for him here this morning, saying that he couldn't be granted a transfer from the Goodwill to Vedanta Place, and from C.O. to Theological Student status. Jim took the news very well; he is good at adjusting to things, very strong, I think. Because he had been looking forward to becoming a monk at once, and last night he told me he felt certain it was going to be all right; so this was a severe disappointment. Now he is going to ask around and find out what the prospects are for making an appeal.

(I'm hiding this diary and my two previous diaries in the closet, lest Jim should find them while I'm away and read something in them which hurt his feelings!)

(The visit to New York, written after getting back to Los Angeles on May 22:

May 14. The flight was uncomfortable, although the plane was empty, because I was on the wrong side of the aisle, where there were only three seats instead of four, so I couldn't stretch out. Maurice Grosser's flat wasn't as dirty as I'd feared, but it was beset by outside noises: a barking dog, belonging to some nuns who ran a children's school; the children themselves; hammering by workmen in the house next door; piano lessons and a mysterious public address system which was apt to produce utterances at any hour—for example, "Someone on the third floor has left his phone off the hook," repeated over and over. I at once forgot my good resolutions and began sulking about this and about the general squalor of New York, which was indeed hideous. Don was very patient, sweet and apologetic. New York is his "property" rather than mine, because of all his previous visits there alone, so he feels responsible for it.

We had lunch with tiny Anita Loos, who always seems to have her life under control and to be amused by it, and who never looks a day older. Her tensions, if any, are all down below the carefully-made-up surface. Then we saw Joe Brainard's show of cutout paper grasses and flowers, framed in plastic, and bought one (five hundred dollars) for Don's birthday present. Also the Andy Warhol show at the Whitney, which is slightly like the Tower of London in its grimness and, for that very reason, all the more striking as a setting for this vast area of blazing flowers and faces; when you come out of the elevator you are staggered by the emptiness of the huge room with its flaming scrapbook walls.[1]

Supper with Glenway Wescott was quite spoilt by the fact that he had invited a bunch of white-collar faggots we scarcely knew; so all was leering sneering politeness.

May 15. We had lunch with Vera [Stravinsky] and Bob Craft and Ed Allen at their apartment on Fifth Avenue, 920, looking out on the park. It was astonishingly like earlier meals at times when Igor was sick in bed, and, after the first five minutes, not in the least embarrassing. Both Vera and Bob spoke freely and in great detail about the last days of Igor's life and about the funeral. Bob seems anxious to define the nature of his grief. The chief impression I

[1] A retrospective, "Andy Warhol," May 1–June 20, Whitney Museum of American Art.

got was that they and Ed now made up a working team which was still full of potential energy but lacked any immediate project. Vera's eyes looked yellow and tired and she seemed a bit shakier and older, but after all she had only just recovered from the flu. She was sweet and affectionate as always.

Julie Harris (who we also saw, in a poor play called *And Miss Reardon Drinks a Little*[1] and then had a snack with at Sardi's) was worried about her son Peter, who keeps running away from school; now he's to be put on a training ship which sails round the world. This was just a symbolic meeting. We hurried away from it to see Mishima's dreadful arty but disgustingly realistic film about a Japanese army officer disembowelling himself and his wife stabbing herself in the throat.[2] The morbidly interesting thing is that Mishima made it not so long before he really killed himself.

May 16. It rained a bit. We went to the Lincoln Center to see the New York [City] Ballet in *Jewels*. Don was in raptures; ballet really gets to him as almost nothing else does. And I was happy he was happy; it left me impressed but cool. A stylish dancer we hadn't either of us seen before, Robert Weiss.[3] Handsome and romantic.

Supper with a man named Ron Holland, who's J.J. Mitchell's latest date. Virgil Thomson came, and Joe LeSueur. I hadn't wanted to come and I rather hated it; and Don rather hated the situation, which was that Holland, who is rich, was being bribed to give us all a meal in exchange for having Virgil and me at the party. Or, to put it in another way, J.J. was working off his "obligation" to us (being taken to Disneyland) by making his lover buy us a meal. I don't much like either Joe or J.J. Joe is so unattractive and J.J. is silly in the wrong way. Furthermore, the restaurant was tiny, French, pretentious, and crowded, and a cap came off one of my teeth.

May 17. Saw Peter Schwed at Simon and Schuster's—also Michael Korda, who looks like a Jewish queen in his mid-twenties. Liked him; felt embarrassed as always by Peter. They showed me the design for the jacket of *Kathleen and Frank* and, lo and behold, side by side with a photo of Frank was a photo of—Emily! That's

[1] By Paul Zindel; Rae Allen won a Tony for Best Featured Actress, and Estelle Parsons was nominated for Best Actress.

[2] *Yukoku* (1966), based on his 1960 novella; the English title is *Patriotism* or, sometimes, *The Rite of Love and Death*.

[3] He danced for Balanchine from age seventeen and later became a choreographer and the first artistic director of the Carolina Ballet.

the sort of mistake which only seems possible in publishing, and it shows a fundamental lack of interest.

I was then interviewed by Daniel Halpern, the editor of a magazine called *Antaeus*; this was arranged by Andreas Brown of the Gotham Book Mart[. T]he two of them seem to be great buddies, maybe lovers, for Halpern is quite attractive, a big fresh-faced Jewish boy with sort of Afro hair. I was slightly but not entirely charmed by him; I suspect him of being a teaser, a climber and a hustler. He has certainly collected a lot of people to contribute to his magazine and presumably for free—Paul Bowles, Lawrence Durrell, Thom Gunn, Tennessee, Gore, Roditi, Kosinski, Burroughs, Böll, etc. etc.[1] He came with me and waited while a nice old-fashioned New York Jewish dentist put the cap back on my tooth for only five dollars. Then I met Don and we saw Garbo in *As You Desire Me*—a miserable adaptation,[2] but she and von Stroheim were marvellous. And then we had supper with Vera and Bob at Romeo Salta's. When we went there, we had the right street but the wrong number; and the cabdriver shattered the myth about the rudeness of New York cabdrivers by cruising up the street on his own, *after* we'd paid him, looking for the restaurant and then running back to tell us when he'd found it! This restaurant visit was sheerly symbolic; we felt obliged to spend as much money as we possibly could. In fact we'd offered them the Colony and Vera had accepted it with pleasure, but thank goodness it was closed, this being a Monday! As it was, our bill came to over a hundred dollars. But Bob was so helpful, talking Italian to show off and ordering everything, just as if he was the host. I forget what we talked about, but it was a snug evening. They struck both of us even more than usual as being innocents. They seem hopelessly extravagant. But maybe if they really find themselves poor they will innocently accept that too.

May 18. We had lunch with Jens Yow, who really is a very sweet good-natured person. We picked him up at the Morgan Library where he works; it is like the inside of a pyramid, so massive and seemingly theft- and dirt- and bombproof, and packed with treasures. Jens still sees Lincoln [Kirstein] a lot, so I complained to him of my sadness at the utter rejection of the two of us by Lincoln and his refusal even to answer the peace note I left for

[1] Halpern (b. 1945), a poet, co-founded *Antaeus* with Paul Bowles. He also co-founded Ecco Press, acquired in 1999 by HarperCollins, which made him publisher of one volume of Isherwood's diaries.

[2] I.e., of Pirandello's play.

him last February. Jens said he couldn't understand it. According to him, Lincoln isn't at all crazy at present.

That afternoon we went over to Brooklyn and saw Paul Cadmus and his friend Jon Andersson. They were both very nice to us, but Don doesn't really like Paul, he finds him cagey and sarcastic and full of bitterness. Paul looked healthy. He had an unbecomingly long hairdo; it's all white now.

Then we saw *Follies*[1] which was ghastly, although Don liked it a little bit more than I did, and suddenly the birthday was a failure, because I hadn't arranged a birthday supper to top it off. We hurriedly tried to call people and couldn't get anyone. It *was* my fault, of course, but Don's birthdays are always tricky, because he will not say beforehand what he wants or doesn't want to do, and he doesn't like it if I tell people in advance. I had been dreading this birthday for several weeks, and I was thankful when it was over. We came back gloomily to eat at home—soup out of a can—this was my punishment. When we got back to the house we found the tenants in a state of fury because, as they claimed, the new owner—who wants to get them all out of the building because their rents are fixed—had sent two men to tear down the door which protects the staircase, thereby exposing them to burglars. They had already sent for the police and were planning legal action. We were a bit worried, wondering if we'd be harassed with stink- or firebombs or beatings-up as the representatives of Maurice Grosser!

May 19. We saw Brian Bedford in *The School for Wives*. He was better than either of us had ever seen him before. As Don said, it was so wise of him to switch from romantic leads to a character part. With his Tony Award, he is very much *Mr.* Bedford, and the matinee audience—largely school kids—ate him up, whistling and stamping. Brian received us afterwards in his dressing room, wearing a very short robe and repeatedly rearranging his bare legs; very friendly and gracious, but he showed no further enthusiasm for *A Meeting by the River*, not to mention the film of *A Single Man*.... In the evening we had supper with Elaine de Kooning,[2] Joe LeSueur and his friend Alan Martel. I like her, hadn't met her before, and it seems she dug me. The utter silence of Alan Martel gradually becomes infuriating and aggressive. I had fantasies of wiring his chair and giving him an electric shock. And, altogether,

[1] Stephen Sondheim's musical, playing on Broadway since April 4.

[2] American painter, teacher, writer, and lecturer on art (1920–1989), married to the Dutch-born artist Willem de Kooning since 1943.

why why why does one spend such evenings? In this case the answer was that we met because Don wants Elaine to arrange a sitting for him with de Kooning. Which is quite sufficient, as far as I'm concerned—except that nothing *was* arranged. When we got back home, we found that there had been a small fire in the house next door. Needless to say, the thought occurred that this might be another broad hint from the new landlord that we should all move out.

May 20. Judge Isaacs, a friend of Virgil Thomson, who is a specialist in housing disputes, assured me on the phone that there was nothing to worry about; he had talked to the new landlord who had convinced him that he was not a crook and had not had the door removed for malicious reasons, but because a large piece of furniture had to be carried upstairs! (This doesn't explain, however, why the corresponding door in the other part of the building was also removed at the same time—or so the other tenants told us!)

I spent the rest of the day running around seeing this person and that. (My disinclination to record any more of this dreary and tiresome visit is now so strong that I'll only mention two or three more items.) In the morning I went to see Pavitrananda. We spent about an hour together, of which I found the first three-quarters extraordinarily trying. I felt like someone who understands the language of the birds, but not well. Pavitrananda's weird vowels and speech rhythms seem at first utterly nonhuman but, if you strain your concentration to the utmost, you find you can *just* make out what he's talking about. However, shortly before I left, we began to communicate, mentally rather than verbally. Once again, the wonderfulness began to shine forth from him, as I have seen it shining many times before. This old silver-haired scarecrow was shining and saying joyfully, his eyes full of tears, "Repeat the name of the Lord, that's all that matters," and I was filled with joy and bowed down before him without any taint of politeness and then went out and wandered about Central Park feeling elated.

Claire Bloom wasn't up to *Hedda Gabler*; she seemed merely vulgar.[1] We saw good old Myrna Loy afterwards.

May 21. Just before having lunch with Virgil Thomson (who I'm slowly becoming quite fond of) I met Clinton Kimbrough in the lobby of the Chelsea Hotel. He was just off to England with the

[1] Bloom (b. 1931, London) won a 1971 Drama Desk award for the role.

girl he is going to marry. He had a bit of a beard and looked tired but pleased with himself and his life. I was glad to see him. I have always rather liked him, or at any rate wished him well.

In the afternoon, Don and I went to see Truman Capote in his apartment in one of the buildings on the United Nations Plaza. He has almost the only tolerable view in New York, because he looks down the river, instead of across it to that junkyard, Brooklyn. You see lots of barges heading seaward, making white wakes in the grey water, and the park is below you, bright green, and then the U.N. building in profile, looking so slender that you can hardly believe it's packed from top to bottom with diplomats, and to your right is lower Manhattan with the chief skyscrapers in the foreground, including the two towers which are to become higher than the Empire State.

Truman looked bulky and unhealthy and seemed somewhat listless and sad. A wise little old child who has stayed indoors too long, with his toys, collections of glass paperweights, jeweled boxes and other treasures, including a marvellous ivory (Chinese?) banana, half peeled. Of this, he remarked that some people who saw it were completely fooled and went away saying that Truman was a slob, to leave fruit lying around in that condition.

He is probably worried about his novel which he says he can't finish; he already has seven hundred pages of it. Also, his life must be full of anxiety—though that's something one can never be sure of, with another person. He is now forty-seven, so when I first met him he must have been twenty-three. There was a photograph in the apartment taken about that time, with the Paleys[,] and he looks so cute and sexy in it, an impudent little blond with a smooth sturdy body, wearing nothing but a check cap on the back of his head and a pair of white shorts.[1]

He told us that Cecil Beaton has written an indiscreet show-off book about his relations with Garbo, including their fucking.[2] Truman thinks it will do Cecil a lot of harm.

I have asked Simon and Schuster to send Truman a set of proofs of *Kathleen and Frank*. I begged him to write me his opinion of it, no matter what.

In the evening we had supper with Vera, Bob and Ed Allen. Bob and Ed were busy unpacking boxes of books which had just

[1] Probably the often-reproduced shot taken beside the Paleys' pool in Jamaica soon after Capote was introduced to them by the Selznicks in 1955.

[2] *The Happy Years* (1972), volume 3 of his diaries, tells about the love affair which began in 1947; further passages about Garbo appear in volume 4, *The Strenuous Years* (1973). The sex is only implied, not explicitly described.

arrived. It seems unwise to do all this work when they may be leaving the apartment so soon, but I could feel that the excitement of unpacking had cheered them all up, even Vera; it was their first project since Igor's death. Bob kept remarking that they had far more books than they could possibly find room for and that they would have to get rid of a lot of them. "So help yourself," he added. But when I noticed and asked for a novel by Ford Madox Ford, *The Fifth Queen* (?),[1] Bob said that Stravinsky had liked it, or Mad*am* had liked it, implying that it was a bit sacred. He and Vera pressed me to take it, nevertheless, but I took the hint instead and left the book lying on the table.

Our flight back to Los Angeles next morning was our first on a 747. We were both nervous at takeoff and held hands, but the huge thing got itself off the ground without apparent trouble. There were lots of empty seats. The hostesses tried to be arch and were merely rude; the food called itself "gourmet" and was merely lukewarm.

We got back to find that Jim Gates had washed his own bed sheets and had not taken the car to work that morning (although I'd told him he could) because he figured we'd be needing it as soon as we arrived. He had also packed his bags and had Peter move them back to the house in Venice. Even Don, who continues to suspect all Jim's motives and actions, has to admit that this was thoughtful.)

[1] A historical romance about Catherine Howard originally published as a trilogy: *The Fifth Queen and How She Came to Court* (1906), *Privy Seal* (1907), *The Fifth Queen Crowned* (1908).

June 3. Yesterday we finished nearly twenty-one pages of the "Frankenstein" teleplay. It is a ghastly chore, we are going so slowly; at this rate we shan't finish before the middle of September. Don is upset because he feels he is a drag on me. Actually he is and he isn't. He is a drag when he's typing because he simply cannot sit quiet and work something out, it drives him up the wall. And yet without him I wouldn't work on the fucking thing at all. And he does very often have good and even brilliant ideas. Oh I wish to Christ we had never started it. But then we have to earn money somehow.

Swami told me Pavitrananda wrote him about my visit and said I had such great humility! This really does amaze me. When Swami says such things I feel I know why—it's because he subconsciously expects all western "intellectuals" to be overtly arrogant. But how could I have conveyed humility to Pavitrananda? Because I took the dust of his feet? Surely not. Don't these two dear saints realize that it is the very height of pride for the proud man to have a few people before whom he humbles himself—as much as to say, behold, even I, in all my greatness, am bowing down!? That is exactly what T.E. Lawrence used to do.

There has been a big pot fuss at Vedanta Place, because of the goings-on of Charlie Mitchell, who runs this so-called First Liberty Church. Charlie and all his disciples are now forgiven, more or less.[1] I do wish however that Swami had made it a little bit clearer that what he is really condemning is the elevating of pot into being a sacrament and an adjunct to meditation. That Charlie Mitchell should set himself up as a guru is almost incredible, anyhow [. . .].

An infuriating David Susskind program on T.V. on the 29th: "What It Means To Be a Homosexual." What David Susskind meant was that he confronted four functioning homosexuals with four who had been "cured" by Eli Siegel's Aesthetic Realism[2]—which teaches, apparently, that you can't be a satisfied homosexual because aesthetics require true opposites (in this case, a man and a woman) and queers are not opposite. What humanly emerged from the show was that the four uncured cases seemed surprisingly "wholesome" and "healthy," while the four cured ones had the twisted malicious faces which are supposed to belong to Boys-in-the-Band-type faggots.[3] Also, the cured ones were desperately

[1] They were Vedanta devotees; see Glossary under Mitchell.

[2] Siegel (1902–1978), Latvian-born American poet, philosopher, critic, and essayist based his personal philosophy on three principles: living honestly, the danger of contempt for the world, and uniting opposites.

[3] I.e., self-haters, suffering from shame and disappointment.

eager to convert the uncured, but not vice versa.

Clement Scott Gilbert is in town, but we haven't seen him yet. And we haven't succeeded in talking to Jennifer since her dramatic marriage to Norton Simon (May 30). A suggested telegram to her: "Sister, can you spare a dime?"

June 16. Chetanananda, the new assistant, finally arrived on the 11th and I met him last night. It turns out that we had previously met at Belur Math, the last time I was there. He is thirty-four and looks younger—quite cute looking, with big dark eyes which he opens wide; he is flirty in that Asian way, shows his beautiful white teeth in endless smiles, poses his face sideways on his hands. He can also maybe be malicious or teasing. He is tall, slim, fairly well built. He wears his hair cut close. He makes mistakes in English but speaks fluently, without hesitation.

This is only a first impression, but I fear Asaktananda may become jealous of him, because he is more at ease socially than Asaktananda. Asaktananda seemed silent and a bit sulky when we were together.

Don and I are still toiling at "Frankenstein," just approaching the making of the Creature.

An offer from the BBC to do a radio play of *A Single Man.* Also, a young actor named Michael Brandon[1] wants to do it as a film. Jim Bridges knows him. The question is, does he have any money and can we get a director we want?

We have seen Clement Scott Gilbert, but there seems no prospect of our getting *Meeting by the River* done. Grete Mosheim[2] showed up out of nowhere—such a sparkling, pretty old thing still—and I gave her a copy of the play for possible production in Germany; but not a word from her, yet.

And I am still waiting for the British proofs of *Kathleen and Frank.* It is sort of a waiting time.

July 13. Just an entry to restart, after this long gap. The British and U.S. proofs of *Kathleen and Frank* are all corrected and returned. No word from Clement Scott Gilbert or from Grete Mosheim

[1] American (b. 1945), he later starred in the 1980s T.V. series "Dempsey and Makepeace" and in the 2003 stage musical *Jerry Springer–The Opera.*

[2] German stage and film star (1905–1986), famous at seventeen in Max Reinhardt's *The Speaking Ape*; she married a co-star, Oscar Homolka, and, later, an American railroad tycoon Howard Gould, then returned to Germany where she appeared in modern American plays.

about *Meeting by the River*. Nothing more from Michael Brandon about filming *A Single Man*.

We have been slogging along on "Frankenstein" and are now only about halfway through. Hunt Stromberg remains in Texas and doesn't even call us; merely sends messages through his secretary at Universal to say that he likes the teleplay so far!

Truman Capote called, really (I think) enthusiastic about *Kathleen and Frank*. He says he wants to review it in *The New York Times*.

I went to see Dr. Ashworth about my hand and he said it must be operated upon; the nodule on the joint is getting bigger. So we have tentatively agreed that it shall be done in September, when Ashworth gets back from his vacation. I'll have to spend two nights in hospital and have a general anesthetic; which is rather depressing.

Don works and works. Since the end of June we have been going on the beach and in the ocean quite a lot. Also we keep up our vitamins and the gym. I can't get below 148 but seldom go above 150.

July 26. This *is* a quiet period, though a happy one on the whole. My one misery is the "Frankenstein" teleplay. I seem to have so little energy for it and it's the sort of job one should finish in a single night-and-day session powered by coffee and Dexamyl. To make matters worse, Hunt has just sent one of his letters of suggestions, wanting us to stop and rewrite past scenes in accordance with them because, as he puts it, "I am thrilled with what you have done to date—I just want it now to be platinum studded." His letter also tries to blackmail us into doing this at once by saying that he hasn't sent the last lot of pages in to the front office at Universal because he is afraid they won't like them quite as much as they did the first lot. However—unless he raises a real fuss—I am determined to go ahead, at least as far as the costume ball scene.

This morning, the original two sets of page proofs of *Kathleen and Frank* arrived from Methuen, postmarked June 14! They had been sent surface mail, just as I suspected.

Don has talked to Irving Blum and they have agreed that Don shall prepare a new show—of big drawings, little ink drawings and paintings. And then Irving will come and see if he likes it and is prepared to put it into his gallery.

Am reading *Henry James at Home*, H. Montgomery Hyde; *The Secret Lives of Lawrence of Arabia*, Knightley and Simpson; Jung's *Memories, Dreams, Reflections*, and the eighth volume of Chekhov stories, *The Chorus Girl*.

Weight at gym today, only a fraction under 150. And Don's weight yesterday was 137, a record low! Am slightly worried about this—but he looks well.

July 27. Don's weight went up a bit yesterday, to 140. Today he has gone to see Ray Unger and Jack Fontan at Ray's late mother's house at San Bernardino. He has been reading Proust—all of *Swann's Way* and then the opening of *Cities of the Plain:*[1] "Introducing the Men-Women." Proust's falseness and his respect for the Establishment both disgust Don; Don really loathes him. I suppose his character will be loathed increasingly by younger generations; in my time we were ready to eat his huge shit-sandwich for the sake of the delicious morsels of truth in the middle. Now I'm beginning to hate him through Don's eyes.

Last night I had gut-ache and we saw *Guys and Dolls*, which is interminable, but the songs are still marvellous, some of them. Today I have gotten on with "Frankenstein" in rough, quite a bit, and have been in the ocean, jogging down there. (I met Madge MacDonald, who cried, "You look wonderful—you and I are going to live to a hundred!") This afternoon, a nice man from *The National Observer* named Bruce Cook interviewed me. He likes Kerouac and has written a book about him and the other Beats,[2] which was a bond of interest. I now make it a point to talk to all interviewers about my queerness. He was just a tiny bit embarrassed, smiling bravely. In the midst of this, Hunt called from Texas, still wanting us to kill off Mrs. Blair the landlady and have the Creature unjustly accused of the crime! He told me that the prefab house he and Dick had put up on their ranch property was burned to the ground only a couple of nights ago! He was wonderfully calm about it. I got him to promise not to show any more pages to Universal for the present. He says he's coming back to Los Angeles in two weeks.

July 28. At the gym, Don weighed 141 and I about 149 and ½, which was a drop, considering that I'd had quite a bit of lunch not long before. But, drop or no drop, I'm in a fat gassy state. I *feel* fat. After the gym, we saw *Peter Rabbit*—such a miserable pale coy film which has the impudence to say it's based on the Beatrix Potter stories. Actually it's based on the Royal Ballet at its

[1] I.e., *Sodom et Gomorrhe.*
[2] *The Beat Generation* (1971), one of Cook's (1923–2003) more than twenty fiction and non-fiction books; he also worked at *Newsweek* and *USA Today.*

most tepid—the boys and girls cavort in animal masks—it's like a dancing class and I found almost no consolation in knowing that, if I could have pulled off the right animal's head, I'd have unmasked adorable Wayne Sleep. Don hated it too so we left in the middle, to eat turkey salads at the Beverly Wilshire drugstore with Mike Van Horn. Then Mike (who'd already been to a ballet class) took Don off to dance with him at The Farm. I hope they are having fun, bless their darling hearts. Old Drump is going to bed and to sleep, perchance to snore.

The termite man came to inspect and we have to pay 170 dollars for sprayings.

Cloudy today but I went in the ocean and the water is warm. There is a "red tide," the worst in many years; it is expected to kill about one hundred tons of fish.

July 29. Lunch with Swami today at Malibu; Pavitrananda and Swahananda (from Berkeley) were there too. Conversation at lunch: how many words had I written, which poets were most admired nowadays—both Pavitrananda and Swahananda came on very literary. Swami was happy because the New American Library has agreed to publish his *Sermon on the Mount According to Vedanta*.[1] But he seemed very tired. He told me he has just heard that Prema (he never calls him Vidya) is being paid by some French devotees to go with them to India and interpret. Swami brought up Vidya's unfortunate remark again—writing to Swami he said he would see him *if the Lord wills*. Swami interprets this as a claim by Vidya to know the will of the Lord! But, anyhow, poor Vidya does have a most unfortunate knack of rubbing people the wrong way.

When I got home, Hunt Stromberg called, to say that the head of Universal (I forget his name[2]) is going to England about the 18th of August to arrange to film "Frankenstein" there and that Hunt wants him to have the script as far as the costume ball—the rest will be filled in from our treatment. He says they want to show the film in movie houses. Hunt himself is coming back here about the 10th or 11th, to see about this. I'm relieved in a way that we have got to get a move on.

Then I ran down to the ocean and took a dip, and on the way back I met Mike Steen, who told me that Tennessee's play[3] didn't do badly in Chicago and that a man from the Ahmanson Theater

[1] It appeared in 1964 from the Vedanta Press and from Allen and Unwin in London.
[2] Evidently Sidney Sheinberg; see below and Glossary.
[3] *Out Cry*.

has gone to see it and decide if it shall be produced here.

July 30. Last night, when we were ready for bed, Doug Walsh showed up, drunk. He has been to see me again today, drunk. Very briefly, one of his daughters, aged seventeen, ran away with a black man, a married black man who is a musician of some kind, and lived with him for a couple of weeks. When she came home, under pressure from the police, she talked "nigger talk" to Doug, which so enraged him that he hit her and broke her nose. Soon after this, she ran away again. Now she has written him from Boston, saying that she left town to save the black man, whom Doug has been threatening to kill. She also said that the black man saved her from the drug scene and that she wants to live with him and bear his children. Doug is now checking up to find if the black man is still in Los Angeles or if he has joined her in Boston. If he is still in Los Angeles, Doug says he won't kill him.

During our talk today, Doug said his father died of cancer and that he is sure he has it too—only in his case it's lung cancer. He was very happy in the army because it gave him "guidance." Otherwise he has been "a miserable sonofabitch." His wife doesn't understand. He sacrificed himself for his country and what is his reward? His daughter runs off with a nigger. Do I think he's prejudiced? I evade the question.

Doug is a living proof that you don't have to be educated or win an Oscar to be like Harry Brown. But, actually, he's much better company. The funny thing is, despite his self-pity, he has lots of drive and is an affectionate dog-type. We both think he has a big queer streak in him. Today he shed tears and then said, "A man of my stature ought not to cry." I thought to myself, thank God I'm not a psychiatrist—imagine earning your living by listening to this sort of thing.... And yet, in my cool Isherwood way, I'm really quite fond of Doug. I would even like to help him. But all the help he wants is something I won't/can't do; write a television story about him and his daughter and the tragedy of parents in our day and age!

At the gym, down to 148 and ½. We went in the ocean, which is brownish red. Very humid. Ninety degrees in town but only seventy-six down here.

July 31. We went in the ocean this morning. I've been in at least twenty times this month! A slightly drunken drip with a moustache came over and told me he'd read *Prater Violet* five times and still didn't know what it was about. He went on to say the same thing about several of my other books. He was a perfect specimen of the Hostile Fan type.

Here's part of a questionnaire sent me by a Miss Jean Ehly from Amarillo, Texas: What is the loveliest thing about your home? What does home mean to you? Can we not teach love, compassion, tolerance to our children, thus enabling them to better cope with the world? How important is beauty to you in your surroundings? Sir, how important is meditation to you (at home)? Today many people do not like to stay home—is boredom evidence of lack of communion with God in our home life?

Gut-ache and gas. Desperate struggles to get on with "Frankenstein." Don absolutely angelic. Somehow or other, whenever I talk to him about the story, we hatch out an idea. So on it crawls.

August 1. Last night it was very hot and humid, even down here. I felt wretched, blown up full with gas. And Jo's sad old life added to the wretchedness, although, I must say, she didn't whine nearly as much as usual; even when she talked about Benjy.[1] She says that he hasn't drunk anything now in a very long time, but that he tells her he still thinks about liquor continually. He has just taken up some form of meditation and goes to classes. Meanwhile, in New York, Anne Baxter has had a somewhat unexpected triumph, having taken over Lauren Bacall's part in *Applause* (the musical based on *All About Eve*).

Today my gut is much better, at least for the time being. And, thank God, I have at last been granted a breakthrough on "Frankenstein." I've written about five times more than my usual stint and it seems fairly okay.

August 2. This heat continues; it's bearable down here but awful as soon as you get inland. I still feel lousy with gas, although I ran down to the beach and went in the ocean. Now I'm trying to cut down on eating. Did quite a big chunk of "Frankenstein" scenes today.

Last night we saw Gavin, who is still convinced that there is going to be a depression very soon. He has got a lot of his money in Switzerland. Has been seeing Merle Oberon and her husband, [Bruno] Pagliai—and says of them, "There's something very comforting about the rich, their attitude never changes." Still no definite news of financing for Gavin's two film projects.

[1] I.e., Ben Masselink.

August 3. More "Frankenstein," more gas—though I've been eating practically nothing. Don with Mike Van Horn to see Billy Al Bengston and to draw Penny Little. I shopped, buying nuts for our breakfast cereal and Makro-Slim, our latest patent chew-food. Then I went in the ocean, which is still red and dirty but warm with delicious waves. Oh, I am such a compulsive old thing, jogging down the road to the beach, sitting for a moment only on the sand alert for dogs (lest they should pee on my towel), then into the ocean[,] alert for surfers (lest they should collide with me) then to take a shower on the beach (hurrying lest someone else should get there first) then hobbling uphill over the gravel and wiping off the sand from my feet on the lawn of the corner house (hastily, lest they should look out and tell me not to). *My* secret life isn't a bit like Walter Mitty's—it's mostly ratlike scurrying to secure myself some tiny advantage.

August 4. Another broiling day. Last night we had supper with Billy Al Bengston and Penny Little at Musso Frank's. It was a caprice of Billy's—everything one does with Billy is one of his caprices—to drive right into town on such a hot night, very slowly and leisurely, in his huge air-conditioned car. Actually he was sleepy and got tired of the caprice and wanted his bed, even before he was halfway through his great chunk of beef. I do like him and all his poses and acts, but I don't feel at ease with him. ("I never feel at ease with any heterosexual man," I told Don, which rather annoyed him; he thought I was just being difficult!) Billy is now approaching forty—from his point of view; actually he's about three years off it. He looks fine, which is probably why he talks like that. He told Penny he has "old arms."

Don has heard (he can't remember from whom) that there is to be another big earthquake on August 9 right here in town.

Lunch with Swami and the others at Malibu today. While we were alone before lunch, Swami told me something he said he'd never told anyone else. (I doubt this—I think by this time he has told *every word* to someone or other!) We were talking about Hindus and Muslims, and Swami, to illustrate the fact that he had been on good terms with Muslims when he was a young man in India, told me how, while he was on the train travelling to see Maharaj at Kankhal (when he was sixteen or seventeen) he sat next to a Hindu priest, with three Muslims sitting opposite to him. Suddenly the priest groped him, right up beside the crotch—Swami demonstrated this—and, as Swami put it, "I felt very narvous," adding, "I didn't know what it was about." So the three Muslims invited him to come over and sit with them, and they offered him food.

Swami also told me that Michael Barrie had been to see him, and that Michael had told him Gerald had had a dream, shortly before his strokes, in which he was at Belur Math, and Ramakrishna was there surrounded by his disciples, and as Gerald walked past them, Ramakrishna pointed to him and said, "That one belongs to me!"

August 5. Another hot day! And we have to drive clear through town to Mount Washington, to have supper with Bob Ennis (Exotica) and the Reverend Troy Perry, whom I've been wanting to meet for some time but am also rather dreading. I am determined not to become involved in his gay church. Still, I can't help respecting him, idiotically as he sometimes behaves. At least he defends us queers at the risk of his own life as well as ours!

To see Dr. Allen this afternoon about my stomach. At the first sight of his good-looking good-tempered face I felt better at once. He prescribed Daricon.[1]

August 6. As hot as ever, and more predicted, but I feel a lot better inside—thanks either to Daricon or to psychological detensioning produced by seeing Dr. Allen.

Our evening was pretty much of a drag, because we talked homosexuality relentlessly for nearly four hours, and dinner was deliberately delayed by Bob in the hopes that Troy's boyfriend Steve would show up—he didn't. Meanwhile, a coconut rum drink was served, followed by lots of wine. I laid off all of this, on Allen's orders. Also, dinner itself wasn't up to much. The spinach salad would never have passed muster at the Casa,[2] and what Bob grandly called Beef Wellington was more of a Napoleon Waterloo.

On the credit side was Bob's new and quite beautiful though nearly silent boyfriend, David Mosley. Also, Reverend Troy really impressed me. He is a queeny southern belle, and vain, but brave, innocently shameless and (I believe) fundamentally without shit. I like Bob too.

Don thinks David Mosley was dipping into a dictionary throughout the evening and looking up words we used which he didn't understand. It is true that once I distinctly heard him repeat to himself the word "sophisticated"!

At the gym today—it was too hot to go there for the past week—I weighed a little over 147.

[1] Oxyphencyclimine, an antocholinergic used for treating peptic ulcers.
[2] La Casa de los Animales, i.e., 145 Adelaide Drive.

August 7. Very hot again. Today I missed going down to the ocean, in order to get on with "Frankenstein." This is the part just before the attempted destruction of the Creature by Polidor and Victor.

Last night, we had supper at the Beverly Hills Hotel, in Tennessee Williams's room. Mike Steen cooked it—meat loaf. And Victor Campbell, Tennessee's secretary and boyfriend, was there. We liked him. He is a very sweet-natured girly-faced rather husky young man from Florida. Tennessee says he used to be retarded and maybe he still is. He read an awful short story out loud to us, at Tennessee's suggestion, without the slightest embarrassment. And he does magic tricks, with apparatus he bought at Disneyland. They are off to Tahiti and maybe other Pacific islands, tonight.

Tennessee seemed fatter, quite sane, though a bit high on something before dinner; he laughed incessantly and said life was like a Pinter play. He told me he had always loved me, even before we met; and Don that he gets more and more beautiful. He hopes to have his two-character play, *Out Cry*, performed here. He gave us an old copy of Stevenson's *A Child's Garden of Verses*—an edition published in 1905, with illustrations by Jessie Willcox Smith. He said this was the one his mother used to read to him from, when he was a child.

Tennessee went to bed early, saying that he was "not a well lady"; but his reason for doing so was also, evidently, in order to hold a levee—or rather, it's opposite—for he assembled us around him, sitting on the bed. There was then a good deal of dialogue with Victor, making him recall, for our benefit, the night of their first meeting, etc. I wondered later if this form of conversation is perhaps particularly characteristic of dramatists, since it is so often used by them on the stage, when one character is forced by another to make some terrible or at least indiscreet revelation containing necessary exposition!

August 8. Have now got as far as the destruction of the Creature—how is it to be attempted? Our latest idea is to involve one of the two Chinese menservants—I mean, let the Creature kill him; we can't afford to kill both, as one is needed for driving! According to this arrangement, the menservants would be doing the dirty work, with Polidor and Victor watching.

Last night we had supper with Nellie Carroll, Miguel[1] and

[1] Not his real name.

Wilbur Flam,[1] who is here on a visit with his two sons. He is now divorced from his wife. He looks older, worried, fat and rather sad. He was very pleasant, and indeed so was Nellie. I have a sort of prejudice against her and yet, often, when I see her (which *isn't* often) I am entertained and feel fond of her.

Not quite so hot today, and the red tide has disappeared.

Yesterday afternoon, I went over to 147 to talk to Donald Hall, who is writing the Charles Laughton biography.[2] Elsa was there too, but withdrew so we men could dig up the dirt. I gathered I was supposed to discuss Elsa's relations with Charles and answer the question: "Granted that he was often unpleasant with her, did she provoke it in any way and was she ever unpleasant to him?" Of course I wasn't about to commit myself on this. So we retreated to the safer paths of homosexuality; I being careful only to lead Mr. Hall up the ones he knew already. It is all very well to talk about being truthful in the interests of history, future generations, etc.—but not with Elsa around; she will never, never forget anything I say. Though she may not fully realize it, she is actually using this project to find out if certain of her suspicions about Charles are correct or not. I'm not sure what I think about Hall himself. He smiled rather too much. I suspect that I terrified him—he was terrified lest he, a square, should say something tactless about my "people," the queers—or even merely use some tabooed word, like calling a Scot Scotch.

August 9. As hot as ever. All records are now broken—this is the longest spell of over-ninety temperatures in Californian history. But, down here, it isn't really anything to complain about. Most of the time we have our breeze. And the ocean is refreshing.

Hunt's secretary called this morning. He is returning today and we are to see him tomorrow. Today I reached the end of the fire scene at the Old House and there's only one more short scene before the ball sequence.

Yesterday, Bob Ennis and David Mosley came over, so Don could draw David. Meanwhile, Bob poured out his emotions, pure intentions and hopes for a beautiful relationship. His silliness

[1] Not his real name.

[2] Hall (b. 1928)—poet, playwright, autobiographer, essayist, educated at Harvard and Oxford and appointed fourteenth poet laureate of the U.S. in 2006—had already published books on Henry Moore and Marianne Moore. He drafted half a million words and roughed out one hundred pages of the biography, but Lanchester didn't like it, and Hall destroyed the manuscript and his notes.

is endearing. But what is impressive is that he has already talked to several members of this white and apparently fairly wealthy family and somehow made them accept the fact that David, who has run away from their home in Georgia, is living with Bob and planning to stay there for some time. Bob says that David is very protective towards him, and this I can believe. He seems like a good boy. Don likes him. As for sex, David claims that he doesn't want any of any kind, at present.

August 10 [Tuesday]. Yesterday at the gym I weighed 149. This is the kind of thing which enrages me out of all proportion to its importance, which is nearly nil. I have been trying hard to bring my weight down and I had set my heart on getting down below 145 for my birthday. And yet, what the hell does it matter? I am not significantly overweight anyhow. No one—not even Don—cares, within reason, how I look. And on top of that, I'm well aware that weight is a freak thing and that probably mine was up temporarily because of too many cups of Sanka that day.

Today I more or less got through the ball scene in "Frankenstein." Hunt Stromberg is coming by tomorrow morning to see us about it. He wants to give Sheinberg the extra pages on the 16th, to take to England on the 18th.

Had supper at Vedanta Place, where Swami told me that [one of the monks] ran away the other day and wanted to disappear—maybe even kill himself—because he is, in his words, "no good for anything." This stunned me. And it is one more reminder that I don't really understand anybody—I only sometimes kid myself that I do. [The boy] is back at the monastery now—seemingly his usual cheerful self, and ready to take brahmacharya with several others at Trabuco next Friday (the 13th!). Probably it was because of this impending change of status that [the boy] got so upset. He even sent Swami a note saying that he wasn't fit for it. Swami replied that he, Swami, was the best judge of that! I saw [the boy] tonight but only for a moment. I wish I could have talked to him alone. I also saw Chetanananda again and am quite charmed by him. He is marvellously campy, rolls his eyes, giggles and does incredible things with his immensely long fingers. Surely he *must* be queer? But let me beware of thinking I understand *him*!

Number 147 was broken into again, last night. This time nothing was stolen except for some liquor. One of the downstair beds was slept in. Elsa has started saying that maybe she will sell the house, or anyhow rent it for a while. This is slightly aimed at us, implying that if we were better neighbors and closer companions for her she wouldn't feel disinclined to come down there, as she does since Ray Henderson got married.

August 11. 147 and ½ at the gym.

Hunt came to see us this morning. Most of his suggestions are fairly easily complied with and I dreamed up an end for Mrs. Blair which satisfied him—she goes snooping in Victor's rooms, comes upon the Creature and has a stroke. Hunt wants the new pages tomorrow or Friday, and the whole thing is to be given to Sheinberg on the 17th. Hunt's ideas for casting—Burton, Taylor, Chamberlain—but Chamberlain wants half of the picture! *And* top billing!!

Renate called to say Ronnie is crazy again and she is afraid he will visit her and be violent. More alarming, for us, is that Ronnie was in some looney bin in Texas and he decided that *I* released him from it by my mystic power—so we can expect a visit from him any time! Poor dear Ronnie—for him to show up just at this moment would be hideously appropriate—like the Creature looking in on Frankenstein; the prospect of his gratitude is appalling. It will use up several work hours.

August 12. Shortly before seven this evening; have just sent off our big batch of "Frankenstein" pages to Hunt by messenger, up to and including the ball. Now we have a couple of days to do the alterations he wants on the earlier pages.

Last night we had supper with Gavin and Mark at a restaurant called the Fiasco in the crowded slummy-grand marina, with its logjam of small boats—what a depressing place. Mark talked for hours and hours but was really interesting about Merle Oberon, who quizzed him regarding his career plans, finances, relations with Gavin, other sexual activities, family background etc. etc., and gave him solid advice, as one who had succeeded in getting exactly what she wanted. Gavin told us that Jim Bridges had made nasty remarks about *Natural Causes*[1] and had said that Mark is just being jollied along—Gavin doesn't really intend to put him in it. Gavin is very hurt and resentful. He feels that he ought to have it out with Jim because otherwise they can never be friends again. He also told us, however, that most of Jim's remarks had been passed on to him by Clyde Ventura, who is a pathological mischief-maker: so maybe they are twisted or exaggerated. Jim must be mad to talk to Clyde about any mutual friends, anyhow. What does he expect will happen?

[1] A screenplay Lambert was working on; never filmed.

August 13 [Friday]. At the gym today I weighed almost exactly 147. And last night, when we went to have supper with Ed and Avilda Moses, she said I had visibly lost weight. That's fine, but I still suffer from this bloated belly—so much so that I have had to loosen my belt one hole!

I like both the Moseses, they are really very ready to be friendly. Billy Al and Penny were there too. Billy was at his most competitive. Whatever anyone talked about, from mescalin to surfing, Billy had to have done more of it and better, or had a worse bummer of a trip or injured himself more severely!

Either Billy or Ed said I ought to be president. Whereupon Don said, "Well, I absolutely refuse to be First Lady!" He laughed when he said it and they all laughed too, but it was a curiously daring remark for Don to make in that circle, because he is always a little afraid of embarrassing them (or rather, just Billy) by bringing up the queerness thing.

Hunt has raised no objections to the pages we sent him yesterday—I guess because he is too eager to get them typed and handed over to Sheinberg. So now I only have to do the alterations he proposed on the earlier pages. Sheinberg isn't leaving till Friday anyhow.

Yesterday I called [the boy who ran away from the monastery] and told him how happy I am that he is taking brahmacharya (today). I think this really pleased him.

I forgot to mention Jerry Byrd (Bird?)[1] who was also at the Moses' house last night. A very husky broad-shouldered boy with spectacles, who looks rather like a homelier Mike Van Horn and is an art student who is working as an assistant to Ed. He appears to have a sweet disposition. We both thought him attractive.

A card yesterday to say that David Hockney and Peter will soon be coming by here on their way to Japan. Also a note from Peter Schwed, saying that his son Greg and a friend are motoring across the country and that he, Peter, had suggested us among various "people I thought might be willing to give them a floor to sleep on if they need one." It is not so much the gall of this suggestion that amazes me—it's that Peter should suggest that his son should spend a night amidst us declared faggots!

August 14. Gerald died today, about 2 p.m. Michael had been

[1]Byrd (b. 1947) got an MFA from the University of California at Irvine, went on to paint in abstract geometric style, and illustrated a text book for Macmillan.

feeding him soup. He stopped breathing. There was no struggle at the end and no apparent change in consciousness. The body was sent away almost at once, to the UCLA medical school—Gerald had willed it to them years ago. Now Michael is tidying up the house. He said he didn't feel like seeing anyone until tomorrow—except Dr. Cohen, who was there when Gerald died. He is very anxious to get in touch with Swami and tell him as soon as possible. But Swami is down at Trabuco and the phone is off the hook!

I am only just beginning to realize that it has really happened, at last. My chief feeling is that I would like to get on the air and tell the great stodgy thick-skinned world that it has lost one of its most tremendous men, one of the great magic mythmakers and revealers of life's wonder.

Talked to Chris Wood, who feels that now he must perhaps go to England. For him it is a terrible dilemma. I think he is deeply shaken.

Talked to Tom Van Sant, whose reaction was that all of us who knew Gerald should get together and exchange our experiences of him. Fine—but include me out!

August 15. To see Michael this afternoon. He looks terribly withered but isn't in the least sentimental or tearful. I think he wants to be allowed by Swami to live at Trabuco, or maybe somewhere around Vedanta Place. He is planning to go to Honolulu and stay with Mrs. Luce for a while. He doesn't seem to want company particularly; indeed I felt he was suffering my visit out of politeness, after the first half hour. He has been tidying up the house. The back part, where Gerald didn't live, is spotlessly clean. I wasn't shown the front part. He said, "My life has been completely changed."

Coming away, I met Madge MacDonald, who told me Jack Jones had told her that Gerald had told him, on the 13th, that he was going to die next day! (*How did Gerald speak?*)

August 16. I forgot to mention that, yesterday, Michael gave me the sloping board on which Gerald used to write. I tried writing on this today, a letter to Bob Craft about his description of Igor's death, and I find that I prefer a sloping surface rather than a flat one to write on—so now I'm kind of physically in touch with Gerald. On the back of the board, in Gerald's handwriting, are the names of stories and books which he wrote on it—*The Creed of Christ*, *The Code of Christ*, *Training for the Life of Spirit*, 1 and 2, *Man the Master*, *Havelock Ellis: Desert*, *Sigmund Freud: Ecstasy*, *Taste for Honey, Reply Paid, Murder by Reflection, The Fog, The Cray Fish, Despair Deferred, A Tide in Time, The Cat I Am, Desert Dialogue, Eden on Ice, Aum, A Prologue to Prayer.* (Michael says that some of these are the original titles of books published under different names.)

Got up extra early this morning and finished revising the "Frankenstein" script, the earlier part as well as the later, and sent it off by messenger to Universal. So now we'll have a day or two of holiday, maybe, before we start on the final section.

Then we went on the beach; the water was cooler but refreshing. Then I went to the gym. Still just over 147.

Nearly every day, Don makes some remark, half-jokingly referring to the fact that I don't find him attractive in a beard. He thinks it is a kind of obstinacy and maybe it is. But at least I am consistent. I find beards, considered as sex adornments, repulsive—no matter who wears them. I find something peculiarly indecent about a beard on a nude. And I am increasingly getting a hate on shoulder-length hair and nearly all moustaches—the thicker the worse. I *yearn* for a new age of crew cuts.

August 17. Have just finished reading a manuscript sent me by Paul Wheeler, the singer and composer whom I met at Cambridge with Mark Lancaster on April 29, 1970, the last time I saw Morgan. It is an almost disconnected narrative, like a lot of entries from a journal, dealing with Paul's life during that year. It's called *Hold It.* Wheeler sent a note with it, saying he had meant to write me and was sending this instead. There is a lot about girls, and bars, and pot, and music, with glimpses of John Lennon and David Hockney and many others. It is dull because he doesn't show his own viewpoint clearly enough and fails to explain situations sufficiently. But he is very good at rendering people's dialogue. After saying of me: "He is precise in a heart-rending way: he unties all the little knots of human tension surrounding him, so patiently and considerately...." (which I don't find particularly illuminating) he goes on: "At the same time there is a puzzled schizophrenia about his reaction to circumstances he is in; as though he knows how to speak the language of the situation but always does so in a very slight accent, in case there is anyone there from home..." (that's brilliantly observed).

Tomorrow I'm to drive down with Michael to Trabuco. It was my idea and my heart sinks at the prospect, just a little. But it was a good idea—Michael is delighted and Swami seemed pleased that I'd thought of it.

August 18. The trip to Trabuco was quite a success. Michael got what he wanted—Swami's permission to come and spend at least several months out of the year at Trabuco, and I got a welcome glimpse of the nostalgic old place—now landscaped and trimmed and pruned and cleaned up as never before—and the boys. Cuddly ruthless Peter [Schneider] was there on a short visit, playing basketball with Mark, Franklin, Doug [Rauch] and a red-haired new boy whose name I forget.[1] Swami seemed rather low, he has asthma. It seems he had darkly hinted to Anandaprana that he had finished his last task—giving brahmacharya—and that he might now depart. (Whereat Anamananda had said brightly, "Nonsense, Swami, you may live for two or three years!") Swami brightened at lunch, however. We left soon after.

Michael has been asked by Claire Boothe Luce to come and live with her, in the event of Gerald's death, and he has promised that he will. He made a great deal out of this promise and how promises *must* be kept—telling me, on the way home, that he had told Swami of this promise and that Swami had said, yes, indeed, he must keep it. This kind of ethical fussing is characteristic of an aspect of Michael which I still find tiresome and basically insincere. Because the thing which he didn't mention is that Mrs. Luce has the power to reward him handsomely for his services.

At the gym today, just a shade *under* 147!

August 19. I didn't have time, yesterday, to write all I wanted to about Michael. Getting back to Mrs. Luce, I asked him right out if she was going to provide for him in exchange for being looked after. This quite shocked him at first, but then he told me that he was certain he was in her will; she had always regarded him and Gerald as being "family." Actually, I think Michael has quite a sound instinct for feathering his nest. He already has some property in the islands which he now wants to sell and then—if he is going to live with Mrs. Luce and/or with the Vedanta Society—he can also sell his Rustic Road house which is very valuable, he says, because it is zoned so that he is officially entitled to lease one half of it to someone else, if he wants to. On top of all his other plans, he has offered to go with Chris Wood to England, if Chris wishes to make a visit of exploration, to see if he wants to live there again—and this, I couldn't help thinking, might very well make Chris decide to leave Michael some of *his* money. Well, good for Michael!

[1] Dan Crane; he later left and married.

Probing to find out what he thinks of Jack Jones, I asked about Jack's relations with Gerald. But Michael was very careful what he told me. He said that Gerald had always regarded Jack as "part of his karma," since Jack had been the one who was determined that they should know each other. ("But I was just the same," Michael added; "I kept writing Gerald letters, and then I practically forced my way into Trabuco.") Michael did however say, in a superior tone, that Gerald had been unable to get Jack interested in "the ultimate goals," so he had decided, in Jack's case, to concentrate on "building a character"—that is, trying to make Jack reliable, truthful and careful to keep his promises. Michael said he was now in a difficult situation because Jack had come to him and said that Gerald had told both of them that they were to practise automatic writing together, after his death, in order that he might communicate with them. Michael says he can't remember that Gerald ever said any such thing and he doesn't want to do it anyway. He has decided that Gerald's personality has already dissolved and that he is now with Brahman and therefore could not communicate even if "he" wanted to. This was another question he asked Swami, and Swami obligingly answered that Michael should not make any attempt to get in touch with Gerald by psychic means. (I know, all too well, how easily Swami can be "worked" as an oracle. This is no criticism of Swami; it's just that you can make him give answers to questions according to the way you phrase them, because despite his fluency, he doesn't understand all the shades of the English language. Michael must surely know this too; so I find his questioning dishonest.) And now here was Michael saying demurely that he didn't know if he could tell Jack Jones what Swami had said, because this would be speaking of the Guru's instructions to someone else, which is strictly forbidden. Oh, Mary! (I'm not writing all this just to bitch Michael. It's just that I am profoundly interested in his state of mind and indeed his whole predicament at this time—it is such a marvellous subject for a novel!)

There are many other details which will perhaps occur to me later. The only one I can think of right now is that Gerald revealed, during his illness, that he had always disliked the way Michael's side hair was kept cropped around the shiny bald patch; so Michael had let it grow. *That* detail I find touching and true. What touches me is Gerald's unwilling aversion and Michael's being able to speak of it.

This morning, he called and thanked me sweetly and with obvious sincerity for having arranged the Trabuco visit.

Hunt called this morning to say that he has reread the whole script and that he finds it marvellous and that the dialogue has the "distinction of Shaw or Wilde"! More impressive, businesswise, was a call I got this afternoon from Robin French, telling me he'd met Sheinberg this morning who'd said, "I've got the best script I've ever had!", meaning ours. But he hadn't read all of it yet.

Meanwhile I am in a fearful flap because we can't seem to find a satisfactory progression for the episodes between the ball and the sailing of the ship which is taking Victor and Elizabeth (and the Creature) to their doom.

August 20. Another good opinion of "Frankenstein" (from Herb Schlosser, whoever he is[1]) quoted to me by Hunt. It is to be Universal's definitely biggest production of the year, etc. Today Sheinberg has presumably left for London with our script. Worked quite a lot on the scenes which follow the ball. They are all wrong, so plotty. When a scene is really and truly wrong, you sometimes find it won't stop, you go on writing and writing and there is always more to say and nothing you say is the end line.

Am reading Leslie Marchand's one-volume life of Byron, *Byron: A Portrait*, Chekhov's *The Schoolmistress* [*and Other Stories*] and Montgomery Hyde's *Henry James at Home*—the last of these bores me, rather. It isn't frank enough—or maybe there's not enough to be frank about.

August 21. We had supper last night with Irving and Shirley Blum, at their house. Jack and Jim were there too. I've probably described the house before, somewhere in this diary. It's exactly like living in an art gallery. The pictures are nearly all gallery pictures; large and needing to be looked at from a distance—Frank Stella, Ellsworth Kelly, Kenneth Nolan, Lichtenstein, Warhol—and, because they are well spaced and strategically hung, the living room seems austerely bare. The pieces of furniture and the few other items lying around are all "objects of value"—just as much so as the German sculptures in shining metal which remind you of Lang's *Metropolis*. [...] The most surprising of the Blums' treasures is their adorable two-year-old son Jason. He doesn't fit into the setting.

Jack talked a lot and was very lively, but we still suspect they are mad at us. Don thinks it's because we didn't like *Carnal Knowledge*. Jack and Jim have become very thick with Mike Nichols.[2]

Don has to buy another car. He was out today looking at jeeps

[1] Wall Street lawyer turned broadcasting executive; he became president of NBC in 1974 and later a media and entertainment analyst.

[2] Who directed it; see Glossary.

but thinks them too clumsy and heavy and expensive. Foreign cars are supposed to become more expensive, because of the weakness of the dollar.

Evelyn has been bothering us again, about getting in touch with the Spenders when she goes to Europe. They are holed up in that house near Les Baux and I don't have the exact address and can think of no way of finding it. Evelyn, who is still hysterical underneath and apt to be weepy, professes to think that they have "forgotten" her! She can be quite tiresome. She also said that *The Advocate* wants to write an obituary of Gerald! I can't imagine that Michael would ever agree to this.

August 22. I was right. Michael doesn't agree. He says that Gerald never declared his queerness in print, and when he wrote articles on the subject he used an assumed name.

We had supper last night with Paul Dehn, who is staying at the Beverly Hills Hotel, writing another Ape film. He can't drive, so he is quite cut off there, just commuting to the studio and sometimes walking into Beverly Hills. He is friendly and full of jokes and worked hard at entertaining us. But, as Don said, he treated him like a woman. He has a tiresome trick of lowering his voice until you simply can't hear him unless he is speaking directly to you. As he sat between us, we shared his conversation, and Don got much less than half of it. So the evening was tense. Just the same, I feel I must see him sometimes while he's here. It's a curious bond, that we both come from Disley—a curiously strong one!

August 23. Back to 148 at the gym. Don said it was only because of all the coffee I'd drunk. For the past two weeks (at *least*) I've given up alcohol altogether. And today I not only ran round two blocks but ran down to the beach as well. Everybody says how healthy I look.

Last night we went with Gavin and Mark to see Ingmar Bergman's *The Touch*. It's really a very poor script, the most ordinary adultery story, just a feminine caprice in the usual Frog style; she wants to do it so she does it, and they both yell bloody murder because they find they are "in love" and "can't live without each other," and meanwhile the husband sulks. Yet Bibi Andersson made it all seem fascinating and subtle and even almost touching, as long as she was on the screen. She is one of the most marvellous actresses in the world and I have never seen her better. But *why* did she have to play opposite that loathsome slob-Jew, Elliott Gould? Because Bergman wanted it!

As usual, Gavin insisted on going to a terribly expensive restaurant, Jack's at the Beach—our half share of the bill was twenty-three dollars—and the food at these places is almost never good. Mark is looking almost beautiful, these days, or at any rate very striking—but there is nevertheless something comic about his appearance and we keep wondering how he will photograph. He told me I looked like a little boy when I was eating my sherbert—that's his idea of a compliment. He asked Don if we touched each other much in bed!

Swami is now up at Santa Barbara and will stay there till his asthma gets better—so I shan't be able to visit him on my birthday.

Michael Barrie leaves for Honolulu and Mrs. Luce the day after tomorrow.

A postcard from David Hockney and Peter. They are in Spain. They say nothing more about coming here on the way to Japan. The message ends: "p.s. sell your dollars." We keep getting panicky about money. In a few days we are having Robin French to supper, hoping to get some financial advice from him.

August 24. Last night I saw Peter Watkins's *Punishment Park.* It disturbed but irritated me. Today I feel that I have had an experience but I am still irritated. The scenes at the tribunal, where the prisoners speak, accusing and cursing their judges, are memorable and convincing, because the actors were (or so I believe) sincerely convinced pacifists and activists in their daily lives. But the pseudo-newsreel technique became all the more phony for that very reason, when it dealt with the manhunt. The Watkins film about the Battle of Culloden *was* convincing all through, despite the anachronism of the newsreel technique, because one knows that everything it describes actually did happen. I have written this down in order to "know what I think" by seeing "what I say"; because I shall have to discuss the picture with Robin French when he comes to supper on Saturday. Chartwell had something to do with producing it.[1]

The Sorrows of Frankenstein. Shall Elizabeth tell Victor she is going to have a baby *before* her interview with the police chief and Polidor? Or wait till she and Victor are safely on board the ship? Shall they go straight on board or hide out at a waterfront inn until it gets dark? How does the Creature, after its escape from

[1] Chartwell Artists produced the 1971 faked documentary in which war protesters must choose between prison sentences or a three-day desert crossing pursued by National Guardsmen authorized to shoot and kill them. "Culloden" (1964) was made for BBC T.V.

the police station, discover where the others are? However, we have made one tremendous breakthrough, entirely due to Don. He has had the brilliant idea that the Creature shall carry Polidor up the mast and that they shall both be struck by the same bolt of lightning—killing Polidor and invigorating the Creature! This is a perfect example of cinematic symbolism. For, as Don at once pointed out, it was always Polidor who hated electricity and Henry (now part of the Creature) who believed in it.

August 25. This is the end of the month of entries I'd set myself. After this, I won't write unless I want to—which is almost never! (Even now, I'm playing games with myself; tricking myself into writing perhaps.) Well, anyhow—

I feel depressed because of the gloomy state of the economy and because of a British T.V. film they showed last night, called: "San Francisco, the City That Waits to Die." It is vulgarly alarmist, but at the same time it clearly makes the point that a major earthquake up there is apparently inevitable before long; and there are schools and whole residential areas which are built right on top of the fault. The film (or rather, an appendix to it) stated that the same situation exists in southern California—the fault hasn't moved in a long while and must release its huge pressure in the near future. Oddly enough, the middle section of the fault keeps shifting and is therefore far less dangerous.

I talked about this to the man at the Duck Blind liquor store. He told me that they don't have any earthquake insurance. And they had some of their bottles smashed even in the last quake!

Yesterday afternoon, a young man from Oxford named George Rumens interviewed me. He came here last year, too. He has a girl named Lindy whom he's going to marry and a friend whose name I didn't get, who is training to be an actor. Rumens says that Oxford students nowadays are only interested in pop music and their own poetry, they care nothing about writers of other generations. He also complained that all the villages in that part of England are spreading out and merging with each other. Still, they have a literary society at which, apparently, they are ready to listen to interviews which Rumens has taped with literary characters, such as me![1]

[1] Rumens (b. 1949) attended Magdalen College; later he made environmental films and was a Green Party activist. The literary society was The Oxford Critical Society.

September 2. As a birthday present, Don gave us a large gym-type scale; he had taken enormous trouble to find one which was secondhand but in good condition. So we have been weighing ourselves several times a day. Of course the weighings in the gym are all with gym clothes on, which means at least half a pound more than your weight nude. As of this moment (5 p.m.), I weigh 147 and ¼ nude; 147 and ¾ in gym clothes. We are inclined to think that the scale in the gym weighs slightly heavier—and *of course* ours is the accurate one!

Now I'm sixty-seven and have my operation ahead of me this month, unless Dr. Ashworth has a very long waiting list. I rather dread this. It will obviously be quite a big production. Never mind. My little finger is getting slightly worse, which is an inducement to go ahead.

Don Howard showed up from Baltimore, exactly the same, except for an unbecoming blond moustache. Paul Wonner has grown a *very* unbecoming beard. Bill Harris also appeared, from San Diego where he is visiting his mother. His figure is still all right and he has neither beard nor moustache, but, oh, his face is so nervous and weary, a sad anxious queeny face. I am still fond of him and felt sorry; he is one of those about whom one says, "*What* will become of him?"

I got the impression that Paul Wonner didn't care much for *Kathleen and Frank*, though Bill Brown had told me that he (Paul) loved it.

Have been reading through the second file of the 1939–1944 journals, to get background for my work on the reconstruction of a narrative for 1945; my only material is the little day-to-day diary which isn't even kept up every day. The 1939–1944 journals seem very self-conscious, and very conscious that they are about to be read by Dodie and Alec Beesley. But there *is* interesting material in them. I keep marvelling that I didn't break with Peggy years earlier than I actually did. She really was intolerable.

Much impressed by the atmosphere of the Australian film, *Walkabout*.[1] Amazing how it captures the weirdness of that desert country, its hills and ridges which always promise something marvellous beyond them, and the astonishing variety of scenery which is nevertheless all basically inhospitable to human life. And how it made you feel that slowly increasing horror of being lost!

Don finally bought a car, a new V.W., light yellow, three inches longer than the standard bug model, a super bug. He had intended

[1] Directed by Nicolas Roeg.

that the new car should be mine and that he should take over the beige V.W. But his father and mother took such pride and delight in the purchase (particularly Jess, of course) that he hadn't the heart to disown the new one. Don hates having a new car and besides, when the older beige one starts to break down, he won't be able to ask Jess to fix it!

We have now at last got Frankenstein, Elizabeth, Polidor *and* the Creature on board the ship—and set sail for the final sequence!

During last month, I was in the sea twenty-four days out of the thirty-one. All that running down to the beach has undoubtedly been healthy, but it (and the acquisition of the bathroom scale) has rather disinclined me to go to the gym.

September 4. Don has gone off this morning to spend the day with Jack Fontan and Ray Unger. I have just heard from Vedanta Place that Swami has a fairly serious blood condition and that he has been put in the Cottage Hospital at Santa Barbara for treatment and a thorough rest. He'll probably stay there a couple of weeks.

Last night, we had supper with Chris Wood. He was very frank about Gerald; Don thinks that he was feeling he must "confess" to someone. He told us he had been unkind to Gerald quite often; that he had once said to him, "You're a very stupid man." He repeated what he has so often said, that he cannot feel that Gerald was really spiritual, "not like the Swami." Chris thinks that Gerald was always trying to convince himself; that he didn't *know*. Chris's general attitude is that Gerald used to be ever so much nicer and more fun in the old days, in England, before he stopped having sex and got religious. Chris says that he loved Gerald, as he was then, more than anyone else in his whole life, except for Mark Palmer.[1] But, when Gerald got to America, he came under the influence of people like Allan Hunter—puritans who disapproved of Chris. Chris says he never wanted to come to America in the first place and only did so because he believed it would be for a short visit; later he realized that Gerald had always intended to stay here permanently. He never wanted to set up house with Gerald, either. Gerald was living at Peggy's, and Maria Huxley persuaded Chris to take a place with Gerald (this was Arlene Terrace)—no doubt to spite Peggy (this is my guess, not Chris's). According to Gerald and to Michael Barrie, Margaret Gage behaved badly to Gerald and more or less ran him out of the house. But Chris

[1] Wood's best friend, who died in England in 1943, as Isherwood records in *D.1*.

thinks quite the opposite; he says Gerald was most inconsiderate in his dealings with Margaret, went away for weeks on end without telling her, etc. Finally, Chris said he didn't at all agree that Gerald is now merged in Brahman, as Michael thinks. Chris feels that Gerald is still very much himself, wherever he is.

Of course, some of this is quite obviously what Chris *needs* to believe. Just the same, I feel and have always felt that anything Chris says must be taken seriously, because he is nothing if not truthful and he gives the impression of seldom if ever kidding himself.

September 17. On the 15th, we finished the "Frankenstein" teleplay and early yesterday morning we sent it off to Universal. Today, Hunt called, saying he is very pleased with it. He wants to send it to Sheinberg just as it is. He said, "We won't consider any revisions until we get a director." So now, quite suddenly, it is out of our hands and we are left without a project. Hunt talks about "The Mummy," but I most certainly do not want to start on that without taking a break.

Glancing through the script I feel really quite pleased with it. There is far too much dialogue and things are clumsily expressed, but that's very easily remedied. The whole thing is 197 pages. It breaks neatly into two sections: the first ends with the confrontation between the Creature and Polidor at the ruined house; the second begins (on page 98) with the wedding of Victor and Elizabeth.

My operation is now set for the 24th, but I have to go into hospital on the 23rd. I spend at least two nights there. I get depressed about this, of course.

On September 5th, we had a nice, mildly drunken lunch with Truman Capote, who was staying with the ex-wife of Johnny Carson.[1] I wish I had written about this, because Truman is very amusing. All I remember is that Truman teased me by announcing he is going to start a story that I gave the U.S. rights for *Maurice* to the National Institute of Arts and Letters only on condition that it must elect Gore a member, even if he got no votes at all!

Swami has been quite sick. But I won't write about this now as I'm waiting for a call from Anamananda to tell me what the doctors said about him this afternoon.

Am in good shape physically; I go to the gym or run down to the beach or both, practically every day.

Nearly three-quarters of our street is parked up solid, every

[1]Joanne Carson; see Glossary.

working day, by the cars of the men who work on the new buildings on the corner. The word is that they won't be finished for another year! Also, they are putting up a new house on Maybery Road, with much early morning hammering. Also, there are more dogs around than ever before!

Don (describing himself, I think—I wrote it down on my pad some while ago because I laughed so when he said it): "A dried effigy on a throne of cardboard."

The doctor is pleased with Swami, so the scare is off for the time being. But I did think he looked both ill and very old and weak when I saw him the day before yesterday. Poor little darling thing, peeping out of a blanket, as he does when sick, and telling me that he had "lost the use of" his mind the day before, while he was sitting in the bathroom; apparently he blacked out. Also, which was very distressing, he had hiccups!

September 23. Swami seemed much better, when I saw him briefly last night. I asked him if he had had any spiritual experiences while he was sick. He said he hadn't, exactly—only a dream which made him very happy. He had been outside a temple and someone came and said to him, "What are you doing outside?" and brought him in and offered him a seat. But the seat belonged to Swami Nirmalananda, so Swami wouldn't take it. Nirmalananda tried to give up the seat to him and Swami clung to his feet to prevent him. Then he woke up.[1]

Then Swami talked about grace and was so beautiful, so shining. He told me that he didn't feel he had ever gone through any particular spiritual struggles; Maharaj had made everything easy for him. "All those visions I had, Chris, I *never* felt I had really earned them." As Swami said this, he positively shone with grace, he was the blessed one, the lucky one, and his luck was adorable. When I left I made an extra long prostration to him, and he said, "Get up now." Maybe he was a tiny bit embarrassed, feeling that I was worshipping him. So I was, but it wasn't *him*.

This afternoon I have to go into that damned hospital. Amazing, how I hate the prospect. It isn't the prospect of pain. It isn't worry

[1] Nirmalananda (1863–1938), or Tulsi Maharaj, is sometimes considered one of seventeen direct disciples of Ramakrishna, but in *Ramakrishna and His Disciples* Isherwood includes only sixteen. He appears in some photographs as Ramakrishna's disciple, but also described himself as a disciple of Vivekananda. His uncertain status evidently symbolizes Prabhavananda's own conflicting feelings of longing and humility as his death approached and he prepared to "join" the circle of disciples.

that this thing may prove to be more serious than I expect. It is basically my horror of The Hospital, the frontier post of death. I hate submitting to their rituals, all the more so because I am otherwise in the pink of health.

And then there is the pain of leaving Don, which is also symbolic; because the parting is actually only to be for two nights. But that too is a symbol of death—and he feels it instinctively, as I do. We don't speak of it but we are gentle together; we are more than ever The Animals on an occasion like this.

October 1 [Friday]. I can't use my left hand properly until the splint and bandage have been taken off, which won't be until next Wednesday (October 6) at the earliest. So I'll write this in longhand now and type it up later.

When we got to the California Hospital, I was terribly depressed for a few minutes, because it was just exactly what I had been expecting. The lobby crowded with people waiting and looking as if they'd been waiting hours and hours, with a terrible impassive acceptance of delay and a stolid readiness for suffering. I wanted to send Don away, so that I could resign myself utterly to this death experience. But he wouldn't go; said he was going to see me settled into my room. His gentleness and sweetness made the experience all the more painful, because they made me *feel*, which was exactly what I wanted to stop doing; and yet, what a comfort it was to have him there!

But then, quite quickly, someone appeared to take me up to my room. And I had one all to myself and it was quite cheerful—very quiet, but with a pleasing view of the freeway viaducts in the middle distance and the traffic whizzing silently by behind the closed windows. The nurses were all real professionals, not the kind of stopgap help they had at the Cedars of Lebanon when I was there. The only bad thing was the food. But Don returned after supper bringing a great bag of fruit and we sat up all evening watching *The Best Years of Our Lives* on television.

I had sleeping pills that night and again next morning—the fact that I almost never take them pays off hugely on such occasions, I'm so beautifully easy to sedate. Also I had a calming shot of scopolamine, after which I felt relaxed enough for execution. This was just as well, for I had to wait quite a long time outside the operating theater, lying on my trolley, and then some more time inside, on the table, before Dr. Ashworth showed up. When he did, I laid my left hand on the chopping block, thinking, with dopey humor, of Cranmer and feeling that I was behaving with remarkable grace and style.[1] The anesthetist, Dr. Musicant, told me I should have sodium pentothal with gas to follow. As he gave me the shot, he said "goodnight," which I thought charming. The operation began at 11:50 a.m. and lasted a little over an hour. It is technically described as a "fasciectomy; palm."

When I became clearly aware of things again, I was back in my room and there was Don, waiting. I felt so happy, seeing him there. I had known that he would be, of course—but at moments like this one you take The Loved One less for granted than usual—it was like returning from a long journey—I saw Don's marvellousness and his presence in my life as the beautiful miracle it always is.

I left the hospital later that same afternoon. Dr. Ashworth had said that I could, if my heart and blood pressure were all right. And they were; I knew that they would be. I felt rather vain of my good physical condition. We had supper at the Fuji Gardens. My hand was in a cast and my arm in a sling.

Not long after my return home, I got a letter from Bill Caskey saying that he was leaving for England from Greece on October 3, and that he might stay several weeks trying to get his photo books published and to arrange a show of his photographs at a gallery. Don at once said that he didn't want to see Billy if we were all in England together. As for me I know that, if he were there, I would have to introduce him around to people who could help him, and that this would establish all sorts of embarrassing links. So on September 28 I called Richard Simon and asked him did he think it was *really* necessary that I should give those interviews in London to help advertise the publication of *Kathleen and Frank*. He said no. So I decided not to go and wrote to Methuen, blaming my decision on the operation. By great good luck, [Richard] Simon[2] has had the operation too—and had a much worse time with it than I am having—so he will be able to convince John Cullen at Methuen that this is a legitimate excuse.

On September 29, David Hockney called from London in tears, having just received a letter I wrote him saying how sorry we both were to hear that he and Peter had split up. (I neglected to record this earlier, but Peter had sent us a postcard from Greece saying,

[1] Thomas Cranmer (1489–1556), architect of Henry VIII's Reformation and Archbishop of Canterbury, was burned at the stake by Henry's Roman Catholic daughter Mary Tudor when she became queen. He pushed his right hand into the flames first for its offence in signing a recantation, forced from him, of his Protestant beliefs.

[2] Isherwood wrote "John," perhaps transposing with the first name, mentioned in the same sentence, of his editor at Methuen.

"I'm sort of convalescing here as I've left David! I've moved into my studio at 5 Colville Square, W11. I don't know what will happen. It just got too impossible.") It now seems that Peter hadn't really had the matter out with David—this doesn't surprise me after the things he told me in London last year—and had simply slipped out from under. So poor David had a shock when my letter arrived. Now I've written to Peter begging him to be absolutely frank with David. Haven't heard from either of them yet.

It was also on the 29th that Dr. Ashworth took off my cast and said he was pleased with the condition of the wound. But he has rebandaged the hand and put a metal splint on the little finger. It does seem much better but I get terrific nerve twinges in the palm whenever I make a wrong move. Psychosomatic note: Just before the operation, my right hand became violently "jealous," because the lazy left hand, which it supports by earning their living, was getting all the attention. So it started an acute attack of arthritis in its thumb—a hundred times worse than any pain I've had from the operation itself!

We have both registered for unemployment insurance. Also, we have had our television put on the cable; a huge improvement.

A very nice visit to Peter and Clytie Alexander on September 30. He wants to trade one of his wax paintings for the drawing Don did of her. I feel Peter really respects Don's work, so I almost love him.

Swami is still not at all well. He leaves for Santa Barbara tomorrow. He will move into a rented house there and rest for at least two months. I saw him last night. He had insisted on going into the shrine that morning and chanting—I think it was because of Durga puja and some intimate memory of Maharaj which was connected with it. When he spoke of Maharaj, it was as if he had seen him only a few moments before; it was *instant*, not the boyhood memories of an old man. I realized, more vividly than I usually can, that, for Swami, Maharaj actually is *present*, quite a lot of the time. He made my eyes fill with tears.

Just before I left, he said in his unaffected childlike way that he was hungrier than he thought, and he phoned the kitchen and asked for a lamburger to be fixed for his supper. What seems childlike is that he never apologizes for showing appetite or otherwise taking pleasure in something "worldly," as puritans do.

October 3. Yesterday we had to use the heating for the first time this year. The mornings are cold but the weather is warm and beautiful at midday.

Don is terribly depressed about his work. He still hasn't shown any paintings to Irving Blum. However he did show about ten of them to Billy Al Bengston when we had supper there on the 29th, and Billy was quite impressed by them, in his cool way. I keep urging Don to bring things to a climax by showing them to Irving; let's at least know where he stands.

Yesterday I drove the car to the market without paining my hand. So I'm independently mobile again, which is a blessing. As far as I can judge, it is healing; but I get occasional stabbing pains right through the palm.

Doug Walsh came this morning to fix the garbage disposal. I had dropped a screw from the meat grinder into it. The disposal bit one of the teeth out of the screw and cracked itself in doing so, but Doug says it will be all right for a while; it is working again now.

Doug made his trip east but could find out nothing more about his missing daughter. He now takes a grim sulky don't-care attitude, saying that she's either in bad trouble or dead. While he was in a small town in Illinois (maybe it was his home town) the daughter of one of his former buddies was dating a black basketball player from her high school. So Doug's buddy called the boy on the phone and the boy apologized and said he had been telling the girl that there was no future in it; he promised not to see her again. Doug told this story to show how simply and sensibly and nonviolently these things are dealt with in the Midwest!

October 7. Yesterday, Dr. Ashworth took the stitches out of my hand and removed the splint. The little finger is bloated and stiff and there is still a zigzag scar like a W running down from it into the palm. He says I can take the bandages off in two days' time and wash the hand and exercise it.

Last night, while we were having the reading at Vedanta Place, there was a big brush fire up around Montecito. At the end of the reading, Ananda announced publicly that the flames had reached Ladera Lane and that the convent was already evacuated and the temple was in great danger. "All we can do is pray," she concluded, in her weepy Jewish voice. I was hugely irritated, because she was so obviously just calling attention to herself and her anxiety—she interrupted Asaktananda in the middle of a sentence—and had neglected to mention that there were two fire trucks standing ready to save the buildings—and anyhow as a Vedantist, how dare she pray for them to be saved? Maybe Krishna wants them burnt! Oh dear, how horribly obvious it is that the women are taking over already, now that Swami is out of the way! They think they can boss Asaktananda. They may well be wrong about that; I suspect that they are. But, when the time comes for him to take over, he will have to smack them down hard.

Today we hear that the danger is over, but that the fire was indeed very alarming; even worse than last time.

Much more alarming is the news that Swami doesn't seem to be recovering. He got sick while driving up to Santa Barbara on Saturday and is now in the cottage hospital there.

Two days ago, I got an advance copy of *Kathleen and Frank*. It looks very nice—but some of my writing seems to me terribly sloppy.

Also a letter from [Richard] Simon, saying he's leaving Methuen to start his own business. This is a big blow. Am seriously considering switching to him from Curtis Brown. But I want Don to meet him first and then to talk this over with him. It would be complicated, leaving Curtis Brown after all these years.

An absurd letter from Glenway, enclosing a copy of the "blurb" he wrote for *Kathleen and Frank* and sent to Simon and Schuster. Most of it is really about himself; the rest is faint praise dressed up in verbiage. Don says Glenway is still envious of me and can't bring himself to write anything really favorable. Some extracts:

"It is a little masterpiece of the head and the heart. Nothing little about it, except that Kathleen wrote half." "The subject matter: As I am a *paysan parvenu*, that is, a farmer's son risen to be a high bohemian, I thought that the concentration of upper middle class life might, so to speak, alienate me. No." "The style: The instant one feels any rebelliousness or boredom, the pleasure of his style takes over. How it gleams, in its bright but plain colors. What a lively movement it has, sinewy and at the same time, simple. How amusing it is, like a little friendly green snake."

October 13. Swami is now about to leave the hospital. He is better.

Dr. Ashworth says that I have healed quicker than any of the three other men on whom he operated, that same day.

October 14. Early this morning, Don had a dream which he described as being "full of love." He and I were at a film made by Fellini. The film was intended to make you feel "as tall as a skyscraper." The rest of the audience didn't realize this. Only Don and I did. "We were so happy, we cried." Fellini came over to us and bent down to peer into our faces; he knew that we were feeling what he wished us to feel and he was pleased.

Also this morning—maybe it was after hearing Don's dream—I dreamt I was up to my waist in the ocean and a tremendous black wave appeared, coming straight toward me. I couldn't escape, except by deliberately waking up. It was a dream about death, but I felt quite relaxed, not frantic.

October 21. Today is the official publication date of *Kathleen and Frank* in England. It is also Jack Isherwood's birthday—I happened to notice this, this morning, while leafing through the book—as I always do for the first few weeks, when a book of mine is just out. (After that, I feel a growing disinclination to open it at all.)

Robin French called to report that Audrey Wood, who works for his writers in New York,[1] has read our *Meeting by the River* play and likes it very much and is eager to do something about it. Also, a man at the Lincoln Center wants to see the script of *Black Girl*, with a view to possible New York production. And, yesterday, Jim Bridges called us to know if he could try to set up a production deal for a film of *Meeting by the River* with himself as director. So things are humming!

Meanwhile, we have met Boris Sagal, who is supposed to be going to direct "Dr. Frankenstein." If they do it. This isn't certain yet. First a budget has to be worked out and the studio has to see if it can find some available stars. We shall be hearing more about this fairly soon. If they do make the picture (either as T.V. or theater movie or both) it will definitely be in England and probably start in February.

Sagal looks a bit like Jerry Lawrence. He is very pleasant, even charming, but he is definitely a second-string workhorse type of director; the type which is called in when costs have to be cut and the show gotten on the road. Meanwhile, Jim Bridges wistfully wants to direct "Frankenstein" himself. He definitely turned this idea down earlier, when we suggested it. Now, his own projects have fallen through for the time being, so he wants to get something started. Admittedly, Jim has far more imagination than Sagal and would probably be better with the actors. Sagal might be better on practical production details and he is accustomed to work in England. Don is very much against him, and I am hardly thrilled. But I doubt if we could persuade Universal to hire Jim, anyway.

A rather touching letter from Peter Schlesinger. He makes it clear that he and David *did* discuss their difficulties before parting.

[1] Wood (1905–1985) represented many celebrated playwrights, including, until 1971, Tennessee Williams. The agency she once ran with her husband, William Liebling, was bought by MCA in 1954.

There is somebody else Peter is involved with, but they aren't living together and David and Peter are still meeting and going to parties, etc. Peter is living at his studio. He says, "I've never lived by myself so it's quite exciting. I read again at night in bed. Little things like that I could no longer do, we led a very busy social life and lots of smoking which also gets me down.... Basically though I still and always will love David...." From David himself we haven't heard anything.

October 27. Just after Don and I had left the gym, two days ago, we saw a small man who looked a bit crazy and who limped. His hands were covered in small bags of cellophane; he was wearing them like gloves. They were very red, so they may have been burnt and therefore sensitive. But the thought struck me that maybe he had a neurotic fear of contamination. This naturally made me think of Dr. Polidori in our Frankenstein script, I said to Don: "Suppose we have an epilogue at the end of the picture? Mary has just finished telling her story and they are all about to begin the picnic. Polidori is there too; he has forgiven them.... Some of the picnic food is sticky, so Mary says, "Dr. Polidori, I fear you will have to take off those elegant gloves of yours, you'll spoil them, otherwise." And Polidori does so. They all watch him—they are still under the spell of the story and half expect to see that his hands really *are* crippled. Polidori realizes what they are thinking. When the gloves are off, he holds up his hands to show them that they are perfectly formed. He smiles. The others all laugh. End of picture."

Now—here's the synchronicity:

Immediately after seeing the man with the cellophane bags, we went on to Columbia Studios, where they were showing *The Shooting*, a Western made by Monte Hellman, who directed *Two-Lane Blacktop*. (It's one of the best Westerns I've ever seen.) In it, Jack Nicholson plays a hired gunman *who wears gloves all the time*; it is said of him that he never takes them off, "even with a woman." At the end, he and Warren Oates have a fight. Oates wants to put Nicholson out of action as a gunman, so *he cripples Nicholson's hand* by beating on it with a rock!

John Lehmann and a nice friend of his came to stay with us for the night of the 23rd. The friend's name is Douglas Stoker. He's a Scot from Glasgow, around thirty, big, homely, with thick glasses, rather attractive.[1] John is teaching down at San Diego again. We

[1] Stoker (1932–1988), who ran away from home at seventeen, was well read,

had been dreading his visit and indeed, he was as inconsiderate as ever, allowing us to wait on him hand and foot, leaving cigarette ash all over the floors and messing up everything he touched. But it's so marvellous now that he's gone—that makes the visit almost worthwhile. And "friendship never ends"; even Don admits to having become fond of him, a little.

John brought with him a review of *Kathleen and Frank* by C.P. Snow, in *The Financial Times* (October 21), beginning, "This is far and away his best work since the early Berlin novels." (I wonder, though, if Snow has actually read anything in between!)

As so often happens, the best came first. Since then I've been sent several others, in which the reviewer spends most of his time making the book sound dull. You could dig quotes out of them for a blurb, but that's all. Roy Fuller is the most favorable, in *The Listener.* Stephen Spender writes one of his half-assed bitchy lukewarm putdowns, in *The Sunday Telegraph.* I must own, I do resent that. Why did he have to review the book at all? It is sheer aggression and envy; he is sick with it—far sicker than he knows.

Altogether, I feel disappointed. I have to admit to myself that I had expected this book to be a real instant success—in England that is, not here. If they don't like this they will never like anything I've written. Well, fuck them. It's their loss.

October 29. Cold, windy, beautiful weather. Yesterday we ran on the beach. I feel very well but my weight is mysteriously up to nearly 150.

Yesterday we got a letter from David Hockney. He is coming here on November 8, travelling with Mark Lancaster, on his way to Japan and around the world. He writes, "I must admit that I am very unhappy, more unhappy than at any time in my life, and I know Peter is not happy ... our situation seems to have deteriorated to a point where we can hardly even be friends, so I thought it a good moment to go away. I am taking Mark Lancaster ... I just couldn't travel to Japan alone as I'm sure I would finish up just talking to myself, and go mad." We have written inviting both of them to stay.

Have just called Vedanta Place and heard that Swami has had a virus infection but is better now. I am getting seriously worried about him. There's a horrible feeling, this time, that he won't

intelligent, a drinker, and never successfully employed. He was Lehmann's sexual partner, cleaner, and travelling companion from the late 1950s until an angry split not long before Stoker's sudden death in 1988.

be able to regain his strength and just keep slipping slowly downward.

I saw him for five minutes (all that was allowed) the day before yesterday, in the late afternoon. He had come down from Santa Barbara to see the doctor and was sitting up in a chair in his room. He had been told not to sit with his legs crossed as that interferes with the circulation—what do *they* know about Bengali circulation, I couldn't help thinking. Anyhow, his feet in their blue socks were down on the carpet, so I could touch them with my forehead as I prostrated. He was pleased to see me but seemed very quiet, withdrawn and deserted. I asked him if he had had any spiritual experiences and he said no—rather forlornly. "You see, Chris, I couldn't eat anything. When you don't eat, you can't think. I realized for the first time what it means in the scriptures when it says, the food is Brahman." He also told me that there were "grey waves" before his eyes when he looked at anything (this is apparently due to his cataract).

Later: a card arrived by this morning's post, postmarked Paris, saying that David will arrive about 5 p.m. on November 8. "Peter and I are sorting it out O.K. but I would still love to discuss things with you if possible."

November 11. David arrived here on the 8th and left on the evening of the 9th, for San Francisco, where he was to join Mark Lancaster and then take off for Japan, via Honolulu. The rest of their tour is Hong Kong, Bali, Bangkok, Rangoon, Calcutta, Delhi, Afghanistan (Kabul), Istanbul, Rome, London.

David has lost a lot of weight and is really slim. He seemed quite cheerful and as full of energy as ever. He was sad that Peter wasn't with him on the trip; just before he took off he had asked Peter again to come along and Peter had refused, saying that he must stay in London and attend the Slade regularly, or he won't get his degree. David admitted that he had been lecturing Peter on sloth, telling him that he must work much harder; so David could hardly give Peter an argument about this.

David told us in detail about his scenes with Peter last summer, and his own despairs and tears. But, underneath, Don and I both sensed a bland assurance that all would turn out the way David wanted it; we don't think he understands, even now, what the fuss Peter made was all about. The Swedish boy, Eric(?),[1] is really

[1]Eric Boman, then an illustrator and fabric designer, later a well-known fashion photographer, became Schlesinger's long-term companion.

no threat, it seems; he is merely a mischief-maker and a situation queen. David says Peter isn't in love with him. According to David, Peter merely has to work hard; then he'll make it as an artist and all will be well. Meanwhile, Peter lives in his uncomfortable studio, where there is no bath. David's beautiful enlarged flat is only a short walk away; a permanent temptation. David has suggested that Peter shall live there while he's away, and we could see that David was expecting that Peter wouldn't want to move out again by the time he returned. So all would be back to position A.

I sound hostile, writing about this, but I'm not. Yes, David is a bit thick-skinned. He has spoilt Peter rotten and given him champagne tastes and then made fun of his extravagance. He tells stories about Peter in public which aren't always in the best of taste; one hears the note of North Country ruthlessness. Oh yes, indeed, David is a monster in the making; before long he will be full grown. But I love him and Don loves him and he is lovable, truly lovable and wonderful and kind and generous and full of life.

This time, he even praised one of Don's paintings!

Last night, I read at Vedanta Place and saw Swami. They are giving him tests to find out if his almost constant dizziness is due to bad circulation, or what. Swami seemed better than before, but he is very quiet and thin. Krishna and Len (Bhadrananda) both sleep in his room, now; Len because, as a former pharmacist's mate, he is able to take Swami's pulse at intervals during the night; Krishna because he jealously refuses to let anyone sleep in Swami's room unless he himself can sleep there too! Swami complains that this overcrowding makes the room become "smelly." He told me that he can nearly always go to sleep by feeling that Holy Mother's feet are on his head.

Asaktananda answered questions at the meeting. Attendance has dropped off a lot, now that people know Swami won't be coming. A cunty fat-legged woman heckled Asaktananda, saying that she preferred "the Zen way." (So very often, it's the references to "woman and gold" in the Gospel which get under their skin.) Asaktananda held his own, but he became a bit sharp toned and schoolmasterish, and suddenly he seemed to be the leader of a tiny aggressive sect, out of step with everybody else, hidden in a corner on the wrong side of the freeway. And I looked at Jim Gates and the other boys and put myself in their place and felt afraid their loyalty would falter and they would soon lose heart under such a leader, and ask themselves in dismay, "Is *this* what I've given my life to?" Asaktananda is admirable, full of courage and conviction—but you don't feel, yet, that he *knows*. Did I always feel that about Swami, that he knew, right from the beginning? I don't remember. I think I did fairly soon. But then Swami could always say, "I have seen The Son of God."

Elsa Laughton has just discovered that Ned Hoopes(?) has finished a book about Charles, after pretending that he had given up the whole project. It seems that he had kept duplicates of all the interviews which he and Elsa did together. When Elsa first heard this, she was so upset that she took to her bed. Now the question is, can Ned be legally restrained from publishing?[1]

My latest worry, a small hard pimple inside my mouth. The tooth Dr. Kurtzman pulled out is still sore. My hand is still tender; I think it's getting better, though.

November 14. Very strong wind, nervous making. Hunt Stromberg called this morning, just back from Texas. Talked about how Oliver Reed[2] is interested in doing "Frankenstein" and Vanessa Redgrave maybe is, but the only cold fact is that Boris Sagal is definitely in, because Universal trusts him not to exceed the budget. Oh yes, and Hunt would now like us to do *Pride and Prejudice*! I called Robin French this morning and asked him to make a last-ditch attempt to interest them in Jim Bridges. But then, I don't know if Jim *really* wants to; I suspect that he might try to wriggle out at the last moment.

We rush about, seeing films; the festival is on. Yesterday, we dashed to Santa Barbara, arriving on campus to find that we had literally only three minutes in which to see a Charles Demuth[3] show. Then to Bill and Paul—but thanks to Bill's social instinct we had to have supper with dull old Wright Ludington, and I drank wine because I was so bored and now my weight is up to 150 or over, nude. The only satisfactory part of the trip was that I slept nearly all the way home and was simply amazed to find myself back in our carport, after what I'd thought was a brief doze.

The only good news, I'm getting along steadily with my work on the reconstruction of the 1945 journal. Sometimes that's really interesting.

The day before yesterday, we had supper with Robin and Jessie French. Their house is so dark and gloomy and Robin has such an air of resentment, when you see him at home. He appears to

[1] The book never appeared; see Glossary under Hoopes.

[2] British star (1938–1999) then at the height of his career after *Oliver!* (1968), *Women in Love* (1969), and *The Devils* (1971).

[3] American Modernist painter (1883–1935), trained in Philadelphia and Paris; he exhibited with Alfred Steiglitz's group at the American Place Gallery.

spend most of his home time fiddling with gadgets, taking them apart to see how they work—with a sort of sulky listlessness. They have two huge dogs; Robin makes rather a thing about their being dangerous, and he tells, with obvious satisfaction, how he was taken to court by the neighbors because the dogs bark so much in the night.

The young man who wrote *The Andromeda Strain* was there: Michael Crichton. He is six feet seven, fresh faced, obviously very intelligent and a bit arrogant. Jessie told me later she thinks he is Jewish, "You can tell by the way he moves his hands." (Don says this is the only interesting remark he has ever known Jessie to make!)

November 17. Yesterday morning I woke to find that the congestion in my lungs had got up into my throat; I had almost lost my voice. Saw Dr. Allen. He didn't seem much impressed but agreed that I had better not go to Claremont and give my talk about film writing, at the men's college. I showed him my hand and he didn't seem much impressed by that, either. He even implied a criticism of Dr. Ashworth's surgery, saying that he hadn't expected so much scar tissue. He evidently doesn't expect it will get much better. When I said that the thickening at the base of my little finger felt like a ring, he grinned and said, "You'd better get a diamond and have it set." But I like his good-humored cynical manner; it is much more cheering than worry and concern.

This morning I had a dream: Either Asaktananda or Vandanananda—definitely *not* Swami—was talking about Vivekananda and how he had switched from Jnana Yoga to Bhakti Yoga, and he added that this showed Vivekananda's "humility."[1] After Asaktananda or Vandanananda had left the room (it was in the Hollywood monastery) I turned to some of the monks who had been listening and said, "Do you notice how he always hits the target right in the middle—he's like a revolver—the way he said *humility*?" This was a vanity dream; I was pleased with myself for my analysis of the swami's statement; I felt that the others would never have noticed the significance of the word "humility" if I hadn't pointed it out to them.... Still and all, this was at least a dream about something religious. I doubt if I've had one in the past six months.

[1] Bhakti yoga, the path of love, is the most natural path to God; Jnana, the path of knowledge, relies on the intellectual power of the mind to discriminate between the real and the unreal.

This afternoon, Hunt tells us, Sheinberg is having a meeting with him and Boris Sagal. This means that Sagal is definitely in. Hunt says that he asked Sheinberg if I should come along and Sheinberg said no. "That's because he's in such terrible awe of you," Hunt said. "That's the way I want it," I said. Hunt also told me that photographs of possible locations for "Frankenstein" have arrived from England and that they are "as if the descriptions in the script had come to life."

December 4. Colin Wilson wrote a good notice in *The Spectator*, to add to the yes votes, and the Beesleys obviously think that I have had a triumph. But I don't, and the cold fact remains that the sales up to the 20th of November were only 1,815! Richard Simon tries to make this sound better by saying that the season has been bad and that there may be a pre-Christmas spurt.

My cough has dragged on ever since the last entry here. But I am due [to] give my talk at Claremont the day after tomorrow.

Am still worried about the little lump inside my lip.

Hunt has told us to rewrite the scenes of Agatha's death and the meeting of Polidor and the Creature. The Creature is to demand to have Agatha brought back to life. Hunt wants the change because it will strengthen the role of the Creature. The idea originally came from Jack Larson, who said he thought we should have stuck to Mary Shelley's version and had the Creature appeal to Frankenstein to create a wife for it. Hunt also wants us to do something to make Frankenstein play a more aggressive part at the end of the story. This because he has now begun casting, or trying to.

Yesterday we saw a young producer at Columbia, Henry Gellis, very square and serious. Gellis wants us to write a religious story for him—by which he means a story in which a very rich man gets disgusted by his riches and turns to God.

I am creeping along with the 1945 journal; sometimes I feel really fascinated by it, other times I have to force myself.

In the mornings I wake up and think of death—not with terror yet but with a heavy heart. Because there is my darling beside me and the prospect of our parting is painful beyond all words. Love is agony, but Proust is a shit for saying so the way he did. A saint is right to speak against attachment. But saints suffer too. Prousts just sneer and whine, so they have asthma instead.

A few mornings ago, I had a dream that there was a little white kitten in bed with us. And it could speak. *It spoke for both of us.* It was sort of a soul of our relationship, our guardian angel.

All this time I haven't seen Swami, because of my cough. I'm afraid he isn't getting any better. Now he's being treated for arthritis.

Bart Johnson called this afternoon, after nearly a year. He has now broken off his second marriage. He said, "I'm a loner" and "I'm a sexual mugwump."[1]

December 11. I saw Swami twice, this week. Both times, he seemed listless and said he felt giddy. But he did also say that he kept being aware of the presence of the disciples of Ramakrishna around him. At the vespers of the puja, two days ago, he came into the temple and bowed down before the relics and gave the devotees his blessing.

This morning, he called to ask me to come to both his birthday celebrations—on the 26th and on the 27th. The reason for this is that, this year, they are having two celebrations, the first for the women and the second for the men, on the two successive days; and Swami wants me to come on the women's day also because Amiya is arriving to stay. I'm sure he dreads her coming and all the hostility it will arouse; and I'm supposed to help cushion the shock!

It has been frightfully cold at nights, which makes the basket extra snug.

Hunt Stromberg left for Texas today and won't be back until after the New Year. Now he is all excited about our next project, "The Mummy"—it is taken for granted that "Frankenstein" is going ahead and that we shall spend months and months with him in England, beginning early in the spring. I don't even think about this yet. No use getting into a flap. Hunt believes we can have Brando as Polidor.

Dr. Allen says the pimple in my mouth is a varicose vein which can be burnt off if necessary. He had my chest x-rayed and couldn't find anything wrong. The cough continues.

Ted is crazy again and has lost his job. This is upsetting Don even more than usual, because he is beginning to foresee the time when Jess and Glade will be dead and he will have to take responsibility for Ted's welfare. He thinks that Ted's attacks will become more and more frequent and finally be his permanent condition.

December 16. Yesterday I saw Dr. Ashworth. He didn't seem at all discouraged by the condition of my hand, said that it will gradually loosen up. (His assistant remarked that this process sometimes

[1] I.e., someone who supports one party, then another.

takes as much as a year!) Dr. Ashworth has given me a splint which I am to wear at night on the little finger, to straighten it up. He says that it is not a question of straining the finger; in time it will yield even to a very slight but continued pressure.

On the evening of the 12th, Jon Voight and Marcheline Bertrand got married at a dreary furnished house in Brentwood which they have just bought. The ceremony demonstrated the psychological confusion of nowadays; one foot in squaresville, the other in hippiedom. Jon and Marcheline's folks were present, Jon wore a tux, Marcheline a veil, and they were married by a judge. But a lot of the guests were dressed "casual," including John Boorman (in a turtleneck) who had to give the eulogy (or whatever they call it) and who referred to marriage as "this battered institution"—a phrase which might well have been intended for his wife. His speech shocked the judge, who then spoke, insisting that marriage is a very serious matter indeed. Then came the "vows," which Jon and Marcheline had rewritten in would-be flower-child style. Jon forgot his lines altogether and finally had to excuse himself, exclaiming "actors!" Marcheline knew all her lines and said "I love you" in a tone of absolute firmness. It was obviously her wedding rather than his.

Later, John Boorman expressed possible interest in filming *A Single Man.* We encouraged him as much as we could.

That was the day Don finally shaved off his beard. I am so happy to get seven-eighths of his dear face back again; he still has a fairly thick moustache.

The day before yesterday we saw Gavin, for the first time since his return from Europe and New York. He has been having dizzy spells due to the effect of the cold on his low blood pressure. While we were alone together, Mark told me that Gavin became terribly upset in New York by fears that Mark would leave him if he became sick and couldn't "keep up." Don is afraid that Gavin is seriously sick, right now.

In London, Gavin had seen Tony Richardson and said he has become very conservative; he has got himself a girlfriend and is making a film written by John Mortimer.[1]

Last night we saw part of Peter Brook's *Lear* film with Scofield. Realism of this kind swallows up Shakespeare in a flood of galloping ponies, fur, snow, Danish extras and storm noises; it becomes all plot and no contents. What a showy Jewish ass Brook is—hardly

[1] Richardson directed Mortimer's stage adaptation of *I, Claudius*, but there was no film. The girlfriend was Grizelda Grimond (b. 1942). See Glossary.

better than Kazan. And how dull Scofield was—not to be compared with Laughton in the part.

We left it in the middle for a wonderful film about Australia, directed by Ted Kotcheff, called *Outback*. You are shown an inferno not run by devils but by thick-skinned, noisy beer-drinking men who drag you down into the mud with them and then behave with unexpected unsentimental group-decency. The film has made absolutely no impression here and is about to close.

December 26. We've had rain on and off for the past four days and more is expected, though the sun is shining this morning. On Christmas Eve, Don had a hideous experience. In the midst of a downpour, while he was sitting in his car at the stoplight at the bottom of the California Avenue incline, waiting to turn onto the Coast Highway, two teenagers whose brakes had gone out crashed into him from behind. He had to wait an hour for a tow truck, leave the car on a garage lot (all garages were already shut for the holidays), find a place to rent a car (nearly all cars were already rented and he could only get an aged Toyota). I was out at the gym and getting my hair cut, so didn't receive his phone calls. He arrived home drenched and in a state of delayed shock, took half a bottle of wine with him into his studio and relaxed. Nevertheless he pulled himself together marvellously and came out with me in the evening. We visited Dorothy Miller who was adorable as always, heard about Don's accident with delighted groans and then described her latest falls and injuries. She is so full of inward joy that her complaints seem purely camp.

From Dorothy's, we went on to a party given by Joe Goode and Mary Agnes Donoghue. Billy Al Bengston and Penny Little and Peter Alexander and a lot of other artists were there. We felt great warmth from Billy and Joe; they both kissed us. But the party got wearisome and I drank too much Greek wine and had a headache all yesterday. Also my back is bad again, probably from the damp.

Yesterday I saw Amiya, who arrived from England a few days ago. There had been much talk about her arthritis and I was surprised to see her looking so well, marvellously fresh faced for a woman nearly seventy. She is staying at the house of a rich youngish gay [. . .], and there was an all-male party in progress. Amiya loves being with "the boys"—her friendly condescension toward them and her conscious broad-mindedness is intensely offensive, but they are seldom offended; after all, she *is* a countess, and that still means a great deal—in Bel Air. Poor old thing, I am fond of her; she means very little harm. But, oh God, how she talks. She told me the whole history of the Sandwich family, about which she is writing an article for the *Reader's Digest.* She stresses the fact that she is (temporarily at least) the *only* Sandwich—now that her stepson's renounced the title and become plain Mr. Montagu, there can be no Earl of Sandwich as long as he's alive.

Last night we had Christmas dinner with Leslie Caron and Michael. Much talk from some European intellectuals about the "evil" in Stanley Kubrick's new film *A Clockwork Orange*. Don and I were disappointed in it and unimpressed by the violence. Polanski's *Repulsion* (which we saw for the second time, some days ago) did, on the other hand, seem frighteningly evil. You got the impression that he had dabbled in the Forbidden and become dangerously involved. [...]

No words can describe how happy I am with Don, just now. But oh how difficult it is to enjoy happiness in and for itself!

1972

January 1. We saw the New Year in at Jack's at the Beach, with Gavin and Mark. We started eating too early and had to sit too long afterwards, not really wanting to stay but not knowing where else to go. Mark, who wasn't even drinking wine, was bored and silent; we would have been much more at our ease without him. The hostess made us wear cardboard hats and gave us noisemakers. I asked Gavin to be my literary executor in the event that both Don and I are killed in an airplane crash or other disaster. I think he was pleased.

For several days now I have had a painful back, right down at the bottom of the spine. I hope it will get better before the 7th, which is the Vivekananda puja (early this year), or sitting in the shrine reading the Katha Upanishad will be misery.

Beautiful bright cold weather, with the sun going down yesterday and today into a hard-line blue ocean.

Don is writing a diary too, which makes me disinclined to write this. His will be far more interesting, because he's in the first flush of composition. He told me he worked two hours on it this morning. He will read this one day but I don't suppose I shall ever read his.

Jim Charlton showed up this week, on a short trip from Hawaii. He wears his hair rather long and keeps saying "right?" in the middle of his sentences. "I saw this man down at the bar, right?, and he told me we'd been at school together ... etc." He talks a great deal about Mark Cooper,[1] still: indeed he repeated word for word his description of meeting Mark which I heard when I was over in Honolulu with him, on my way back from Australia. I felt that he had called me chiefly to see if I would have him to the house and if Don would receive him; he still challenges you to reject him. I couldn't be at my ease with him. To relax toward Jim I have to be alone with him for at least one whole day. But I still feel some of that old charm. He's a bit like an aging actor, with all his tricks, including his much-too-deep "masculine" voice.

Two things about Christmas I forgot to mention. One is that Glade had a most alarming attack during dinner with Jess, Ted and Don—this was not so long before I arrived. She began to gasp and couldn't breathe and was terribly scared. Ted was so shocked that he became quite sane and concerned; but Glade absolutely refuses to see any doctor.... The other thing is that the kimono which I bought for Don in Japtown was a great success, which relieved me enormously; I had been afraid that he would be furious with me for paying so much money, nearly a hundred dollars. Our relations between then and now have been blissful, until this afternoon when I was unintentionally rude to Avilda Moses. I had called to tell her that she and Ed should come to us later for dinner, and I added—with what Don calls my fatal habit of supplying superfluous information—that Irving and Shirley Blum were going to be late, too. Don heard me and blew up, saying that I'd suggested that we couldn't endure their company alone, even for a short while, before the Blums arrived. You would never be rude like that, he continued, to one of *your* friends; and he said how utterly selfish I am.... But now I'm forgiven.

A man from Simon and Schuster's publicity department rang from New York to say they want me to come there at the end of this month and appear on the T.V. show called "Today" and maybe some other shows. He implied that this is because *Kathleen and Frank* is going to get good notices and be talked about; but I don't know how they know this. Still and all, I'm inclined to go, and so is Don. Simon and Schuster will pay my fare and expenses, and anyhow we can probably stay with Vera and Bob.

January 8. Yesterday was the Vivekananda puja and once again I read the Katha Upanishad in the shrine. Asaktananda did the worship but Swami was able to sit in on it. Don was there too. I got very moved, but it was a mixture of spiritual and theatrical

[1] Not his real name.

emotion, as usual. Still and all, it is a great day in my year, and I did feel a momentary slight brightening in the midst of the Cloud of Unknowing.

Asaktananda is very much in control, nowadays, and he isn't at all shy about it; he gives orders right and left. He was impatient because breakfast wasn't served quickly enough. Swami has now authorized him to give interviews. (Which reminds me that Larry Holt called me up in a flap a few days ago because he had read an announcement about the interviews and was afraid that this meant Asaktananda would also be allowed to initiate people. I asked Swami about this and Swami said no. But Swami *would* like Asaktananda to have that power as soon as he dies. Belur Math, however, decrees that only older swamis can initiate; and this would mean that some senior swami from another center would have to come to Vedanta Place and initiate a batch of devotees who were strangers to him.)

At breakfast, Asaktananda gave Don and me a cigarette each from the pack which had been used in the shrine. So we gave them to Jim Gates and Peter Schneider, who promptly smoked them. Peter's face is changing, from prettiness to homeliness. Don says it is the compulsive need of the Jew to be ugly. He is still as lively as ever, busy reading masterpieces for his college work. He is finding *Don Quixote* a terrible drag.

On the 5th, I had quite a long talk with Swami, who seems much better. (Yesterday, after breakfast, he was sleepy because he'd slept badly the night before.) I began asking him about his life. Had it been difficult? No, he said, it had been easy—except for the material difficulties when he first came to Hollywood and they had no money: Maharaj had taken care of everything. "I made only one big mistake in my life," he said. This was in Madras, while Maharaj was still alive. Maharaj was leaving the monastery there, to go elsewhere. He said to Swami, "I shall miss you." He had given Swami instructions to go to various places of pilgrimage and meditate there. But Swami feels that he ought to have told Maharaj, "I'm coming with you." He believes that Maharaj would have agreed, if only he had insisted, and that then they would have been together for the rest of Maharaj's life—more than a year. As it was, he never saw Maharaj again. Swami also believes that, if they had been together, Maharaj would have taught him some special spiritual disciplines which had been given to him by Ramakrishna and which he had never taught anyone else.

This conversation with Swami made me feel that I should have some sessions with him when he gets stronger, using a tape recorder, and asking him questions about his life, right from the beginning. He has agreed to this. But, of course, he would want to talk almost entirely about Maharaj. I want to find out about *him*. What fascinates me is the mystery of a vocation. What were the very first manifestations of it in Swami's early life.

Yesterday was also Jo's birthday. She came to supper with us, bringing her daughter Betty Arizu. We got to talking about Jim Charlton, whom she saw during his last visit, and Jo told Don, quite without being prompted, how much Jim admires him, how he spoke about Don's beauty and his great talent. I was so delighted that Jo said this, and I think it made at least some impression on Don. Jim is one of the people Don is hostile to, partly out of jealousy and partly because he feels Jim is jealous of *him* and bitches him. And yet he would really like to like Jim and be friends with him. Jim Gates is another of these. But, in the case of Jim Gates, I'm afraid he is right. Jim really is quite a bitch, a demure bitch. Such a pity, he would have made a charming grandson. Let's hope he grows out of it.

January 24. Tomorrow we leave for New York, where I'm to do interviews on T.V. and radio, to promote *Kathleen and Frank*. It has been a sick month. We've both had coughs, on and off, throughout it. And on January 14 my back was so bad I called in Neil Saker, an Israeli masseur from the gym, whom Gavin employs and recommends. Saker is very funny and quite a charmer and maybe even a first-class masseur; but he was unlucky with me, and turned the stiffness at the bottom of my spine into a really agonizing sciatica-type pain darting from my left buttock right down my left leg to the ankle. This has continued ever since. I tried treating it with acupuncture; went to a Japanese downtown, who gave me some funny little jabs all over but didn't alter the situation. So going to New York is in the nature of a desperate shock treatment.

Don has been such an angel as words can't describe. But Don, as usual, demands action. He too feels that kill-or-cure New York is better than sitting here waiting for the pain to stop.

Well, anyhow, our wills have been drawn up, by charming sad John Kerr,[1] so if we're killed on the plane Gavin will be reading this as our literary executor.

No news of Boorman's effort to become the director of

[1] American actor and lawyer (b. 1931), educated at Exeter and Harvard; he came from an acting family and appeared on Broadway and in films before studying law at UCLA. He still worked occasionally in T.V.

"Frankenstein" as a feature picture. The studio still hasn't replied to the letter he wrote [Lew] Wasserman.

Enough for now.

February 7. We got back here on the 4th. New York turned out to be much as I expected. So, I regret to say, did my behavior; I sulked. But there was some satisfaction in speaking out about my queerness on two T.V. shows. ("Today" and "Dick Cavett") a radio show (Arlene Francis) and a video tape (for Arthur Bell of *The Village Voice*). And we saw Wystan and Vera and Anita Loos. And Don was the greatest imaginable support—except for a few hours when I managed to fray through his patience.

My back and leg are as before, but my cough is almost gone, also a mysterious rash which broke out while we were in New York. And I feel good today because we got through most of the rewrite of the scene in "Frankenstein," when the Creature brings the body of Agatha to the old house. We can finish the rest of it tomorrow to give to Hunt, when we go to Universal to see three of the mummy films.

February 16. Yesterday we dashed up to the vespers of the Ramakrishna puja before going to Barbara Poe's show at the Rex Evans Gallery. Got touched by the relics—I get priority now right after the monastics, but the difficulty is to establish Don's right to follow me, so I don't have to wait for him—then saw Swami, who looked wonderful, quite his old self; that morning he'd initiated Anandaprana's daughter and someone else the morning before. He was phoning to Amiya, whose sister Joy has just tried to commit suicide, perhaps because she thinks she has cancer.[1] Amiya is still up at Carmel(?) with her but is coming down on the 21st to spend a month driving everybody crazy at Vedanta Place. Swami was trying to get her to hang up. Meanwhile Don and I knelt and took the dust of his feet and he put his hand on our heads and blessed us. Then we were sat down in front of him and made to take prasad, a big fat-making slice of rich cake each, and coffee for Don and tea for me and two tangerines each to take away. This morning I weighed 150, nude.

I forgot to say that Swami was still in the shrine when we arrived. As he left—Asaktananda was handling the relic ritual—he spoke to us all. He told how, in 1914, right after he had joined the monastery, he had been present at the Ramakrishna puja with

[1]Joy Palmerton survived to old age and enjoyed a second marriage to "Mitch" Mitchell, with whom she settled in Monterey.

Brahmananda, and Brahmananda had said that anybody, anywhere, who took the name of "my Thakur"[1] on that day would be liberated. Swami added that he was sure this was still true, and he made us all chant Ramakrishna's name. Later in his room he told us that this was the only occasion on which he had heard Brahmananda say "*my* Thakur."

As I was leaving the shrine after being touched by the relics, I was able to slip Jim Gates the draft-board notice (which they'd mailed to him at our house) saying that he was to be released from his service obligation next month. When that happens he'll immediately become a monk. So this was a sort of Ramakrishna birthday present for him.

This morning, Lenny Spigelgass called to ask me if I would help him with the Academy Award show, and, specifically, with the wording of the presentation of the special award to Charlie Chaplin. So I had to tell him the whole tale of the accusation that I peed on the Chaplins' sofa when drunk and how, for this reason or for some other which he never explained, Chaplin thereafter refused to see me.[2] Lenny agreed that we had better not take any chances, lest Chaplin should make a scene when he saw me and maybe refuse the award in an outburst of senile temper. How truly astounding the workings of karma are, sometimes! I'm sorry, because I felt really honored that I had been asked.

March 1. I forgot to record one important detail of the story Swami told in the shrine on February 15. During that puja in 1914, Brahmananda had a vision of Ramakrishna while a musician was playing the vina. So he asked [the] musician to move his instrument a little, because it was touching the robe of the figure which Brahmananda, but not the others, could see.

There is much to record, mostly confusion about "Frankenstein." Boorman has left and is in New York, on his way back home to Ireland. Hunt is now on his side, or says he is; Wasserman and Sheinberg haven't seen Boorman's film yet,[3] so don't have an opinion. And meanwhile, Warner's is trying to buy the project away from Universal!

The Phoenix Theatre in Leicester definitely wants to do *A Meeting by the River*, almost at once. They would start rehearsals

[1] I.e., Ramakrishna.

[2] In *Lost Years*, Isherwood urges that he did not pee on Chaplin's sofa and that Chaplin may have had some other reason for ending the friendship, around April 1950.

[3] *Deliverance*.

on April 17 and open from May 10 to May 27. But Bob Chetwyn (who called from London to tell me all this) won't be able to direct the play because he is tied up with BBC work. And now it seems unlikely that "Frankenstein" will take us over there that soon, if at all. So we'll probably have to say no. It is sad, but we can't afford to make the trip on our own. And a performance without Chetwyn and without us seems like suicide.

Don, poor angel, has a terrible cold. It infuriates him, because he hates to think of himself as the sort of person who catches colds or other infections. Usually, he is able to say that he got it from old Dobbin, so at least Dobbin is guilty. But not this time. So Dobbin has to sleep either in the studio or in his workroom, so he won't catch it—because that would be the last straw, having a cold yourself and having old Dub sneezing and coughing all over you.

March 11. Despite all precautions, old Drab got a cough which has grown worse and worse. Also there are now sciatic twinges in *both* his legs. It's really a bore—I long to get my health back and be able to run properly and take exercise again. I do go to the gym, but only jog round a single block and do very little else.

Poor old Jo now has to have an operation on her foot.

Don is going through a very bad patch with his work. Being Don, he keeps on working and doesn't moan about it. But I know he is miserable and I can't help.

When we were up at Vedanta Place on the 8th, Anandaprana asked me to help them do something to discourage "the personality cult" of Swami which is getting more and more rabid and ridiculous, according to her. Young girls weep whenever they see him, and people stand around outside his room, hoping to get a glimpse. Ananda wanted me to bring up the subject during question time, after the reading. She said all this in Swami's presence. Don, who was in the room too, told me later that he agreed with her; also that he felt Swami hadn't liked Ananda's attitude.

My own feelings are very mixed. On the one hand, of course, I can understand Ananda's point of view perfectly. Certain people have indulged in playacted devotion for Swami ever since I first came to the center and it has always nauseated me—especially when the pseudo-devotees were frumpy middle-aged women. Furthermore, Ananda is right in seeing a certain danger in the possibility that a cult of Swami might make things very difficult for Asaktananda after Swami's death; the act might have become too difficult to follow and the Vedanta Society might therefore lose most of its membership and fall apart. This happened at the

La Crescenta Center; after the death of Paramananda (or whatever his name was) the devotees refused to accept another swami from Belur Math to replace him, and so La Crescenta became a little shrinking cult group.[1] (Not that I seriously believe that could happen in our case, because we already have Asaktananda firmly established as second in command.) And then again, as Don pointed out, this *is* supposed to be a Vedanta Society—even the cult of Ramakrishna is secondary to the impersonal philosophical teaching, at least on our brochure. For Vedanta, in the ultimate sense, negates all personality cults.

On the other hand, I personally am a devotee of Swami first and a Vedantist second. I flatter myself that *my* devotion is, in the last analysis, not to Swami himself as Abanindra Nath Ghosh but as "the vessel through which Le Sacre passed," the living proof that spiritual enlightenment is possible. I flatter myself that I can see Swami as "the vessel" and also as my adorable but quite human and fallible little Bengali friend, and keep the two separate in my own mind. I can bow down to the God which is sometimes manifest in him and yet feel perfectly at ease with him, a minute later, on an ordinary social basis. My religion is what I glimpse of Swami's experience of religion. But I still firmly claim that it isn't a personality cult.

Nevertheless, it is very possible that I have encouraged others to practise a cult of Swami. It is all very easy for me, after thirty-some years, to be able to distinguish between The Guru and Abanindra Nath Ghosh, but, when I talk about Swami to others, I am nearly always talking about The Guru. Therefore I am apt to say, for example, that I believe (and I do believe it) that it is a tremendous privilege to set eyes on Swami even once and that a single meeting with him might have incalculable effects upon an individual in later life. And that's apparently cult talk; it's nearly certain to be misunderstood as such.

Ananda made a reference to "the Venice Group," with obvious hostility—and that too I can understand; from her point of view, such people are most unsuitable members of a congregation and scare away the respectable. I tend to be sentimental about "Venice" types, provided that they are cute and young and male. And I am particularly impressed by their instinctive understanding of *darshan*. Why *shouldn't* they want to sit and gaze at Swami, I feel, especially

[1] They followed Paramananda's charismatic niece, Gayatri Devi, thereby ending their official relationship with Belur Math; and after her death, they followed a longtime American nun, Sudha Ma. Prabhavananda and, later, Swami Swahananda tried—so far without success—to help them rejoin the Vedanta Society.

since they know they won't get many more opportunities of doing it? I don't know if Ananda blames me for bringing Jim Gates and Peter Schneider into the society. I doubt if she objects to Jim but she may well have taken against Peter. He *does*, more and more, seem to be playacting. I say "seem" because I do believe that he sincerely wants to be initiated—though that may be partly out of his Jewish competitiveness, to catch up with Jim; it's his methods of keeping himself in Swami's eye which appear like clowning. He is always lurking in the shadows, after dark, when I come out of Swami's room. Last time, when I saw who it was, I involuntarily said "Peter Quint!"[1] Also, lately, he has written Swami another of his letters—the most outrageous, so far:

You have smeared me with sandal paste
And splashed me with water
And laid me here on a silver plate with other flowers,
Unfamiliar and strange.

You didn't just uproot me, keeping me in some of my native soil so I could be planted again.
No; you broke my stem, so I'm separated from the earth,
And even if my broken stem were planted again
I would not grow, but droop and rot;
My original fragrance (!?!) and the sandal paste's, too,
Would mingle with the world's winds, leaving me
Raging ridiculously, thinking it must have been a wishful dream to ever have been on a silver plate.

But, why won't you offer me?
Have you only now seen my blemishes,
Eaten-away bad spots,
Or am I just not pretty enough?
Or have you finished your worship now,
Leaving a few unlucky unused flowers, having
Picked too many?
(I'm not surprised if you're ashamed to put me at His feet, and I know I'm closer to Him now than before you picked me.)

Please put me at His feet, before I wither and blacken myself with my jealous restlessness, and become
Too ugly at last to even stay on the waiting plate.

As a psychological exhibit, I find this enormously interesting. Even if it was written primarily as dialogue for the role Peter is

[1] In Henry James's "The Turn of the Screw," the dead valet who haunts Bly.

playing, two things in it seem to reveal more about him than he intended—the question "am I just not pretty enough?" and the phrase "jealous restlessness."

That reminds me of another demonstration of the personality cult which I took part in a few weeks ago, I forget which day. I had gone to Vedanta Place during the daytime when Swami was about to take one of his invalid walks; a few turns round the temple and then back to his room. So we all set out, Swami ahead, me a few paces behind, then Krishna, then about ten devotees who had been waiting around for this moment. After we'd started, Peter sidled up and got into step with me, and as we circulated he asked me if I thought he should concentrate on taking his degree at college or rather study subjects which interested him, even if they weren't part of the required credits.... Thus he established himself as a specially favored cultist, practically leading the procession. And the only person who remained isolated throughout the walk was Swami himself—an onlooker would have supposed that he was just too sacred to be talked to!

March 20. At 11 p.m. on the 16th, Ed Parone called to tell me that the Mark Taper has decided to do our *Meeting by the River* play. Jim is to direct it. And—wouldn't you know—this overlaps the period of rehearsal and the opening of Jack's Byron opera in New York! Jack got into a terrific flap about this, the other evening, and Jim told us he raved all that night and said he was leaving Jim and going off somewhere on his own, because he was being rejected by Jim and slighted by Virgil Thomson (who won't let him publish his version of the scenario, and won't allow him to have a queer scene with Lady Caroline Lamb). So Jim, who rather loves all this fuss, has sworn to dash back and forth between here and New York and hold Jack's hand and be at the opera opening, even if it means missing our rehearsals!

For a while, it looked as if we'd get Jon Voight to be in the play. Now he is backing down. Maybe that's just as well. He's what Jim calls "a talking actor," meaning that he has to discuss every move.

Am feeling hounded, right now. Letters, manuscripts, people who have written theses and want to interview me. (One of them, David Geherin, did a whole tape of conversation with me and then somehow goofed and erased half an hour of it—so I had to redo it over the phone![1]) Then Hunt keeps plaguing us with "The

[1] "An interview with Christopher Isherwood," *The Journal of Narrative Technique*, vol. 2, Sept. 1972, pp. 143–158, published at Eastern Michigan

Mummy." The only thing which doesn't budge is "Frankenstein." No glimpse of a director and those two farts, Wasserman and Sheinberg, are still in New York. Boorman is in Ireland, writing his own film,[1] and that seems to be that.

March 25. Hunt Stromberg called me yesterday, saying that he will never employ Boorman now because Boorman has lied to him. He has discovered that Boorman's commitment is and always has been firm, and that he can't possibly get out of it to do "Frankenstein," even if asked by Universal. We don't know whom to believe [. . .]. But [the] fact remains that Boorman is almost certainly out of the picture. Now Hunt talks about Polanski! He also fusses us to get something done on "The Mummy." He is going back to Texas today. If it weren't that we need money and would be silly to break with Hunt, I would say that we don't want to do "The Mummy." The whole subject bores me.

We are having a good deal of fun, auditioning people for our play. We now have a really sexy and amusing Tom (Gordon Hoban), two possible mothers, a good Rafferty (Jason Wingreen), Laurence Luckinbill for Patrick—except that he's [urging] us to take his wife, Robin Strasser, for Penelope, [though] we think we have a better one, Susan Brown. (Jim Bridges, who has gone off to Harvard to look over locations for a film he may be directing, will probably be seeing the Luckinbills on his way home.) The big problem is Oliver. Jon Voight mulls and ponders and keeps putting us off, but we're pretty sure he won't do it. The two other actors we have seen are wrong and utterly wrong, respectively.

Amiya left for England yesterday. I had quite a long talk with her, the day before—or rather, I assisted at a tape-recorded interview she was having with herself, recalling her memories of the society in the early days. She is such a blowsy drunken old bag and yet the thing she got from Swami is still apparent, nearly all of the time, and often she makes remarks of great perceptiveness. For example, she said that Swami doesn't love us, he lets us find the love that is in him, if we need it. (I've expressed this badly—can't remember how Amiya phrased it; but I think it is very true. And it explains why one cannot feel jealous in Swami's presence. Jealousy arises when a person sends his love out toward other people on a personal level; because, if he does that, he is bound to favor

University where Geherin became a professor of twentieth-century British and American fiction; he got his Ph.D. from Purdue in 1970.

[1] Probably *Zardoz*; see Glossary.

someone. Did Swami do this in the past? Yes, perhaps. I am thinking of Sarada.)

Jimmy Barnett (now called Sat) is going to help Jim as a technical director, with the music, costumes, etc. I want to keep this a secret from Swami and the others as long as possible, because of all the fuss and excitement there'll be, as soon as the news about the play leaks out.

April 2. He is risen, as I wrote to Dodie and Alec today, telling them about the forthcoming production of *Meeting*. And indeed this Easter does seem unusually cheerful, thanks to my happy life with Don, the nice prospect of rehearsals starting on Tuesday (though we'll get sick of them very soon, I realize) and the breakthrough, mostly thanks to Don, on "The Mummy." We can now almost glimpse a complete continuity, to be produced as soon as Universal pays us some money.

Hunt is in Texas. I greeted his arrival there by reading him another communication from Boorman, over the phone, in which Boorman repeats that he is ready to make himself free and do "Frankenstein" if ever or whenever he is made an offer. This sent Hunt into a flap. He asked me to read the letter to his agents. So I did. I also asked the agents to call Hunt back. I don't know if they have. We still can't figure out who is lying—Hunt or the agents or Boorman. Probably all of them, a little.

Last night, we had supper with Gavin and Mark. Gavin has definitely decided to leave California, probably to settle in Hawaii. His house is already up for sale, $85,000. His reasons: everything is so expensive here, and there are too many people, and he likes the idea of living on an island. If he and Mark get a picture to do, then they can come back and live in a motel. Gavin said approvingly that we were the only people who haven't been horrified by his decision. Well, we didn't express our horror, but we are horrified, rather. We both remember the glimpse we got of island loneliness, on our trip through in 1957. And I remember how isolated Jim Charlton seemed. It is very hard for me to imagine Gavin and Mark together there. But Mark has lived there before, for a number of years; and Gavin, as Don says, is very resourceful. He will be able to keep himself occupied.

April 7. Our four days of rehearsal have gone by quickly and already the play is coming together. I must say, it seems most awfully good. Even Don is greatly impressed by it. Jim Bridges directs in an easygoing but assured way. Sam Waterston has real power as Oliver and Larry Luckinbill is very funny. He shows signs of bitchery however and Don thinks he'll make trouble before we're through. Sam is a very nice boy and a really dedicated actor; his only danger is that he'll work himself too hard, he is playing in *Volpone* every night! Gordon Hoban is also nice and sexy and sweet, but a bit stupid. He can't grasp the moments of camp. Florida Friebus is really excellent as the mother—not great as Gladys Cooper would have been. And Susan Brown, though not at all Penelope, gets enough of the part to make everything work, except, I fear, in her final speech to Oliver. All in all, we got a cast far and away beyond what we might have expected, and we feel that the play will get as good a presentation as it could have, anywhere in America. Certain British nuances will be lacking, but there will be an American freedom of emotion which we might not be able to get in England. Only Gordon as Tom has a couple of hangups, he won't say "darling" at all, and it bothers the hell out of him to say "faggot"! But when he declares his love for Patrick he is true and moving—you feel something old-American, Whitmanesque.

Am reading right through Thomas Hardy's poems.

April 22. Jim Bridges got back from New York yesterday morning and we had a rehearsal on stage in the Mark Taper. There was bad feeling brewing from the beginning, probably because Jim went off to New York and left them, in the midst of rehearsals. Anyhow, no sooner had we started the run through than Jeremy Railton (who is the sloppiest and most inefficient art director)[1] came on the gallery at the back of the set with Donald Harris, the light designer, and a carpenter and proceeded to talk and take measurements. Larry Luckinbill objected. Jim ought to have thrown them off the set at once, but he didn't; so Larry's mood got nastier. And then he and Sam Waterston were playing their scene up on the gallery and they couldn't be heard and Jim told them so, and also called on me to confirm it. Whereupon Larry got really nasty and declared that his whole performance was ruined if he had to shout. (Actors!) There was quite a fuss. Jim seemed weak and upset by it, which alarmed us; we had been thinking of him as a tower of strength—but no doubt this was partly due to his exhaustion after the trip. I was careful not to mix in—lest I should

[1] Since then, winner of four Emmy Awards for production designs, and production designer for the opening and closing ceremonies of the 2002 Winter Olympics in Salt Lake City.

get annoyed and tell them What Every Author Knows: that the lines come first and that the most exquisite acting is useless if you can't hear them.

Four days before opening, and I do feel that, by and large, we are in good shape. Larry is very good—as right for the part, physically and psychologically, as any American actor could be; Sam ditto. (But Sam is a much nicer and more serious person—which is as it should be.) Gordon Hoban is nearly perfect in every way, and he no longer seems stupid. After Julie had played Sally Bowles, nearly all the other actresses who took the part seemed like whores. In the same way, the actors (if any) who follow Gordon will seem like hustlers. In the scene where he unbuttons his shirt, offering himself to Patrick, he is truly beautiful and noble. (Admittedly, I have a slight crush on him; so does Jim.) Florida Friebus is all right, though irritating and too strident at moments; Don and I both particularly dislike [the] way she trots across the stage with the photographs of Oliver as a swami. She's a bit of a bitch too, and she blows up on her lines. Poor Susan Brown is awful; she has a whining voice, and a disgusting middle-class sweetness. We fully expect, however, that the audience will like her. Jason Wingreen is perfect as Rafferty; he never varies, knows all his lines and can speak more clearly than anybody else. Logan Ramsey[1] isn't a swami, but he is a very good actor and is fat and has authority. Sirri Murad isn't a swami or an actor and you can't understand a word he says, because he is Turkish; but he looks very distinguished and is charming and courteous and friendly. The two friends of Tom, John Ritter[2] and Jack Bender,[3] are very sweet boys, and I think they both understand and appreciate the play better than any other member of the cast (with the possible exception of Sam); no doubt they have talent, too, but it isn't called for in their tiny parts. They were both in Jack Larson's *Cherry, Larry, Sandy, Doris, Jean, Paul*, at USC, a few years ago.

Speaking of Jack Larson, Jim talked to him this morning on the phone in New York and Jack says that the Byron opera got a bad notice in *The New York Times*, written by an enemy of Virgil. It said that the opera was too bland and that the libretto lacked wit. Jim commented bitterly that they didn't like it because it isn't "heavy with German-Jewish angst." He said he had realized himself that the opera wasn't a hit. Virgil is said to be philosophical,

[1] American character actor (1921–2000), mostly on T.V.

[2] T.V. star (1948–2003), later, in the 1980s series "Three's Company"; he had numerous subsequent stage and film roles before his early death.

[3] Actor and, later, T.V. director.

Jack is upset. Jim thinks the opera has now no longer a chance of being performed at the Met or elsewhere in New York.

Hunt Stromberg called a few days ago and said that "Frankenstein" is now declared a feature film—after Wasserman finally got around to reading it. We at once cabled Boorman, asking him was he free to direct it. Boorman hasn't replied yet.

On the 16th, at Jerry Lawrence's, there was a reading of Tennessee's *Two-Character Play*.[1] It was at least an hour too long, and Tennessee was drunk and bawled out Jerry for using the phone and Mark Andrews for yawning during the performance. And then he tried to strangle Oliver Evans and to throw him over the balcony. He came to see our rehearsal of act 1 and went out in the middle to pee and get a soft drink; then to see act 2 on another day and didn't arrive until it was nearly over. We are both fond of him but oh dear, he is tiresome.

April 30. The play opened on the 26th and now we have had four of our six public performances; the last two are today. It has gone very well—considering that Larry and Sam lose us a great many laughs and serious lines because of their poor diction, and considering that Hindu religion and homosexuality are still sticky subjects for many of the audience. The Hinduism bores some; the homosexuality makes many uncomfortable, though they are visibly reassured by Gordon Hoban's healthy appearance and grin and big teeth and masculine army-surplus clothes. (At the dress rehearsal, he played the first of his scenes naked to the waist, in pajama bottoms, but turned out surprisingly to have a bulky, rather unattractive body; so now he's in a T-shirt and jeans and looks charming.) Sam gets more and more wailing-wall Jewish in his scenes of spiritual anguish; what his soul needs is a stiff upper lip. Larry Luckinbill is marvellously right for Patrick, if only he could speak better. Susan gets better and better and the audiences like her a lot, but her voice is still horrible.

The theater has been packed every night, because there are so many regular subscribers, and this in itself helps the play a lot; it creates an atmosphere of success.

Don and I are already at work on rewrites, because we know that if we don't do them now we never will.

May 2. Only two days since it ended. It seems strange and sad, not to be going down every day to rehearsals (though it was a drag),

[1] Which he had revised and retitled as *Out Cry*; see Glossary.

not to be able to come in through the stage door, to be greeted by the doorman (who remembered me from the days of *Black Girl*), to stroll around backstage or sit on the steps during a performance, feeling you almost owned the theater. We had champagne with the cast after the evening show on Sunday. (We'd given each one of them a signed copy of one of my books, with one or two photographs of drawings of me by Don, pasted into them; an egomaniacal sort of present, but the only one we could think of.) The future didn't look bright for most of them; most were going back onto unemployment pay, for as long as it would last. Only Larry Luckinbill had an aura of prosperity. He was returning to New York with his wife and child and its nurse, with the prospect of various parts and the performance of something he has written. His coming out here had [been a] gesture in the grandest manner, an affirmation of his belief in our play, for it had cost him (he said) $3,500! Gordon Hoban seemed suddenly much older. He is the eldest of a family of eight; the only one of them who is an actor. And what are his chances? He thinks he may go to New York soon and try his luck there. The Clyde Ventura production of *Who's Afraid of Virginia Woolf?* has been cancelled, because Albee won't permit them to perform it with a male actor in the older woman's part.

May 5. The reviews in the *Hollywood Reporter*, the *Los Angeles Times* and the *Herald Examiner* are all condescending toward the play, saying that it needs a lot of work. The *Examiner* says the play "obscures itself in a murky drizzle of verbiage," but does call it an "otherwise important work." Larry and Sam and Florida are praised, but poor Gordon is described as "soggily out-of-sorts in the role of Tom." Dan Sullivan in the *Times*, after putting the play down for being "just a bit too pale for the theater" and its letter-reading technique for being "static and strained," adds, "even so this was one of the most civilized evenings New Theater for Now has given us, the sort of play about which one defiantly says: 'Well, I liked it.'" Jim feels that this won't ruin the play's chances of being put on again for a longer run at the Taper; but the worst of it is, we're in competition with the play which has followed us, *In a Fine Castle* by Derek Walcott—and that's so bad that Gordon Davidson is sure to prefer it, especially as it's being directed by Ed Parone.

Jim Bridges and Jack Larson are both of them eager to feud with Gavin, because I incautiously told Jim how Mark Andrews had admitted to us, without the least embarrassment, that he had talked to Sullivan, who was sitting next to him during the performance, and made it clear that he didn't like it. Mark calls this "being honest." Personally, I feel sure that Sullivan would have given us a lukewarm review under any circumstances.

Now I have a cold, which is probably my reaction to the strain of rehearsals and hopes. Weight just 150, nude. Don is sleeping out in the studio, so as not to catch cold too. Jim Bridges has one, and Jack has been in bed with one ever since his return from New York. Virgil Thomson arrives here today or tomorrow. Apparently they no longer regard the Byron opera as a flop, but I don't know what its prospects are.

Jim Gates moved into the Hollywood monastery and officially started his life as a monk, on the day our play opened!

The day before yesterday, when we saw Swami, he told us that he had been unable to sleep the other night so he had made japam for a while—and discovered later that he had been doing it for two and a half hours! (He says he never uses his beads now, because he can't sit up properly with his legs crossed.)

May 23. Now we're back on "Frankenstein" again, cutting it so it can be budgeted—after which Hunt Stromberg will condescend to return from Texas, and a director will be found and casting will begin. What we fear is that, with all this time lost, we shall end by losing not only Boorman (who still holds out some dim hopes), Jim Bridges (who may well have to take on another directing job), and Jon Voight (who may well get a part in another picture). And who knows, we may even end up with Boris Sagal!

No news of any future for our play. A copy has gone to the Royal Shakespeare Company. And Gordon Davidson is supposed to be considering if he'll do it again here. But at least we do have a rewritten version which we are very pleased with—so much so that we hate to think it wasn't the one that was performed. We finished the rewritten version on the 18th, Don's birthday, so that seems a good omen for it. And indeed Don said that he had enjoyed this birthday of work much more than most of his birthdays of celebration. In the evening we went to St. Germain for supper, inviting Billy Al Bengston, Penny Little, Joe Goode, Mary Agnes Donoghue and Mike Van Horn. The super-Frog food made a great impression upon all, as the bill did upon us; it was $135. But it seemingly poisoned both Don and Mary Agnes; anyhow they were both vilely sick for several days, vomiting and shitting water. Don, poor darling angel, even shit in the bed several times. Don had had sweetbreads and had given some of his to Mary Agnes to taste. However, I had sweetbreads too and didn't feel a thing.

On the 15th, I talked to John Rechy's class at Occidental College; such a sleepy old-world oasis, I'd never visited it before in all these years, though Aldous and Gerald used to lecture there often. One of the students was a cute English boy from Manchester, so of course I told him that was practically my hometown too; and when the talk was over, he mumbled bashfully to John Rechy, "He made me feel very proud." (To balance this bouquet, I must record a brickbat, thrown by someone named Richard Toscan, reviewing our play on KPFK; he referred to me as a "has-been novelist," sneered at me for contributing to New Theater for Now at the age of sixty-eight, and said I ought to write a gay series for T.V., as that was all I was good for.)

Next evening, the 16th, I was given an award by the Hollywood Authors' Club, for "A lifetime of distinguished contribution to literature"; it is a cast of a statuette by Henry Lion, called "The Book," and shows a nude woman apparently overcome with disgust after reading what looks more like a filmscript open in front of her—or maybe it has just sent her to sleep. The award was engineered by faithful little Fred Shroyer, bless him. But it came only at the end of a four-hour evening, spent mostly in celebrating the fiftieth anniversary of the club and the one hundredth anniversary of the birth of its founder, Rupert Hughes, that nasty old horror who said that "we send a gangster to the electric chair but we do not treat these pacifists as traitors" (July 5, 1940).[1] He was described as "Christlike" and "almost a saint." The speakers also bragged about the old days of the Authors' Club, when only men were admitted and the conversation was "ribald and raunchy." The one bright spot in the evening was when a cute seventeen-year-old violinist appeared, his name is Endre Balogh. But he only played stunt pieces, very fast, with buzzy noises and saw strokes.

Saw Mrs. Maltin at last, on May 13. Felt awfully sorry for her, but I had to ask her for all of our tax papers. We were very polite. I also had to tell her that a man from the District Attorney's office had called and asked about Arnold, who has apparently been reported for fraud of some kind. This didn't seem to worry her unduly. She is faithful unto death; her huge ass seems somehow to express this. She said quite seriously of Arnold that he was the most sensitive man she has ever met. She obviously accepts his view of himself as a victim of society.

On May 19, Truman Capote called, from Long Island. He just

[1] Hughes, a writer and film director who had served in the military, was speaking at the Rotary Club; Isherwood recorded this in *D.1*.

wanted to talk. He was very bright and friendly. But he told me that he had had two operations for cancer in the rectum; he had thought it was piles for a long time. The pain had been terrible, and then, after the operations, he had a nervous breakdown and had to go to a clinic in Switzerland for three months. I had just the slightest impression that this was a signal for help. Maybe he is sending out a lot of them. It is peculiarly shocking to see this terrible thing happen to a whiz kid of success. I wanted to take him in my arms and somehow reassure him. And there was absolutely nothing to say—at least, not over the phone, not to Truman. I am glad that he has Jack Dunphy with him.

June 3. Mary Lazar, whom we saw at a party on the 28th, given by Jennifer and Norton Simon, made light of Truman's cancer to Don, saying it was nothing at all serious. (Is there such a thing as an unserious cancer? The Lazars are in disfavor slightly, anyhow, because we invited them to the dress rehearsal of our play and they excused themselves at the last moment through a secretary.)

The Simon party was given for Lauren Bacall; it was also a first-anniversary celebration of their wedding. Their marriage seems to be going well and I began to like Norton; he does seem to be sensible politically, and being sensible is all one can expect of Republicans. Bacall was terrific—like a campy alligator, very like Beatrix Lehmann at her funniest, but not so bitter. Senator McGovern[1] also showed up—thus causing the house to be immediately surrounded by secret service men. He was probably exhausted. He thrilled none of us but did his best to be pleasant. He's all we've got to vote for—that was the attitude of the Democrats present. I'm sure we can't win, anyway.

The "Frankenstein" script is now cut down from 199 and ½ pages to 139 and ½, which is exactly what Hunt required. I think we did a very neat job. Now there is no excuse left for Hunt—he must leave Texas and come back here and do something about getting a director. We have decided to start work again on "The Mummy" the day after tomorrow, but at slow speed.

Anecdotes and miscellaneous information:

John Gielgud told us this story about Mae West. She was asked, "Do you ever smoke after you've had sex?" She answered, "I never looked."

Tennessee (who's here again) told us that his sweet but

[1] George McGovern (b. 1922), Senator for South Dakota, then campaigning for the Democratic presidential nomination.

simple-minded friend Victor went with him to Italy but hated it. The first morning, Victor went out on the street among the Italians and came back at once, exclaiming in dismay, "I can't understand what they're saying!"

Both Virginia Pfeiffer and Gavin have been bitten by black widow spiders. Neither one of them felt the bite. But Virginia's leg swelled up terribly and she was in bed for four weeks. Gavin's symptoms were much less violent. His doctor gave him cortisone and vitamin C.

Swami told us, the day before yesterday, that several people ask him for a Brahman mantram when they are initiated, but that they always want to exchange it later for a mantram containing the name of a chosen ideal—the Impersonal God proves unsatisfactory for meditation. Three of the monks have Christ mantras.

June 16. An almost incredible absence of news. Nothing from Universal, or from Hunt, or from Ed Parone in London, or from anybody in New York. Meanwhile, we keep plodding on with "The Mummy"—now provisionally called "The Lady from the Land of the Dead," a title I think I like very much, though Don has doubts. It does sound a bit like an operetta.

My left-hand little finger, after getting straighter, is now curling over. I daren't show it to Dr. Ashworth, because he said, "If you don't wear it in a splint at nights, don't come back," and I haven't been wearing it in the splint because the splint keeps coming off.

Have just been out shopping for a Father's Day present for Swami. Got a silk scarf I'm not sure about; on these occasions, one takes refuge in spending a lot of money, $18.38.

Robin French just called to say how much he likes the cut version of "Frankenstein." Word has lately got around that Jessie is pregnant again. [...]

Paul Millard called me this morning to say that he wants me to ask his friend Rob Matteson to photograph me *and* to offer to pay for the pictures. This is because Rob is very insecure—although a marvellous photographer, according to Paul. Because of his insecurity, Paul says, they split up for a while and are now back together again on trial. If Rob's confidence isn't restored, Paul fears Rob will leave him again. I agreed, of course, and yet the whole thing irritates me a bit. It seems such a cheap trick to play on Rob, and, if he ever does get to know about it, he almost certainly *will* leave Paul. I told Don I felt inclined to praise the pictures to the skies and then say all my friends must have them and order six dozen. Because Paul has assured me that he'll pay for the pictures, *no matter what they cost.* (I couldn't resist telling Paul that I *never* pay for photographs taken of me; which is true.)

When I saw Swami the day before yesterday, I asked him if he had had any dreams about Maharaj lately. He said smiling, "No, I think he is waiting till I die." Swami got annoyed with Anandaprana, who had said at the board meeting, "Why don't we sell some of the Trabuco land, so we can start the new buildings for the nuns?" Swami is determined not to do this; he feels it would be a betrayal of Gerald's memory. (Swami always thinks of Trabuco as being Gerald's *personal* gift, which, technically, it wasn't.) Ananda, whiny voiced but obstinate, told me, "I can't help thinking of the poor girls, living next to that terrible freeway noise, and the boys at Trabuco have far more room than they need." Swami thinks they can raise the money by getting a property-tax exemption which, in its turn, will release some of their invested capital to be used for building. They want to reconstruct the entire Hollywood center.

According to Swami, Jim Gates is getting on very well at the monastery. He works at the bookshop and everybody likes him. But Peter Schneider is out of favor. Swami thinks him lazy. He is being given money to look after the apartment-house garden, and he doesn't. Also, Swami doubts that he really wants to become a monk.

June 24. Dennis Altman just sent me this from Australia. It appeared among the want ads in a Sydney newspaper:

"Single man, an Isherwood fan, seeks another, 28 plus, to share snug Sydney northside flat."

Paul Shaw, our new tax accountant, is arranging for us to share some of the money and thus avoid death taxes. At present, I feel boundless confidence in him and imagine we'd now be rich if we'd met him ten years ago.

Nothing will budge Hunt from Texas, it seems. We hear that Sheinberg is shopping around for a cheap television director to do "Frankenstein."

I dreamed that I was walking with Stravinsky on the seashore, talking about my coming death.

The worst time of the day or night for me at present is immediately after waking, because Don insists on having the alarm ring, again and again from 6:30 onward. We are supposed to get up in turns and stop it. But I always am the first to do this, and it wakes me up, and I get back into bed and worry. But another of our basic disagreements, about Don's fast driving, has had the sting taken out of it, because I now lie down on the back seat of the car. I rather enjoy this, except that it is so uncomfortable, because from that position you can see almost nothing but tree-tops, only occasionally a tall building.

Michael York is really quite adorable and incredibly young looking for thirty. (We went to a party there, last night.) When he sits down beside you, he projects a very strong feminine appeal. We think him capable of deep resentments, beneath his beautiful manners. (As we were leaving, I said I liked to leave a party without saying general goodbyes, and he said, "You're just like me," and then corrected himself, "I mean, I'm just like you." Who else in the entire Los Angeles area would have said that?) He kisses us in public, too. Pat his wife is a power lady, pop-eyed with ambition.

Martin Hensler told Don that John Gielgud often begins to cry when he thinks about his death. Martin was furious yesterday because Truman Capote had said in an interview that John was stupid. He wanted to write Truman an abusive letter.

July 1. Don keeps drawing Michael York but can't get a good likeness. He has done several good portraits of Pat.

Ed Parone has returned from London. He hasn't called us, but he told Jim Bridges that he had no luck there with our play; perhaps because it has been overexposed by Clement Scott Gilbert. Several people said they weren't even interested in reading the new version. However there is a very favorable notice in July's issue of *After Dark*, complete with a picture of Larry and Sam in their parts. Gordon Hoban is also praised, which pleases me because he got slighted in the local notices. The *After Dark* critic, Viola Hegyi Swisher[1] (can this be a joke name invented by a drag queen?) writes: "The New Theater for Now season struck a beautiful opening chord, under Jim Bridges' direction."

Last night, we went with Lauren Bacall to hear Lee Roy Reams sing at the Little Club. He's in *Applause* with her and she is trying to promote his career as a night-club singer. He has a nice voice but is hopelessly shy and a nonwinner. Bacall is all power, but we both like her. Got a headache from drinking half a bottle of wine—so low is my tolerance, nowadays—so went down to the beach and dunked it in the ocean. Don drank two vodkas and felt awful.

Yesterday, I took $33,000 out of three of our savings accounts, and Don deposited the money in his personal checking account. Now I have to live three years (I believe) or else Don will have to

[1] Los Angeles dance critic who regularly reviewed ballet, musicals, and films.

pay death duty on it. Mr. Shaw is very disapproving of the joint tenancy grant deed—which Albert Spar drew up, years ago. He fears it will land us in trouble, later. But Don absolutely refuses to spend money on consulting a lawyer to find out how we can annul it. He bitterly resents the fact, and so do I, that I shall have to pay $3,170 state gift tax on the $33,000.

For the moment—or to be more exact, for the past week—we are stuck in "The Mummy." Can't think how to end it.

July 2. This morning we sort of fudged an ending to "The Mummy." We now think we'll sketch it out as quickly as we can and then throw the whole burden onto Hunt, if he ever comes back. Mummy talk has become a terrible time waste.

Meanwhile, I get on very slowly but not too badly with my reconstructed 19[4]7[1] diary. There's a lot of good material in it.

Bad news/good news, from the *New York Mattachine Times* of May–June:

> After three months, the new penal code enacted by the Idaho legislature, legalizing homosexual activity between persons of sixteen years and older, has been repealed. Now they are back to the old law: "Every person guilty of the infamous crime against nature is punishable by imprisonment for not less than five years."
>
> A new criminal code for Hawaii, effective next January, contains no penalties for solicitation and lewd conduct and makes fourteen the age of consent for "deviate sexual conduct."
>
> Other relatively free states are now Colorado, Connecticut, Illinois and Oregon.

July 3. The usual holiday jitters. How one longs for them to be over! Last night we had supper with Stathis Orphanos and his friend Ralph Sylvester. Stathis is a sweet boy, a good person, but his photography takes longer than anyone else's and tomorrow he plans to spend most of the day photographing Don—who will then later be photographed by Pat York! Luckily, Stathis now has nearly all the desirable editions of all my books, so I don't have any more to sign for him.

This morning a letter came from James M. Foster who is the only gay delegate from California to the Democratic National

[1] Isherwood typed 1957.

Convention at Miami Beach this month. He is also a friend of Troy Perry's. He tells me they have set up a foundation called the Isherwood-Radclyffe Foundation which is to lobby for gay rights, etc. The women on the committee which set it up voted for "Radclyffe" in memory of Radclyffe Hall; the men voted for me! I'm going to talk them out of this, if I can. Foster is up in San Francisco and I haven't been able to reach him yet.[1]

July 4. Both yesterday and today we were in the ocean, big waves and beautiful, but still very cold. We got away before the big beach crowd this morning and I have spent the day working on my 1947 diary, while Stathis takes photographs of Don. He has already spent three hours on this and shows no signs of having finished. Don groaned to me, as he changed clothes for the third or fourth time, "This'll punish me for all the times I've complained because I wasn't photographed!"

Nothing will get my weight down; it fluctuates around 149–151.

Last night we saw Truman Capote's friend Rick Brown. He is living in Hollywood, sharing an apartment with a boy named Michael who got stabbed in the buttocks on a parking lot by a guy who followed him out of a bar because the guy thought Michael was making a pass at his girlfriend. When the police arrived, they checked up on Michael and found he had never appeared to answer a drunk-driving charge, some while ago. So, disregarding his wounds, they shut him up for the whole of one day in the tank before they let him lie down on a hospital bed. Now his wounds are infected. A big nice-looking good-natured boy who could easily be terrifically sexy, except that he's a hopeless slob. Both he and Rick want to be actors. They don't appear to have a chance in a million. But then, how much chance does someone with real talent, like Gordon Hoban, have?

July 5. A singularly pleasant harmonious dinner party last night. Mrs. Leavitt cooked and we had Michael and Pat York, Billy Al Bengston and Penny Little, Jack Larson and Jim Bridges. Michael got good marks again, he was charming to Penny and suitably macho (in a restrained British way) with Billy Al. Pat came on strong about Don's drawings and paintings, compelling him to bring out two albums, which everybody looked at, even Billy. And

[1] Foster (1934–1990), founded the lesbian and gay political group called the Alice B. Toklas Memorial Democratic Club and was being paid to organize what was eventually called the Whitman-Radclyffe Foundation.

then Jack and Jim came on even stronger with their new craze for skiing, which pleased Pat. And then we had the Fourth of July fireworks with our dessert, not so plentiful as usual, but enough.

Don has done three very interesting paintings, painted over drawings.

This morning we really have more or less found an ending for "The Mummy."

My weight, just a fraction under 150.

Gluttony; after a marvellous run down to the beach and plunge into the ocean, I grabbed all the remaining cookies from last night, three cherries, a slice off the sweetmeat which Ralph, Stathis's friend, made for us and three Bath Oliver biscuits, two of them spread with (very old and stale) honey, one of them with cheese!

July 6. Woke up this morning feeling very strongly that I must not allow my name to be put on to this foundation. If it remains there, it will compromise me; I shall have all the disadvantages of being dead without any of the advantages. I'll be like a president of the Soviet Union. I called Foster and at once felt that my hunch had been correct. I realize that they only voted for me because they think of me as a stuffed limey. And when I told Foster, he was horrified, because they have already sent the papers to Sacramento in my name—they did that before they even asked me! Foster immediately said that I must talk to his colleague, David Goodstein(?), who's a lawyer[1]—in other words, someone who would be able to bully or wheedle me into changing my mind. Oh, they little know Dobbin! I do hope I shan't have to go to law about it.

Last night we saw *Morocco* with Gavin—what a noble masterpiece of high camp it is, and how incredibly Dietrich rises above her dumpy kraut figure! Once again, Gavin showed up without Mark. We are now nearly sure that he doesn't want to see Don or me again before they leave for France. Gavin totally uncommunicative about his home life, as usual. But neither Don nor I can believe that Mark will want to stay long in France.

July 7. Leslie Laughlin doesn't believe Mark will want to stay in France, either, though she admits it's just possible he might make a hit in French films as "a tall Texan beauty." (For some reason, Leslie had supposed Mark was from Texas.) We had supper with her last night—alone, because Michael was down in La Jolla,

[1] Goodstein hired Foster to run the foundation; see Glossary under Goodstein.

selling the idea of a hotel project in Corsica to some money man; Leslie says Michael is an amazing salesman. This dinner was a delayed birthday dinner for her and it was a bit of a bore, because Leslie has to pretend to be interested in things like my visit to China or my friendship with Stravinsky, instead of getting down to some good relaxed gossip. We noticed how her face twitches. But how adorable her face still is, such sweet little-girl prettiness. And she does try hard—it's almost heartbreaking, sometimes.

This morning, James Foster's colleague, David Goodstein, called me. He didn't attempt any pressure, said that if that was what I wanted, they would take my name off the foundation—they might not be able to do this at once, but they promised never to use it. So that's that. We were quite pleasant and I got the feeling that he is sensible and efficient. It seems that he and the other San Francisco organizers take a dim view of the leaders of the gay movement down here. He blamed it on them that I hadn't been informed properly before they had sent the documents to Sacramento, so that I could have taken my name off then. This may have been just an alibi, but I don't think so.

On the beach this morning and yesterday and every day this week, in fact. Yesterday, I heard a queen bragging about the man she'd met who wore a chain round his neck: "He took it off and put it around the two of us—*well*—I'd never been through *that* before, and let me tell *you*—!" Nearly everybody has big whiskers, most have moustaches, many have beards. I yearn for a new Razor Age. Never mind, the water is getting warmer at last.

July 8. Last night I called Hunt. He was drunk. He assured me that all is well. "Just let me play the game in my own way. You'll see, we're going to get Boorman.... But, please, *do* finish 'The Mummy' as soon as you can—that's all part of it—" This time, I didn't even heckle him, it seems so hopeless. Meanwhile, Jon Voight has written Sid Sheinberg a very pompous letter, saying that he won't do "Frankenstein" if it's going to be for television, and continuing: "Had you suggested a strong director I would have done all I could to accommodate him and try to make it possible to include the very best actors for the project ... so it is all the more distressing to me that I must terminate my interest at this time...." He ends by saying that he would like to buy the script if Universal ever wants to sell it. He's a good boy and a loyal ally, but Jesus what a stuffed shirt he'll be when he's older!

Also last night, I talked to Bob Ennis and found out that his house was burned down last December. He lost nearly everything in it, including his dog. But Don's drawing and painting of him were saved because they were out in the hallway. Troy Perry told him the burning of the house was a blessing in disguise, because he had gotten himself married to it!

July 9. Last night, Rick Brown called with a message from Truman Capote; Truman had told him to tell me that Tony Bower has been murdered. I had heard this already—from Paul Sorel, of all people—and had been meaning to mention it here as soon as I knew more details. But Rick told me more or less the same story as Sorel did. Tony was found sitting in a chair in his apartment with two bullet holes in his chest. Both Paul and Rick take it for granted that this was done by a hustler. It is "shocking" as they say, but I'm not particularly shocked. Surely such a death is lucky—like a sudden unexpected fatal heart attack—when you compare it with deaths by painful disease, long weary sufferings in a hospital? Of course my objective attitude is due to the fact that I hadn't seen Tony in a long while before this and that we'd anyhow lost touch with each other. There was a time when he had a crush on me and a time when I wanted (not violently) to go to bed with him. We never did. And I never felt more for him than the affection which goes with having known someone for more than thirty years. I always thought of him as being very unhappy, underneath a show of cocktail-party high spirits, and of being a bitch because he felt afraid of nearly everybody. I was fondest of him during the war because I identified with the miserable time he had in boot camp. I made him one of my characters (Ronny) in "Paul,"[1] not merely because he had known Denny but because he seemed to me to be a perfect specimen of the denier, the person who is roused to fury by any mention of the spiritual life and those who try to practise it. The denier's fury is caused by fear and Tony was full of it. There was something childlike about him. Poor Tony—the more I think about him, the more I feel my fondness for him coming back. But I still say he was comparatively lucky.

There is a very good notice of our play in the July issue of *Coast*, by Sam Eisenstein.

This morning we went with Mike Van Horn to a studio show of Billy Al Bengston's work. We would like to have bought one of his paintings or even one of his tables, as a gesture of friendship, but they are so expensive; hardly anything under $2,000. Billy and

[1] Fourth part of *Down There on a Visit*.

Penny put on a wonderful breakfast snack of Chinese cookies and puddingy cakes; they both wore Chinese hats. Whenever I go to his place, I always admire the style he lives in; it is all contrived, down to the last detail, as artificial as a palace but with fun—that's the difference between him and, say, Tony Duquette.

July 10. We went on the beach again this morning and in the water, although the ocean was dirty and the sky full of thin fog which kept breaking open and closing in. We met Bill Bopp, grey haired but with the figure of a muscular young man. Don says he still finds Bill attractive but can't forget what a bore he was to be with. He really is a dog person in the best sense, so friendly and willing to be amused and eager for companionship.

Last night, John Gielgud and Martin Hensler had a goodbye party; they leave the day after tomorrow, as John's part in the picture is finished.[1] Bumble Dawson the stage designer is staying with them, a sick old woman in agonies of arthritis with a huge bent heavy-bellied body who nevertheless manages to create an effect of extreme elegance by wearing beautiful full robes which cover her right down to the ground. We both like her. Don has done a subtle sympathetic drawing of her which Martin says he wants to buy; but now Don is worried because he asked $200 for it and feels that was too much. Sweet kissy Michael York was there and Pat, who has become intense and potentially tiresome since reading *An Approach to Vedanta*. Also Leslie, being bitched by [her] Michael; he really does behave like a petulant child. And Roddy McDowall, with his curly-haired sexy boyfriend Paul Anderson. We both think Paul is terrifically cute. He keeps looking adoringly at Roddy. Don says he's stagestruck.

July 11. Don had a bad day. His sitter in the morning—an actress we had met with Jack and Jim—showed up and calmly announced she could only stay one hour! And Leslie, who had promised to sit in the afternoon, called up *two hours late* to say she wasn't coming! I went to the beach alone but unfortunately ran into Jess Flood, who is going to Stella Adler's acting school because they told him in France or Italy that he really must learn to act a tiny bit before being in any more of their pictures. He's a nice boy but I would have enjoyed the ocean much more without him.

In the afternoon I spent two hours telling about Huxley to a

[1] *Lost Horizon* (1973). Michael York was also in it.

young man named Jeff Sadler who just hadn't done his homework.[1]

We went later to say goodbye to John Gielgud and Martin Hensler and Bumble Dawson, and that was sad. My opinion of Bumble rose even higher because she suggested getting a London gallery for Don to exhibit in and because she ran out after us to thank him extra, having suddenly realized that she was getting one [of] his actual drawings of her as a Christmas present from John and Martin, and not just a photograph of it. Martin paid Don $200 for the drawing, in cash!

July 12. Last night we saw a saddening Japanese film called *Lake of Dracula*—saddening because every glimpse of modern Japanese life in it seemed so hopelessly and drably Americanized.

This morning, after doing our stint on "The Mummy," we decided to tape an interview with Don about his work methods and his attitude to his sitters, to be used for a magazine article. We were both amazed to find how alike our voices sounded. If bits of this tape had been played to me, I couldn't have been certain who was speaking.

July 13. So sad to read in *Newsweek* that Brandon de Wilde,[2] that beautiful boy, was killed on July 6 in a car accident.

Last night we had supper with Gavin, after seeing Ozu's film *The End of Summer*—so clear, so haunting, so full of love and noble irony, and so unhurried that you have to slow down for the first half hour in order to get in step with it. We ate on the pier at Sinbad's, which is one of the snuggest places I know on a foggy night, with all the weird pier folk wandering past the windows. Lenny[3] doesn't work there any more. And before long it will be closed and the pier torn down, condemned for being old, like Dobbin.

Gavin came back with us later and astonished us by asking our advice on "a very delicate subject"; he said it was like Hitchcock's *Suspicion* (or was it *Shadow of a Doubt*?)—well, anyhow, he is nearly

[1] He submitted a doctoral thesis, "The Politics of the Margin: Aldous Huxley's Quest for Peace," to the University of Wisconsin English department in 1973.

[2] Child actor (b. 1942), he debuted at nine on Broadway in *The Member of the Wedding* and starred in the 1952 film; later movies include *Shane* (1953) and *Hud* (1963). He also appeared in Speed Lamkin's short-running play *Comes a Day*.

[3] Lenn*ie* Newman, a boyfriend of Jay de Laval and assistant chef in his restaurant during the 1940s; he appears in *Lost Years*.

certain that Mark is stealing from him, and what shall he do about it? It turned out that he has a place where he keeps money and he has been noticing that some of it was missing; and then he happened to look in a drawer in Mark's room (for an address book) and there were some dollar bills and he knew that Mark didn't have that much. Then came the real surprise—as far as Don and I were concerned; the bills in question were tens, just a couple of them, I think. So I asked tactfully how did Mark get money and Gavin said, "Oh, I give it him when he needs it." Then I had to get very very tactful and suggest that perhaps it embarrasses Mark to have to ask for money in this way (I didn't of course add, and for such very small sums) and wouldn't it perhaps be better to give him a lump sum (I told him you can give $3,000 a year without paying gift tax) and thus make things easier for him. Gavin thought this was a good idea and is going to act on it, I guess. But what an odd light it cast on their life together! (Don reminded me that I used to say Gavin was a miser at heart.) How odd, too, that we should be in the position of defending Mark's behavior!

July 14. Saw Swami last night. He is all excited because Mrs. (Xerox) Carlson has given the society $25,000.[1] With that and other funds, the work on the girls' dormitory can definitely go ahead. Swami and Pavitrananda leave for their Malibu holiday tomorrow; Pavitrananda is thinner than ever, but seems fairly well.

Swami says that Gerald, just before he got sick, had a dream that he was at Dakshineswar, and that Ramakrishna pointed to him and said, "He belongs to me." Can't remember if I've heard this before. (Looking back through this diary, I came on it, quite by chance: August 4, 1971!)

Later in the evening, we sat up watching McGovern and Thomas Eagleton making their acceptance speeches.[2] Both were fluent and Eagleton was even quite witty and youngish and full of energy. But, oh dear, they do not for one minute look like winners. And all the groovy young people, who'd been working their asses off and who were cheering and vowing victory in November, had the heroic pathos of a lost cause. When Ted Kennedy appeared and spoke he was like a veteran actor at high-school theatricals. The

[1] Her husband, Chester Carlson (1906–1968), once president of the New York Vedanta Society, invented xerography; the Carlsons appear in *D.2*.

[2] At the Democratic National Convention; but Eagleton (1929–2007), Senator for Missouri, gave up the vice-presidential nomination on August 1, when the press disclosed that he had undergone electro-shock therapy in the early 1960s.

whole building felt this and roared for him. He was so conscious of his power, but not ready to use it yet. I suppose he thinks another four years of Nixon will produce a backlash, and then he'll stand up and America will hug him.[1] But he shouldn't be too sure. There may have been somebody in that crowd last night at the convention—a gay black abortionist with a Mexican mother, who has been married by Troy Perry to a Chinese Maoist boy. The Mexican mother would be women's lib, of course. Our first truly all-American president.

July 15. Hunt called from Texas yesterday to say that Boorman's agent had told him that Boorman's science fiction story has been turned down by Columbia. The agent wanted to make a deal with Hunt; Boorman will direct "Frankenstein" if Universal will buy the sci-fi story as well. Hunt said he'd read the sci-fi but would make no promises. Hunt also said he had talked to Sheinberg who had told him to "keep the faith." Whatever that may mean. I do hope I shall one day get the chance to tell Sheinberg what I think of him.

We had supper with Paul Wonner and Bill Brown. Once again they are planning to move. Probably up to San Francisco. But they did say they wished they had had a chance to buy Gavin's house and that they might be interested in Elsa's, next door to us, if it's for sale. Bill ran on about the great Stravinsky ballet memorial festival, which he had been in New York for,[2] and Bob Craft's new book of his collected diaries; he had told Bill that he has only mentioned famous people because they "describe themselves" whereas the unfamous have to be described. Bill, not having been mentioned, was a shade sour about this; and he claimed that Bob had quoted something he said without giving him credit for it. (I think it was a parody title for a Stravinsky biography.)

Bill also talked about Vera, how marvellous she is; gets the breakfast in the morning and fixes the last post-midnight snack and goes to all the necessary parties and performances and manages to paint in between, and drinks and smokes and entertains people. Meanwhile, Paul sat glum, wearing an unbecoming patchy sparse beard; he really is very sulky. Don said he is sure Paul dislikes him—and then added that most people do. Whenever Don makes a statement of this kind, I am apt to get irritated and contradict

[1] Edward Kennedy (1932–2009), Senator for Massachusetts since 1962, ran for president in 1980.

[2] During the last week of June; Balanchine and his New York City Ballet staged thirty-one ballets to Stravinsky's music.

him, and then we have an argument and he says I hate him when he complains, because it is "rocking the boat." Last night, I was careful not to contradict him too much. I admit it, monster that I am, I do *not* want to rock the boat. Why in hell should I? The seas I am sailing on are so beautiful, most of the time, and I know I can't expect to enjoy them very much longer. Maybe I should be sadder and more thoughtful and more mindful of my latter end. But there is some virtue, surely, in just not being a miserable old man and polluting the emotional atmosphere?

July 16. This morning, on the beach, Don accused me once again of taking him for granted, and a long discussion followed. I'm not going to write anything about this now. For reasons—

A note came from Michael Barrie yesterday. He is at Trabuco and wants to find Edward James's address, in order to give him back some letters he once wrote Gerald. In his note, Michael also refers archly to an exhibition of temper by Swami, when he was at Trabuco for July 4: "Swami P. sent back his chicken, saying that it was undercooked and referring darkly to the fact that if there were no one here who could cook for him he would be forced to return to Hollywood." Perhaps I misjudge Michael, but I sense in that remark the moralism of a Peggy Kiskadden. Why *shouldn't* Swami complain and get angry? Even from a Kiskadden point of view, monks need discipline. I must confess, I don't believe I could live at Trabuco as long as Michael was there. And yet I am really quite fond of him. What makes me vomit is his smug brand of piety.

Last night, we saw Franju's film of Cocteau's *Thomas the Imposter.* It isn't well cast or acted, but it did seem to me full of haunting atmosphere; amazing that they could have made it in the sixties. Even the ruins seemed characteristically World War I. A terrible glimpse of a dobbin with his mane on fire! We went with Pat Faure. I can't help it, I don't warm to victims—especially a woman victim of a Frog who leaves her for a boy.[1]

Sunshine at last—but that means the Canyon is crammed to the top with cars.

July 17. Ken Anderson lives in a house on the Sherman Canal in Venice, only a few doors along from where Jim Gates and Peter Schneider used to live. These little wooden shacks seem holy and

[1] Faure (b. 1928), an American model, photographer, and later, art dealer and gallery owner, ran Nicholas Wilder's gallery, 1972–1979; her second marriage, to Jacques Faure, with whom she lived in Paris from 1959 to 1970, was ending.

pure, because they have nothing to do with the culture of the marina, whose highrises and cranes are visible in the distance. We spent a couple of hours deciding which painting and which drawings of Ken's to buy.[1] Ken asked us a thousand for the painting but came down instantly to five hundred; the two drawings cost us another hundred. Ken is bearded, mild eyed, nonchalant and absolutely sure of himself. When he speaks of painting he is apt to use art jargon. He is very cool but could become friendly, I think. Later, we had supper at a French place called Puce, where they make pancakes. Kind of hippie. Boys tend to have two girls apiece and wear open waistcoats over bare bodies. The service very slow because all the cooking was being done by a little woman single-handed, though there were two women to wait and a proprietor who did absolutely nothing but nag the help. Don thought the pancakes good; I didn't. While I was eating mine I felt like a spy trying to chew and swallow a very tough secret document before the police arrive.

Am reading, or rather rereading, Cocteau's *Thomas the Imposter*. Between the leaves, I found a photo of Larry Paxton and me, mugging and wearing derby hats, which must have been taken in San Francisco only a short while before his death.[2]

July 18. Have been dipping into my old journals of the early sixties; a mistake. Now I feel sad as shit, but must admit things are much better nowadays, at least from my point of view. Is it really good to keep a journal? I loathe doing it at the time and I get depressed when I read it. But it's such a marvellous treasure trove. I have vowed to make an entry a day throughout July, so I'll stick to this, but I protest, I protest.

Actually, the current blue phase with Don is clearing up. We had a beautiful beach morning with big waves. Don has typed out our "interview" of him for the Wyatt Cooper magazine[3] and really it is awfully well expressed: he talks with such authority—not at all like a downtrodden kitten, as I told him.

Books I am in the process of reading—some of them for many months, already: *The Way We Live Now*, Trollope; *Forbidden Colors* and *Spring Snow*, Mishima; Nijinsky's *Diary*; *Inside the Third Reich*,

[1] Bachardy had previously bought a large drawing from Anderson's UCLA graduate show.

[2] Paxton recommended the book around the time the photo was taken, as Isherwood tells in *D.2*. He was diabetic and died suddenly in May 1963 when a Christian Scientist persuaded him to stop taking insulin.

[3] The article never appeared; see Glossary under Cooper.

Speer; *She Knew She Was Right*, Ivy Litvinov; *Mysteries*, [Knut] Hamsun; *Thomas the Imposter*, Cocteau.

July 19. Yesterday afternoon another representative came around from Occidental Petroleum; this one looked like a Jewish college professor and smoked a big pipe. I shut the door in his face without saying a word; he didn't seem surprised, no doubt the others had reported on their reception. This aroused all the paranoid fury I try to suppress; I wanted to order a wholesale massacre. Only a day or two ago, we got a letter from [James] Covington, who is selling his house and moving to Miami; he is still interested in getting us and our neighbors, Elsa, John Hardine, Donald Fareed and Dr. Christian Herrmann, to sign a group agreement with the corporation, claiming that it might in the future bring us in $10,000 each. But how loathsome to take their money![1]

Another spurt of fury today; for the second time lately, there's been a girl in the gym.

Last night we saw *Dr. Strangelove* again. It is brilliant. That endless flight over the arctic regions. Those shots of the war room at the Pentagon, with the great circle of faces. Sterling Hayden's mouth and cigar. Strangelove's obscene mechanical arm.

Faride Mantilla came in this morning to clean for us. We like her so much, better than any of the others. We really pay her less than we ought, two dollars an hour and bus fare, so we've had bad consciences. Today we eased them considerably by giving her fifty dollars toward a dentist's bill; she owed $117.

July 20. Cloudy morning; sunshine too late for beachgoing. Last night we had supper with Swami, who is with Pavitrananda, Bhadrananda and Krishna at the house on Malibu Road where they stay for their holiday. Swami's eyes are getting worse—he strained them working on a new translation of Saradananda's book[2]—and his doctor is against letting him have the cataract operation. He told us how some young girls who were devotees of the Baby Krishna once insisted on feeding Brahmananda with milk like a baby, which caused him to go into *samadhi*. Swami

[1] Occidental offered financial incentives for a neighborhood promise not to oppose drilling off the coast at Pacific Palisades; community objections eventually prevented the oil rigs.

[2] *Sri Sri Ramakrishna Lilaprasanga*; Prabhavananda did not get far with it. Swami Jagadananda translated it as *Sri Ramakrishna, the Great Master*, and later Chetanananda made another translation. Saradananda was a direct disciple of Ramakrishna.

said he couldn't put this story into *The Eternal Companion*[1] because people in the West wouldn't understand. The long lines of the evening breakers rolling into the bay—our eyes kept being pulled away from Swami's kind of beauty to theirs.

There's a new group going around town: Jews for Christ. And senior citizens are being referred to as Gray Lib. Incidentally, I save quite a bit of money at movies now; a lot of theaters have special rates for senior citizens.

I've been reading three long extracts from S.N. Behrman's *People in a Diary*, in issues of *The New Yorker*. Here's an abridged version of a story he tells about Siegfried Sassoon. Behrman and he had been to visit Sassoon's mother, during a visit by Behrman to England. As they were driving back to London, Behrman noticed that Sassoon was in a "dark mood." Sassoon finally confessed that he was upset because he had been thinking about Edna St. Vincent Millay. Behrman knew that Sassoon had met Millay and liked her but sensed that now something had gone wrong between them:

> There was an immense interval before Siegfried said anything more. We were approaching London. Siegfried went on,
>
> "A friend of mine told me ... someone was praising my war poems—perhaps excessively. Miss Millay said, 'Yes, yes, I agree. But I wonder whether he would have cared so much if it were a thousand virgins who had been slaughtered.'"
>
> "I don't believe she said it."
>
> "She did." There was a pause. His cheek muscles were twitching. He was suffering.
>
> "What's the use?" he added, "What can one do?"
>
> Did he mean against a concentrated malice and venom of the world?—of even a fellow poet—a nice creature like Edna Millay?
>
> "It couldn't have meant much to you," I got out finally, "or you'd have told me."
>
> "It meant so much to me that I didn't tell you. It all came back while you were telling those stories at dinner."
>
> We drew up before the Park Lane.
>
> "I'll call you in the morning," he said. "Good night."
>
> "Good night, Siegfried," I said.
>
> He shook my hand. My heart ached for him.
>
> I went upstairs to my room. I sat on the bed thinking of him. Everyone sang, I thought, except the poet.

[1] *The Eternal Companion: Brahmananda, His Life and Teachings* (1948) by Swami Prabhavananda.

This story—or rather, the way Behrman tells it, makes me dislike Behrman, Sassoon and Millay equally. Which is quite a feat. And how odd that Robert Graves made much the same remark, later, about both Sassoon and Owen!

July 21. Another semivictory is announced in *The Advocate*: a revised code for Delaware, but the lewd conduct and soliciting clauses still stand.[1]

Last night we had supper with Jack Larson, Jim Bridges and a young director from New York, Joe Chaikin, who has been giving classes here for directors, on the invitation of the Mark Taper Forum. Gordon Davidson and Ed Parone, as well as Jim and Jack, have been attending the classes. Jim and Jack say Chaikin is amazing. He's a little stocky plump curly-haired red-blond, a Russian Jew by descent, with light blue eyes. He has a lot of charm and seemed rather impressive, though prone to sententious art talk, which may be explained by the fact that he worked with Peter Brook in England. He reminded me a bit of Richard Dreyfuss, the actor.

A heavenly morning on the beach with perfect waves. We are very near the end of "The Mummy" outline.

July 22. Tomorrow is Gavin Lambert's birthday. We celebrated it by taking him to dinner at Raphael's yesterday. Before dinner, we went to see *The Stranger*, with Orson Welles. The sound system in the theater was in such bad shape that you could scarcely hear the dialogue. The drama is corny, anyhow; and I am temperamentally against all movies set in small American towns. Welles was at his best.

Mark, once again, wasn't with us, but this time he had a convincing excuse, he has sprained his ankle. Gavin talked casually and without much sympathy about this, saying how tiresome it was to have to cook for him and to have him laid up just now when Gavin needs his help in getting ready to move. We felt he was a bit bored with Mark. I asked him if he had brought up the subject of Mark's alleged thefts, and he said no, the right moment wouldn't be until after they had settled in France, because whatever money he decided to give Mark would be in francs or pounds, not dollars.

We have a scene in "The Mummy" in which Kay dreams that

[1] The revision to the Delaware criminal code (enacted July 6, 1971) redefined sodomy to exclude consensual acts and made sixteen the age of consent. Loitering was still criminal and included soliciting for sex of any kind.

the mummy kisses her; she wakes and finds white dust on her lips. This suggests a title for our film, "Put your Mummy Where Your Mouth Is."

Gavin wants to use one of Don's drawings of Selznick as an illustration in his book about the making of *Gone with the Wind.* And Bryan Forbes, who had tea with us yesterday, wants Don to draw him so Bryan can use the picture on his next book, I think it's his autobiography. Bryan didn't seem to have changed a bit, full of bounce and good nature as ever. How kind he was to Don around the time of Don's London exhibition in 1961. He took our play with him, saying he wanted to send it [to] some theater in London he's connected with.

July 23. This morning we finished "The Mummy" outline in rough; it still seems to need something doing to the ending.

Last night, supper with Rudi Gernreich[1] and Oreste Pucciani;[2] Shelley Winters, Bill and Peggy Claxton and an actress named Diane Ladd were there. Rudi wore a thick red burnous which made him look absurd. Oreste looks exactly like Radclyffe Hall. Oh yes, and their precious stud Frog, Jacques Faure,[3] the defaulting husband of Pat Faure, was there too; I forgot him, which shows, I guess, what a racist I am at heart. Shelley did all the talking. Peggy still had on her dead white and black geisha makeup. God, what an idiot bore evening. Then we went on for a short dip in Bryan Forbes's birthday party. I do like him. And at least there were some groovy boys. Also an English rock star, Elton John—who, Bryan had previously assured us, is a real respectable married gay, twenty-four, made over a million last year and shares it with his lover-manager, also young. I find something particularly touching and charming in the abandon with which an English gay boy dresses up to the teeth and throws himself into the role of a Californian. Saw one of them doing it at that party.

July 24. We still haven't quite finished "The Mummy," but almost. Don had a bright idea for the ending, while we were on the beach

[1] Viennese-born costume and fashion designer (1922–1985), in California from 1938; known for his topless bathing suit modelled by Peggy Moffit Claxton in 1964, the thong, and futuristic designs in plastic and vinyl. He was a founder of the Mattachine Society.

[2] Professor of French at UCLA (d. 1999), raised in Cleveland, educated at Harvard; Gernreich's longtime companion.

[3] Art director for French *Vogue* and other Condé Nast publications; he settled in Santa Monica in 1970.

this morning—a revelation that the nurse is now working for the mummy. But it will take some fixing.

Yesterday we saw Dan Luckenbill and talked about the two stories he gave us to read. Later, Don told me that he thinks Dan regards me as having a closed mind; that I have my own standards and apply them to every story I read and if they don't fit, that's too bad. Dan is right, but I don't apologize for that at all; it's the way I have to be. Real broadmindedness is for noncreators. What did disconcert me was hearing from Don that Dan had told him he was once in a class of mine on some campus and had shown me a story and that I'd made much the same criticisms of it as I made this time. I mean, I was disconcerted to find I had forgotten this so completely.

Later we went to see *Butterflies Are Free*, ghastly stuff, thick sentiment sprinkled with Jewish jokes. But Edward Albert, whom we used to see at the Palisades gym, gave a really moving and noble performance as the blind boy. He is quite beautiful but not slickly so and he makes you believe he is in love with the girl—how seldom can one say that about an actor! His emotion is very Latin in a good sense, and when he plays anger he is warm, not nasty. Considering the dialogue he had to speak, this was a genuine personal triumph. He got absolutely no support from anybody else in the cast.

July 25. We still haven't finished "The Mummy," because we can't quite decide on an ending. I really wish we could have a cloudy morning, tomorrow. These brilliant warm mornings lure us down to the beach after we've worked less than three hours, which isn't nearly enough.

Michael and Pat York came yesterday afternoon; he to be drawn, she to photograph me. The drawings didn't turn out well. Don says Michael sat all right but somehow didn't project. Indeed, he later fell asleep! Pat took about a hundred pictures of me, talking all the time. And then, at her request, I showed her some of my albums, which she loved seeing. I think it jolted her, though, when I told her that, for me as a queer, the difficulty wasn't to be able to fuck a woman but to be able to fall in love with one—which, I said, had been out of the question throughout my life. She would have much preferred it if I had said the opposite, because then my falling in love with a woman would have been touching and romantic and my impotence something to be saddened by—in other words, I should have presented homosexuality as a disability, which is how women like to think of it.

But I do find her sympathetic, in her desperate way. Don still thinks she is a bit crazy. [...] When I showed her the ridiculous snapshot of Mrs. Lanigan[1] and told her that Mrs. Lanigan was my first heterosexual fuck, she hastened to say that she thought Mrs. Lanigan really quite attractive—as if to boost my morale by assuring me that I'd been to bed with at least *one* attractive woman! Heinz's pictures, before and after the operation on his nose, interested her enormously, for obvious reasons—we both agreed that Heinz looked far better with it broken. Michael broke his nose—perhaps the most beautiful broken nose of his generation—when he was around three.

Michael kept saying how sad he felt to be leaving California. Pat is going to the Vedanta Center in London, to seek guidance. I would love to see her interviewing Buddha (Yogeshananda)! We probably shan't see them again before they leave. Michael made a stiff little speech, thanking us for our friendship. Darling as he is, he embarrasses both of us. I feel I could only break the ice by covering him with kisses all over, until he screamed and giggled. The kind of kisses we do exchange are mere culture-bridging gestures—like a white man exchanging the correct ritual greetings with an African tribal chief.

July 26. A very hot but beautiful day, enough to make anyone regret leaving California. We had supper with Swami at the Malibu house. He asked, "Do you think I've gained weight?" It seems he now weighs 105 pounds with his shoes and pants on. When they dehydrated him at the hospital some time ago, he weighed 93! Dub's weight this morning, just under 150; Kitty's just over 139.

Today we really did *almost* finish "The Mummy." If we fix a couple of tiny things, we can send it in tomorrow. We've also been working on Don's magazine article. Now he's picking drawings to illustrate it; and he has to write comments on each one!

Suddenly Don is getting commissions. His goal is to earn enough to pay for all the paintings he's bought.

July 27. Am writing this listening to an Elton John album (*Honky Château*) which Don brought back this morning inscribed to us from Elton John and Bernie Taupin, who writes the lyrics. At present I like the lyrics much better than the music, especially "Honky Cat"—"They said get back, honky cat, / Better get back

[1] Isherwood's employer for a month in 1919, when he tutored her son Wallace at the hotel she ran in Bettyhill, northern Scotland.

to the woods ..." Also "Rocket Man." Don got the record because he was over at Malibu at the house where Bryan and his family are staying with Elton John, drawing Bryan's daughter.[1] One of the drawings is marvellous and Bryan is going to buy it, as well as Don's drawing of himself (of Bryan, I mean).

After I wrote my entry for yesterday we actually did finish "The Mummy" last night. This morning I had it xeroxed and a messenger from Universal is to pick it up. Talked to Hunt this morning in Texas. He says he has heard, in some backstairs way (the way he seems to hear everything) that Sheinberg has set the beginning of shooting on "Frankenstein" for October 15. And that they are "going in the direction of Boorman."

Another glorious day. But the beach is really too crowded and the sand is too hot.

July 28. Robin French talked to Sheinberg and was told there's no truth in Hunt's story that a date has been set for "Frankenstein." It seems that what has been holding things up all this while is NBC, which can't decide if it wants "Frankenstein" for T.V. But now Robin thinks it will probably be made as a feature film, because even the head of NBC has advised them to do this.

This morning, right after breakfast, we went down to the beach—partly because it looked like being a very hot day later, partly because Don had to go to Santa Barbara and see Bill and Paul. Don has left now, but the hot day isn't so hot down here, after all; there's a wind. In town, it's been breaking records, going up into the hundreds.

On the way back from the beach, we met Gordon Davidson outside his house. He told us he hasn't read the revised version of our play yet but he's just about to—no doubt he decided that when he saw us. At the bottom of the steps leading up to the Casa, somebody had dumped a dead opossum, a few days ago. It began to stink horribly, but this morning it has been nearly devoured by maggots; the ground all around it was swarming with them.

Last night, with Gavin, we saw *Touch of Evil* and it was as wonderful as ever. It makes a special appeal to me because it's about a frontier.[2] It almost makes me want to have another try at writing about my mystique of The Frontier, as I did in *The Forgotten*.[3] But first I must see if I can't understand better what it is.

I had meant to write a great deal today in this journal—particularly

[1] Sarah (b. 1959), later a fashion journalist as Sarah Standing.
[2] The border between the U.S. and Mexico.
[3] An early version of *Down There on a Visit*.

about my day-to-day life with Don, and thoughts about his future and my death and attempts at meditation. But the day has flown by, and now I must fix my supper.

July 29. Very hot again today, and humid. After going to the gym, I was too tired to do anything but lie on the couch in my workroom and read the rest of Bryan Forbes's novel, *The Distant Laughter,* and Gavin's book (in manuscript) about the making of *Gone with the Wind.* He's called it *GWTW,* but isn't sure if the publisher will like the title; they may object that the book will be mistaken for yet another exposure of one of the big corporations! Gavin's book is journalism of the very best kind; it could hardly be improved on. Bryan's novel could have been a quite entertaining story about moviemakers, if only he hadn't dragged in this embarrassing would-be macho love affair. But I shall have to say something nice about it, maybe even write a blurb, because he has done so much for Don. Don has earned a thousand dollars, just in the last few days!

July 30. Hunt called this morning from Texas, having read "The Mummy." His first reaction is that it is better than "Frankenstein" and "on a very high literary level"! His only request—before having it typed—that we should state that the coffin of the Princess Naketah is made of silver, not wood. Hunt also wanted us to delete our note that the roles of Laura and Naketah are to be played by the same actress. But I persuaded him to let it stand.

Another very hot day, but not so muggy. Thunderheads are piling up inland; we'll probably have a storm tonight. Don and I had a very nice swim but the beach was already terribly crowded by noon. They say as many as 325,000 may come on a Sunday like this one. I suppose that would be more people than I've met during my entire life! Don and I talked about his early days of drawing and going to art school. He said, "You were the only one who encouraged me, I'd never have gone on without you." We were very close to each other.

Evelyn Hooker called me a couple of days ago. She said she had had all kinds of ailments and had finally gone to the UCLA hospital, and they had feared a tumor at first but had found nothing wrong except a microscopic ageing of the ventricles of the brain. So now Evelyn is somewhat hysterically determined to be well. I saw her this afternoon. She wants to go to Vedanta Place and get instruction in meditation. That means she'll have to see Asaktananda, I expect. But then her sister came in. She said she'd met me before; I didn't remember and was surprised how Jewish she looked, rather like Shirley Booth![1] Anyhow, Evelyn then said that she wanted to do exercises as well as meditation, and I told her that Swami is against breathing exercises. Whereupon the sister said that *she* had done breathing exercises for years and they hadn't harmed her. So I sensed hostility developing and shut up.

July 31. Don got an infuriating traffic ticket last night, because he drove the wrong way down a deserted one-way alley in Santa Monica, late in the evening. We'd been seeing *The Lady from Shanghai* and *The Third Man.* I think Welles's scene with Cotten on the big wheel in Vienna is one of *the* classic demonstrations of hamming *with the eyes*; Welles doesn't seem to need his other features at all. Also, the square in which the final sequence begins, part baroque architecture and part bomb ruins, lit by flashlights, is a magnificent Piranesi theater set. Just now, for some reason, I'm finding the end-of-war period in Europe tremendously nostalgic. This morning I read Speer's description of Hitler just before his suicide—the best thing in the whole book.

We went to see the films with Gavin. When we're with him (always without Mark) we say very little about his move. It seems tactless to mention it—as though he were about to undergo surgery.

Some specimens of art jargon, from a brochure put out by The Underground, a gallery at Los Gatos: "Geoff McCormick... constructs small, container-like kinetic sculptures.... Some change shape or make sudden thrusts.... One is left of a silent impression of sped-up geologic forces.... Larry Rief ... forces art happenings from the ordinary and sometimes dismal environment.... Ernest Mathews colors medium-sized canvasses employing an original method using capillary action."

August 1. This morning, at breakfast, just as I was telling Don he really should do some more drawings of Swami while he's at the Malibu house, a white cockatoo appeared and started flying around the Canyon. It reminded us of those mornings at Tony Richardson's house in Australia.

Don's earnings are now up to eighteen hundred dollars for this recent period! We went and deposited some of the money in a

[1] American radio, stage and screen star (1898–1962); she won an Academy Award for *Come Back, Little Sheba* (1952) and became familiar as "Hazel" on T.V. during the 1960s.

different savings bank, called American Savings. Somehow the place seemed totally unreliable and unserious, but that was probably because we were taken care of by a very campy black queen and because everybody in the office was in pseudo-Hawaiian drag—it's part of a publicity drive by Western Airlines. When the deposit had been made we were given an orchid each. We left them in the gym, to which we went right afterwards.

Harry Rigby, for whom Gielgud is directing *Irene* in New York,[1] is one of the people Don has been drawing. He came around with Scott Schubach to pick up the drawing. (Don wasn't there because he was drawing Elton John and his collaborator Bernie Taupin; this morning he had to dash out to Malibu to deliver the drawings to them before they took off for England.) Rigby is a mixture of Van Vechten[2] and Frankenstein's monster. He kind of drifts up to you and makes gestures all around you, like an octopus. Scott alarmed me by saying that Hunt was "a prisoner" of Dick Shasta and Dick's mother, and that Hunt is terrified that Dick will murder him for his money. He doesn't think Hunt will be allowed by them to return to Los Angeles or go to England for the film—unless, of course, he is heavily guarded. This is just the sort of hysterical half-truth which one can't dismiss altogether. Hunt certainly did believe his life was in danger from Dick; that was when he had that bodyguard.

Surveyors have been working a lot on the road below. Fear that means they are going to widen it at last.

Another beautiful morning on the beach. We went to the ocean twice.

August 2. Talked to Jim Bridges on the phone this morning. He is now more or less committed to directing *The Paper Chase* and is off to look Harvard over and then go to Toronto, to decide if the buildings there will do as a double for the Harvard Law School. He has just found out, to his disgust, that the production of *Streetcar* which he's to direct at the Ahmanson next spring isn't the only production opening at that time; there will be another at the Lincoln Center. Jim feels this will take away a lot of his publicity and prestige. Also, he has begun to realize that Tennessee isn't really interested in the project; he's busy preparing to write his autobiography. Also, Jim has had a change of heart about the

[1] In the end, Gielgud did not direct it; see Glossary under Rigby.

[2] Carl Van Vechten (1880–1964), American novelist, poet, critic, photographer, New York bohemian, and promoter of the Harlem Renaissance; he appears in *D.1*.

play itself; he now feels that it (hush, hush!) isn't all that good.

Robin French called to say that he has put us up as possible writers of the script of Galsworthy's *The Apple Tree*, with Bogdanovich directing. Also that Sheinberg is leaving for Europe, where he will talk to Boorman and/or Lindsay Anderson about directing "Dr. Frankenstein." Also that he still believes he can sell *The Vacant Room*[1] to T.V.

From first indications, *Deliverance* is going to be a huge success.

August 3. We had supper with Swami out at Malibu again last night. He had been discussing spiritual experiences with Pavitrananda. He told us that his vision of Vivekananda in the shrine of the Hollywood temple had come after Ashokananda had attacked him for allowing me to write (in the introduction to *Vedanta for the Western World*) that "if he (Vivekananda) had never visited Dakshineswar, he might well have become one of India's foremost national leaders." Ashokananda found this insulting to Vivekananda's memory. Swami didn't agree but nevertheless was very upset. And then the vision came, apparently to reassure him. All this I've heard dozens of times already. But yesterday evening Swami remarked, almost casually, that, just before the vision, he had been going through a mood of self-blame in which he considered suicide. "How could I live, if I had insulted Swamiji?" This was what amazed me and Don. I asked him, "Surely, Swami, you couldn't have considered killing yourself and leaving your work unfinished—that isn't like you at all, you always have such a sense of responsibility?" Swami didn't seem to get my point. But we both remain astonished. Was this just a bit of Asian hyperbole? I find it hard to dismiss it like that, when I remember how matter-of-factly Swami went on to speak of his three great experiences, this vision of Swamiji, his vision of Holy Mother in his room at the center and his experience of God without form in the temple of Jagannath.[2] It was so convincing that it gave me goose pimples. Then he added that the "I" that witnesses these experiences is not the ordinary egotistical "I"; it simply records and remembers.

Don asked Swami if he would sit for more drawings, but he screwed up his nose and said, "It takes too long—and this body is so old!'

[1]The ghost story Isherwood wrote with screenwriter Lesser Samuels in 1950.
[2]Isherwood tells more about Swami's three visions in his entry for June 6, 1963 in *D.2*. The earliest was in the temple of Jagannath, in Puri, Orissa, eastern India, the next in his bedroom in Sister Lalita's house before the Hollywood temple was built.

Rob [Matteson] sent prints of his photographs yesterday. They really are surprisingly good, of both of us. Only I must give up smiling at the camera altogether. My assorted mouthful of false and real teeth looks grotesque!

August 4. I forgot to say that I asked Swami which of the direct disciples—other than Brahmananda—made the greatest impression on him, personally. He answered without hesitation Premananda and Turiyananda.

Last night we had Chris Wood to supper at the Bellevue. His back is better but still bad; he is stooped and shrunken. He weighs 112 pounds—only a little more than Swami! He still talks of going to England, but maybe the back is a psychosomatic obstacle; Don thinks so. Chris talked about Peggy, who has obviously been bugging him by trying to run his medical care. He says that Derek [Bok] is now "like a famous actor." Then, obviously quite forgetting who he was talking to, he decided that only people of the same age should have sex together—otherwise "it isn't pretty"!

This morning, Evelyn Hooker went up to Vedanta Place and had an interview with Asaktananda. I spoke to her on the phone just before she left and knew at once that she was "disturbed"; her voice was weepy and she told me, "Things are happening which I'm not ready to have happen—samadhi." I thought it best not to ask her to describe what she thinks samadhi is. Later, having seen Asaktananda, she called me to say that he was "a man of great wisdom." She was vague about the interview itself, but apparently she got "samadhi" again just as it started and was unable to speak for some time. Asaktananda gave her some instructions; "and now," she said, "I know where I belong."

August 5. Last night we went with Jack Larson and Jim Bridges to hear Tom Hayden talk about the war in Indochina. Jon Voight arranged this series of talks and he wants us all to get others to come, so we'll know the facts and be able to use them—in the presidential campaign, for example. Hayden spoke lucidly and in great detail. He seemed terribly weary and no wonder; he has been doing this for years now. It is right and necessary that we should be reminded that all this is going on, just over the horizon of our fat-cat lives. But what *exactly* is one to do about it? Making Nixon responsible seems petty and even cynical, in the face of this ghastly suffering. Are we to use it merely as a weapon in our party politics? Our entire economic system and its philosophy is involved; and what are the alternatives to it? China? Russia?

Cuba? God save us! So, practically, what does one do? Retire into an anarchist commune. Or a monastery. Work for birth control. Go to Indochina with the Friends and help in a medical unit. Yes, those are ways of keeping your hands clean. I suppose, for me, the only way I could be even minimally useful would be to teach again. I did manage to tell a lot of people a few home truths, in those days, and one or two may have been influenced by them. It's far more effective to give a little prod unexpectedly, now and then, in the midst of what's officially a course in English literature. When you speak at a peace rally everybody knows in advance what you're going to say. It makes no difference if they're for or against you; at such times their minds aren't open.

Before going to the talk, we went to a restaurant called The Corkscrew which has been opened fairly lately on San Vicente. The sort of place we never ordinarily visit—hostessy, your "wine girl" is introduced to you by name, and you go to the counter and mix your own salad; the object being, I guess, to give the sheep the illusion of making individual contacts and choices. Jim objected to it strongly, calling it "bourgeois," but Jack obstinately likes the food. (This morning, both Don and I had gained weight; I'm over 151.)

The talk was at the Company Theater.[1] Two of its members told me they want to do *Dogskin* soon, the original version. I'm wondering if Elsa will raise a stink about this. Don says that, as usual, I'm manufacturing obstacles.

August 6. Elsa, to my surprise, took the news about *Dogskin* and the Company Theater quite calmly, saying that it would make no difference, as far as our musical version is concerned. She's now getting ready to pursue John Houseman, who is coming here soon to direct a replay of *Don Juan in Hell*.

Don, waking up this morning: "'I think I'm going to have to drop my ban on Jesus, eventually'—that's the first line of a movie we wrote about Stephen Spender." Don quite often begins the day like this, by describing a dream he has had. He is still half asleep while he talks to me. "Did Stephen say that?" I asked. "No," said Don. "Was the movie produced?" "Not yet."

I have kept meaning to record a discovery I made, last time I was up at the Malibu house, seeing Swami. In his bedroom is a shelf of books belonging to the hostess. Most of them are religious or at any rate seriously instructive in tone. But, between the New

[1] Founded in 1967 by American director and activist Steven Kent.

Testament in Modern Speech and St. Francis de Sales's *Treatise on the Love of God*, I saw *Myra Breckinridge*!

Last night, we had Mrs. Leavitt in to cook and had Roddy McDowall, his friend Paul Anderson, Dick Gain the dancer, Jack Larson and Jim Bridges to supper. It didn't quite work. I think Roddy is nervous unless he has some other stars in the room with him and preferably some women. And maybe he was embarrassed by the flower-child behavior of Dick, who kept repeating how much he loves Don and me. But Dick's a sweet creature, and his figure, in his late thirties, is still terrific. He's up at Marymount Convent, teaching children dancing, for two weeks. I still find Paul attractive, though he is very much Mr. Showbiz Junior. He's a real farm boy, with a thick trunk and sturdy legs—"workhorse legs," Don calls them. Roddy has a weird hobby, he makes candles. He brought us one, or rather a sort of wax embryo containing three wicks and many lumps of colored wax embedded in wax. Without my glasses, I took it for some sort of fruit dessert and was about to put it in the icebox.

August 7. As we ran down to the beach this morning, we talked about one of Don's paintings which is hanging in his studio. Don described how he prepares the background first, covering the whole area of the paper; then he paints the head and shoulders over it, allowing the background to show through in places. That's to say, he has to adapt the figure to the background. I asked him to explain why he does this and he answered that it's because he has a temperamental horror of making plans. He feels that plan making destroys his artistic impulse. If he tries to plan, he loses all desire to paint the picture. So he confronts himself, as it were, with an emergency, a fait accompli, which demands to be dealt with.

We live, many people would say, entirely for pleasure—that is, we mostly do only what we want to do, and even when we do something we don't want to do it is as a result of our own mistakes; for example, we go to somebody's house against our better judgment and then have to invite that person back. We aren't wage slaves, we don't have to conform, we impose our own lifestyle on others rather than vice versa. Yet our lives are curiously rigorous. We go to bed late, get up early after what seems like too little sleep, feel guilty if we haven't finished breakfast by eight, put in our stints of reading and work, run down to the beach and go into the water but don't stay long, hurry off to the gym, rush out to movies and sometimes see three of them in a row, often sitting through them grimly and almost hating them, then dash home and drop stunned into bed.... And yet this is, without doubt, the most continuously happy time of my life. The nearness of Don, the sharing of experience with him—that's what makes everything worthwhile.

August 8. Last night we went with Gavin to see *Rebecca* and *Jamaica Inn*. The print of the latter, or rather the way it was projected, was so bad that we left. In many shots, the heads were cut off. But it was obvious that Laughton was awful, anyway. This is Gavin's last week in his house, then he'll move to the Marmont, then leave. We still haven't seen Mark or even discussed the situation with Gavin. I feel we can't let him go without having it out with him, otherwise the whole thing will fester into a feud. At the Golden Bull, where we had supper later, Evelyn Hooker was sitting with two guys at another table; either she didn't see us or she decided not to come over. She was drunk and talking with maximum audibility. I heard her telling the others that she was Lebanese. I never heard this before. Was she kidding, and, if she wasn't, does this explain why her sister looks so Jewish?

August 9. This is a period of intensive moviegoing, partly because several theaters are running revivals of old films made by some particular director or starring some particular actor. Last night we saw *To Have and Have Not*, which I thoroughly enjoyed and Ozu's *Tokyo Story*, which is perhaps the best of all his films and tremendously moving. Even his boring passages—in this case the death of the old woman—seem artistically intended; he slows the action down until it nearly stops and this in itself is a most daring kind of realism, a way of making you feel the drag of Life.

The moviegoing keeps us up very late and we wake exhausted. This morning Don had a dream "that we bought a lot of furniture and when we brought it home we didn't like it as much."

Rob Matteson came to be drawn yesterday afternoon. He absolutely refuses to be paid for the photographs, which means (a) that I shall have to get one of them into a magazine when next I'm interviewed and (b) that Don will have to give him one of his drawings. He really is a sweet boy though.

Richard Haydn[1] came by and had tea with me. After reading *Kathleen and Frank* with great enthusiasm, he wanted to see Frank's

[1] British comedy revue star (1905–1985), in Hollywood from the early 1940s; his films include *And Then There Were None* (1945), *Forever Amber* (1947), *Please Don't Eat the Daisies* (1960), *Mutiny on the Bounty* (1962), *The Sound of Music* (1965), and *Young Frankenstein* (1974). He also directed.

and Kathleen's watercolors. He is an odd lonely creature, proud of being able to live alone. He says that when he gets sick he recovers much more quickly than most people because there is no one to encourage him to stay in bed. I can't quite picture how he uses his time.

August 10. Two or three days ago we discovered, with some anxiety, that Hammer Films of England have made a movie based on Bram Stoker's *The Jewel of Seven Stars*, which is one of the books I read while we were concocting our mummy story. Last night we went to see it. It is called *Blood from the Mummy's Tomb.* (Don and I were talking about the nuances of meaning in "*A* Lady from the Land of the Dead" and "*The* Lady" etc. "Blood from *a* Mummy's Tomb" would be like "Elegy Written in a Country Churchyard"—campy romantic, not a horror title.)

There *are* resemblances between our story and theirs, but I don't think they're serious enough to matter. The emphasis is quite different. The biggest resemblance is that both stories have a girl in them who's a reincarnation of an ancient Egyptian. But theirs is an evil witch, ours is a perfectly harmless princess. Their incarnation takes place more or less arbitrarily; ours is motivated. Their story is cluttered with minor characters, ours uses very few.

I shall call Hunt about this, but largely to tease him as a punishment for not having been in touch with us the whole of the past week.

August 11. Don says not to tell Hunt about *Blood from the Mummy's Tomb* because if I do I shall start something. What we ought to do, he says, is say nothing unless somebody raises the question, and then say casually, Oh, that thing—we went to see it, of course, but it was so obviously quite different and inferior that we didn't think it worth mentioning.... So that's what I'll do.

We saw Robin French and one of his colleagues yesterday, to discuss T.V. projects. But, as always when you do this with an agent, the attitude is, You write something marvellous and then we'll see if we can sell it. *Their* idea of something marvellous was a story of Germany in the thirties. Robin quoted a *great* story someone had dreamed up, about a very ambitious dissatisfied young World War I vet who can't get work and everybody laughs at him and he's full of rage, and then he turns out to be—Hitler! Actually, I believe *Jean-Christophe* would make an excellent four-part T.V. film, if only someone would buy my screenplay from Fox.

We had supper with Swami at Malibu yesterday evening. In

addition to the regulars, Swami Prabuddhananda, the head of the San Francisco Center, is staying with them now. He hardly spoke at all and seemed stodgy and almost Germanic; no doubt he has had to assume an authoritarian persona in order to control the rebel "white swami" elements in his congregation which so nearly succeeded in taking over before he arrived! Swami was full of love. He seemed really pleased to see us both.

August 12. Hunt has never returned my call. I wonder if Dick Shasta (who answered the phone) ever told him about it? Ever since Scott Schubach made those remarks, I've been half expecting some terrible drama. A little paranoia goes an awfully long way, especially since we have been seeing so many Hitchcock movies lately. How easy to imagine Hunt gagged and strapped in the cellar, because he refuses to sign a deed of gift, making over his fortune to Dick and Dick's mother!

Last night, we took Jo to supper and then to hear Tom Hayden speak again on Vietnam at the Company Theater. The talk itself was less interesting than last time, because it was the past history of Vietnam. It was all based on the proposition that the Vietnamese are sure to throw the invaders out because they always have done so in the past. I kept falling asleep. But I am being much instructed by one of Tom Hayden's books, *The Love of Possession Is a Disease with Them.*

Jo seems a little "better," I mean wiser. I felt this quite definitely yesterday. Glade Bachardy, alas, isn't "better." Today, Don tells me that she has refused to go to Europe with Jess, saying that she doesn't want to "come back in a box." Don says it is all spite; she has to do something to spoil his enthusiasm for new experiences. Don is wondering if he ought to speak to her. But he is afraid that, if he does, she will simply fly off the handle and refuse to see him again.

This morning I got a letter from Calder Willingham, asking me to write something about his new novel, *Rambling Rose.* He complains that critics have never known what to make of his writing: "They have persisted in regarding me as a naturalistic slice-of-life writer who somehow went wrong. *End as a Man* was no work of Dreiserian naturalism, but rather was a gothic comic-horror nightmare, a work of pure invention from start to finish.... I consider myself one of the most miscomprehended writers on the scene. The literary crowd for the most part has either ignored my work or misrepresented it.... For over twenty years now they have killed me in hardcover and I do not say it lightly." He ends,

"Others have written me wonderful letters.... but you are one of the few who ever went on public record, and thereby you stand almost in a solitary splendor."[1] Luckily for me, I've already read enough of the new book to know that I like it.

August 13. This morning Don and I went to a place in Santa Monica and he bought a Fiat with four doors, so Dobbin can get in and out without gasps and contortions. It is bigger than his V.W. but still fits easily into our carport. I mean, it will—it hasn't been delivered yet; they promise it for tomorrow, but we're sceptical.

Jess Bachardy has already sold Don's V.W. for us. For $1,750!

Steven Arnold, who directed *Luminous Procuress* which we saw on the 9th at Universal, is about to start work on another film, *Monkey*, based on the novel by Wu Ch'eng-en which Arthur Waley translated. He gave the outline to Don, asking for our opinion, and Don thinks this is a good opportunity to get into "underground" moviemaking. So it is, or would be, if only we could make any suggestions. But Arnold's version is so hopelessly unambitious, just a series of tableaux, strangeness for the sake of being strange and "oriental" and low-campy. We are trying to figure out how the original conception of the monkey symbol could be introduced into this version.

August 14. According to a statement from Simon and Schuster, we sold 8,848 copies of *Kathleen and Frank* for the period ending March 31.

After four dull hot days, today has been glorious. We ran down to the beach and went in the water. Waves big.

Last night, we watched the first of a five-part series on the life of Leonardo da Vinci. Dignified, deadpan, unthrilling stuff. An idiot bearded commentator in modern dress kept appearing amidst the scenes; no doubt he was supposed to be like Kenneth Clark in *Civilisation.*[2] He calmly informed us that, when Leonardo was arrested in 1476, it was on an unknown charge—probably heresy! Durant says unequivocally that it was homosexuality.[3] If Durant's

[1] Isherwood discovered Willingham in 1948, as he tells in *Lost Years*, and praised his novels on the radio, in personal interviews, and in print, pronouncing *Geraldine Bradshaw* (1950) a "comic masterpiece."

[2] The narrator was played by an Italian actor, Giulio Bosetti, rather than an art expert as Clark was; the series was directed by Renato Castellani who worked mostly in Italian T.V.

[3] *The Story of Civilization*, vol. 5, *The Renaissance: A History of Civilization in Italy 1304–1576 A.D.* (1953), p. 200.

information has been proved wrong that could only mean someone had discovered what the real charge was. How could it be unknown? Must look into this.

August 15. Oscar Levant died yesterday. I talked to June today and she told me that he wasn't particularly sick. He had been expecting a visit from Candice Bergen who was writing about him and he went upstairs in the middle of the day to rest before the interview. When June came into his room a bit later she found him dead. She began to cry and I wished her daughter Amanda hadn't made her speak to me. But I did learn to my great relief that they are having a strictly family funeral; both Don and I had been dreading going.

Steve Black came to see me this afternoon. He hasn't changed much since the Cal. State days and I guess he *is* still fairly young.[1] He is settled in Vancouver now, with a different wife, and is perfectly happy in his job and may become a Canadian citizen. He has nearly finished a book on Whitman. He read me "As I Ebb'd with the Ocean of Life," very badly but with deep emotion—so deep indeed that he tore the page quite seriously as he turned it and didn't even apologize. I feel warmly toward him but he is hopelessly square. I cannot imagine him understanding Whitman. That's a dumb remark, though. What I mean is, I can't imagine him understanding *my* Whitman.

Don got his four-door white Fiat yesterday. It is much more convenient for me to lie in than the V.W.

August 16. This afternoon, parking on Santa Monica Boulevard near the freeway bridge, I backed into a plump Japanese lady named Mrs. Grace Mori. It was one of those tiny bumps which might be expensive. She was cross for a minute, then remembered her Japanese manners just before I delivered a line I'd prepared: "I am sorry to see that you have adopted our Western rudeness—we look to your race to teach us better behavior."

Rumors that Nixon may be settling up the war.[2] The awful thing is that one doesn't want him to, because then he's certain to be reelected.

[1] He was a student at Cal. State in the late 1950s when it was called L.A. State; he appears in *D.1*.

[2] Henry Kissinger, then national security advisor, was conducting secret talks with North Vietnamese representatives in Paris; on August 15, one representative returned to Hanoi and Kissinger visited the president of South Vietnam, suggesting the talks had reached a breakthrough phase.

In a booklet which contains everything Vivekananda wrote about Ramakrishna, I found the original Bengali–English translation of Vivekananda's Bengali hymn which I later Englished as "Breaker of this world's chain. . . ." It contains the marvellous line: "Samsara's bondage-breaker, taintless Thou"![1]

Last night we had supper with Swami at the Malibu house. He seemed better than ever. Vividishananda was there, from Seattle. He is nearly blind. I do like him. He is an example of a really lovable bossy person. He cross-examined me exactly like a police officer, just for the sake of asking questions. "When exactly did you first come to the United States?" "On which years did you visit India? And you saw Calcutta only? Why was this?"

August 17. That was our last visit to the Malibu house, at least for this year. Swami and the others are leaving it early next week. Seeing him there is a special experience; the ocean creates an atmosphere of massive calm, despite all the noise it makes, and you are secure from the apologetic but bossy interruptions of the women at Vedanta Place. Whenever Don and I visit Swami now, we make *pranams*; and I am beginning to realize what a bond this has formed between the two of us, both being his disciples and bowing down to him together. The fact that we meditate at the same times, though in different rooms, is also part of this bond. It has formed very slowly, imperceptibly almost, and yet it is perhaps the most important feature of our whole relationship.

We talked to Jo Lathwood yesterday on our way down to the beach. What a rugged old thing she is! She remarked how big the waves are just now, and said that she was nervous swimming out through them that morning; "But, of course, once I'm outside them, then I'm all right, I'm launched, I'm like a ship."

Don did some very good drawings of Paul Millard's friend Rob [Matteson] yesterday, and then Paul came down and we all had dinner together. While Paul and I were alone, he told me how Rob left him for six weeks and went to live with a man named Graham in the Valley, in a single room. Graham is a hypochondriac, in his fifties, very poor, very stingy and incredibly ugly, but he seems to have temporarily bewitched Rob; it's all over now. Paul also told us that Speed Lamkin's aunt has died at last, leaving him a huge chunk of Delta Airlines stock. (Delta recently had to pay out a million dollars to a hijacker, which is so far a record!) Speed says

[1] The booklet was probably Vivekananda's *In Search of God and Other Poems* (1947).

he is going to settle in London, presumably to annoy Marguerite!

Mrs. Grace Mori called very early this morning, to tell me that I owe her thirty-four dollars and ten cents for repairs to her car. She was quite apologetic about this, but I was hugely relieved it wasn't more.

Wyatt Cooper has called at last to say he wants to put five of Don's drawings—of Dali, Virgil Thomson, Margaret Leighton, himself and me—in the second issue of his magazine. We had begun to fear he was going to back out of the whole thing.

August 18. We had supper with Roddy McDowall and Paul Anderson last night; they live in the house George [Hoyningen-] Huene used to have. Trashy oddments of furniture have destroyed all its charm, but the patio at the back is still attractive. Paul fixed cheeseburgers in extremely short shorts, his legs aren't as sexy as I'd expected. Then we saw Lubitsch's *To Be or Not To Be*, which I simply hated. I loathe all films about the Nazi–Jew confrontation, although I have to admit theoretically that a good one *might* be made. This is sheer schmaltz.

Because Roddy wants to meet Mary Miles Minter O'Hilde[r]brandt, I called her this morning and got a full hour of talk, chiefly about her lawsuit against Rod Serling. (He showed her picture on T.V. in a documentary about famous murderesses!) All she would finally say was that Roddy had her permission to call her and talk (ha!) to her. M.M.M. also kept on telling me that we should put up a sign opposite her house, to show people where we live, so she won't be disturbed.

August 19. We're just off to Santa Barbara, to stay the night with Bill and Paul and then I have to give a reading at the Vedanta temple.

Last night we went to another of Tom Hayden's talks on Vietnam. This one was the best we've heard. What interested me most was his explanation of the way the South Vietnamese troops make an arrangement with the North Vietnamese, so that the South can seemingly "occupy" and "pacify" a district when in fact the North is still controlling it. The South agrees not to rock the boat and in return the North agrees to allow its presence in the area and not move in and destroy its forces completely, as it easily could. Jane Fonda was there, looking older, but still good. We left during the question period because it was so hot and so many people were smoking.

Incidental information, obtained from the radio: in one of the national parks, the park signs used to be made of redwood. But the bears ate them, so the signs were made of plywood instead. It was then discovered that the plywood signs contained glue which the porcupines simply love; they got eaten too.

Have at last reached the end of 1947 in my reconstructed diary.

August 20. When we got to Bill and Paul's house we were told, to our amazement, that they are definitely going to live in New Hampshire. Bill found a house for them there while he was away in the East. Don thinks that Paul doesn't really want to go—although one would think that it would be Paul, not Bill, who would be more apt to take to life in the country. (They will be surrounded by woods.) Paul had already done something to his back, that very day; it got worse during our stay. So maybe that was the first shot being fired in a big psychosomatic offensive.

My reading at the temple went off all right, and afterwards Don and I had lunch with the nuns before returning home. The chief topic of conversation (and no wonder!) was that Mick Jagger has bought a house right across from the convent.

Have just had a call from Dorothy Miller. She has had pneumonia, but what she really wanted to talk to me about was Oscar Levant's death. I'm sure she hoped I would have some gruesome details which hadn't been reported by the press.

August 21. Very hot this morning, almost too hot for the beach and much too hot for the gym. However we went in the water. We also tried to work out a storyline for *Monkey.* Not much success.

Paul Millard called to say there may be an offer coming up for our Hilldale property. But Arnold Maltin told Dick Dobyns that we wanted $75,000, which is surely far too little considering that there is still between thirty-six and forty thousand to pay on it.

We have a heavy dinner party tonight, John Houseman, Merle Oberon, her boyfriend Robert Wolders (who was with her in her last film), Joan Didion and her husband John Dunne, and Gavin and Mark. It is Gavin's farewell party. And it'll be the first time we've seen Mark since the quarrel. Well, anyhow, thank God we have Mrs. Leavitt to cook for us!

August 22. The party went quite well. Merle, looking amazingly like her famous self (at sixty), was being as gracious as she knew how. Don thought her exactly like British royalty. Her friend Robert Wolders is good-looking in an entirely uninteresting way, with nice manners and a slight accent (Scandinavian?)[.][1] He did something very tiresome and inconsiderate, however—arrived in a car with a boosted battery which went dead on arrival *and* drove said dead car down our ramp till its fender touched our cars, nearly. It was a miracle that he managed to back it out again but he did—after a repairman had been phoned for, and the battery recharged and the engine of the car left running all through supper.

John Houseman was also royalty of a sort, very grand and gracious. He has a languid, dogmatic tone when he talks about the theater which I find disagreeable; I don't think I'd enjoy working with him. But he seemed much more sympathetic when he complained that his book *Run-Through* hadn't sold nearly as well as he had hoped. This started Joan Didion off and also her husband John Dunne. We all deplored the ruthless commercial tactics of the bestseller writers, who tour the country getting to know the owners of all the key bookstores and send agents to buy up dozens of copies at the shops which furnish the sales figures on which *The New York Times* bestseller list is based. I said I always feel that Jesus was referring to them when he said, "Verily, they have their reward." Did he?[2] I'm always dubious about my own quotations. But this one got a big laugh.

Joan, as usual, spoke in that tiny little voice which always seems to me to be a mode of aggression. Or an instrument of it, anyhow; for it must be maddening in the midst of a domestic quarrel. She drinks quite a lot. So does he.

Mark, whose hair had been inexpertly rinsed and was therefore pinkish, seemed most anxious to be friends. And Don recognized this by promptly kissing him when he arrived. Nothing whatever was said or hinted about the quarrel. I guess we shall see him again, and certainly Gavin, on Friday afternoon. They leave for New York next day.

It was very hot yesterday and has been again today. Almost too hot for the beach. I ran down, jumped in the water, didn't lie in the sun at all. This morning we worked on Steven Arnold's *Monkey* story; we have now worked out quite a charming continuity and only need to fill in some details.

August 23. More work on *Monkey* today. I really wish this film

[1] Wolders (b. 1936) is Dutch; he appeared in *Interval* (1973) with Oberon and in the T.V. series "Laredo." They married in 1975. After her death, he was Audrey Hepburn's companion.

[2] Cp. Matthew 6.2 and 6.5, the Sermon on the Mount.

could be made properly, because I think we could produce an excellent script for it.

Incidental information; at the barber's: Human hair used to be mixed with mortar to help bind it together. The barber (at the shop next to the Tudor House) told me that he has a customer who is a doctor and who asks for all of his cut hair; the doctor takes it to his lab and analyzes it to find out if he has enough of certain minerals in his system.

Still very hot. There is a big fire up near Ojai. You can't see much smoke in the sky but the sunlight is reddish.

Two boys, Michael McDonagh and Jeff Bailey, came up from Long Beach to see me yesterday. They were both fairly attractive, quite well read and almost certainly queer, maybe lovers. But all they wanted to talk about, seemingly, was had I known So-and-So, what was So-and-So like?

Saw Jess and Glade with Don, because his car needs a radio putting into it. Jess and Glade seem to be definitely going to Europe—despite the fuss Glade made about it and her threats of dying during the trip. But oh my God it is heartrending how worried they are about details. The passengers on this tour have already been given a briefing session, and, from the way Jess describes it, the briefer must act like a marine sergeant addressing recruits before the start of boot camp: "This is it, men!" He has obviously succeeded in terrifying all these elderly people at the thought of all the problems they face. One big question Jess asked me, "They tell us that the ladies may have to use rest rooms with coin machines on the doors; now, we'll be going through a lot of different countries with different currencies—shall we be able to turn in the coins we don't use and get them exchanged for another currency, or will they only change bills?"

August 24. A talk with Peter Schneider on the phone yesterday, about the poems he sent me. He really is astoundingly thick-skinned; took my faint praise of some of them as a matter of course, asked me to send one of them on to Swami and to send him a copy of *Prater Violet* to give to a friend. Woe unto him when he ceases to amuse me! However, this time he did amuse me because I got him onto the subject of Jim Gates. Peter said, "Now I'm surprised I could have lived with him for three years.... I hoped, when he joined the monastery, he would become forceful and quiet and serene, but it makes him more jumpy and obsequious.... We've always been very close but I've always been disgusted by him; he tries to please everyone and he agrees and says he likes everything.

I used to be brutally frank with him, when we lived together. His eyes used to fill with tears and he'd say, Do you want me to move out?"

And yet I can't just dismiss Peter as "a bad boy" (which is what Don calls him). I like him when he says how he hates religious people and how he doesn't even like Vedanta philosophy. I think he is sincerely impressed by Swami. He says that his feeling for Swami is quite different from what he would feel for a father, "or a young lady" (he always calls girls "young ladies," which sounds so oddly prim).

Supper last night at Jo Lathwood's, with her daughter Betty and Anne Baxter and a middle-aged male friend of hers (not a lover, Jo told us in advance, and indeed he seemed to be probably a closet queen). Anne talked our heads off. She is a true demon—even her looks are demonically well preserved. She could have taken on Peggy Kiskadden in her prime. Much of her talk was about playing the lead in *Applause*, and we had to be very careful what we said, lest we should reveal that we saw Lauren Bacall in it here, an unpardonable crime, since we didn't see *her* in it while we were in New York!

August 25. This morning at breakfast, Don told me about a conversation he had had with Mark Andrews, the night he and Gavin came to supper, the 21st. (Don was surprised he hadn't told me before and it's just possible he had, because I was a little bit drunk and that always brings amnesia.) Don had asked Mark, "I suppose you're excited to be going?" and Mark had made a grimace and said, "I'd be excited if I was in a picture." And then he had gone on to tell Don that he is dreading leaving, that this is "the only place in the world." "As soon as we get there, Gavin will start working on his film" (Galsworthy's *The Apple Tree*, which Bogdanovich is to direct, and which Robin French tried to get for us to write) "and then what'll I do?" Don remarked how we all jump to conclusions and how wrong we usually are; we had all taken it for granted that it was Mark who had insisted on moving away from Los Angeles because of his disappointment over the screen roles which didn't materialize.

Yesterday we went to see Steven Arnold and told him our suggestions for his *Monkey* film project. He liked everything except the proposed ending—that the princess finds the holy man and becomes his disciple and remains in his cave high up in the mountains, while the demon takes on the form of the princess and returns to torment the wicked uncle. At this point, Steven became evasive and said he would shoot it "both ways," adding that he always shoots several endings. The meeting gave both of us doubts, especially Don, who feels we shan't be able to work with him because he will always end by doing what he wants. He is very graceful, silly-pretty, gentle, passive seeming, with his pale eyes and absurd little moustache—and it is obvious that he has a "whim of iron" and unlimited energy. He has already made up a thick pack of what he calls storyboards—cards on which every shot is described; and for each shot he has made a drawing, in his more than somewhat Beardsley style.

The problem is, how can we gracefully back out of this project now, if we decide to do so?

Last night we went to another talk about Vietnam by Tom Hayden. This one was supposed to analyze the Pentagon Papers but was actually rather vague and often repetitive. Jack told Don that last time, after we had left, a speech was made by a man who Jack and Jim decided must be an FBI provocateur. Jack said that when they told this afterwards to Jon Voight he was terrified.

Tonight there is a final meeting to decide what should be done to spread information about Vietnam. We aren't going. If there *is* anything we can do, we shall hear soon enough, and every meeting is more and more crowded. Last night, the place was bursting.

August 26. Last night, as a birthday request, I asked for the alarm not to be turned on. Don agreed, on condition that he shall have *his* birthday request next time, which will be for *both* the shades to be left up at night, instead of one. When he woke, he said, "A morning without an alarm is like Dobbin without a belt." I dreamt that we were rehearsing a new play of ours. The word "skinflint" came into the dialogue. Don objected that "no actor would call himself a skinflint." So we took it out.

In the night, the dogs from the house right across the Canyon barked for a long time. They haven't done this in months. The owners must have gone away and left them.

I weighed a hundred and forty-nine and three quarter pounds, this morning.

Just before breakfast, Don shaved off his moustache, as a birthday gesture. He hates himself without it and says he is now sure he must always have one. I prefer him without it, because I hate to see his (or anybody else's) mouth hidden. Don claims that the moustache hides the lines around his mouth. I love it, I mean his mouth, with or without lines. But I had to admit that his mouth looked funny for the first few hours because he was so terribly conscious of its bareness.

He gave me a beautiful long scarlet velour bathrobe, which makes me look like a cardinal. And a new billfold, the same model as my old one, which is coming apart, but with the more practical feature that you can remove the money clip and carry it separately.

A cable of birthday wishes from Richard. Calls from Larry Holt and Mary Herbold—and from Paul Anderson (Roddy's friend) yesterday. Birthday letters from Joan Elan, Harry Heckford, Lee Prosser. Jennifer and Norton Simon have given me a very good quality leather shirt which looks terrible on me; I'm going to exchange it.

Yesterday afternoon, Gavin came in to say goodbye. He left this morning. Mark wasn't with him yesterday; Gavin told us, "He's sleeping," with a slightly ironic smile. We didn't know what to say to him, and made a great deal of protest that this wasn't really a parting, the South of France is so near, etc. Yeah, so is the Vedanta Center at Gretz. Gavin will stay in a hotel near Toulon and from there they'll house hunt. And he will start work at once on *The Apple Tree*. The way he described it, it sounds corny and dreary beyond words, Galsworthy at his worst. But Gavin is obviously determined to make the best of it. Never have I heard louder whistling in the dark. He even told us with a perfectly straight face how the lover, twenty years later, finds the little unmarked grave of the Welsh girl who killed herself for love of him! We both felt so thankful that we missed out on this assignment.

Driving to supper with Nellie Carroll and Miguel, yesterday evening, I told Don that I saw a story about the Animals which at the same time would make it clear that the Animals are also human beings. And I do, still, today. But the objection is, as always, that I feel it is a kind of sacrilege to write about the Animals at all, except privately.

The supper was quite [brutal][1] because Nellie wanted to pressure Don into recommending Miguel and his lamps to Irving Blum and getting Irving to come and look at them. Don doesn't really admire the lamps all that much and doesn't want to do this and is furious with Nellie for the pressure. Although the supper was therefore for their benefit, not ours, Nellie had the gall to call Don and ask him to bring the beer—Mexican, because we were having Mexican food, from the Mexican delicatessen on Mississippi; it was tepid and cardboardish. Jack and Jim were there too. They

[1] Isherwood typed "brutally."

had stayed on at the Vietnam meeting on the 24th and heard the discussion which followed Tom Hayden's lecture. Some tactless oaf, meaning to be on the right side, had stood up and quoted a remark made by an enemy of Jane Fonda, "She's the only dike we haven't bombed." Jane had said nothing and everybody had been deeply shocked, and then someone else had said, "That remark is not to be repeated." Aside from this, there had been much gloomy foretelling of what will be done if Nixon is reelected and therefore becomes the president who presides over the second centenary celebrations. It was said that he proposes to issue stickers to be carried on the cars of all "real" Americans: presumably the rest of us will be refused gasoline. And there is great fear of a possible Agnew presidency to follow.[1]

Well, that's all for this volume of diary. Had intended making some remarks on being sixty-eight but can't think of any right now.

[1] Spiro Agnew (1918–1976) was still Nixon's vice president. He resigned October 10, 1973; see below.

September 2, 1972—December 14, 1974

September 2. Now I'm just one week into my sixty-ninth year. On my birthday I made a resolution; henceforward I'll try to make three acts of recollection every day instead of two—I don't know what else to call them, meditation is much too grand and even bead telling may not always be possible in the middle of the day, if I'm away from home and in the midst of doing something. The important thing is to *recollect.*[1] I feel this more strongly as I get older and nearer to the great test; will "this thing" be true for me and with me when everything else fails, when I'm on my own and it is my only hope and support? Well, at least I am glad I'm thinking this way and doing something, however feeble, to get myself ready.

I worry more and more about Don. How am I going to avoid involving him, if I get sick and die slowly and expensively, like poor Igor? I think of this every morning, nearly, when I wake up.

But this doesn't mean that I'm depressed. Most of the time I am very content. And I have been through another period of excellent health. Yesterday, one of those sorenesses in the muscles of the abdomen started up again, but it may not be anything much. Every day we either run down to the beach or I go to the gym and jog around the block.

Have just finished Philip Roth's *The Breast*. I only read it because the publisher sent me an advance copy and because it is short. It stirs up my worst prejudices. I find it "disgusting" because it is a heterosexual fantasy and because it is so Jewish in its boring gallows humor and its delight in misfortune. But why doesn't Kafka

[1] I.e., to feel the presence of the Lord, as against more formal meditation or making japam. Prabhavananda and Brahmananda both emphasized this kind of remembering.

make me feel the same?[1] Because he's Kafka and Roth is Roth. Even so, the Kafka thing isn't really my thing and never has been. I don't delight in him.

Have also been reading Bob Craft's *Stravinsky: Chronicle of a Friendship*. Whenever Igor speaks, it is marvellously alive; everything to do with Igor and Vera and their world is fascinating. But Bob's travel sketches are too painstakingly cute, and a classic demonstration of how not to use similes. The sun going down is likened, not to a sinking ship which suddenly dives and disappears, but specifically to the *Titanic*. Why the *Titanic*? And, a few lines later, "a herd of buffalo oozes to the neck in the mud, like overweight bankers in Turkish baths." Why bankers?[2] But at least I never want to stop reading.

September 9. Don will be away for the day. He left at eight to pick up his parents and take them to the airport for the beginning of their trip to Europe. They are miserable with worry, of course, and probably won't have a single happy moment until they're safe home again. I talked to them on the phone this morning, told Jess to disregard the "sights" and just look around at the people in the different places they visit. I don't think he understood what I meant, and now it seems to me to have been one of my typically simplistic remarks. (Reading Wystan's *Epistle to a Godson*, which I have just bought, makes me feel how stupid and platitudinous I usually am.)

After the airport, Don will visit Beverly Baeressen in her mechanical lung.[3] They decided to make a reading list and then each read the same book on it at the same time and later discuss it. The trouble is, poor Beverly has all the time in the world and Don nearly none. The books he and I decided on were: *Under Western Eyes*, *Emma*, *Dangerous Acquaintances*,[4] *Jacob's Room*—I can't remember what else.

After Beverly, Don will go down to Laguna Beach and see Jack Fontan and Ray Unger. He isn't altogether looking forward to this because he knows Ray will fix an enormous meal and he wants to

[1] In "Metamorphosis" Kafka's hero wakes to find he is a giant beetle; Roth's hero is a giant female breast.

[2] Craft cut the bankers in his revised edition.

[3] A friend introduced by Jack Larson and Jim Bridges. She had had polio and lived with her parents in Torrance, an hour's drive away. She died a few years later, in her early forties.

[4] A translation from the French of Laclos' 1782 novel; another English title is *Dangerous Liaisons*.

starve himself; his weight this morning was up to 141, which for him is gigantic. (I was 149 and ½, which for me is average.)

No interesting letters this morning but a large nuisance mail. Am irritated by a letter from a Betsy Crockett who calls herself the Program Officer of the Los Angeles World Affairs Council and wants to arrange an appointment for me with Mr. Martyn Goff. He's a publisher from London, I think,[1] and Angus Wilson sicked him onto me. What makes me mad is that this Crockett, who has my address, calmly writes, "I could not locate your phone number." My first instinct was to tear up her letter but I tell myself that's just senile spite, so I'll call her Monday instead and bitch her about her incompetence.

For the record, I thoroughly dislike Mark Spitz, a gloomy pretentious self-satisfied Jewish drear, with unsexy legs. But let's not say anything more about the Olympics. The massacre releases nothing but negative emotion—a plague on both....[2]

I was talking to Laura Huxley on the phone a few days ago. Every time I do this, I feel affection for her, yet we so seldom meet. I'm afraid Jinny [Pfeiffer] isn't well at all. They had been up to British Columbia, to some ashram, and Laura said, "The air was so *marvellous*, and then we come back here to all this filth. I love that in the Book of the Dead, where it says 'Oh, Nobly Born—.' If *I* am nobly born, why do I breathe this un-noble air?"

Saw Swami last Wednesday (6th). I asked him if he had any desire to go back to India—four of the girls are leaving in November. He was absolutely positive: *No*! I am so glad he feels like this. So often, as people get old, they long for the Homeland—and, in Swami's case, such a longing would mean that Maharaj's presence here, and the shrine here, hadn't been enough to sustain him. It would mean, to a certain extent, that the inner life had *not* paid.

Don still hasn't started to let his moustache grow again, as he threatened to. He is so beautiful without it—and so infinitely much more *revealed*. Why does he have to mask himself? The irony of it is, most people simply haven't noticed that he shaved it off; I am the only one who really minds and is frustrated by that silly superimposition of hair.

[1] Goff (b. 1923), a novelist, ran the Booker Prize for thirty-six years and was a director of Henry Sotheran's, the book dealer.

[2] Spitz (b. 1950), the American swimmer, won seven gold medals at the Munich Olympics before Palestinian terrorists broke into the Israeli compound on September 6, killed two athletes, and took nine hostages who died the next day in a gun battle at the airport. For "A plague o' both your houses!" see Shakespeare, *Romeo and Juliet*, 3.i.

September 17. Gavin sent us a postcard saying that the South of France is off—high-rise buildings everywhere and bumper-to-bumper traffic; it reminded him of Marina del Rey. So now they're trying Rome!

We saw Swami the day before yesterday. (I love the moment when Don and I bow down and touch his feet simultaneously and get blessed together.) Talking about Maharaj and Premananda, he said, "When you were with them, you were in another world." This was at supper, just after we'd eaten. He was full of joy, in-drawn yet eager to talk, as he relived the experience. His words were vibrations. I thought, he talked to God and now he's talking to us. He told us how Maharaj would call for his water pipe, take a puff and then "go away somewhere else." Much later, he would exclaim with surprise that the pipe had gone out.

This was recorded. One of the girls had switched on the tape recorder which has a mike hidden in a flower vase on the dining room table. But then Swami got onto the history of the centers in the U.S., telling which swamis had been sent to which centers and how some had flopped and how others had succeeded—which was on a lower plane, as they say, but funny and charming, like a veteran actor talking about show business.

Before supper, while we were in his room, Swami told us that he can read very little now, because of his cataracts. I asked, "Don't you miss it?" and Swami answered, with the sweetest significance, "I don't miss *anything*."

Utter silence from Hunt in Texas, and from Sheinberg, and from London and New York about our play—except that Edgar Lansbury, after talking to Jim Bridges (who was in New York making arrangements for his *Paper Chase* film), did ask to see a copy. We have talked to Henry Hathaway about his project, *Praying Mantis*. He is a dear, a bit like King Vidor, but the film story (by Ben Hecht) seems to us old-fashioned and dreary-dirty. Tomorrow or the next day at latest, we must decide what to do about this.

I run every day, either down to the beach or while I'm at the gym. My weight this morning was down to 148. Chiefly because we had a supper of stilton cheese and melon.

Have kept up midday bead telling—except that it sometimes doesn't get done until late in the afternoon—every day but one. The first thing is to establish this as a habit.

September 23. It was reading the paragraph immediately above which reminded me, at about 3:30, that I hadn't made my midday japam for today. That's the way it still goes, most days. Certainly there is some part of my mind which mischievously distracts my attention around noon, so that I forget. Is it the same "part" (if one can use that expression) which reminds me, later? It does seem like a kind of game, played with mischief but without malice.

What I keep working on is this thought, that I must keep reminding myself of my *only refuge*. And now, past sixty-eight, I have no difficulty in understanding how desperately important this is; even though I feel wonderfully healthy at present. Meditation on the shrine, or on Swami in his room, is becoming easier. When I imagine myself in either of these places, I feel courage and protection. (Also, with closed eyes, I feel a strange upward turning of and pressure on the eyeballs. I must try to remember to ask Swami about this.) Imagining myself in Swami's room has an extra advantage, because I can think of Maharaj's photograph over the mantelpiece at the same time, and remember that I am not only in Swami's presence but also, through Swami, in Maharaj's, because Swami is always in his presence. Sometimes, the sense of safety which this brings is quite strong. I feel that I would not be at all afraid to die, if I might be carried to Swami's room to do it, or even put in front of the shrine. But I must never forget that this faith is a grace; I can't command it and daren't give myself the least credit for it, or I shall probably lose it. And there's a tremendous added joy in the thought that my darling beloved Kitty isn't excluded from any of this. I don't have to bring him into Swami's room, he sits there in his own right as Swami's disciple, and I can picture him sitting across from me, also at Swami's feet, as he so often does.

Swami, when we saw him three days ago, didn't seem as lively as he had been the week before. I think the heat tires him, and he is still waiting to hear if the doctor will agree to his having his cataracts removed. However, we had our first post-holiday Gospel reading in the temple and he answered questions.

In the late afternoon of the 17th, we drove up to the top of Las Tunas Canyon, to visit Peter and Clytie Alexander, who are getting ready to build a house at the head of an adjoining canyon which winds down to give a glimpse of Malibu Colony on the shore. At present, the Alexanders and their children are living in a box-shaped semitransparent tent and cooking out of doors on butane stoves and hibachis. (It seems that the Forestry Service hasn't been after them yet about the possible fire danger; as for the danger of being swamped, they expect to have the house at least partially built before the rains come.) Larry Bell, Chuck Arn[o]ldi, with a pigtail, Guy Dill and some other art men accompanied by art girls were there, drinking saki and, a few of them, sniffing.[1] (That at least isn't a fire hazard!) Two of them were playing catch with baseball mitts—I can't help it, it always embarrasses me to see grown men doing this in a social situation; it seems sheer macho showing off.... The place is still wildly beautiful and almost uninhabited, though constantly overflown by planes and helicopters. We watched the sun set and the stars come out. It seemed incredible that we were only twenty minutes' drive from Santa Monica. Talked mostly to Clytie. Don says he likes her better than any other woman he knows and I can understand why. We ought to visit her sometime when she's alone. The children go to school in Topanga Canyon by bus and Peter spends a working day at a studio in Venice.

Almost the only house visible from the Alexanders' homesite is called Sandstone. It is a nudist colony; Peter describes it as a "fuck club" and says that the members have sex according to the latest techniques of group therapy, curing each other's impotence and other hang-ups. Sometimes the whole school of them will sun themselves on a mound near the house. Peter then watches them through a telescope. (Today, on the beach, Don and I were discussing a possible film—Sandstone, as it really is and as it appears to the Alexanders and their friends, contrasted with the Alexanders as they really are and as they appear to the nudists.)

We have sadly told Hathaway we don't want to work on his story. I'm afraid he was terribly disappointed. Wrote him a note later, suggesting he might like to consider making a film out of *Mr. Norris*. No answer yet.

Dr. Kurtzman says I have got to have two more of my upper teeth pulled. If I lose another after these, I shall have to have a whole mouthful of falsies, I fear. Was bitten by a dog in the Canyon, while visiting a rather charming Englishman [who] had a crush on, and probably an affair with Vernon Old, in 1940. The dog was only a puppy and wanted to play with my red socks but I flew into a sick rage because its owner flatly denied that it had bitten me. I yelled at him and felt later as if I'd had a sort of anti-orgasm.

Michael Barrie came to see me yesterday. He wanted information about the La Verne seminar, for his book on Gerald. In the spring he is going to England, to interview people there. What Michael really wanted me to confirm was that the La Verne seminar was

[1] Cocaine.

basically Gerald's idea and that Gerald had led the discussions—also that it was Gerald who founded Trabuco. Michael has read somewhere that Aldous was the leader of both projects and he resents this, enormously. I assured him that Aldous was incapable of that kind of planning and leadership. Then Michael told me something I'd never known before—the money which paid for the building of Trabuco, the hundred thousand dollars, had all been saved by Gerald in the course of his life. Gerald use[d] to hint that it had been given him by some rich woman. This means, of course, that the ten thousand dollars Gerald gave Denny [Fouts] were also Gerald's own money.

September 28. Dr. Kurtzman pulled out my two teeth today and now I have a much better-looking partial, full of bold broad white choppers.

The day before yesterday, I went to see Dorothy Miller, on my way to Vedanta Place. She has lost a lot of weight but looks well, quite recovered from her pneumonia. I realize that Dorothy's form of vanity is that she can make absolutely anybody afraid of her. When she tells stories to illustrate this, she bugs her eyes at you like a tiger. Her beloved church is a paradise of gossip. She says, "I like to find where the juice is dropping from the grapevine." She told about a friend whose doctor gave her some medicine saying that she must not take it until she was lying in bed. "But when that she got home, she couldn't resist. *Oh—just a taste!* And, Mr. Isherwood, when that she drank that medicine she fell right on the floor. She's a mulatto, and, I'm telling *you*, she turned purple!"

Dorothy repeated that in her opinion it was Jo who drove Ben away. Dorothy believes that, if you know something bad about a person who is about to get married, you are bound to tell the other party. Recently, she told a girl that her fiancé was still seeing his former wife (and screwing her, Dorothy implied but didn't say). The man was furious, but later Dorothy was glad she'd done it.

When I told Swami about the pressure I sometimes feel on my eyeballs while meditating, he couldn't explain it. But when I went on to tell him that after this I nearly always find myself taking a very deep breath which gives me a feeling of great calm and happiness, Swami was very pleased. "That's a very auspicious sign," he said, and added that "two of our new girls" have been experiencing the same thing.

Yesterday, we had a visit from Bruce Chatw[i]n, a blond blue-eyed but somehow not really attractive friend of Peter Schlesinger.

He is an anthropologist[1] and has spent some time with native groups of hunters in the lands south of the Sahara; Mali, Niger and Chad. He maintains that hunting groups aren't religious; religion only begins when people settle down and have individual possessions. (I didn't want to get into semantics so didn't challenge this, because it was obvious that Chatwin attached a different meaning to the word "religion".) But he was extremely interesting, describing how the boys between thirteen and sixteen wear a sort of drag and are regarded as girls. The whole hunting group is perfectly adjusted to its environment; even young children know what stars are rising and setting and when the migrations of birds take place and what habits the various animals have. As Chatwin put it, they differ from us in that they never try to interfere with nature in any way. They also think that our preoccupation with possessions is crazy; according to their way of thinking, you share everything you have, so they "steal" from tourists, only it isn't really stealing because they don't want to keep what they take.

September 30. I had some more talk with Chatwin yesterday morning on the phone—I think he has now left Los Angeles, on his way to see some of the pueblos of New Mexico. He repeated some of the things he told me when he came to the house—that he regards the hunting groups as being fundamentally unaggressive; that "much of what passes for aggression is a response to confinement" and that there is no confinement in the way of life led by the hunting groups; that, in their case, gift giving replaces aggression—when two groups are in the same territory they don't fight over it, they exchange gifts, one group leaving its gift at a certain place and the other accepting it only when the gift seems sufficient; if the gift is not accepted, the giver adds to it until the recipient thinks it adequate and takes it away, leaving another gift in its place. Chatwin is very scornful about Konrad Lorenz's [*On*] *Aggression* and says that his philosophy is derived from the same sources as that of the Nazis. (He mentioned a book called *Die Weltraetzel* by Heckel(?).[2]

[1] Chatwin (1940–1989) worked at Sotheby's and studied archaeology at Edinburgh University; later he became known for his travel writing and novels, including *In Patagonia* (1977), *On the Black Hill* (1982), and *Utz* (1988).

[2] *Welträtsel* (1899), by German zoologist and evolutionist E.H. Haeckel (1834–1919), translated into English as *The Riddle of the Universe.* Haeckel's racial account of human evolution became part of the "scientific" basis of Nazism.

Alan Searle came to supper last night, with George Cukor, Roddy McDowall, Paul Anderson, Gary Essert and Gary Abrahams (who run the Los Angeles Film Festival).[1] Cukor was charming and amusing—Don says that he feels in almost one hundred percent agreement with everything Cukor says about films and actors—Roddy was lively, Paul was cute, Gary Abrahams quite dazzling, Gary Essert rather silent but in a sympathetic way, in fact the evening was a success. Except for poor Alan, who seemed out of it, vague, dazed and gaga. He stared dully at our faces. He is as round as a barrel. He talked of some boys they had seen at UCLA but it seemed impossible that he could still be capable of any sexual feeling. Is this the result of the famous youth cure he and Willie took? It seems to have turned him into a prematurely old man.

Hunt Stromberg called this morning from Texas, for the first time in many weeks. Still nothing definite from Sheinberg. Hunt is all excited because Julie Christie has said she'll play in "Frankenstein"—maybe sometime. Meanwhile he has heard that a first-draft screenplay has been completed by our rival, Coppola!

Gavin writes that he has found "a terrific villa," forty-five minutes from Rome, which he has taken for a year. "Rome itself seems as seductive as ever; am beginning to feel the boost of a European injection, after the downer of France." He doesn't mention Mark.

October 9. Gavin wrote (dated October 2): "Mark, alas, is unhappy here.... So he'll be going back to California pretty soon. It saddens me, but there's no alternative. In a way I have not been good for him—postponing (without meaning to) the necessary day when he stops being automatically dependent. It's the old story of someone having to learn that you can't be attractive and naughty and funny and totally selfish for ever."

Both Don and I are slightly staggered by this. Don says he's sure that Gavin consciously or unconsciously made this move to Europe *in order* to get Mark to leave him. That would certainly explain why he sold his house here, instead of renting it. Maybe Mark suggested to Gavin that he, Mark, could stay on in the house while Gavin was away.

When I last saw Swami, on the 4th, he asked about my meditation. I said I was finding it helpful to keep reminding myself how near my death may be. Swami then told me that Vivekananda had

[1] Known as "the two Garys," they founded Filmex in 1971 and ran it together until the early 1980s.

said: If you are trying to know God, you must imagine that death is already gripping you by the hair; but if you are trying to win power and fame, you must imagine that you will live for ever.

October 22. Toothache because my new bridgework is pressing on the gum, depression because Nixon is going to win and because oil is coming to the Canyon and maybe high rises. But a deep happiness because of my life with Don. He said today, "Dobbin is so lucky to have Kitty, and sometimes Kitty loves Dobbin just because he's so lucky."

Swami, Krishna and I have now made two recordings—an hour and a half each—of Swami answering questions, about his childhood, chiefly. We taped them on the 11th and the 18th. And though there are many repetitions of old anecdotes, I think we have some valuable new material. Lately, Swami has seemed eager to tell us as much about the past as he can, and he says he can remember more than he used to. Larry Holt, who follows all such Vedanta Place doings feverishly from his flat—by means of telephone calls which drive the bookstore staff nuts—is of course pessimistically convinced that this attitude of Swami's is an omen of his approaching death.

Lately, when meditating, I've been using the photograph of Holy Mother which I got not long ago. It turns me on, like none of the others, because it shows her deeply absorbed. It is a marvellous glimpse of her. This is the first time that I have really wanted to worship Holy Mother. At least for a very long time.

With Jo, last night, we looked at her still photos of our Mexican trip in 1954 and then at our movie film of it. Astonishing, how dead the stills are—and then the figures jump out of them into the movie and are as alive as they were then: boyish-modest obliging Ben, bossy self-confident Jo (surprisingly plump), Drub far sprier than nowadays but just as sulky-suspicious when caught off guard, just as painfully smiling when he knows he's on camera—and black-haired delicious Don, a wild, defensively hostile kitten, the sexiest imaginable jailbait—but oh, how different and how much less dear than my darling Angel of nowadays!

October 27. Swami told me yesterday afternoon that he now feels dizzy if certain people touch him. (I think this included a married couple who he initiated the other day.) I wanted to ask him if this applied to me, but I hadn't the courage. He says this has never happened to him before.

We now have the go-ahead to do the screenplay of "Lady From

the Land of the Dead." Universal wants it to be long enough for two and a half hours but I don't think we ought to worry about that, for the time being. Our first big problem is to find a good opening.

Meanwhile I crawl along with the reconstructed diary; am now nearing the end of 1948. The work bores me but I know how valuable it will be when finished.

The day before yesterday, I went back to Dr. Kurtzman because it still hurt me to bite on my denture. He discovered that a bit of bone had broken off from my jaw and worked itself up through the gum; he says this isn't unusual. He stuck a needle in my jaw to give me a shot of novocaine and it was very sore and I yelled. When I came out through the waiting room, I noticed that two small children looked quite pale; they had heard the yell and were beginning to fear that they would get hurt too.

Another ailment: the pad of flesh at the base of the second and third toes of my left foot is swollen and it hurts me a little to run.

November 2. Just before eight in the evening. I'm at home alone because Don has gone to draw and have supper with a friend (Dan Price; he's a film historian, quite cute). Hunt called about half an hour ago. He has arrived here, he claims, to get to the bottom of the "Frankenstein" mystery at Universal.

Meanwhile, we are slogging along on "The Lady from the Land of the Dead" which we started October 28. We have made quite a good start, but it will be a long time before we know how much of a story we actually have. Meanwhile, I must skim through parts of Frankfort's *Ancient Egyptian Religion* to find out just what the Egyptians did or didn't believe about reincarnation. What an odd working of providence, that Gerald should have given me—or advised me to buy—this book, years and years ago. I don't think I have ever opened it before today!

The grim shadow of the election is on us. I just want it to be over with, and Nixon to have had his obscene little hour of triumph and be immersed once more in his own political karma.

One little blow today, a letter arrived at last from Bryan Forbes—he wrote it on September 18 and sent it surface mail—to say that he couldn't persuade the National Youth Theatre to accept our play.

November 5. A chilly breeze but glorious clear weather. Don has been away drawing, so I ran down to the beach alone and went in the water, which was, as they used to say in England, "bracing."

I also ran along the beach—the tide being out. When I started to run, a strange vigor possessed me, I felt almost weightless and able to sprint all out; I don't suppose this lasted for more than a couple of hundred yards, however.

Yesterday, I saw Dr. Allen about my foot. He gave me some ointment called Synalar Cream,[1] a tiny professional sample tube. I think he meant it merely as a placebo. He simply didn't know what's wrong with me and didn't much care. And yet I get such an odd impression of his goodness—a sort of Dostoevsky innocence, which wouldn't necessarily be incompatible with a readiness to commit murder. He complimented me on my clothes—the wide Jerry Magnin pants and the blue soft leather Mexican jacket from Ohrbach's: "That's a nice outfit you've got there." And he told me that, in the past ten years, he has noticed that his senior patients seem to be behaving in a much less "elderly" way. He ascribes this to (1) better nutrition and (2) a changed psychological attitude; they no longer accept the role of ancients so easily, or maybe the young don't force it on them so ruthlessly.

Hans Mechtold, Elsa's gardener, is running for assemblyman in the 63rd district. He hasn't any money for publicity, so he has had printed up a lot of fake dollar bills with his name and picture on them. The bills are marked 250, because that is the amount he promises to get taken off our property taxes or rent if he is elected.

Yesterday, I saw *Cabaret* for the second time and liked it much better than before. I still don't think it adds up to anything much, but Michael York this time seemed not only adorable and beautiful but a really sensitive and subtle actor. Liza Minnelli I liked less, however; thought her clumsy and utterly wrong for the part, though touching sometimes, in a boyish good-sport way.

November 7. We voted this morning, with heavy hearts. It was estimated by a T.V. speaker last night that balloting would take each voter about ten minutes, because of the many propositions to be voted on. Don said afterwards that he felt frustrated and I know what he meant. One longs to be able to stand up in public and yell one's vote aloud—"One vote against the liar and crook Nixon!" etc.

Mark Andrews is just back from Rome. He called to ask if we could loan him $1,600, because he needs the money immediately for a car and an apartment. He must make the down payments at once or he would lose both, and Gavin's money order wouldn't

[1] Steroid cream, to reduce inflammation.

arrive for three days. Tried to reach Gavin in Rome to check if Mark is speaking the truth, but couldn't. Don said not to trust him, remembering how he had stolen from Gavin. So I told him we couldn't let him have the money. He took this quite calmly. Incidentally, he told us that Gavin has had yet another robbery—after his three in Santa Monica. Someone stole nearly all his clothes from his Roman villa! Mark said he cried, telling this over the phone.

A couple of nights ago, a man named Philip Chamberlain called me to ask if I would introduce one of the films at the Filmex exhibition. I asked him how much they paid. He said nothing—and then made the tactless mistake of going on to give me a list of all the distinguished people—like Norman Corwin[1] and Ray Bradbury!—who were doing it for free. This was supposed to be a rebuke and it made me mad, especially since the invitation was obviously a last-minute need for a substitute and they felt they were scraping the bottom of the barrel. So I told Chamberlain that I thought it was "deeply dishonest" to organize an exhibition and not put aside any money to pay speakers. This shocked him. Don says I did wrong, and I did, I should have just refused. Now I have probably offended all the other organizers, including his cute friend Gary Abrahams. It's just that I hate this type of academic public relations fart.

Today we had lunch with Hunt Stromberg at Universal. He's back from Texas to get "Dr. Frankenstein" on the road, ha ha. When we met, he told us that ABC has announced a Frankenstein film of its own, but this didn't seem to worry Hunt a bit. He says we have a big start on them! To me, he seemed dazed and not with it, like a dope taker. Still, at least he's here and answerable to us.

November 11. The Forces of Hell won by a landslide, which I hope will carry them downhill to disgrace and Nixon's impeachment, during the next four years. There's much talk of the upcoming scandals, which will involve the White House.

Every single person we voted for lost—except that Bugliosi, the Queers' friend, still claims that he will upset Busch when the damaged ballots have been counted.[2] And the death penalty was reinstated. And the marijuana initiative was defeated. *But*, for our

[1] American writer, screenwriter, producer, teacher (b. 1910), best known for his radio dramas during the 1930s and 1940s.

[2] Vincent Bugliosi (b. 1934), public prosecutor in the Charles Manson case, ran against his boss, District Attorney Joseph Busch; the race was close, but Bugliosi lost.

side, the obscenity initiative was thrown out and so was the labor relations initiative, *and* Proposition 20, to protect the coastline, won, contrary to all expectations.

Mark Andrews says that Gavin has now sent him the money. He got the car he wanted to buy but lost the apartment.

Peter Schlesinger came to see us on the 7th, looking seedy and pale. He says that David still makes scenes from time to time and tries to get him back. But he has no intention of going. He won't even live with his Scandinavian boyfriend, says he still loves living alone. Somehow, Peter charmed us less, this time. For one thing, he kept saying about nearly everybody he mentioned, "He's so stupid." And he seems so languid and cold-blooded and unenthusiastic and sloppy and ungracious. Still—he *is* doing a lot of painting.

On October 10, for the first time, I started recording my weight in my daily diary. Perhaps because I had hit a longtime high, 152 and ¾ pounds. During the month that has followed, my low has been 148. This morning I was 150 and ½.

Hunt now declares that the studio has given him the go-ahead and that he wants to get a director this coming week. Don says this means we get a hack. We're going to try calling John Boorman in Ireland tomorrow, but he never did answer our cable; and now Hunt claims that Boorman's agent here has told him Boorman is no longer interested in the "Frankenstein" project. Jon Voight says languidly that he'll still do "Frankenstein" if he can fit it in, maybe. Meanwhile, he is eager to get us to work on a film with him, says he'll raise the money and we'll be independent of Universal. He wants to do *The Idiot*.

Today we got to page 30 of the rough screenplay of "Lady from the Land of the Dead." It's sort of fun, as long as you do it in a carefree spirit.

Another refusal of our play, by Ronald Bryden for the Royal Shakespeare Company.[1] Well at least it was considered "very seriously" by Trevor Nunn[2] and David Jones,[3] but they decided they couldn't do it unless they could get Alec McCowen (he's far too

[1] Bryden (1927–2004) left his job as theater critic for *The Observer* to become play advisor to the RSC in 1972. Later, he became a drama professor in Toronto.

[2] British stage and film director (b. 1934), later world-famous for *Cats* (1981) and *Les Misérables* (1985) among many others. He was artistic director of the RSC from 1968 to 1986.

[3] British stage, T.V., and film director (1934–2008), then RSC artistic controller; in 1973 he took over RSC Aldwych where he became known for his productions of Pinter and Gorky.

old, *we* think) to play Oliver. And he turned it down, choosing to do *The Misanthrope* instead.[1]

November 23. We are having our Thanksgiving dinner with Joe Goode and Mary Agnes. Much to give thanks for: life with Don, work accomplished and good health. We are through the first part of "Lady from the Land of the Dead" in rough, started the second part today. I have very nearly finished the big thick manuscript volume in which I'm reconstructing journals from 1945 to 1952 and maybe onwards. Am now halfway through 1949. Mike Montel, who has a repertory theater in New York, The New Phoenix Repertory Company, is doing *A Meeting by the River* for one or two performances at the beginning of his season on December 18, and we have got Larry Luckinbill and Sam Waterston to play in it again.

Not to give thanks for—that Hunt Stromberg has double-crossed us and signed Jack Smight, a C director, behind our backs, and this will almost certainly mean a C grade cast for the picture. They swear they will shoot in England in February.

Maybe to give thanks for—Pancho Kohner[2] wants to make a movie out of *A Meeting by the River*, but he doesn't want to pay us for developing a screen treatment. There is also a danger that he may want to direct. He's away right now, so this is in the mulling stage. He's a nice young man, sensitive, intelligent but no humor.

December 3. This morning Hunt called, from Texas—and lo and behold, we are even more completely double-crossed. He is going direct from there to England, sailing from New York on the 10th, and we are *not* asked to join him for the shooting of the film (which will be either in February or March). How he puts it is that the studio won't pay our fares unless we have rewrites to do. Well, *of course* we shall have rewrites to do—indeed, they already exist. But Hunt doesn't want us there rocking the boat.

Our only resource now, it seems, is to make friends hurriedly with Jack Smight and try to influence him to insist on our coming over.

I can't say that I feel a great rage against Hunt. It's as if a snake

[1] McCowen (b. 1925), celebrated for roles in Shakespeare and Chekhov, played in Tony Harrison's adaptation of *The Misanthrope* for the National Theatre in 1973. In 1990, he was George in the Greenwich Theatre production of *A Single Man*.

[2] Screenwriter, director, and, mostly, producer (b. 1939) of Charles Bronson films and *Madeline* (1998).

bit you. You know that you shouldn't handle such creatures without precautions.

We have set our trip to New York for the 11th, hoping to have finished the rough draft screenplay of "Lady from the Land of the Dead" by then. I doubt if we shall have quite, but it should be near enough. We have kept up a quite commendable record of five pages a day—and all this in the mornings only, with often a run to the beach or a visit to the gym also fitted in before lunch! We've been fairly good about getting up at seven.

Don has been adorable beyond all words, lately. I can honestly say that the passages in our life together which are like the one we've just been living through represent my idea of "the earthly paradise."

Am delighted that he has at last got the official invitation to have a show at the Municipal Art Gallery, from July 10 to August 5. This will be his biggest—and he will have, for the first time, a proper illustrated catalogue!

December 8. We did finish our rough draft of the "Lady" screenplay—on the 5th. It is terribly slapdash, but we won't look at it until we return from New York. This may be on the 20th—or later, if we decide to spend Christmas there. We definitely don't want to go to Bill Brown and Paul Wonner in New Hampshire, even if they ask us. (It looks like they aren't going to, for Bill called Don just now and didn't mention it. They are in New York at the moment and want to meet us there.)

Pleasant excitement over the prospect of working on the play. Unpleasant travel jitters. However, we have found an apparently trustworthy (and cute) house sitter named Charles Hill. He's an artist, a friend of Billy Al Bengston. He will stay here with his girlfriend. Michael Montel has called again from New York, friendly and pleased that we are coming. Gordon Hoban has gone over there to play Tom, and Larry Luckinbill's wife, Robin Strasser, will play Penelope and Jacqueline Brookes the mother. Rafferty = Steve Macht. Swamis = Barton Heyman, Tony Manionis and Charles Turner. They are short of one actor, so Rafferty[1] will have to take his own photographs.

On December 5, we met Jack Smight at Universal. We both liked him and doubtless he is competent, but Don feels sure he has no talent. However he is more than ready to be our ally and said at once that *of course* we must come to England and work with

[1] The reporter who interviews Oliver about his vows.

him there on the script. He called Hunt about this, while we were in his office, and Hunt agreed—*if* Universal would agree. I still suspect he may manage to find some out. He says he'll let us know definitely by the end of this month, after he gets to England. They still plan to shoot in either February or March. It is definitely to be a T.V. picture and apparently they are going to cut it up into lengths and serve it to the viewers two or three nights in a row!

The day before yesterday we saw Swami. He was in a strange mood—that's to say, he seemed *other* than himself. We were talking about the fragmentary poem by Vivekananda, scribbled on a piece of notepaper while in the United States. There is one line about age having "no hold," and Swami said, "I can tell you this from my own experience—I have seen that there was *no difference* in ages between Thakur and Holy Mother and Maharaj and Swamiji." What was strange wasn't just the absolute authority with which he said this but also the look in his eyes. He was looking at me as he said it, fixing me with his look, and yet I felt he wasn't seeing me. He looked *inseeing*, withdrawn, closed. I thought, as I have so many times, that his look was like a rock face; it inspired awe. And then, a few moments later, he was himself again and full of such sweetness. He asked Don why he wasn't sitting on a cushion and he looked at him really lovingly. And later, when Don was helping him on with his shoes, he placed his hand on Don's head, obviously blessing him. Don was well aware of all this; he spoke of it later.

I must repeat it again, I have been so happy with Don lately. Even my rides with him in the car, which I used to hate, are snug now. He drives more recklessly than ever—everyone who rides with us notices it and gets nervous—but I lie down on the back seat, with a pillow under my head, wrapped in a blanket, and we have long intimate talks, punctuated only occasionally by a terrific jerk as he stands on the brake, having just avoided a smash! And the wonderful intimacy of sitting with him in the movies and feeling his closeness. And the joy of waking with him in the basket—the painful but joyful tenderness—painful only because I am always so aware that it can't last forever or even for very long, Kitty and Old Drub will have to say goodbye.

I still try to keep up my midday sits, but wow what resistance! The Opposition is bent on making me forget to do them, and often I do forget, or only remember hours later. What *do* I think about, moon about, worry about, when I'm not thinking of God? Plenty, apparently.

December 22. We got home the day before yesterday. The plane was crowded and two noisy parent-indulged moppets were sitting near us. However they had their function, as far as we were concerned; they made us so nervous that we started working on "Dr. Frankenstein" to distract ourselves. Don wrote and I dictated dialogue, my voice getting louder to drown out the screaming until I was giving a performance which must have been audible to all our immediate neighbors. (No one complained either of the children or of us; most of them were elderly, so perhaps they were also deaf.) By the end of the flight, we had roughly reconstructed the entire Prima-Elizabeth-Fanshawe sequence, insofar as it needed reconstructing. This was all the more satisfactory because we hadn't done one lick of work on it all the time we were in New York.

Charles Hill met us at the airport and drove us home. He and his girlfriend Paula Sweet appear to be the ideal house sitters; the place was spotless and they enjoyed their stay, into the bargain. (However, in justice to former house sitters, I must add that they are the only ones we have paid for being in the house. When I suggested this to Charles, before we left, he asked for twenty-five dollars.)

Yesterday, Jim Charlton called; he has just arrived from Honolulu on a short visit. I had to go the UCLA library so I took him along. He looks suddenly very much older—or rather, his youthfulness looks far more damaged; he is still incredibly young-looking for a fifty-some-year-old. He was more affectionate than usual; he seemed really glad to see me and anxious to show me that he was. I was happy to see him, as always, but this is a most awkward time for visitors. There's so much to do.

Jim struck me as being perhaps a bit less lonely than when I last saw him. His work is going well and he has a capable assistant. He still lives alone but there are lots of boys around; he says he falls in love about three times a year and that he usually falls in love with a boy only after the boy has left him and gone off to some other place. He also has a lot of friends. Does he really want anything more than this? I doubt it. He wants to wander. He wants to be the stray boy who finds shelter with a succession of families or individuals but stays nowhere long. He told me he has no intention of visiting Hilde at Christmas time [...]. Jim is still stingy. But why shouldn't he be, as far as [she] is concerned?

I was going to the UCLA library because I wanted to look up an article on Gore Vidal written by a man named Gerald Clarke, who has been told by *Esquire* to interview me. I want to find out if he's a bitch or not, before I say yes. I couldn't discover the article, because I was looking too far back in the files. But Jim found a copy of a book he's been hunting for for years, an anthology complied by Dashiell Hammett, *Creeps by Night*, containing a story called "The Red Brain."[1]

What about New York? I really cannot go into all that right now. Maybe I'll keep slipping in memories of it, day by day, if I keep this journal going continuously till the end of the year. Just for openers: It was icy, windy, wet, snowy, warmish by turns but always so miserably dark—coming back here was like a return to paradise; yesterday was so warm that Don went in the ocean. I tore a ligament in my right ankle. More of that another time. The play was very well received by the audience—much more about that, another time. Of the two notices we have seen so far, one was goodish, one poorish. The goodish one is in the *New York Post*, December 19. It is by Jerry Tallmer, titled "One Saint in Two Acts." He starts with a little bit of bitchery, noting that the play is by Christopher Isherwood (born 1904) and Don Bachardy (born 1934) and continues: "It is a witty, intelligent, provocative drama, and though not long in minutes it's still perhaps a trace too long near the end.... this sketched version of a full production was as invigorating as the material, with its overtones of E.M. Forster.... I think the drama if mounted in full, though modestly, could find an audience here." Richard F. Shepard in *The New York Times*, December 20, says: "....the new work deals with Hinduism and homosexuality, and each, although not related, comes out a moral winner.... These constantly shifting triangles and relationships make an over-all interesting but wordy evening, one that slumps into troughs of tedium.... But it is a play worth more work and refinement." (Arnold Weissberger, whom I talked to yesterday on the phone, doesn't think *The New York Times* notice is really bad from a business point of view. It dashed our spirits a good deal when we read it, because we had begun to feel that we had had a success.)

December 23. This morning I weighed 153, the all-time high since I started recording my weight. This is probably due to supper with Jack Larson and Jim Bridges last night; we drank a lot of wine and ate handfuls of sugar-coated peanuts. Jim was sulking because (as Don discovered today from Jack) they had had a quarrel about Jim's cameraman on this film. [...] Jim wants to make him his

[1] By Donald Wandrei (1908–1987).

partner in a film company. (The cameraman's name is Gordon Willis.[1]) Early this morning, Jim left to spend Christmas with his family in Paris, Arkansas.

Don had lunch today with Gary Abrahams and they finally discussed my quarrel with Philip Chamberlain (see November 7) and the friction with Gary and his friend Gary [Essert] which has followed it. Don gave them to understand that I hadn't really meant what I'd said and would probably have agreed to speak if I'd been asked nicely. So it is all smoothed over. That's okay, as long I don't have to apologize to Chamberlain. That I will never do, because I did mean exactly what I said.

I had lunch with Jim Charlton on the pier. I always enjoy listening to him talk—he describes things so well. Today he talked about the weird derelict island which belongs to the millionaire (Paley?)[2] whose life Bill van Petten is writing. Then about Japan, to which he would like to return, and Mutsuo(?) the boy he deserted there. Jim said, with a significantly affectionate look at me, "He was the only other really good person I ever went with." Pretending not to recognize this compliment, I said, "Oh, you mean Mark Cooper?" (Jim had just told me that he thinks Mark killed several people, while running around with the Hell's Angels!)

Gerald Clarke came around in the afternoon. I have decided to let him go ahead with the *Esquire* interview. Not because I'm sure he's not a bitch; I think he probably is. But what do I care? The interview won't be for several months, anyhow.

About our New York visit: this was the first time that Don and I have ever meditated together. We decided to do it as long as we were staying at the hotel, because the alternative was for one of us to go into the bathroom while the other meditated, which created a pressure situation. I think it worked very well and even made another bond between us. But we haven't kept it up since our return.

On the 12th, the day after our arrival, we met Michael Montel and went with him to a rehearsal of our play. It was at the Edison Theater, where *Don't Bother Me, I Can't Cope* was playing in the evenings. (Our single performance there was set for Monday the 18th because the show was dark on Mondays.) We had seen *Don't Bother Me* in Los Angeles—it was produced by Ed Padula,

[1] Cinematographer for *Klute* (1971), *The Godfather* series (1972, 1974, 1990), and a number of Woody Allen films, including *Annie Hall* (1977), *Interiors* (1978), *Manhattan* (1979), and *Zelig* (1983).

[2] Bachardy thinks it was not William Paley, whom they knew slightly, but another millionaire whom they had heard of but never met.

who is interested in making *The Adventures of the Black Girl* into a musical—and had hated it. To us it seemed a very usual sort of "sincere," "challenging" freedom play, in which the white audience is insulted by the black performers and just loves it.

Michael Montel is nothing much of a director and his only virtue was that he let Larry Luckinbill and Sam Waterston follow the moves worked out by Jim Bridges. Larry was as good as ever and seemingly quite unchanged. Sam was as good as ever, but his success (in *Much Ado About Nothing*) did seem to have changed him for the better. He had more confidence and he actually looked a lot better, almost handsome. Dear Gordon Hoban was as good as ever, but *he* seemed definitely less attractive, beginning to fade. Jacqueline Brookes as the mother was utterly wrong; a squashy-faced mushy-mouthed woman, incapable of irony or bitterness. They all say she is a great actress, but she wasn't in this part. Robin Strasser, Larry's wife, was wrong too. She didn't even seem professional. She looked ugly and mean. Don said she was just a spoilt rich girl. She made a tremendous fuss over Larry, holding hands, stroking his sleeve, kissing him; it was embarrassing. I think he was embarrassed, too. Stephen Macht, who played Rafferty, was quite a young man, and sexy. He had a slightly demonic air and a big-toothed, rather devilish smile. You could think of him as a sort of familiar who had fastened onto Patrick, a symptom of Patrick's deplorable spiritual condition. Maybe, before long, Patrick would start going to bed with him and be controlled by him entirely—unless he could be rescued by his good angel, Tom. . . . The three swamis all learned the Sanskrit chant from us marvellously quickly. The brightest was a black actor, Charles Turner, Anthony Manionis was good too. Barton Heyman, who was a fairly important actor, showed that he thought the part was beneath his dignity. We decided that he might be a troublemaker.

The set (which belonged to *Don't Bother Me*) looked like a miniature golf course. Down the middle of it was a short slippery ramp, which outraged Sam; he kept losing his foothold and skidding, in protest. Later he had some kind of stickers put on the soles of his sandals. But the set remained tricky. The actors kept having to cross the ramp on their way to one or the other of the little putting-green platforms on which they had to play their scenes. And the ritual had to be performed right on the ramp itself. It looked like an uneasy picnic.

December 24. A brilliant day with a strong wind. Don, trying to catch the low-angled sun, lies in a hollow on our pie slice of land, his head lower than his feet, in trunks. I lie beside him with my clothes on. I can now walk, even trot, but my ankle warns me that it would just love to twist over and throw me, preferably down some steep steps. My left thumb still has that little lump on it—growing(?)[.] And there is the hard little inflamed thing on the side of the calf of my right leg which Dr. Maxwell Wolff says he will have to remove, after the holidays. My weight is down half a pound from yesterday.

Stathis Orphanos found out from catalogues that they are selling postcards I write to fans. So I got mean and sent the following to some applicant, a short while before leaving for New York:

> Because dealers solicit autographs and then sell them, I have decided that I will only send autographed photographs to strangers if they will first send me:
>
> Ten dollars, which will be given to a charity of my choice, and a stamped and addressed envelope.

I got my note returned by a Mr. B. Berger, with the following rebuke:

> Mr. Isherwood, my son Jeffrey is fourteen years old and he has been collecting autographs for going on seven years now. Among his most famous are Harry Truman, President Nixon, Allen Drury, Leon Uris, William Manchester, plus many Nobel prize winners. When he received this, he brought it down and gave it to me to read—we both smiled at each other—you may keep your autograph sir—he will surely not miss it! Have a joyful holiday season—so sorry you've been so taken advantage of.

I was just about to send some money to the Gay Community Services Center—the outfit run by Kight and Kilhefner—when something made me call Bill Legg at One Incorporated. (Actually, I had a reason for calling him, I wanted to know how to make the check out to One so that it is tax deductible.) In the course of our conversation, he told me that he and Troy Perry have lately been seeing Yorty[1] and also the District Attorney, and have arranged to consult with them on gay problems. "Some people tell me, 'They'll use you politically,' and I answer, 'Why not, we're going to use *them*,'" Legg said. So then I asked him about the Gay Services Center, and he said that, whenever anybody applies to

[1] Samuel William Yorty (1909–1998), mayor of Los Angeles 1961–1973, a conservative Democrat—anticommunist, anti-Equal Rights Amendment, antibusing; he ran for president in 1972 and, when McGovern was nominated, became a Republican.

them for help, they try to make him into a revolutionary activist; they are radical about this and declare that anyone who isn't with them is against them. Of course Legg is prejudiced. He sees all other gay groups as johnnys-come-lately, trying to grab the credit away from Veteran Warrior One. But what he said was just what I suspect about Kilhefner—Kight I'm not so sure about, but he's certainly slippery. So the result of talking to Legg was that, rightly or wrongly, I didn't send the Gay Services any money, after all. I must go down there, though, and see for myself.

To return to December 12 in New York: That evening we went to a party, given by some rich guys for a mural which had been painted for them by a friend of Mary Louise Aswell[1]—I can't be bothered to go into all this more explicitly. Our reason for going was entirely that Vera Stravinsky, Bill Brown and Paul Wonner were coming also. Vera had to leave soon—she was as wonderful and stylish and gaily worried as usual—but we stayed with Bill and Paul and had supper with them later. From them we gathered that their retreat to New Hampshire is already a disaster, at least in Paul's eyes. They are cut off from their friends, the snow is a bore and the neighbors are hostile and dishonest. While I was out of the room, they asked Don if we would come up there for Christmas. Don replied tactfully that I simply couldn't stand the cold. Paul later told Don that he hates Vera. Don says this is quite natural, because he feels that he isn't welcome when they entertain Bill Brown. Probably Paul's resentment is really directed against Bill, but he finds it easier to vent it on Vera, whom he doesn't live with.

The next day was the day I hurt my ankle. We had a lively wordy lunch with Glenway [Wescott], full of compliments to both of us and Proustian digressions which would have been pages long in print. Really, he seems unable to stop talking and perhaps hardly aware of what he is saying. In the afternoon, there was another rehearsal, at which we learned that Barton Heyman, the potential troublemaker, had been obliged to give up his part and fly to the Coast to see his sick mother. The substitute was an unimpressive-looking balding actor named Tom Tarpey,[2] who turned out to be excellent and a true professional; he had mastered the whole of his part including the chant by the next morning.

From the rehearsal we went to the Oak Room at the Plaza,

[1] American novelist (1902–1984), fiction editor at *Harper's Bazaar*.

[2] He had two small parts on Broadway in the early 1970s, followed by others on T.V. and in films, including *Nine to Five* (1980) and *Lost in America* (1985).

where Truman Capote was waiting for us, planted in a corner with his stomach in his lap. He didn't look well but he was very cheerful. He told us about his great problem, should he give a reading of part of his new novel in Carnegie Hall? If he did, all his enemies and all the people he had put into the novel would be sure to be there, and perhaps he'd be involved in libel actions.... But we felt that he had already decided to give the reading, for he already knew exactly what he would wear, a dark grey sweater with a black bow tie, tuxedo trousers and black pumps. The real object of the reading, as he presently admitted, was to prove to the world that this novel does indeed exist, or at any rate large bits of it.

We came out of the Oak Room and left the building by the side door which opens onto Central Park South. I don't think I was the least bit drunk; I had had two dry vermouths, nothing more. But I slipped on the bottom step. (People told me later that it is well known to be dangerous because too narrow.) I fell quite tidily, without dirtying my suit or cutting my hands or scraping my knees. But there was this horrible feeling—more than just pain—as if I had twisted my foot off. I felt I was going to faint. I nearly did. Then I was all right. Don held me, and Truman came out of the Plaza and made a tremendous demonstration, "Oh, poor Baby!" etc. He wanted to take me to a hospital but I wouldn't go. We went back to the hotel and the house doctor was phoned to. He advised us to get a bag for ice cubes to put on the swelling, and I lay there and felt the pain drain away, and the adrenalin start to give me a big positive lift. That night, I slept perfectly, without having to use any pills.

(It occurs to me that I left out two very important things, when writing about our lunch with Glenway. First, that he told us he believed he had had a stroke, a few days previously—this *might* account for his slightly incoherent ramblings. Second, that he said he had had to face the fact that his journals weren't nearly as comprehensive as he had imagined they were. One of his prospective publishers had announced quite brutally that they weren't interesting enough to print. So poor old Glenway, heroically, is sitting down to go all through them and fill in the gaps.)

December 25. A very dry hot Santana[1] is blowing. We went down on the beach and the sun was nearly too hot to lie in. Don went in the ocean. I paddled. It was absolutely beautiful, except for the little gusts which blew handfuls of sand over us every few minutes.

[1] I.e., Santa Ana.

Last night we went to a movie, *The King of Marvin Gardens* (wonderfully photographed, but that was about all we could say) and two parties. The first was given by Garys Essert and Abrahams. There were very few people there when we arrived, because it had been going on since midday and was to continue till midnight. But peace was tacitly made, which had been our object in going; and I had a long conversation with an attractive boy in spectacles named Rick San[d]ford, who told me that one of his youthful daydreams was to be drowned in the *Titanic.* He is going to send me some stories he has written about his earliest sexual experiences. The other party was at Joe Goode's, with the usual gang—Billy Al, Penny, Bob Grah[am] etc. It was intimate and snug and Joe and Mary Agnes served a Fortnum and Mason Christmas pudding and Dom Pérignon. Like Billy, Joe is rather fond of these little displays of luxury living.

I asked Don this morning where he would most like to spend Christmas. His unhesitating answer surprised me a bit: with Salka, Peter Viertel and Deborah Kerr. Second choice, Marguerite [Lamkin], David Hockney. Third choice, Dodie and Alec Beesley.

My weight is now down to 151.

Early on the morning of the 14th, the hotel doctor, Dr. Joseph Pincus, finally paid us a visit. His chief concern was with my medicare coverage—he merely glanced at my ankle. He was like an insurance agent. He made me feel I was about to be processed—ambulanced, x-rayed, put into a cast and reduced to invalidism. Don became desperately depressed and suggested that we leave for Los Angeles instantly. He saw the accident as the start of a repetition of my sickness during our visit here last January. "We're even in the same room," he said grimly. From his point of view, the spraining of my ankle was psychosomatic; I did it because I hate being rushed and hate New York because it is the city of haste. I wanted to slow everything, including Don, down. As for myself, I couldn't confidently tell him this was nonsense. I think it may well have been true, or at least as true as any other explanation. However, as I pointed out, we had to go to the orthopedist first and get me fixed up; I couldn't travel as I was. And the orthopedist, Dr. Fred Hochberg, was young and efficient. He x-rayed my ankle, said it wasn't broken—at, least not in a major way; a small bone *might* be cracked, but he didn't think so. He taped it up in a compact manner, so I could get a shoe on. (The ankle was swollen and discolored because a ligament was torn.)

So the psychological clouds lifted. We decided to stay. In the evening, Don went to see Joe Brainard and I hobbled across the street to see sex movies. The hotel porter provided me with a weird aluminum contrivance, fitted with four small feet. It took the weight off my ankle but threatened to put so much strain on my wrist that I could easily have sprained *that.* (The next day I exchanged it for an ordinary walking stick.)

The two sex movies were *The Boys in the Sand* and *Bijou.* I had seen the former already, at that same theater, on February 2. I found the blond boy in it more attractive this time and I jacked off while watching him being fucked by the black telephone linesman. The only really imaginative moment in the film is in one of the other episodes, however. The blond boy is lonely. He sits beside his swimming pool playing with his cock or runs with his dog on the beach. Then, one day, he sees an ad in a newspaper—we aren't shown what the ad says, only that the newspaper is gay. He comes home, writes a letter. Then he waits. Days pass. He calls at the local post office and picks up an envelope containing a small box. He returns to his swimming pool, opens the box and takes out a large pill. He throws the pill into the pool. The water foams violently. Then a big black-haired boy with a muscular brown body pops up out of the foaming water, grins at the blond boy and swims over to him. They have each other, every which way, beside the pool. Later, we see them walking along the road to the post office, fully clothed, with their arms around each other. As they pass the post office, a rather absurd-looking queen comes out of it. He is eagerly reading the same gay newspaper. The blond boy and the dark boy grin at each other. I wonder if the idea for this came from something in *The Arabian Nights*?

Bijou has an extraordinary opening. We see a young construction worker returning home from his job. He is being cruised by a man in a car—at least, it looks that way at first. Then it begins to seem that the driver of the car is really interested in a girl, whom we pick up walking in the opposite direction, toward the construction worker. At length—and all of the opening of this film is told at length, with the slowness of an Antonioni—worker and girl arrive on the opposite sides of the same street; they will pass each other as they cross it. But now the driver's car swerves into the intersection and knocks the girl down, apparently by accident. The driver gets out and bends over her. She is unconscious, maybe dead. The shock of collision has flung her purse some distance away from her. The worker picks it up surreptitiously and hurries off, hiding it under his jacket. He gets to his room and opens the purse. Among the various objects in it is a card with the name *Bijou* on it and an address. The card says that one must go there at a certain time. After this, there is at least one reel of pointless dawdling, while the young worker showers and plays with his cock; it is established that he is thinking entirely about fucking girls. Then, finally, he goes to the club Bijou. An old woman, probably a man in drag, signs to him to enter. He gropes his way along dark passages, sees a lighted notice telling him to take his clothes off. He takes them off, walks farther until he finds himself in a room full of guys, all naked. They handle his body. He doesn't seem to mind a bit. And soon he is taking part in a slow-motion stylized homosexual orgy which lasts until the end of the film. You never see what happened to the girl and the driver of the car, and you never know why she had that card for the Bijou in her purse.

December 26. Two more parties last night, one given by Jennifer and Norton Simon, a catchall affair, in which his relatives and friends, rather than hers, seemed to predominate. At least two of Norton's relatives appear to be retarded, one of them probably dangerous; she went around clutching a small red box. Ben Weininger, Jim Charlton's analyst-guru, is a [...] false-innocent, in my opinion.[1] Jennifer was in a party daze and took very little notice of us. Norton ditto. Later we went on to see Leslie and Michael Laughlin, where we were much more warmly welcomed. But oh thank God Christmas is now over. One should always spend it in China.

On the 15th, we had supper with Vera Stravinsky, Bob Craft and Ed Allen, at the apartment. I didn't enjoy it very much. There is a bad atmosphere. Bob was bossing Vera, telling her to have the drinks at one table when she wanted to have them at another. She told him, "I shall do what I like, this is my apartment." She laughed at Don and me as she said this, but I felt friction, perhaps even some hatred. Vera I shall always adore, but I feel myself becoming estranged from Bob. He is full of cleverness and contempt. And, when Vera had been showing us her paintings and we came back into the living room, we found Bob and Ed whispering together like two sneering courtiers of an aged queen.

On the 16th, there was a frighteningly strong, icy wind. We saw Julie in *The Last of Mrs. Lincoln.* Julie was just about as good as she could possibly have been, considering the play. Actually, the play isn't nearly as bad as the critics have said; it is very informative, and the tragedy and comedy of it are well balanced. What is lacking

[1] Psychiatrist, teacher, and founder of the Southern California Counseling Center; he published several practical books, including a guide to aging.

is the admission that Lincoln must have been largely responsible for his wife's madness. He looms in the background, but merely as the usual political plaster saint. In the evening, we had one of those duty dinners with Virgil Thomson. The food was delicious. And I am sort of fond of Virgil. But being with him is such a terrific production; you have to work at it every instant. Among the guests was Lorna Levant, one of Oscar's daughters, who has grown up to be a rather attractive bachelor girl in the music world; she works for John Houseman.

December 27. Don drew Jim Charlton, yesterday morning—a gesture which pleased and touched me and also impressed Jim, who later told me that he had felt very flattered. Then Jim and I had a goodbye lunch at the Bellevue—goodbye for a long time, probably, since Jim says he isn't coming back here next Christmas; he doesn't want to see Hilde and he feels that his ties with former friends here have weakened. We talked of his getting me an invitation to speak at the university, so I could come and visit him over there. He has definitely opted for Hawaii, or maybe for Japan. He feels that he has a life in Honolulu, but he declares that he would much rather not live alone. I still feel that there is strong affection between us. It was sad to see him go.

Later, Don and I went to the vigil of the Holy Mother puja, at Vedanta Place. The singing was wonderfully sweet. Jim Gates played his violin and Peter was at the hand drums. Didn't get to speak to either of them. Asaktananda, for all his seriousness, conducted the ritual with complete informality. There was a lot of laughter from the monks and nuns inside the shrine; I don't know what about. When it was my turn to be touched with the relics, Asaktananda said, "Chris, I thought you were still in New York!", in a tone which seemed more suitable for a tea party. This sort of behavior on sacred occasions is so characteristic of the Hindus at their best and most lovable. Swami came into the shrine, but only briefly. We saw him later. He was complaining of giddiness again.

On the 17th, Bob Regester took us out to lunch at a restaurant called The Monks' Court, with George Cukor and three others. I guess the lunch was chiefly in honor of George, who has been touring the lecture and T.V. circuit, talking about *Travels with My Aunt* but chiefly trying to boost the sales of Gavin's book on him; he is indignant because they have been so poor and because, he says, his publishers have done so little to help. Bob was breathlessly aware of the importance of this social occasion and anxious to be a perfect host. He had rented a limousine and provided champagne for us to drink on the way to the restaurant (where the waiters were all dressed as Catholic monks). Bob himself looks ill, and admits that he has been told by his doctor to lay off liquor; but he did take some, nevertheless. I found his eagerness and anxiety far more touching than it was tiresome; we just had to do our best to be appreciative. It was a hellishly cold, miserable day, and the restaurant was plunged in gothic darkness but otherwise quite snug. When the sun shone weakly into it for a few moments through a grilled window, it looked like a Piranesi dungeon.

In the afternoon, we went on to see Sam Waterston in *Much Ado About Nothing*, which was played in the style of a musical, with a brass band and 1910 costumes. A car drives on stage and backfires. Balloons fall from the ceiling. Sam played Benedick, coming back from the wars in a moustache and a Teddy Roosevelt uniform. He was loud and hammy and confident and the audience just loved him and the whole show. I think it would seem like a miracle to Londoners, after the respectfulness of the Old Vic.

In the evening we had supper with Hugh Wheeler. I find him fascinating, chiefly because he seems so enigmatic, always withholding something, and yet he is warm, not in the least cagey. He talked a lot about this musical, *Irene*, which Gielgud is directing and which may or may not be a thundering disaster. Hugh's indignation, as he tells these stories, is peculiarly British. He revels in it.

On the 18th, we had our dress rehearsal in the afternoon. Anita Loos and Paulette Goddard came to it and loyally laughed and applauded. No one else did (I think many of them were from the New Phoenix Repertory Company, so they were being *dis*loyal.) Jacqueline Brookes and Robin Strasser were even worse than usual. The lighting had been done only that morning, very carelessly. Sam Waterston, after the speech which ends, "I'm all alone," was left grovelling in front of the seat, brilliantly lit. We waited, sweating, for the light change. It didn't come. So poor Sam had to hoist himself up onto the seat, sit there with egg on his face, and count up to a hundred before starting to describe his vision. We were in despair.

And then the classic thing happened. The evening performance was—well, not marvellous but infinitely better than we could have hoped. Larry and Sam were in their best form. Gordon Hoban was excellent. So was Stephen Macht. So were the Swamis. Jacqueline Brookes managed to get a few laughs—though she was still wretchedly inferior to Florida Friebus. Robin Strasser didn't altogether disgrace herself. We had only one bad moment. In order to correct the afternoon's goof, we had asked for a total blackout when Oliver gets up onto the seat. But the audience took this as a signal that the play was over, and many of them applauded. (This incident was referred to by the critics.) Nevertheless, the play was very well received, with laughter in all the desired places and real enthusiasm at the end.

The evening audience included Julie Harris and Jim Murdock (who now calls himself by his real name, David Baker), Virgil Thomson, Paul Cadmus, Glenway Wescott, Myrna Loy, Hugh Wheeler, John Houseman, Hal Prince. We had recklessly invited between forty and fifty people. (A bill arrived today from the Phoenix Theater for half the amount of the tickets, seventy dollars; we are splitting the cost with them.)

Talking of the Phoenix reminds me of a coincidence I have only just noticed, while rereading my diary for this year: the theater at Leicester, England, which offered to put on our play was also called The Phoenix!

We heard later, through Hugh Wheeler, that Hal Prince had been tremendously impressed by our play—which is very important because he is one of the chief people in the Phoenix Company's organization. So, undoubtedly, was John Houseman. So, I really do think, was Virgil; he remained awake throughout. (Through Hugh, I leaked an explanation of my offended feelings about *Cabaret* to Prince. He affected to be amazed. I don't know if this will lead to a reconciliation. It probably won't, unless Prince decides to put our play on for a run.)

Mike Montel, having blandly and blindly declared from the very beginning that all would be well, was sort of vindicated—which we both regretted, because our nondisaster was due to Sam and Larry and Gordon with their memories of Jim Bridges' staging, not to Mike. I don't begrudge him his fool's luck but I do hate it that he has been praised in print, instead of Jim. In fairness, though, I must add that Jim's name was on the program as the director of the original Los Angeles production.

When we went backstage to thank the cast, Gordon Hoban said rather pathetically that he hoped the next performance of our play wouldn't be delayed very long, or he'd be too old for the part. So we joked and told him that in that case we'd change the lines; when Patrick says, "I only wish we'd met ten years ago," Tom will answer, "You mean when I was twenty?"

The evening ended delightfully with a sort of victory party given by J.J. Mitchell's handsome and nice friend Ron Holland, at a restaurant called Ma Bell's, where they have telephones on all the tables which you can use for free, anywhere in the New York area. Ron told a story about a boy he picked up at a gym he goes to. He brought the boy to this restaurant and told him he could call anybody he liked. The boy was delighted. He called his mother and started telling her what a wonderful place he was in. Then his face fell. He turned to Ron and said apologetically "I've got to split—she says my father's dying."

December 28. It was quite hot yesterday, and we went down to the beach and Don went in the water. I can now run, more or less, and my ankle gives only occasional twinges.

Talked to Jim Bridges on the phone this morning. Yesterday he saw an assembly cut of his picture. He seems very pleased with it, particularly with John Houseman's performance. But he admits that some of the photography is "pretentious and dark." Having told me this, he added hastily that it was "just between the two of us—and Don, of course," so I suppose he doesn't want to concede anything to Jack in their argument about Gordon Willis.

On the 19th, we got up early and went to the T.V. studio to be interviewed by Larry Luckinbill on ABC's [morning] program. Sam Waterston was there too. None of us had slept much, but Larry was amazing. Before our interview began, he was dressing a Christmas tree on camera, with viewers calling in to make suggestions; their voices could be heard on the show and Larry replied to them as he worked, suavely sending them up. One lady suggested that he should use fresh grass. Larry raised an eyebrow, for the benefit of sophisticated viewers, "*Grass*? *Fresh* grass? How do you *get* it? I wouldn't know where to get *grass* in *our* neighborhood ... etc. etc." The funny thing was, Larry's behavior and patter seemed like an extension of Patrick's carryings-on in our play, and Sam, watching him with admiration and, at the same time, a certain horror, seemed just like our Oliver. The two of them played a bit of a scene together—when Oliver is explaining *namaskars* and pranams to Patrick, just before the lunch party with the swamis in act 1. Larry asked Don quite a lot of questions, about his art work, which pleased me, and him, I think.

The rest of the day was spent running around—to lunch with Anita Loos, to see the orthopedist again, to look at Maurice Grosser's paintings of Morocco (one of which we considered buying), seeing Michael and Pat York, who had arrived in town just too late to see our play. Michael is to be in Tennessee's *Two-Character Play*. He hinted strongly that he would be interested in playing Oliver in a possible film. Our day ended by seeing *Pippin*, a musical with

John Rubinstein,[1] which I thought a waste of money. Rubinstein has got so thin, and the script was even thinner. But Don liked the dancing. And then we had a boring goodbye drink with Irving Drutman and Mike de Lisio.[2] And next day we flew back here.

Truman Capote never showed up for the opening of the play. Next day he called and made some unconvincing excuse; there had been a flood in the cellar of his country house, or something. This is the second time he has failed us this year; the first time was when he promised to review *Kathleen and Frank* and then didn't, and didn't even contribute a blurb.

The total cost of our New York trip, including seventy dollars we had to pay the New Phoenix Theater for tickets we'd bought for our guests, was $1,152.04.

1973

January 1. We saw the New Year in at Jack and Jim's, eating too much cake and drinking too much wine—with the result that my weight is up to 1[5]2 and ¾.[3]

For the first time that I can remember, Swami called, this morning, to wish us both a happy New Year.

1972 certainly produced some satisfactory accomplishment: our play presented twice and rewritten once, the mummy screenplay finished in a rough draft, and a whole big volume of reconstructed journals filled right up (though there are a few more bits and pieces to be added to it). I'm afraid Don wouldn't say that *his* year has been satisfactory, though. He did (as far as I know) very little painting and he is dissatisfied with his drawing—until a few days ago, when he did some drawings of Renate which pleased him.

What we now have ahead of us are four possible projects:

To go to England and work on the script of "Dr. Frankenstein"—*if* Jack Smight can persuade Hunt to agree to having us. We now take it for granted that he doesn't really want to; he thinks we are boat rockers. But Smight does want us, I'm fairly certain. Don doesn't much want to go anyhow. I only want to go in order to see friends.

[1] Son (b. 1946) of pianist Artur Rubinstein; he originated the title role in *Pippin*, had later Broadway successes, appeared on T.V. and in films, composed film scores, and directed.

[2] Drutman was a friend of Capote; he advised Cocteau on his film *Beauty and the Beast* (1946), wrote songs, and later edited a book of Janet Flanner's Paris essays. De Lisio was his longtime companion.

[3] Isherwood typed "172 and ¾."

To finish the screenplay of "Lady from the Land of the Dead." This is a plain bore.

To get the screenplay of *A Meeting by the River* written. Very difficult but challenging and exciting.

To work with Lamont Johnson on a screenplay of *Adventures of the Black Girl in Her Search for God*, to be directed by him in Africa. This is exciting because of the chance it gives us to see Africa—and maybe also to see the total eclipse of the sun which will take place on June 30. Lamont will be in Africa then, shooting another picture, into which he wants to introduce an eclipse episode; so maybe, with a lot of luck, we could arrange to join him there for a "consultation," just as we joined Tony Richardson in Australia.

The little lump in my thumb persists. And the sole of my left foot still has this swelling inside it. I must see Dr. Kafka, who promises to correct this with a support inside my shoe. Otherwise, I'm pretty well. My ankle gives me very little trouble.

Books read in 1972. Very few made a strong impression. But—Castaneda's *Journey to Ixtlan*, with the two earlier books,[1] seems to me to form a masterpiece of magic comedy or comic magic or whatever you can call it. These books seem even better if you don't altogether believe that what Castaneda writes is literally true. If Don Juan is invented by him (as some say) then Don Juan is one of the greatest characters in the whole of modern literature. Knut Hamsun's *Mysteries*, which I reread, seemed truly marvellous Mortmere writing, far better than I'd remembered it. Halfway through *The Way We Live Now*, I was in raptures and thought it superior even to *The Eustace Diamonds*; now I like the latter more. But Trollope really is one of the most readable of all writers, at least in his later period. Rolf Hochhuth's *Soldiers*[2] affected me lastingly—but as journalism, not as a melodrama; the disgustingness of Churchill is uncannily conveyed. Susan Hill (*Strange Meeting* and *I'm the King of the Castle*) is a good writer but I don't feel she has rung the bell yet. Mishima's *Spring Snow* bored me, but when I had finished it I felt I had visited a world and understood something about it; it haunts me. I shall try to read the other volumes as they appear.[3]

[1] *Teachings of Don Juan: A Yaqui Way of Knowledge* (1968), *A Separate Reality: Further Conversations with Don Juan* (1971), *Journey to Ixtlan: The Lessons of Don Juan* (1972).

[2] *Soldiers, An Obituary for Geneva: A Tragedy*, translated by Robert David MacDonald from the 1967 German play.

[3] *Spring Snow*, translated from Japanese by Michael Gallagher, was the first volume of Mishima's tetralogy *The Sea of Fertility* (1965–1970).

Some good films seen (including a few old ones): *Deliverance, Cisco Pike, The Criminal Life of Archibaldo de la Cruz, Sunday Bloody Sunday*, Pasolini's *Decameron, Taking Off, The Kremlin Letter, Get to Know Your Rabbit, Thomas the Imposter, Performance, That Certain Summer, The Discreet Charm of the Bourgeoisie, A Tear in the Ocean, Bad Company, Fat City*, Fellini's *Roma.*

January 5. The weather has been incredibly cold lately. Another dazzlingly clear morning today. On the opposite side of the Canyon, while the sun is still over behind us, you now see a mysterious-looking shadow—cast by that abomination, the high-rise towers at the end of our street.

David Hockney arrived here yesterday, full of the news that he has fallen in love with Celia, the wife of Ossie Clark the dress designer. So Celia is coming out here, with her children. David doesn't know what he is going to do about this situation. They haven't even been to bed together yet; Celia says she is afraid that might spoil everything. David not only loves her but loves Ossie and Ossie loves him. Ossie may love Celia too but he beats her and runs around with boys, so perhaps part of Celia's reason for coming out here is to take a holiday from her marriage. Don remarks that David now is showing another side of himself; he seems surprisingly innocent and vulnerable, like a country boy. David has a commission here with Gemini, the printmaking place;[1] he will be working here for several months. He says that Peter's attitude toward him has changed since he got involved with Celia; Peter seems "insecure."

Before we heard from David, we had arranged to meet Jon and Marcheline Voight at the projection room on the Strip and see a Russian film version of *The Idiot.* So we did—it was actually only part 1 of the film, but it ran just over two hours; it was literal, plodding but quite well acted and gave you a strong sense of the original mood and atmosphere. After seeing the film we all met up with David for supper. Jon and he got along fairly all right, especially after Jon had asked David how much his big paintings cost and had been impressed by the figure $22,000. Also, of course, they could agree on the subject of the Vietnam War. Later, we went back with the Voights to their apartment[. . . .] Marcheline is now several months pregnant, after one or two miscarriages.

In the afternoon, we went to Universal and briefly saw Jack

[1] Gemini Graphic Editions Limited, an artists' workshop and lithography publisher in West Hollywood.

Smight, who leaves for England and "Dr. Frankenstein" today. He has heard nothing from Hunt and is somewhat worried. All he knows is that Richard Chamberlain is definitely not going to be in the picture; he has another commitment. We came away feeling that maybe, even at this late stage, the whole project will be suddenly dropped.

January 13. Only eight days since I wrote the above, but there's already an absolute logjam of news.

We are definitely going to England, it seems. On the 22nd of this month, in a Pan Am jumbo, leaving at noon, first class. The first class part of it is our great inducement; Don has never flown first class anywhere. Universal is being unexpectedly generous with our expense allowance, in addition to the first-class ticket we get $600 a week *each.* (Robin French found out from George Santoro, the business manager, that Hunt himself is getting $750 a week, but with that he has to pay for Dick Shasta and Dick's mother.)

Now we are trying to find an apartment we can borrow. Bryan Forbes, who offered us his, now says it's being taken apart because damp has been discovered in the building. Marguerite won't be there; she has left already for India. Other prospects are Michael York's flat or David Hockney's.

Yesterday we heard we have won our dispute with Gert Macy about the earnings from *Cabaret*; this means that I shall get about $25,000. What riles me is that I have had to pay between $2,000 and $2,500 in order to win back what was, after all, always my own money. It is hateful to have to pay lawyers *anything.*

The lump in my left thumb has now been x-rayed. It is not a growth on the bone and therefore, apparently, not significant. So I'll try to forget it and perhaps it will go away. Meanwhile, Dr. Kakfa is having some supports made which I can slip into my shoes and thus take the pressure off the ball of my left foot. I'm definitely flat-footed.

A virulent kind of flu, known as "London," has arrived in this city from San Francisco. Jack Larson is down with it and has a fever of 103. It often leads to pneumonia—"the old man's friend." Jim Bridges has a sore throat but hasn't succumbed yet. He is still cutting his film and also preparing for his production of *Streetcar.* He hopes to show us the film before we leave.

This morning, an advance copy of the paperback of *Kathleen and Frank* arrived from Curtis Books. It has a very old-fashioned cover, a painting of Kathleen and Frank faked from their photographs, with both faces distorted and dramatized, so that Kathleen looks "imperious" and Frank brutally stupid. I don't altogether dislike this; it has an arresting pop-art quality. But the print is painfully small—considerably smaller, for example, than Speer's *Inside the Third Reich*, which is a much thicker book of the same size.

January 21. Tomorrow morning we take off.

Omens: This evening the sunset had a green flash, very bright and distinct; we both saw it.[1] Yesterday evening, we had supper at a Chinese restaurant recommended by Chris Wood, called "Lucky"; and Don's fortune cookie had no fortune in it.

David Hockney, at a party given in his honor by Ron Davis, an abstract artist: "I dislike abstract art more and more."

We have seen Jim Bridges' film, *The Paper Chase*, and are both much worried; it doesn't seem to add up. The photography is beautiful, though Jack was quite right when he feared it would be too dark. And John Houseman gives what is maybe the best amateur performance ever. He could play Julius Caesar, easily.

Yesterday we went to say goodbye to Swami. We talked for five minutes about the fact that he hates prune juice and that it disagrees with him and that nevertheless they had made him drink it, with the result that his stomach became bloated. He seems much better. When Don and I bowed down and he gave us his blessing, I felt that his love is the strongest of all the bonds that bind us together. We didn't comment on this, but, when we got into the car, Don smiled at me so sweetly and took my hand.

[1]An optical phenomenon caused by the refraction of light through the atmosphere as the sun drops below the horizon; most easily seen in clear air across an unobstructed view like the ocean. Isherwood also mentions it in *D.2*.

As usual, I failed to keep a diary during our trip. So what follows is written retrospectively:

We took off at noon on January 22. The fact that we were travelling first class gave our journey a honeymoon atmosphere. We sat right up forward in the nose of the plane, and it felt as if we were all alone there. (Actually there were very few first-class passengers.) The hostesses seemed eager to unload their entire allowance of drinks on us. They kept bringing Campari, and later wine. We got drunk and giggly over the pretentious Frog menu, wondering what a "cascade of shrimp" would be like—it was just a few of them on a plate. We held hands and kissed and Don said it was fate that we had met and that he adored me for being so lucky. I felt an intense bliss of togetherness—a distillation of my combined feelings when I'm sitting beside him in a movie, eating a snug meal alone with him, lying beside him in bed—and indeed we were more or less doing all these things at once. The bliss of being where you would most like to be, at this moment and always.... However, when we did get sleepy, we found that the partitions can't be removed from first-class seats. So Don lay down on the floor. So much for luxury travel!

In the morning—or rather, in the latter part of that immensely long night—we found ourselves walking miles and miles and miles along the uncannily clean deserted corridors of London Airport[1]—to be met by our production manager, Brian Burgess, with two envelopes full of pound notes, the first installment of our expense money. And then a studio-provided limousine drove us into London. We were well aware that our driver was shocked by the squalor of Powis Terrace—even the residents agree that it is a bad neighborhood—and besides, what were *we* doing there, when that bulging envelope of notes entitled us to stay at the Connaught or the Dorchester?

(On closer acquaintance, Powis Terrace seemed to be squalid rather than "bad." No doubt one could get mugged there—where *can't* you get mugged, these days?—but the atmosphere is unalarming, peeling houses, trash cans spilling over the sidewalks, seedy shops run by thin pop-eyed Pakistanis who appeared to be far more afraid of you than you could be of them. You see racist slogans written up against them on walls: "Kill Asian shit," for example.)

[1] I.e., Heathrow.

David's flat—which opens improbably off a dingy slummy staircase—is all elegance, much larger and grander than when we last saw it in 1970. Indirect lighting controlled by rheostats, a shower with horizontal jets of water, a dining room big enough for a banquet, a library, a streamlined kitchen. We were given a snug little nest of a bedroom with a bed almost as inviting as our own basket. It was only when you opened the Moorish shutters that you discovered you were looking out into an airshaft across which washing hung on wires; slumland was all around you, with its uninhibited noises. A blaring radio woke us regularly at 7:15, except on the one morning when we had wanted to get up and had counted on it; then it was silent.

Mo McDermott and his friend Mick Sid[a][1] were looking after the flat. Mo was at art school with David and has been a close friend of his ever since, but I hadn't met him before.[2] He is small and squarely built, a blond with a very white skin, going a little bald on top, but still cute and sexy. Indeed, it is easy to imagine that someone could fall in love with him, because of his warmth and anxious intensity and a peculiarly British kind of sturdiness. He is the faithful-unto-death type of retainer; you feel that, if the flat were burgled, his corpse would be found stretched across the threshold. He speaks with quiet fury against people who take advantage of David's generosity. He had recently had a scene with Ossie Clark, after demanding that Ossie should give up a doorkey which he had taken away with him. And we both felt that Mo resents the casualness with which Peter Schlesinger shows up and entertains us there, offering us David's wine.

Mick is a fuzzy-haired skinny freak with an awful complexion; he is shy or rather standoffish, like an animal. He helps Mo make the cutout wooden trees which are now becoming a successful business venture; they are selling them to the States. Mo tells us that Mick never gets home until 4 a.m. We don't quite like to ask if this is because of a job or because he prowls the clubs. In the mornings, when we peep into the big bedroom, they are both in bed. But it is always Mo who gets out to answer the phone or cope with any emergency.

Mo has a very beautiful white cat who is also standoffish. He isn't used to living in David's flat, so he won't use his box and shits on the floor every day, on exactly the same spot.

[1] Isherwood typed Sider, as he would have heard it pronounced—with an English accent.

[2] Isherwood is forgetting an evening at the ballet; see April 11, 1970, above.

That first morning, the 23rd, we had a meeting with Hunt Stromberg, who is installed in an office on Piccadilly, quite near Hyde Park Corner. I suppose these houses all used to be family homes. Their vast rooms have now been partitioned off, with a crudity which creates a sort of war-emergency atmosphere; you can't believe that these cheap plywood walls can be anything but temporary. Two or three doors down the street the ground floor of another house has been converted into a hippie restaurant, The Hard Rock Café. The noise is terrific, which excuses you from talking. They serve you sandwiches so huge that you can hardly bite into them. The customers are mostly long-haired and there seem to be many Americans—yet again you feel the essentially British grin-and-bear-it air of emergency.

We found Hunt interviewing actors. His manner with them is timid, flattering and yet rude. He seems vulgar, insecure. Dick Shasta is playing the role of co-producer. While we had supper with them and Dick's mother that evening, at the flat they have rented on Grosvenor Square, Hunt and Dick brought forth their suggestions for rewrites. The chief one was that the Creature shouldn't be created by means of light[n]ing, with a kite used for a conductor, because that is in the original Frankenstein film. Hunt and Dick suggested solar energy. This seemed an idiotic notion for England, where the weather is so unreliable, but we said, Sure, sure. Then Dick said that the Creature couldn't carry Polidori right up the mast; no actor could manage such a feat. Therefore, he should be hoisted up on a hook, which happens to be hanging down from the mast at the end of a rope. Okay, okay. Then Hunt said that the Creature shouldn't throw Elizabeth's clothes all over the road, otherwise she wouldn't have anything to wear on the voyage. We agreed to this but privately decided to get around it, because cinematically a monster throwing beautiful female clothes all over a road at night is better than throwing pots and pans or anything else.

We spent most of January 24 at Pinewood Studios. Hunt and Dick were there and we met and had lunch with Ian Lewis, who is the production head of the studio, and a couple of his assistants. It is much grander than Hollywood; the dining room has quite a baronial air and they have actually used it for a set in a costume picture. But Lewis and the others were much less grand and much much less full of shit than our bogus Hollywood executives; they actually seemed to know the business and had been cameramen, cutters, directors themselves. In the afternoon we talked to Wilfred Shingleton, the art director, and—such is the power of misplaced ingenuity—at the end of two hours we had recklessly and shamelessly invented the whole process of conveying solar energy to storage batteries by means of mirrors, and also the accident which causes the mirrors to burn each other up and crash down upon Frankenstein and the Creature. After which Mr. Shingleton said he would produce sketches of all this—we never saw these—and we parted with the utmost politeness.

Later we saw Peter Schlesinger and went with him to have supper with Tony Richardson. Tony seemed very much as usual. We talked a lot about his latest film project, [H. G.] Wells's *Tono-Bungay*. But all this was just table conversation; nothing intimate was said. Later we found out that the situation in the house has its depths, or heights—that is to say, Will Chandlee(?) the boyfriend I met at Tony's place in the South of France in 1970, is living upstairs, while Grizelda [Grimond], the girl who has just had a child by Tony, is living in the basement. We saw Will that evening but not Grizelda.

January the 25th was the first of our stay-at-home working days on the rewriting of the script. Luckily, David has a Smith Corona electric typewriter of the same model as ours, so Don felt at home with it. We had supper at Patrick Woodcock's house. Hal Buckley's brother Peter was there with his friend Kevin McKormick.[1] Patrick described them as "dishy." He seemed older and stiffer—he had had an operation on his knee—but looked very healthy. He is planning to retire soon to the South of France and build a house there, to which he will invite his elderly wom[e]n friends to stay—his "White Queens" he calls them. They are Rosamond Lehmann, Hester Chapman and Gladys Calthrop.

As so often before, Patrick held forth on the advantages of dying of lung cancer; he has become quite aggressive about this. It was *the only* way to die. It didn't hurt a bit and you had a whole six months in which to put your affairs in order.

On January 26, we went to see Peter Schlesinger. He has a large front upstair room in a house which is only a minute's walk from David's. Peter's bed—or rather, mattress—lies on a wooden platform which you reach by a ladder. His paintings stand stacked against the wall. He cooks on a hot plate and shares a bathroom with another lodger. The room overlooks the street and is much lighter and less shut in than the Hockney flat. And Peter himself

[1] Hal Buckley was an American actor (1937–1986), mostly on T.V.; he had small movie roles in *Kelly's Heroes* (1970) and *Shampoo* (1975). Peter Buckley also had show business aspirations. Kevin McCormick produced *Saturday Night Fever* (1977) and became a studio executive.

seems different in this setting—freer, more individual, more cheerful, more admirable—no longer a spoilt little boy, nested in luxury. He keeps repeating how much he likes living alone.

After lunch we went to the Universal office and met Nicola Pagett, whom Hunt and Jack Smight fancy to play either Elizabeth or Prima. She is impressive and not, thank God, very nice. Then I realized that she is the actress I so much admired in 1970 as Blanche in *Widowers' Houses*.

We spent the rest of the afternoon working with Jack Smight at his flat in a mews in the Sloane Square neighborhood. This was furnished just as you would expect a shrewd decorator-landlord to furnish a flat which he wanted to rent to Americans. The accent was heavily on British tradition; to intimidate them. "Good" furniture, coats of arms, engravings of eighteenth century fighting ships, etc.

We had firmly made up our minds to be as agreeable to Jack as we knew how, and I think we succeeded. Actually, he is very easy to get along with. Indeed he accepts our ideas and objections with a readiness which is alarming. It suggests that he has no opinions of his own. His wife kept popping in and disturbing us; she is one of those women who hate to see her man busy with other men. And there is a slobbish but agreeable teenage son who was going through the ordeal of adjusting to a London school for the sons of Americans working in England.

In the evening, we saw *I and Albert*, a musical directed by John Schlesinger. It has failed and is soon coming off; and the first act is certainly rather dull. But, in the second act, Lewis Fiander as Disraeli does a number with Victoria which is one of the best things of its kind I have ever seen on the stage. Disraeli plays the scene as a conjurer who is trying to keep Victoria amused and at the same time singing a song to the audience about the best way to handle old ladies. He produces coins out of the air and drops them into her empty treasure box, he fishes the Suez Canal out of a top hat—it is a long blue ribbon, he turns his handkerchief into a Union Jack and then discovers an earl's coronet in it for himself. It was like a series of political cartoons from *Punch* brought to life. The effect was both surrealistic and realistic, because it gave you such a vivid impression of what Disraeli must actually have been like.

After the show we had supper with Bob Regester at Wheeler's Braganza. A supper at Wheeler's was one of the treats we had looked forward to when thinking of this trip. But, in actual fact, we only really like the Wheeler's in Old Compton Street. The

Braganza is too grand and the one off Jermyn Street is so uncomfortable. After supper, we went back to the house with Bob to see Neil. He seemed lonely and depressed and sick. Bob says he gets pneumonia every year. But Bob is nevertheless going back to the States in a short while. The two of them have more or less separated.

January 27 was another day of indoor work—until Stephen Spender came around to pick us up in his car and drive us back to his house for supper. To my surprise, he wasn't nearly as bitchy as usual. When I asked him about Wystan, he was positively reticent. Lizzie Spender was there and, on an impulse, I told her I would try to get her a part in "Frankenstein"; she belongs to Equity and she really is very good-looking. Then Francis Bacon arrived—that powerful star—and we were at once drawn into his orbit and scarcely talked to anyone else all evening. He was most affectionate and kissy. He had dim sour Luci[a]n Freud with him and also a nephew from South Africa, a farmer, named Harley Knott. Knott was quite young—not unattractive but seemingly dead square. He kept addressing Francis as "Uncle Francis." I couldn't make out what his attitude was to Francis's flamboyant behavior; sometimes he seemed disapproving, sometimes impressed. He didn't even seem convinced that Francis was really an important figure in the art world; maybe he thought that the rest of us were merely flattering him and putting him on. I solemnly assured Harley that Francis was every bit as great as Van Gogh. We were all of us pretty drunk.

The 28th was another indoor workday. In the evening, we went with Mo and Peter to see *Lady Caroline Lamb* and then on to supper at Odin's. This was to have been primarily a bit of return hospitality to Mo and Mick Sida. Then Mo told us that Mick wouldn't be coming to the film but would join us later at Odin's. But Mick never showed up. Around midnight, Mo finally contacted him, back at the flat. Mick said vaguely that he'd gone to see some people and had lost track of time. No doubt they'd all got high on something.

Lady Caroline was charmingly dressed and staged, but the chief character (played by Sarah Miles) is a capricious bore sentimentally presented as a martyr, and Richard Chamberlain as Lord Byron looked like the Queen of the Gypsies. Odin's has moved into the house next door and is now a very elegant establishment, more like a drawing room than a restaurant, its walls hung with paintings. But it is no longer snug. And I'm sure no Wayne Sleep will ever dance naked on its tabletops. [Peter] Langan came over to talk to us, drunk as usual and full of semihostile blarney. He insisted on giving me a painting, which he said I had raved about when I used to come to Odin's in 1970. It is late eighteenth century; of a rooster (or maybe hen) surrounded by chickens. The rooster-hen is a very odd-looking fierce bird with a crest of feathers like plumes on a helmet. Indeed, the creature has an arrogantly human air; it might be the portrait of some cruel old general. Its look gives the whole picture a sinister quality, a little in the manner of a Goya—which, I suppose, had moved me to overpraise it when drunk. Langan tried to bully me into saying I didn't want it and hadn't meant what I had said about it, three years ago; so I ended up accepting it. I think he just wanted to get it off the premises. He was full of suppressed resentment against David Hockney and Kasmin—Kasmin chiefly—and told some involved story I didn't listen to properly, of how he had wanted to reproduce some of the Hockney lithographs he owns on the covers of the restaurant menus, and of how Kasmin had demanded an outrageous fee to allow him to do this. I have probably got the facts quite wrong.

Eric Boman, Peter's Swedish lover, arrived to join us for dinner. (Peter had wished him on us, uninvited, but of course we were very curious to meet him.) He is a big, strikingly pretty blond boy, extremely self-sufficient, with cold good manners, who speaks perfect English almost without an accent. We got the impression that he is in love with Peter, but not that Peter is in love with him. Peter plays it very cool and passive. He tells us that he and Eric still go to bed together. (My first, slightly unfavorable impressions of Eric improved a good deal after two more meetings.)

Another indoor workday on the 29th. Fierce pale-faced little Mo! I could hug him for being so ferocious and harassed and dauntless. He must have given Mick a terrific bawling out for his behavior yesterday evening. Because Mick apologized to me, embarrassing me greatly. Mo is also enraged because of the many boys—David's hangers-on—who keep appearing and saying they want to take a bath; David no doubt casually told them that they could. So now Mo keeps the door of the flat locked at all times. If we go out in the mornings, he asks us to lock him in. The first morning we were here, he went out and locked *us* in. We didn't know the trick of the lock and couldn't get out of the flat for nearly half an hour!

In the evening we went to the Royal Court and saw two Beckett plays, *Krapp's Last Tape* and *Not I*, followed by John Osborne's *A Sense of Detachment*. I'm getting to be a chronic evening napper at the theater, especially if I've eaten supper beforehand. I napped through a lot of *Krapp's Last Tape*, although I admire it very much. Don says it was spoilt by Albert Finney's affectations. At the end of it, Finney didn't acknowledge the applause or even rise to his feet; he sat motionless in his chair. This pretentious stunt may not have been his idea, however; perhaps the director wanted it. For Billie Whitelaw—the voice behind the illuminated lips in *Not I*—didn't make an appearance at all. I shan't know what I think about *Not I* till I've read it; in performance it seemed mere clever patter in the manner of Joyce. Beckett (and/or his directors) always puts so many theater tricks between you and your understanding of his work.

The Osborne play seemed almost intolerable to both of us; it's a kind of intimate revue, or performed interview with him, presenting his opinions, prejudices and fantasies. But that sort of thing has to be done with immense art and care; this was sloppy. The satire was wide of the mark. The indignation was sentimental. It all projected the Osborne pose—that he's the one truly just man, incorruptible and fearlessly outspoken, who has outlived his period of fashionable success and now scorns it and curses the corruption around him, like Timon of Athens. It reminded me of Hemingway's drunken peevishness during his last period. And when Osborne dragged in the Vietnam War—as if he had *fought* in it on the side of the North—and dared to quote from one of Yeats's most tremendous passages on the revolutionary heroes of Ireland—his bad taste, in this trivial context, was obscene.[1] The actors worked bravely, trying to win over a bored hostile audience, but they failed. And, when two of them had to come out at the end with the earthshaking news that Osborne nevertheless believes in *Love*, my stomach turned in sympathy for them. Only Rachel Kempson (Lady Redgrave) managed to remain uncontaminated and almost noble as she read from a porno catalogue with which Osborne was trying to shock us.

(I don't mean what I have written to be against John personally. I'm actually trying to describe his failure as a warning to myself. Because all of us writers (nearly) are capable of lapsing into this tone of oracular complacency and making silly pricks out of ourselves.)

On the 30th we worked most of the day with Jack Smight at his flat. Then saw Deborah Kerr in *The Day After the Fair*, during which I did *not* doze. The play is absurd—based on one of Hardy's "ironic" jokes—but curiously absorbing.[2] And Deborah, that

[1] In the published play, Vietnam is mentioned, but the conflict in Northern Ireland figures more. The Yeats passage is from "Easter 1916"; the play also has lines from the related poem, "The Rose Tree."

[2] By Frank Harvey, adapted from Hardy's short story "On the Western Circuit."

queen of good sports, entertained us, heart and soul. She has only one bad habit; she stops every so often to mug at the audience, as actresses used to do in bedroom farces when one lover is under the bed, another in the wardrobe, and the husband is banging on the door. The mugging means: "How would *you* get out of this fix, if you were me?"

On the 31st, we had lunch with Stephen Spender and Cyril Connolly. Stephen had warned us that Cyril liked to have lunch at a certain very expensive restaurant (I forget its name); he told us to insist on going to Bianchi's, which was cheaper. (Actually, all the restaurants we went to in London seemed hugely expensive; the cost in pounds was sometimes nearly equal to what the cost in dollars would be in Los Angeles.) Cyril had agreed to Bianchi's, over the phone, but with a somewhat bad grace.

Stephen arrived first and explained to us that Natasha refuses to see Cyril because he stole one of Stephen's rarest books, an Auden first edition, and later refused to surrender it, saying that books were more important to him than friends. (I suspect Stephen made up this line.) Stephen also told us that he had had an accident and had very nearly been arrested for drunk driving on his way home from driving us back to Powis Terrace on the night of the 27th.

When Cyril joined us, he made quite a production out of greeting Don and then sat down beside me without greeting me at all or even looking at me—maybe this was a gesture of intimacy, for his manner toward me was otherwise most friendly throughout lunch. He started bitching Wystan without delay, saying that everybody in Oxford is already bored by Wystan's stories about Yeats and by his demand to be given his supper punctually at 7:00 p.m. He has also given offence by criticizing Oxford in an interview, saying that it is noisier than New York![1] Cyril had brought some of my books for me to sign. He said that Wystan's blow job poem[2] and my story "Afterwards" were the two best pieces of homosexual literature. But the compliment was dropped so carelessly and unconvincingly that it was nearly an insult. You felt that Cyril just doesn't give a damn about anybody or anything outside of his close domestic interests. His flat blue eyes have no mercy in them and hardly any life, though his talk is still lively. He said that, if we go to Africa this summer, he will tell us the best places to visit. I

[1] Auden was given the use of a cottage in Christ Church, moved there in mid-November 1972, and joined the Senior Common Room.
[2] I.e., "The Platonic Blow," mentioned above.

greatly enjoyed seeing him again and even felt an affection for him; I like him to be exactly as he is.

In the evening we went to a hateful catchall party at Ron Kitaj's. Romana Bouverie McKuen (the daughter of Gore's friend, Alice Bouverie)[1] wanted to know *why* I had been a conscientious objector during World War II. I tried to explain my feelings, instead of snubbing her; I wish I had snubbed her. She has an infuriating "some of my best friends are" attitude to queers.

On February 1 we had supper with Bob Chetwyn and Howard Schuman. It was a happy evening. Don took to them at once, and they to him. Bob and Howard now seem much more definitely a pair.

On February 2, we finished our work on "Frankenstein" and took the final batch of pages to the Universal office, where we ran into Hunt and Dick Shasta; we have seen almost nothing of them during this visit. Then we went around to the dirty little flat in Reece Mews near South Kensington station where Francis Bacon lives. He has a beautiful old house on the river in Dockland but says he can't paint in the studio there, it's too big. His studio here is tiny and dark. Fruit-juice cans full of brushes; a shambles of squeezed paint tubes; the walls smeared with blazing brush-wipings (which Francis calls "my abstracts"). The disorder is extraordinarily violent, like a battlefield. On an easel, two blurred figures which might be trussed-up demons. Francis has tied their legs into knots: they are writhing furiously.

The studio seems as intensely "charged" as a shrine, but its atmosphere isn't in the least shrinelike; it doesn't calm you. This *is* the battlefield; the awe-inspiring scene of Francis's desperate victorious struggles to "get down to the nerve"—I can never forget that phrase of his.[2] What he must have gone through! You see it reflected in the wildness of his eyes. And yet his youthfully fresh skin makes him still look strangely pretty, despite his pouchy cheeks.

He told us about his gangster friends. Some of them once stole one of his paintings. He was ready to prosecute them. They came to visit him. Nothing was said about the painting. No threats

[1] Romana von Hofmannsthal Mc*Ew*en (b. *circa* 1935), daughter of Alice Astor and her second husband, Austrian writer Raimund von Hofmannsthal. She was educated at Brearly School in New York and married Scottish folksinger and botanical painter Roderick (Rory) McEwen (1932–1982). Alice Astor's fourth husband was architect David Pleydell-Bouverie (1911–1994).

[2] Recorded in Isherwood's diary entry for February 10, 1956 and used about Isherwood's own writing on November 26 the same year; see *D.1*.

were uttered. Indeed, they kept asking him reproachfully why he didn't come to see them as he used to. But Francis understood that he would have to drop the charge or else get beaten up. So he dropped the charge. (I think these gangsters may well have been Ronnie and Reggie, the Kray twins.)

Francis appears to live in a no-man's land between the criminals and the police, persecuted by both sides. The police he regards simply as legalized crooks. Once they raided his studio and planted a packet of hashish there. He was only able to clear himself after expensive court proceedings, in which he brought medical evidence to prove that his health made it impossible for him to take the drug.

We had supper at the Old Compton Street Wheeler's. On the way there, Francis told us about [a friend] who has just returned [home]. As they were driving to the airport, [the friend] burst into tears and confessed that he is homosexual. Francis said, "Why on earth didn't you tell me sooner? I'd have found someone for you." He thinks [the friend] is sure to become an alcoholic if he stays on his [...] farm—out of sex frustration.

Francis said of David Hockney's work, "She's no good."

On the 3rd, we went with Mo to have tea with Celia and Ossie Clark. Celia was leaving next day for Los Angeles, to join David Hockney. Perhaps it was just because I know that Ossie sometimes beats her, but her eye shadow seemed exaggerated to me, as if she had two black eyes. Ossie seemed gentle as a lamb, very frail and skinny and stoned out of his mind. But I saw cruelty and rage in his face. Don agreed with me.

Later we saw Leo Madigan and his friend Sean [O'Brien] at James Pope-Hennessy's flat, where Leo is staying and working on his new book. Leo is Leo, not Larry, since he published *Jackarandy*; he seems much more assured and more "literary" than when I saw him in 1970;[1] but his charm has increased too. We both liked him very much. Sean used to be in the Irish Guards but resigned from them when the troubles began in Ulster, lest he should be sent there and have to fight against his countrymen. His skin is very white and his hair blond. I find him much sexier than Leo. We went out to supper with Leo and Sean and Waris Hussein the film director and then on to a bar, where Sean gave me a great big wet-tongued long-lasting kiss—maybe to assure me that I wasn't too old to be still in the running. This was a drunkenly pleasant evening. But the food at the (gay) restaurant, the Masquerade, was bad and expensive.

[1] See April 6.

On the 4th, we went down to see the Beesleys at their cottage. Alec met us at Audley End station, looking as ruddy faced as ever. But he has had pneumonia and wasn't allowed to get out of the air-conditioned car, the weather being damp and chilly. The pneumonia came on quite suddenly after he had had a cold. He became breathless and finally just sat down and gasped, until the ambulance arrived bringing an oxygen mask. The doctor told Dodie later that Alec would certainly have died within a few hours if they hadn't got him under oxygen. Nevertheless, Alec made an unusually quick recovery, of which he is proud.

As for Dodie, she's a little old lady. We felt that our visit was almost more than her nerves could bear but at the same time quite welcome. This latest dalmatian causes more distraction than any of the others. When leaving a room, it makes a beeline, jumping straight over any furniture which happens to be in the way. So there are very few places where it is safe to leave anything. And tension is permanent.

Aside from telling our news about "Frankenstein," our play, etc.—which took up most of the time—we talked about Ada Leverson, because I am reading one of her novels, *The Limit*, which I found in David's library. Dodie knows them all and is about to reread them. She also still reads Henry James every day. And, as usual, she is at work on a book which is a deadly secret—in fact, she and Alec are already discussing possible revisions. We left early, so Alec could get us to the station and return before dark; he isn't allowed to be out at night. Dodie's face, as we said goodbye at the cottage gate, was suddenly tragic. I wondered if she was fearing that we would never see each other again. If she was, it can only have meant that she thought I looked sickly. I'm sure she can't, daren't believe that *she* will die, poor darling.

In the evening, we visited Amiya, whose present flat is much larger than her others, though in a dreary neighborhood north of Kensington High Street. She seemed almost or entirely sober, didn't ramble on, didn't indulge in her usual display of jewels of her spiritual and worldly wisdom. Instead, she told us about a man who is staying in her flat because he is completely broke and about his handsome boyfriend who sends him money from the States, where he is on a singing tour. When we left, Don said that he had never before realized what a good woman Amiya is.

On the 5th, we had lunch with Ivan Moffatt at Don Luigi's. He looked florid and a bit older but was as entertaining as ever. His entrance line (does he make them up in advance?) was, "I feel very bad because I've just done something I never did before in my life—I smacked a girl this morning—" (a chuckle) "not hard, though—" (chuckle, chuckle) "but she cried." Later he revealed that he had smacked the girl because she had irritated him by asking what he was going to do that day—that she was Jewish, and therefore sensitive to ill-treatment, even if mild—that she had been available for smacking because she had been spending the night with him at his flat. Throughout the rest of lunch, Ivan took occasion to mention two or three other girls he was going with—as though he wanted us to be reassured that he still cuts the mustard.

Aside from this, he talked about the film he has been working on—the *Last Days of Hitler*,[1] with Alec Guinness as Hitler. Without ever saying anything damning, he contrived to tell us what an utter disaster this had been. He was wildly funny, imitating the Italian director, hardly able to talk English and hopelessly at sea in this Kraut drama, and Guinness himself, arrogant and sensitive, feeling that his great performance wasn't being appreciated, and the technical adviser, a former officer in the German army, even more arrogant than Guinness and despising the whole lot of them. Oh yes, I forgot—right at the end, Ivan did make one damning admission; that they had had to keep cutting chunks out of the film because they were so boring!

Around teatime, we saw Peter Viertel and Deborah at their flat. Peter had already been urging us to come and stay with him in Switzerland, so that we could see Salka. Now we told him definitely that we'd go. Peter had obviously felt certain that we would. He at once got his secretary to work and within twenty minutes we had a complete schedule drawn up for us—the plane to Zürich, the train to Landquart (where Peter will pick us up in the car and drive us to Klosters), the train back from Klosters to Zürich, a choice of train or plane to take us to Rome (where we'll stay with Gavin), the plane from Rome to London which will connect with the plane that takes us back to Los Angeles. I had never realized before, or had forgotten, what a manager Peter is. While we were with him, we put through a call to Adelaide Drive; I'd been getting mildly worried because a couple of efforts to contact Charles Hill had failed. Didn't get him this time, either. Meanwhile, Deborah was being interviewed in the sitting room by a woman journalist. We were both impressed by the ease and naturalness with which she handled this; never for a moment did she make like a star.

[1] *Hitler: The Last Ten Days* (1973), directed by Ennio de Concini.

Then we went to see Robin Dalton, who is the agent for our play in London. (She gives an impression of efficiency but I doubt if she's much good.)[1] On the way there, I was nearly hit by a car which came at me as I was crossing the street—neither Don nor I can ever quite get used to their driving on the "wrong" side. Because I was startled and also because I wanted to startle and worry the driver—he didn't give a shit—I let out a great yell, which merely had the effect of first startling and then infuriating Don. So he yelled at me. Our nerves are badly strained by the tension of all this work under uncomfortable alien conditions and the wild running around pursuing our social duties in the interim.

Later we had a dullish drunken supper with well-meaning Paul Dehn and his friend Jimmy Bernard at a bad pretentious restaurant, The Minotaur. But it was truly thoughtful of Paul to buy back for me a watercolor by Frank which his sister (I think) had seen in a shop at Disley; and it was in Frank's best manner, well worth having.

On the morning of the 6th, I went alone to the Ramakrishna Vivekananda Center. Phil-Buddha-Yogeshananda opened the door to me. My first impression was that he was lopsided, perhaps crippled by arthritis, very skinny but jolly. However, as we talked, he gradually straightened himself and when, later, I questioned him tactfully about his posture he said he was maybe a bit stiff because he and a few of the others have weekly classes with Bhavyananda on yoga *asanas*. Yogeshananda said it was astounding how limber Bhavyananda is, although he's so fat; he absolutely refuses to go on a diet. They ate their lunch early, so I sat with them before I left. I think most of the boys were the same as in 1970—anyhow, I got a very good feeling from them and from the whole place. Yogeshananda seemed quite radiant with joy—much more so than when last I saw him—a bit crazy too (as Swami always says he is) but in the right way. "I never want to leave here," he told me, "I never want to go back, I'm so happy here." I left the house feeling genuinely uplifted, calmed, reassured. Yes, thank God, it is we, the worldlings, who are mad. People like this are the only sane ones. I am mad, but not altogether, because I know this all the time—and sometimes, as today, absolutely for certain.

I felt uplifted by Wayne Sleep too, in a different way; Don and Peter Schlesinger and I had lunch with him at the Old Compton

[1] Dalton (b. 1921), an Australian, was a prominent literary and theatrical agent, an author and, later, a film producer, notably of *Oscar and Lucinda* (1997).

Street Wheeler's. How marvellous and admirable he is! He was radiating *physical genius*, I don't know how else to describe it. Not long ago, he was being interviewed on a T.V. show, describing and demonstrating various ballet movements. The interviewer explained what *entrechats* are, adding that Nijinsky had been able to do ten of them during one jump, an *entrechat dix*. (Actually, I forget what numbers Wayne mentioned when he was telling the story; the number ten is reported by Lincoln Kirstein in his *Ballet Alphabet*, with the caution that this is "contemporary myth"—but that's unimportant.) The interviewer then called upon Wayne to demonstrate the step. Anxious to please, Wayne took off with the biggest jump he could manage—and executed an *entrechat douze*. Sensation! Since this had been televised, his feat was incontestably confirmed. The press publicized it as a sporting event: Sleep breaks Nijinsky's record!!

Nevertheless, Wayne seems to regard himself already as an almost-has-been dancer. He is busy transforming himself into an actor, so that he will be able to support himself in his old age. He had just been appearing in Goldoni's *The Servant of Two Masters*. He was very funny about this, telling us he has learnt that, from a dancer's point of view, stage acting consists in overacting. He was training himself to over accent his lines.

After lunch, Peter, Don and I went to the British Museum. First, we looked at the mummies—from a sense of duty, since we have just shown Hunt Stromberg our "Lady from the Land of the Dead" script, and he vows that he will get the Egyptian government to finance the production. They (the mummies) were both impressive and exceedingly depressing. What a *dead* culture! I felt such a fatal, inhibiting obsession with death—which sounds like a naive remark, considering that we were looking at coffins. Perhaps I'll react differently if we ever do go to Egypt and work some more on the picture. Hunt's latest notion is that he wants the entire story set in Cairo, instead of London—just because he finds that the Egyptians are all excited at the prospect of having a Hollywood film set in modern Egypt; it would be a smack at the Israelis and a betrayal of them by the Hollywood Jews, I guess, from Egypt's point of view!

Also at the museum was an exhibit of master drawings through the ages. A wonderful collection, but Peter and I could take them only in small doses; we kept retiring to a bench to rest. (Peter questioned me with great interest about my visit to China in 1938; I forget why.) Meanwhile, Don moved slowly from drawing to drawing, tirelessly and minutely inspecting them. It is at such moments that I realize how deeply artistic he is—how he reacts to and compares different techniques, making them part of his personal graphic knowledge. I also realize that this wouldn't surprise me in the least, if it weren't for Don's endlessly repeated declarations that he isn't really an artist at all. This self-denigration makes me furious with him sometimes because I feel its power—perversely trying to destroy my belief in him. Not that it ever could. There is too much evidence to the contrary.

From the museum we went on to the Hayward Gallery, where there was an exhibition of paintings of England done by French Impressionists. They had Monet's marvellous *Waterloo Bridge* series, dating around the beginning of this century. Sometimes the bridge seems brutally dark and heavy, laden with toiling traffic, a ponderous lifeless heartless commercial contraption, seen before a hideous background of factory chimneys and drab clouds—like a political cartoon directed against Victorian capitalism. Then, again, the scene is transformed into weightless magic, flashes of light on the water, golden gleams on the bus tops, the bridge and the factories melted into mysterious indications amidst fiery vapors.

Finally we all three went to the Roundhouse, where Stomu Yamash'ta's Red Buddha Theater was performing. The Japanese actresses and actors were lithe and cute and eager to entertain us—even a bit too eager, they seemed more like airline stewards and stewardesses. The best parts of the show were the least modern and even these weren't remarkable; an uninspired reference to the Kabuki tradition. Oddly enough, I was only really turned on by the skullsplitting noise of the orchestra, though I usually hate electronic music. Maybe this proves that Stomu Yamash'ta is something special. The program said he is a supergenius.

Eric Boman joined us. We met John Houseman, with a girl. I was glad for him to see me escorted by three such striking and varied examples of the beauty of the tribe. The Roundhouse has strangely contrived to preserve the squalor of its railroad days. You get the impression that everything—the chairs, the tables and the food in the restaurant—must be filthy—though actually it isn't.

On the morning of February 7, John Lehmann came by to see me. (I neglected to mention that he'd already given us drinks at his flat on the evening of January 29.) He wants very much to publish the letters we exchanged about the contributors to *New Writing*, during the late thirties. I agreed to write notes to be added to the text, if the publishers decide to accept it. Now I'm sorry I did this. I fear that the letters must be a thundering bore and that all comment on them will be superfluous. John is also writing a sex memoir, in the third person and under an assumed name. He showed me part of it. It is commendably frank but not really all that interesting. John has a besetting blandness in his writing which persists even when he becomes pornographic. Each boy gets his compliment—he is attractive, handsome, muscular, or what have you—and I feel that this is essentially the same tone which John has used so often in his autobiography, when he bestows literary compliments on a group of writers, making them all alike and equally mediocre.

In the afternoon we saw *Crown Matrimonial*,[1] a play about the abdication of Edward VIII. At first, I was astonished that the royal family would permit such a play to be performed. Then I understood why they did; it is most ingeniously inoffensive, the candid camera has a filter attached to it. You don't see either Mrs. Simpson or Baldwin,[2] and anyhow Mrs. Simpson and Baldwin are lavishly excused and even praised behind their backs. Edward is impetuous, boyish, rash, but he never behaves like an arrogant fascist cad. Queen Mary is marvellously absurd—Wendy Hiller impersonates her stunningly—but she is never revealed as a domineering old cunt. We both found the whole thing immensely entertaining and could have gone on watching Hiller for hours.

After this we had a drink with my new agent at Curtis Brown, Peter Grose, an Australian. He seemed energetic and intelligent and made a very good impression on both of us. In fact, I think I'm going to like him better than Richard Simon—who, incidentally, has been rude and cool during this visit. Obviously he isn't interested in seeing me, now I'm no longer a client.

Then we saw the first act of Tom Stoppard's *Jumpers*—pretentious double-talking shit which everybody here is lapping up. We left at the intermission and had a delightful supper with James Pope-Hennessy, Leo Madigan, Sean [O'Brien] and Waris Hussein. We both think James is a really good person. Leo and Sean are fun to be with—especially now that Leo has gotten over his star-struck attitude toward me, which was probably partly playacting anyhow. Waris is complicated, hypersensitive and a show-off, full of oriental jealousy and indiscretion, but he can be very entertaining.

February 8. I left Euston at noon, got to Stockport at 2:30. As a matter of fact, I think we arrived a bit earlier. Richard wasn't there. Outside the station, it was drizzling. The taxis were being taken,

[1] By Royce Ryton (1924–2009).

[2] Stanley Baldwin (1867–1947), the prime minister who gave Edward the choice of renouncing the divorcée, Mrs. Simpson, or renouncing the crown.

one after another. On an impulse, I grabbed the last of them, saying to myself that maybe Richard had got the day of my arrival wrong. As we drove away from the station, we passed a car in which I was nearly sure I saw Richard and Dan Bradley—but I wasn't *quite* sure and it seemed less trouble not to stop the taxi driver. (I didn't mention this when I saw Richard and the Bradleys later.)

As we reached the end of High Lane and turned off the Buxton Road up the road that leads to Wyberslegh (Carr Brow?),[1] I saw that the hillside was already built up with rows of houses—not the old solid brick villas but the flimsy glass and plaster bungalows of nowadays. Knowing from Richard's letters that his bungalow and the Bradleys' were both at the top of the hill, I told the driver to go up and around, so that we entered the "development" (it's called "Wybersley Rise") from the Ridge road, only a few yards from the drive gate of Wyberslegh Hall. This area was still under construction and deep in mud. The way through it was blocked by a cement mixer and several trucks. The men working there told us we would have to drive back down the hill and enter by the lower road. No one seemed to know exactly where number 61 Thornway (the Bradleys' bungalow) was. So down we went and around, and up past the already finished—fatally, finally finished—houses, until we reached the same goddamned cement mixer from the other side! And this time I discovered that number 61 was a mere dozen hop-skip steps away, across puddles and along duckboards.

Evelyn Bradley welcomed me warmly, telling me how well I looked and herself looking like a big plump apple-cheeked girl, about thirty years younger than her age. As she served me tea—a tea which never quite ended but became "high" and merged, several hours later, into supper—she told me that Dan and Richard had probably missed me because they had both been drinking. She clicked her tongue over this—not that it shocked or worried her—she was merely deploring my bad luck in choosing the wrong day to come here; as though their drinking were like the weather, something beyond anyone's control. On the whole, she said, Mr. Richard was "much better" than he had been in the old days, when he had spent hours shut up inside the Hall, getting depressed and downing bottle after bottle of beer. Now he only went in there a couple of times a day, just be sure that everything was all in order.... This led Evelyn on into psalms of praise of their bungalow. "It's like being in paradise," she said,

[1] Carr Brow runs from the Buxton Road to the turning with Wyberslegh Road, where it becomes Jackson's Edge Road.

and "sometimes, we can't believe we're really here!" I could only suppose she meant that the living room was big, far bigger than the one they had had in the Disley house. Certainly, the view from this bungalow is much inferior to the view from Bentside Road. The hill slope is only gradual, up here at the top, and so you look straight at the bungalow opposite, not over its roof. But maybe the real source of Evelyn's joy is that she feels that "Wybersley Rise" has *class*—no doubt it *would* rank socially above the Bentside Road area; office-worker commuters in the one, factory operatives in the other. And then, of course, these bungalows are separate buildings, while the Bentside houses are semidetached. A very important distinction! At the same time, Evelyn was on the lookout to anticipate and counter any criticisms my roving eyes might be implying. She pointed out that the electricians hadn't fixed up some of the lighting and that the plasterers had left some cracks in the wall—all that would be attended to when they came around again on their monthly circuit.

Then Evelyn's daughter Susan came out of the bedroom where she had been lying down. She had spent the day teaching—it was some kind of a test for her student-teacher's diploma—and now she was suffering from migraine. (I suspected that the migraine might disappear forever if she could escape from the Bradleys' bungalow and be on her own in Manchester and get herself fucked.) But she is obviously a bright, interesting girl. I tried hard to project sympathy but we had no chance to have a proper talk.

Finally, Richard and Dan returned. They had missed me by only a moment—it *had* been them I'd seen in the car as I drove away from the station. Richard was pretty drunk and remained so. He showed this chiefly by his air of preoccupation. He had set his controls on automatic, saying "yerss, yerss" at appropriate intervals but rather too loudly to be convincing. Dan talked obsessively. Since his accident he has been homebound and all his energy goes into talk for talk's sake. He described, in repetitious detail, exactly how the heating system of the bungalow works and what legal difficulties he has had to contend with in making his claim for accident compensation. He was drunk too but only moderately.

I was to sleep in Richard's bungalow, number 63. Mrs. Bradley had fixed a bedroom for me but the rest of it is still only partially furnished. I realized that Richard himself had never slept there before that night and that he had been pressured into doing so now as part of the ritual of hospitality toward me. All this time he has been sleeping in the Bradleys' spare room, because he doesn't like sleeping alone in his bungalow, even though it's only a few yards away from theirs. Next morning (February 9) Evelyn Bradley told me that Richard had in fact not slept at all or even got into bed; he had stayed up the whole night, drinking.

The morning was showery, windy, with gleams of miserable sunshine and brief flurries of snow. Richard refused to eat any breakfast and sat gulping brandy as though it were beer; I've never seen him do this before. Throughout my visit, we had almost no contact, and yet I didn't feel that his drinking was a hostile demonstration—it was simply, as Evelyn had said, that I'd picked the wrong day.

He did tell me that he has switched from the Rosicrucians to Unity. And he made a great point of giving me—pressing me to take—a watercolor of Wyberslegh Hall painted by the Mr. Standen who had given painting lessons to Frank. Was this a guilt offering? I only suspect so because I found out, quite accidentally, from Mrs. Bradley that Richard sold Frank's entire portfolio of watercolors to a dealer, some while ago! This explains why Paul Dehn's sister found the one he gave me, for sale in a Disley shop.

Mrs. Bradley's only other indiscretion was to point out to me, among the workers on the building project, a young man whom Richard fancies. Her smile, as she did this, was full of loving amusement; absolutely unbitchy. She really is an adorable woman.

The Hall is shockingly close to this bungalow townlet; from the Bradleys' back window you look right into its wilderness of a garden. There isn't even a fence. Yet this is a frontier, between two worlds, two utterly alien time rhythms. I find their confrontation inexpressibly spooky. The Hall seems ghastly, abandoned, betrayed, dead—yet overpoweringly *there*. The bungalows are evidently alive yet scarcely present—mere shoddy momentary things.

I went over to the Hall about midday, with Richard and Dan. The roof is leaking badly, now, and won't be repaired till the spring. The water has run down into the Stone Parlour, which reeks of damp and rot and dirt. The shutters were closed. From this dim death chamber, I put a call through to Santa Monica and soon heard Charles Hill's chirpy young voice speaking from the land of sunshine. He explained quite casually that we hadn't been able to reach him because he had been spending his nights at Ed Moses's house, looking after the Moses children. Paula, Charles's girlfriend, had had to go to Chicago, and Charles had decided that he didn't like sleeping in our house alone.

The Bradleys told me that Thomas Isherwood now owns the field which is directly below the house. This is due to a deal whereby Richard gets a larger sum of money guaranteed during his life and Thomas gets some part of his inheritance in advance, before Richard's death. The Bradleys seem to think that Thomas will sell this field to the developers, who will then fill that section of the hillside with bungalows and block the view from the Hall almost entirely. Thomas has already made an enemy of the Bradleys, by some condescending remark to the effect that they are only occupying their bungalow by gracious permission of the Isherwood family—a permission which may be withdrawn. The Bradleys are terribly afraid that they will be left unable to pay their rent and will be evicted, after Richard dies. Dan spent the last hour before I left in alluding to this problem with much circumlocution and repetition. No doubt he hoped I'd promise to help them out if necessary. I didn't, but I suppose I probably would.

The fields below the Hall are part of Wyberslegh; if they are built over, its view taken away, it will no longer be able to breathe or see. And yet the farm, which is its Siamese twin, will remain almost unaffected. The farm's view, out toward Cobden Edge and the High Peak, hasn't changed—though, of course, it could easily be cut off by building two or three houses on the opposite side of the road.

It isn't surprising that the Bradleys are pleased with their bungalow townlet, but it amazes me that Richard is. With his intense underlying obstinacy, he can hardly have let himself be brainwashed; this must be a case of subconsciously willed acceptance. Since he has agreed to build a nest for himself and the Bradleys here, he must have made up his mind to like it. He keeps suggesting, quite seriously, that Don and I should come over and settle here. As for Evelyn Bradley, she says with conviction that Kathleen would have been so comfortable and happy in one of these bungalows. This statement staggered me so that I couldn't help showing my astonishment. I think Kathleen would have had a stroke at the mere suggestion.

After lunch (another meal he refused to eat) was over, Richard began to fall apart. He couldn't even stand up. Evelyn used all her powers of coaxing to get him to go to bed and at last she succeeded, just as it was time for me to leave. He was helped into his little bedroom in the Bradleys' bungalow, still protesting that he wanted to come with me to the station. I caught the 3:38 train and was back in London at 6:12. An hour and a half later, washed and changed, I showed up with Don at a party given by Alain Bernheim[1] for John Houseman. My sense of psychological

[1] French literary and film agent (1923–2009), he worked in Hollywood after W.W.II and eventually became a producer.

transportation couldn't have been stronger if we'd just flown in from California. Here was Ingrid Bergman, still beautiful and still radiantly smiling as she was when I first set eyes on her, more than thirty years ago. (The last two times we met, she seemed hardly to remember me, but now she did, vigorously—she is the adorable queen of all the moo cows.)[1] From the party, we went on to a happy gossipy supper with John Schlesinger and his friend Michael Childers at their newly acquired and reconstructed house. Each room was like a fresh advertisement, executed in breath-taking, boldly contrasted colors, for their successful, with-it way of life. Nowhere to relax, be untidy or display unfashionable objects of sentimental value.

On the morning of the 10th, we phoned Wystan, who had just returned to Oxford from Brussels, where they'd been performing [Nicolas] Nabokov's opera, *Love's Labour*['s] *Lost*, for which Wystan and Chester wrote the libretto. This was the only possible day for us to have seen him. Chester answered the phone and there was a feeling of awkwardness between us. Then Wystan got on the line and said quite firmly that he couldn't see us that day. Perhaps he didn't want to, with Chester around. I don't blame him.

We filled in our more or less wasted last day in England by going to see the Futurismo and Dante Gabriel Rossetti shows at the Royal Academy, eating a cramped lunch at the Jermyn Street Wheeler's and having a drink with Kate Moffat. She was unexpectedly unfriendly and obviously only received us as a duty. Ivan arrived while we were there, but couldn't warm up the atmosphere. He still seems very much at home in her house. Didn't see Kate's lover.

Our last evening was spent at the Powis Street flat. Peter Schlesinger and Eric Boman cooked supper. Patrick Woodcock was there and Mo—but not Mick Sida; he was off, I suppose, on his mysterious nocturnal wanderings. My relations with Patrick weren't as pleasant as usual, I'm sorry to say. Maybe all his talk about settling down in France irritated me, after a few drinks, for I launched into one of my tiresome cantankerous Francophobe tirades. Also I declared that, as a writer, I needed all my life to master the English language—implying that Patrick and the rest never had and never would—and that I therefore had no time to waste in dabbling in foreign tongues. Patrick rightly found this statement pretentious. It was also rude to Eric, who speaks at least three languages fluently.

[1] I.e., large, overpowering, totally female females. And see Glossary under Bergman.

We packed before going to bed.

Next morning, February 11, Tom Pugsley (the driver who had brought us here on arrival) came round early with a limousine—Universal's last service to us—and took us out to the airport where he refused a tip. We left most of our baggage in storage, so as to travel light to Switzerland and Italy. All went according to Peter Viertel's schedule, except that the plane to Zürich was delayed nearly an hour. I think there had been a last-minute bomb scare, because the baggage had been put on the plane and taken off again, and each of us had to identify his own pieces. But, when we did get to Zürich, the Swissair representative who was a friend of Peter's saved us from missing our train by unlocking doors with a private key, thus bypassing passport and customs officials, then personally changing money for us and getting us a cab to the station. This was dirty and gloomy, not Swissbright as I had expected, and crowded with skiers, many of them just riding up to the snow for the afternoon. We must have been almost the only nonskiers on the train. When we got off the train at Landquart, we found ourselves marching inside a compact formation of them; all around us, the dark skis, held upright, looked and clattered like antique weapons.

Peter was there to meet us, as he'd promised, all briskness and wisecracks. He drove us up a winding road into the Klosters valley. (The train goes there too, but you have to change from the Zürich train into another one, so it's quicker to drive.) Light snow fell, the peaks were hidden by clouds; you felt the claustrophobia of a forlorn grey day in the Alps. The chalet belonging to Peter and Deborah seemed forlorn and shut-in too when we first saw it; dark woods of conifer all around, the branches extending their heavy paws of snow. Near it is an immense old barn which used to be part of the original farm, Wyhergut.

The inside of the chalet is richly comfortable, furnished in a mixture of styles, plutocratic peasant and Kensington dainty. Local craftsmen carved its ceilings and put stained glass windows in its doors.

Peter put us to work at once, shovelling snow off the steps, lest we shouldn't be able to get back up them when we returned from Klosters later in the evening. He didn't ask me to help, knowing that I would anyway. With Don he was apologetically bossy. "Don't make a lifework out of it, as my sergeant used to say." Peter's conversation keeps reverting to his army memories. As long as the light remained, skiers would keep shooting out of the woods and whizzing by, in twos and threes, uncannily silent. The massive snow-hush was broken only when Peter started up his miniature snowplow. Once, we heard a faint rumbling as the train passed, improbably high above us, on its way to Davos.

Wyhergut is about a mile beyond Klosters. We drove down there to have supper with Salka, who has a pleasant little flat in the middle of the village. She is still strikingly handsome but now she is old, really old. You feel how bitterly aware she is of her increasing isolation from the outside world. No doubt Klosters, with its gothic, incurably Swiss villagers and vulgar international millionaire ski-bums, seems a place of exile to her; the local dialect must jar on her theatrically trained ear; and the place itself, though actually so accessible, has an air of hopeless remoteness, tucked away in the very back of Switzerland near unfindable Liech[t]enstein. Also, Peter told us that Salka had been terribly depressed by a recent visit from her sister Rose, who is younger than she is but already gaga, with a failing memory. Nevertheless, Salka is probably better off here than she would be anywhere else. Everyone says how kind Deborah is to her—when Deborah is there. And we all have to make our last stand somewhere.

After supper, we went round to see Irwin Shaw and his girl friend Bodil Nielson.[1] He was very drunk; Peter says he always is. Shaw has turned into the usual type of heavy-duty huffing and puffing Jewish writer, oracular but unsure of himself, and expecting a pogrom to start any minute. (The most ghastly punishment imaginable for him would be to condemn him to live in Israel.) I liked him. I always have.

February 12. The sun shone and everything was instantly transformed. The whole valley seemed to have opened up. Peter cooked breakfast, refusing to let us help, and announced the orders for the day. We were to go up with him into the mountains by cable car. He would ski down and we would ride down. After breakfast, he made and received long-distance phone calls—talking to London, Paris and Rome as if they were neighboring villages, and switching without effort from English to fluent French, German or Spanish. Driving down to Klosters, he told me about a girl he had known who was killed by an avalanche. This was one of his beloved Hemingway-style guilt stories; Peter claimed that he felt responsible because the girl had started out with him and several others and had gone off on her own, accompanied by Irwin Shaw's son; Peter, as the most experienced local resident in the

[1] Bodie Nielsen, American writer of Danish background, raised in Connecticut and educated at private boarding school and Vassar; she was managing editor of *Interiors* in New York until she met Shaw and moved with him to Europe.

party, felt he ought to have stopped her from doing this because they were in an area of possible avalanches.[1] If there are a lot of you together and an avalanche does fall, it's more likely that a few of you will escape and be able to dig the others out of it.

It was only after I had questioned him that he had to admit that the girl herself was no beginner but a ski instructor. And the avalanche only killed her indirectly, by knocking her head against a tree. Shaw's son, who was also buried by it, was able to struggle out of the snow and could have rescued the girl, if she hadn't been dead already. The fact is, Peter is carefully cultivating his guilt because he is planning to turn the story into a novel.

Later in the morning, Don and I went down to Klosters on foot, to enjoy the exercise and the intoxicating pre-pollution air. We took a walk with Salka before lunch—she is still surprisingly active—and watched people setting out on a langlauf, which is skiing on the more-or-less flat. (Peter has decreed that, when we next come to Klosters, I shall be sent on a langlauf while he teaches Don to ski. In Switzerland, if you can't manage a langlauf you are ready for burial.)

After lunch, it was time for our ascent. Although I had quite made up my mind to go through with this, I mentioned my proneness to vertigo, chiefly to give Peter the pleasure of feeling that he was putting me through a test of my manhood. No day in the life of a Hemingway devotee is complete without one. However, Peter decided that, as a concession to my weakness and advanced age, we should definitely go up the Gotschnagrad on the cable car rather than the ski lift to another peak. Even the cable car gave me a breathless moment but I really enjoyed its vast upward swoop and the exciting joggling with which it passes the towers. (The downward swoop, as I found later, is far more breathtaking, but by then my nerve was stronger.) The upper snowfield and its view over a vast havoc of mountaintops and gorges, and the whiteness of the snowshine—that was beautiful beyond description. And I was so full of joy to be seeing it with Don. We rejoiced together in the wonder of the light, after London's dank dim days. The snow itself was so dry that you could plunge your leg into it and brush it off again without getting the least bit wet.

Then Peter left us to return on the car while he skied down to Klosters. He makes this (I believe) quite dangerous run nearly every afternoon, by way of a short constitutional. He arrived at the inn which was our rendezvous only a few minutes after we did.

[1] Adam Shaw was then in his early twenties; see Glossary.

While it was still light, Peter drove us up to Davos, an uninteresting hotel-village, with scenery inferior to that of Klosters and a heavy odor of Swissrichness. He pointed out to us the clinic which is the original of Mann's Zauberberg sanitarium. I was disappointed because I'd imagined (quite arbitrarily) that it would be perched on the edge of a precipice with a tremendous view below and snow peaks above.

We had supper at a restaurant which is halfway between Davos and Klosters, called the Landhaus Lantet, as the guests of Count and Countess de Chandon, champagne aristocracy.[1] The Count had beautiful manners and a sulky blonde mistress named Susie—or was Susie the mistress of the Countess? Anyhow it was all highly Frog and civilized and suave and I had to fight off an acute attack of Francophobia all evening. Meanwhile, Peter chattered and Salka looked a bit dazed and deaf and Don, as always, did his charming social best to pretend to be entertained.

February 13. Since the dollar was still on the skids in Europe, Peter decided that we'd better change our great bundles of remaining expense-account pounds back into dollars while they were still so cheap. So we followed him on a paperchase around the Klosters banks; it was necessary to visit several, no one bank had nearly enough. We stuffed the money into our pockets as we ran back and forth across the street—it was getting late. I imagined that people were laughing at us, mad Yanks in a rush to buy back their unwanted dough. Salka came to the station and said goodbye with the tears of an old woman whose every parting from friends may be her last. Actually I do hope to see her again, because we are both eager to visit Klosters in summertime.

We started at 10:40, got to Zürich at 1:00, took off by Alitalia plane at 2:25, landed at Rome airport at 3:45. Luckily we didn't have to drive through Rome to reach Gavin Lambert's villa. It's on a side street called the Via Lugari, off the Appian Way. One might say that it is in the Campagna—some sheep were wandering about in the drizzle—but that part of the Campagna is being encroached upon by buildings and billboards. No doubt it was more romantically isolated when it was new. Its rust-red walls, within a courtyard, still suggest the romantic melancholy of a nineteenth-century watercolor. Our driver (after treating us to a positively surrealistic display of recklessness) exclaimed on arrival: "It's a

[1] Comte Frédéric Chandon de Briailles (b. 192[7]), vice-president of Moët et Chandon Champagne and later chairman of Moët Hennessy Group, and his first wife, Francesca Sanjust di Teulada (b. 1929), a painter. The restaurant was probably Landhaus Laret.

paradise!" Gavin, emerging from the house in a caftan, seemed amused by his enthusiasm; and when we later complimented him on the elegance of the interior said, in his inimitable tone of campy surprise: "Oh, do you *think* so?" We didn't really think so. The library had a certain old-world air of meditative calm, but most of the rooms seemed merely cold and empty. We were given a big naked bedroom with beds too small for it and a bathroom which was out of order. (There were two others which did work, however.) Of course, the wet weather did nothing to brighten one's spirits. I got the impression that Gavin had stranded himself here, neither in nor properly out of the city, in a situation which—if you didn't keep reminding yourself that this was Roman—was merely suburban and dull. Gavin must have felt as we did. Anyhow, he told us he was leaving very shortly to take another look at Tangier and perhaps find a place to stay there.

Gavin lives with a very handsome big Yugoslav lover whose first name is George, or rather, Georges, since his only other language is French. Gavin is very proud of Georges and of their affair but, at the same time, characteristically, he kept giving us the impression that Georges is lazy, passive, fatalistic. Georges is having trouble with his passport; it must be renewed almost at once, otherwise he will have to return to Yugoslavia. Renewing the passport is one of the things he has been passive about.

Georges had been to the dentist, so decided to stay at home. Gavin drove us into Rome to have supper at a restaurant, the Bolognese, on the Piazza del Pop[o]lo. This was an extraordinarily happy choice, because it was the restaurant Don most warmly remembered from our stay in Rome in 1955. He had often said to me since[,] how he looked forward to going back there and eating their delicious risotto with peas. So the risotto was ordered—but, alas, it was a huge disappointment.

The inner city wasn't a disappointment, what we saw of it—fantastic glimpses of buildings and columns and arches in all their authority of fame and history. At night they are as improbable as designs for a vast imaginary theater which, you think, could never actually be constructed. But the traffic makes any drive miserable. Seeing all those cars converge from a relatively broad street into one of the narrow alleys, you know in advance what a maddeningly long wait is in store for you. The Italians must have a totally different attitude to this kind of frustration; otherwise they would all ride bicycles or horses or reintroduce sedan chairs.

February 14. It rained in the morning but stopped later. We drove into Rome with Gavin. Gore Vidal had invited us to lunch.

We went up to his apartment first, for drinks. Howard Aust[e]n was there too, amidst dogs. They showed us pictures of a spectacular hillside villa they have just bought at Ravello. Gore talked about his nearly finished historical novel; the central character is Aaron Burr. Gore claimed to have "debunked our founding fathers"; he also indulged in the usual anti-Capote anecdotes. He took us to lunch at Passetto's, unshowy but very grand. When I told him that today was Don's and my twentieth anniversary, he ordered Dom Pérignon. Afterwards, he showed us the inside of the nearby Pantheon, which, oddly enough, Don and I neglected to visit in 1955. It really is overwhelming. Don thought it the most impressive building he had ever seen in his life. Then Gavin left us and we wandered around the streets, windowshopping. We had been into Bulgari's that morning, trying to find a chain like the one Truman Capote had bought there, a few years earlier. They told us that they no longer make that kind. But we returned in the afternoon and asked to be shown bracelets, and finally chose one which both of us liked enormously—pale gold with heavy links which looked rather like the links of a bicycle chain,. When it was time for supper, we met Gavin and Georges at Nino's. Georges and I managed to converse in French.... I have written all this down with deliberate flatness, not knowing how to convey the day's essential quality; it was a strangely joyful anniversary, simply because we were both of us happy that we had met each other and that we were still together.

February 15. We left Rome by Alitalia plane at 9:05 a.m. On the way to the airport, I had noticed a Russian Aeroflot poster, announcing that you could fly from Rome to Moscow and thence across Siberia to Tokyo, in fifteen hours. That would also be approximately the time span of our flight today. Crossing the Alps, we got a perfect view of the Matterhorn; I had never seen it before. We arrived in London at 10:20 a.m, where we collected our baggage and took off by Pan Am plane at 1:20. We got to Los Angeles at 4:10 p.m, Pacific Time. The flight was one of the most boring I've known. *Judge Roy Bean* on the screen. Having given up our first-class tickets in order to pay for the Klosters–Rome side trip, we had to endure tourist class and its swarming children. And never has the Canadian wilderness looked more dreary tha[n] it did that afternoon. I wish it could at least have been Siberia, to make a bit of a change.

As a postscript to this record, here is Richard's comment on my visit to Wyberslegh, in a letter I received about two weeks after my return home:

Your visit was terribly short but *much* better than nothing, and if it wasn't perhaps quite all it could have been it was entirely my fault, and it shan't happen again. Sometimes I think when it is a case of one's nearest *and dearest* who one only sees occasionally one is apt to feel a little nervous, specially when they are famous. It was *very* nice of you to travel up to Stockport to see me....

No. 63 does seem ideally planned, don't you think, as regards the central heating, there is something positively exotic about it. But it makes me feel very remiss that poor Mum never enjoyed such luxury at Wybersleigh.... Of *course* No. 63 hasn't the charm of Wybersleigh and it would be ridiculous to compare them—so near and yet so far apart. However it has got a lot of individuality for a brand new residence, don't you think. M. would have liked it I think somewhere else but as a modern dwelling so close to Wybersleigh, *never*.

February 23. Last night we went to a farewell party given by the Laughlins. They have already sold their Bel Air house and are going to live in France. Leslie looked so charming, in a very feminine flouncy soft grey dress. She told Don, "I thought of you when I put it on," meaning that she knew he'd like it, which he did.

Jack and Jim were there; Jim has now begun rehearsals of *Streetcar.* He is very pleased with Voight and Faye Dunaway[1] but worried about the play and about whether Tennessee will like his direction. David Hockney was there too—without Celia, with whom he's now living in a house at Malibu; she had decided to spend the evening at home. I had a talk alone with him out on the terrace, from which I gathered that David (apparently) isn't having sex with Celia and that (certainly) he is still carrying a torch for Peter. I had to be careful not to say too much about our meeting with Eric Boman; David is very jealous of him.

The day before yesterday (21st) there was an earthquake centered near Oxnard; 5.75 on the Richter Scale. When it happened, at 6:46 in the morning, Don and I were both in the front bathroom weighing ourselves. (I seem to have lost some of the extra weight I am sure I was carrying while I was in England; am now down to around 150.) At first there was the usual mild unalarming jarring. Then, suddenly, the house seemed to be rocking about on a bed of tapioca pudding; it was the same horrid feeling we'd had during the 1971 quake, though less intense. Don said, "Oh dear—"

Later that day, I went up to Vedanta Place and talked to Vidyatmananda, who had been staying at the monastery during his short visit to Los Angeles. (He left yesterday, to go back to Gretz, after a stopover in Boston.) It was seven years since we'd last met. He didn't look any older, only heavier and much healthier; he is nearly sixty. The healthy look was no doubt due to farmwork. They have what he describes as a commune there, including a guest-house for people wanting to make a retreat and a fully functioning farm. The guests all have to pitch in and help milk cows, etc. And there are several young brahmacharis of various nationalities; some of them cute, no doubt. I could just see Vidya bustling around, supervising everything. (Larry Holt's friend Tom had described him as "regal." Don, who saw him later in the evening, and who is quite favorably disposed toward him, nevertheless thought that he made a bad first impression; prissy, sour and unctuous. I noticed

[1] Dunaway (b. 1941) was already a film star—*Bonnie and Clyde* (1967), *The Thomas Crown Affair* (1968), *Little Big Man* (1970)—when Williams asked her to play Blanche in this twenty-fifth anniversary revival.

that his mouth twisted into an ugly grimace whenever he spoke disapprovingly (but with an evident obsessive interest) about sexy films, books, etc.)

Still and all, I felt a great deal of affection throughout our talk, on both sides. He made a point of speaking of himself to me as Prema, not Vidya—seemingly because he wanted to get back to the atmosphere of our earlier relationship. On the whole, he appears to be happy at Gretz, although he is still having hard work with the French language. He still writes out his lectures in French, then gets someone to correct the French, then memorizes it. He said he had no plans to return to the States. He would rather die in India than here. He described, with obvious satisfaction, a recent visit to India during which he had had a red-carpet welcome everywhere—as if contrasting this with his rejection by the Los Angeles and Chicago centers.

He asked me if I regretted growing older, admitting that he did. And he told me that he thought I was the luckiest person he had ever known—there was a slight hint of reproof in this remark, I thought.

Swami was visibly upset by Vidya's presence in the house. And he was eager to be told exactly what we had talked about; I suppose he expected that Vidya would have said something against him or the Hollywood Center as a whole. Swami even brought up the old accusation that Vidya had been indirectly responsible for giving him the heart attack which he had had soon after hearing the news that Vidya had been conspiring with Gambhirananda in Belur Math, to undermine his influence and take away his authority to give sannyas to American nuns!

His final interview with Vidya had apparently been quite friendly. Vidya had asked Swami if he was still angry and Swami had replied that, "The guru in me was never angry with you, only the man—if the guru had been angry, then nothing in the three worlds could have saved you." This reply startled and shocked me for a moment, until Swami explained that "the guru in me" wasn't himself at all, it was *Satchitananda*.[1] He added, "I never feel that it is *I* who am initiating a disciple, it is The Lord."

During the interview, he had discovered that Vidya wasn't repeating the mantras which he had been given on taking sannyas; he ought to have been using them every day. Vidya had to confess that he couldn't remember them. He had written them down and put the paper away in a box. Swami told him that this was like

[1] The triad of absolute consciousness, absolute existence, absolute joy.

buying airplane tickets and then locking them up instead of using them to go flying. So he coached Vidya in the mantras, and this had made him dizzy.

Later, not referring to Vidya, he said that his body had become "very subtle" and that it therefore always made him dizzy to be with people who were insincere and untruthful.

After supper, when we were about to leave Swami's room for the reading in the temple, Swami remained sitting in his chair instead of getting up and letting Krishna help him on with his coat. Then he roused himself, saying, "I was daydreaming." I asked him what he had been thinking about. "About Maharaj—that's my daydreaming."

He also remarked about Vidya that his face showed he had suffered a great deal.

Mother Hubble[1] died at last, a few days ago, aged 102.

Robin French has heard from Ken Tynan, sending us his love and saying how much he likes our play and deeply regretting that he can't get his colleagues on the National Theatre board interested in it.

Since getting back from Europe, I've had a bad cold; now it has turned into a cough.

March 4. Beautiful weather but am feeling depressed. Chiefly because Don has had ulcerlike pains for the past three weeks, if not more. He went to Allen, who seemed unimpressed. Now he says he wants to find another doctor, one he can feel more at ease with. We have heard of someone, through Evelyn Hooker, but Don still hesitates to go, because he doesn't want to be x-rayed. Besides this, he is still much worried about his forthcoming show. How will he ever get people to come to the opening party? Shall he show paintings as well as drawings? He thinks not. I think yes. And then, last night, having supper at Nellie Carroll's, Michael Laughlin held forth, as he frequently does, on the probability of galloping inflation. He himself is about to take off for France to join Leslie forever, and he advises us all to take our money and put it into Swiss francs and a Swiss bank. This kind of talk makes us both nervous, Don particularly.

Another reason for depression is that the Screen Writers are almost certainly going to declare a strike against the studios, when

[1] Yogaprana's mother, who lived at the Vedanta Society in old age. Yogaprana, formerly Yogini, became a disciple of Prabhavananda during W.W.II, along with her husband, Walter Brown (Yogi), who later left. They appear in *D.1.* and *Lost Years.*

they meet tomorrow evening, and this will mean that we shall have to picket, I don't know how often. Jim Bridges, being a "hyphenate," a writer-director, has been threatened with exclusion from his own cutting room if the strike comes; the studio says that someone else will finish the work on *The Paper Chase*. Meanwhile, he is in a state of exhaustion because of long gabby rehearsals with the *Streetcar* company. He claims that Jon Voight talked for four hours yesterday without stopping. We both think Jim overplays his role of exhausted artist, however. Yesterday, just before supper started, he left the table to lie down—and then let Nellie feed him.

On the other side of the picture, such a happy glimpse of Jim Gates and Abedha (Tony Eckstein) cooking something for the Shiva Ratri celebration, the day before yesterday. They were beaming with joy. How that kind of happiness gladdens my heart. I had a nice long visit with Swami too, much of it in silence; me in Swami's presence, Swami in Maharaj's. No depression can exist in that atmosphere, no anxiety, no sorrow.

March 8. On March 5, the sun set behind the headland for the first time. On March 6, the writers' strike was declared.[1] We went to the vespers of the Ramakrishna puja. Never have I felt less spiritual. Swami was dizzy and weak; he had initiated someone that morning. But I felt his affection for Don, when we went to see him afterwards.

Yesterday, Don had to picket Universal Studios. But, luckily, the woman in charge of his picketing group owned one of his drawings and knew all about him, and she said that she didn't feel he need do any more picketing, since screenwriting wasn't his real profession. So we're hoping that she will be able to persuade the strike committee to agree with her. As for me, I'll have to wait until I'm called and then figure out what the best policy is for strike dodging. Picketing is a really grim stint—three hours on end!

Yesterday evening, we had supper with David and Celia Clark at the Malibu beach house. Joel Grey[2] and his wife Joey[3] were

[1] The Writers Guild struck against major T.V. and movie producers for the first time since 1960; see Glossary under Writers Guild Strike.

[2] Broadway star Joel Grey (b. 1932) won a Tony Award as the Master of Ceremonies in the stage musical *Cabaret* (1966), and he was about to win an Academy Award for Best Supporting Actor in the film (see below).

[3] Jo Wilder, actress and singer from New York, appeared on Broadway and on T.V., including commercials, then ran a gift gallery on Melrose Avenue and later became a real estate agent.

there—don't like them much—and Nick Wilder and his friend Gregory Evans—am getting to like them both increasingly. Ossie Clark is supposed to be coming out here almost at once. Don has drawn and talked to Celia. It seems clear that she and David haven't had sex together; she told Don that David is like a little boy. She doesn't altogether want Ossie around and yet she misses him.

Rain most of today. Hunt rang up very drunk to tell me that the cast had just read through the script and that David McCallum is sensational as Henry Clerval and that James Mason had said that Leonard Whiting is so great that he'll steal the show and that every time Michael Sarrazin (as the Creature) had said "Victor" he (Hunt) had felt tears come into his eyes, etc. etc. They have got an incredibly beautiful girl named Jane Seymour to play Prima and the part of the foreign lady at the opera is to be offered to Simone Signoret[1]—wow! Hunt had the gall to suggest that we should write an extra scene for Mason, in which Polidori denounces Henry. I told him we aren't about to strikebreak—there are too many spies around.

March 15. Hunt called today to tell me that they have just finished their first day's shooting—Nicola [Pagett] and Leonard Whiting and David McCallum in the scene of Elizabeth's visit to the lodgings and her hostility toward Henry Clerval. Everybody had been terrific, and McCallum terrific-terrific. (Hunt was drunk again; it's very tiresome that he calls at eight in the evening, their time.) However, he let slip some ominous hints. They have altered the period of the costumes quite a bit, because Hunt thinks the Regency clothes look unattractive on women. They have also altered "a few words" in the script. Also, on this first day, they were ahead of schedule by midday—which sounds like quickie filming.

Am getting worried because of the utter silence of the Writers Guild strike committee. No one has called to tell me when I'm to picket. I can't help wondering if maybe I am supposed to contact them, and if I shall be fined for not having done so. According to some bossy woman who called Jack Larson about Jim Bridges (who can't picket because he is rehearsing *Streetcar*) the fine is one hundred dollars a day for picket dodging!

Ananda and the other girls are back from India. We saw them last night for the first time. Ananda can't stop talking about their

[1] It was played by Margaret Leighton.

experiences. Perhaps their greatest scoop was seeing a very old lady who had met Ramakrishna. She was in her high nineties, and she has since died! But Ananda said, "Just the same, it makes you realize, we have everything right here." I repeated this remark to Swami. He said, agreeing, "How many places are there where you can live all the time with your guru?" This was one of his utterly impersonal uncoy remarks which never fail to startle me.

Ananda brought us back a leaf from the mango tree which Ramakrishna planted at Kamarpukur.

March 19. This morning I'm going for the first time to picket, at MGM: noon to three o'clock. Tomorrow we are going down to Palm Springs to see Truman Capote. I got excused from picketing that day but forgot to get let off the next day too, which means that I am supposed to picket from six to nine in the morning! The only hope is that the strike may possibly be ending, or at any rate reaching a settlement with some of the independent companies, thus reducing the number of pickets needed. It is all a dreary nuisance, but I'm playing it very cooperative until I see a way of wriggling out of the whole thing.

On the 16th, we went to see one of the previews of Jim's production of *Streetcar.* Faye Dunaway was excellent, about as good a Blanche as you could expect. But Jon Voight is all fucked up by his own methods. He got tapes of dialect from New Orleans and very carefully tries to imitate them, thereby inhibiting himself from playing Stanley! It is infuriating, the way these actors tie themselves in knots—when all they have to do is *speak up*; many of Tennessee's best lines were lost because Voight was trying to speak with his mouth full of spittle, or some such idiocy. Jack Larson said later, "All that generation of actors, they're done for, there's nothing, no way of saving them." Jim has fits of despair. But no doubt they will all perk up on the press night—tonight, in fact. Tennessee himself is coming, which is regrettable; one never knows how he will act up, nowadays. We haven't seen him yet.

March 25. This morning, Swami rang up to tell me that Mokshada (Meta Evans)[1] died at midnight last night. She was in her nineties. Abhaya (I can never spell that name, but she's the hospital nurse devotee who looks after Swami[2]) was with Mokshada and so was

[1] A devotee for many years; in *D.1*, Isherwood records that she worked as a laundress at the Vedanta Society in the 1940s.

[2] The spelling is correct; the name is pronounced Avoya; it means "fearless."

Krishna. At about 11:30 p.m. Mokshada told them that she had seen Brahmananda. She asked them to prop her up in bed, so she could be in a posture of meditation with her beads. Then she cried out, "Raja! Raja!"[1] She didn't want anybody to sit with her; she said she wanted to be alone with Maharaj. Swami was obviously enormously impressed by this story. I had unworthy doubts, remembering what a playactress Mokshada used to be and what a conniving old Jewess, full of schmaltz and bitchery. But then Don said, "Does one playact when one's dying?" I thought maybe yes, but on second thoughts had to admit that it was at least remarkable not to want to have people with you at the end. Swami said, "And that will happen to every devotee."

The local notices of *Streetcar* in the papers, including the Free Press and the two trades, were very bad. That is, they took it for granted that the play is a masterpiece and that Jim's direction had ruined it, and they disliked Voight. But they praised Dunaway greatly. (And Tennessee even said she was the best Blanche ever.) Tennessee also told Jim that he loved the direction and wanted to work with him again. Jim was shattered at first, stayed in bed the day after, but has perked up since the play and his work were praised on T.V. Voight is said to be very upset. We haven't seen him since the preview. Meanwhile, you can't get a seat for love or money. It's the biggest hit they've had. And, often, there are standing ovations at the end of a performance!

Tennessee has arranged to leave for the Orient—Hong Kong and then Thailand—on a cruise ship from Honolulu, with a fierce-looking blue-eyed blond young man named Robert Carroll.[2] But today we hear that they may split up, because of some awful quarrel. In fact, Ten is very much himself. He has recently given a deplorable interview to *Playboy* with indiscreet personal stuff about Frank Merlo's sex life and death.

Also, on the 20th, we drove down to Palm Springs and saw Truman. He seems very sick, says he vomits constantly. He's swollen up enormously and looks very old, rather like Churchill.

My picketing on the 19th was rather fun, except that I wore myself out talking to my colleagues. Since then, I haven't done it. I got excused by the picket captain, in order to go to see Truman, and then, on the 23rd, because I'd lost my voice. It has more or

[1] A nickname for Brahmananda given him by Vivekananda for his regal bearing; he is often called Raja Maharaj by Ramakrishna devotees in India, where Maharaj is used generically for most monks in the order.

[2] Vietnam veteran and would-be writer from West Virginia, ninth child (b. 1942) of a coal miner; Williams's companion 1972–1979.

less come back now. But we only picket alternate weeks, so this coming week we'll be free.

A heaventime with Don. But Kitty caught Drub out yesterday, he had eaten a whole bag of dried dates. So the sentence of the cat court was No Supper. The sentence was revoked at the last moment, and they had eggplant, lots of it. With the result that Drub weighed 153 and ¼ this morning (up a pound and ¼) and Kitty weighed 144.

March 29. Triumph of *Cabaret* over *The Godfather*. They got only three awards, we got eight.[1] Excitement over Brando's refusal to accept the award, a silly show-biz stunt which was probably nevertheless of some value because it made people remember the Indians. We minorities are like a lot of tiresome (to the majority) children jumping about yelling, "Look at *me*!" But Brando should have appeared personally.[2]

This is a jittery period, as far as I'm concerned. What with the prospect of the writers' strike dragging on maybe for months and the fuss of having to make several public appearances after all this long while, and the feeling that we're not getting on with our *Meeting by the River* screenplay and Don's continuing uncertainty about having his show and about his future as an artist altogether and my wonderings about what to write next. Well, jitters are better than inertia.

Tennessee finally did go off with Robert Carroll to the Orient.

A very interesting interview with Truman Capote by Andy Warhol in *Rolling Stone* (Jim Charlton gave Don a subscription to it as a present). He says that, once he has finished *Answered Prayers*, he won't do any ambitious writing again. He says:

> I can't count ... five people [...] of real importance in the media who aren't Jewish. [...] If those people could have done me in, they would have done me in like nobody's ever been done in. But they couldn't do me in. They would have done me in because not only wasn't I Jewish and wasn't in the Jewish clique, but I *talked* about not being part of it.

He says that he gets lots of poison-pen letters from a man who signs himself "U." They are becoming increasingly threatening.

[1] Including Best Actress, Liza Minnelli, and Best Director, Bob Fosse. *The Godfather* won Best Picture.

[2] Brando's Award for Best Actor was refused on his behalf by Sacheen Littlefeather; see Glossary.

The writer keeps changing the style and color of his envelopes, so Truman won't recognize his letters and simply throw them away. He thinks Ackerley's *My Dog Tulip* is one of the greatest books ever written by anybody in the world. He says: "I know everything there is to know about success, so how can success change your life? I know everything there is to know about failure, too, so nothing can affect me on that score. I mean professionally ... from the beginning I always attracted a lot of attention, because—well, really—there really isn't anybody else like me." He says: "For me, every act of art is the act of solving a mystery.... There's a technical mystery to be mastered, and there's a mystery of human nature. Art is a mystery."[1]

Truman impresses me enormously and I feel that he has a great deal of worldly wisdom. I feel a mature weariness in him; to me, he seems much older than I am. And yet at the same time I'm aware that this stance of his is partly a pose and that, probably, when he seems most assured he is whistling in the dark.

This morning a postcard arrived from Tom Wright; I can't read the postmark but it's somewhere in Colombia. "I had a splendid adventure in Brazil, in the wilds above Manaus. A (*name illegible*) guide, an American youth and I took a skiff along a deserted and densely forested river tributary of the Rio Negro. After a grueling fourteen-day trip we got to the Waimiri Indians (eluding the Indian Protection service), a tribe who massacred thirteen missionaries four years ago, and who killed three road builders just last month. There was no danger at all. I found them charming and hated to go back to Manaus!" Dear Tom! I feel such a warmth of love, thinking of him, that plump white porky-pig, sticking his inquisitive nose into darkest Amazonia. I love his genuine but somehow ridiculous daring. This news brightened up the day for me.

March 30. My day was further brightened up when someone from the Screen Writers Guild called during the afternoon to say that the whole picket schedule has been revised. My group doesn't have to picket again until April 9, and then the picketing will be from 12 to 3 every day for five days, after which there will be a two-week interval before the next stint begins!

David Hockney came around to talk about Peter. He has decided

[1] From "Sunday with Mr. C.: An Audiodocumentary by Andy Warhol Starring Truman Capote," April 12, 1973, pp. 28–48 and Jann Wenner, "Coda: Another Round with Mr. C.," in the same issue, pp. 50–54.

to write him a letter saying that they mustn't meet again for a long time. It seems that Peter has been dropping into the Powis Terrace flat and using it like a club while David has been away—making Mo furiously indignant. Meanwhile, David has been flirting with Gregory Evans, Nick Wilder's friend. Gregory has a crush on David and would like to go back with him to England. David is really very tough, but he keeps saying how much Peter has "hurt" him.

April 6. We have just heard that Virginia Pfeiffer died on February 24 of cancer, after a very painful illness. I talked yesterday to Laura, who is still quite shattered.

When I saw Swami last Saturday, the 31st of March, Abhaya told me quite firmly, in his presence, that Mokshada did *not* claim to have seen Maharaj, though she did cry out, "Raja! Raja!" Anandaprana (not in Swami's presence) told me that he had been talking about dying, a few days before. Ananda had said, "I'm sorry, Swami, but *I* don't believe your time has come yet."

Early in the morning of the 3rd, I had the following very vivid dream. With the greatest of ease, Don picked up a large full-size oak tree. (I think this happened in Coldwater Canyon.) In the branches of the tree, an old Chinese woman was sitting. I was terribly afraid that she'd fall and that he would therefore get into trouble. Later, the oak tree became tiny and was inside a leather bag, like the camera bag he uses to carry his paints and brushes in. I guarded it, afraid that it would be found.

Was this a dream about big worries that become small and absurd ones?

Yesterday, Faride Mantilla (who has decided she won't come to us any more because she's got a better job) sent us a Texan Mexican named Vangie or Angie Bravo. This seems worthy to be called a synchronicity because, a few days ago, Don wanted me to suggest a name for the chihuahua he has just bought for Glade and I suggested "Olé"!

Angie's enormous advantage to us is that she speaks fluent English. But she is also a good cleaner and very agreeable. She calls us Don and Chris, not out of what Kathleen would have called "familiarity" but because those names are easiest to say. She has a very big ass.

When she saw our Shirley Temple doll, she picked it up with an extraordinarily moving expression of tenderness. (She has two children of her own.) She looked at it for a moment, then became self-conscious, murmured, "A grownup playing with dolls!," put the doll down again and straightened herself, letting go a loud fart.

April 9. This in haste, because I have to picket today, from 12 to 3. The new schedule only requires you to picket one week out of three and the dreaded 6 to 9 a.m. shift has been eliminated—because, apparently, the people who go to work at the studios that early are not worthwhile picket-victims.

In my last entry I neglected to say that on April 4 I made two public appearances in aid of the *Kathleen and Frank* paperback; one downtown at the Alexandria Hotel, talking to the Los Angeles County Public Library breakfast; one in Berkeley that same evening, at which I didn't have to talk, merely mingle with two hundred and some booksellers from northern California. The one truly memorable moment of that day was when I was about to leave for the plane and Don reminded me to take my beads with me—I hadn't even thought of doing so. But I've noticed, whenever we're on a plane together at take-off, that Don tells his beads. (I merely make japam.) Don's reminder moved me very much. It seemed to express what is essential in our relationship, what makes it different from and dearer than any other I have ever had in my life.

When I saw Swami on April 6, I asked him, did he fear death. He said no, not at all, but he then repeated a story he has told me before, how, when he was going to have a double hernia operation and had to fill in a form relieving the surgeon of responsibility in the event of his death, he had felt afraid and had prayed to Maharaj for release from his fear and had been instantly reassured.

On April 7, Swami initiated Peter Schneider—at last! I have been trying to get in touch with Peter ever since, to hear his impressions, but without success.

This reminds me that Allyn Nelson told Don that Jim Gates has told her he mustn't see her any more, because he's a monk. This seems to Don and me a piece of almost incredible hypocrisy. Don thinks that Jim made Swami forbid him to see Allyn by asking innocently, "Swami—ought I to have a friendship with my old girlfriend, now I'm a monk?" Swami could only answer this question with a "No." It's the same thing as programming a computer to get the answer you want. The only mystery is, *why* did Jim want an excuse to stop seeing Allyn?

Oh yes—and on March 30 Dr. Maxwell Wolff removed a little growth from the tip of my nose. This seems to be all right, now. I don't think he was seriously concerned about it.

April 22. Easter Sunday and a truly glorious day—lately we've been having too much wind for beachgoing. This afternoon, we're driving down to Laguna Beach to see Jack Fontan and Ray Unger.

Yesterday, I finished a weary chore: twenty pages of reconstructed diary about our trip to England. I forced myself to go on writing it, but I know most of it is mere typewriting. Now I can go on with the reconstruction of my 1949 diary. That's an effort, too, but I know it's worthwhile. I would love to push right through to the end of 1952, as soon as possible. Then I want to sit down and read the whole bunch of my diaries. I have a feeling something will come out of this. I keep playing with the idea of a version of *Specimen Days*—nothing like Whitman's, but that's its "secret" title. Kind of prose poems, each one quite short, following certain trials or scents through the jungles of memory and fantasy and dream.

On April 18, when last I saw Swami, I asked him, "If Gerald Heard had stayed with you, would you have arranged for him to take sannyas?" Swami said yes and then added, "And you too, Chris—even now—why don't you come back to us?" I said, "I'm not fit to," and he answered, "Only the Lord can judge who is fit." We looked at each other and he giggled. He giggled again at supper when he told the girls, "Chris is going to become a monk!" I laughed and the girls laughed too, rather uneasily, watching my face to see if this was a joke. It was very odd and disturbing. Doesn't Swami accept my present way of life, after all? I had begun to believe that he did. Oddly enough, during that same conversation, Swami had said that it had been all right for Amiya to leave the convent because she had only taken brahmacharya, which was only a resolve, not an absolute vow. Sarada's case was different, because she had taken sannyas.

Talking to Mrs. Norman Lloyd, the actor's wife,[1] in the market the other day, we got onto the subject of Don, who recently drew her. She said, "While he's drawing, you feel that pure spirit—the marvellous grace with which he handles your immortal soul!"

Don and I keep slowly but steadily on with the draft of the screenplay of *A Meeting by the River.* Fifteen pages to date. It seems pretty solid so far.

[1] Peggy Lloyd (1913–2011), actress and T.V. director. She appeared on Broadway in the 1930s, joined Orson Welles's Mercury Theater Company, and directed T.V. shows for Alfred Hitchcock. Norman Lloyd also worked with Welles and Hitchcock and appeared in *Saboteur* (1942), *Spellbound* (1945), *Dead Poets Society* (1989), and *The Age of Innocence* (1993).

April 26. Am just off to talk to a group at Claremont. This was arranged by a young man named Adrian Turcotte,[1] "Butch," I've heard people call him, who comes to Vedanta Place for the readings, etc. Peter Schneider is to drive up with me and we're to have supper with his mother afterwards.

At Vedanta Place last night, much talk about the Watergate scandal and Zeffirelli's *Brother Sun, Sister Moon*, which the boys seem to like better than the girls—to my surprise. As for Nixon, Swami frankly hopes he'll be impeached.

Talking of Swami—at the question period last week, he was asked, "Should one regard the guru as being the same as God?" To which he unhesitatingly answered yes, going on to say that he speaks sometimes to Ramakrishna and sometimes to Maharaj. Some of those present were a bit bothered, wondering did Swami mean that we should regard *him* as God. He realized this—how often he does realize the inwardness of a question just when I'm worried because I think he hasn't! He said, "But I'm not that kind of guru." At which someone asked, "What kind of guru are you, Swami?" Swami answered, "I am the dust of the feet of Maharaj." A pause. Then he smiled: "There is some holiness even in the dust."

On the 21st, I was in Hollywood. Turning off the boulevard, I passed a group of young hustlers. One of them looked exactly like Wayne Sleep, only much taller. The others were going someplace, and this boy said he didn't want to, "I've got to sleep." Can you call this a synchronicity?

May 6. Last week, I picketed at MGM, until Thursday, when I got an attack of sciatica (or something similar) and had to quit. My back has been bad since then—despite two treatments of DMSO, recommended by Laura Huxley. (It's *great* virtue, in her eyes, is that it is banned in this country, although used in Germany and elsewhere.)

We keep on steadily with our *Meeting by the River* script. And I creep on with the reconstruction of my 1949 diary.

Jim Bridges' *The Paper Chase* was sneaked at Goleta, right next to the U.C. Santa Barbara campus, and greatly applauded by a student audience.

Last Wednesday (the 2nd), when I went to see Swami, I got a very powerful spiritual vibration from him. It began when I asked

[1] Later a success in the computer business; he married three times and settled near Seattle.

him what he does all day, now that he can't read. He said that he feels the Lord's presence. I asked him if he meditated. He said no, not in the daytime. When he meditates, he meditates on the Guru in the center of the brain and on the Lord in the center of the heart. And then he began to recite the meditation on the Guru, partly in Sanskrit and partly in English. This recitation was broken off by the arrival of Krishna to get Swami dressed to go to the temple for the reading. (This is one of Krishna's prerogatives, and I know, without Swami's having told me, that I must never help him, even if Krishna comes in late.) Then Swami took my arm (this is one of *my* prerogatives) for the walk to the temple. I always try to meditate on the Guru while we're doing this, but my vanity at being Swami's attendant and making the entrance with him usually spoils my concentration. This time, just as we were about to enter the temple, Swami suddenly continued his recitation, with the words: "He is calm and he is pleased with you, smiling, gracious...." I don't know if he said "you" instead of "me" intentionally, but I couldn't help taking it personally—feeling in fact that the Guru was speaking to me through Swami. It was an extraordinary moment.

May 7. My back is better but the lump on my left foot is more painful than ever before. We went to the gym but I couldn't run. Put DMSO on it, and on the lump on my left thumb, which still hasn't gone away and is maybe even a bit bigger. (This isn't a moan, just a routine health report.)

This morning, we went to watch Clinton Kimbrough directing his film at a tiny café on Washington Boulevard in Venice called Rick's—one of the snuggest places I have ever seen in Los Angeles. Clint seems calm and businesslike, but the film sounds like the most improbable kind of quickie; when he is through[,] it will have taken him three weeks to shoot. It's about three nurses and one of them (at least) gets murdered. In this bar sequence, Tom Baker[1] was playing the barman. There was also a stunningly beautiful black girl and a campily handsome and elegant black man. Clint's wife Frances is the script girl. Don thinks he has already turned her into his slave.

Maurice Jarre, who composes music for films and T.V., called to say he had been in England and had seen an hour and a half

[1] American actor (1940–1982), he appeared in Andy Warhol's film *I, a Man* (1967) as a last-minute substitute for his friend Jim Morrison of The Doors. He died of a drug overdose.

of "Frankenstein" footage, because Hunt wants him to do the "Frankenstein" music. He said that he had been greatly impressed by what he had seen, but that he didn't want to do the music if the film is to be only for T.V. This is to be decided within a week. I think Jarre's real motive for calling me was that he isn't sure if he can trust Hunt and wants my advice.

Yesterday afternoon, we drove up Las Tunas Canyon to visit Peter and Clytie Alexander. Their house is now more or less built; that's to say, a big awkward wooden structure like two packing cases on top of one another has been somewhat insecurely erected on a platform of land cut out of the hillside. Clytie is just loving her life there, she says. She is working on a vegetable garden and gave us homegrown chard and lettuce (delicious, we ate it last night). They want to avoid using the public utilities, except water, and are researching into the relative advantages of solar energy and a generator driven by a windmill. They are also shopping for the best type of filter, so as to be able to make use of all the waste from the shower, washing machine, etc. to irrigate the land. Clytie was anxious to show us how ecology-minded she and Peter are—yet they seem absolutely unable to understand why the building and zoning authorities are giving them trouble; they can't see any reason against spoiling the appearance of the canyon with this ugly dump they have planted in the midst of the landscape!

To live as they are living is an absolutely wholetime job, and you have to be quite rich to do it. For example, during the winter, they had appalling windstorms. Clytie said that a pile of heavy lumber being used by one of their neighbors was picked up by the gale and scattered on the hillside "like a pack of cards." And then the top part of their own house keeled over and had to be rebuilt—which meant that they had to hire a big trailer to live in, until the storms subsided. Don shuddered at the starkness of the interior; the completely public shower which is their only washing facility and the thronelike chemical toilet which is the most exposed place in the house. But Clytie spoke of the joys of lying in bed and looking right down the valley to the ocean and seeing the lights of ships out at sea. She is a truly sweet girl and it's obvious that she's thriving on this discipline. Peter looks very gaunt—maybe from too much sniffing and too little eating. I like him too. And I liked large Guy Dill, the young artist, who was helping him dig a hollow which may become a swimming pool or a fishpond.

One of the reasons why I like Peter is that he's such a good fan. He and Clytie read our "Frankenstein" script aloud to each other.

Yesterday I really loved him when he suddenly asked Don about the paintings of his which used to hang in my room. (Don later took them down in a rage because I let Diebenkorn see them.) Don was really warmed and pleased by Peter's gratuitous praise.

May 8. Suddenly it seems as if Don's show at the Barnsdall gallery may not take place. They have just told Don that they won't pay more than five hundred dollars of his framing costs, which would mean that he would have to pay about a thousand, maybe more, out of his own pocket. Don is furious and is inclined to cancel the show. But he's going to ask Irving Blum's advice first.

The latest news is that the writers' strike may soon be over.

Today I finished Hardy's *Collected Poems.* Am now reading Thomas Pynchon's *Gravity's Rainbow*, Quentin Bell's life of Virginia Woolf, Mishima's *Runaway Horses*, Steegmuller's *Flaubert in Egypt.*[1]

May 10. The latest news is that the writers' strike will *not* soon be over. Mort Lewis,[2] who said goodbye to everybody on our picket shift last Friday, now seems to think we'll certainly be seeing each other again on the 21st.

Don still hasn't quite made up his mind about his show. Joe Goode advises him to economize by having his drawings "bound" in plastic instead of putting them into frames. The plastic can be thrown away later, whereas frames would have to be stored and would cost hundreds of dollars more. Joe says spend the extra money on the catalogue and the party.

We try to get up at six every morning and generally succeed in getting up at six-thirty. Our old alarm clock is absurdly capricious; sometimes it makes a noise like a lawn mower, sometimes it doesn't ring at all but just faintly whirrs. A couple of days ago, I spilled some water while watering the plant which hangs beside the window. Don: "There's an old Dub for you, starts the day with a splash."

Hunt called again today, full of schemes to cut the script radically. Also to rewrite it, making Elizabeth be the one to decide to leave for America. As usual, he was drunk. Don wants me to call him some time when he's likely to be sober and try to reason with him.

Today, John Mitchell and Maurice Stans were indicted for

[1] I.e., Steegmuller's 1972 edition of Flaubert's travel notes and letters.

[2] Radio and T.V. writer; he scripted Burns and Allen in the 1930s, and in the 1950s and 1960s he wrote for "Bonanza," "Rawhide," "Combat," and "Bewitched."

conspiracy, obstruction and perjury.[1] Also, the House, for the first time in its history, voted to refuse the administration funds to bomb Cambodia. But Nixon will go on bombing Cambodia anyway.

May 13. This morning I weighed 154 and ¼—all-time high except for my fat period in England in 1970. This was probably due to two cooked dinners at home, yesterday and the day before, with lots of wine. But my fluctuations are remarkable. The day before yesterday, I weighed nearly four pounds less!

Don's insides, like mine, are upset as the result of last night. This morning, as we were working on our *Meeting by the River* filmscript, he farted. "If Kitty had known that it would come to this—lying farting helplessly on an old couch!"

John Dunne and Joan Didion came to dinner last night, with Irving and Shirley Blum. The talk was mostly about Watergate of course, and the dismissal of the Ellsberg case. John Dunne told us that Anthony Russo had wanted to insist on a jury verdict, but that Ellsberg and their defence attorneys were afraid to risk it. According to journalists who have now interviewed the jurors, the jury would most probably have acquitted them.[2]

Don has now decided to go ahead with the show. Irving Blum has written a foreword for his catalogue which I don't like the tone of; it's patronizing and it doesn't mention Don's previous exhibitions, except for the one at Irving Blum's gallery. Also, it's sloppily written. Don is in an awkward position, because it was Irving who volunteered to write this foreword and because Don feels grateful to Irving for his past support. Another awkwardness is that Irving has delegated the job of organizing the opening party to hateful Brook[e Hopper]! I see breakers ahead.

I forgot to record that Jo Lathwood has recently had a long talk with Ben about Ben's father, who is now over ninety and getting frail, and who would like to come out and live here. Ben wanted to have his father come and live with him and Dee, but Dee refused. [...]

A recently conducted poll of people in Cleveland suggests that the real lasting result of the Watergate scandals will be greatly increased public cynicism in its view of *both* parties and *all* politicians.

Cukor has suggested that we should work with him on a

[1] Both were under investigation in connection with the break-in at Democratic headquarters in the Watergate Hotel complex in June 1972; see Watergate in Glossary.

[2] Ellsberg and Russo leaked the Pentagon Papers to the press; see Glossary under Pentagon Papers.

script of *The Aspern Papers*, for Katharine Hepburn and Maggie Smith.

Jon Voight called last night to say that Marcheline and he have a son, born on the 11th.

And Lamont Johnson called to say goodbye. He's off to Africa to shoot his other film, without even finding time for a talk about our *Black Girl* project. All his big talk about taking us on a location hunt in Africa and giving us a chance to view the eclipse of the sun!

May 24. Watergate hearings on T.V. are a constant temptation to waste the best part of the morning watching; they start at 7 a.m. our time. Most of us are rather charmed by James McCord.[1] He is so positive in his replies; he speaks rather ornate, old-world FBI English. He turns his face away from the camera and looks at it with one eye, which is like the bright wily eye of a whale. Don even thinks he's sexually attractive.

The most frequently used phrase in the hearings is "at this point in time." The burglary is referred to as "the Watergate intrusion."

The other day, while I was sitting in the steam room at the gym, in burst an absurd wouldbe-macho queen, yelling "Son of a bitch, cunt, mother fucker, fucking asshole—!" I thought for a mad moment that he was speaking to me, but he continued, "Impeach the bastard, he fucked up the stock market!" After which, he spat, three times, on the rocks of the heater—an action which was made ten times more melodramatic than he had intended because, an instant later, the automatic shower turned on, raising clouds of steam.

Up at Vedanta Place, they have been having a garage sale of unwanted clothing, furniture, kitchen utensils, books etc. which have been donated to the society throughout the years. The boys made a lot of placards to advertise this. One of them was: "Watergate Sale—everyone's got to go!" But it was vetoed by Asaktananda.

Swami told me, when I was there yesterday evening, that Dr. Kaplan[2] had "gone sex mad" and had even propositioned the nuns. Having also talked to Anandaprana, I do think that he must be having some sort of nervous breakdown. He is now saying he wants to go up to Santa Barbara, presumably because he wants to have a go at the nuns there! But Swami's real anxiety is not that Joseph Kaplan will be able to seduce anybody but that he is longing

[1] Who led the break-in; see Glossary.
[2] A distinguished physicist; see Glossary.

to make a speech at our Father's Day celebration (which, this year, is also the celebration of Swami's fifty years in the States). Kaplan's speeches are always intolerably wordy and long and egotistical, and this one will probably be crazy as well.

Swami now refers to Sarada as "Ruth."

Everyone now says that the strike will go on indefinitely. Yesterday we had a mass picket at MGM, over sixty of us. But we don't picket regularly.

Don's show keeps changing its aspect. Billy Bengston now says he will give a party, but this is to be only a small one, for the inner circle. So the question remains, what about Irving's party? The ill feeling between Billy and Irving complicates matters. Shall there be two parties, or will they merge? And then, again, who is the show *for*? Don had taken it for granted that it was for celebrity-conscious establishment squares. But now his attitude has been greatly changed by Nick Wilder, who has come out with a most handsome vote of confidence in his work, telling him that he now realizes that the portraits of celebrities are only a small part of Don's work and that he, Nick, wants to show Don's work in his gallery—first with other artists, this fall, and then in one or more one-man shows, later. Nick's encouragement has greatly boosted Don's morale. He now feels that Nick is really appreciating him as an artist. To be appreciated as an artist, not jollied along as a portrait drawer with interesting social connections, is what Don needs more than anything else. For years he has labored under this terrible defeatist sense of inferiority, wondering if he is really an artist at all, etc. I have been able to do nothing to undermine it, because I can't speak with authority. Billy Al, Joe Goode, Peter Alexander and some of the other art friends Don has made lately, have already done quite a bit. But Nick has the supremely convincing quality of being able and ready to put his money where his mouth is. Nick says he thinks Don has been poorly handled by Irving, but that he couldn't make Don an offer until Irving decided to move to New York and thus ceased to be Don's dealer. Oh, if only Nick likes Don's paintings as well as his drawings! (Irving never did, and that, too, made Don doubtful about them.) Right now, I'm prepared to love Nick dearly for the rest of my life. As for loving Gregory Evans, that's no sweat.

May 28. The party problem is still unsolved, because Don hasn't had a chance to talk it through with Billy Al. He is afraid that Billy may be tiresome about it, because Billy hates Irving and would love to spite him—even, maybe, if it inconvenienced Don.

Billy was talking the other day as if he wants to force Don to choose between the two of them by opting for the one party or the other. Irving has said definitely that he won't give a party with Billy and that, if Billy wants to give a party, he must do it earlier, before the show, or later, after it.

Yesterday afternoon, Irving came to look through the pictures Don had chosen tentatively for exhibition in the show. He firmly nixed all the paintings and said Don shouldn't consider showing any paintings whatsoever, because he was only just beginning to find a style. He also nixed all the little ink drawings and all the nudes, but this was chiefly because he felt that they didn't go with the big drawings and that the show should be all of a piece. The only drawing he really seemed to like greatly was the one of Montgomery Clift. He told Don not to show the ones of Katharine Hepburn and Norman Mailer.

I can't help feeling that Irving doesn't really think much of Don's work in any category—because, Don says, he was so terrifically surprised that Nick Wilder would want to give Don a show in his gallery. Now, Irving's even beginning to hint that perhaps he can arrange a New York show for Don! This is all to the good, of course. But it doesn't endear Irving to me.

Meanwhile, Nick gets dearer by the minute. Or rather, to my fondness sympathy has been added, because he has just had a horrible shock. The day before yesterday, Gregory Evans took off for London. He was going to stay with David there and then spend the whole summer in Europe. Yesterday, Nick got the news that Gregory had been stopped at London Airport because the immigration officials saw that he was dopey (on Valium) and then found needle marks on his arms (he used to shoot heroin) [...] After being questioned, poor Gregory was sent back to the States, where he was again questioned. We saw him last night, after he had gotten back here, very late. He and Nick are determined to appeal and somehow force the British to let him back into the country; but this looks like being a long campaign.

Amidst all the strains and stresses connected with the show, Don has been at his terrific best. He works tirelessly, takes all responsibility upon himself, never despairs when things go wrong. After driving down to San Diego to see Rex Heftmann,[1] who was to design the show catalogue, he made up his mind that Rex wasn't up to it and phoned Rex and called the whole thing off—a horribly

[1] Graphic artist and designer, later an instructor at Miramar Fine Arts College in San Diego.

embarrassing scene to have to play. And how marvellous he looks! At thirty-nine he has the figure he had in his early twenties, after he'd started going to the gym. And how adorable and sweet he is to his old Plug (whose weight is again this morning up to 154 and ¼)!

This is Memorial Day and I'm working away, sitting out the holiday. Don has gone to draw Norton Simon. I fixed this up, after a scene of (if I do say it) Kissingerlike diplomacy. I asked Jennifer if she thought Norton would agree to it. Jennifer said she would *make* him agree, because she wanted the portrait for herself. While they were arguing, I said, "Norton, I can promise you one thing—if you say no, Don will never ask you again—he never nags at people—I'll give you an example: years and years ago, he asked Jennifer to sit for him, and she said no, and he's never once suggested it since then, has he, Jennifer?" To be frank, this was a lucky shot, not a carefully aimed one. But it hit. Norton grinned and said, "I'll make a bargain—I'll sit for Don now if Jennifer will promise to sit for her portrait in time for me to have it as a Christmas present." And Jennifer agreed! The interesting thing I noticed was that Norton was genuinely eager to have his portrait hanging there amidst all the other celebs in Don's show. *If* Don can do a portrait that pleases him—a huge if—I really believe Norton and Jennifer will come to the opening party, which would delight Irving and all the gallery snobs.

May 31. Don did an excellent drawing of Norton—so good that he'll probably include it among the eighteen which will be reproduced in his catalogue. After talking to Billy Al, he feels that Billy is going to be reasonable, after all, and that both parties can be held at different times, without creating friction.

Triumph over Yorty, yesterday. Bradley is mayor[1] and Burt Pines City Attorney. I have considerable doubts about both of them, but hope for the best. At least we're rid of Yorty, who's as slimy as oil and as crooked as his friends the oilmen. Bradley is pledged to protect the coastline and Pines to protect the gays.

Was at Vedanta Place yesterday evening. Swami seems very frail, but he gets out for his two daily walks and continues to initiate people. He told us how Brahmananda reproved him once for not being sufficiently interested in the work of the Math. Brahmananda told him that he must love everything about the Math, including the trees and the flowers, because everything there was dedicated to God.

[1] Tom Bradley (1917–1998), lawyer and former policeman, first black mayor of Los Angeles, held the post for twenty years.

This reminds me that I talked to Jim Gates on my previous visit, May 23. I asked him how he now felt about being a monk and he assured me that he was very happy. He said that, when he left the monastery and went into other people's houses, the atmosphere seemed "burnt up." (I wondered if he was consciously referring to Buddha's Fire Sermon.) He's a bit too pure, and quite a bit too malicious. We were standing outside the office in the temple building and Jim suggested that I should ask Krishna for copies of his photographs of the garage sale placards. "He's in there now," Jim said, and before I could object he unlocked the office door with a bunch of keys and there was Krishna working on the tape recorder and really cross for a moment at the intrusion—cross chiefly with Jim, but with me too. I felt terrible. And then I knew instinctively that Jim had done it to tease Krishna.

June 4. The writers' strike has again been prolonged and Lenny Spigelgass (whom we met at a party given for Gottfried and Silvia Reinhardt, the day before yesterday) seems to think there is now no hope of ending it for a long time. Gottfried looked just the same but fatter; Silvia a bit shrivelled. All the guests were old, nearly all were film Jews. I spend very little time with old people and almost none with big groups of old people, so this party depressed me—not because the old were old but because they were so desperately competing with each other as show-off survivors.

Last night, Joe Goode, Mary Agnes Donoghue and Peter and Clytie Alexander came to supper. Joe is teaching at Irvine and he was very interesting, telling us his ideas on how to teach art students. One of the most important things was, he thought, to have carpenters, plumbers and builders explain to them how to build their own studios or convert existing buildings into studios. He got Nick Wilder to come and tell them all about the relations between the artist and the art dealer.

Joe loves talking about nature. He and Mary Agnes had gone up Canyon Drive in Hollywood, right to the top, in the hills of Griffith Park, where you can picnic. There had been lots of rattlers; this is the season when they breed. Joe told us how he had sat on a rock, with the snakes rattling all around him. "I enjoyed the feeling of danger."

After talking to Ted on the phone yesterday, Don is afraid that he may be starting another of his attacks.

Today, Don has gone to the printers in Pasadena to see about the printing of his mailer for the show and of his catalogue. When he drew Muff Brackett the other day, she told him that she had always thought his portrait of Charlie was too grim, but that, after Charlie had had his stroke, she had *recognized* the grimness in his face. Another instance of Don's uncanny gift of seeing potentialities in people—like the madness he saw in Sarada.[1]

June 7. On the 4th we had supper with Chris Wood, just returned from England and wildly enthusiastic about it—except that, being Chris, his enthusiasm is expressed in terms of anti-American aversion; he kept saying how ghastly the food is here—you can't get eels, or proper cold meat, or Melton Mowbray pies. He had spent most of his time in London walking around, especially in the parks. He had seen John Gielgud and Raymond Mortimer, both of whom had been very kind to him. He hadn't seen Dodie and Alec, or Joe Ackerley's sister Nancy, or Patrick Woodcock. He still can't make up his mind if he really wants to go back and live in England. Maybe the charms of living here and bitterly yearning for England are greater.

Gavin writes from Tangier that Georges has vanished into the depths of Yugoslavia. He couldn't get another passport and he was so eager to rejoin Gavin that he altered the date on his old one, and was detected by the Italian authorities, with the result that they handed him over to the Yugoslavs. The Yugoslavs didn't arrest him but took away his old passport and gave him "a rather sinister document good for one journey to Yugoslavia only, to be used within two weeks." Gavin also says that Tom Wright has returned to Tangier and is busy on his Amazon book. Gavin seems fairly happy otherwise; says, "Tangier is growing on me."

Last night, Swami told us that one of the devotees had an initiation from him in a dream. She came and told him about it and he agreed that the mantra she had dreamed that he had given her was an authentic one, and that the instructions that went with it were the right ones, also. I couldn't quite figure out what Swami's attitude was to all of this, but I got the impression that he was pleased, satisfied, maybe a bit honored. He said, "Well, it's happened at last!" And then he told a story of how a devotee received a mantra from Maharaj in a dream, but a word was missing, Maharaj was already dead, so the devotee went to one of the other direct disciples (Shivananda?[2]) and he went into Maharaj's old room and

[1] Sarada was outwardly serene, but Bachardy sensed neurotic disturbances within and, unlike Prabhavananda, was not surprised when she decided not to be a nun.

[2] Isherwood's guess is correct.

meditated and presently returned to the devotee and gave him the complete mantra.

My left foot is getting, if anything, more swollen, despite the "Earth Shoes" which Don and I bought, a couple of weeks ago. They have very low heels and raised soles, so that you walk on your heels which is supposed to be better for your posture. They are comforting to wear.

Poor wretched old Larry Holt called this morning and moaned because, after having at last arranged an interview with Swami through Anamananda (behind Anandaprana's back) he got colitis and wasn't able to see Swami after all.

June 8. Well, we're off tomorrow to Yosemite, with John Schlesinger and his friend Michael Childers. We are both of us dreading it just a tiny bit, but that's chiefly, I think, because it's an interruption of our beloved routine and because we don't usually care to spend much time continuously with other people. We are to fly to Merced, then take a rented car.

Don has decided to include a batch of photographs of drawings in his show. These will be of drawings which are of celebrities but not really up to standard. Don says they will look better reduced. And they will be of interest, anyhow, as celebrity portraits. Irving Blum approves of this idea.

Gottfried and Silvia Reinhardt came to supper last night. They arrived drunk and got drunker. They squabbled together like children, and they both talked to us at once, so we couldn't attend to either. Then Silvia fell asleep on the couch. But we enjoyed their visit. They are both of them warm and affectionate, in their different ways. Gottfried asked me to speak at a Max Reinhardt memorial concert which is to be held in the Hollywood Bowl on August 30. I said yes because I want to see what it's like, speaking at the Bowl. It appeals to my show-off side. What struck me again was Gottfried's extraordinary vitality. Despite his physical grossness and his Austrian air of indolence, he seems so vital, so relatively young. And under his easy good humor there is real strength, and behind his jokes real compassion and serious concern. I think he is a very good man.

June 13. I've kept meaning to write in this diary because such a lot has been happening. Now all I can do is to put it all down briefly in a single lump.

On the 9th, we left for Yosemite as planned. We'd promised to meet John Schlesinger and Michael Childers at the airport at 7:10

and didn't arrive till after 7:30 because that angel Kitty had been fluffing his dear fur. But John and Michael were late too. They had just started to get into a flap and, as we came in, they had me paged—the only time I've ever heard myself paged at an airport. Otherwise all went smoothly. The rented car was actually there when we got to Merced, and we duly navigated the twisty back roads and got ourselves onto the highway that comes up from Fresno and turned off at the Chinquapin turnoff which leads to Glacier Point. I was terribly edgy, hoping that Glacier Point would come up to my memories of it and that it wouldn't disappoint John and Michael. We made the mistake of stopping at another viewpoint first which *was* disappointing because the view was only partial. And, when we did get to Glacier Point, it was swarming with schoolchildren. I don't think John and Michael really thought it worth the effort, but Don was caught at once. I knew it, when he squatted down and gazed, motionless. There was Half Dome. There was the Yosemite Fall. There were the two great falls away on your right. There were the ridges behind, with snow still on them. A chipmunk ran across the face of the precipice and a little bird launched straight out over the abyss. You saw the hotel and the meadows and the river and the tiny cars, far far below. Whenever the children stopped screaming for a moment, there was the silence and the far faint purring of the falls. Don said to me in a low voice, "It's unanswerable," and I knew that we shared yet another secret experience.

Then we drove down to the valley and had a late lunch at the Ahwahnee Hotel. People everywhere, but that didn't matter so much. The Bridal Veil Fall was splendid. We walked right up to it and there was a rainbow in the spray. All the falls were still full from the melting of the snow water.

There was nowhere to stay. Everything booked up. So we drove on to San Francisco. It was beautiful, coming out of the mountains onto the plain, and then the sunset lighting the clouds. I hated the Fairmont; the guests seemed to be all Watergate folk but in fact they are probably little provincials out to splurge on some special occasion and therefore eager to be overcharged. They will brag about their bills later as people do about their illnesses when they are let out of hospital. Don and I snapped at each other and had a brief quarrel, which merely made me realize how beautifully we have been getting along for the past month. Next day we saw a painting by F.E. Church, *Rainy Season in the Tropics*—a marvellous romantic landscape, rather like the Andean plateau. (Later we found that Paul Wonner, whom we saw last night, is a big admirer of Church.) And then we got a plane back to Los Angeles, in the afternoon. That night on T.V. we heard that Bill Inge had killed himself. After two unsuccessful tries with pills, he gassed himself in his garage. The obscene ruthless obstinacy of despair. Now, of course, we all feel we should somehow have prevented it. But we couldn't have.

The day before yesterday, I spent the whole day downtown at the American Booksellers' Association convention, being interviewed and autographing copies of *Kathleen and Frank* in paperback. While there I had to endure lunch with several of the Curtis representatives. They were all married men, talking about the girls they were seeing here in Los Angeles in a way which took you back to Dorothy Parker. It was depressing. I kept very quiet and I think they were aware of this, but at least they kept off fag stories—to my relief, because I didn't want to have to speak up and make everybody uncomfortable and yet I knew I would have to. The real literary star of the day was the actress who had played in *Deep Throat*![1] There was a line right down one side of the gigantic convention hall to get her autograph.

June 14. Swami has had another setback; he very nearly developed pneumonia while up at Santa Barbara, but managed to give his lecture and then insisted on being brought back to Hollywood, so he could see his doctor. He's better now but weak. The Father's Day lunch has been postponed until June 23. I saw Swami briefly last night. When I left, he said, "God bless you, Chris." I said, "I'd much rather have *your* blessing, Swami." I don't know what made me say that; it wasn't just a corny compliment. And I don't know exactly how Swami interpreted my meaning; anyway, he laughed.

I forgot to record that Swami told us, last time we were with him, that he hadn't slept more than two hours, the night before. I asked, "So you made japam all night?" And he answered, almost comically, "What else should I do?"

I ate with the boys at the monastery yesterday, before the reading. The two who most impress me at the moment are Bob Adjemian and Abedha (Tony Eckstein). I feel a very strong current of love from Bob, which is all the more impressive because he doesn't demonstrate it by looks, words or general behavior. And Abedha, that sour, moody little Jew, is now, it seems, almost uncannily happy. Either he's gone a bit nuts, or something essential has happened to him. But I don't think he's nuts. Last night he told

[1] Porn star Linda Lovelace (1949–2002).

me that two of the other monks—Jim Gates is one of them—are doing the worship every day. "And that makes it wonderful here," he added, or words to that effect. It was as if he'd said that the monastery had been moved to the ocean and that therefore they were now getting wonderful sea air.

When I meditate nowadays, I try to conjure up the shrine with all of them sitting in it, and I say the meditations Swami told me to say as if they were being said by different monks and nuns in turn, or else speaking in chorus. In this way I can join in with them and lose myself for a few moments. It is much easier to say "we" than "I" when trying to avoid distractions. (In doing this, maybe I'm reverting to Kathleen's preference for group worship, after all!) Bob is one of the people I chiefly think of. Also Beth [High], Abedha, Gauriprana.[1] And then I remind myself that, at this very moment (7 a.m. to about 7:45) Swami, Krishna and Abhaya are meditating in Swami's room. So I mentally join them.

I talked to Lenny Spigelgass this morning, asking him if I should risk going to see Mike Laughlin's film at a projection room at United Artists studio.[2] He warned me earnestly not to. "They're longing to get us, Chris. You and I are too well known. Don't do it!" Lenny thinks the strike may be over quite soon, however.

Lenny said of Bill Inge, "In the last few months, there was nothing there."

One thing I forgot about my visit to Swami last night. When I came in, I prostrated as usual and touched his feet with my head. As far as I was concerned, that was it. But, as I started to get up, I saw his hand extended over me. In my dumb vague way, I thought he was holding out his hand for a handshake, so I tried to take it. Immediately, he raised it a little, out of my reach. So I reached up for it. He raised it higher. And then I realized that he was blessing me! I was quite embarrassed, but he didn't laugh.

June 17. On the 14th, I had lunch with Dr. William Melnitz[3] at

[1] American Vedanta nun at the Hollywood Vedanta Society, also known as Sushima, then in her late forties or early fifties, vivacious, a good singer, and a good cook. She left the order around 1980 and died a few years later.

[2] *Nicole*, starring Caron; see below, July 15, and Glossary under Laughlin.

[3] German-born stage director (1900–1989); he joined Max Reinhardt in Los Angeles in 1941 and became Professor of Theater Arts, Dean of the School of Fine Arts, and a founder of the Professional Theater Group at UCLA. He was director of the Max Reinhardt Archive at SUNY, Binghamton, New York 1969–1973 and co-author, with Kenneth MacGowan, of *The Living Stage: A History of the World Theater* (1955) and *Golden Ages of the Theater* (1959).

the UCLA faculty center, to talk to him about Max Reinhardt, in preparation for my speech at the Bowl. He says that Max was just like Gottfried to look at (before Gottfried got so fat) and that he resembled Gottfried in many ways, but that Gottfried didn't love his father as much as Wolfgang did—at least, not in the early days. I got the impression that Melnitz doesn't like Gottfried very much. And disapproves of his book about Max.[1]

Melnitz thinks that Max's greatest theatrical talent was for working with actors. It didn't matter who they were. He treated the students at the Reinhardt School in Los Angeles just as he had treated his biggest stars in Berlin. I said I'd heard that the Los Angeles students performed so impressively because Max had trained them like animals, teaching them every move and every intonation. Melnitz denied this absolutely, saying that Max very often left an actor free to develop his performance in any way he chose. He was a wonderful audience and enjoyed the rehearsals more than anybody else.

After seeing Melnitz, I went with Don to visit Guy Dill and Charles Hill in their studios. Guy lives in Venice, on the corner of San Juan and Main Street, in a building from which boats used to be launched into the canal which once ran along Main. Now, all the interior walls have been knocked down, and the boarded floor of a former room is now just a low platform in the midst of the one great bare barnlike enclosure. Here Guy sleeps and eats in the presence of his concrete pillars, tension pieces and other artworks. Some of the tension pieces (he showed us photographs of them) are huge contraptions in which heavy blocks of concrete are involved and strips of metal are kept bent by wire cables bolted down to the floor. Guy says that he is considering switching from concrete to pumice because it will be so much easier to handle in bulk. The whole place is kept as tidy and clean as a showroom.

Guy is a tall very well-built boy with a rather puffy red face—the kind one describes as "meaty" and associates with football players. He wears his hair longish. Though not exactly handsome, he is extremely attractive and also intelligent. He projects quiet confidence and an air of intention; you feel that his works express just what he wants them to. Among the most self-revealing of them, perhaps, are some early "objects": these are old faded adventure books such as you would find on the twenty-five cent tray outside

[1] *Der Liebhaber: Erinnerungen seines Sohnes Gottfried Reinhardt an Max Reinhardt* published that year in Germany; it later appeared in English as *The Genius: A Memoir of Max Reinhardt* (1979).

a used bookstore; Guy has tied them up like parcels with black insulated wire (which has cruel-looking spiky knots in it, suggesting barbed wire) and has thrust jagged pieces of glass between their pages. They are astonishingly ugly, and they really made me react to them. I felt curiously outraged and almost scared. (Would the effect have been stronger if the books had been valuable, and/or sacred (Bibles, for instance)? I'm not sure. Maybe that would have seemed merely corny.) On the surface, Guy is nice mannered and friendly. He was really pleased to see us, I think.

Charles Hill and Paula have their studio in the building which is rented by Billy Al Bengston; it used to be a (semigay) bar. It is now a very tall room because the bar's false ceiling got burned in a fire and had to be taken down. Charles and Paula sleep on a platform which is about three-quarters of the way up the wall. The place is untidy because Charles is using every inch of it. His tattered paper paintings—most of them at least as big as flags—look like mummy wrappings. Charles buries them in the earth, so that the dye runs and they turn pale and moldy; if they show signs of falling apart—and they all seem about to—he stitches them onto a paper backing. Lately, however, he has switched to making tangled cutout designs like winding creeper-stems within an oval outline. To us, the most interesting items in the studio were a collection of photomat pictures for which Charles has posed—he somehow got the use of a photomat machine in privacy for a week or two.[1] In many of the pictures he is naked to the loins, camping in dance postures, sometimes with a female wig and makeup. He has a terrifically sexy lithe smooth brown body, and a whole new aspect of him seems revealed. He also makes films in which he is the only actor, but we didn't see any of these.

Then we went to have supper up at the Alexanders' house. On the way back, just as we got to the bottom of Las Tunas Canyon, the brakes of Don's Fiat began to seize. He had barely enough mobility to get us across the highway—otherwise we might easily have been hit and killed. Then the car wouldn't budge an inch. Don walked to the Las Tunas Isle Motel and phoned the Auto Club. The repair truck came quickly; it had been in the neighborhood. Two big youths, one of them sweetly pretty, jumped out and put a dolly under the car, since it couldn't be towed otherwise. Both boys seemed to be enjoying their job, although they are on the all-night shift and sometimes get called out of their beds. They kept laughing and joking and clowning, but they knew exactly

[1] I.e., the coin-operated booth which produces instant automatic portraits.

how to handle the situation. The pretty one's name was Morgan. They drove us to Adelaide, and there we found that the wheels had somehow come unlocked! The boys asked for ten dollars and we tipped them an extra five, for making this tiresome experience so pleasant. Later we realized that they had never asked Don for his club card or given him the usual form to sign. Were they going to pocket all the money? If so, bless them and preserve them from being found out.

On the 15th, I got a cable from Stephen Spender, telling me that Jean Ross died last month and giving me the address of her daughter Sarah Cockburn. It was truly considerate of Stephen to do this. Have written to Sarah.

That afternoon, an Englishman named Winston Leyland came to interview me for the San Francisco paper, *Gay Sunshine*.[1] He is maybe Jewish, and looks like Gavin Lambert somewhat, but I fear he's a slob. He used to be a Catholic priest but has given all that up and is now deeply into gay lib. I warned him, before we even met, that I would not discuss my sex life in relation to any other named person, and he agreed. But today, when he came again, he announced that he had been disappointed when he listened to the tapes of our previous talk—in a word, he wanted names or nothing; specifically, he wanted me to discuss Auden's sex life in relation to his poetry. I said absolutely not. He pouted, but stayed on to waste another of my hours. When he left, he still wasn't sure if he would use the interview in his paper or not. The gall—especially since Dobbin had waxed eloquent on Love ("Love is tension, not fulfillment, not some sort of an insurance policy") and actually brought tears into the slob's eyes! The insulting suggestion behind his behavior was that old Drub alone wasn't interesting; you have to serve him garnished with gossip and newsy names!

June 26. Two days ago, on Sunday, the guild met and voted to end the strike, except against the networks, ABC, CBS, NBC. Which reminds me that we've just heard, from someone at NBC, that our "Frankenstein" is definitely scheduled to be shown on NBC television in two parts, Saturday November 17 and Monday November 19. So that's the end of our hopes.

Tomorrow, the house is to be fumigated for termites. We shall

[1] Leyland (b. 1940), who emigrated to the U.S. in boyhood, founded the Gay Sunshine Press, also publishing gay erotica and translations. He reprinted the Isherwood interview in his *Gay Sunshine Interviews*, Vol. 1 (1978), which includes interviews with Allen Ginsberg, John Rechy, Gore Vidal, Tennessee Williams, and others.

have to spend tomorrow night in the studio and won't be able to get back in until the next afternoon. They can't put a tent over the house, it's too big, so everything must be sealed up.

[Don's London friend] is in town and has written to both of us, asking if he may see us. Don doesn't want to see him but is torn by doubts—wonders if he ought to and then tell [his friend] frankly how he feels about him. We both feel [the friend] will make mischief if he possibly can. What he actually wants, as of now, is the German version of the Wedekind Lulu plays, edited by Wedekind's daughter. But I'm certain I sent the book back to him years ago.

Howard Kelley called last night, slightly drunk and sentimental, to tell me how much he had loved *Kathleen and Frank.* He also told me that Wallace Bobo has had his leg amputated just below the knee. The bone has been infected a long time and there have been several operations on the leg already. I felt really pained by this news but much of the pain was guilt because I don't see them and maybe I never will again—Manhattan Beach is psychologically farther than Australia and the welcome they would give me would be more embarrassing than a reception given by Reagan.... (Having written this, I have almost succeeded in shaming myself into visiting them!)

Talking of embarrassing receptions—on the 23rd we had Swami's Father's Day lunch, which had been postponed from the 16th because of his health. This time, Swami was well enough to go through the whole ceremony, from the Grand Entrance to his speech. The entrance out from his room into the garden was particularly solemn this year. First came Swami, then Pavitrananda supported by Krishna, then Chatanananda, then Asaktananda. And we had to pause, as first Swami and then Pavitrananda stopped to face the massed cameras of the devotees. I was standing respectfully to one side, but Asaktananda took my arm as he passed me and firmly included me in the procession. I felt exactly as if we were getting married. At last we were all seated amidst the flowers, under the colored parachute-silk awnings. Ananda coyly reminded me, in Chatanananda's presence, that I should get on with the recording of the Upanishads.[1] Swami reproved the monks for starting to eat before he did. Then came some very tough meat, and some of the usual blush-making songs by Sudhir, Sat and their ensemble.[2]

[1] Seven Upanishads translated by Prabhavananda and Isherwood and read by Isherwood, produced as a double L.P. in 1976.

[2] Sudhir was a Hollywood devotee who lived nearby in Vedanta Terrace; he sang in the choir and wrote folk songs. Sat was Jimmy Barnett.

And then Swami spoke. It was astonishingly long and rambling—a sort of history of his doings in America, largely in terms of the sums of money donated or paid out for the building of the temple, Trabuco real estate, etc. Swami also talked about Huxley, Heard and me, saying that, "although I am nothing," these famous men had sat on the floor in his presence. There was a vanity in all this which I (and Don) minded not because it was vanity but because it seemed so senile. I have never heard Swami talk quite in this tone before. When it was over he was obviously exhausted. He went to his room saying that he wanted to be alone. (But since then he has seemed fairly all right.)

Then I went to see poor Mary Herbold (in a nursing home on Vermont). She was attacked from behind by a burglar. He crushed her ribs and hit her on the head. She was knocked unconscious and bled so much that he must have feared he'd killed her, for he ran out of the house without stealing anything. (He had been asking where her money was and she'd been telling him that she'd show it to him if he would let go of her.) But already she seemed quite recovered and was so pleased to see me and so lively and gay that I loved her and didn't even mind her torrent of talk. Most of it was about her wonder-sons, John, who's the greatest high-school baseball coach in America, and Tony who has seven children and has taken a poorly paid post at a college in Maine because the fishing and hunting is so good in that neighborhood. What interested me much more was her speech about the beautiful love which has sprung up between David Nelson, the slightly cute Finnish boy who was a monk for a short while, and an older guy named Joe.[1] They are now living together and Mary seems to think this is most inspiring. I must try to find out more about them.

July 12. Today is Don's big day. This morning we went down to the Barnsdall Park gallery to see how his pictures look on the walls. They look splendid—truly an oeuvre, a human comedy ranging from Norton Simon to Gregory Evans, from Mrs. Reagan to Candy Darling.[2] And Opliger[3] has lit it intelligently, not using what he calls "the garage lights," the big glaring vertical lamps in

[1] Joe Cooper; Nelson is also known as Devadatta. They settled in Santa Barbara.

[2] New York drag queen (1944–1974); she starred in Warhol's *Flesh* (1968) and *Women in Revolt* (1971), and had bit parts in *Klute* (1971) and Tennessee Williams's play *Small Craft Warnings* (1972).

[3] Curtis Opliger, art critic and abstract film maker, then in charge of the gallery.

the roof, and thus keeping the middle of the gallery relatively dark and the drawings featured. Oh, I felt so proud of my darling. How happy we are together now, happier than ever before.

Opliger, who was so stuffy and grand with Don at the beginning, now treats him quite respectfully. Irving Blum's also provisionally back in my favor because this morning he told Don that he has definitely arranged a show for him in New York next spring.

Chris Wood had nearly all of his prostate removed at the end of last month. When they got it out, they found cancer in it; but now they say that the cancer hasn't spread. Chris was marvellously brave and calm while he was waiting for the result of the tests. Now he's at Peggy's, much against his will, because he has to keep quiet for a while and isn't allowed to drive a car.

The other day, a woman phoned to tell me that Guttchen is dead. She was his landlady. It seems that he had left the names of two friends who were to be called in case of emergency. I was one of them. The landlady kept saying that she hoped he had repented and accepted Jesus. I got the feeling that she was afraid I would claim some of his belongings; it seems that he had some Chinese pictures which she thought were beautiful. Since then, I've heard nothing from her.

On the 6th, we went to see Rita Hayworth. We had arranged to take her to supper at the Chianti and were a bit worried that she would be drunk when we arrived. She was drunk, quite, but, to our delight, she admitted she hated going to restaurants. So we stayed with her in her rather spooky house (Don, who went into the back part of it, said that it had a terribly sinister atmosphere) and drank wine and ate cheese and listened to a record which had been sent her from England, made by the New Vaudeville Band. The words of the song began, "Dear Rita Hayworth; I just saw your picture in the daily press. . . ." and went on to say that the singer had "loved you every way I can" and wanted her to come and visit him in England. Rita played this over and over, and danced to it. She also played the castanets, with extraordinary coordination, so that they purred, or clicked like rattlesnakes. Rita's dancing got progressively less steady, but she was charming and touching and even oddly beautiful.

(I've just realized that I left out a most significant detail: Rita didn't say immediately that she didn't want to go to the Chianti. First, she told us that she had lost her key and therefore couldn't leave the house because she wouldn't be able to get back into it. She let us hunt around for the key a bit, *then* she said that, in any case, she would rather not go. She never admitted that the losing of the key was just playacting. Maybe she really had lost it.)

Rita says she often spends weeks in the house without seeing anyone, except her two dachshunds and (I suppose) her maid. She made like she was delighted to see us, indeed she kept saying so and hugging us, but still, behind the drunkenness and the little-girl clowning and little-girl pathos, there was a suspicion of hostility. When she saw us to the door, she was anxious that we shouldn't stumble in the darkness on the way to our car. "Are you all right?" she called, from the doorstep. We called back that we were. Then, as she shut the door, we heard her voice from inside, strangely ironical and mocking, saying to herself and/or the dachshunds, "*I'll bet!*"

July 15. Don's invitation to his show called forth one of Joan Crawford's memorable letters. She is truly The Last Great Lady: "Don dear, thank you so much, etc. etc.... I wish I could be there but I am in New York.... The drawing on my invitation shows such beautiful strength in your lovely face, Don. Bless you, and my best wishes to you.... My love to you and Christopher. As ever, Joan."

The opening was certainly a success. Hundreds came, including a few stars—Julie Harris (who scuttled in like a mouse as soon as the doors opened), Elsa Laughton (less bitchy than usual), Joan Didion (who had to be pressured and made a martyred face when she was photographed), George Cukor, Roddy McDowall, John Huston (who all behaved perfectly), Glenn Ford (full of reproachful glances, like a jilted lover), Virgil Thomson (who came down especially from San Francisco and was the life and soul of everything). There were probably others I've forgotten. No-shows included Jane Fonda (excused because she's just had a baby), Lance Loud (excused because of lack of transportation from Santa Barbara), Jennifer and Norton Simon (not excused). Don was at his best—he blazed with manic brilliance, greeted everyone, knew everybody's name and submitted to much photography by a fumbling amateur. Later, there was a supper party up on the open-air roof of Ceeje's Upstairs Restaurant, hosted by Billy Al Bengston, Joe Goode and Nick Wilder. This was delightfully relaxing—except that I had to talk to Leslie Caron (another of our stars, whom I'd forgotten to mention) and Michael Laughlin about the picture which Michael produced and Leslie played in; we had seen it that afternoon, in a wild rush, just fitting it in before Don's opening. Well, at least we could both say truthfully that Leslie's performance was probably the best she has ever given, a sort of Lady Macbeth. "That's what I'm really like," Leslie said.

The picture itself has some beautiful photography but it is arty and hopelessly muddled and adds up to nothing. A dramatic event of dinner was the behavior of Charles Hill. Perhaps because he was nervous about his show in San Francisco next day, he began to eat and drink immensely. After a while, he went to the men's room and passed out cold. They had to get an ambulance and take him to the hospital. But next morning he insisted on leaving for San Francisco by car on schedule!

William Wilson, who has always praised Don, had been to see the drawings early on the morning of the show. On July 13, the *Los Angeles Times* published his piece about it. It is the strangest mixture of panning and praise, self-contradictory and double edged. Irving Blum says that the nastiness in it is probably due to his not having been invited to the supper party. Here are some extracts:

> One of the cheapest shots at artistic fame is the flattering society portrait.... Today the genre is so debased that its mere mention is enough to make the cognoscenti nudge each other's padded ribs. Last week an exhibition of such portraits opened at the Municipal Art Gallery.... If it were Bachardy's purpose to arrive on the wings of social maneuverings, his selection of subjects could scarcely have been cagier. Many luminaries of Venice's art mafia are represented ... etc. etc. ... Igor Stravinsky and Virgil Thomson stare from Bachardy's characteristically smooth pencil and brushstrokes.... The entire exercise borders on the sort of inbreeding that used to make idiots out of Egyptian pharaohs. There are serious technical flaws in Bachardy's basically academic contour drawings: shapes of heads get lost in masses of expressively messy hair (a device used successfully by Ingres in his portrait of M. Bertin). Fussy concentration on internal shadows around eyes, noses and mouths threatens the premise of dimensional, modelled drawings. Yet the Bachardy exhibition remains among the most absorbing socio-psychological documents I have ever seen in an art gallery. The artist flatters his sitters with a suavity of style often resembling that of the great fashion illustrator René Boucher. Bachardy is mesmerized by glamor but he is not blinded. One perceives his sitters with a shock of recognition that has nothing to do with his regular achievement of simple physical resemblance. He responds with sympathy, detachment and an unnerving stainless steel insight. We see a handsome young artist being slowly engulfed by satanic madness" (*Peter Alexander?!*), "a millionaire collector who has

seen too many tragedies" (*Norton Simon*), "the shrewd stare of a hip dealer" (*Nick Wilder?*), "the wistfulness of a faded actress" (*too many candidates to permit a guess*), "all presented without a hint of sentimentality or satire. As a total experience the large exhibition is a panoramic view of a slightly faded aristocratic subculture as fascinating as Huxley's *Point Counter Point*.

What emerges from all this verbiage is Wilson's desperate fear of seeming to be unsophisticated. *Of course* you can't praise "flattering society portraits," so his praise of Bachardy, such as it is, must be severely qualified. However, by the time he is through, he has tacitly admitted that Bachardy's sitters aren't "society" and that his portraits aren't "flattering"—you can't flatter your sitter *and* show him being "engulfed by satanic madness"! The word "suavity" is also used in a pejorative sense, yet Wilson says it is a quality of Boucher, whom he calls "great." The "device" of covering the heads with "messy" hair is said to be a serious technical flaw; yet Wilson declares that Ingres used this device successfully. (Admittedly, this contradiction is just sloppy writing, but Wilson nevertheless fails utterly to prove his point.) I have no idea what the next sentence means. Nor do I understand the distinction he makes between the metaphors of being mesmerized and blinded.... I suggested to Don that he should send Wilson a drawing of some great movie star, completely bald!

July 28. What a radiantly happy time I have had, these last months! Such great joy, being with Don. I am certain that he feels it too. We live in unbroken intimacy and love, day after day, from waking till sleeping. This is most certainly one of the happiest periods of my entire life.

We finished the first draft of our *Meeting by the River* screenplay on the 26th—the the day that Jim Bridges got back from New York, where he has been showing his picture (now definitely retitled *The Paper Chase* and purged of its offensive song). Jim has read our script and doesn't like it much. He feels that most of the tension between the two brothers has been relaxed and that the other characters upstage them. We have no opinion because we haven't either of us read the script through properly, we were in such a hurry to let Jim have it. We are both ready to do more work on it—if only Jim will continue to believe in the project and not lose faith and ditch us for something else.

Today is glorious—at least, this afternoon is, after a morning of sea-fog. (This has been one of the foggiest summers I've ever known.) I was in the ocean for nearly ten minutes. Meanwhile, Don went off to Laguna Beach with Lance Loud, who is so lively and friendly and looking really sexy; he has a good figure and he has modified the amusing but too drastic henna-red of his hair to a darker, more becoming shade. Lance is to be introduced to Jack Fontan and Ray Unger; he is eager to have his horoscope read.

August 3. Jack Fontan and Ray Unger said that Lance Loud is inhibited by his Catholic upbringing—hence the frenzy of his camping; he feels he has to go all the way or none of it. But they think he will be successful in life. Lance has a thing about *The Day of the Locust*, which Don has just read and loathed, so we are planning to introduce him to John Schlesinger, when he returns from England. Meanwhile, Michael Childers has told us that John is already greatly interested in Lance and will certainly give him at least a walk-on part in the picture.

John Lehmann has already signed a contract (or at any rate a "memorandum of agreement") with Methuen; Curtis Brown has just sent me the papers. So I've had to write to John saying firmly that I will not sign until I've seen copies of my letters because I still haven't made up my mind if I want them to be published at all. Don says John is trying to railroad me, and he's right, of course.

On the 31st, I went with [Bronislau K]aper to the Hollywood Bowl, to meet Ernest Fleischmann, its director, and talk about the Max Reinhardt memorial evening on the 30th of this month, at which I'm to speak. It was a lovely though slightly smoggy morning. Alfred Brendel was rehearsing with the orchestra for a concert that evening. To my great joy he was playing Mozart's D Minor Concerto when we arrived, with that utterly blissful passage for the piano at the beginning of the second movement. Brendel's gestures are extravagant yet not ridiculous, because they are obviously for himself rather than for the public. He will suddenly rip his right hand off the keyboard, as though the keys had become red-hot and the flesh of his fingers had begun to stick to them. At the end of a solo passage, he will seemingly throw the music at the orchestra, so they can take over. He will hold his left hand high in the air, drooping from the wrist ("like a lily," as Kaper said). When both hands are free, he will pat them together, the fingers crooked and helpless looking, like a cripple trying to applaud. Then, again, he will "fish" with his right hand inside the open piano, as though he were a cat trying to catch goldfish. We met him later, a charming, slightly stooped man with comic eyes. [...] Fleischmann was very smooth and very polite. Kaper says he is terrifically vain. I do like Kaper so much, whenever I see him. He lives all alone with a housekeeper in a big house on Bedford Drive. I have a feeling that he pursues girls like a maniac. We talked about Gottfried, whom we both love. Kaper says that, when he and Gottfried meet, they repeat the punch lines of stories which they both know by heart—"just to reestablish contact."

August 6. Yesterday afternoon, I went alone to see Don's show again. (He naturally finds it embarrassing to hang around the show himself, as if waiting to be recognized and complimented.) Yesterday was the last day.

I was terrifically moved—far more so than I expected I'd be. The whole thing—it appeared to me, for the first time, not just as something done by a person I love but "in its own right," existing quite apart from Don. And it is tremendous. All those faces, famous, infamous, unfamous—each one doing his or her "thing," trying to impress, to charm, to intimidate, or not trying to project anything, just being—all of them were made equal, they were all part of the family that Don's art has conjured up. I kept thinking of Edward Thomas's lines: "All of us gone out of the reach of change—/Immortal in [a] picture of an old grange."[1] The tears ran down my cheeks and I felt so marvellously joyful.

August 14. Am not so joyful today, because yesterday we saw an untitled undubbed print of "Dr. Frankenstein" (the cut they're going to show on T.V.) and today I had to tell Hunt Stromberg in no uncertain language what we thought of it.

God knows we went (to the projection room at Universal) with no great expectations, but this was worse than anything I'd expected. Smight has no imagination, no fun, no sense of style, no feeling for atmosphere. Which also means that he can't direct actors. Even James Mason wasn't nearly as good as usual; he missed nearly all the campiness of his part. Leonard Whiting is just a fat young man with an expressionless face and no emotion beyond a certain petulance; he could say a line like, "Why the devil can't the buses run on time?") but that's all. Nicola was a disappointment. David McCallum could have been really good if he had had more humor; Smight is probably to be blamed for this. Michael Sarrazin was physically wrong. When he looked in the mirror and said, "Beautiful!" it was terribly embarrassing because he wasn't. And the makeup they gave him was totally wrong, too mild to begin

[1] From "Haymaking."

with, so that you couldn't believe it could have scared Mrs. Blair into a fit, and later merely disgusting—sores and carbuncles—not frightening. The best of them was Jane Seymour. She would have been very very good in the part if Smight hadn't fouled up her scenes. He has no sense of timing whatever.

And then Hunt and he had put in words and made cuts which spoilt much of our dialogue. And they had written a whole scene for Gielgud and made nonsense of the escape to the ship. And ruined the ball scene. And made the drowning of William meaningless and the avalanche ending ditto. No, I don't want to write any more. I feel really sick to my stomach. This wretched pair have taken months of our work and destroyed it.

Don was so wonderful last night, while we were having supper. He said (in effect), "Imagine if, twenty years ago, we could have seen ourselves as we are tonight, talking about this disaster—we'd think we must be unhappy." Meaning that we aren't. And truly we aren't—because the life we have together now makes all such disasters unimportant, even funny.... Still and all, we are mad at Hunt!

August 21. On the 19th, Don went up to Carpinteria, where Ken Price has taken a beach house and was giving a party for The Gang—Billy Al, Joe Goode, etc. etc. Before this, he drew Howard Warshaw and found him very hostile. This seems perfectly natural to Don. He reads it as a typical heterosexual reaction; making a heavy out of the junior partner in a homosexual relationship—"Chris was our friend until Don came along and poisoned him against us" etc. But I'm puzzled. It doesn't seem to me that simple. Nor do I understand the very much more evident dislike which Fran Warshaw now feels for me—has felt, I'm sure, for a long while. No doubt she hates Don too, but she hasn't had as many opportunities to show this.

While I was writing the above, Robin French called and told me that George Santoro at Universal told him NBC is delighted with the "Frankenstein" film and is planning to show it on two consecutive nights—November 30 and December 1—instead of on a Saturday and a Monday night, November 17 and 19, as previously planned. I haven't told Don this yet because he's out in the studio drawing Randy Kleiser (who has just signed a contract to direct at Universal and is interested in doing our "Mummy" script—if he likes it when he's read it!).

To get back to Sunday the 19th. Ben Underhill unexpectedly showed up—he was in town staying with an old friend, [...]. The two of them came down and took me to lunch. Ben looked incredibly healthy, his eyes so clear and his skin so pink. But he had a stare of switched-on brightness which seemed a trifle uncanny. I had no chance to speak to him alone, so couldn't find out anything about his life. He seemed very glad to see me. And indeed I felt happy to see him.

In the afternoon, Rudi Amendt came over to talk to me, at my request, about Max Reinhardt. He is such a weird creature. He worked himself up into a frenzy of gestures, stage whispers, writhings, eye rollings and impersonations. At moments he looked like a reanimated skeleton—his head is skull-like and his eyes become eyeholes. I have no doubt that he can talk *exactly* like Max himself. From this point of view, our conversation was as good as a séance; Max was conjured up. But I don't feel that I got all that much material to use in my speech at the Bowl.

I'll try to note down what I *did* get.... Reinhardt was shy, he only came to life in the theater. He didn't understand business. His brother Edmund looked after that, until he died in 1929. Meanwhile, Reinhardt had become accustomed to borrowing large sums of money and expecting people to have confidence that they'd be paid back, sooner or later. That worked all right in Europe, where he was an internationally famous person, but it didn't work in America when he came there during the late thirties and ran up against the psychology of the Depression. He was in great money difficulties when he died (seventy years old, in 1943).... When he was going to direct a play, he made elaborate plans in advance. He imagined himself playing all the parts and he knew exactly how he wanted them played, but this didn't mean that he imposed his ideas on his actors if they had ideas of their own. If one of them was at a loss, however, Reinhardt didn't leave him to work it out, he showed him how to do it. When he was moved by good acting he wept, no matter if it was tragic or not. And his laugh was famous. If you could make him laugh, that was the highest honor. He never lost his temper with his colleagues. His attitude to the theater was fundamentally religious; he instinctively presented plays as religious ceremonies. But he never forgot that the public must be entertained.... At Salzburg, he learnt style and became an art collector. His sense of style made him pay great attention to detail. Amendt described how, in one play, he installed a specially heavy well-made door, so that it would shut "like the door of a Cadillac" when one of the characters left the house with a lover.... There's lots more, but I forget—

August 24. More from Amendt, by telephone, yesterday: When Reinhardt was working with his students at his school in Los Angeles, he was "gentle, reserved, polite," he treated them "with infinite courtesy." He made the most of whatever talent they had and even overrated them. He gave them "something to stir their imagination".... I asked, was it difficult for him to stage *Sister Beatrice* in a small room, after having presented it (as *The Miracle*) in a building like Olympia?[1] No, said Amendt, because he never compared what he was doing with what he had done before. He never considered any project as being finished. He couldn't stop working on a project—which caused him to keep postponing openings, to the despair of his business advisers. Amendt said that, when he read in *Ramakrishna and His Disciples* that the atmosphere of Ramakrishna's life could best be described by the one word *now*, he had thought of Reinhardt. (I was well aware of the characteristically Amendt compliment implied by this—slipping in a little flattery of me along with Reinhardt.) Reinhardt's life, when he was working, was all *now*. He didn't have any sense of time. He would work for days on end and his stamina (until shortly before the end of his life) was extraordinary.... Reinhardt had a perfect ear for dialogue, and this was demonstrated when he supervised the translation of foreign plays into German. Not only this, he saw how to tighten them up and improve them in detail. (Am a bit dubious about Reinhardt's talent in this respect, however. I remember the heavy-handedness of his adaptation of Maugham's *Home and Beauty* (called *Too Many Husbands or Viktoria*) and Gusti Adler describes approvingly how Reinhardt had the actress who played the king's mistress in *The Apple Cart* (*Der Kaiser von Amerika*) do gymnastic exercises throughout the second act in order to liven up the dialogue. Adler comments with true kraut smugness that this was "a wittily inserted circus act which fascinated the public—an example of how Reinhardt could help a writer out, when one of his scenes was not very successful dramatically."[2] I wonder what Shaw thought of this bit of wit. Probably he didn't care

[1] In West Kensington, London, the largest indoor exhibition hall in England when it opened in 1886 as a venue for agricultural shows, circuses, motor, and air shows.

[2] Adler (1890–1985), born in Vienna, was Reinhardt's personal secretary and later worked as a researcher at Warner Brothers. Her memoir, *Max Reinhardt, sein Leben*, was published in German in 1964. Bachardy recalls that Isherwood possibly had a typescript of an English translation (published later); otherwise, this is Isherwood's own translation.

too much, since Reinhardt made the play into a hit, after it had more or less flopped in England.).... Amendt says that Reinhardt never made comparisons between actors who had played the same role.... When Reinhardt was working with you, you felt that you were very close to him. But, to Reinhardt, it was only the work that mattered. If another project came along and you weren't the right person to help him with it, then he dropped you. And so, as Amendt puts it, "He left frustrated lovers behind him".... His absolute dedication, his singlemindedness, inspired awe in other people. Amendt says, "One couldn't pee with him".... Reinhardt always attempted too much. He was unable to supervise all his productions and some of them thus slipped out of his control and became second-rate. He was very extravagant and accustomed, in his middle life, to live in luxury. His great house, Leopoldskron, was in itself a kind of theatrical production. But he never became addicted to his possessions as ordinary rich men do. His great attachment was to a dog named Mickey (after Mickey Rooney) and it was a terrible blow to him when Mickey was killed by a rattlesnake in the Hollywood Hills. And on Fire Island in 1943, the excitement and exertion of trying to defend another of his dogs against two others caused him to have his first stroke.

August 26. Sixty-nine. A very quiet birthday—partly because I wanted it that way, partly because Don has a cold. We had a talk this morning with Jim Bridges about the *Meeting by the River* film. He now wants to use most of our script—rewrite some sections and then send it around to possible backers. We feel that he is much hotter on the project than he has ever been before, but his enthusiasm is fickle, and he'll cool off if he gets something else which offers a chance of immediate employment. He is very hard up.

One of our chief problems; what is Tom's role in the story? Jim would like to reduce it, because Tom scares off possible buyers. But he wants to keep Tom in the picture because he loves him. We all agree that the relationship between Patrick and Oliver must be strengthened in every possible way.

Don brilliantly found a present for me which I really needed and yet wasn't expecting: a briefcase.

The thickness—it doesn't seem to be actually a lump—on the ball of my left foot persists and sometimes hurts, especially when I get out of bed in the morning. My feet are altogether quite painful when I run. I notice a slight shortness of breath, nothing serious. The lump in my left thumb persists, neither smaller nor larger. My memory is very bad—am constantly drying up on people's names. My weight this morning 151 and ½.

Chris Wood, whom we had supper with last night, has just started cobalt treatment. He feels well, though complains of being tired, and doesn't seem unduly worried.

Few birthday greetings: Stathis Orphanos and Ralph Sylvester (who made me some sweetmeats out of dates), Peter Schlesinger, Richard (a cable), Marguerite [Lamkin] (ditto). Have just called Rod Owens to wish him a happy birthday, but got no answer.

September 3. We gave a goodbye dinner party for Mike Van Horn last night—Penny Little and Billy Al Bengston, Jim Crabe,[1] Jack Larson and Jim Bridges came. Mike and Jim Crabe are splitting up, quite amiably, and Mike is going to try his fortune in New York. Mike is worried about this and regards it as a test of himself and his talent and character—though Don has been trying to persuade him not to take such a dramatic all-or-nothing attitude. Don thinks Jim Crabe's attitude to Mike is still condescending; that Jim treats him as a quaint little boy. Jim is certainly not heartbroken over their split-up. He wants to live on his own because then he won't be embarrassed by Mike's presence when his square studio friends come around. I'm sure he has never really appreciated Mike for what he is and probably can become. Jim's closet is terribly terribly square.

Jim Bridges has just been told that he has been awarded the grand prize for *The Paper Chase* at the Atlanta Film Festival. So he's going there on Thursday and so is Randy Kleiser, who also has an award for *Peege*. And Randy, who is signing a contract with Universal, seems quite interested in "The Lady from the Land of the Dead," if they'll let him! Not another word from Hunt Stromberg.

My appearance at the Bowl was a disaster. When I asked Ernest Fleischmann how long my talk about Reinhardt should be, he said, "Just use your judgment." So on I went. The Bowl was two-thirds empty—that's to say, the audience count was just over 6,000 out of an available 17,000 seats—but still and all that's more people than I've ever spoken to before, outside of India. When I got to the mike, it was too high, because it had been adjusted to suit a tall British actress, Rohan McCullogh, who was later to speak some passages out of *Midsummer Night's Dream* against a background of

[1] American cinematographer (1931–1989), his many T.V. and feature films include *Save the Tiger* (1973), *Rhinoceros* (1974), *Rocky* (1976), and *The Karate Kid* (1984).

Mendelssohn's *Midsummer Night's Dream* music. Also, I found that I couldn't see anything but a great blazing spotlight shining down at me out of a blackness which was like outer space. It was quite overwhelming for a moment—like having to address God; you knew the blackness was conscious and could hear you. I felt a sinking feeling of dismay but launched out into my speech and wasn't bad, I think. Very soon, however, there were flurries of clapping out of the darkness and more clapping with increasing murmuring from the orchestra behind me. At first I disregarded this, but finally the murmuring became so loud that I asked, "Am I running overtime?" "Yes," they said, "twenty minutes." I asked if I could finish with an anecdote. They said, "No." The clapping got louder. So I said, "Sorry, goodnight," and walked straight off the stage. I may have spoken for twenty minutes in all but I doubt it. (The anecdote was to have been about Jean Ross—how she claimed that she and her fellow extra used to fuck every single night on stage during the party at Guilietta's Venetian palace in the second act of *Hoffmanns Erzählungen.*[1])

As I was on my way back to my seat (in a box with Don and a guy from the gym named Jim Shadduck) a lady named Mrs. Olive Behrendt ran after me and told me I had been marvellous, etc., and not to feel bad about the rudeness of the audience. She is one of the most important people on the Hollywood Bowl Committee.[2] I met her again today at Roddy McDowall's and she repeated what she'd said, but she wouldn't at all accept the idea that Fleischmann was really to blame for the whole thing. After the show was over—and Rohan McCullogh had disgraced herself by giving a hissy unctuous performance of her lines—quite a batch of people came up to say how sorry they were that I'd been cut short. And then there was an article by Martin Bernheimer in the *Los Angeles Times*, pointing out that the tribute to Reinhardt had been underannounced and that I wasn't introduced. And then a member of the orchestra wrote a letter apologizing for his colleagues but excusing them by saying that they had none of them known about my appearance in advance or why I was going to speak or for how long; and that the concerts are timed precisely because of union regulations. He also pointed out that, since members of the audience are allowed to eat and drink throughout the performances,

[1] Reinhardt staged Offenbach's *Tales from Hoffman* in Berlin, November 1931–April 1932.

[2] Socialite and fundraiser (d. 1987), wife of George Behrendt; she became President of the Los Angeles Philharmonic and later was on the board of the Norton Simon Museum.

many of them are drunk! (The writer of the letter was named Dennis Trembly.[1])

Don, that avenging angel, was furious of course, but we decided to show the flag by going to the party afterwards. However, when we arrived, and asked for wine, we were told that we couldn't have any because it was needed for supper. So we left!

September 14. Letters about the Bowl incident keep arriving. And I keep answering, in effect; don't apologize to me, apologize to Reinhardt—don't blame the audience, blame the management.

On the 6th I sent a cable to John Lehmann, and next day a letter, telling him that I cannot agree to the publication of my letters to him. They are dull, mechanical, false. Don was horrified by their insincerity when he read them—he hadn't believed that even old Dobbin could be capable of such falseness. I haven't heard a word from John yet, but I surely shall soon.

Robin French has at last admitted that he is giving up being an agent. He has got a job as Vice President in Charge of Production at Paramount. We are rather pleased and excited, because we have practically decided to switch to Irving Lazar. If this is a mistake, it will at least be an adventurous one. We sounded him out through Peter Viertel and he was enthusiastic to have us. (Peter is in town because Deborah is here in *The Day After the Fair.* She opened last night, boldly overacting because of the oversize of the Shubert Theater. Her courage is adorable and I think the audience really liked her.)

During the last few days, I quite suddenly decided to make a stab at my autobiographical book about America—or maybe just the first volume of it. And yesterday I actually started. I am basing the opening on my diary, but already I'm expanding the material and, I dare to hope, beginning to see how I can work away from mere narrative. My inspiration is Jung's resolve "to tell my personal myth."[2] Therefore, I shall try to dwell only on the numinous, on the magical and the mythical. I shall try to avoid much reference to characters who lack these qualities and arouse negative emotions in me. I don't want to gossip and bitch. I don't want to complain and denounce—unless I am denouncing mythological demons. I aim for a tone which is positive, good-humored, tough, appreciative—but not goody-goody or mealymouthed. I want to

[1] Bassist, trained at Juilliard; he joined the Los Angeles Philharmonic in 1970 and became a Principal Bass late in 1973.

[2] In the Prologue to his autobiography, *Memories, Dreams, Reflections* (1963).

refrain altogether from blaming myself or indulging in guilt. When I mention famous people, it will never be merely to debunk them. Unless I can see something marvellous in them, I'll leave them out altogether if possible. I want to write about my sex life very frankly and without the least hint of self-defence.

The kind of book I do *not* want to write—the kind which is full of sly debunking put-down gossip—is exemplified by S.N. Behrman's *People in a Diary*, which I am just reading. (I was told by someone that he died the other day,[1] but that's no excuse for him.) On page 250 of this diary, July 20, 1972, I've already referred to one of his stories about Siegfried Sassoon, which was part of an excerpt from the book published in *The New Yorker.* Here is another, equally nauseating—

Behrman has a talk with Sassoon's ex-landlady and friend in London, "Burton." She hints darkly to Behrman that Sassoon ought not to stay at his country cottage near Salisbury—it's no use his staying there, the longer he stays, the unhappier he'll be. When Behrman goes down to visit Sassoon, he finds that Sassoon is lingering at the cottage because of an obsessive involvement with someone who lives at a nearby country house—a very beautiful but somehow sinister house; Sassoon himself describes it as being like "The Turn of the Screw." The house has a tower and in this tower lives a young man who is the nephew of a famous British public figure. When Sassoon comes to call, he is told that the young man isn't well enough to see anyone. Sassoon knows that this is a lie and says so, bitterly. This embarrasses the two nice middle-aged ladies who are looking after the house. They are even more embarrassed when another young man arrives, "flamboyantly dressed and arrogant":

> The women knew him, evidently. He treated them with condescension.
>
> "'Ow is 'is Lordship today?" he asked, with a jocular wink at them.... He nodded up toward the tower room.
>
> "'E expects me an' I 'ave to make the six-thirty to London. Apologies."
>
> He turned abruptly and walked into the house.... Siegfried had gone white. He turned and walked toward his car. I followed.
>
> We rode in silence. Siegfried's face was set; he kept his eyes on the road. I wondered. Had the invalid contrived this visit from

[1] On September 9.

> the disagreeable young man in order to humiliate Siegfried? ... The incident was ugly, grotesque.... The visit had left an impression of evil, intensified somehow by the surrounding beauty ... a poisoned beauty....
>
> About a year later, in New York, I had lunch with a famous English stage star. He said that there was no more pernicious young man in England than the occupant of that tower room.

Here, Behrman seems out to prove that homosexuality is a kind of madness. Sassoon's obsession is so pathological that he lays himself open to a snub from his lover and to humiliation by a hustler who drops his h's! This, Behrman describes as "grotesque." The only grotesque feature of the story is the British upper-class snobbery affected by Behrman—an American Jew born in a ghetto (his word, not mine). And to think that the most pernicious young man in England was poor silly crazy touching Stephen Tennant! What an abject masochist Sassoon must have been, to expose his feelings to a bitch like Behrman.

One more Behrman anecdote—this time about Maugham—

Behrman is staying with Maugham at the Villa Mauresque. After supper, Maugham and Behrman go into the drawing room. Behrman catches Maugham off guard:

> He was standing, his back to me, before a sunburst mirror hanging on a wall.... I saw him make an extraordinary gesture. He stretched his arms out wide, flexed them, and stretched them out again.... It was a gesture of despair, a gesture of trying to burst unbreakable fetters, to escape from the unescapable. His arms fell to his sides. He stood there a moment, limp. It came over me that Willie was the most miserable man in the world. I felt for him....

Boy, what compassion! Behrman *feels* for Maugham!

Now, Maugham tells how he wanted to go to India with Gerald Haxton to do some research before writing *The Razor's Edge.* At first, Haxton wasn't permitted to go (no doubt because the British objected to his homosexuality) but Maugham got the ban lifted.

> "We were received everywhere, unofficially—by the maharajahs and so forth—but there was no official recognition of our presence in India." There was a freighted pause and then in a voice that stabbed like a knife thrust, he explained the official neglect. "The Viceroy was Lord Linlithgow."

There was, in the last words, an ultimate concentration of hatred. Lord Linlithgow was Liza's father-in-law....[1] Did Willie think that Liza was an accomplice?...

The vituperation continued. He was embalmed in hatred....

Finally, sensitive Behrman can endure no more. He excuses himself and goes up to his room. The Villa Mauresque turns into another Turn-of-the-Screw house. "You felt it was malign, that you had to get out of it to escape what it might at any moment reveal." But innocent Behrman is still *puzzled*, hatred is something he just cannot *understand*.

Sitting in a chair, trying to ravel out the heart of the mystery, I remembered suddenly a novel by Christopher Isherwood, *A Single Man*. It tells the story of a teacher in a California university whose male lover has been killed in an accident before the story opens. He is driving along the freeway to his class when he passes a block of newly built apartment houses. As he thinks of all those un-gay squares who will occupy those buildings, he gives vent, in his mind, to a sustained aria of applied sadism.... Is this how they really feel about us? I couldn't help asking myself—is this what they'd really like to do to us? And then I thought of Mr. Isherwood himself, with whom at one time I had a very pleasant acquaintanceship. This personal knowledge did not, at any point, cohere with the aria.

The joke of it is, I wrote that hate passage as a parable for the members of all the other minorities as well as mine. Because we are all so unwilling to admit that our own dear little injured minority can ever feel hate—except, of course, when it's one hundred percent *justified*. I hope that Behrman the Jew at least got my point subconsciously, that the vast majority of all minority members sometimes give way to a paranoia which makes them temporarily insane. I believe Behrman did get the point, though he didn't dare to admit it to himself. That is why he's so disturbed, and writes all this gooey shit.

September 20. No word from John Lehmann. Is he so shattered that he simply can't bring himself to reply? I doubt that.

[1] Liza Maugham's second marriage was to Lord John Hope (1912–1996), 1st Baron Glendevon, the younger of Linlithgow's twin sons; see Glossary under Maugham.

Jim Bridges keeps having reasons why he can't see us and talk about our *Meeting by the River* film. Quite acceptable reasons. He had to go to Atlanta twice, to be honored at the film festival. (Not only did *The Paper Chase* get the grand prize, but Jim got awards as best director and writer and the cameraman got an award and so did John Houseman!) This weekend, Jim has gone off to another festival, at Stratford, Ontario. And of course he has every right to be too busy or too exhausted in between these trips, or to tell us, "I just haven't gotten around to working on it." All the same, we are worried. Jim is a terribly sloppy, trial-and-error type of worker, we realize that now. He throws out suggestions without thinking them through, "spitballing" as he calls it, and then, if we make some slight technical objection, instantly abandons the whole idea, instead of trying to find a solution. I fear that it's only too likely that he'll accept some ready-made commercial film which the studio wants him to do. His success has made him all the more vulnerable in this respect.

Good news. R[o]y Strong, of the National Portrait Gallery in London, is at least interested in buying some more of Don's drawings.[1] Irving Lazar, as I noted before, is definitely interested in being our agent. Deborah has had wonderful notices for her play. For the past eight days, I have been working with enthusiasm on my autobiographical book; I really do think it is going to hatch out!

Terrible weather, the worst summer anyone can remember, fog day after day. When you go in the water it's cold but bearable. I've been running quite a lot—but also overeating. Gluttony is truly the vice of old age. The latest temptation, Doritos taco chips.

About a week ago, poor old Jo woke up in the middle of the night and felt thirsty, so she drank some orange juice, thought maybe it was a bit tainted but finished it anyway. A short while later, her tongue began to burn and swell. It got bigger and bigger until she couldn't speak, so wasn't able to phone for help. Terrified, she rushed out in her nightgown, got into the car—it was now around 3 a.m.—and drove to the emergency ward at the Santa Monica Hospital. As she told me, "I was snow white, wet and shaking like an old car!" They told her it was some kind of an allergy, resembling in its effects the bite of a snake or a spider, and that she would probably have died before much longer, if she hadn't been treated. They gave her adrenalin and cortisone. In a

[1] Strong (b. 1935), then Director of the NPG, had purchased Bachardy's drawing of Auden in 1969. See Glossary under National Portrait Gallery.

few hours she was nearly all right again, and then the adrenalin made her terrifically energetic for several days!

September 25. John Lehmann has finally written, in a tone of restrained pathos, accepting my decision not to publish our letters. "Very well," he begins. End of paragraph. Massive pause of reproach. Then: "As you remind me, I liked and continue to like your letters very much, indeed I treasure them, and I hate to be told that the writer of them thinks they are false. That is painful and I mind it a lot." Next paragraph repeats that John *does* think they tell a story—of our relationship. Next paragraph moans a little because of all the work his agent and Methuen put "into their side of the scheme" (whatever that means). Then: "We shall just have to wait until someone makes a fuck-up of it after our deaths. 'We!'—our ghosts." And then another bit of pathos: "Of course I shan't be coming to Texas/California this autumn. No point now." Then, in case he's gone too far, he announces that he'll be coming to Berkeley in mid-March and looks forward very much to seeing us both. Signs himself, "love, John."

This irritated me so much that I wrote quite a stinker to him, which Don wisely restrained me from mailing. The reply I did mail is as follows: "It pains *me*, and *I* mind it a lot that you have decided to take this personally and make up your mind that my friendship is false because I tell you that my letters are. We'll talk it over before long. I won't write any more on the subject—it's too silly. Treasure the letters, by all means—but one day try to read them through objectively. If you can, you'll see what I've saved us from."

This is weak and evasive, but I absolutely could not be bothered to go into the whole thing and point out to John how he tried to bulldoze this project through and commit me to it before I'd reread the material, and how he always resorts to pathos and being pained when he doesn't get his way. Once started, I might have found myself cutting much deeper and telling John *why* my letters to him during the war were so false—namely because I knew he wasn't on my side, I knew he didn't believe I was serious about Vedanta or pacifism and I knew he would disapprove, on principle, of any book I wrote while I was living in America. I was false because I didn't want to admit how deeply I resented his fatherly tone of forgiveness for my betrayal of him and England—"England" being, in fact, his magazine.... Well, I probably shall tell him one day if he provokes me enough, but I'll do it face to face, not in writing. The stupid thing is that I'm fond of him in a way, and that

I've often defended him, even. I think, as everybody in London thinks, that he's an ass and that he has almost no talent. But I am fond of him, which is more than most people are.

September 30. Wystan died yesterday—or anyhow sometime during the night of September 28–29. A man from Reuters was the first to phone and tell me, then a man from the BBC. He died at a hotel in Vienna, after giving a talk to a literary society there. According to the BBC man, he was quite himself during the talk; no one suspected he was sick. The BBC man said it was a heart attack, but didn't know if it happened while Wystan was asleep or not. His body is being flown to England to be buried.[1]

I have written all this down so I can read it. This is still so uncanny. I believe it, I guess, but it seems utterly against nature. Not because I thought Wystan was so tough as all that. He seemed to have been ruining his health for years. And then, whatever he may have said, he was awfully lonely—isolated is what I mean—he made a wall around himself, for most people, by his behavior and prejudices and demands. Perhaps he deeply wanted to go. His death seems uncanny to me because he was one of the guarantees that *I* won't die—at least not yet. I think most of us, if we live long enough, have such guarantee figures. On the other hand, the fact that he has gone first makes the prospect of death easier to face. He has shown me the way.

The thing which seemed particularly poignant about his death—which made me start to cry—was that he didn't get the Nobel Prize, after all. He did so want it—and, God knows, it isn't much to ask of life.

An odd thing: That night he died—or rather, in the afternoon here, which might well have been the exact time of his death in Vienna—I started a sore throat, the first I've had in a long long while, and it got so bad during *our* nighttime that I couldn't swallow. And today, despite doses of penicillin, I still have a fever and headache and feel lousy. This makes me glad. I like to believe that he sent me a message which got through to me.[2]

October 7. Now the awful Arabs have attacked the intolerable Israelis and all is tension and fighting speeches.[3] The only way to

[1] Auden was buried in Austria; see below, October 7.

[2] *Lions and Shadows* tells how Isherwood was stricken with fever and tonsillitis whenever he met "Hugh Weston," the Auden character: "these psychological attacks became one of our stock jokes" (pp. 216–217).

[3] On October 6, Yom Kippur, Egypt attacked across the Suez Canal and

keep from giving way to depression is to work hard. I finished a rough draft of the first chapter of my autobiography a few days ago, and now we are trying to get our screenplay revised for Jim Bridges while he's still interested in the *Meeting* project. We haven't even told him that Clement Scott Gilbert just wrote with another of his false good tidings—Simon Ward, who played the young Winston Churchill,[1] wants to play Oliver and Basil Hoskins wants to play Patrick. (On the stage.)

At least Chester was with Wystan in the hotel the night he died, although they had separate rooms and Chester knew nothing about it until next morning. Now (on the 4th) Wystan has been buried, at Kirchstetten.

Heinz wrote about it in a way that made me laugh, because I could just hear Heinz saying the words. After saying (in German) that he was sad about Jean Ross's death, he went on: "Weil wir schon mal beim sterben sind, W.H. Auden hat es auch erwischt in Austria ..."[2]

John Lehmann writes the usual Lehmannese: "I was just going to write to you about William's death when the second hammer blow fell ... Wystan was much closer to you than to me, but his death made me feel as if a chapter, our chapter, was coming to an end, and an icy chill seemed to be in the air...." He goes on to say that William Plomer had his fatal heart attack at night in the middle of a thunderstorm and that Charles [Erdmann] had to rush out and phone for a doctor, who didn't arrive until two minutes before William died.

Upward wrote too, saying that he felt sorry he had never written to Wystan praising his *City Without Walls*—or rather, that he had never sent the letter: "I suppose I was inhibited by the unhappiness I felt about the remarks he was reported to have made on the Vietnam War, though I never checked up on the reports and half tried to hope they weren't true or that if true they had been made in some semi-ironic or deliberately provocative way. And also I had never quite recovered from his having removed, in a subsequent edition, his dedication to me of that poem beginning "What siren zooming." I was happy to see he had restored the dedication in the recent reissue of *The Orators*." Another letter which is just as characteristic of its writer as the two above!

Syria attacked along the Golan Heights trying to recapture territory occupied by Israel since the Six-Day War in 1967.

[1] In Richard Attenborough's *Young Winston* (1972).

[2] "While we happen to be on the subject of death, W.H. Auden has also caught it in Austria ..."

Incidentally, John adds, "I would like our last exchange of letters to be forgotten altogether"—which merely means that he reserves the right to go on sulking till the end of time.

October 9. Tonight the sun made its first complete descent into the ocean—last night, the tiniest smidgen of Point Dume was visible against its rim—and so the blessed work season of autumn, according to our private calendar, begins. There's certainly a lot of work in prospect: my autobiography of the first years in this country, the reconstruction of my diaries (I'm still only at the beginning of 1950, but I am quite pleased with what I've done), our screenplay of *A Meeting by the River* to be rewritten (before Jim Bridges gets back from a tour of personal appearances with his *Paper Chase* film, in two or three weeks), and also possibly the beginning of a treatment for George Cukor—either of the life of Virginia Woolf or of Schnitzler's *Casanova's Homecoming.*

I would be feeling excited and happy about all of this, if it weren't for the cloud of Wystan's death and the growlings of the Israeli–Arab war, which is obviously going to create an international crisis, if nothing worse. Already the kids on the UCLA campus are fighting about it.

We had supper with John and Cici Huston on the 7th. He's like a sly old king with his women all around him. Don says he's sure Huston can only live with women who fight each other and have bad characters—and the domestic animals are subtly being encouraged to act up, too. A drunk woman named Stephanie Zimbalist[1] went on fussing with the chicken belonging to Cici's [...] son Colin until it shit all over Don's jacket. Then the two dogs—who are adorable—suddenly attacked and very nearly succeeded in killing the adorable kitten belonging to Cici's archenemy Mrs. Hill, John's secretary. And somehow, although he never ceased talking to me with the utmost charm about Kipling and *The Man Who Would Be King* (which he is about to film), John was aware of all this and was sanctioning it. After supper, he retired to his bedroom and watched T.V. reports on the war. Cici told us that he is passionately on the side of the Israelis, but he didn't say one word about this. John is so much of a sultan that he makes you feel you are part of his harem too—a eunuch at best—and so he can ignore you if he's in the mood to do something else. And yet he's polite like no one else in the world.

[1] Stephanie Spaulding Zimbalist (193[4]–2007), second wife of actor Efrem Zimbalist Jr. (b. 1918) and mother of actress Stephanie Zimbalist (b. 1956).

October 11. Yesterday was Resignation Day. The immaculate Agnew had to stand up in court and admit that he had been guilty of income tax fraud. He had said he would never never resign, and he has resigned. He had assured the Republican women that he was innocent, and he has admitted that he lied. He had also assured them that he would never let his lawyers make a deal with the prosecution, and he has let them make a deal; in exchange for his resignation and his admission of guilt, he is not to be prosecuted on all the other charges against him.[1] These creatures, with their sickening patriotic mouthings, are all turning out to be crooks and perjurers, and yet, even after all that has happened, Agnew's representatives are trying to make his resignation sound like an act of patriotism, he is doing it "in the national interest"!

We saw Swami yesterday evening. Last weekend, he was up at Santa Barbara; they had the Durga puja there on the 6th. While Swami was offering a flower at the shrine during the puja he was suddenly overwhelmed, thinking how gracious Mother had been to him. He burst into tears. Chetanananda had to help him out of the shrine into the little office room at the back. Chetanananda said to Swami, "This is a sign of the great grace Mother is showing you"—at which Swami began to cry again and couldn't stop for some time. He begged Chetanananda not to mention Mother's name. When he had finally got control of himself he went back into the temple and blessed the congregation, lest they should think that he had suddenly been taken sick. After this, he was "very jolly." The nuns told him that his face seemed changed. It was flushed. Swami seemed very well when we saw him, though perhaps a bit tired. Now they are getting ready to celebrate the Kali puja, here. Anandaprana, that compulsive fusser, spoke to Swami about the hiring of a boat at Newport Beach for the immersion of the Kali statue after the ceremony. Swami immediately got angry: "I don't want to hear about that. It's like arranging the baby's funeral before it is born." It seemed to me that Swami's indignation was due to the intensity of his experience at Santa Barbara. I suppose, to him, it actually is apparent that the Divine Mother enters the statue.

October 28. The sun has just gone down on one of the most perfect days we've had this year; the whole bay visible out to the headland, the ocean cold but warm enough to dip into (I did) and

[1] Evidence suggested he had accepted bribes in office, but he was charged only with tax evasion and money laundering for failing to report the income.

the mountains standing out clear, nearly smogless, all around the city. But my darling Don has gone to New York. Today at noon.

Ten days at least he'll be away; this year with him has been so beautiful and near and tender; we seem to grow together. And he himself feels this, I know. He went off in his usual style from the airport, late and heavily laden. He had written a letter to Bette Davis (about drawing her while he is back east) and he dropped it three times and had to have it picked up for him, because he was trying to hold a portfolio, a couple of drawing bags, a suitcase and a copy of Nigel Nicolson's book about queer Harold and Vita and her girlfriend.[1]

And now old Drab is alone and full of good intentions; chiefly to get on with his autobiography.

Last night, at about 11:30, we finished our revision of the *Meeting by the River* screenplay. Scene stretched into scene—I thought it would never end. And Don, who practically retyped the entire manuscript, told me this morning that he felt sure we wouldn't make it before he left.

Don's objectives: to see about his show in New York—which is now to begin next February—and to bargain for at least some money for its catalogue; to draw Davis; to meet Alice Faye at a ball to which he has been invited and if possible draw her; to go to Avon and find out what version of our "Frankenstein" screenplay they are using for their forthcoming paperback—if they are using the Stromberg–Smight butchered version, then we will take our names off it.

Swami had a slight infection in his lungs this week, so I didn't see him. When I saw him last, on October 19, he told me that he had had a repetition of his experience at Santa Barbara (see my last entry) but not such a powerful one. It was one morning, when he was preparing to meditate in his room at Vedanta Place. He was walking around burning an incense stick before the various sacred pictures. When he got to the picture of Maharaj over the fireplace, he felt "a wave of love" and began to cry. Krishna didn't at first know what had happened. He helped Swami to a chair.

Swami had just found out that married couples indulge in partner swapping! This sincerely amazes him. One of the girls who comes to Vedanta Place has admitted that she used to take part in it. And Swami says that—before he knew this—he saw her, just after his return from Santa Barbara when his experience at the Durga puja had made him "become very subtle," and felt at once that there

[1] *Portrait of a Marriage* (1973), about Nicolson's parents.

was something wrong with her. (I often wonder, when I help him on with his shoes or otherwise make physical contact with him, if he doesn't have similar feelings of uneasiness and have to conceal them out of politeness!)

October 29. Was up soon after six and now it's nine-thirty and I have finished fussing with chores—phoning around, getting Don's bathrobe in the washer, shaving, making the bed, taking vitamins etc.—and now there is absolutely no reason not to begin thinking about my problem; how to start the autobiography.

What I have written so far is no good at all. It's the wrong tone and the wrong rhythm. It is dull, prudent, cagey. (One can be as frank as Allen Ginsberg and still sound cagey if the rhythm of your narration is cagey, if you seem to be playing with your cards held tight against your vest.)

My difficulty is that I want to have this book start with our departure for America. But I have now realized that I can only put our departure in perspective if I begin with Germany—why I went there—"to find my sexual homeland'—and go on to tell about my wanderings with Heinz and his arrest and the complicated resentment which grew up out of it, against Kathleen and England, Kathleen *as* England. This is such a long story in itself that it seems absurd to begin with our departure and then immediately go into a long flashback.

I'm also much bothered by the first-person–third-person problem. I thought I'd tell the story as "I" when I'm speaking as the narrator and as "Christopher" when I'm talking about myself in that period. That worked in *Kathleen and Frank*, because Christopher in that narrative is just one out of many characters and not an important one. But this is a different situation. "Christopher" is such a cumbersome name to keep repeating, and the use of it has a dangerous cuteness. I think I'll probably end up by writing "I."

Later. After going to the gym and doing quite a bit of thinking while driving there and back, I have come to a different idea about the autobiography. I feel that it must start with my going to Berlin—not with my first trip out there to see Wystan, or with my visit to Wystan in the Harz Mountains that summer, but with my real emigration sometime later in the year. (I can't even remember the date but maybe Richard can find it in Kathleen's diary.)

So this book will be a record of my wander years, taking up more or less where *Lions and Shadows* left off, and carrying on, with jumps over the periods which I've covered elsewhere, to a time in the States when, for one reason and another, I had accepted the fact that California had become my home, and that I was therefore no longer a wanderer. The provisional title of the book will be *Wanderings*.

Jack Larson made me laugh a lot this morning, as he described how he had to take charge of the fifteen-year-old son of a friend of his who had run away from home (Lake Tahoe) and had to be sent back there. This little boy, who Jack says is "adorable and sexy," had gone to Palm Springs and become involved with a pair of male lovers, in their early twenties, who were painting a house. Having (presumably) had sex with him, they decided that he should be restored to respectability; so Jack was contacted and Jack contacted the boy's father, who said that he would only pay for the boy's bus fare home. The two housepainters were generous, however. They volunteered to make up the difference between that and plane fare. The only difficulty was, they were going to a drag ball last night and they couldn't very well take the fifteen-year-old along. So Jack had to pick him up outside the drag ball—trembling with fear that he and the drag queens would all be arrested for corrupting a minor. And then, since there was time to kill before the plane left for Lake Tahoe, Jack decided that the boy should suffer for his tiresomeness by coming along to a concert of baroque music which Jack had been planning to attend!

October 30. This afternoon, a fire somewhere back in the opposite hills is rolling a cloud of purplish golden smoke right out to sea and across the sinking sun. I hope the Alexanders are all right. We were threatened with strong Santana winds, but there isn't a breath at present, luckily.

Altogether a beautiful day. I was up at five-thirty. Vera Fike appeared, a bit weary, but quite up to cleaning the house and doing the laundry. She is so sweet and uncunty; very proud of her son who came down here to play against UCLA for Berkeley and got all the applause although his team lost. I talked to Don twice, once very early, about our strategy with Avon and the publication of the "Frankenstein" screenplay and once this afternoon, when he called to say that Avon has agreed to print our version. He thinks they checked with Universal in the meanwhile. But I'm not so sure. I called George Santoro about this but he says I must talk to a man named Steve Adler, and I can't get to him till tomorrow.

Today is Dostoevsky's birthday. So I decided to restart my autobiography—the version I call *Wanderings*. I went down to the beach and went in the cold but bracing water, biggish waves, and dozed in the sun, feeling so thankful that this book is in my mind and ready to be coaxed out of it onto paper. Then, sitting on the deck in the hot midafternoon sunshine, I wrote the first two pages with the gold ballpoint pen Don gave me. Of course it will be weeks before I'm really sure that this *is* the right approach. But I'm very optimistic.

November 2. Of course it isn't as simple as all that. I still don't know how to narrate the book. Today I'm going to try beginning in the present tense and the third person singular, to tell how I met up again with Francis Turville-Petre in Berlin. But isn't this third-person present tense a stunt? Shouldn't I be better off just telling the whole thing from my point of view? And admitting that I don't remember very much?

The whole project does excite me, though. It seems to contain everything I want to say—far too much perhaps for one book. And my chief worry is that I may be unequal to the job. When I reread my earlier work, I feel that perhaps my style may have lost its ease and brightness and become ponderous. Well, so it's ponderous. At least I still have matter, if not manner.

Don called this morning from New York; he's probably going to stay until the end of next week. Problem: shall he stay on at the Chelsea and spend money or shall he stay with Mario Amaya (as invited) where there may be tiresome leather parties? Since Amaya is the boss of the gallery, Don feels that it would be unwise to refuse his invitation.

Don has drawn Bette Davis and feels that they have become friends;[1] but he doesn't like his drawings.

The night of the 30th, the fire in Topanga Canyon got really big and there were flames a hundred feet high, jumping up the hillside. A huge crowd of onlookers gathered along Adelaide Drive, which made me nervous because a lot of them were smoking, and our bushes would burn like fire bombs. I find I am much more inclined to paranoia when living here alone.

Jim Bridges seems to like our new draft of the *Meeting* screenplay. He is worried because several lymph glands have become swollen in his armpit. He goes to the doctor today to hear if they have to be cut out or not.

Betty Harford and Alex [de] Naszody were evacuated from their house during the fire; they went to the evacuation center but

[1] They had already met three times: at the Bracketts' in 1954, as Isherwood records in *D.1*; again in 1965 when Tennessee Williams introduced them backstage before a preview of the New York opening of *The Night of the Iguana*; at a recent dinner party at Roddy McDowall's.

only stayed half an hour. The astounding thing is that no house was actually destroyed, despite the huge range of the fire. Betty is greatly relieved. The tumor they took out of her uterus had been found to be malignant, but now they feel fairly sure that it was completely eradicated. They aren't even giving her cobalt treatment.

The night before last, getting home from Vedanta Place, I had worked myself up into a state of acute tension about all the letters I had to answer—twenty-five at least. So I tore up nearly all of them.

Paul Anderson went to see Jack Fontan about his horoscope, not long ago. This is what Paul told me on the phone about their meeting. Jack said that Paul shouldn't try to hurry but "poke along at the rate you're going"; that Paul has a personal magnetism which draws people of power and influence to him; that Paul would be drawn into writing, but that that didn't mean he shouldn't go on with his singing; that Paul "is a show-off but it's very deep inside"; that Paul has always had to have people support him, first his grandmother, then his aunts, and now Roddy; that Paul should never refuse any gift because "no one will ever give you anything which you haven't earned." When I repeated this last remark to Don, he said he thought it was very good. I told Don that I have just the faintest suspicion that Jack and Paul went to bed together or came near to doing so.

November 6. I got up at 5:30 this morning—not deliberately but because I happened to wake then; while Don is away, I don't use The Bee, I sleep as long or as little as my body decides.[1] Did my sit—I'm ashamed to call it meditation—for the best part of an hour. I have less concentration than ever, it seems. The only thought I can anchor onto at all is of Swami. Occasionally I can get something by imagining myself in front of the shrine, but I can't hold it. I keep reminding myself how soon I may have to face death. (But my expectation of life is ten more years; I just looked it up!) I feel dull headed but fairly healthy. Well, I say to myself, at least I am making this act of recollection, and isn't that what really matters? I think of Bob Adjemian and Jim Gates doing the worship in their rooms up at the Hollywood monastery. Perhaps they're thinking of me, and sending me helpful thoughts.

A beautiful day, but cold. The hills opposite are burned elephant gray by the fire.

[1] Bachardy was the one who set the alarm clock.

All yesterday and again this morning I have been looking through Wystan's letters and manuscripts—that tiny writing which I find I can, almost incredibly, decipher. He is so much in my thoughts. I seem to see the whole of his life, and it is so honest, so full of love and so dedicated, all of a piece. What surprises me is the unhesitating way he declared, to the BBC interviewers, that he came to the U.S. not intending to return to England. Unless my memory deceives me altogether, he was very doubtful what he should do when the war broke out. He loved me very much and I behaved rather badly to him, a lot of the time. Again and again, in the later letters, he begs me to come and spend some time alone with him. Why didn't I? Because I was involved with some lover or film job or whatnot. Maybe this is why he said—perhaps with more bitterness than I realized—that he couldn't understand my capacity for making friends with my inferiors!

Yesterday I did quite a lot of work on *Wanderings*. Now I'm going to take a look at it.

November 11 [Sunday]. Yesterday I worked all day long, from about nine to seven, in my bathrobe like Balzac, eating nothing but one orange with several cups of Sanka, and I finished the first chapter of *Wanderings*. What a great grace it is, to have so much work to do! I was hurrying to finish so I could show it to my darling when he got home today—but, just as I was getting up from the typewriter, he called and there's another postponement, because he *may* be able to catch Alice Faye today. He says of her, "We're like two bulls—both Taurus—and now it's right down to the line"; because, on Monday Faye starts rehearsals for her show.[1] I say I think Don and she are like Sherlock and Dr. Moriarty when they face each other at the Reichenbach Falls![2]

So Don's returning on Tuesday, not tomorrow because I have to go to USC in the morning and talk to a class there.

After eating so little during the day yesterday, I had supper with Gavin, with the result that my weight is up again, 153 and ½. On November 3—after supper with Gavin right after his return from Morocco—I hit a ghastly high, 155!

Gavin is delighted with Tangier, doesn't want to live anywhere else, doesn't regret leaving this place. He went to see Didion and Dunne and thinks they've gone Hollywood—at least he has and

[1] *Good News*; see Glossary.

[2] Holmes and his nemesis, master criminal Professor Moriarity, reel into the falls fighting in Conan Doyle's "The Final Problem" (1893).

she is going along with it. They plan to make a new version of *A Star Is Born*, with Elvis Presley as the male star who is setting. Gavin also told me that he loved Clinton Kimbrough more than any of thc others. He's working on a book about thriller writers, including Conan Doyle, Eric Ambler, Raymond Chandler.

Now I'm going to UCLA to research Dr. Hirschfeld and his institute, because my next chapter will be about them.

November 17. Kitty got back safe and sound—although he took such a long time leaving the plane that I thought someone must have frightened him and driven him to hide under one of the seats, so I went on board to look.

He did catch Alice Faye! The drawing was kind of a token, done in haste. But she has seen his catalogue and been greatly impressed so probably another sitting can be arranged much more easily. (Since he got back here, Don drew Jane Powell[1] and she told him she thinks Faye may crack up and leave the show before they open, because she's so terribly scared. Powell had been shocked by Faye's appearance when they met again, but Don wasn't.)

Now it's dripping from moist fog, very saddening, if Dobbin didn't have Kitty to cheer him. Also, I have the following ailments. A nasty circular red rash on the left thigh, up against the scrotum, about the size of a quarter and apparently growing. (I showed this to dear Dr. Wolff who said it was either herpes or something else and probably due to something I ate, and not to worry and to rub it with Cetaphil lotion[2] and it will go away in about two weeks.) The toe next to my right big toe is very sore, as though I'd banged, even perhaps broken it, which I don't remember doing. I have the usual lump on the ball of my left foot—it seemed better for a while, now it's back. I have sharp arthritic pain in my right thumb. I have a painful sore lump in my mouth, lower jaw, right side, perhaps caused by my bridgework; can't see Dr. Kurtzman about this as he's away till the 27th.

Don likes my first chapter of *Wanderings* very much, I think. I shall try to get on with this in a rough draft, although I need to do much more research on Hirschfeld. All I could find so far is a journalistic book of interviews of G.S. Viereck with one chapter in it on Hirschfeld calling him "the Einstein of Sex"!

Gore is in town, with Howard. And, at the end of this coming

[1] American singer, dancer, and actress (b. 1929); she was a child radio star, a lead in MGM musicals including *Seven Brides for Seven Brothers* (1954), and later performed in stage musicals and on T.V.

[2] Non-irritating moisturizer.

week, Truman will be arriving. Gore says he's ready for a confrontation! "By that time," he said, "*Burr* will be at the top of the bestseller list."

We got the proofs of our "Frankenstein" screenplay from Avon and it really is a more or less adequate version, although Hunt cut out the episode of Byron wanting to kill the butterfly and Shelley saving it, from the prologue. They can't put this back into the first printing, as the book has gone to press, but it can probably be added later, also the scene between Polidori and the captain, when the schooner is being chased by a British coastguard cutter and Polidori says, "God bless America!"—which now seems amusingly topical, because it sounds like that other crook, Nixon, in his first statement about Watergate.

November 21. I forgot to mention that, when I went to the UCLA library to see what they had on Hirschfeld—not much, no biography—I ran into Marion Hargrove who told me that two of his sons have become disciples of the Maharishi, in Switzerland, and have consequently given up beer and pot, and taken to meditation. One of them had described himself as having become "a bliss ninny."

On the 18th, Gore, Howard and Gavin came to supper. Gore is very fat (for him) at the moment; he has been travelling around promoting *Burr*, and Howard has come out to join him. They brought a tiresome little dog "Rat," which was distracting. It seems to me that Gavin and Gore don't go well together. Not that there was the least friction but the combination of them made for low-level conversation; Gore was alternately dogmatic (Nixon is *certain* to be forced out) and career conscious (talking about a new attempt to vote him into the Institute,[1] and getting me to read him a list of its writer-members and concluding from it that they represent Jewish, East-coast, Wasp influences and prejudices—predominantly Jewish).

We have told Mark Andrews that we will not go to his wedding or the reception afterwards. I suppose I am being a bit cantankerous, or worse, pretending to be cantankerous because the simple truth is that I loathe all such gatherings, even slightly preferring funerals. But I do honestly feel that for Mark and Lydia[2] to pretend that they are pledging their troth in the solemn sacred sense is a rather blasphemous farce. Instead of going off and having a five-minute civil ceremony they are camping through some kind of

[1] I.e., the National Institute of Arts and Letters; see Glossary.
[2] Not her real name.

Zen ritual or whatnot. How *dare* they—when Mark was actually offering to run off with Gavin, right up to the last moment. But then, Mark is an irresponsible baby. As for Gavin, he has told Mark that he never wants to see him again. It seems that Gavin is particularly furious because Mark has been trying to window-dress the party by inviting people like Merle Oberon. Don observes that it's all very well for Gavin to be furious now, but that he never restrained Mark's indiscretions as long as they were living together.

The day before yesterday, we both saw Irving Lazar and decided to have him for our agent. Yesterday I wrote to Perry Knowlton at Curtis Brown and told him I don't want Curtis Brown as my U.S. agent any more. I imagine there will be a stink about this.

We are being threatened with gas rationing and/or the closing of gas stations at weekends.

About the ailments I listed on the 17th: the rash on the thigh has not disappeared by any means—Dr. Wolff predicted it would go in two weeks, today it's just one week—but it does look paler and much less "angry." The toe seems quite recovered. The lump on the ball of my foot is slightly less evident. My thumb is almost okay. (These two symptoms probably respond to the weather; we've been having cold and rain and high wind, but today is beautiful.) I have relieved the lump in my jaw by not wearing my lower bit of bridgework.

Irving Lazar is certainly no asslicker. He didn't say one word of praise—or blame either—for our *Meeting by the River* script. He just told us that he was sure one couldn't raise the money for it. And he also said that Cukor won't be able to raise any money to do his pictures, because he isn't box office any more. People think he's had it. Don likes Lazar's frankness. So do I, at present.

December 6. A long lapse and a lot to report.

Jim Bridges claims that he has got John Calley—who worked with us on *The Loved One* as an associate producer or something and is now head of Warner's—interested in *A Meeting by the River*. He was to read the novel over the weekend. (Jim was determined *not* to show him our screenplay, which makes me think that Jim must be secretly intending to rewrite it drastically.) Well, we've heard nothing from Calley[,] and Jim has gone off on another of his *Paper Chase* promotion trips and will be away for a couple of weeks—leaving Jack to receive Salka Viertel, who is due to arrive here almost immediately and maybe stay through Christmas!

Irving Lazar is in New York and scheduled to see Curtis Brown. Perry Knowlton answered my letter and urged me to reconsider leaving them, particularly for Irving Lazar, who, he says, is all right for film business but how can he represent me in New York, where I need "a full time representative who specializes in literary representation and has the staff and expertise to do the job." I haven't answered this letter because I don't want to get really nasty and point out to Knowlton what kind of representatives he and his expert staff have been. Even if Irving turns out to have been a mistake, I do trust him to produce some action.

Already, he has called Universal and they have agreed in principle that Don and I should get some money from the sales of the paperback of our "Frankenstein" screenplay, which is just out. The two parts of the T.V. film were shown on NBC, on November 30 and December 1. Many people called us about it, all professing to have enjoyed it despite its faults—no, not quite all, Cukor was frank, so was Billy Bengston. We thought it even worse than it had seemed in the projection room at Universal. Aside from all the faults I've already mentioned, another, a quite basic one, appeared; the picture hadn't been shot for television—that's to say, it was hopelessly lacking in close-ups and therefore the players' reactions were lost; in many scenes, they were far far away from you, wandering about in the backgrounds of the big sets. We've been cursing Jack Smight for being "a television director"—would that he were! Hunt Stromberg called from Texas, trying to placate me by quoting from various "rave" notices, including one by Cecil Smith in the *Los Angeles Times*. I told him he had betrayed us, but I didn't raise my voice. He just kept repeating that he was sure we would get an Emmy!

Gavin left for Tangier on the 26th. We haven't heard from him yet.

For a long time now, Jack Larson and Jim Bridges have been raving about The Open Theater Company. Having met Joe Chaikin, its director, we were inclined to think we'd hate it, because he's so pretentious when he talks about the drama. But at last we felt obliged to see it because the company is breaking up—chiefly, as far as we could gather, because three or four of the top people in it are such prima donnas that they couldn't bear to be with each other any longer, having stuck it out for ten years! So, on the 26th, we drove down to Huntington Beach and saw a performance at the Golden West College, and then again, on the 29th, we drove to Costa Mesa and saw them, in another show, at the Orange Coast College. The first performance was called *The Mutation Show* and the second *Night Walk*. Nearly all of it is movement, partly dancing, partly miming, partly tableau, partly just mugging at the audience—accompanied by animal noises or noises imitating the rhythms of human speech, also there are drum beats and other percussive sounds. Very little intelligible dialogue. These are not improvisations; indeed, it is obvious that they have been very carefully rehearsed. They hold you for quite a while; then the lack of structure—or perhaps I'd better say quite brutally "plot"—begins to be felt. All the more so, in a way, because the actors are really good at what they're doing. Talent minus structure equals cuteness; the greater the talent the greater the cuteness.

On December 1, we went to a farewell lunch for the company at Jack and Jim's house. Seen offstage, they nearly all seemed charming, unpretentious, friendly, really lovable. A big blond young man named John Stoltenberg, who manages the show and is also a writer, was dangerously affectionate, a true dog person who would have moved into the house and our basket, it seemed, if he had received one pat too many.[1]

At Vedanta Place last night, Swami asked me if I had had any experiences. That's the word I always use when I ask him the same question and so it staggered me for a moment, but of course he didn't mean it in the same way. Anyhow, I found myself instantly in a state of emotion. I told him that if I hadn't met him my life would have been nothing, that I knew this now, and that I like best to meditate on his room because I know that Maharaj is there. My voice was shaking and tears ran down my face. Swami didn't say anything but his face became aloof in that way it does, when he is in a spiritual mood—"deserted," one might call it, because you feel that he is "out of himself." We were silent for a long time.

After supper at Vedanta Place and the reading, I went to Nick Wilder's gallery to put in an appearance because a new show was opening and I want to declare my solidarity with Nick on every possible occasion—tomorrow he is to come down here and look at Don's work and discuss showing it in the gallery in the near future. (Don is worried because he has nothing new, he says, to show Nick and of course I am urging him as usual to give Nick a chance to react to the paintings.)

This opening was for Bruce Nauman.[2] Nauman had surrounded

[1] In the spring of 1974, Stoltenberg met the feminist Andrea Dworkin, became her life partner and, like her, an anti-pornography activist. His books include *Refusing to Be a Man: Essays on Sex and Justice* (1989).

[2] American conceptual and installation artist (b. 1941); Wilder staged Nauman's first solo show in 1966, and Nauman rapidly became known for his experiments with sculpture, video, performance, neon, hologram, photography, printmaking, and interactive environments.

the walls with white doorless cubicles. The rest was up to you. You were supposed to go into one of them and take any position you liked, accompanied by as many people as you wished. Nick told me that one of Nauman's other shows had featured a long narrow corridor down which you walked toward a television set. As you approached it, you saw yourself in the set, getting smaller and smaller. I've left out a lot of other details, but that was the general idea.

I have now finished the second chapter of *Wanderings* in a very rough form. It deals with the Hirschfeld Institute period of my stay in Berlin and my preliminary attitudes to the city and its boys and to Hirschfeld's sexology. And now comes the problem of covering the rest of my time in Berlin. How to deal with the characters from my books? Should I simply quote from my own descriptions of them? I don't want to make this merely a handbook and guide to my fiction. I want to cover the ground as quickly as possible because, after all, it is fundamentally an introduction to the American material, an answer to the question, why did I go to America.

But how deeply all this interests me! I don't think I have ever felt so challenged and turned on by any other project. I'm merely in a flap because I feel I can't possibly get it all "in." But of course the thing I must remember is that I can "get in" anything that I really want to get in. What I have to do is just write all of it down and then take a good look at it and then begin "packing."

I see this book essentially as a study of successive attitudes to my life. It is true autobiography, not memoirs.

December 8. Nick Wilder came down yesterday morning to look at Don's work and talk about the show Don is to have at his gallery. Before he arrived, Don and I had an argument about what to show him. I begged Don to show some of his paintings and small ink drawings as well as the big portrait drawings. Don was very dubious about the paintings, afraid that Nick would be put off by them—which was only natural, after the negative reactions of Irving Blum. But I argued that a dealer is like a lawyer, you can't afford to have secrets from him if you want him to represent you, and Don agreed and we finally picked out about a dozen paintings—that's to say mostly the blotty watercolors.

Well, to Don's amazement and to my much smaller amazement but huge joy and relief, Nick loved the watercolors and was altogether impressed by Don's versatility and said that he wants to give Don a show in which the whole front room of the gallery is full of the watercolors with a few drawings in the back room. And, when we met Nick again, yesterday evening, at the opening of a show of Charles Hill's work, Nick told Don that he had nearly called him that afternoon, because "I can't get your paintings out of my mind."

And Kitty was so joyful and said that he would always listen to Drub's advice in the future, and he exclaimed—thinking of all the painting he would do for this show—"Now I can begin to live!" And honestly I believe, if I had been told I could have one wish granted to me (within the bounds of possibility) this would have been it.

Another bit of good news, though hardly in the same category, is that Calley is really interested in our *Meeting by the River* project and wants to go ahead. Jim is back here, just for today, because Salka arrived yesterday. She plans to stay with them through Christmas and maybe longer, if Jack and Jim will accompany her back to Switzerland early in the New Year, to go skiing.

A rather drunken fan on the phone, a few days ago, told me, "I've read several of your books." I asked which ones. This baffled him for a moment, then he marvellously retorted, "A man like you—you're a man who's *beyond titles*!"

To return to Nick. He said that he wants Don's show [to] be very important—either in May or early next fall, "in prime time." What's so marvellous about Nick is that he is quite consciously out to alter Don's image—from that of the celebrity-hunting portraiteer who got the chance to draw famous people through his association with Isherwood to that of a serious artist who has to be respected for his talent. I get the impression that Nick thinks Irving Blum doesn't really take Don seriously.

December 12. This afternoon, Julian Jebb is due to arrive here with his assistant, Rosemary Bowen Jones, and we are to be in the grip of the BBC for a week.[1] Am at present sulking about this, wishing to Christ I'd never agreed to it, even wishing I'd agreed to go to Berlin because maybe once I was there I'd have remembered something interesting. Now it seems to me that Berlin was one of the least important episodes in my life, which is nonsense of course—but it does bring home to me that my life in those days was a pretty shabby little affair in comparison with what I have had since.

[1] Filming for an "Arena: Cinema" segment called "Hooray for Hollywood"; see Glossary under Jebb.

The day before yesterday, Don and I went down to Palm Springs to see Truman Capote and John O'Shea; they had two other men with them, sort of business types, and Truman insisted with his devilish gleeful whim that we should smoke pot, and I got very aggressive, feeling I was being pushed around. However, we emerged from the situation—thanks to Don—without getting into a quarrel. Only result next day was that I weighed 155 (not due to pot but cracked crab!) and today, after abstaining from practi[c]ally everything, as I fondly imagined, I weigh 155 *and ½*!

Last night, Salka came to supper here, with Jack Larson. She is so shaky and deaf and it is sadly dreary and exhausting being with her. You have to shout and she takes forever understanding what you're saying. I had to bear the brunt of this, naturally, but I really wish I had been quite alone with her. Don detected in her face a look of obstinacy and I can understand this—when you get in that state you are all the more obstinately determined to be part of the scene, so you instinctively make it impossible for others to talk about their own affairs in your presence; at least you try to. Actually, I was distracted from my efforts to cope with Salka by hearing snatches of a fascinating conversation Jack was having with Don. Jack was telling how Jim suffers when an actor puts him down for being gay. Jon Voight did this during the rehearsals of *Streetcar*, telling [Jim] (or anyhow implying) that he couldn't possibly understand the play because he couldn't understand how a man feels about a woman! Jack says that Jim sheds tears on these occasions and smashes things when he's alone, later, in his room.

(The above sentence is misleading, as I now see, rereading it. I meant to say that Jim will shed tears in Jack's presence—not in the presence of the actor—while he is describing what took place, and then, later, go into his room alone and smash things.)

December 20. The filming is all over, thank God. I quite enjoyed the first day of it, the 17th, because we were out of doors, on Muscle Beach, and I clowned around, jogging and sitting on the swings and then under a palm tree, reading from *Goodbye to Berlin*. But the day before yesterday and yesterday they filmed in this house, mostly in my workroom, and disturbed everything and made me feel jostled out of my nest. (To restore my morale, I've made a point of restarting work on *Wanderings*, *and* doing a small entry in the 1950 journal, *and* writing up this diary.)

Just the same, I think the interview may turn out quite well, and I feel Julian handled it not only professionally but tactfully, as far as I was concerned. I like his assistant, Rosemary Bowen

Jones; she is one of those smiling unflappable British treasures who can manage everything without being bossy. I also thought that the sound recordist, David MacMillan, was sweetly sympathetic as well as being powerfully sexy; I felt quite dizzy while he was attaching the mike to my shirt. He is—small world!—the boyfriend of Ann Gowland, Peter and Alice's daughter—and, less creditably, a member of the Francis Ford Coppola film colony in San Francisco.[1] The cameraman, Bryan Anderson, I liked less; he acted a bit over-virile, maybe because I was saying a good deal about being queer in my answers to Julian's interview questions. Bryan has an Asian wife named Tamiko—I mean racially Asian—she is technically an American and speaks perfect English. Tamiko is a palmist and an astrologer. She knew at once that Don belonged to one of the earth signs—by looking at his hands, she claimed. She asked Don—not while I was there—"How long have you and Christopher been married?" Don firmly informed her that we don't regard ourselves as married: "We want to avoid the mistakes the heterosexuals have been making."

I pointed out to Julian that it would be absurd to show me at home and not show Don. It might even look, I said, as if I were trying to conceal his existence. Julian was very sensible about this, and agreed, and so Don and I were filmed going up to the mailbox and getting our letters, and then sorting out which ones were for which of us!

December 25. This year, I've celebrated Christmas Day by working quite a bit—on my *Wanderings* book and on the reconstructed diary for 1950. Also, I've been out jogging on San Vicente. Don is with his parents. Later, we're to go to the Norton Simons' catchall party for their families and other hangers-on. I write this nastily, because I am still mad at Jennifer for ignoring Don's show at the Barnsdall Gallery last summer. I'm mad at Norton too, come to that. But Don, who so rightly puts business before grudges, says that he'll have a last try at getting Jennifer to sit for him and maybe even remind her of her promise to do so *before Christmas*, if he drew and exhibited a portrait of Norton.

Yesterday, we drove down to Palm Springs, to spend the day with John Schlesinger, who has been lent a house there by Robert

[1] Coppola, George Lucas and Walter Murch founded American Zoetrope Productions in 1969 to escape the commercial control of the Hollywood studios. Isherwood disliked Coppola and thought his films overpraised. MacMillan later shared three Academy Awards for sound work on *The Right Stuff* (1983), *Speed* (1994), and *Apollo 13* (1995).

Wagner and Natalie Wood. On the way, I asked Don a lot of questions about himself—about how he is feeling nowadays. Of course he is in a flap about this forthcoming show in New York, but less so, I think, than ever before—because this time he has got all the pictures he needs for it and because he has the prospect of this quite different kind of show (of his paintings), and of the work he must do for it, ahead of him. In other words, this is the end of a period with the satisfaction of knowing that it isn't a dead end. He can see his next move. I asked how he feels about his meditation and he said that it is now definitely part of his life but that he doesn't at all share my reliance on Swami as a guru. "If anybody's my guru, you are." Well, that's okay, as long as he merely believes in my belief in Swami. Then I asked him about sex. He said that he doesn't mind our not having sex together any more; he agreed with me that our relationship is still very physical.[1] The difficulty is that what he now wants is a sex object, not a big relationship, because he's got that with me. But no attractive boy wants to be merely a sex object; he wants to be a big relationship. I suppose I knew all this, kind of. But it was good to talk about it. Our long drives in the car are now almost our only opportunities to have real talks. As Don himself says, he is obsessed by time and always feels in a hurry, unless he is actually getting on with doing something. He says that there are now quite often moments, while he is drawing, when he feels that this is the one thing he really wants to do and experiences a great joy that he is actually doing it. But, even during the drawing, he says that he also feels harassed because he isn't drawing as quickly and economically as he could wish.

The house party at Schlesinger's was crowded and often noisy with children and dogs. John, perhaps with the friendly desire to show me off to the other guests, took me through a long cross-examination about my pacifism, Hindu beliefs, etc. etc., which I found embarrassing. Also, John wanted to argue: how could one be a pacifist in Israel, when it is fighting for its existence? At this, an extremely cute young man named Roger—who is the boyfriend of the daughter of Jim Clark, John's film editor[2]—spoke up unasked and said that the Jews always exaggerate and he challenged

[1] Isherwood and Bachardy still slept together, and Bachardy recalls that during his infatuation with Bill Franklin (see below), he and Isherwood had sex several times; see Introduction, p. xxi.

[2] Clark worked closely with Schlesinger, beginning with *Darling* in 1964, and is credited with salvaging *Midnight Cowboy* although he was not officially the editor. They were then shooting *Day of the Locust*. He was accompanied by his wife, Laurence, and two daughters.

the statement about Israel's position. This of course deeply offended John[1]—and indeed the young man went on to make a lot of uncharming, right-wing remarks, although he had previously been attacking Nixon. By the end of the evening, he seemed rather sinister. He is planning to become a lawyer.[2]

Also at the party were a nice musician named Andrew,[3] who is a cousin of Schlesinger's, and a rather snooty man named Michael Oliver, who is Schlesinger's lawyer.[4] Oliver said that Palm Springs was "unreal." So I went after him, saying I'd much rather live in an "unreal" place than in a *seemingly* "real" one, like for example, Oxford or Cambridge. Both Andrew and Oliver are queer and British.

We had a big Mexican dinner at a restaurant in Rancho Mirage. All these desert-fringe communities have run together into a single roadside shopping alley, which is truly depressing. But the mission of our day was well and truly accomplished: Don did a drawing of John which John liked so much he wants to buy it. This was also another triumph of Don's iron nerves over interruptions and barking dogs. No, no—his nerves aren't iron—they are terribly sensitive and yet capable of being strained to an almost incredible extent without snapping, like some kind of super rubber.

December 27. Yesterday was Swami's eightieth birthday. We both went to see him, but Don skipped lunch; he finds these functions even more painful than I do and he isn't obligated, as I am, to attend them.

However, Swami was "well worth the visit." Before lunch, talking to the boys from Trabuco, he really gave forth power. After he had retold that spooky story of Brahmananda coming up behind him and saying, "Lovest thou me?"[5] he added, "And I didn't love him." Despite all my years of indoctrination, my

[1] Jewish by background, though not a religious man.

[2] He was a builder. Kate Clark, then about eighteen and still a student, married and divorced him; later she married tenor saxophonist Hart McNee and settled in New Orleans.

[3] Andrew Raeburn (1933–2010), Cambridge-educated musician and conductor; an administrator for the Boston Symphony from 1964 and for Detroit from the mid-1970s. Later, he directed the Van Cliburn International Piano Competition in Fort Worth and the Honens International Piano Competition in Calgary.

[4] And also his business partner; Schlesinger later split with him.

[5] Christ's words, repeated three times, to Simon Peter in John 21.15–17; Swami published the story in his *Eternal Companion: Brahmananda, His Life and Teachings*.

responses are so conventional that I was puzzled and rather shocked, and I asked almost indignantly, "What do you mean?" To which Swami replied, "How many of us can love God—how many of us can return that love? It's too big for us. We can't understand it. He loves all of us, everybody, and without any motive. How can we understand that?" What is so overwhelming about Swami at these moments is that he speaks with such absolute, matter-of-fact certainty *and* at the same time with awe, not the familiarity which breeds contempt. It's as if he were talking about Mount Everest. God's love is *there*, I do believe, for him just as obviously as Mount Everest is for all of us. And he is awed by it as we are awed by the mountain.

After lunch, he signed to me to take his arm and to help him walk back to his room. When he lets me do this, I always feel it is a special grace and I try to meditate as we are walking. But this time, on his way to his room, he had to stop and receive the pranams of every single monastic member who was present. My first instinct was to draw my hand away, but he held it firmly under his left arm. So there I stood, like an inferior Siamese twin, attached to this being through whom Brahmananda's blessing was being conferred by the touch of Swami's right hand upon all those who came and bowed down before him! All I could do was to keep my eyes lowered and try not to seem to participate outwardly in the giving of the benediction. It was an absurd and embarrassing and beautiful situation to be in.

1974

January 4. Have been too busy to write anything here until today. This is a busy time. Not only are Don and I both working to a deadline—Don getting ready for his show in New York, I trying to finish a rough draft of the opening chapters of my book before I go east—but there are these out-of-towners who have to be coped with; at present, Salka Viertel, John Collier and the Boormans.

Salka really is a trial, poor thing, because she is so deaf. It's a tremendous effort talking to her. She is scheduled to leave at the beginning of next week and at present it seems that Jack Larson and Jim Bridges will fly with her to Switzerland and spend some time there skiing. But this—I only yesterday discovered—depends on Jim's commitments as a director. Jim now has a deal to direct three pictures for Warner Brothers—it may not be firm but it's been proposed. One of these pictures is to be *Meeting by the River*, another Peter Viertel's *White Hunter, Black Heart.* However, there is still some question as to whether Jim will be paid for writing a script of *White Hunter* with Peter Viertel—this is called a "development deal" in the jargon. Jim had planned to do this in the intervals of his Swiss skiing, but he says that he won't work for free. If he gets no money he may not go to Switzerland at all. Jim makes plans and abandons them like a child—well, that's show biz. But he does seem utterly unrealistic when he says things like, we'll do Peter's film in Africa this summer, then we'll do *Meeting* in the fall. At such moments he talks as though he were a complete amateur. How can he imagine a picture like *White Hunter* could be prepared for in a couple of months? And, before *Hunter,* he wants to whizz off to India with us!

John Collier, whom I had lunch with the day before yesterday, looks older but seems exactly the same person. I do like and respect him very much. If we lived in the same town, we should become close friends at once. Alas, he's leaving almost immediately.

The Boormans we are to see for the first time tonight, at a preview of his film, *Zardoz.* I have fears about this. It's something he wrote himself.

Heavy rain all night and this morning, but clearing now. My disgraceful weight, 154 and ¾, due to gluttony. My worst weight last year was 155 and ½, my best (August 8) was 149 and ¼, which isn't nearly low enough.

January 6. More heavy rain, yesterday and today. *Zardoz* was awful beyond description. I could hardly bring out even a flattering lie to John Boorman. We are afraid this picture will make it hard for him to raise any money for another.

A big dinner party last night. We had invited Salka, Jack and Jim for a farewell. Then John Houseman showed up in town, so they asked if they might invite him. Then Don, who drew John Collier in the morning, asked him and his wife Harriet and his son John to come to dinner on their way to the airport, to take a midnight plane to England. It wasn't a success. John is less than himself when his wife is around. And Salka is so deaf. And Don dislikes her—I see exactly why; she is awfully arrogant. And Houseman was tired and didn't bother not to show it; our company didn't thrill him. Collier's son is cute, will perhaps be beautiful. He has a kind of animal fixation on his father, keeps touching him. This embarrasses John and makes him very British; he puts on an air of disowning John Junior, of whom he's obviously very proud.

Today I've stayed at home working. Every day I get some more of the book done. It's hopeless, almost incoherent, but that doesn't matter; one *never* regrets writing anything. The jungle is being cleared.

I suppose Salka and the boys are taking off tomorrow. Nobody has called. I called twice, got no answer, decided not to call again. For one thing, Peggy Kiskadden was to see Salka this afternoon and I don't want to get entangled in that. Salka irritated me by urging me to make it up with Peggy. She simply does not understand. I don't have a quarrel with Peggy. I have a peace treaty, which is based on our never seeing each other again.

Today they started daylight-saving time, in the name of power conservation. More and more, people are saying that the oil companies have lied and that there is no real gas shortage.[1] I think there will be a big stink soon. Houseman told us last night that in the East they still expect that Nixon will be forced out, one way or another. I'm not so sure.

A dream last night: Don and I were going into people's houses and Don was looking at their letters, diaries, photographs, etc., to find out what they were like. He was caught doing this by one family; I was standing in the porch and didn't become involved. They weren't really mad at him and I knew they wouldn't do anything serious. But they demanded that he should apologize, and he did—making it into a marvellous parody of their way of talking and their general philosophy. I realized what a great actor he was. Then I said "Ganga!"[2] (to astonish the others) and Don and I began to fly. I said, as we flew, "The Animals skimmed low over the meadows," but, in fact, we were skimming over a freeway.... This was a very happy and propitious dream.

January 17. Am working away on *Wanderings*, hoping to get a very rough draft of the introductory (pre-1939) chapters finished before I go to New York to join Don at the opening of his show. It now looks as if he will go on ahead, around the beginning of February, and that I shall arrive only a few days before the opening party on the 14th. (Incidentally, Jack Fontan has predicted, on the basis of Don's horoscope, that his show of paintings at the Nick Wilder gallery will open on October the 8th.)

Tony Richardson has shown up here again, saying he will stay at least a month. He seems to have a project going but is

[1] Arab oil-exporting countries imposed an embargo in retaliation for U.S. involvement in the Arab–Israeli conflict in September. See Glossary under Energy Crisis.

[2] Sanskrit and Bengali for Ganges, the river sacred to Hindus.

mysterious about it, as usual. He was very amusing about Vanessa, and altogether on his best behavior, when we had dinner together last night. This may be because he has found out what Derek Dietz[1] told Don, when they met on the 8th. (Derek was Tony's constant companion on his last visit here.) Derek misunderstood something Don had said and thought it was, "Tony likes me," to which he replied, "He doesn't like you, he says you're nothing but a mimic." (This kind of unmotivated malice—malice for the sake of malice—makes me sick with rage; I could kill Derek for it.) Of course this upset Don a lot. At first he didn't want to meet Tony again; then he said, "No, I'll play it cool." Then Tony called, yesterday afternoon, and asked Don if he thought Derek was dishonest. Don said he hadn't suspected it. Tony said, "Well, I'm pretty sure it was he who stole quite a large sum of money from me." So then Don said, "Well, that surprises me—but I do think he's very malicious, he said all kinds of things about you which I didn't want to listen to." After this, we felt sure that Tony would cross-question Don at supper, and thus Don would have a perfect cue to repeat to him what Derek had said he'd said. (Derek had also described Tony's masochism, in and out of bed.) But Tony didn't refer to Derek once throughout the evening. This was perhaps because he had another companion with him, Buddy Trone, a boy who used to be around in the days when we were writing and filming *The Loved One.*[2] He's fat now and looks rather like Margo,[3] but is still the same amiable easygoing partly Latin character. When we knew that Buddy was coming, we invited Mike Van Horn, whom Tony rather fancied.

Tony said that Vanessa has become a sort of Trotskyite—I forget what it's called nowadays. Whenever she gets a job in a play, she tries to indoctrinate the entire cast. The young impressionable inexperienced actors and actresses are terribly flattered at first, to be asked to attend meetings after rehearsals by this great star. But soon she bores them to death and they try to wriggle out of it. According to Tony, Vanessa is completely under the influence of her brother Corin who is even more fanatical than she is.

On January 9, Swami wasn't feeling well, so Chetanananda talked at the reading—mostly about Vijnanananda, who was the

[1] Not his real name.

[2] Trone was an aspiring actor; he later changed his name to Buddy Ochoa to take better advantage of his Mexican-American ethnicity. He had a few small T.V. parts in the 1970s and then worked in radio marketing, selling advertising.

[3] Mexican-American actress wife of Eddie Albert; see Glossary.

last of the direct disciples to become president of the order. He told how Vijnanananda came to Belur Math on a visit but only wanted to stay one day. When they urged him to stay at least three, he said he would do it if they gave him a fountain pen. So they did, and he did. Later, they asked him to be vice-president and he said the same thing and got another pen. Chetanananda giggled wildly: "Those knowers of Brahman—you never know *what* they will do!" I think quite a lot of people in the audience were slightly shocked and more were puzzled. This kind of talk makes them uneasy. They want to understand *everybody*, even if it's Jesus or Ramakrishna. What they can't understand, or dismiss as crazy, seems to them a bit unwholesome, sinister. Surely there must be a *motive*?

On January 12, we saw an almost complete rainbow in the morning; one end in the hills, the other in the ocean. And at about 7 p.m. (*energy-saving time*, as they call it) we saw the comet Kohoutek at last. It was low down over the sea, below Venus. When you looked at it with the naked eye, it seemed to be just a very bright star, as bright as Venus or maybe brighter. Through binoculars, you could see a reddish-orange short tail to one side of it, like exhaust from a plane. I saw it again on the 14th and again yesterday, but the tail wasn't visible; it looked like a tiny crescent moon, lying on its back.

On the 14th, I read the Katha Upanishad and the Hymn to Brahman from the Mahanirvana Tantra. Asaktananda did the worship but Swami came into the shrine and sat on a chair. When the breakfast tray was brought in to be offered to Swamiji, Asaktananda made *mudras*[1] in the correct ritualistic style over the food, before pouring the coffee and lighting the cigarettes. Swami objected to this. He got down on his knees behind Asaktananda and reproved him, telling him that this isn't the ritual offering like the others—one shouldn't make the mudras, because "he is alive." When we saw Swami after breakfast, he said that Sister Lalita never made mudras, she simply offered the breakfast as she would have offered it to Swamiji when he was alive. But Don felt that Swami shouldn't have corrected Asaktananda in public like that—it embarrassed him in front of all of us. I didn't agree, remembering how Swami himself always regarded Maharaj's scoldings as blessings. But I think Asaktananda himself may have been a bit hurt. He was silent and rather aloof during breakfast. He became very

[1] Symbolic hand gestures, connecting external actions with spiritual ideas in order to focus the mind on God.

Asian, wearing his scarf over his head like a snood under his cap. This made him resemble women in the fashions of the mid-forties.

I didn't get as much emotion out of Swamiji's breakfast puja, this time, as I usually do. And I have never read worse. My sits have been so dull lately, too. I keep thinking of my death, and saying to myself, "Suppose It isn't there for me, at the end?" But that's not for me to worry about.

February 4. Bad news tonight. Don called me from New York—he went there yesterday—to say that at least twenty-one of the twenty-four frames that have so far arrived are damaged. And the firm that sent them was recommended by Billy Al Bengston and Nick Wilder! And then it looks as if there'll be delays in getting the catalogues delivered from the New York airport—they may not arrive in time for the show!

There is much else to tell, about the appearance of James Ivory and Ismail Merchant as possible partners in the deal with Warner's to get *A Meeting by the River* filmed and about our fears that Jim Bridges may not go along with this because he has other irons in the fire—but I'm [in] no mood to go into all that, at eleven thirty-five, after a long day.

Swami has been sick again. I was told this by Anandaprana on the 30th. The previous night—it was actually on the morning of that day—I had a dream. I was in a place like the bookshop at the Vedanta Center, standing under a high shelf and protected by it. Swami was there too, but he was standing away from the shelf, near the middle of the room. Suddenly, about half a dozen very heavy packages (of books probably) crashed down from the shelf. They fell in slow motion, but I was so taken aback that I did nothing to kick them aside as they fell. They landed all around Swami. He was a bit shaken but quite unhurt. I blamed myself for my lack of presence of mind, but this wasn't a dream about guilt. The point was that he had been in danger of actual death and had escaped.

First pangs of Kitty-loss were felt this morning, when I found I'd put out two cushions on chairs on the deck, when I sat down to eat my breakfast there. It was such a beautiful day, which made things sadder.

February 11. Tomorrow I am to leave for New York, with Nick Wilder and maybe also the Moseses. Well, good luck—I dread it, the snow and the parties and the dirty grim old Chelsea Hotel. But my angel is there. I only do hope I shan't get sick and be a nuisance like the last two times.

I saw Swami yesterday. He is going to take a holiday, up in a house in the hills near the Montecito convent. Bhadrananda, who is going with him, described him as "testy." He only seemed rather remote. He is beautiful when he is like this, like some animal, just experiencing his body and waiting—until it either gets better or stops living. Then, with nervous precision, he instructs huge-assed Bhaktiprana[1] how to pack his bag. She is to put in the winter socks and the black and white scarf—it was black and gold, but she didn't dare say so. Krishna was there too, unflappable, getting scolded and taking it all as a grace. Why should he ever worry about anything any more? He has it made. He is one of *los Felices*, the Blessed. Sitting for a few minutes in this atmosphere, I couldn't for one moment stop worrying about getting Don's catalogues which I must get from the printer today, and about the problems of depositing the *Cabaret* check and then taking money out of it to deposit in Don's account—his yearly three thousand, all I am allowed by the income-tax law to give him. Swami then asked me to see Swami Adiswarananda, who is running Nikhilananda's former center. I remarked that I have never yet been to that center. And Swami smiled, seeming pleased, and said, "Nikhilananda was always jealous of me, because of my disciples." This touched me, too—the memory seemed to come from such a long long way off—like an old woman remembering some satisfaction of her girlish vanity. And then he said, "Now I have to go to the bathroom," indicating that I should leave him. And I bowed down. It's odd, I feel that our relations are so much less personal than they used to be. I am just another householder devotee. But I don't really mind. There is no shit here, no sentimentality, absolutely none. And I do feel "in the presence" when I am with him—the truth is, "he" has very nearly disappeared.

I did not get a rough draft of all the pre-1939 chapters of *Wanderings* finished before leaving. But I have done 104 pages, which only brings me to the end of 1935 and our departure for Portugal. It seems to me that I could easily make two volumes out of this work—one of them entirely pre-American. But I don't like the idea of that; it is incomplete when the first one is published, by itself. This project is all of a piece.

[1] Gwendolyn Thomas (1922–2007), born in California, educated at San José State and Julliard, where she studied violin. In New York, she was a devotee of Swami Bodhananda, then she joined the Hollywood convent in 1954, took brahmacharya in 1959, and sannyas in 1965. She performed at Sunday services, directed the women's choir, and taught Vedanta classes in Spanish. She also helped establish the Vedanta Society in San Diego.

February 23. I got back here the day before yesterday. Don is still in New York, partly because he is still hoping to get another chance to draw Alice Faye. I haven't heard from him yet.

When I left, to catch the 7:00 p.m. plane to Los Angeles, Don came with me in the cab to the bus terminal. When I got on the bus, he waited until it drove off. Then he ran quickly ahead of it and surprised me by appearing at the corner, just outside the building, as we drove out. This somehow moved me deeply. I thought, "... they do but part as friends cross the seas, they live in one another still"—a misquotation of Penn[1] which doesn't even quite make sense, but that was how the words came into my head; and they made the tears run down my cheeks.

Don's opening was crowded and, as one says, a success. And he appeared as an item of T.V. news, commenting on some of his sitters, next evening but one (February 16), and looked so beautiful and distinguished and I was so proud. But no notices yet, and probably there won't be any—Mario [Amaya] the gallery director is so slobbish and incompetent.

I want to write more about my visit but am not in the mood. Don's car-insurance agent, or a spokesman for him, called me and asked a lot of bureaucratic questions which were none of the company's business, like how old I am—just because I had admitted that I lived in the same house! So I flew into a towering old man's shrill rage and, when he kept mispronouncing Don's name and I corrected him and he said *he wasn't interested* in how it should be pronounced, I told him I wasn't interested in his questions. And he said, I'll tell them you were uncooperative—as though this were a threat. So now I'm still shaking with fury and rehearsing, too late, all the things I should have said to wither him.

Well, I've just seen the new moon through glass. And now I must stop and fix Dobbin his nice meal—fillet of sole with carrots and a tiny Boston lettuce, and then the low-fat yoghurt pineapple dessert, and meanwhile, *Twenty Thousand Leagues Under the Sea*, or *Alexander Nevsky* to watch on television.

February 25. Heavenweather, brilliant, hot, nearly smogless. Am in a dither, not knowing what work to undertake first—restart the book, the 1950 diary, or get on with letters, etc.

The day of Don's New York opening, my right upper front tooth-cap came off, breaking the stump. I rushed to a dentist

[1] "Death is but crossing the world as friends do the seas; they live in one another still." William Penn, *Some Fruits of Solitude* (1693).

recommended by Virgil, Dr. Seymour Feinberg, got it put on again, then raced to get photographed with Liza Min[n]elli, who was already late for her plane to Rio. I rather liked her. She has a complete manner for such encounters and occasions; she genuinely enjoys them, so it doesn't seem false. We were photographed with her sitting on my shoulders, or peeking around me from behind.[1]

The tooth cap held throughout the New York visit and appeared in many more photographs and interviews; Don and I are now beginning to appear publicly as a pair, before long we'll be giving advice to newlyweds on How To Stay Together For Twenty Years. But as soon as I got home, it fell off again, and now they are adding a new fang to my bridgework. Today I have nearly nothing in my upper jaw and must have a soup dinner tonight with Jo.

Joan Crawford's secretary called to ask me why Joan's picture wasn't in the show. So now I have to ask Don to call her and explain. The real reason is that Joan signed it all over with loving words and a huge signature, absurdly and fatally upstaging the drawing. But Don probably wouldn't have exhibited it anyhow, he wasn't pleased with it. I can't imagine what he will think of to tell her.

Yesterday I went to the vespers of the Ramakrishna puja. Swami came to have the relics touched to his forehead, before it began. He is very frail. Seeing him shuffle back to his room and totter up the steps—that fragile vessel containing the boy who met God sitting disguised as Brahmananda at Belur Math—I was quite overwhelmed and began to cry.

Rereading the revised first draft of our *Meeting by the River* screenplay, I really like it. (The first draft, in which he goes to California, wanders too far from the point.) We are to send it, or maybe both drafts, and also the text of the stage play, to James Ivory. If he likes it, he may even be the one to direct it—because Jim Bridges told me flatly, when we talked on the phone between New York and Klosters, that he is planning to direct Peter's *White Hunter* before he does *Meeting*—that's to say, this summer, fall, winter, who knows how long it will take. When I protested, saying that after all I am getting along and I want to see our film done while I'm still alive (this was said in a joke tone but with underlying indignation) Jim at once suggested that we should get

[1] For *T.V. and Radio Times*, promoting the T.V. broadcast of *Cabaret*; one photo appeared on the cover. The previous year, a month before she won the Academy Award, Minnelli (b. 1946) appeared dressed as Sally Bowles on the February 28 cover of *Time* and, during the same week, on the cover of *Newsweek*.

Ivory to direct it. So I feel he has already resolved to ditch us. Don is furious with him—well, yes, and I am too. He is so fucking self-indulgent.

February 27. Got up at six this morning. It is amazingly quiet here, early. Which reminds me of the incredible volume of noise produced in the early morning on 23rd Street, outside our room at the Chelsea Hotel. As if they were dragging trains without wheels over the surface of the road.

But even at this hour, still dark, I couldn't meditate. Even though I knew that the boys up at the Hollywood monastery must be doing their individual worship right at that very moment. All I can say to myself is, I am exposing myself to this thing, exposing myself with all my worries and absurd senile trivial preoccupations. It's like a treatment which you have to believe is good for you, since there are no perceptible results whatsoever. Oh yes, there was one thing. I started by chanting Om, and the noise came out of me very loud and resonant and strange, as though it were a sound being made *through* me, not by me. Well, that's not the point. That's just theater. I do believe. I do really quite nearly totally believe; something is present and hears. Only, isn't it that something which makes me make these efforts? Is it praying to itself? Well, why not? As always, one comes back to the paradox. You must make the effort and yet you cannot make the effort unless it wills. The fact that I can accept this idea is perhaps the only sign that I have achieved a certain tiny degree of—no, I won't say wisdom—sophistication.

Reading Byron's letters. How he keeps debunking the very myth of himself which he has created! Right in the midst of the *Childe Harold* craze, he writes, "I am grown within these few months much *fatter* ... and I can't think of starving myself down to an amatory size."[1] Which reminds me that I became truly enormous in New York—all because of those mandatory three meals—the story of Sodom ought to have been about a city of gluttons which was destroyed for the unnatural sin of eating "business lunches." Even now, I am 153 and ½.

But, in contrast, how beautiful those breakfasts with Kitty, fixed with his own paws and eaten in our snug little double bed! The manager (Mr. Bard) told Don that he regarded it as an honor to have Mr. Isherwood staying here, and he wouldn't charge Don anything extra for the room. But surely, he said, Mr. Isherwood

[1]November 10, 1812, to Lady Melbourne, *Byron's Letters and Journals*, ed. Leslie A. Marchand, Vol. 2, 1810–1812, *Famous in My Time* (1973).

would want a larger double room? Don couldn't very well explain that the larger rooms have two beds.... As for the cockroaches, they are everywhere in the hotel—except maybe in Virgil Thomson's apartment, and they are nothing to fuss about. They don't run over you.

March 1 [Friday]. I had supper with Bill Brown and Paul Wonner last night. I suddenly felt like seeing them and they gave me the meal they were anyhow going to eat themselves, which was so heavy that I wonder they're not both as fat as hogs—great lumps of pork and then a rich apple pie. It was quite nice, being with them. I noticed how Bill keeps upstaging Paul, however. When I asked to see Paul's paintings, Bill at once produced his. When I talked to Paul, Bill started playing the piano. Bill's paintings seemed inferior to and imitative of Paul's. Paul calls his paintings illustrations—there is a series supposed to illustrate poems about the moon, and another which is a comic strip, or so he says. They are romantic, often grotesque, enigmatic. Figures are strangely interlocked with each other or with mythological creatures. The landscapes which surround them are in pale fresh colors. I saw that they had been inspired by the Indian paintings on the walls of the apartment. Bill and Paul say that their Indian collection is now worth more than $50,000!

We talked about poetry and quoted bits. But altogether I felt a remoteness. They are already elsewhere; I am out of touch with them. I probably bore them. They don't bore me, but the things which interest me about them—chiefly questions about their relationship—how can two such characters endure to live with each other—can never be even hinted at while they are both present. It's odd to think that I've been to bed with both of them. But then, of course, "I" haven't and "they" haven't. That was three other guys, long long ago.

Don isn't coming back till Tuesday at the earliest. Meanwhile a brush-off letter has arrived from Jim Bridges in Switzerland:

> I don't want to stand in the way of a production of *A Meeting by the River* this year and since you've gotten together with Ivory and Merchant ... I think I should step aside and let him direct it. I don't want to be responsible if something should happen and by the time I can get around to making the film the money would not be in India any more. As we all know, the economics of film money is precarious and constantly changing and also studio heads roll with a strange and frequent rhythm...

I have a feeling that part of this was dictated to Jim by someone else. Irving Lazar? He has been in Europe lately, hasn't returned here yet. Did he double-cross us? It's quite possible. I'm all set to break with him. But all of Don's fury is directed against Jim.

Who could we get to play the brothers—*if* James Ivory likes the script, which I mailed to him yesterday? While we were in New York we saw Michael Moriarty in a terribly dull play about a youth in love with a middle-aged married man, *Find Your Way Home*. I think he is a truly great actor. I mean by that that he can transform himself, without giving an impersonation or using makeup. He could play a hard charming blue-eyed ruthless Watergate conspirator and he could play The Idiot. In this play he seemed so innocent and vulnerable that we could see him as Oliver. But he could easily do Patrick too, if he weren't so young. Because of the dullness of the play, I was in a strangely ambivalent state, sometimes thrilled, sometimes falling asleep. (The same thing happened, for a different reason, while I was visiting Swami Pavitrananda on February 18. When I first came in, he was so full of calm joy, really shining with it, that I *knew* he was a saint and felt quite overwhelmed. But then his very calmness relaxed me so much, after all my frantic running around, that I found myself dozing!)

I should add that we saw Moriarty after the performance (on February 20) and talked to him about *Meeting*, without actually saying that we wanted him. Since then, I have sent him a copy of the novel. And Don has drawn him. Don says he has a French wife, older than he is. This suggests a similarity to Michael York. But actually Moriarty looks much more like Jon Voight; they *could* be brothers. Oh, if only Jon had more fun in him!

Stephen Spender was in New York while we were there. Also David Hockney with a French boyfriend whom he refers to as Yves-Marie from Paris, or rather, Yves-Mar*ee* from Par*ee*.[1] Yves-Marie is quite attractive and intelligent and nice altogether, but I don't think David is really hooked. For one thing, he really cannot be bothered to learn French properly, so he and Yves-Marie can't communicate beyond a certain point. On February 16, Stephen, David, Yves-Marie and I went across on the ferry to Staten Island. It was David's idea of course, one of his whiz tours. But I was eager to see the new monsters, the World Trade Center's twin towers. They are only one hundred feet higher than the Empire State, but, planted down there near the Battery, they have effectively fucked

[1] Yves-Marie Hervé, then studying art history at the École du Louvre. Hockney drew him several times.

up the marvellous effect of Manhattan as first seen from the water. The rest of its skyline is now dwarfed and looks out of scale and insignificant. As though this were an island of quite small buildings with just these two giants. I said to Yves-Marie, "American architecture is the architecture of selfishness," which was playing to the Frog gallery, but he loved it. David was his wonderful breezy self—how uplifting he is! Stephen seemed sly and worried, but he too has amazing vigor. We are very friendly at present, chiefly because of this book I'm writing and the fact that he is helping me with it. I think he is enormously relieved that Wystan is dead and can now be both bitched and honored without one's feeling either indiscreet or envious. Maybe Stephen would like to get me out of the way, too. Meanwhile, I feel that I am promoted to Senior Old Man, an official figure who has to be flattered. Stephen said, "Christopher's the only one of us who hasn't changed at all."

March 8. My angel got back safe on the evening of the 5th. Mario and his assistant Sue are behaving badly, trying to get him to pay for all sorts of things they had agreed to pay for; so letters have to be written. Nick Wilder has been helping us with his advice. That wretched pair are both dishonest and incompetent. They didn't get Don a single notice from any critic and they didn't sell a single one of his pictures. As Nick says, Mario's treatment of the show was as a social, not an artistic event. As long as there was a big crowd at the opening and Don got interviewed on T.V., he felt that was all Don could possibly expect or want.

It is pouring down rain, in violent storms. I am pounding away at the manuscript of *Wanderings*, which, in this first draft, means chiefly copying out letters and diary extracts; I'm just getting the outline of the narrative, not yet writing the narrative itself. My chief concern is that it will be so awfully long. I fear those two volumes.

Nick Wilder told us that, a week ago, he had a dream. In the dream he felt wonderful and he knew this was because he had given up smoking altogether. Up to that time, Nick had been smoking three packs a day; but, since the dream, he says he has not only stopped smoking but has absolutely no desire to do so. If this continues, it will be next door to a miracle. I see it as a sudden intervention by the Deep Will. It gets impatient and tells Nick to cut it out. This makes sense, because Nick does have heart trouble and may very well be in danger of losing his life. Nick goes right on drinking, however. He said he'd been drunk every night this week. He is a really lovable character.

March 14. Nick Wilder is still off smoking—or was, at any rate, when last heard from on the evening of the 12th, when we went to the opening of Joe Goode's show at the Wilder Gallery. Joe's new work is called *Vandalism.* The pictures are torn in places; the effect is a bit like that of posters which have been exposed to the weather. I like them better than his earlier work.

Afterwards, there was a party at Ceeje's; the upstairs restaurant where Don had his after-opening party on July 12, last year. Joe Goode has another girlfriend now. Mary Agnes appeared briefly, just to show that she didn't mind, but she did. We sat with Billy Bengston and Penny and Robin and Jessie French. Billy gave me a very beautiful rose pink silk scarf he was wearing, and then announced he was going to get drunk, and did, and passed out. It was understood that we were no longer mad at Robin—for having deserted us as an agent without warning us in advance—and Robin has begun to talk vaguely about hiring us for some job at Paramount.

Last night we saw Swami, briefly. He complained that his pulse was too rapid but admitted that the doctor hadn't been able to find anything particularly wrong with him. Almost for the first time, he asked us to leave after only a few minutes, saying that people tired him. I could see that it was a strain on him, just trying to attend to what we said. When Don had said something and then I said something, his head jerked around painfully, like an old animal which is being teased by two people competing for its attention.

Ananda and some of the other nuns take the attitude that this is the beginning of the end; Swami isn't gaining any ground, as they put it. Ananda said, "He's failing." But Chetanananda—who really cares about Swami, I feel—I mean, really *values* him in a way that the nuns, with their cunty "oh, he's just a cute little boy, inside" attitude, never never could—said, "Swamiji is becoming more and more indrawn." In other words, Chetanananda sees Swami as preparing himself spiritually for his own *mahasamadhi*, not just passively "failing" physically. Chetanananda answered the questions at the reading, last night. I felt that we were consolidating our relationship; we get along together far more easily than I do with Asaktananda. This morning, on the phone, Chetanananda said to me, with bursts of giggles, "We feel that you belong to us, Chris—you are our very own!"

I talked to Bob Adjemian, who enthused about their last Shiva Ratri, at which the worship had been broken up into a number of small groups in different parts of the temple, instead of taking place in the shrine only. A few people brought their own *lingas*,[1] as well as bells to ring and other implements for worship. Bob told with some satisfaction that the bells used by Jim Gates and himself had made an ugly noise and had annoyed many of the nuns. I sensed that this was one of those occasions on which the monks had asserted themselves. At the Hollywood Center, the monks are always having to do this because, as latecomers, they are second-class citizens. Nun hating is one of the great dangers there for the young male aspirant. And yet, I envied Bob's utter involvement in this scene. However parochial they may appear to be, they are all going somewhere or trying to.

Abedha volunteered to drive me home, because the devotees from Venice hadn't shown up for the reading, knowing that Swami wouldn't attend it. We talked about his approaching sannyas—he has been there twelve years. Abedha astounded me by saying that he wanted "more than anything" to become a swami, "because then I'll know that they really accept me, they don't want to throw me out." He said that he had no idea what the others thought of him. I got a frightening glimpse into his utter self-seclusion. I'd always known that he had had black moods, but this confession was so odd and shocking. Did it somehow refer to his Jewishness? As for Abedha, I could tell that our conversation had done him good. And I'm not really worried about him. I feel fairly sure that he'll make it.

Chiefly because of Don, but also because of the work ahead of me and of the health and strength I still feel, I am marvellously happy. I don't think I have ever been so *conscious* of happiness as I am now. That's because I'm aware that my happiness is threatened on every side. I mean, from the material angle, by all the approaching weaknesses of old age. But is it *really* threatened? How strong is the foundation of love and faith beneath it? That's what I'm going to find out, one day soon.

March 28. All this time I have been plodding along. *Wanderings*, as of yesterday, is on its 149th page; I have now reached the period just after Heinz's arrest in 1937. So it looks like I shall reach page 210 by the time we sail for the States. That would be about seventy thousand words. Of course a lot of that would be cut. But, even if it's cut to fifty thousand, I would need at least another hundred

[1] Symbol of Shiva, shaped like a pillar or sometimes rounded like a doorknob, and often set inside a basin to catch offerings. The phallic element embodies the male principle, the container, the female principle.

thousand for the rest of the book. Can it possibly be done in one volume?

What I'm writing now is entirely worthless, except as notes, quotes, reminders. It's the kind of writing I always do when I'm driving against a great stolid mound of tamas. Rewriting now seems infinitely attractive but I won't let myself begin until the first draft—I mean, up to our sailing for the States—is finished. I have firmly abstained from rereading the first chapter, the only one which is more or less as it should be.

Also, I keep on with the reconstructed diary; am now nearing the end of 1950. A dreary period. But I would like to record the winter of 1951–1952, even if I go no further.

Also, I'm slowly writing some stuff for the expanded *Essentials of Vedanta*,[1] as requested by Anandaprana. The only interesting aspect of this task is that I am reading all sorts of forgotten or hitherto unknown passages in Vivekananda's *Collected Works*. His thoughts about the importance of dwelling on the prospect of death. His image of the cab horse which says that human beings must be very immoral because they are not whipped regularly. He is so marvellous.

My angel is sick. He has severe muscle pains probably caused by some strain at the gym. And he has broken another tooth, which Kurtzman is seeing to right now. And we are supposed to go out tonight to a party at Nick Dunne's[2] to see Marguerite [Lamkin], who [has] just arrived from England. (No, I've just realized that that's tomorrow, thank God.)

As for me, I've got an almost complete upper plate, anchored to the back teeth. It is very light and rocks when I eat[,] like a house in an earthquake[,] and may also have a tendency to stink if it isn't kept clean, but Don says it looks much better than my former collection of odd fangs. Toward the end, two of them began to project forward, like Dracula's.

Swami was very much better, when I saw him last night. Last Sunday, the 24th, he even gave a lecture and didn't collapse afterwards. But he complains of not being able to sleep.

James Ivory and Ismail Merchant are coming here on April 6. Ismail told me this on the phone from New York this morning. Ismail says he has talked to Calley at Warner Brothers and that he and Ivory will in any case do something about getting *Meeting*

[1] Originally published by Isherwood in 1969.

[2] Dominick Dunne (1925–2009), then a film and T.V. producer, later a novelist and contributor to *Vanity Fair*. Brother of John Gregory Dunne and brother-in-law of Joan Didion.

by the River produced and directed. Ismail says he likes the script. Ivory so far has only read the play; according to Ismail, he likes it very much. Meanwhile, Jim Bridges is apparently all set to go ahead on *White Hunter*. He called me on the 18th. I told him we considered his letter from Switzerland a "Dear John." He denied this, hotly but unconvincingly. He said he would come down to see us and discuss the whole thing. He hasn't been down and we have heard nothing from him or Jack either. They are at the top of the shit list.

On March 23 we had supper with Jo Lathwood. Paul Wonner and Margit Fellegi[1] were there. Bill Brown leaves for San Francisco soon, to find them yet another new home. Paul is staying on here to teach, for a while. Poor old Margit keeps having pains from her cancer (or whatever it is) but she tries so hard to be bright. Jo was telling a story about a visit she recently paid to Palm Springs. She arrived by plane and there were three paraplegics on board, also getting off there—"three *cripples* in *wheelchairs*!" Jo exclaimed, her face screwing up with aversion, "Why do they send them there?" Margit answered soothingly: "Perhaps they want to live a little longer."

Of course, I realize that Jo's aversion is every bit as touching as Margit's compassion. Jo is scared of death and just as conscious of its nearness as Margit is, so she doesn't want to be reminded of it. Margit tries to ignore the signs of her own death, but she isn't really scared. That's the only difference between them.

April 14. I'm determined to write something today because it's Easter Sunday and I've written nothing in ages. I celebrated the day by finishing the first draft of part one of *Wanderings*, up to our arrival in New York in January 1939. It ends on page 201, a bit less than I'd expected. Shan't read through it for a day or two.

Don is in New York, packing up the show. He's due to return tomorrow night. I won't report on his doings until he's back.

Before he left, he went (after consulting me) and made a scene with Jack and Jim about Jim's failure to reveal his plans to us; his decision to do *White Hunter* before *Meeting*. This had a good effect. Not that it made him change his mind but it put him in a bad strategic position. Jack was obliged to side with us and agree that Jim is cowardly and cagey. Of course it's by no means certain that they will be able to get *White Hunter* cast and started. What does

[1] Hollywood costume designer until 1936,when she collaborated with silent screen star Fred Cole to found the bathing suit company Cole of California.

seem certain is that Calley will have no part of Ivory–Merchant. He doesn't have a deal with them and isn't about to. This he told me himself, on the phone, a few days ago. Am going to see Merchant tonight. Ivory is in San Francisco. He told me definitely, on the phone from New York, that he is prepared to direct *Meeting*, although he doesn't altogether like our present script.

A horrible shock: my income taxes came to over thirteen thousand dollars. This was the result of getting the money which was in dispute with Gert Macy, all in one lump!

Jack Larson admitted to Don and me that it was he who advised Jim not to direct *Meeting* until after he'd done *White Hunter*. I don't blame him, really. But the fact remains that they could have told us much sooner.

Now we really don't want Jim to direct the picture anyhow.

The ground at Vedanta Place has now been cleared for the building of the new convent. It is rough, of course, and there are trenches in it; no more than that. But Swami seems to regard it as a potential deathtrap, a kind of miniature Grand Canyon. He phoned me specially on the 10th, when I was coming up there to read, to warn me to approach his room through the kitchen. "Oh, I'm so glad I caught you!" he exclaimed. I only saw him for about ten minutes. He seemed much better. After describing his new diet in the most minute detail, he suddenly began talking about Rama.[1] The whole room was filled with his joy. Tears of joy ran down my cheeks. I forget everything he said. I came out into the kitchen and Krishna was there. All I could say was, "Oh, I'm so glad to see you!" But Krishna understood instantly. We beamed at each other in delight.

Ed Wilson, Jan van Adlmann's friend,[2] came with him to dinner here on March 21—I forgot to record this—and told us that the fossil shell we use to stop the door belongs to the Topanga formation from the middle Miocene and is about twenty-five million years old! I believe Charles Laughton told us something like this too, but Ed is a professional geologist. The shell was there when we bought the house and it was being used for a doorstop then. We are still using it as one, since hearing what Ed said. It seems sort of silly not to.

On April 11, I had supper with Bill Roerick, who is here playing in *The Waltz of the Toreadors*. He really is a vain old goose. I hadn't

[1] Hero of the Ramayana; he was prince of Ayodhya, an incarnation of God, and a perfect model of humanity.

[2] Probably Jan von Adlmann, an art critic and curator.

realized how much his looks—his past looks—still mean to him; and this is all the more surprising because I think he is very modest about his acting. He told me that he and Tyrone Power used to be regarded as "the two great beauties." He also told me that Joe Ackerley had told him he was so like Forster, which sounds so incredible that I wonder if Joe wasn't making fun of him. Bill also quoted from his own sayings: "Concentration is falling in love with the present moment. There are so many pretty little moments in life which I simply can't resist." And Bill added: "After I'd said that, I thought it was so good that I memorized it."

April 23. This is a real exotic situation. I am sitting in a stylish green-pink-white room at the Saint Louis Hotel, Bienville Street, New Orleans, at 10 a.m. having just breakfasted on bits of cheese, lettuce, apple and two glasses of Saint Louis Beaujolais. The reason why I had the wine and snacks was that the manager sent them up, to greet me on arrival. Also, I remembered how Jay Laval used to recommend wine with breakfast and I am in a piss-elegant mood which wine seems to suit.

This hotel, into which I was booked by Philip Dynia on behalf of Loyola University,[1] is a quite beautiful old mansion with an interior courtyard, complete with fountain. It has rather chilly air-conditioning. The weather here is steamy and Gulf-coastish; yesterday, when I arrived we had a tropical rainstorm. Today is beautiful and I shall wear my white suit.

Last night, in my green corduroy[2] with red socks (which were remarked on by someone in the audience during question time) I gave a reading from *Goodbye to Berlin*, *A Single Man*, *Lions and Shadows* and *Kathleen and Frank*, which was followed by questions. I was pretty good but I did show off outrageously, even for old prancing Dobbin, and made a gay lib declaration which brought them to their feet, clapping. The audience was too small, though. Afterwards, Philip Dynia and his exceedingly cute and flirty friend, Patrick Dunne, a plump little blond, gave a party in their atmospheric slum-elegant apartment in the Quarter. (Patrick deliberately has part of the ceiling paper hanging down in tatters.[3]) Philip Dynia, who teaches political science, is an anxious,

[1] Dynia (b. 1944) had been an undergraduate at the Georgetown University School of Foreign Service and got a Ph.D. in Government from Georgetown in 1973; he had just begun teaching at Loyola.

[2] I.e., his green corduroy suit.

[3] Dunne became an interior designer and ran a local antique business, Lucullus.

quite sympathetic young man with an Afro hairdo. I think he was humiliated by the poor turnout and by the fact that he parted with so much money for it; two thousand dollars for my fee, plus plane fare, plus the bill at this hotel, forty dollars a night. Never mind, a paid Dobbin will always prance his best, even for a dozen faggots.

Other luxuries of this hotel: a light which shows you you have a message waiting for you at the desk, if you come back and forget to ask for messages; a mint left on your pillow with a note, "We at the Saint Louis wish you a very good night."

There is an attractive view from my window, looking down the street. The Quarter is much bigger and more beautiful than I'd remembered it. Last night, they were having a party on one of the balconies opposite. Every house has balconies, nearly. The streets seem strangely quiet at night.

Before I get up, I'll just note briefly that I spoke at the Cal. State University Honors Convocation, on April 19. I used to regard the place as a rough and ready teach-factory. But on this occasion it put on the dog, ludicrously. I had to wear a gown and mortarboard along with the others and walk in a procession beside the President, John Greenlee. And someone ahead of us carried a silver mace! The ceremony was held in the gym in front of a huge captive audience. I was allowed to take my mortarboard off before speaking; it was balancing uneasily on my head. I gave a twenty-minute version of the "Last Lecture" I gave in Santa Barbara, back in the early sixties.[1] But this time it included a gay lib statement. It was very well received, though it probably shocked Fred Shroyer, my sponsor.

Now to get up. I'm to talk at a class this morning. I'm rather on their hands, but they urged me to stay this extra day. I leave tomorrow morning.

May 1. Rather a flurried First. So much to be done. So much I ought to record here but can't now because other chores are more urgent.

Poor old awful tiresome John Lehmann is coming on Friday for the night. Also Ed Mendelson, to start looking through the Auden letters.[2] He'll be on our necks for several days. We have to see Jon Voight this afternoon with Ivory and Merchant, although we're now convinced that he would be fatal for *Meeting*, even if

[1] The sixth lecture in Isherwood's UCSB series "A Writer and His World," given in the autumn of 1960; see *D.2*.

[2] He was preparing a selection of Auden's work; see Glossary.

he wanted to do the film and even if they like him. Lazar has gone off to Europe just when needed; we shall have to get rid of him. A producer at Paramount, Emmett Lavery Jr., wants us to do a T.V. film of [Poe's] "The Murders in the Rue Morgue"; and a very good director, Daryl Duke, who made *Pay Day*, wants to make a film out of *Prater Violet*. And I have got a polished version of the first chapter of my new book ready. I showed it to Don yesterday and he likes it.

Feeling rather senile, and I really must go to the Japanese doctor (Roy Ozawa) about my foot. He told Don that he doesn't have cancer in his, and Don says he is adorable.

Swami called me the other day, again worried lest I should have fallen and hurt myself when crossing the rough ground where the building site is, at Vedanta Place. He also told me Peter Schneider is to become a monk within a few months. Peter himself wasn't aware of his, but seemed delighted when I told him!

May 19. Don marked his birthday yesterday by having most of his hair cut off and by shaving off his moustache. He looks much more handsome, interesting, formidable and aggressive without the hair. Also younger. The truth is that the long hair, beautiful as it was, seemed a bit absurd and unworthy of Don as a person. That elaborate coiffure would have been quite suitable for someone of Don's age or far younger, provided that that someone was no one in particular, just a plain face that needed an amusing frame. Don's doesn't.

Today he's depressed. Perhaps just a little bit because the hair is gone—it takes him a long time to grow it that long—but chiefly because Nick Wilder said to him at dinner last night that he wanted to come down and see some more of Don's paintings. And Don says there aren't any more that are any good. He is rattled and talks of putting off the show.

The birthday dinner, at Chianti, was with Nick and his new friend Raymond and Mike Van Horn. The food was inferior and they kept us waiting a long time for a table.

Last time I saw Swami, on the 8th, I asked him if he had had any experiences and he said no. "If I have another one, it will be the end of me." He said this quite casually, with a certain amusement. "I couldn't stand it," he added.

On the 12th, Elsa Laughton and I were part of a Tribute to Auden at USC. We both read some of Wystan's poems. That is to say, I read; Elsa wriggled, camped, half sang, rolled her eyes and was generally embarrassing. We both got very good notices in the press. Don says I really was good and I felt it went well. I read "O What Is That Sound," "A shilling life," "Danse Macabre," "The Ship," part of "St. Cecilia,"[1] part of the Yeats poem,[2] "At Dirty Dick's,"[3] "Song of the Devil" and "Since."

Now I'm making resolutions to work harder and be more recollected during the three months between our two birthdays. Enough for now.

May 24. First, some afterthoughts and additions:

On May 17, there were the following misguided attempts to pre-celebrate Kitty's birthday: the FBI's massacre of the six Symbionese Liberation Army members in South Los Angeles, the bombings in Dublin, the Israeli raids on Lebanon, the earthquake in Peru.[4]

On the 12th, at the Tribute to Auden at USC, Mae West was present, sitting like an idol, dumpy, expressionless, with her creamy hair piled up on her head and great sausage-curls hanging down over her shoulder. When I read "Danse Macabre," I had made up my mind to look straight at West as I said the line, "He would steal you and cut off your beautiful hair"—but the bright lights made the audience almost invisible and I can't be sure that I beamed it to her or that she was even aware of it. We didn't meet afterwards.

Speaking of hair, Don has received almost one hundred percent of compliments on his haircut. Only one person, I forget who it was, failed to say that the cut made Don look younger; this one said it made him look older—but handsomer.

The day before yesterday, I went to Vedanta Place to read. Don came by at the same time, just to see Swami. We both got the impression that he was in a mood to tell things—particularly to tell them to me, so I should record them.

He spoke of the time he had had the vision of Holy Mother. This was while he was still in his room in the old house. He had gotten out of bed (I think this was all part of his dream) and sat down to meditate and Mother had appeared to him. First his hair stood on end and then his surroundings disappeared. "And I went into samadhi." Swami's using that word about himself gave me quite a shock, I'd never heard him do it before. He added, "There is samadhi in sleep also—there are many kinds of perfection."

[1] "Song for St. Cecilia's Day."
[2] "In Memory of W.B. Yeats."
[3] Spoken by Master and Boatswain in *The Sea and the Mirror.*
[4] See Glossary under Kitty's birthday.

He also told how he had seen The Big Four[1] in the shrine, and added to the story—I think I've heard this before, but I'm not sure—that he had been terribly upset at the time because Ashokananda had taken exception to what I wrote about Swamiji in my introduction to *Vedanta for the Western World*, and how he had said to himself, "If I have offended Swamiji I shall kill myself." So this vision was, so to speak, an assurance that he hadn't offended Swamiji—and that I hadn't, either. I have to admit, it bothered me when Swami said that about killing himself. It seemed hysterical and unlike him. I want to ask him more about this.

He also talked about Ramakrishna and Girish Ghosh[2]—how they had had a competition to find who knew the most risqué words—*risqué* was the word Swami used. At the end, Girish bowed down and said, "You are my guru in this, also."

There is a great deal of flurry about a book of selections from Swamiji on the subject of meditation which Chetanananda is preparing.[3] Anandaprana, who seems to get bossier and bossier, is trying to influence me to insist on certain changes. She is patronizing in her attitude toward Chetanananda; she keeps cooing, "He's such a *good* little man." Then, when I was up at the monastery, Chetanananda himself talked to me, evidently wanting to get me on *his* side. It is a ludicrous situation. I must be very tactful.

Well, and then Bob Adjemian rather secretively called me upstairs into his room, shut the door and made me sit down on his bed while he showed me a printed pamphlet which contains a very outspoken appeal by Swamiji to young people to renounce everything and become monks and nuns. Obviously this is something Bob identifies with strongly. He wants to get it into the book, and I guess Anandaprana has already said it isn't suitable because this book is primarily for householders. (I think I might sneak it into the preface I've promised to write.) Anyhow—while Bob and I are sitting there on the bed, so compromisingly close together, Jim Gates throws open the door without knocking and says Chetanananda says it's time for me to go to the reading. He obviously did it just out of nosiness.

As I was driving back home, I saw Jim going down to the market at the corner of Franklin and Cahuenga. I stopped the car and he got in and we talked. He said that he is getting along all right but

[1] Ramakrishna, Sarada Devi (Holy Mother), Vivekananda (Swamiji), and Brahmananda (Maharaj).

[2] Bengali actor, playwright, songwriter, and Ramakrishna devotee (1844–1912).

[3] *Meditation and Its Methods*; see Glossary under Chetanananda.

there is a lot of hate in the monastery and he doesn't have any real friend. He doesn't know if Peter Schneider will be able to stand it. He told me that Beth [High] (D[ee]pti) nearly left the other day, to go back to her husband Gib [Peters]. She told Swami she planned to and he said all right and he told the girls not to argue with her about it. But, on the very day she was to go off with Gib, she decided she couldn't. Somehow, I keep getting the impression that Jim is becoming more and more of a conniver and getting less and less spiritual. He leaves me with a bad taste in the mouth.

This morning we are to drive over to Riverside to see Ivory shooting *The Wild Party*. Poor old Jo just rang up in tears to say that the doctor told her she has Paget's Disease and that she must keep her foot off the ground. She is terribly worried and doesn't know what will happen to her.

May 27. Memorial Day, the last of the most dreadful weekend in the year—with the menace of the crowded summer beyond it. Yesterday we had thick fog and the beach was swarming. Today it's beautiful but the road below us still has very few cars parked on it.

I am getting slowly ahead with chapter 2 of my book. I keep feeling that it's all too abstract—all about my feelings and attitudes, with hardly any description of my surroundings. The trouble is, the descriptions are written already, in my Berlin books.

Don is drawing in the living room, with Mike Van Horn and Paul Wonner. They have a girl for a model because the groovy guy who'd promised to come didn't show.

Old Jo seems a bit better. She came out to dinner with us, the day before yesterday. And now she's discovering that "everybody" has got Paget's Disease!

Paul Wonner is leaving this week to join Bill Brown in San Francisco. They're going to live up there for a while.

Swami called me yesterday to say that I should come and see him earlier next Wednesday because he has something to tell me, "something I never told before," which I'm to take notes of.

Our visit to Riverside was unpleasant because of the smog but otherwise quite fascinating. I have never before explored the Mission Inn which is quite as mad in its own way as anything Ludwig of Bavaria can have built—the top half Germanic, the lower half Spanish, a bar with a huge Buddha behind it, a chapel like one of the golden churches in Quito,[1] a sort of *bierkeller*, and

[1] Ecuador, where he visited with Bill Caskey in November 1947; see *The Condor and the Cows*.

some magnificent twent[ies]ish apartments with period furniture. It's in some of these that Ivory is shooting the film. There are very few exteriors, some of them to be shot in the patio at night. Other than that, there's a beach scene, I think, and one scene in Los Angeles.

Michael Childers and his bitchy little friend Bruce Weintraub[1] are both working on the picture. It was surely a mistake to have hired them. They both put on the airs of professionals who are forced to consort with amateurs and keep referring to the professional way things were done during the shooting of *The Day of the Locust*. Weintraub bitched Ivory and Merchant continually and kept adding, "I know they're your friends," until I nearly said, "Yes they are and you sure as hell aren't."

Obviously, there is a lot of confusion. This situation is entirely different from shooting in a proper studio. Big crowds of extras have to be brought out from Los Angeles and dressed and made up. The day we were there, their wigs had just been stolen! Also Ismail *seems* confused, because that is his Asian way; probably in fact he has things quite well under control, considering the fact that he is being the producer, stage manager, costume master etc. etc. all at once. He has immense energy and good humor. But he scurries, instead of strolling with the dignity befitting his position, and that makes underlings like Childers and Weintraub despise him.

Their chief problem is the bitchery of Raquel Welch. (The other leads, James Coco and Perry King, are very professional and charming too.) But Welch nearly walked out on them once already, and tried to get Ivory fired. Now she has brought her own dramatic coach with her and consults him about every take—a good deal to his embarrassment. This is of course an outrageous insult to Ivory. We both thought his self-control was impressive and, altogether, his behavior on the set confirmed our belief that he would be a splendid director for *Meeting*. He has a first-rate cameraman on the picture, Walter Lassally, who has worked a lot with Tony Richardson. Lassally has lived a long time in England but he's still full of German arrogance[2] and it seems much more likely that he might tell Welch to go fuck herself than that Ivory would.

[1] Set decorator (1952–1985); he worked on a number of Hollywood films, including *Scarface* (1983) and *Prizzi's Honor* (1985), sometimes as production designer or production assistant.

[2] Lassally, born in Berlin in 1926, refugeed to England in 1939. He worked for Richardson on *A Taste of Honey*, *The Loneliness of the Long Distance Runner*, and *Tom Jones*, and he won an Academy Award for *Zorba the Greek* (1964).

We pray that the picture will be a success, despite all of these problems. Then Ismail and James will be able to get the money for *Meeting*.

We've heard nothing more from Jim Bridges.

June 16. On May 28, Hunt Stromberg called from Texas, wanting us to do another script for him, *A Tale of Two Cities*. This would be T.V. It is really unthinkable that we should work with Hunt again. We can't trust him and the only way we could more or less protect ourselves would be to demand director approval, which Universal would never grant. However, our talk was quite friendly. No use getting nasty. It's too late for that.

On May 29, we saw Swami and he recorded some memories of Maharaj, but there really wasn't anything new—except for a couple of details: Swami said how, when Premananda had bathed in the Ganges, he, Swami, had taken the mud from his footprints and smeared it on his own chest. Also how, on one occasion, he massaged one of Maharaj's legs and a very beautiful girl massaged the other; because of Maharaj's purity, Swami, who wasn't a monk then, hadn't felt the least desire for the girl.

On June 3, Don flew to New York, to work with Bryan Forbes on his film, *The Stepford Wives*. Don had to do some drawing on camera. His hand was supposed to be the hand of William Prince, who is playing an artist in the film.[1] Don returned on June 7. Bryan had been very nice, he said, but the whole job was a bore and a waste of time and he was sure the film would be awful.

On June 8, I went to a banquet at the Century Plaza Hotel, given by the Institute for the Study of Human Resources. I had to introduce Evelyn Hooker, who was the guest of honor. And whom should I meet there but Bud Mong, who has settled down with a friend and become quite prosperous in business![2] He was still flirty. The friend is a psychiatrist; that was why they were at the dinner.

On June 10, I got around to telling Irving Lazar that we don't want him as our agent any more. He was very pleasant about this; I didn't feel he wanted us either. But now the search is on for another agent, or rather, two, since we need a literary agent in New York.

[1]Prince (1913–1996), American actor of stage, screen, and T.V. soaps, played Christopher Isherwood in the original Broadway production of *I Am a Camera*.
[2]Mentioned in *D.1* as typifying the Tragic American who is broke yet has a fancy car which symbolizes the success for which he longs; see December 1939, pp. 59 ff.

On June 11, I finished the second chapter of *Wanderings*. Fairly pleased with it.

On June 12, Don got his hair cut even shorter, by Mike Van Horn. Mike is an excellent barber but he works like a snail. It took him two whole hours, in the middle of the night! Alas, Mike is going away to New York to try to get himself a job there. This saddens Don who says Mike is his only real friend and that he really loves him. And I more or less love Mike myself. What a sweet passive inhibited creature!

Yesterday, we went to Vedanta Place for the Father's Day celebration (one day early). This year there was no lunch, only a kind of darshan with fruit punch and cookies in the puja hall. Swami told several of his favorite stories about Maharaj. He seemed very very old and rather like a tiny black man in his West Indian-style straw hat.

This morning, Swami called me and came as near as he has ever done to giving me a bawling out. He said he had been "shocked" by two questions I had asked him while he was speaking in the puja hall yesterday. "You were not yourself," he said, rather as if he were accusing me of having been drunk or high on something. The worse of my two offences was that, when Swami told how Maharaj had jokingly asked someone at a gathering of women, "Which one do you like?" I had asked him: "Did Maharaj say that to you, Swami?" My question shocked Swami because it was a suggestion that Maharaj could possibly say such a thing, even in joke, to a boy who was already a monk and his disciple. Actually, he had been speaking to Ramakrishna's nephew, Ramlal, who was a married man. Swami says he made this clear and I'm sure he did. But the heat was stupefying inside the puja hall and I guess I wasn't attending. Also, I frankly couldn't see anything very terrible in the idea that Maharaj could make a joke of this kind, even to a young monk. Surely, purity such as his gives one a great deal of license? My other offence was that, when Swami told how Maharaj was able to inhibit his sense of smell completely, I asked, "Do you mean he did it permanently?" This was sheer silliness on my part, admittedly. Well, I asked him to forgive me and of course he ended by laughing and saying, "How should I not forgive you? You are my disciple and my child." "A very silly child," I said. "Oh, no, Chris, you are the most intelligent of all my disciples." After this conversation, I felt that his scolding had really been a true blessing. But now I can't resist a little resentment—not toward him—because I suspect one of the nuns of having mentioned my mistake to him and thus put the idea of scolding me into his head.

Fred Shroyer, on the phone this morning: "I'm a boy who could never get away from under the Christmas tree." He was talking about the presents he had been given for Father's Day, including a print of the original *Phantom of the Opera*,[1] and how he loves to read romantic horror stories. Le Fanu is one of his favorites.

Don is drawing Michael Moriarty. We had supper with him last night and gave him the *Meeting* play to read. We are hoping to lure him into the film.

June 22. Today, Michael Moriarty told me on the phone that he wants to play Oliver in *Meeting.* But that he wants to get the play put on and do that before he does the film. He wants to do the play either in New York or London; he mentioned London first. Says he feels that he can only prepare himself for the film by acting in the play. He is a bit grand about it all but he does seem genuinely enthusiastic about the whole project. I have already told Ismail and Jim Ivory about this. I think Ismail is chiefly worried lest he and Jim get pushed out of the play-and-film deal by middlemen. They have finished their work on *The Wild Party* and are now off on vacation. They were both very pleased and excited about Moriarty. So am I.

Don doesn't know yet, because he went back to Connecticut the day before yesterday. Bryan Forbes wanted him to draw more portraits and also to do some more drawing on camera. When he called me after arrival he was very disgusted because they had put him in a motel, miles from anywhere. His only amusement: reading the first volume of Dodie Smith's autobiography.

There are lots more little details but maybe I'll put them in in a day or two, because right now I should gradually get ready for a dinner party at which Tony Richardson, Neil Hartley and Hugh Wheeler will be present.

June 25. The only "little detail" I can remember now is that shortly before midnight on June 20 there was an apparent earthquake shock; not a big one, but the house gave quite a shudder. Later, Cal. Tech. stated that this shock had *not* registered on the seismograph. So the papers called it a mystery shock. There were two others I didn't feel.

As for the party referred to above, it was given by Alan Shane and Norman Sunshine at their house on the top of a hill... overlooking Rock Hudson, which gives you a rough idea how high it

[1] I.e., the 1925 film, starring Lon Chaney, of Gaston Leroux's novel.

is. I think Shane and Sunshine are a pair of gracious livers and ash-tray emptiers; they start tidying the place up almost before you've left the room. Hugh Wheeler had come to Los Angeles for the opening of *A Little Night Music.* He was a bit greyer but as charming as ever. I find it impossible to have an intimate conversation with him unless we're alone together. It's something to do with our both being English. Tony was a little bit absent. He's worried about himself. The doctor—to whom he went supposing he might have rectal clap—told him that he didn't have clap but did have a cyst which may have to be cut out.

The day before yesterday, I got up early and left in the car at 7:10 a.m. It was a perfect cool blue morning. The road was fairly empty and the dreamy flat ocean had that beautiful Californian look of remoteness, far-stretchingness. I made japam nearly all of the way to Montecito. The mantra soon began to say itself aloud until I became quite hoarse and had to switch to saying it mentally, lest I should lose my voice before my afternoon talk. I was in a good mood, no hate fantasies, no anxieties, nothing but happy thoughts, chiefly about my darling, including memories of him—as for instance when I passed the Point Mugu state beach (now unrecognizable) and remembered that first outdoor lovemaking in the cove.

Reached the Sarada Convent just around nine, too late for breakfast with the nuns but not too late to be given some, as I'd hoped. Saw Swami briefly. He was constipated and uneasy in his room—he now dislikes spending a night anywhere except his room at Vedanta Place. He gave his lecture sitting down and in his western clothes; he says that he is apt to trip over his gerua robe while walking from the car into the temple. He looked marvellously remote and carven and Chinese, one eye nearly closed. But then he was funny and we laughed. At the end, after pronouncing the formal blessing, he blessed us all and then made a gesture toward the shrine, saying, "Who spoke through me." I didn't find this particularly mysterious but several people did and asked me what exactly *had* he said. I think they supposed it was a question—that Swami was asking, as a ventriloquist's puppet might, which of his manipulators had made him speak. *Honestly*! Swami didn't feel up to seeing me afterwards. But Krishna sent me away happy by giving me such a deep smile and warm handshake. I felt at that moment that he really loved me.

Then I went to speak at the writers' conference which was being held at the Cate School, up in the back country, looking out onto the mountains of the Los Padres forest. Ray Bradbury was there, acting youthful in shorts; I'm fond of him. Also James Michener,[1] rather professiorial but trying to be friendly. We embarrassed the living shit out of each other, and I think my talk shocked him. It was questions and answers. I have the impression that I was quite good, an A minus. It was very hot. One of the disadvantages of being so frank about one's queerness is that everybody expects you to leer at attractive boys, so you try not to, out of perversity. There was one in the front row. He sprawled right back, wearing almost nonexistent shorts, with his legs naked to the crotch and wide apart. Whenever he asked a question, I had to keep my eyes high by a conscious effort of will.[2]

The drive home was unpleasant as I neared Los Angeles. The San Fernando valley was oven temperature, with a fierce hot wind ruffling the golden teddy-bear hills. Scarcely was I home when Don called from Connecticut and then Michael Moriarty arrived with his girlfriend, Anne Martin(?). She was friendly but struck me as a bit arrogant, with a look of Princess Kelly; she's a career girl, an executive in the telephone company. Much talk about *Meeting*. Moriarty is fairly silly and self-absorbed but I still like him and, what's infinitely more important, still believe in him.

This morning, for the first time this summer, I went in the ocean. First I had to clear the path below the house, it's so overgrown that you can hardly get through; but I had to be careful to leave it still looking jungly, as a deterrent to trespassers.

In a letter from Rolf Ekman, my Swedish professor fan, a perfect specimen of the Lord-giveth-and-Lord-taketh-away type of compliment: "In my opinion they should have given you the Nobel prize—many who got it were not so worthy of it as you. Or perhaps share it with some other British or American novelist. Such as Angus Wilson?"

June 28. Angel is coming back today, I hope. He should be on the 2:25 plane from New York. This morning I got up early and inscribed all the uninscribed books of mine that I have given him; he was complaining that there were so many I hadn't written in. But how impossible it is to write anything that doesn't sound

[1] American novelist and short story writer (1907–1997); he won a Pulitzer Prize for *Tales of the South Pacific* (1947), which became the Rodgers and Hammerstein musical. Other best-selling works include *Hawaii* (1959), *The Source* (1965), *Kent State: What Happened and Why* (1971), and *Centennial* (1974).

[2] In *Lost Years*, which Isherwood was writing at this time, he describes how Jack Fontan was required to dress in shorts and sit front and center exactly like this for his role in *South Pacific*. See Glossary under Fontan.

stupid or false or indecent. Inscriptions are for The Others.

In Ezra Pound's *Cantos*, there is a reference to the green flash; Canto 100: "a green yellow flash after sunset." So he saw it too! I've nearly finished reading right through the *Cantos*. I only read them on the toilet seat. They are perfect for making you shit, because they give you a feeling of pleasant apprehension. It's the way I can imagine the Ancient Mariner talking.

I hear from Jack Larson that Jim Bridges went off to Spain to join Peter Viertel, either yesterday or the day before. Jack was rather bitchy, saying that *White Hunter* has nothing about Huston in it, only about Peter. And implying that the script Peter wrote is unusable.

Bob Adjemian told me that Swami's dismissal of one of the nuns—I think I heard all about it at the time, but I forget—had "turned the society upside down."[1] We were discussing (the day before yesterday) the peculiar behavior of the enlightened. Pavitrananda is at the center now and not looking nearly as ill as I'd feared, from the accounts; he has had two operations. Thank God, Swami has decided that it will be too hot and tiring to go to Trabuco on the Fourth, so I won't have to either. Swami said that, when he told Pavitrananda this, Pavitrananda "jumped with joy."

Went to supper with the Hustons on the 25th. Tim Durant[2] was there. At seventy-four he had just won a horse race and he was talking John into entering a race in Dublin and telling him how he had to get into training, doing a squatting exercise. Somebody stepped into the fish pool in the living room, trying this out. I got rather drunk (with no ill effects) and Cici got me to recite poetry. So I spoke it to the cat.

July 13. Since writing the above, Cici told me that Tim Durant rode in the Grand National. He didn't win it but he was one of the few who completed the course. That was a sufficiently astonishing feat, at his age. Cici also said that it was out of the question for John Huston to ride the race in Dublin. Right now, he and Cici are down in Mexico, at Puerto Vallarta. Cici wants us to come down and stay with them there, later. This might be a good idea, because Don is determined not to have Jack and Jim in the house and it will be difficult to avoid inviting them, when Virgil Thomson comes here in August. It would be amusing to see John in a Mexican setting but I dread the great heat.

[1] Mona Meredith, once married to Greg Meredith; her Sanskrit name was Chinmayi, and later she became Mona Reddick. Devotees took sides strongly.

[2] Former tennis star, mentioned in *D.1*.

I keep meaning to write in this book and failing to. Mostly because of work which seems to go more slowly than ever before. I feel uninspired and stupid. Also I am so fat from gluttony and my eyesight seems dimmer and I have this chronic thing in my left leg and the ball (I never know exactly what to call it) of the foot. It feels as if it is something wrong with the veins. I hesitate to go to a doctor about it for fear it will mean some kind of operation.

I think a great deal about death, nowadays. With apprehension, of course, but much more as a reminder. I mean, I try to think about death because that is the best way I can keep reminding myself of "our only refuge," Holy Mother particularly. So the thought is really an inspiring, invigorating thought—as long as I don't dwell on the negative aspect of it; separation from my darling and the probable dreariness and pain of dying.

I'm at present rereading Carlos Castaneda's *Journey to Ixtlan*, that marvellous masterpiece. I just picked it up and then couldn't stop. There is a chapter in it called "The Last Battle on Earth," in which Don Juan says: "Focus your attention on the link between you and your death, without remorse or sadness or worrying. Focus your attention on the fact that you don't have time and let your acts flow accordingly. Let each of your acts be your last battle on earth. Only under those conditions will your acts have their rightful power."

What Don Juan means by "power" isn't what I'm looking for. But that passage is still valuable for me as a meditation.

Michael Moriarty has gone back to New York. Now he is going to show the play script of *Meeting by the River* to Edwin Sherin, who directed him in *Find Your Way Home*. He says Sherin is his guru, which I suppose means that if Sherin doesn't like the script he won't want to play it. It's more or less taken for granted that Sherin will direct him in it if Sherin does like it. Am somewhat pessimistic about this, because I read Sherin as the kind of Jew who only admires "gutsy" material.

Have just started a revised rough draft of chapter 5, which is to be the third and last chapter about Berlin, in *Wanderings*. Have also more or less finished the foreword to Chetanananda's Vivekananda anthology on meditation. I do hope I don't have to write any more of this stuff. I simply cannot help ringing false.

Swami has been down at Malibu with Pavitrananda and Bhadrananda and Sat for about a week, now. No word from him. I wonder if he's displeased with me about something.

Don is painting quite a lot, I'm delighted to say. I think the work is excellent, but who cares for old Dub's verdicts? I do wish he'd let Nick Wilder see it.

July 26. Just beginning the last whole month before my seventieth birthday. I have a nagging soreness in the right side of my throat and altogether feel shaky and old, but not really unhappy at all. Vivekananda says we should think of death always, so I am trying to do that, and I know it will help when I get the hang of it.

Meanwhile Don, journeying at my side, seems the very embodiment of life, alive every instant in his work and his anxieties and intense nervous strain and the warmth of his love. Despite his worries about the future, he seems all *now*, he's got that instantaneity Gerald was so fond of talking about.

Poor Nick Wilder has been sentenced to five days in jail for drunk driving and may get a further sentence later. So his inspection of Don's paintings is delayed. Don continues to paint, however, and that's all that really matters.

July 30. Poor old Jo has been having shingles, in addition to her bad leg and the nervous tortures she goes through because of the dance hall which has been opened in what used to be Ted's Grill. All the kids from miles around come to it and there is a band with singers blasting away from ten to two-thirty in the morning. Dozens of residents have complained but the police will do nothing. The place has been zoned as a dance hall, they say, and they can't interfere until the law is broken. They would bust the place quick enough if it was queer, and plant some pot in the kids' pockets, to make the bust stick.

Jo complains so insistently that I feel it's part of her psychological self-treatment. And today she admits to feeling better. But the pain must be truly awful.

Steamy tropical weather. I wish my throat wasn't sore. I long to go in the ocean. Dr. Allen says all my tests are okay so far.

I keep slugging on at the book. Now I have finished the third chapter in its third draft. I feel so uninventive. I keep being reduced to quoting from Stephen's autobiography—which has some marvellous descriptive phrases in it—or from *Goodbye to Berlin*. Richard is helping a lot, sending me extracts from Kathleen's diaries. I love one of her exclamations: "That hateful Berlin and all it contains!"

Swami and Chetanananda are both said to be pleased with my foreword to the book of Vivekananda's teachings on meditation.

Today, Elsa has gone to spend the night at St. John's Hospital where she is to be x-rayed some more and the decision is to be made, operate or wait. Poor miserable old thing. How awful her remnant of a life must be.

Am reading *The Canterbury Tales*, Pound's *A.B.C. of Reading*,

Andersen's *Fairy Tales*, Claud Cockburn's first volume of autobiography (*In Time of Trouble*). Have just finished the five Dashiell Hammett novels. I like *The Dain Curse* and *The Maltese Falcon* the best. The others are so densely plotty. And all of them I find displeasingly sentimental, macho-sentimental. I keep feeling what an unpleasant man Hammett must have been.

August 11. Old Jo is in the orthopaedic hospital. They have taken scrapings from her bones and don't think it is cancer but aren't sure yet. Elsa is in St. John's Hospital and will be operated on tomorrow. It may be cancer or it mayn't. I'm going to see her this afternoon. What upset me much more is that Larry [Miller], the sweet little new monk up at Vedanta Place, has cancer in his back and only a fifty-fifty chance of getting over it. When I had supper at the monastery he seemed so cheerful and looked so young. It was heartbreaking.

Only three days since Nixon's resignation speech. Most of the commentators praised it. But he didn't show the least awareness of his guilt. He talked as though his resignation was a big patriotic act of self-sacrifice—of "binding up this country's wounds." But he *is* The Wound. And he is being nicely dressed and bandaged by being given huge sums of money every year for the rest of his life, at the cost of the taxpayers.

Don is in a great phase of painting. Now he doesn't want Nick Wilder to see the work, lest he should make discouraging remarks. He's right, of course. Not that I expect for one instant that Nick won't admire the new pictures. But the great thing is to get them painted while Don is in the mood.

I am gradually nearing the end of the Berlin section of my book. God, it is hard to put together and I have all manner of doubts and fears, but at the same time I know it is worth doing—oh, far more than that: this book creates a situation in which I can say things I have never—could never have—said in any of my other books.

At present, I don't look beyond what I now think of as the first volume: March 1929 to January 1939.

I admire the first part of Claud Cockburn's autobiography very much. But it hasn't helped me at all in my researches. I can't find the faintest allusion to Jean Ross,[1] or indeed to myself either. I must get the later volume.

[1] Ross and Cockburn were in Berlin at the same time, but did not meet until after she moved back to London; see Glossary.

August 19. The day before yesterday, I finished this draft of my Berlin section, and now I really have to get down to *writing* it, not just assembling the bits and pieces.

I feel as if we were passing through a sort of minefield of disease. Elsa has had her operation and, yes, it was cancer. They tell her that they have got it all out but Elsa is rightly suspicious; she has heard all this before, when Charles was sick. She is very strong and calm and "British" about it. Old Jo is not. She keeps hysterically weeping and complaining. Her daughter Betty has got her back home in the apartment, but she is as miserable as ever. They haven't told her definitely if the test shows cancer or not.

A few days ago, Don and I were both being tormented by the noise of a toy motor horn on a child's tricycle. So I walked down to the house where the noise was coming from and talked to the lady who lives there. It wasn't her child but she promised to talk to the child's mother. And then she told me that her husband has just died of cancer and that the husband of a lady across the street has just died of cancer. And then, when I was coming back from the beach after a swim, I met Madge MacDonald, looking the picture of health, and she told me she had had a colostomy two or three months ago—"I've got no rectum, thank you very much"—and laughed loudly and said she is going around reassuring everybody that it isn't the end of everything. And then Roddy McDowall was here for supper on the 16th and told us that Paul Dehn is dying of cancer in London. Oh yes, and poor Swami got up in the middle of the night and felt dizzy and fell and cut his head, not seriously. So Sudhira has been spending the night in his room, so he shan't be left alone. They are giving him oxygen to sniff from time to time, because the doctor says his dizziness was due to lack of it.

August 23. When I went up to Vedanta Place the day before yesterday, Swami didn't feel like seeing me, said he felt so weak. He is said to be in a strange state, partly with his mind in his early life with Maharaj, partly cantankerous about his food. He said on one occasion that he mustn't talk about spiritual things and he turned on the radio news. Anandaprana told me that he came out of the bathroom and said to her, without any preamble, some words which Brahmananda had said, quoting from the Gospels; I think it is when Jesus addresses Philip and tells him that he has seen the Son of God, meaning himself.[1] And then he told one of the girls, "I don't scold you for now but for after I'm gone." To hear

[1] John 14.8–11.

Anandaprana talk, the girls are really acting in the most wretchedly self-indulgent manner. One of them is "shattered" because she had to cook three meals for Swami in one day, and the rest are so upset because he finds fault with their cooking. Good God, any faithful housekeeper of the old school would put them to shame. And then Avoya (I can never remember how her name is spelled and I have just looked right through this journal in vain to find it written[1]) in her matter-of-fact hospital-nurse's dry tone told me: "Yesterday morning, he said something that didn't sound so good—he said, 'I've been dancing with Maharaj'"! Despite the seriousness of all this, I have to find the situation funny. It's so like the Hostess in *Henry V*: "I hoped there was no need to trouble himself with any such thoughts yet."[2] Only, in this case, "The Hostess" is a devotee and "Falstaff" is her guru!

Last evening, Don had a slight collision on the way to the gym. Later we had supper with George Cukor, who is quite enthusiastic about the prospects of working with the Russians in Leningrad on the filming of Maeterlinck's *The Blue Bird*.[3] But he admits that the food is awful and that he will have to take lots of books because there is nothing to do but read when you aren't working. He finds Leningrad beautiful and nostalgic but also squalid and run-down and sad. He met a lot of Jews and doesn't feel that they are all of them being persecuted. The other day, Cukor asked us if we would be interested in writing his biography. He seemed to think that Don'd be willing to do all the research—because of his interest in films—and also write a rough draft of the book itself, which I would then touch up! We declined politely. But we would be interested in touching up the *Blue Bird* script, if necessary. Partly because that might also earn us a visit to Cukor in Russia.

August 26. Yesterday, Nick Wilder looked through the fifty paintings which Don had picked out for possibles, for his show. On the whole, he approved of them and was just as enthusiastic as ever and the show's opening has been set for November 5. So Don is enormously relieved. He had been so afraid Nick would cool off. As for me, as I told Don, Nick's approval of the paintings was the best birthday present I could possibly get.

[1] Pronounced Avoya, but spelled Abhaya, as in the entry for March 25, 1973 (above), or Aboyha as in the entry for January 24, 1975 (below).

[2] When she tells how Falstaff cried out "God, God, God!" and then died, II.iii.

[3] The 1976 film of Maurice Maeterlinck's play *L'Oiseau Bleu* (1909) was a U.S.–Soviet coproduction.

I wasn't there during Nick's visit, because I had to go to Vedanta Place and read one of Vivekananda's lectures to the congregation (it was "Maya and Illusion") so that the assistant swamis could get a bit more of a holiday. This time I saw Swami. He seemed weak but he gave me such a sweet smile. No talk of Maharaj. He told how he soiled two pairs of pajamas every day in the bathroom . . . I suppose he meant his sphincter muscle is weak.

WE INTERRUPT THIS REPORT FOR A NEWSFLASH: 5:35. EDWIN SHERIN, MICHAEL MORIARTY'S DIRECTOR AND ADVISER, HAS JUST PHONED FROM NEW YORK TO TELL US THAT HE LIKES *A MEETING BY THE RIVER* VERY MUCH AND WANTS TO DIRECT IT AS SOON AS POSSIBLE.

So there's Dobbin's second super birthday gift. But now I must get ready to go out with my darling, so the rest must wait till tomorrow.

August 31. The first of the three days of the Labor Day weekend—and then the beautiful autumn begins. I have just got off the phone talking to Paul Millard ("Jemima Puddleduck") who quacked away about the kites which he and Rob Matteson fly and sell on the beach. They drive around with them in a royal hearse bought in England and painted yellow.

Every day I have meant to write something here and every day there have been other things to do. I keep on with my book, but at such a crawl. It seems that I simply cannot write a sentence straight until I have done it over half a dozen times. And I have so little energy. The "shutter" of my brain is only open for about three-quarters of an hour in the morning—I mean, really open. Nevertheless, I continue to feel that this book is one of the most challenging and potentially marvellous projects I have ever undertaken.

To get back to some of the things I have been meaning to write about—

On August 25, we had a very nice, really well-planned supper at Jennifer and Norton Simon's house—Hope Lange, Nick Wilder, Leslie and Michael Laughlin. Nick arrived informally dressed—as we'd all been told to do—whereas Michael seemed dressed for a wedding and Leslie was every inch la Parisienne in a long white formal. Hope was jolly, plump and a bit boozy. Michael got into a big dialogue with Norton about gold—Michael rejoicing that he had bought so much and Norton kind of dubious—as far as I could hear; it seemed funny to hear someone bragging about his wealth to a multimillionaire! Afterwards, Jennifer told me that

Norton had said he wanted to talk to me about politics and get my point of view, as something I had said had interested him. I can't imagine what.

My chief birthday present from Don is a silver bracelet, of very good design; he found it at Tiffany's. He wants me to wear it always, as he does the gold one I got him at Bulgari's for our twentieth anniversary last year. At first I didn't want to; I find I have an absurd prejudice against bracelets as being faggy—for me, not for anybody else! But now I have worn it since my birthday without taking it off and already I'm so pleased with it that I keep letting it slide down my wrist onto my hand, to look at it. The only time I do feel still faintly embarrassed is when I go to the beach in it.

Nick Wilder gave me a brass paper knife, kind of art nouveau with a floral pattern, French, signed by the maker, whose name looks like "A. Terler" or "Tezlet." It is handsome and nice to hold but so blunt that I sometimes can't cut paper with it! Leslie and Michael gave me a big illustrated book called *Farm Boy*, which is a reportage about three generations living on a farm in Illinois, by an ex-*Look* photographer named Archie Lieberman, over a period of twenty years. I'll try to remember to write about this when I have finished it.

Ismail Merchant has called to say that he has talked to Edwin Sherin and that Sherin wants to do our play in London in April. We await confirmation of this.

After an unusually long spell of sanity, Don's brother Ted seems to be flipping again, and Don of course is terribly upset. But I think perhaps not quite so much as he has been before, because he has gained in self-confidence as an artist. He says that he now no longer wonders if he's one or not; he knows that he is and that he'll go on working, whatever other people may think of his work. That's my great cause for rejoicing, now. Whatever happens to me, I know he's "all right."

The day before yesterday, I saw Evelyn Hooker. She has been in a lot of pain because of inflammation all around the vagina, due to something wrong with her blood. Because of the pain she has been drinking a lot of wine, and this has made her fat. But now she has met a doctor she believes in; his name's Thomas Hodges and he lives at Malibu. So she has stopped drinking because he told her to. She kept repeating that she needs discipline. She wants to go to Vedanta Place and have another interview with Asaktananda. (Which reminds me that Tom Wudl is having an interview with him too, on the 5th.)

Evelyn gave me a cushion with late-nineteenth-century embroidery on it; Santa Claus carrying some toys. It looks very German. Her reason for giving me this gift was that she had misremembered something I say in *Kathleen and Frank* about my name—that I hate it in some combinations, like it in others: "Christopher Marlowe yes, Christopher Robin *no*!" Evelyn thought I had written that I like Christopher Robin, and the cushion had reminded her of him!

I saw Swami yesterday, still weak, but he does get out a little. He had told me at first that he couldn't see me, then sent a message that he could. So I drove right into town for ten minutes with him. He asked me if I was still thinking a lot about Holy Mother and was very pleased when I told him I keep praying to her. He said how Brahmananda had bowed down before her "and trembled," and how he had said that she was the gateway to knowledge of Brahman. About his health, he said that he sometimes gets discouraged. But then he said that he hopes to take the question period when we restart our Wednesday night readings. Also he mentioned the lecture he is due to give in September. And for the second time he told me how, after Sudhira had sat up with him for two nights in a row, he had discovered that she had been working throughout the day on those two days. Sudhira had had to sit up in a chair because she was in such pain from her arthritis.

September 11. This diary keeping is being done in a mood of desperation, nowadays—or perhaps I should say intense nervous irritation. There seems to be no time for it, and yet here I am sitting at my desk for at least five hours out of the day, often more. There is always something else; the book, now in chapter 4, and then a huge batch of letters which is being added to faster than I answer them, and then a sort of symbolic task—turning Swami's taped reminiscences into an orderly narrative—the value of the task is symbolic because all of this material has been written down already and much of it is in print; I am only doing it because Swami asked me to and because I can guess how frustrating it must be for him in his old age to keep worrying about such matters and being put off again and again by the lazy, excuse-making nuns. Oh yes, and there is also the reconstruction of my 1945–1953 diaries; a job I really enjoy but haven't worked on in a long while. Right now, I've reached January 1951. I would like, at least, to get the rest of that year recorded, particularly the production of *I Am a Camera*.

Well, now Nixon is pardoned[1] and Evel Knievel is alive after

[1] By President Gerald Ford on September 8 for any offences committed in office.

failing to jump the Snake River Canyon on his rocket.[1] *Newsweek* says that, in eight years, when all the planets align themselves on the same side of the sun, there will be major earthquakes, particularly in California.

Last night we had supper with Anita Loos, who is in town to promote her new autobiographical book, at that horrible restaurant in Century City, Señor Pico. Her big silly niece[2] was there with a rather dreadful decorator friend, Dewey Stengel(?),[3] who was loudmouthed and drunk and quarrelsome. Anita is truly amazing. She had had four interviews that day and had also appeared on a T.V. talk show. She didn't seem a bit tired. She said, "The only thing that really impresses them about me is that I'm over eighty and that I get up at 4 a.m." She also told us that "Miss Moore"[4] has become boy crazy and is a bore. Anita is very successful and very very tough and seemingly quite unsentimental. She said that she wouldn't dream of living anywhere except New York. Her niece Mary disagreed, saying how horrible New York is. Mary's example of New York's horribleness was that she had seen, from the window of Anita's apartment, a huge black pimp wearing a fur coat and a diamond brooch drive up in a limousine to check up on his girls, who were working someplace opposite Anita's building.

Nick Wilder came down yesterday afternoon and took four of Don's paintings away with him, so he can try them out in various frames to see what kind of frames will be best for the show. Don has been working without models, this last week; and he has produced at least two paintings which seem to me as good as the best of his model work. Altogether, he does seem to be making a breakthrough. My only prayer is that Nick doesn't get himself jailed, doesn't have a heart attack, doesn't do anything to upset the applecart before November 5.

September 22. This morning, we had gotten ourselves pledged to go to Vedanta Place, hear Swami lecture and have lunch with the nuns—which I almost never do, nowadays; and meet Sudhira there. Don got so upset at the thought of all this that his stomach became upset and I begged him not to come. He said that he has

[1] The motorcycle stuntman (1938–2007) used a jet-powered sled launched from metal tracks for the three-quarter-mile leap, but his parachute deployed prematurely, and he fell into the canyon.
[2] Mary Anita Loos (1910–2004), screenwriter and novelist; her first novel, *The Beggars Are Coming*, was published that year.
[3] Dewey Spriegel.
[4] Loos's adopted daughter.

no role at Vedanta Place and that even I don't want him with me when I'm there. This is untrue, but I understand absolutely what he means about the role. I wouldn't have a role either, if I didn't read in the temple and do literary odd jobs for them. Probably even my visits to Swami are resented by some people.... Well, anyhow, I said okay and Don turned the car around and we started back home for me to get my car, and then Don turned around again and we went there and it was even worse than either of us had feared. It was a blazing hot day. And Swami gave a long talk about Holy Mother, and was heartbreakingly sweet at moments but often incoherent, not finishing the stories he told and leaving out the point. And then we were informed that Sudhira had called off coming to lunch because her old patient is sick. And then one of the nuns (Dipeka?)[1] took us around the half-built convent compound and, as Don rightly said, she was so cunty that it was almost unbelievable. She kept saying over and over again that the nuns would have privacy and that they wouldn't be intruded on by visitors and that the new arrangements would save them from having to carry things—it was all so self-pampering, as though this place were a health resort instead of a convent. She sounded almost like Merle Oberon.

Larry [Miller] has left the monastery and gone back to his parents to be treated there. No doubt it's his mother's influence—anything to get him away from the monastery. Jim Gates wrote me a letter about it which was more or less a confession that he was in love with Larry:

> Then he came, and without realising how deep it was, that kind of friendship grew, so easily that I hardly noticed it in a way. Now I'm really amazed and appalled at the grief that strikes me almost whenever there is a pause or quiet moment.... I was alone in the puja hall trying to do something, suddenly I actually started to cry. I haven't done that for a long time.... Well, obviously no one (maybe not even Swami) would understand this unnatural behavior of one monk upon the departure of another; I think you will and do and I already feel relieved to have been able to tell you this.

Bhadrananda says that, in his opinion, Larry won't recover.

September 28. Since then, I have had a few words alone with Jim

[1] Dipíka; she later became Pravrajika Vivekaprana.

Gates—on the 25th, when I went up there to do the first reading of the fall season. Jim says he is feeling better about the situation. Nothing has been heard from Larry, yet.

Nature notes: On the 22nd, that irritating girl Dip[i]ka (I'm still not sure how to spell it) told us that the snakes have been very bad up at the Montecito convent this year. One rattler actually attacked two of the nuns, running them up onto a wall where they had to wait until help arrived. This sounds improbable but possible, I guess. She also said that rattlers can strike their whole length, which surely isn't so. They lie out on Ladera Lane, in the evening.

On the 25th, I finally heard from Michael Moriarty, after I'd called his answering service and left a message. He told me that he doesn't want to do our play until either the fall of 1975 or the fall of 1976, because of *Richard III*, filmwork and a musical based on *Merton of the Movies*. Furthermore, he doesn't want to promise to do the film after he's done the play. I fear this means a breaking off of relations between us. We'll just have to try to go ahead with the film. Ismail now says *The Wild Party* will be released in December.

Am now over a month into seventy. I am very much aware of this. At the same time, my health remains good—except for my bad left foot, which I don't want to have examined, for fear it proves to be something serious—and I am in good spirits, most of the time. Great happiness with Don. Steady work on the book. Today I finished chapter 4—a tiresome one, because it is just studies of the principal characters in my Berlin books. I'll be glad when the part about Germany is finished, but that's two more chapters ahead.

Elsa Laughton seems almost well again, after her operation. Old Jo grouses but gets around on a crutch. The weather is grey, foggy, dead. So is my meditation.

October 3. On October 1, Don went to teach his first class (life drawing) at the Art Center College of Design. They offered him the job and I urged him to take it because I thought—and still think—that it was an opportunity for him to realize a whole side of himself he doesn't believe in; the side which could teach and be instructive and inspiring and marvellous. Well, he bitterly opposed me for getting him into the situation; and, as it turned out, the situation was impossible, because they had lied to him that he would have experienced students and then they gave him beginners. So his faint faith in himself was destroyed and he gave up the job. This was a disaster, from my point of view. But it was entirely their fault. I doubt if he will ever try anything like that again.

Now I'm worried about the Nick Wilder show because he has heard nothing from Nick and the framer has told him that Nick owes him a great deal of money and Don doesn't want to tell Nick that he has been told this. Nothing makes me more scared and savage than the thought that someone may destroy Don's self-confidence. It's like what Jesus said about causing the little ones to stumble.[1] At the same time I'm sure that, as far as Don's painting is concerned, he really has come to believe in it, at last—but, oh my goodness, what we went through, and how Dobbin was blamed for being insincere and mealymouthed because he kept praising it!

Today, Edward sent me a cassette, with his voice reading the first chapter of his new novel, *Alan Sebrill.*[2] I haven't a machine to play it on, so I must rent one and then record it on tape. But my little Uher recorder doesn't work properly, it seems, although Don's father tinkered with it. I have been experimenting with it all afternoon.

Strange half-mad Abedha (Tony Eckstein) has finally blown a fuse. He kept saying that the Hollywood monastery was a fake and full of hypocrites and that Asaktananda was "a bastard" (I don't know exactly why). So it got to Swami's ears and he came up to the monastery and called all the monks together and talked to Abedha and told him that he must go and live at the Vivekananda House in Pasadena and that the society would go on supporting him until he gets a job. Jim Gates says that Swami was most impressive; very firm and very sweet and full of love. He said to Abedha: "What I am doing is the best thing anybody could do for you. But you won't understand that until you reach the moment of death." The remarkable thing is, Swami doesn't seem at all exhausted by the effort of conducting this showdown. Yesterday, when I saw him, he talked for half an hour or more, correcting the transcripts I had made of his tapes.

Jim Gates has heard from Larry. He has made himself a shrine and is meditating three times a day and feels very bad about leaving the monastery; but he didn't say anything about his physical condition.

A letter from John Lehmann, today, says, "Alas, I'm going to be away when Wystan enters the abbey next week." I don't know if

[1] "And whosoever shall cause one of these little ones that believe in me to stumble, it were better for him if a great millstone were hanged about his neck and he were cast into the sea." Mark 9.42; see also Matthew 18.6, Luke 17.2.

[2] An early title for *No Home But the Struggle.*

this means that he'll be reburied there, or have a memorial tablet, or what.[1]

October 11. I borrowed a Japanese tape recorder from Bill Scobie (an Hitachi) and played Edward's cassette on it. Edward's voice sounded weird, and yet I'd have known it at once. There is a bit of British waffling—vague wah-wah sounds—and then Edward announces the title and the author. The way he says "Edward Upward" is so deprecatory, so charged with irony, that he might be that actor in the television commercial who says, "I'm Granny Goose."[2]

I find it extraordinarily hard to get the flavor of the writing as I should while reading it. In one sense, Edward reads very badly; that's to say, apologetically. You are so conscious, throughout, that *he* has written this. On the other hand, his voice sometimes rings with the consciousness of the beauty of a phrase which makes it marvellously memorable. For example, the way he reads the words "my everlasting death."

What is wonderful about this style of his is its reticence. He tells you everything in his own time. He isn't one bit worried about your possibly getting impatient. His deliberation is remorseless. He builds the structure of matchsticks with maddening patience. But it gets built, and what's more, the matchsticks are magnetic. They can't be blown over—no, not by a hurricane. They are locked together.

And then there is the tremendous, rigidly repressed excitement of the *mental journey*. Edward, in this Victorian drawing room in Sandown, is setting forth into the outer space of his own mind. Edward conveys the excitement of that in a way which makes me think of Zen.

Two days ago, we took Tom Wudl up to Vedanta Place. He had supper with us at the monastery and then came to the reading, where Chetanananda answered questions. I think Tom was interested, but we both suspect that he was also a bit put off.

Old Age notes: After a day of answering letters, writing checks to pay bills, etc., I am more than ever conscious of my incompetence. This could be described as senility, but is basically due to an enormous resentment at having to do these chores at all. Maybe old age is expressed in resentment. My left foot is always bad now.

[1] A memorial stone was unveiled in Poets' Corner on October 2.

[2] Philip Carey as a secret agent who rescued women in danger—for Granny Goose Potato Chips.

And my left eye seems to be clouding over. Yet I run every day. And I see well enough. My eyes don't smart any more than they always have done at the end of a day.

October 14. Another name to add to the Cancer List: Larry Holt has had a malignant tumor taken out of his lower intestine. I haven't talked to him or seen him yet. And Dorothy Miller, she's also in hospital; because she has at last agreed to have her leg surgically straightened.

Don has listened to Edward's tape of *Alan Sebrill* and thinks that Edward is very near to madness. I think he *lives* very near to madness and perhaps always has; but I don't expect him to cross the line now. I believe that danger is over.

My meditation has never been worse. Sometimes I spend the entire period just rattling away to myself about chores or anxieties or my book. What I must learn to do now is live in the present—with one exception: I must dwell continually on my death until I am absolutely accustomed to the thought and not one bit scared by it. Otherwise I must live in my love for Don and in the work I am doing. And I must fill in every empty moment with japam. Never mind if I can't think of Holy Mother or Maharaj; I don't believe that that really matters, as long as I never cease to make the effort.

The novel—just look at me, I call it a "novel" because most of my mind simply isn't attending to writing this—the book goes very slow but I am progressing. I keep feeling that I am leaving out "the point," however. This book, more than any of the others, is like a net. I keep letting it down into the mind water and failing to bring up the real right fish. Don's suggestion for its title is *A World of Difference* or maybe *Worlds of Difference.* I have an uneasy feeling that these titles have been used already.

After all this time, I have taken to drinking coffee again; it seems to blow the nearly dead embers of my imagination into a faint red glow. Also I occasionally increase the effect with a Dexamyl tablet.

Old Jo now walks without a crutch or even a stick and her complaining is cut down to a minimum. Now it's chiefly because there are so many kids everywhere: "It's like they let loose a swarm of beetles." She says they come down to the beach while it's still dark and begin to surf at dawn. I suppose this will be a marvellously romantic childhood memory for those of them who grow up capable of having romantic childhood memories.

October 16. Suddenly an overpowering heat wave. Even down here, at 9 p.m., it is hot outside. Don has gone to draw Alice

Faye at the theater during her performances—a wild undertaking which has really no relation to art but to his private sport of star teasing. The more she tried to put him off, the more determined he became to trap her.

Yesterday a letter from Leo Madigan arrived, with a postscript to the news of James Pope-Hennessy's murder which is almost more upsetting to me than the murder itself: namely that the murderer was a sexy blond Irish boy named Sean O'Brien whom we met twice in London last year with Madigan, on February the 3rd and 7th, and who gave me that memorable kiss in the pub. Madigan writes:

> He got a seventeen-year sentence a few months back though what good that does after the event I can't tell. Miss James sorely but, in his innocence, it was partly his own doing. He had a way of putting people on towering pedestals till they thought they were confirmed in grace, then cutting them dead. Aren't trying to justify or excuse Sean but that's the position he suddenly found himself in and for a bloke who used to go out and hustle to bring home a nigh-on bankrupt James the takings to find himself abandoned just when all the papers were screeching about him, JPH, getting a 75,000 pound advance for a book on Noel Coward, it was understandable treachery. I believe Sean wanted to humiliate James. He, Sean, had been drinking for twenty-four hours beforehand and the whole thing ran amok.[1]

I wrote back asking Madigan, if he was in communication with Sean, to tell him how sorry we were and to give him our love. Don was amused and a bit thrilled, saying that what I had written was "daring," but it was simply how I felt and he agreed that he really felt that too.

October 30. Some young Mexicans, employed by Elsa, are trimming the eucalyptus which tends to overhang this house. Looking at it more closely, we find that another tree, much smaller but already quite big, has grown up near the parent tree but on our side of the fence. So we shall have to pay to have that one pruned. While watching the Mexicans at work on the big tree, I saw something extraordinary happen. A limb of the smaller tree suddenly cracked and hung down, nearly broken off, as if in sympathy. There was no wind blowing, at all.

[1] O'Brien beat Pope-Hennessy so that he choked to death; see Glossary.

I'm sort of stuck in my book or at least dragging my feet[.] I think I am being too picky, trying to make this a final draft, which it can't possibly be, because I can't see where to put the emphasis on certain characters and events until I can read through the whole thing.

Now the big event is Don's show, on the 5th.[1] Several people, including Diebenkorn and William Wilson, have seen the few paintings of his which are already framed at the gallery and have been greatly impressed, saying that the paintings made them revise their whole opinion of Don as an artist. What Deibenkorn has evidently forgotten is that he actually saw four of these same paintings when they were hanging here in my workroom, and that he praised one of them, without knowing who it was by *and* (after I had told him) without repeating his praise to Don.

We've been having a cold spell, with rain. The nerves and muscles in my hip promptly started giving me pain jerks. But enough of that—complaints are for Old Jo—from whom I've heard nothing for quite a while.

November 15. Henry Seldis, this morning, in the *Los Angeles Times*:

> Don Bachardy, long famed for his revealingly drawn portraits of celebrities, has also created some absolutely fascinating watercolor portraits on paper using anonymous friends and acquaintances for sitters. All but one of these remarkable paintings is frontal. It is in the directness and in Bachardy's sure and expressionistic use of color that we seem to be able to plumb the mood of each of his subjects as it looks back at us. There is a tension here akin to Egon Schiele—though both terror and eroticism identified with that great Viennese master are nowhere to be found in these Bachardy works. Rather he comes closer to Kokoschka's probing of the psychological aspects of the people he painted rather than stay[ing] content with their outward appearance. To me, Bachardy's line portraits have always had too self-conscious an artifice. Here he demonstrates a [far] deeper, more subjective aspect of his considerable talent.

This notice is considered very good by Nick Wilder and of course I'm delighted. But I wish that there had been a picture accompanying it and that it had appeared on Sunday. The show is a success, though—no question about that. Several of the paintings

[1] At Nicholas Wilder, 8225½ Santa Monica Boulevard, until November 23.

can only be called masterpieces. They are so absolutely Bachardy and no one else. And you could feel how sincerely surprised many of Don's acquaintances were by them. Diebenkorn, talking to me about them, got so enthusiastic that he seemed to become quite boyish.

No sales so far, though.

Last night we had supper with Nick Wilder. He got a bit drunk and displayed a rather alarming side of himself, saying that he hardly knows how he'll get through each day, and that he simply has to get drunk toward the evening. He also told us that he is making no money and that, if he doesn't do better in the next four(?) years, he may give up the gallery and go into some other profession. But we both think that this kind of talk, coming from him, isn't to be taken too tragically. He is obviously a very strong, self-reliant person, and a very shrewd businessman.

On the afternoon of the 13th, when they were getting the temple ready for the Kali puja that night, Swami came in to look at the image and said, "My hair is standing on end." Then he prostrated in front of the image, lying down full-length. Then, coming out of the temple, he told people, "You are all Shiva." "But," said Ab[h]aya, telling us about this an hour later, when we arrived to see him, "he's quite all right now."

Dorothy Miller died the 12th, of a heart attack. She had been discharged from the hospital and was in a convalescent home, learning to walk. Her sister Rose had seen her a short while before and she was cheerful and seemed quite all right. The funeral is tomorrow.

November 21. Dorothy's funeral was more white than black, because the members of her church—the Monroe Street Christian Church—attended. The minister [...] was obviously "a good little man" and even a closet queen, but oh dear how terribly he sang "Nearer My God to Thee" and how desperately he affirmed his belief that Dorothy is now with Christ. He had to thump and wave his hands to *make* it true. What a contrast to someone who *knows* it's true, like Swami!

Now Vera Fike is in hospital, having just had an operation for gallstones.

Nature note: a few nights ago, near midnight, we heard a very loud chattering, scolding noise from the road below the house. Don said he thought it must be the squirrel, being attacked by one of the dogs—a dog was barking—but, when we looked down, there were two raccoons, right out in the middle of the road, fucking! We were afraid a car would hit them but there was no traffic.

Don, describing someone: "One of those Jews that live under rocks and grow noses in the moisture."

At the gym: The big noisy male impersonator, named Dick, I think, says to the bleach-blond plump queen: "Shit, I though[t] you was starving—and today, godammit, you drive up in a fucking Cadillac!" The queen, very grandly, as he walks out of the locker-room, displeased: "I *never* starve."

Today, for some reason, I got the impulse to use Gerald Heard's writing board, writing on my lap on the couch. I seldom do this. Instantly, I wrote quite a good passage on Forster. It was a bit spooky.

November 30. Now that Don is (as far as I know) keeping a diary fairly regularly, I can't help feeling "excused" from doing the same. But what that really means is merely that I should stick to the strictly personal, or to things which he didn't experience.

He did experience the evening with Gore Vidal and Howard Austen, at which I told off Sue Mengers during dinner, on the 27th. She said the Jews were better than anybody and, because I was drunk, the words flew out of my mouth and I said their art was fundamentally vulgar and second rate and that, having made themselves the boss minority, they did absolutely nothing for homosexuals, etc. etc. I felt terrible next day, but I must admit largely because I'd expressed myself badly and hadn't been calm and superior and bitchy. Gore swears they didn't mind—she and her husband and Howard and a producer named Howard Rosenman—three and a half Jews in all—but I feel that Howard Austen did mind very much, which I'm truly sorry for.[1]

Next day I went down to Trabuco for Thanksgiving lunch. Despite all the horrors of "development" around El Toro, the view from the monastery is still nearly unspoiled—and the silence, it's so odd; I'd forgotten what it is like. I mean, the silence there is the permanent state of affairs with occasional interruptions. Here,

[1] Mengers, Hollywood super-agent (she represented Vidal), was born in Germany in 1938 and escaped with her Jewish parents to New York, where she grew up. Her husband, Belgian-born screenwriter and director Jean-Claude Tramont (1934–1996), was half-Jewish. Producer Howard Rosenman—whose films include *Father of the Bride* (1991), *Buffy the Vampire Slayer* (1992), and a 1996 documentary based on Vito Russo's book about homosexuality in the movies, *The Celluloid Closet* (1981)—was born in New York in 1945 to Israeli parents. Austen was also Jewish.

the silence itself is very very occasional. Jim Gates drove me home; the two of us alone in the car. He'd fixed this because he wanted to talk, and he did—I think he thought he was pouring out terrible secrets, but all he told me was about very natural resentments and feelings of loneliness. Also, that he doesn't feel Asaktananda and Chetanananda have real reverence for Swami. Particularly Asaktananda. Jim thinks his feelings have been terribly wounded by some of Swami's rebukes, and that he can never get over this. Jim also told me that he has swellings in his lymph glands and that he doesn't want to go to the doctor lest he should turn out to have cancer and be a financial burden on the monastery, as Larry Miller was until he went back to his family. (He is apparently getting along quite well now, but he never mentions the cancer in his letters.) Jim wants to wait until it has become very serious and then die quickly. He wasn't melodramatic about this. I felt much more warmly toward him than I have felt, recently.

On the 14th of December, we're leaving for Chicago by train. Now I'm hurrying to get chapter 7 finished and with it the whole section on Berlin. Aside from accidents, I can do this easily before we leave. Meanwhile, tension. Also irritation with Dick Dobyns who has gone off and left us stuck with this boy Corderman who has failed to send us the rent checks.[1] Trivia, trivia. But how it all rattles me! I am in a very strange state. My relations with Don are as close and loving as possible; we hardly seem separate people. We both have our work and are engaged in it constantly. Thus we have the two greatest kinds of worldly happiness. But my prayers are empty; the line is dead. I try every day to keep my mind in God and I have less control than when I first started meditating. The only difference is, in some irrational way I feel that all is well.

A sudden idea—to try meditating while typing. I wonder if it'd work. I'm not sure exactly what I mean. But I'll try it soon. Not enough time, now, before supper.

December 14. I never did try it, and now two weeks have slid by and it's the day of our departure for Chicago—where Tony Richardson no longer is, because he has resigned from the film.[2] But Leslie Caron is there in a play and Neil Hartley has paid for us to be able to stay in his former apartment, the two nights we are in town, which was truly generous and thoughtful.

[1] Corderman was living in one of the Hilldale Avenue apartments owned by Isherwood and Bachardy and managing the building while Dobyns was out of town.

[2] *Mahogany*; see Glossary.

I'm happy to be able to record that chapter 7 is finished, and with it my whole account of the Berlin period. I have even written a few pages of chapter 8, getting Christopher, "Luis"[1] and Erwin Hansen to Greece. Oh God, this book is going to be long!

When we went to say goodbye to Swami on the 11th, he was in good spirits. He told us scornfully that there were some new regulations about brahmacharya made by the Belur Math (I think). One was that each candidate must be recommended by his guru as being devoted to God. "What nonsense!" said Swami. "As if one can say that anyone is devoted to God! It's only when you feel that you are not devoted that you are devoted." However inconsistent and illogical this remark might seem to an outsider, it was beautiful and true because Swami made it. I asked him to bless us on our journey and he told us to say Durga, Durga,[2] as we started.

Well Durga, Durga. I am signing off, and this is the end of this volume.

P.S. After watching the dreadful T.V. production of *After the Fall*, Don said, "Arthur Miller shone a searchlight into an empty tin can."

[1] Evidently a proposed pseudonym for Heinz Neddermeyer.

[2] Departing Bengali travellers repeat the name of the protectress of the universe so she will watch over their journey.

January 1, 1975—December 31, 1975

January 1. This is just to make the symbolic act of beginning a new volume. I have already symbolically contributed a couple of pages to *Wanderings* (chapter 8) in celebration of New Year's Day. Also, if I have the time, I'll rough out an opening for the preface I'm supposed to write for the new British omnibus volume of my Berlin novels—to be called *The Berlin of Sally Bowles.* The whole point of the preface will be to make it clear to the reader that it *wasn't* her Berlin.

We got back from New York safe and sound on December 29; I may feel like writing about this later, but not now. I'll only record that Dale Laster,[1] who'd been house-sitting for us, dismayed us by having put up a lot of the corniest Christmas decorations. He actually has a supply of Christmas tree decorations, pine cones, etc. which he uses year after year; and he'd brought in a lot of Christmas-tree branches which he had made into a sort of bush, sticking out of one of our iron balcony chairs. Other branches he stuck into the railing of the balcony, outdoors. I had to get him to take them down by telling him that we had a superstition in our family: Christmas decorations must be removed before New Year, or there'd be a death!

Dale confessed that, before inviting his mother down to see the house, he had moved some of the pictures around, hiding the Hockney nudes of Peter! But he's a good reliable soul and we are certainly lucky to have him when we need him.

January 24. This morning, I hear from Bill Gray, a professor at the Randolph-Macon College in Virginia, that Chester died in Greece on the 17th. He got ill during dinner and died that night. (The curious thing is that he was having dinner with James Merrill the

[1] Photography assistant and lab man for Peter Gowland.

poet, about whom I've just been corresponding with a tiresome woman who was at the Modern Language Association in New York and heard me speak there; she is writing about Merrill's work and wants to discuss his closet homosexuality and he doesn't want her to and she expects me to advise her how to get around him, which I can't and wouldn't even if I could.)[1]

I am glad that Chester is dead, both unselfishly and selfishly. He was obviously terribly unhappy—partly no doubt because of typical Jewish guilt feelings about his behavior to Wystan—and a misery to himself and a nuisance to others and an incurable unjoyful drunk. So it was best for him to go and go quickly. I am also glad because he might just possibly have made difficulties for me, out of mere meanness, when the time came for me to ask permission to include some of Wystan's lines in my book. And anyhow it is unpleasant to have someone around who is actively hostile to you and talks maliciously about you to people who don't know you. Indeed, I had come to regard him with powerful dislike—so much so that I found myself, for the first time, praying for him the other day, as I do for Peggy Kiskadden on the too-rare occasions when I say my prayer For Those I Love and For Those I Hate.

Talking of Bill Gray's letter reminds me that the mail this morning brought several other letters worth mentioning. Calder Willingham wrote one to accompany a copy of his new novel, *The Big Nickel*, which is the third volume of the trilogy, with *Geraldine Bradshaw* and *Reach to the Stars*. (I reread *Geraldine* a short while ago and was just a bit disappointed; it seemed to go on too long. But some of the dialogue is marvellous.) Willingham writes that he saw our "Frankenstein" and liked our conception of the story, thought "it was undoubtedly, by far, infinitely the best variation on this theme, really an original work." At the same time, he says that he felt Jack Smight's direction "was awry vis-à-vis concept and performance." He once worked with Smight himself: "He really didn't have the vaguest idea what he was doing."

[1] Isherwood spoke at the MLA Forum on Homosexuality and Literature, December 27, 1974, then had a drink with William S. Gray (d. 1992), born and educated in Louisiana, at Exeter, and briefly at Harvard, a friend of Tennessee Williams, Truman Capote, Gore Vidal, and once employed by *The New Yorker*. That evening Isherwood attended an MLA symposium at which Alan Wilde and Carolyn Heilbrun discussed his work. On January 14, he wrote to Judith Moffett who was working on *James Merrill: An Introduction to the Poetry* (1984), deflecting her questions about Merrill's "sexual nature" and telling her he didn't know Merrill, though he admired what poems he had read.

Then there is a charming letter from Dan Brown, who wrote *Something You Do in the Dark* under the name of Dan Curzon. We met him at the convention in New York and then he came to see us here on the 19th. We both liked him so much. He wants to come down again and be drawn by Don. Also, Julie Harris sent me a newspaper clipping describing Liza Minnelli giving a triumphant concert in Berlin. *Cabaret* has been running there for going on three years. Then there is a weirdly pompous letter from a Doctor Jeffrey Elliot, who is "Dean of Curriculum" (a new one on me) at Miami-Dade Community College. "I have found your writing perceptive, thoughtful and engaging. In every case, it exudes a special warmth and humanity, qualities which reflect your own devotion to the ideals of love, nobility, courage, service and truth. In addition to my fondness for your writing, I have marvelled at your lifelong commitment to those ideals which reflect admirably on the human condition. Indeed, everything that I have read about you suggests that you make a conscious effort to live and love as though life and love were one. Such a commitment is really precious. Guard it, please. It might even grow!" He then goes on to ask if he may come and interview me—at his own expense—for the *Community College Social Science Quarterly*—which is he says the nation's largest circulation journal serving two-year colleges; as he has been commissioned by them to "conduct a feature interview with a prominent individual whose life is a testimonial to humane ideals." I am writing to ask him if he knew that I'm queer and if, in the event that he didn't know till now, he still wants to interview me.

Don had a fan to visit him yesterday, a young artist named John Sonsini,[1] who had seen the Nick Wilder show of his paintings. [...] Nick tells us that the man who used to be head of the Marlborough Gallery in New York has just seen Don's paintings and was very impressed. He wanted to see the work of artists who don't have a representative in New York. But, so far, he hasn't got in touch with Nick again.

Wednesday last, the 22nd, I was at Vedanta Place for supper and the reading. A new major crisis situation is rapidly approaching. Some weeks ago, Swami definitely made up his mind that Asaktananda isn't up to taking over the center after his death. So he has written to Belur Math asking for Swami Lokeswarananda, the Swami who used to be in charge of the Narendrapur College or Boys' Home or whatever it's called. (I met him when I went

[1] American portrait painter (b. 1950).

there on December 31, 1963.)[1] Lokeswarananda now runs the Cultural Center in Calcutta. He's about sixty-five and very well thought of by everyone.

So far, Asaktananda knows nothing whatever about this. At least, we are told so. It seems a miracle that this huge secret can be kept when at least six people already know it. Swami's plan is to wait until a favorable reply has been received from India. Then he will call an emergency board meeting to which the letter will be read, in Asaktananda's presence. This seems to be about the most shocking way possible of breaking the news to poor Asaktananda. Indeed, Abhaya (whose name I at last am able to spell, because I just called the Vedanta bookshop—it seems you can write it either as Abhaya or as Abhoya; the latter is Bengali spelling, I think) took me aside when Swami had gone into the bathroom and asked me if I couldn't persuade Swami to tell Asaktananda privately before the meeting. But I don't want to get mixed up in this.

Swami himself tacitly admitted that it would be unwise to break the news in this way, for he said that Asaktananda couldn't run the center because he was "emotionally unstable." As an instance of Asaktananda's instability, Swami said that he had wept while lecturing on Holy Mother at Santa Barbara. I nearly said, "Well, what about you, when you broke down in the Santa Barbara temple and had to be led out—was that merely instability?" but I thought better of it.

I smell some sort of plotting and pressure by the nuns, behind all this. But Abhaya told me one curious thing which doesn't seem to fit into this theory. She says that Anandaprana doesn't know that Prabha knows about the Lokeswarananda plan, and that Prabha doesn't know that Ananda knows—they haven't been told, according to Abhaya, because they are often jealous of each other's influence in matters of policy.

Jim Gates told me that Larry—I forget his other name*—the one who left because he had cancer—has written to say that he's coming back and that he's so happy about living in the monastery again. Larry is coming back, according to this letter, because the doctors who have been treating him in the East now say that the UCLA hospital could do a better job. (This sounds horribly like buck-passing; they know the case is hopeless and want to give someone else the discredit of having failed to save him.)

*Larry Miller.

[1] See *D.2*.

But Swami had just told me that Larry *couldn't* come back, because his parents aren't prepared to support him here; they'll only do it if he stays with them. And because the society can't support him, having spent all its money on the new convent buildings. When I told Jim this, he replied that that was odd, since Michael Barrie had already offered to pay whatever expenses there might be and that they wouldn't be great, because Larry's treatment at UCLA was mostly for free—it's regarded as experimental research. Jim said, "Ah well, it's probably something to do with Anandaprana—she's against Larry's coming back." Those nuns! Don can't even look at the new buildings without raging against their selfish cuntiness.

Krishna has been having a severe cold, and he absolutely refused to take medicine or even to eat. He merely drank water. Swami called him on the phone and threatened to send for Krishna's brother from the East unless he consented to see the doctor. Krishna growled that Swami was "going to extremes" and that he was "quite ready to die, if it's so easy."

January 31 [Friday]. Larry Holt called this morning to tell me that the doctor thinks he may have cancer of the throat. I had tea with him last Wednesday, the 29th—this was the first time I'd seen him in ages, since long before his operation for the malignant tumor on his colon. I always had the feeling that he was very sick and that that was why he didn't want to see me. But on Wednesday he actually seemed better, not nearly so heavy-gloomy. I asked him if his religion had helped him through the operation and afterwards and he said yes, absolutely. He had been up to the center, for a meal with the boys at the monastery, but they won't let him see Swami—I suppose that really means, Anandaprana won't. However, Larry says that that isn't so important to him any more, because he feels Swami's presence all the time. He thinks of Swami as a great saint. It was quite easy being with Larry, except just at the first. He noticed that I wasn't sitting close to him on the sofa and asked me, not altogether jokingly, if I was afraid I might catch his cancer.

After seeing Larry, I saw Swami. I told him about Larry and he immediately told me a story I have heard before—that Larry and Ben Tomkins[1] once played a joke on him, sending him a fake letter from a girl, saying she was in love with him and asking him

[1] A Hollywood devotee in the late 1940s; he helped with the magazine *Vedanta and the West.*

to call her at a certain number, which was actually Larry's. Swami says he recognized the number and knew this was a joke. (Larry says that the whole story is untrue as far as he was concerned, it must be about someone else.)

Referring to this story, I said to Swami: "I suppose you must have had a lot of trouble with women while you've been in Hollywood?" "Oh yes," he answered, "terrible—you have no idea—" Then he added, with truly adorable simplicity: "You see, Chris, I used to be very handsome."

No news about the possible coming of Swami Lokeswarananda has arrived from Belur Math, so far.

After telling me about the doctor's suspicion of throat cancer—which will be either confirmed or not confirmed by next Tuesday—Larry said to me, "You may think this is very absurd, but I can't help remembering that Thakur had it. I guess I'm a hopeless romantic."

February 9. Larry Holt hasn't got cancer of the throat, but Larry Miller has had a recurrence of his cancer and it now seems that he's doomed. Jim Gates is very bitter because Larry Miller isn't to be allowed to return to the monastery, as he wants to. Surely one of the functions of a monastery is to be a place for people to die in, if they are believers and will get spiritual support by being there at the end? Jim blames Anandaprana.

No news yet about the possible coming of Lokeswarananda, except that Belur Math has sent Swami a cable, saying that they are writing him. I personally believe that they are going to excuse themselves and say that Lokeswarananda can't be spared.

Swami says that Asaktananda couldn't be head of the center because he has an inferiority complex and would therefore, if he got power, become so bossy that he would drive Chetanananda and most of the monks and nuns away and ruin the center altogether.

We hear that Yogeshananda (Buddha) has left the London center and come to live at one in Chicago. (It seems that most of the monks at the London center have left, too. Some of them have gone to Gretz.) The amusing thing is that the Hindu swami who is head of the Chicago center—I forget his name—is going back to India on a visit and so Buddha will be left in charge, thus becoming the only white swami to be the head of a Ramakrishna center anywhere.

The Wild Party, Jim Ivory's film, was a disaster when they previewed it in Santa Barbara, we are told. Not a word from them about *A Meeting by the River*. Or from anybody.

Michael Laughlin and Leslie have been having a serious falling-out. Leslie has announced that she loves David Lonn, the producer of the play (*13 Rue de l'Amour*) which she is starring in in Chicago—we saw it and met Lonn while we were there in December. She has told Michael that she wants a separation. Michael made a scene with Lonn [. . .]. Leslie became motherly and sympathetic and gave him Valium. Michael called me this morning for advice. I told him to refuse to agree to a separation at this time, to leave Chicago and come back here and let Leslie make up her mind—assuring her that he loves her and is determined to save the marriage if possible. Don thinks Lonn is just an operator.

We had lunch today with Truman Capote. Don thinks he is very sick. But he seemed much more cheerful than last time we saw him. We had been told in New York that he was abandoning his novel. He assured us that part of it will be coming out quite soon in a magazine. Don isn't sure that he's telling the truth. It was nice seeing him, as nearly always. But today, as far too often, there were a lot of other people around.

February 19. Am writing this around ten in the evening. We are sitting up waiting until it's time to go out to a midnight supper at Marti Stevens's house. The supper is late because she is playing in *The Constant Wife* and she will be bringing Ingrid Bergman back with her. Bergman is at present playing her part sitting in a wheelchair, because she has broken her toe (or foot, I'm not sure which). I'm sure the audience loves this; indeed, that it's an added attraction. That's what being a star is all about.

Talking of being a star, old Dub is about to prance in public again. I have been awarded the Brandeis Medal in literature for this year. They want me to go to New York to receive it on April 6, but I am holding out for travel money, in addition to the thousand dollar honorarium.[1]

Meanwhile, we have practically committed ourselves to taking on a T.V. assignment: Fitzgerald's *The Beautiful and Damned*.

I keep plugging on at the book. At present, not joyfully. I feel it is somehow flat—that I'm failing to give it the sparkle of life. One thing that keeps bugging me is that I have covered so much of the material in my fiction and what's left for me to write is just—leftovers.

[1] Awarded by Brandeis University in recognition of lifetime achievement; Brandeis is in Waltham, Massachusetts, but until 1990, the ceremony was held in New York.

After the grand lunch with Gore Vidal in 1973 and Don's New York show last year, our anniversary celebration this year lacked public recognition. So we made a virtue of that; spent the evening in bed, watching T.V. and drinking champagne. I really liked it much better that way. I think Don did too. With us, domestic bliss isn't just a phrase; it's an exact description of a mood we often experience.

Larry Holt called me tonight and asked me to collaborate with him on the story of his love affair with the young chief on Maupiti.[1] I said yes, I would record it all for him and write it up later. I have often urged Larry to record it and I don't want it to be lost.

A strange dream about Swami—four or five nights ago. Like an idiot I didn't write it down at the time. Now I've forgotten part of it. To begin with, Swami and I were walking in a landscape which was very like the Derbyshire Peak District—say, the pass above Castleton. Green turf over limestone, with grey rocks showing through. A steep smooth rock path, leading downhill. Swami was old as he is now and weak and I was helping him. Suddenly he lost his balance and slipped from me and slid rapidly down the rock path on his bottom into a very small cleft in the rock. I tried to get him out again but couldn't. I knew that he was sitting inside a tiny cave which I was too big to enter.

So I ran down into the neighboring village and began calling to the villagers for help. I was addressing them in Spanish because that most resembled their native language, which was Italian. But they didn't understand me.

The bit I have half forgotten was a sort of X-ray view of Swami inside the tiny cave. My impression is that he was making a terrible fuss and show of panic, but that I knew all the while that this was some sort of deliberate playacting.... The explanation which presents itself is that this was a dream about Swami's death, and about the absurdity of our concept of death. Certainly, it wasn't an unhappy dream. But it wasn't particularly joyful, either.

Leslie has just returned from Chicago, but she is staying in a hotel. Michael, who came to dinner last night, said she was terribly touchy and difficult. He doesn't yet quite know what the latest score is.

February 27. Yesterday, having bought a new mike, I tape-recorded the first part of Larry Holt's narrative of his love affair with the son of the chief, on Maupiti. Larry really dictated very clearly and

[1] Maupiti is in the Leeward Islands, northwest of Tahiti; the youthful love affair is described October 19, 1967 in *D.2*.

logically and I feel pleased to have got some of this down, and eager to get more. He was so moved that he shed tears. Later, he said it was probably bad for him to bring these memories back. But I'm sure he loved doing it.

Before I saw Swami, Anandaprana told me that the doctor had said that Swami is "beginning to fail," according to all the physical signs. He seemed much as usual, however. He said with his usual, rather satisfied aggressiveness that there was going to be a fight with the Math over Lokeswarananda's coming here. The poor man has got to have a gallbladder operation first, and no doubt that's why they're unwilling to let him travel—quite aside from their general unwillingness to let any of their good men come over here.

Anandaprana had also told me that Swami had another spiritual experience when he was up at Santa Barbara, the other day. So I angled around until he told it to me. This time it was a vision during sleep. He was feeding Holy Mother and he began to weep. "I could have wept myself to death," he said. "When the doctor examined me, he said, 'You have had a shock.' It was like a heart attack.... When I wish to die, I can die. Whenever I wish. But I don't want to die yet—not until this place is saved." (By which I suppose he meant, when he has got Lokeswarananda over here.)

He began to talk about the lower samadhi, and how doubts still remain, after you have had it. "Do you still have doubts, Swami?" I asked. "Oh *yes*—well, not doubts, exactly. It's what you call divine discontent. You know, Maharaj used to tell me, 'What's wrong with you is, you're too contented here. You shouldn't be contented in spiritual life. You should always want more.' Well, now I always want more—*more*."

Deepti, I am sorry to say, has left. This time, she didn't give anybody the chance to dissuade her. She disappeared, saying good-bye by notes. They don't know where she is. Swami says Bhuma (Cliff Johnson) is "so restless—he keeps going out shopping in Santa Anna—he stays away all day." And Jim Gates has rather turned against Swami. He went in to see Swami and begged him to at least write a letter to Larry Miller. Swami refused. But even Jim admits that Swami probably misunderstood exactly what Jim was asking him to do, and thought Jim wanted him to write that he'd take Larry back. So Jim is getting restless, too.

April 7. No use apologizing to myself for the huge gap here. The truth is, I am slowing down; I simply cannot get through all the jobs I set myself to do. And so I develop a masochistic attitude toward myself as my own taskmaster.

Don went off to New York on the evening of the 5th and I said to myself that I will accomplish marvels while he's away. But today I have merely puttered, wasting all these hours of precious solitude on a few letters, telephone calls, shoppings, jogging at the gym. Well, anyhow, Don has already accomplished something. He called me last night to say that he had made the speech I wrote for him to the Brandeis people and received the medal. And I could tell that he had been a success. My darling, he had told himself he couldn't do it and he had done it—as usual. It seems that Glenway was almost entirely responsible for my being awarded the medal, as I suspected. And now Don will stay on until the beginning of next week, so as to draw Ingrid Bergman after *The Constant Wife* has opened in New York.

I have said this often before, but I'll say it again: I think this latest period of our life together has been as happy as any happiness I have known in my life. And yet it's not by any means a smooth calm. Now and then Kitty storms, shows his claws, screams furiously about the tiniest trifles. Sometimes I worry about this—for his sake, not mine—but not seriously. This lack of control may get him into some difficulties, especially after I'm dead and he has to vent his nervous fury on others. But it surely is better than being bottled up. The thing I do worry about is the possibility that he'll go deaf. He often seems deaf because he simply doesn't listen when he is absorbed in his thoughts, but such absorption can have psychosomatic consequences and I foresee a time when he will cease *to be able* to hear. I must find the proper moment to speak to him about this. Not easy.

I don't exactly think about Don while he is away. I don't have to. He is with me. He is part of me and I am part of him. Some of the inner rage he feels against me is because of the fact that I am going to leave him. He feels that this is a trick which I shall play on him—have, indeed, already played, by involving us so with each other. Any sign I show of illness, even of fatigue, makes him intensely nervous; he behaves as if it were a kind of bitchery on my part.

Enough of that for today.... There are two things I want to put down, and that's all, till next time.

April 8. I didn't put them down because, almost at that moment, Don called from New York. He told me that Lincoln Kirstein has had a massive heart attack. My immediate thought was, will he now feel he wants to make it up with us?

Here are the two things:

Paul Sorel called yesterday, to tell me that the doctor says that Chris Wood's X-ray pictures show that there is cancer in his pelvis. According to Paul, Chris doesn't know this yet. The doctor must think that it will develop slowly because he still says Chris can make the trip to England which he has planned for next month. Meanwhile, Chris has no more pain from his hip and is becoming more and more able to walk around.

A short while ago, Paul felt he had to have a holiday from the strain of looking after Chris, so he arranged to go to the San Francisco Bay area for two days. Other friends promised to look in on Chris during this period. Chris was perfectly agreeable, and indeed everything went off without a hitch. But, when Paul called Peggy [Kiskadden] to tell her what he was about to do, she told him he had no right to go, and she slammed down the receiver. "It was then," Paul told me, "that I understood for the first time what you have against Peggy." Paul also told me that he and Peggy had spoken about Don and me, quite recently, and that Peggy had said to him, "How was I to know that he (Don) would stay—the others didn't?" Poor wretched woman, she really is incredible in her arrogance. Her attitude is exactly like that of a government. She is prepared to "recognize" a new regime in a foreign country, but only after a suitable interval and only after the new regime has proved that it is worthy of recognition—as Don has now done by becoming a known and exhibited and praised artist!

Three days ago, Stephen Spender called me from New Orleans. He was there on one of his lecture tours; but the tour doesn't include southern California. He wanted to tell me that he and Ed Mendelson are about to write a life of Wystan—well, not a life, exactly, but a book about Wystan in various aspects—as a poet, a teacher, a Christian, etc.[1] He wanted to know if I would help them in this project—by which I think he actually meant, would I endorse it? He is well aware that he is going against Wystan's expressed wish, not to have his life written. But then so am I, on a much smaller scale. I am writing little bits about Wystan in my book. And, I must say, I can't help feeling, wishes or no wishes, it is better if those who knew Wystan write now, instead of leaving it to those who didn't know him, a generation or two later. So I didn't discourage Stephen—not that it would have made any difference if I had.

[1] Spender had already edited *W.H. Auden: A Tribute* (1974), a collection about Auden; in the end, Mendelson wrote a two-part literary-critical biography of Auden on his own; see Glossary.

I then asked Stephen if it was true that he was rewriting *World Within World*—I had heard rumors that he was. He said yes, he was going to enlarge it, expanding the parts about Berlin, etc., and also bringing it up to date. He was very anxious to see the manuscript of my book. I assured him that of course he should see it before it was given to the publishers. In the midst of this conversation, I got a most uneasy feeling that history was repeating itself; Stephen is attempting to scoop my material, as he did back in the thirties. He doesn't even quite know that he is doing this. It's just that he's so competitive.

April 9. Last night I went to Billy Al Bengston's party for the Academy Awards. He had set up T.V. sets in several different rooms. The rooms in Billy's apartment keep changing, because he keeps knocking down and putting in walls and partitions. At present he has a very tall, very narrow door leading into one of the rooms which looks wonderfully sinister.

There was a lottery on the awards. Each entrant had to fill out a form of predictions and it cost you five dollars. Don had done two different ones for the two of us. I hadn't even looked at what he had written. The one he did for me won the lottery easily; no one else came near to the correct answers. So we won $290, as well as getting our $10 back. Such an odd coincidence: Don just received a prize which I had won and now I received one which he had won! My speech much amused the guests, it seems, but alas I can't remember what I said, as Dobbin was somewhat juiced. I do loathe these parties—although I am really beginning to be fond of Billy. I used to feel ill at ease with him. And I like his new work far more than his earlier.

Yesterday, John Preston, the editor of *The Advocate*, called about the piece I wrote for him in lieu of the awful ass-licking interview Don v[o]n Wiedenm[e]n did on me. Preston was very embarrassed and I had to say his lines for him, telling *him* that he didn't really like my piece, thinking it too simpleminded. I urged him to show it to as many people as possible and then make up his mind—assuring him that I wouldn't be offended if he decided not to print it. What I didn't add was that, in that case, I wasn't going to give them another interview, at least certainly not with Wiedenmen.[1]

As a matter of fact, I feel I want to add to this interview with a paragraph about gay literature and its difficulties. More about this

[1] Neither piece was used. Von Wiedenmen had already published an interview with Bachardy, and W.I. Scobie contributed a piece about Isherwood a few months later; see below, April 19, October 23, and December 31, 1975.

some other time perhaps. Right now I must go out, having spent the whole day working on our "Beautiful and Damned" outline. I fear I shall only just get it finished, if at all, before Don returns on Monday.

Talking to my darling on the phone this morning, I mentioned that we are having so many rainstorms and that the wet front turned around and has come right back on us for the second time. And I described that supermelodramatic forecaster who acts out all the weather movements, miming this one. He's on channel seven. We usually watch channel two. So Don said: "Faithless Dobbin! As soon as Kitty's gone, Dobbin watches another channel!" I quote this because it's so typical of our jokes.

April 19. Today I hit upon what I believe may be a valuable psychological trick for dealing with my literary sloth. Yesterday I finished the war section of the "Beautiful and Damned" outline and said to myself that the job was practically finished. This morning I realized that there was a lot more to be written; a fourth act, as it were. And I felt a panic that I wouldn't be through it before Don gets home. I sat down at the desk with all the paper spread around me and simply couldn't make myself start and began wasting time. And then I got the idea of dictating it, in rough, to the tape recorder. Amazing how much easier this was! When I had finished, in about two hours, I made the recorder dictate it back to me and I typed it. This was an amusing game and a relaxation. I had recorded rather too fast, other[w]ise I could have easily kept up with the recorder on the typewriter as long as I didn't worry about typos. I shall do better in future. As a result, I have a rough draft of the last part of the outline, seven pages, which shouldn't be too much trouble to revise. This afternoon, I followed this up by dictating some of chapter 10 of *Wanderings*. I doubt if this will be as valuable, because I really do have to think hard about this; the narrative is so deadly flat at present. Still, I am pleased with myself.

Two fan letters. One from a gay couple in Long Beach who have been together ten years, saying how our example encourages them. The other from an eighteen-year-old Canadian boy in Toronto, telling Don he is really beautiful, as well as a great artist. He encloses a drawing (terrible) which he made from one of Don's photographs in *The Advocate*.[1]

[1] Bachardy was on the cover of the March 12, 1975 *Advocate*, No. 159. Two more photos of him appeared inside along with the earlier von Wiedenmen

Last night, after having supper at the monastery and reading (but not seeing Swami) I went by to visit Chris Wood. He now knows about the cancer in his pelvis; Paul Sorel burst into tears and told him. Chris was wonderfully and I think genuinely calm about this. He was so sweet and cheerful. But he is terribly shrunken and thin; his arms like sticks. He doesn't look at all "ill" though; he is the cutest little old man you could imagine. He said that his life at present is lived from day to day and is "most enjoyable—and when it ceases to be enjoyable, I shan't want to continue it anyway, so that's all right." It seems that he probably will go to England with Paul, as originally planned.

April 11. 9:15 p.m. Just finished the rough outline of "The Beautiful and Damned." An odd day of gluttony and energy. I worked out at the gym, then ate two Toblerone bars and a bag of Turkish delight, then worked for about three hours. Now I'm going to fix some sausages and a steamed carrot for supper.

Michael Laughlin called to tell me that Leslie had phoned him from Paris, asking him to come over and join her. She had been very depressed, disenchanted with Paris, even. Michael is delighted, of course. But he'll probably annoy her all over again. He doesn't seem to have the slightest idea of how to deal with her.

May 4. The names of the senators who voted *against* the Willie Brown Bill (A.B. 489) are: Democrats: Ayala, Collier, Holmdahl, Kennick, Presley, Stiern, Wedworth, Zenovich. Republicans: Berryhill, Carpenter, Cusanovich, Deukmejian, Grunsky, Nejedly, Richardson, Russell, Schrade, Stevens, Stull, Whetmore.

The vote was twenty-one to twenty approving the bill. The tie was broken by Lieutenant Governor Mervyn Dymally.[1]

Two jokes by Ivan Moffat: I said that I thought Frank Langella had overacted his performance as one of the lizardlike creatures in Albee's *Seascape*. "In other words," said Ivan "his performance needs scaling down." Then he said that everybody seems to want to forget about Vietnam—"to let Saigons be Saigons."

My back has been bad for a week, now. I suppose I shall have to see a doctor soon.

May 26. Memorial Day—three months to my seventy-first

interview, a 1953 Arthur Mitchell shot of Isherwood and Bachardy together, and three Bachardy drawings—of Isherwood, Warhol, and Bette Davis.

[1] Assembly Bill 489 removed criminal penalties for adultery, oral sex, and sodomy between consenting adults; see Glossary.

birthday. Watchword: The night cometh, when no man can work. So hurry up, Dobbin. I have been criminally slow, getting on with my book. Still fussing with chapter 10. But the *Beautiful and Damned* screenplay should be ready on time—the end of June.

Causes for anxiety. A little black spot which my darling may have to have removed from the top of his skull. It could be something bad. He sees a doctor—recommended by Billy Al Bengston—tomorrow. This is for a second opinion. Dr. Maxwell Wolff has already said it should come out. Other causes? Well—none, now I come to think of it. My back, after being very painful, is now much better. The only other cause is political. Police Chief Ed Davis's almost certain success in getting together a petition to put an antigay proposition on the ballot next year, in order to stop A.B. 489 from becoming law.

Otherwise we are so happy and both working well. But Nick Wilder must come and look at Don's painting soon, or Don will begin to lose confidence in the new ones he is doing now.

June 10. Good news: The surgeon says Don's black spots were benign—that infuriating word; as if they had done him a favor by costing him money and both of us anxiety!

More good news: Harry Rigby and a favorite director of his, Marshall Mason, are wildly enthusiastic about our *Meeting by the River* play. Now they are trying to get Michael Moriarty to be definite. But can they?

We are getting along pretty well with the "Beautiful and Damned" script but this is only the rough draft; we have to get it revised and typed up before the end of the month.

I have been crawling slower and slower on my book; still only two-thirds through chapter 10. So today a resolve. I will raise my morale—or come to some kind of definite decision—by trying to do another rough draft of the remainder of book 1 (the first half). According to the existing rough draft (from which I've been working) that would be about another 120 pages—making about 280 pages altogether for the first half of the book.

Last time I saw Swami (on the 4th) he said that he had refused the application of a lesbian girl who wanted to be a nun. The next day he called me specially to assure me that he hadn't refused the girl *because* she was a lesbian but because she had attacked one of the girls at the convent (Santa Barbara?) and was also over age! A sort of apology to me, as senior gay-lib representative!

June 26. On the 24th, we finished our draft of "The Beautiful and

Damned." We were both really pleased with it. Don is fixing up some last-minute alterations. Then we'll send it off to Cramer and Baumes,[1] tomorrow morning. We have asked Perry King if he'd be interested in playing Anthony Patch. He got wildly excited, so much so that I feel a bit nervous, for we have no voice in the matter and he may easily be turned down for all kinds of movie-political reasons.

Poor Asaktananda! It now seems that he will receive a dishonorable discharge, instead of being merely downgraded by the arrival of a senior swami from India. One of the girls at Santa Barbara accuses him of having taken her in his arms and a married woman in the congregation has complained of his behavior during a private interview. I suppose this must be true and yet I doubt it, or rather, I feel that it is somehow not his fault but due to the psychic power of those ball-cutting religious cunts.

Some miscellaneous notes[.] A dream, a few nights ago: Don and I had spent a night in Swami's room, looking after him. (It wasn't literally Swami's room because we had been sleeping one on either side of his bed which wouldn't be possible there, his bed being against the wall.) When we all woke, we were spattered with tiny dots of gold which was Swami's shit. (I immediately remembered that Layard used to maintain that shit signifies gold, in dreams.) Anyhow, it was quite odorless and not at all disgusting in any way. I woke feeling that we had been blessed—and that Don had been much more blessed than I, because there was much more of Swami's shit on him than there was on me.

On March 17, Mort Sahl, on his T.V. show called "Both Sides," made antihomosexual remarks, against which Cici Huston, who was one of his guests, violently protested. Here are four of Sahl's remarks (addressed to Cici and some other women): "They despise you because you have the real thing," "They dominate classical music," "Do you know a *poor* faggot?" "They're your enemy."

Two kinds of death reminders I often have. A realization that I shall never be able to learn something now (e.g. Spanish). A poignant memory of the pleasure and *the hope for the future* with which we bought some item of household furniture (e.g. a chair) which is now, like me, nearly worn out and ready to be discarded.

Sometimes, not often, I get strong chills of senile depression. But they can—so far—always be countered by making japam.

[1] Douglas Schoolfield Cramer (a producer of "Love Boat") and his partner and companion Bud Baumes planned to produce the program, but it was never made.

However, I still have no idea what it will be like to begin to "fail" physically. I was a bit scared, today, when my lower lip had fits of trembling. It has stopped now.

August 21. Another attempt to restart this. So much has happened that I won't bother to try to tell it right away. This new beginning is chiefly because we're now free again: we sent the revised script of "The Beautiful and Damned" in to the studio two days ago. They—that is, the NBC people—wanted narrations. So now they've got narrations.

On August 4, Dr. Dayton told me that I have a small cataract in my left eye. He added that about thirty percent (I think) of people over sixty-five have them, and that this one might not have to be removed before I die. (He had, as it later came out, actually seen this cataract the last time he examined me, but hadn't mentioned it.) At present, this doesn't bother me much. But I can tell that it's there.

A note, which arrived this morning, evidently by hand: "To Mr. Isvewood, from a kid on Mabery Rd. From you complaining your a sourbud of the year!!!!!! If you don't like the noise down here on Mabery, than go away! You can't stop a whole street of kids, just cause of you! Take your books and move away, far away."

I can't help feeling that the kid must have been encouraged to write this by his parents. When I first read it, it rattled me in a curious way. One yells at those kids, or rather, at their toy dog with its tormenting bark, but one is amazed that they answer back—because, I suppose, they are part of what seems a purely subjective, psychosomatic environment. For example, I have nearly come to believe that, whenever we go to the beach, ball-players and dogs are psychosomatically compelled to come and annoy us. Often, they seem to appear out of *nowhere.*

August 28. I read the note to Judy Davidson—Gordon Davidson's wife—since they live on Mabery Road, in the Viertels' old house. She too thought one of the parents must be involved. Her curiosity was really aroused but she wanted to see the note itself, hoping to recognize the handwriting; and I doubt very much if I shall ever have the energy to take it down and show it to her.

Now, today, we see that the neighbors immediately below us, who have several kids and entertain all the others on the street, are about to move out. It's like asking for new cards in a card game; the new ones may well be worse. These people, at least, had a relatively nonirritating dog and they did their best to keep it quiet.

Krishnananda was suddenly overcome by a stoppage of the bladder—which, characteristically, he had been trying to ignore—and was sent to St. John's Hospital. I saw him there on the 21st, and Don and I saw him, very briefly, on the 26th. On the 21st, he was lying doubled up, in great pain, but being with him was marvellous. Not in the least because he was noble or sweetly suffering. He was grunting and muttering to himself like any worldly impatient patient. But the vibrations which he gave off were beautiful and exhilarating. After being with him for about ten minutes, I felt so uplifted that I went into the hospital chapel—since any sacred place is better than none—and knelt before its hideous arty modernistic altar.

Krishna has now been operated on, for enlarged prostate and/or stones in the bladder. We haven't heard the results and the prognosis yet.

Larry Holt is just home from St. Vincent's Hospital and in his case the prognosis seems negative. They found a whole lot more cancer in his intestine.

On the 22nd, I went to Malibu to see Swami at his holiday beach-house retreat, and told him how I felt about Krishna. We also talked about the ultimatum from the Belur Math—chiefly inspired by Gambhirananda, no doubt—that they won't send us another swami until we accept the Math's old-fashioned monastic setup, in which the monks and nuns would be almost totally separated and the nuns would probably have to move out of their just-built convent (what poetic justice!) and give it to the monks, and would also not be allowed to meditate with the monks in the temple and would have to have a shrine of their own, and live in what is now the monastery. Swami seems to take all this relatively calmly. Maybe because he guesses that it won't happen, maybe because he is sick of having the nuns fussing over him and bossing him. However, he remarked to me, almost playfully, that it was such a pity I wasn't a monk, because I would have been able to help out with lectures. So I said, "Perhaps in my next lifetime." "What do you mean—next lifetime?" he exclaimed indignantly. "You will not be reborn! You will go straight to the Ramakrishna loka!" "Well, Swami," I said, "if I do, I'll be sitting right at the back."

I had almost forgotten this conversation when, yesterday, Swami phoned me and said—starting in the midst of the previous conversation as he often does: "In God's eye it is all the same." He then went on to assure me that, in God's eye, there is absolutely no difference between me and Krishna and himself. What I find so adorable about Swami—one of the many things—is that he keeps going over and over something which has been said, days or weeks before, and then suddenly feels he must talk about it. It gives me such a sense that he is watching over us. I can say "us" now, because I really do feel that he accepts the relationship of Don and myself absolutely and loves us as a pair.

My seventy-first birthday was passed quietly and sweetly alone with my darling. There was even one of those little spats which are a feature of the tension of loving each other—but it was forgotten in the bliss of eating Kitty's homemade kedgeree in bed, watching T.V.

September 26. All right, now we're going to get this record restarted, despite all internal opposition.

Jonathan Fryer, my biographer-to-be, left Los Angeles this morning. A big pudgy boy, slightly pretty and slightly feminine with a moustache. He describes himself as bisexual, has B.O., a rather charming smile and a very low, almost inaudible voice which may have something to do with the fact that he is subject to a mild form of epilepsy. (He becomes altogether unable to speak just before the fit starts.) Other points of interest are that he is illegitimate, with parents who adopted him; has never met his mother. That he doesn't get along with his foster parents and doesn't want to live in England, so he is about to become a Belgian citizen; he already lives in Brussels. That he speaks Chinese, Japanese, Spanish, French, Italian, Russian (I think) and that he has just published his first book, nonfiction, about the Great Wall of China. The Book Society in England has recommended it. That he has been in Hong Kong, where he studied at the University and became a Taoist, and in Vietnam, where he became a Quaker.... Don began by not altogether trusting him and I wasn't sure, either. We ended by rather liking and respecting him, and we both kissed him when saying goodbye.

On the 23rd, Jim Gates left the monastery and went to stay with Peter Hirschfeld, the ex-monk, who is now married to Deepti, the ex-nun. Jim kept talking to us about his unfitness for being a monk and about his lust. I never really took this seriously but it now appears that he meant it quite literally. He was terribly bothered because he found Larry attractive and also Peter Hirschfeld and also a young Belgian who has been around the center lately. No sooner had he settled into the Hirschfelds' apartment than he went out to a neighborhood bar, The Rusty Nail—ironically enough, the Hirschfelds live on Hayworth, just south of Santa Monica

Boulevard, right around the corner from the sex section. Jim found The Rusty Nail absolutely delightful, not one bit sordid, and he went off with a guy who propositioned him. The guy wanted to be tied up and beaten and Jim obliged, and Jim even enjoyed that, although it isn't his thing and the guy didn't particularly attract him.

What does seem absolutely fascinating to me, and to Don, is that Jim, who was always just a bit too holy to be true, is now reacting in what must surely be regarded as an absolutely sincere no-shit manner, instead of moaning that he's a sinner, as one might have expected. With the result that he seems far more spiritually "advanced" than he did while he was a monk.

Swami, of course, was terribly upset, so were Krishnananda, Chetanananda and Asaktananda, who heard of it somehow and phoned Jim, begging him to go back to the monastery. Incidentally, Jim said that he feels Krishna is very different since his illness. "He's letting it all come out." He talked to Jim at length, very seriously. And he allowed Jim to take the dust of his feet, when they said goodbye. Nobody has ever done this before, except by some trick or other.

October 5. Since I last wrote, I've been struggling doggedly on with chapter 12 of my book. Today, I don't exactly know why, I seem to be aware of the first signs of that breakthrough which nearly always comes when a book approaches its end. But I do work much more slowly nowadays and my mind isn't as sharp, except for a very short period in the morning, usually during meditation. My meditation consists chiefly of worldly thoughts, not sexual, not of hatred against anyone, just of trivial chores. But I think I maybe make more japam each day than I used to. I keep reminding myself whenever I can of my approaching death and calling upon Mother to show me that she is with me and to be with me in that hour. I would say, in general, that this is certainly one of the happiest times of my life. I suppose more people than one would suppose have the experience of happiness in old age and also the experience of happiness with another human being, but I can't say that I know many such people. I wouldn't say that I am enormously confidential with Don, that we talk everything over all the time—no, it isn't like that. But we are *continually exposed* to each other; we are no longer entirely separate people; and I do suspect that we communicate deeply during sleep.

Jonathan Fryer went away somewhere but returned here and we discovered that he had been staying with Bill Scobie or rather with

Tony Sarver, in his spare bedroom. Jonathan called Bill in the first place because he had got the clap and wanted to know where there was a free clinic. Then he called me because he wanted to contact John Rechy. And then he called Don apparently to interview him. So Don said okay, and I'll do another drawing of you. But when Jonathan came, he didn't ask Don any questions at all. Well, at any rate, Don's drawing was brilliant. He does seem a strange secretive creature, though. Scobie simply calls it being shy. As for me, my distrust isn't absolutely dispelled.

October 23. Am getting into an unhealthily manic state about my book, just because it goes so slowly. I worry about it all the time, especially during morning meditation. I tell myself that it is dull and that my mind has lost its edge. And that—least valid excuse of all for worrying—I may suddenly get sick and never finish it. Now I must absolutely pull myself together and resolve to plug on calmly, set no deadlines, let the work take its own time—without regard for my anxieties, or the demands of the publishers (who aren't as yet making *any* demands), or the progress of the books by Messrs. Finney[1] and Fryer (who have both promised to wait for me).

After all, I'm not doing so badly. Have just begun chapter 13, and there are only two more after that. True, there are inserts to be written into the earlier chapters, but not so many.

The day before yesterday, I went to a question-and-answer session at Cal. State University, Long Beach.[2] It was part of their Gay Pride Week. Bill Scobie and Tony Sarver came with us; Bill is to use it in a profile of me for *The Advocate.* I was quite okay, I think. But what mattered much more to me, Don was drawn into the discussion by a would-be provocative question about our domestic life and he answered so clearly and sensibly and amusingly, explaining how, in the early years, he was struggling against his image as "Mrs. Isherwood." I can see that he could become a really good speaker, if he had to.

Jim Gates has now got a job looking after Rudolf Sieber, Marlene Dietrich's husband, who has had a stroke.[3] It is very hard work, all around the clock, but he is to get fifty dollars a day. He wants to save this and go to Europe and India with Toby Pollock,

[1] Brian Finney, see Glossary.

[2] Called "GayThink."

[3] Sieber (1898–1976), a Czech, married Dietrich in 1924, after giving her a small role in *Tragedy of Love* (1923), a German film for which he was a production assistant. They lived together for only about five years (and had a daughter), but the marriage was never dissolved. He was dying of cancer.

one of the Vedanta congregation, and two of his friends. In the meanwhile, poor Jim, with beginner's unluck, has got clap—or anyhow a discharge from his penis and a sore spot in his throat. He is to see a doctor as soon as he has a day off. I had to cross-examine him about his symptoms—in order to be able to reassure him—and he told me that he has fucked and been fucked, several times. He certainly didn't waste any time, after leaving the monastery, getting into the thick of the action!

November 14. Today I finished what was originally chapter 13. I had to break it into two chapters because the new draft of it was becoming so long—forty-two pages instead of twenty-six. But that's all to the good. So now there will be sixteen chapters in the book instead of fifteen. I do feel I am really getting someplace at last, because this next chapter, about the China journey, has a much more clearly defined form and the material will arrange itself, I hope, more easily.

Also, I will try hard to make more entries in this book. Diary writing always seems to give me a perspective on my work; it's a form of meditation. But one's always mysteriously unwilling to meditate, in any fashion.

Have just finished Tennessee's *Memoirs.* A lot of it is silly and/or sloppy but he impresses me greatly. He seems so nobly vulnerable. It's strange how this career seems a failure story, despite his extraordinary triumphs. I do think he is brutally underrated, nowadays. Even if you maintain that he has only written, say, six really good plays, is that so little? Who, now living, has done better?

November 27. Happy Thanksgiving—I hope. We are going to a party given by Truman Capote, always a dubious proposition because he knows so many awful people. But, with us, he is very sweet and warm; even more so, nowadays.

Poor old Larry Holt is in hospital again, because he kept falling down in his apartment and hurting himself. I begin to think he can't last much longer. He said, "My painkiller pills are pacifists—they don't want to kill the pain."

Belur Math has at last promised to send over a swami. I think his name is Swahnanda.[1] But he is only to come for ten years. Then another swami will replace him. Swami is furious about this.

[1] Spelled Swananda; in the end, Swananda became head of the Vedanta center in Berkeley, and Swahananda, head of the Berkeley center and with a confusingly similar name, took over in Hollywood. See Glossary.

Jim Gates is still nursing Rudolf Sieber but thinks he will probably quit soon. Peter Schneider gossiped to a woman member of the Vedanta congregation about the awful haunts of vice which Jim is frequenting. So now everyone at Vedanta Place knows. Jim wasn't mad at Peter. He said he was glad that it was all out in the open.

Peter Pears has been over here. He gave a concert at Schoenberg Hall on the 20th which we went to. We also had him to lunch. He seemed just the same as ever, only rather magnificently older looking. Don said he liked Peter better than anyone he's met for a long long time. And I was so happy to be able to end this apparent feud. I only say "apparent" because Peter maintained that there never was any bad feeling on Ben's part against me, only against Wystan. But Ben did refuse to see me, when I wanted to introduce Don to him in England. I think Peter was just covering up for Ben.

We also much admired the harp playing of Osian Ellis, Peter's co-performer at the concert. Ellis was obliged to stay with two queens at Marina del Rey because he was using a harp which belonged to one of them. He didn't like them and longed to escape, but couldn't. Don said, "Home is where the harp is." Also at the concert was Leslie Wal[l]work, about whom Don said, "He knows the facts about everything and the truth about nothing."

Next week, he leaves for Seattle, to be there for the opening of his show.[1] I'm staying here.

Am at least halfway through the second draft of chapter 15. One more to go, after that. I suppose it's still possible that I might get the final draft more or less finished by March.

December 15. Dialogue between the Animals, late at night, both of them rather drunk, tearful.

Dub: I can't bear to think that I must leave you.

Kitty: Neither can I—that's what makes it so exciting ... But if we could ever be parted, we should never have met.

I don't exactly know what Kitty meant. But one thing I do feel, very strongly. *This was no shit.* At times like this, I get quite a spooky sense of his spiritual insight.

It now looks like I shall finish the second draft of the book before Christmas.

The party Allan Carr (an agent)[2] and John O'Shea (Truman's

[1] December 4 at the Dootson-Calderhead Gallery, 311½ Occidental Street. Bachardy flew up with Nick Wilder, December 3, and did two commissioned sittings during the trip.

[2] Also a manager, promoter, and producer (1941–1999) of the film musical

friend) gave for Truman at the Lincoln Heights Jail. We were finger printed and mug-shotted on arrival. The jail isn't used any more and is quite clean. You had to go into the cells to pee. It was all subtly in bad taste—the young actors dressed up as cops and inmates. You were very conscious of the bars in the background, everywhere. O'Shea wanted to have a fake electric chair, but Truman vetoed this as going too far. He was probably the only person present who has witnessed a real execution. How I hated it all. We left early, after having been dumped down at a place-carded table with bores.

Tennessee, whom we spent the afternoon with yesterday, was quite admirable. Such a dynamo of energy. He was here for a production of *The Night of the Iguana*. He shot off to San Francisco to see one of his new plays which is in rehearsal; then he goes to Vienna to see the opening of another.

December 31. This is just to round off the year—a very poor diary year and a hard but slow work year. I finished the second draft of the last four chapters of *Wanderings* on the 19th; which means that I still have a great deal of revision to do. There's no question—the existence of Fryer and Finney and the prodding knowledge that they are getting along much quicker with *their* books is a depressing rather than a stimulating influence upon me. I get the jitters, think "I'll never finish" and give way to anxious sloth.

When we last went to see Swami, on the 26th, to bring him a crystal drinking glass for his birthday, he amazed us by at once peering at Don (his sight is now very poor) and saying, "There's something different." He had noticed what the vast majority of people have entirely failed to notice—including Don's parents—that Don has cut his moustache off. I still think it's a great improvement and makes him look younger.

Which reminds me of a remark of Don's which made me laugh a lot—and which could have been said by no one else I know. He had been to see the very bad film *Fanny* (with Leslie Caron and Horst Buchholz). When he got back he told me: "I sat through it out of sheer morbidity."

Am now reading the fourth volume of Byron's letters. I really cannot say why I keep on at them. I hardly remember anything I have read—any phrase or any fact—but I somehow like to breathe in his ambiance. Particularly while I am shitting.

Grease (1978) and the stage musical *La Cage aux Folles* (1983), among others.

Jim Gates tells me that a copy of *The Advocate* with Bill Scobie's piece about me talking at Long Beach[1] has fallen into the hands of the nuns and that Abhaya, in particular, was terribly shocked. What she so much deplored was that I had talked about Swami and Vedanta. And one of the reasons she deplored it was that she was afraid all the queers would now start coming to the temple! As if they hadn't started thirty years ago! I felt very self-righteous about all of this and quite prepared to defend my position—in the temple itself, if necessary. But Abhaya, when we actually met her next, on Swami's birthday, couldn't have been friendlier. What hypocrites they are, even the better ones.

Jim is still looking after Rudolf Sieber, but now he shares the work with another ex-monk, Clark(?)[2] and only does three days at a time. This gives him a lot more opportunity to go to The Rusty Nail and make new contacts. He still seems very happy about this and not in the least repentant. He looks much more cheerful and healthier and is free with hugs of affection, but he has grown a straggly moustache which doesn't suit him.

At the end of last year, I wrote about my great happiness with Don and my consistently poor standard of meditation. Both of these have continued throughout 1975. I ought to be more concerned about the meditation, I suppose; and yet I do keep at it and I ask Mother every day for more devotion and at least some sense of her presence. I do feel, I think, quite sincerely that she is my "refuge." And I suppose this is good, since it is balanced by the quite other sort of "refuge" I find in Don. What I am trying to say is that it is doubtless easier to feel that God is the only refuge when you don't have any human being to love and be loved by. But I do. And yet I can still say that even our relationship is within the refuge of Mother. More and more I see that it would be unthinkable to have such a relationship unless the other partner in it shared this belief.

[1] Published December 17, 1975.

[2] Clark Waterman, a monk for a year or two; his wife, Elin, was renamed Kamala by Prabhavananda.

Day-to-Day Diary, January 1—July 31, 1976

[*Isherwood spent the first part of 1976 finishing* Christopher and His Kind, *and he gave up writing in his diary for seven months. Printed below are the entries from his day-to-day pocket diaries for this period.*]

January 1. 154.[1] We went to parties at Billy Al Bengston's and Tony and Jean Santoro's. Ted Pebworth and Claude Summers came to see us. (Claude had just read a paper on *A Single Man* at the MLA convention in San Francisco.) Don and I ate at home.

January 2. 153¼. Started revising *Wanderings.* We saw Swami. We had supper with Jess, Glade, and Ted Bachardy at El Coyote. We went with Ted to see Kubrick's *Lolita* and part of Kazan's *Splendor in the Grass.*

January 3. 151¾. Worked on revision of *Wanderings.* Ran on San Vicente. We ate at home.

January 4. 154. Drove with Dharma Das and Urbashi to Santa Barbara. I read Swamiji's "Practical Vedanta—1"[2] at the temple and we drove back. We had supper with Jo Lathwood—Roger Rosenberg and Marylee Gowland were there.

January 5. 155. Revised *Wanderings.* Ran on San Vicente. Tom Shadduck and his friend Eric Moore pruned trees with power saw. The people opposite (de Cuir) sent for the police. Tom and Eric had drinks with us later. We ate at home.

January 6. 152 ¼. Revised *Wanderings.* We had Tony Richardson and Francis Fry[3] to dinner; Don cooked. Tom Shadduck worked

[1] His weight.
[2] For this lecture and others mentioned in the day-to-day diary below, see *The Complete Works of Swami Vivekananda.*
[3] Son (b. 1955) of Richardson's friend Jeremy Fry; educated at a Rudolf Steiner School in England. He had a small part in the Anthony Shaffer thriller *Absolution* (1978), directed by Anthony Page, but never turned up on the set.

on pruning. Saw John Kniest of Pelican Records about doing an album for them, of readings from my books.

January 7. 154. Don painted Rick Sandford. Tom Shadduck worked on pruning. I saw Dr. Wolff about the inflammation on my leg. To gym. Worked on *Wanderings.* We had supper with Vincent and Coral Price. Roddy McDowall and Paul Anderson were there.

January 8. 152¾. Worked on *Wanderings.* Don painted Ralph Williams. Christian Gollub[1] came to see me. Don and I saw John Huston's *The Man Who Would Be King.* We had supper at Pancho's Family Café.

January 9. 152¾. Worked on *Wanderings.* Stephen Miles came to see me. Don had supper with his parents. I ate at home.

January 10. 151½. Tom Shadduck and Eric Moore worked on pruning. Talked to Mrs. O'Hilderbrand[t] about removing bush. To lunch given by Olivier. (Maggie Smith, Warren Beatty, Dustin Hoffman, Coral Brown[e], Roddy McDowall, etc.) Then we had drinks with David and Beth Schneiderman and met Oliver Stallybrass.[2] We ate at home.

January 11. Worked on *Wanderings.* We ate at home.

January 12. 152¼. To Dr. Allen for cholesterol test. To gym. Worked on *Wanderings.* Don's car being repaired. We had supper with Chris Wood and Paul Sorel at the Bellevue.

January 13. 152. Jogged on San Vicente. Worked on *Wanderings.* Taped readings for Pelican Records. We ate at Casa Mia.

January 14. 153. Worked on *Wanderings.* Stathis Orphanos came by, worked on my bibliography, drove me to Vedanta Place. Don and I saw Swami. I ate at monastery, read in temple with

[1] Christian-Albrecht Gollub, German-born American scholar, translator, poet, and, later, jewelry designer; he published an interview with Isherwood in a German periodical *Litfass: Berliner Zeitschrift für Literatur*, December 1978.

[2] English literary scholar; he edited the Abinger critical edition of Forster's works with Elizabeth Heine and translated Knut Hamsen from Norwegian assisted by his wife, Gunvor.

Chetananananda. Don picked me up. We went to Little Club, heard Paul Anderson sing.

January 15. Warm. 153¼. Leak in V.W. gas tank, took it to Pico Auto House. Worked on *Wanderings.* Picked up V.W. With Don and Ted Bachardy to see *Lucky Lady* (Liza Minnelli, Robby Benson) and part of Ingmar Bergman's film of *The Magic Flute.*

January 16. 151¼. Warm. Natalie Leavitt cleaned house. Tom Shadduck worked on pruning. Worked on *Wanderings.* Don had supper with his folks. Ate at home. Started sore throat.

January 17. 151¾. Very warm. Throat sore. Worked on *Wanderings.* Natalie Leavitt cooked and we had Michael and Pat York, Hope Lange, and Ralph Williams.

January 18. 153½. Throat better. Don drove me to Santa Barbara. I read "[The] Methods and Purpose of Religion" by Vivekananda at the temple. We had lunch with Brad Fuller[1] and Jack at their house. Jaime Green, Margaret Mallory. Drove home in fog. Ann McCoy came by with Klaus [Kertess][2]. We had supper with Tony Richardson and Bob Newman at Patrone's.

January 19. 152¾. Worked on *Wanderings.* We saw Hitchcock's *Frenzy* and part of John Schlesinger's *Midnight Cowboy.*

January 20. 152¼. Roger Rosenberg came to talk to us about our Hilldale property. Don drew and painted Mark Andrews. Worked on *Wanderings.* Jogged on San Vincente. We ate at home.

January 21. 152¼. Worked on *Wanderings.* We had supper with Jo Lathwood at Casa Mia, then saw *The Hindenburg* (George C. Scott, William Atherton). Don drew and painted Ian Whitcomb. Lee and Grace Prosser came to see me with their daughter Mandy.

January 22. 151¾. Worked on *Wanderings.* Jogged on San Vicente. We ate at Pancho's Family Café then saw *The Duchess of Malfi* at the Mark Taper Forum (with Eileen Atkins).

[1] Husband of Romney Tree and former companion of Hugh Chisolm with whom he appears in *D.1.*

[2] New York art critic, curator, author; he wrote books, catalogues and introductions on de Kooning and others. In 1966 with John Byers, he founded the Bykert Gallery, which he ran for a decade.

January 23. 151¼. We went to Vedanta Place for Swamiji's breakfast puja at 7:00 a.m. and then had breakfast there. We picked up Michael York on our way back. Don drew him. Don went to see Joe Chaikin's production of *Electra.* I worked on *Wanderings.* Natalie Leavitt cooked and we had Tony Richardson (who leaves for New York tomorrow) and Bob Newman to supper.

January 24. Worked on *Wanderings.* Jogged on San Vicente. We ate at home.

January 25. 152. Long phone call from Glenway Wescott. Worked on *Wanderings.* We went to a goodbye party for Roland and Janet Flamini, given by Phyllis Nugent,[1] and then had supper with Hal, Peter and Michael Buckley at Hal's house. Don drew and painted Don Cribb.

January 26. 152. Finished revising chapter 13 of *Wanderings.* Jogged on San Vicente. Don drew George Raya.[2] Natalie cooked and we had Stephen and Jean Garrett and Roland and Janet Flamini to supper.

January 27. 152¾. Worked on *Wanderings.* We saw *The Story of Adele H.* (François Truffaut). We ate at home.

January 28. 152¼. Worked on *Wanderings.* We met Mark Andrews on Hilldale and talked about our property and what to do with it. To Vedanta Place, saw Swami, had supper at the monastery, read with Chetanananda. Don drew Bill Bopp and fixed supper for him. I saw Bill when I got home.

January 29. 151¼. Worked on *Wanderings.* Don out with his parents. I ate at home.

January 30. 151¾. To see Dr. Thomas Sternberg about my skin. Jogged on San Vicente. Worked on *Wanderings.* Natalie Leavitt cleaned house. We ate at home.

[1] Wife of journalist and documentary maker, John Nugent, a reporter for *Newsweek.* The Nugents lived in Santa Monica.

[2] Gay activist (b. 1950); he lobbied for A.B. 489 in Sacramento, and in 1977, with a health brief, was part of the first-ever gay and lesbian delegation admitted for meetings at the White House.

January 31. Worked on *Wanderings.* We ate at home. To gym.

February 1. Worked on *Wanderings.* We went to a party given by Paul Millard and Rob Matteson for Estelle Winwood's ninetieth birthday.[1] We had supper with Nick Wilder at Au Petit Joint. Fell and hurt my back.

February 2. 153. Worked on *Wanderings.* Mrs. Leavitt cooked and we had John and [Ch]ristel Boorman, Peter and Clytie Alexander, Rospo and Barbara Pallenberg[2] to supper. We went to the vespers for the Brahmananda puja. Saw Swami afterwards.

February 3. 154½. Worked on *Wanderings.* Arthritis in back. Ate at home. Don out to openings.

February 4. 153¼. Rain. Worked on *Wanderings.* Back bad. Supper at Le Petit Moulin with Billy Al Bengston, Penny Little, Robin and Jessie French. Then with Penny to see Jan Michael Vincent in *White Line Fever.*

February 5. 153½. Rain. Worked on *Wanderings.* Natalie Leavitt cooked and we had Joan Houseman, Renate Druks, Perry and Karen King, Jack Larson and Jim Bridges to supper.

February 6. 153½. Rain. Saw Dr. Sternberg. Worked on *Wanderings.* We went to the gym. We had supper with Avilda Moses—Sam Francis,[3] Ann McCoy, Lynda Benglis[4] and Douglas Bergeron were there.

February 7. 154. Rain. Worked on *Wanderings.* Don saw Alex

[1] British stage actress (1883–1984), perhaps celebrating a few years late. She debuted on Broadway in 1916, made a small number of films—including *The Misfits* (1961), *Camelot* (1967), *The Producers* (1968), and *Murder by Death* (1976)—and often appeared on T.V.

[2] Rospo Pallenberg wrote the original story for Boorman's *Exorcist II: The Heretic* (1977) and later worked on the screenplays for *The Emerald Forest* and *Excalibur*; his wife, an actress, published a book about making *Exorcist II.*

[3] American colorist (1923–1994), born in San Mateo, California, educated at Berkeley and at the School of Fine Arts in San Francisco. He worked in Paris in the 1950s, then settled in Santa Monica.

[4] American sculptor and video artist (b. 1941), from Louisiana; notorious for her 1974 ad in *ArtForum* in which she appeared naked holding a dildo at her crotch.

Garcia.[1] Then Ken Handler[2] came with David Hillbrand and Cindy Fusco[3] to show his photographs. Talked on the phone to David Hockney just arrived in L.A. Don and I had supper at the Dar Maghreb with David Goodstein, Christopher Stone,[4] John Moya, Seymour Black.[5]

February 8. 152¼. Rain. Worked on *Wanderings*. We had lunch with David Hockney, Gregory Evans, Nick Wilder at the Bellevue. We went to Tom Wudl's opening at the Mizuno Gallery. We had supper with Newt Dieter[6] and Tom Morton[7]—Joel Wachs,[8] and Paul Revere, Mark Rockland, David Dear. We saw Wayland Flowers with his puppets[9] at Studio One with Rick Sandford.

February 9. 154. Rain. Worked on *Wanderings*. We had supper at George Cukor's—Maggie Smith, Lauren Bacall, David Lean and Sandy,[10] Beverl[e]y Cross.[11]

February 10. 153. Worked on *Wanderings*. We had Philippa Foot[12]

[1] Their gardener, from Mexico.

[2] Musician, playwright, film director (1944–1994), son of Ruth Handler, co-founder of Mattel, who invented the Barbie doll, named after her daughter Barbara, and of Barbie's boyfriend doll, named after Ken.

[3] Hillbrand was directing plays at U.C. Berkeley and UCLA; he and Fusco later married. He sat for Bachardy several times; she sat once. Eventually he settled in San Francisco.

[4] Journalist and author, born in the Bronx and raised in Fresno, then working for Goodstein at *The Advocate*. He later wrote the bestseller *Recreating Yourself* (1988).

[5] A doctor.

[6] Director of the Gay Media Task Force, founded in 1972 to monitor gay themes on network T.V., and an *Advocate* contributor.

[7] Dieter's lover and partner; he sat for Bachardy in 1977.

[8] Lawyer (b. 1939); Los Angeles City Councilman 1971–2000, mayoral candidate in 1973 and subsequently. He wrote the first U.S. law prohibiting discrimination against AIDS sufferers. In 2001, he became president of the Andy Warhol Foundation in New York.

[9] Flowers (1939–1988) was famous for "Madame," a bawdy old woman puppet. They appeared on T.V. in "Rowan and Martin's Laugh-In" and "Hollywood Squares" and eventually on her own show, "Madame's Place."

[10] Sandra Holtz, fifth wife, 1981–1984, of British film director David Lean (1908–1991). His sixth wife, from 1990 to his death, was Sandra Cooke. His award-winning films include *The Bridge On the River Kwai* (1957), *Lawrence of Arabia* (1962), and *A Passage to India* (1984).

[11] Maggie Smith's second husband; see Glossary.

[12] British moral philosopher (1920–2010), educated at Oxford, where she was a tutor for twenty years. She was a professor at UCLA from 1969 until 2002.

to tea, after Don had drawn her. Natalie Leavitt cooked and we had David Hockney and Don Cribb to dinner.

February 11. 154. Worked on *Wanderings.* With David Hockney we went to the Forum in Inglewood to see David Bowie. Met him and Angie Bowie after performance.

February 12. Finished chapter 14 in revision of *Wanderings.* Don painted Bob Dootson.[1] Jeff Bailey and Jim Hansen came to see us. We went to see Paul Anderson in Pinter's *The Homecoming.*

February 13. 151½. Worked on *Wanderings.* We went to gym. We had supper with John and Christel Boorman at the house they have rented—141 North Bristol. Natalie Leavitt cleaned house.

February 14. 153¼. Worked on *Wanderings.* We went to gym. We saw Truffaut's *Two English Girls* and *The Soft Skin.* We ate at home.

February 15. 153½. Stu Brandt drove me to the Santa Barbara temple, where I read [Vivekananda's] "Hints on Practical Spirituality." Worked on *Wanderings.* We went to John Schlesinger's birthday party—Roddy McDowall, Paul Anderson, Elton John, Charlotte Rampling, Warren Beatty, David Hockney, Henry Geldzahler.

February 16. 153¼. Worked on *Wanderings.* We had a birthday dinner at the Century City Clifton's Cafeteria for Glade Bachardy. Ted and Jess were also there. Then Don and I saw Lelouch's *Happy New Year Caper* and *And Now My Love.*

February 17. 152¼. Worked on *Wanderings.* Don drew Renate Druks. Jogged on San Vicente. Took tape of my reading from my books to Pelican Records. We ate at home.

February 18. 153½. Worked on *Wanderings.* Don drew Werner Pochath.[2] Jogged on San Vicente. We had supper with Jo Lathwood and all went to see a rehearsal of *The Belle of Amherst* (Julie Harris).

[1] Art collector and curator, part owner of the Dootson-Calderhead Gallery in Seattle, where Bachardy exhibited in December 1975.

[2] Austrian actor (1939–1993), in German-language films, and in many small Hollywood roles, including *Venus in Furs* (1968) and *Skyraiders* (1976), and on T.V.

February 19. 152¼. Worked on *Wanderings.* We were photographed by Mary Ellen Mark.[1] Don drew Jean Santoro. We had supper with Billy Al Bengston and Penny Little. We watched *James Dean* (Stephen McHattie) on T.V. Strong wind in the night.

February 20. 153¾. Worked on *Wanderings.* Saw Dr. Thomas Sternberg. Went to gym. Don drew and painted Rick Sandford. Saw *Jim, the World's Greatest.* We ate at home.

February 21. 152¾. Worked on *Wanderings.* We had supper with David Hockney, Gregory Evans, Henry Geldzahler, Don Cribb.

February 22. 151½. Worked on *Wanderings.* To UCLA library to do research for *Wanderings.* Don out to see films. We had supper around midnight.

February 23. 152¼. Worked on *Wanderings.* Dar[y]ll Marsha[k],[2] Bobby Littman's assistant, came by to pick up "Sally Bowles" contracts.[3] We saw *Casque d'Or* (Signoret) with Scott Velliquette[4] and Ted Bachardy; then had supper with Scott at Pancho's Family Café, then all went to a party given by Paul Kantor.[5] David Hockney, Henry Geldzahler were there.

February 24. Rain during night. Worked on *Wanderings.* Jogged on San Vicente. We went to a show of film star photographs arranged by John Kobal[6] at the Barnsdall Park Gallery ("Dreams for Sale"). We had supper with Roddy McDowall and Paul Anderson—Nina

[1] American photojournalist and portrait photographer (b. 1940), educated at the University of Pennsylvania; she contributed to *Life*, *Rolling Stone*, *The New Yorker*, and *Vanity Fair.*

[2] Later a powerful agent, branching out into management, production and packaging.

[3] Evidently old ones, for Littman to review so he could understand the rights situation.

[4] A 1970 graduate of U.C. Riverside, where he majored in history, starred at water polo, and attended Isherwood's lectures; he also wrestled and weight lifted competitively.

[5] Art dealer (b. 1919, New York); he ran a Los Angeles gallery until 1966 where, as early as 1952, he showed Motherwell, Rothko, and de Kooning. He was Diebenkorn's first dealer, from 1950.

[6] Austrian-born Canadian film critic, author, collector (1940–1991); he mounted major exhibitions from his personal archive of images of Hollywood stars and drew on the archive for his more than thirty books about American entertainment.

Foch, her husband Michael Dewell, Sidney Guilaroff.[1]

February 25. 152. Worked on *Wanderings.* Saw Larry Holt at St. Vincent's Hospital. Saw Swami, had supper at the new monastery, read with Chetanananda in temple. Don and I went (separately) to party given by Stanley and [Elyse] Grinstein.[2] David Hockney, Gregory Evans, Don Cribb, Jack Larson, Jim Bridges, Chuck Arnoldi, Ed Moses etc. were there.

February 26. 151¼. Worked on *Wanderings.* With Don, Bill Scobie, Tony Sarver to UCLA, where I spoke in the upstairs lounge of Kerckhoff Hall as part of Gay Awareness Week celebrations. We ate at home later. At 6:00 p.m. we went over to talk to Mr. and Mrs. George Caldwell about the building of Don's studio.

February 27. 152¼. Worked on *Wanderings.* We went to gym. Natalie Leavitt cleaned house. We had supper with Michael and Pat York, Rick McCallum, David Hockney and Don Cribb at Ma Maison, after having been refused service at L'Hermitage because David didn't have a jacket.

February 28. 152¾. Sea-fog. Worked on *Wanderings.* We went to see Ted Michel (Tat Twam Asi) who gave me a ring.[3] We ate at home.

February 29. 154. Worked on *Wanderings.* Charlotte Rampling, her husband Br[y]an [Southcombe], her son Barnaby, David[,] Barnaby's nurse and Mary their secretary came by for drinks. We ate at home.

March 1. 152¼. Rain. Worked on *Wanderings.* David Hockney came and drew us. Don had lunch with him. Don and I ate at Pancho's Family Café, then saw Pabst's *Pandora's Box.*

March 2. 151½. Rain. Worked on *Wanderings.* George and Phila Caldwell came by for drinks. Paul Chin and Chris Gaynor drove me to Fullerton State University where I talked to the gay students,

[1] English-born Hollywood hairstylist (d. 1997), long at MGM.
[2] Co-founder, with Sidney Felsen in 1966, of Gemini Graphic Editions Limited, and his wife, an architect. Isherwood wrongly wrote her name as "Aileen."
[3] Vedanta devotee (1928–1981) and cellist, one of the "Venice gang"; he ran a jewelry store in Santa Monica with his wife Lavanya (Felicia Michel).

met Kurt Yount, was driven home by Bud Schindler. Ate at home with Don.

March 3. Worked on *Wanderings.* We went to gym. We went to the temple for vespers of the Ramakrishna puja, saw Swami. We had supper with Rick Sandford at El Adobe. We went with him to Northridge State University and saw part of *Hamlet*, with Jon Voight.

March 4. Worked on *Wanderings.* Natalie cooked and we had George Cukor, Whitney Warren,[1] Tom Crema, David Hockney, Mark Lancaster and Nick Wilder to dinner.

March 5. Heard that Chris Wood had a stroke. Broke my upper partial and had it repaired by Jim Kearin. Don and David Hockney drew Nick Wilder at Gemini. Don and I had supper with Jess, Glade, and Ted Bachardy at Clifton's Cafeteria. Don and I got naval uniforms from Western Costume Co[mpany] for Nick's party tomorrow. We saw *Nashville* at the Academy, sat with Harry Brown and Barbara Poe.

March 6. 151½. Worked on *Wanderings.* We had supper at Casa Mia. We went to Nick Wilder's party at the gallery.

March 7. 152¼. [Jon], Anna, and Rachel [Monday] took me to Santa Barbara. I read Vivekananda's "Bhakti or Devotion" in the temple. We drove back. Don and I ate at home.

March 8. Worked on *Wanderings.* To gym. We had supper with Chris Wood and Paul Sorel at the Swiss Café.

March 9. Worked on *Wanderings.* Carl Day[2] and Norm Haley[3] came to talk to us about the Coastal Commission hearing on the studio. We ate at home. Don drew David Dambacher.

March 10. 152¾. Finished chapter 15 of *Wanderings.* We saw Swami at Vedanta Place; I stayed for supper at the monastery and read with Chetanananda.

[1] New York art collector and philanthropist (1898–1986) settled in San Francisco, a long-time friend of Cukor; he was a cousin of the Vanderbilts, and his father was an architect of Grand Central Station.
[2] Isherwood and Bachardy's architect.
[3] Evidently assisting Day; he was later a local real estate agent.

March 11. 151¼. Began chapter 16 of *Wanderings*. Jogged on San Vicente. We talk[ed] to Frank Hotchkiss[1] about the permit to enlarge the studio. Ate at home.

March 12. Jogged on San Vicente. Worked on *Wanderings*. We had supper with Michael Laughlin at Mr. Chow's, with Jim Katz. Natalie Leavitt cleaned house.

March 13. 152. Worked on *Wanderings*. We had supper with Ivan Moffat and Hope Lange Pa[k]ula at Le Petit Moulin.

March 14. Carl Day and Norm Haley to see us about the hearing tomorrow. Worked on *Wanderings*. Natalie Leavitt cooked and we had Lauren Bacall, David Hockney, Billy Al Bengston, Penny Little and John O'Shea (very drunk) to supper.

March 15. We drove with Carl Day to the City Hall at Torrance for the first hearing of our application to the Coastal Commission for a permit to enlarge Don's studio. Norm Haley and George Caldwell spoke. To gym. We had supper at Pancho's Family Café. We saw *Sandakan [8]*[2] at the Academy.

March 16. 152¾. Worked on *Wanderings*. Don drew Paula Sweet. Ate at home.

March 17. We flew to Catalina (by Catalina Airlines) at 9:30 a.m. with David Hockney, Nick Wilder, Joe McDonald,[3] David Dambacher, Gilbert Haacke,[4] and Lee Brevard. The flight takes twenty minutes. We had lunch at Michael John's.[5] Don and I flew back at 3:30 (by Air Catalina). The others returned later. We ate at home.

March 18. Worked on *Wanderings*. Jogged on San Vicente. We ate at home.

March 19. 154. Worked on *Wanderings*. Jogged on San Vicente. We

[1] A lawyer neighbor.
[2] The 1974 Japanese film, sometimes translated as *Brothel 8*.
[3] New York model; he met Hockney in Paris and was a subject of "Friends," a series of prints Hockney made at Gemini that year, and of other portraits. He died of AIDS.
[4] A young, blond friend of Dambacher; he occasionally sat for Bachardy.
[5] A restaurant.

had supper with Charles Hill and Paula Sweet at her studio—with Dorothy and Elizabeth.

March 20. 153¼. Worked on *Wanderings.* Don went to Gemini, drew with David Hockney. We went to a party given by Peter Alexander at his studio, then we had supper with Greg Harrold and his friend Gordon Myers.

March 21. 153¾. Worked on *Wanderings.* Stu Brandt drove me to Santa Barbara and back. I read Vivekananda's lecture on "Discipleship" at the temple. We had supper with Gore Vidal at the Luau.

March 22. Worked on *Wanderings.* We went to the gym. We had supper with Bill Scobie and Tony Sarver at Musso and Frank, then saw *Boy Meets Boy* (David Gallegly).[1]

March 23. 154. Worked on *Wanderings.* Jogged on San Vicente. Natalie Leavitt cooked and we had Leslie and Michael Laughlin, David Hockney, and Don Cribb to supper.

March 24. 155. Worked on *Wanderings.* Digby Diehl took me to the *L. A. Times* offices, where I received a Woman of the Year Award for Anaïs Nin, who is in hospital. Don and I saw Swami. Had supper at the monastery. Read at the temple, with Chetanananda. Don drew David Dambacher and [Gilbert] Haacke.[2]

March 25. 154¼. Worked on *Wanderings.* Don drew with David Hockney at Gemini. Jogged on San Vicente. We had supper with John Schlesinger and Michael Childers.

March 26. 153½. Worked on *Wanderings.* We saw *Taxi Driver* (Robert de Niro). We ate at home. Jogged at gym.

March 27. Worked on *Wanderings.* Don drew with David Hockney. Michael and Pat York came by to see us, with Rick McCallum. We ate at home.

[1] Gallegly played Cinderella in the musical comedy by Bill Solly and Donald Ward.

[2] Isherwood mistakenly wrote Vincent Haacke.

March 28. 153½. Worked on *Wanderings.* We saw *Night Moves* (directed by Arthur Penn).

March 29. Worked on *Wanderings.* To party at Billy Al Bengston's, to watch the Oscar awards on T.V. We talked to Carl Day about the studio.

March 30. Worked on *Wanderings.* Natalie Leavitt cooked. We had David Hockney, Gregory Evans, Nick Wilder, David Dambacher, [Gilbert] Haacke, Michael Kaplan[1] and Don Cribb to supper. Showed them the film of our trip to Europe in 1955.

March 31. Worked on *Wanderings.* We saw Swami. I had supper at the monastery and read in the temple, with Chetanananda.

April 1. We applied for unemployment insurance. Worked on *Wanderings.* We went to gym. Said goodbye to David Hockney and Gregory Evans who leave for Australia tonight. Went to Jo Lathwood's show of flower paintings at the Staircase. We had supper with Rick Sandford and then all went to see Hal Holbrook in *Mark Twain Tonight.*

April 2. 152¾. Worked on *Wanderings.* We had supper with Jo Lathwood. Betty Arizu, Janet Gaynor and Paul Gregory were there.

April 3. Worked on *Wanderings* (now to be called *Christopher and His Kind*) and finished the final chapter, 16. Don drew Greg Harrold. We ran a reel of hom[c] movie films which Don had spliced together. We ate at home. Rain in the night.

April 4. 153. Started revising *Christopher and His Kind.* We went with Harry Brown to see *All the President's Men* at the Academy then we all had supper at the Aware Inn.

April 5. 153¼. Revised *Christopher and His Kind.* I saw Mary O'Hilderbrandt. We had supper at Ma Maison with Harry Rigby and Terry [K]ramer.

April 6. 153½. Revised *Christopher and His Kind.* We went to gym.

[1]Costume designer, from Philadelphia; he won a BAFTA with Charles Knode for *Blade Runner* (1982).

We saw *Lipstick* (Margaux Hemingway) and *Conduct Unbecoming* (Michael York). We ate sandwiches from the delicatessen.

April 7. 151¾. Revised *Christopher and His Kind.* We saw Swami. I had supper at Vedanta Place and read in the temple with Chetanananda.

April 8. Met George Caldwell at his house and talked to him about the studio hearing on Monday next. Revised *Christopher and His Kind.* We went to gym. We had supper at Mr. Chow's with Leslie and Michael Laughlin, Jim Katz, and Michael Webster.[1]

April 9. Worked on revising *Christopher and His Kind.* Jogged on San Vicente. Natalie Leavitt cleaned house. Talked to Candida Donadio and Michael di Capua and dictated copy for catalogue listing of my new book, by telephone. We ate at home.

April 10. 153. Worked on revising *Christopher and His Kind.* Don collected signatures for the studio hearing. Natalie Leavitt cooked and we had Harry Rigby and Bobby Sanchez to dinner.

April 11. 152¼. Worked on revising *Christopher and His Kind.* Don collected signatures. I saw Elsa Laughton in town to get signatures. Don and I to Howard Warshaw opening, then to see Hitchcock's *Family Plot.* We ate at home.

April 12. 152¾. We drove down to Torrance with Carl Day for the public hearing about Don's studio before the Coastal Commission. We had lunch with Carl Day and Norm Haley at Peppy's, having waited all morning for our hearing, which finally took place around 3:00. We won 8–0. Then we talked to Mrs. Jeanne Dobrin[2] about managing the Hilldale duplexes. Showed her Tom Warner's apartment. We saw *Robin and Marian.* We ate at home.

April 13. 151. Worked on *Christopher and His Kind.* We went to the gym. We had supper with Vincent Price and Coral Browne. Hugh Wheeler was there.

[1] Leslie Caron recalls this was close friend Guy Webster (b. 1939), photographer of rock musicians—The Rolling Stones, The Beach Boys, The Doors—film stars, and other celebrities.

[2] A different person than Nellie Carroll, whose real name was Jean Dobrin.

April 14. 153. Worked on *Christopher and His Kind.* Saw Swami. Had supper at the monastery. Swami Swahananda from Berkeley was there. I read with him in the temple.

April 15. 151½. High wind. Worked on *Christopher and His Kind.* We took Greg Harrold and Gordon Myers to supper at the Bellevue.

April 16. Worked on *Christopher and His Kind.* Jogged on San Vicente. Don cooked. We had Gloria Stuart[1] and Catherine [Turney][2] to dinner. Don was sick to his stomach in the night—probably a bad mussel.

April 17. 152¾. Worked on *Christopher and His Kind.* We ate at home. Don drew and painted Scott Velliquette.

April 18. 151¾. Worked on *Christopher and His Kind.* We saw Chris Wood at his house, with Paul Sorel. We picked up James Fairfax and took him to a birthday party at Le Restaurant, given by Barry Krost for his friend Douglas Chapin. One of the presents was to be a portrait of Douglas by Don.

April 19. Saw Dr. Allen. Worked on *Christopher and His Kind.* Alan Wallis (just arrived from England to make a film about us) came by with Chris O['D]ell, his cameraman, and Jane Burkett (his production assistant). [Platypus Films.] We had supper with Alan Wallis at La Cabaña.

April 20. We started filming, in my workroom—Alan Wallis, Chris O'Dell, Jane Burkett, Roger Reid (camera assistant) John Page (sound recordist) and Len Emery (electrician). They filmed Don drawing me. We saw Swami. We had supper with Don's parents and Ted Bachardy at the Tic-Toc. We went with Ted to see [Robert Patrick's play] *Kennedy's Children.* Worked on *Christopher and His Kind.*

[1] American stage and screen actress (1910–2010); she was a film star in the 1930s—*The Invisible Man* (1933), *Rebecca of Sunnybrook Farm* (1938), *The Three Musketeers* (1939)—retired in the 1940s and learned to paint, then returned to T.V. and appeared in *Titanic* (1977).

[2] Broadway playwright, screenwriter, novelist, biographer (1906–1998), long at Warner Brothers; her films include *Mildred Pierce* (1945), *A Stolen Life* (1946), *The Man I Love* (1946) and *Winter Meeting* (1947).

April 21. We got our first unemployment checks. Mine was $104. We filmed at UCLA and 333 East Rustic Road (talked to Bud Hammer, who lives there). Natalie Leavitt cooked and we had the Platypus unit to supper. Worked on *Christopher and His Kind.*

April 22. Filmed at MGM. They filmed Don on the beach. We had supper with Chris Wood and Paul Sorel at the Bellevue. Worked on *Christopher and His Kind.*

April 23. We were filmed in the pose of the Hockney portrait. They filmed my books, photos, Don's posters, etc. Read part of *Christopher and His Kind* on camera. Don cooked and we had Virgil Thomson, Billy Al Bengston, and Penny Little to supper. Worked on *Christopher and His Kind.*

April 24. Filmed on the pier. Natalie Leavitt cleaned house. The Platypus unit finished filming in the afternoon. Said goodbye to them—they leave for England tomorrow. Natalie cooked and we had John Schlesinger, Michael Childers, Barry Krost, Douglas Chapin, Mo McDermott (just arrived from England) and Nick Wilder to supper. Worked on *Christopher and His Kind.*

April 25. Don drew Carol Hines. Worked on *Christopher and His Kind.* We saw David Hockney and Gregory Evans, just back from Australia. We saw *Ping Pong*[, the play] by Rick Talcove; John Williams, Nick Wilder's assistant, was in it. Then Nick took us to supper at Au Petit Joint with John Williams, Mo McDermott, Barry Krost, Douglas Chapin, Susan Scott.

April 26. Worked on *Christopher and His Kind.* To gym. Spoke on first day of Gay Week at USC. Don, Bill Scobie, Tony Sarver, Jim Gates and Warren Neal were there.

April 27. Jogged on San Vicente. Worked on *Christopher and His Kind.* Don painted Rick Sandford. We went to Billy Al Bengston's show at the Corcoran Gallery, then to his supper at the Twin Dragons.

April 28. We collected unemployment checks $104. Worked on *Christopher and His Kind.* We saw Swami. I had supper at the monastery, read in temple with Chetanananda.

April 29. Don painted Don Cribb. I worked on *Christopher and His Kind.* Jogged on San Vicente. We had supper at Casa Mia.

April 30. Worked on *Christopher and His Kind.* Mrs. Leavitt cooked. We had John and Christel Boorman, David Hockney, Nick Wilder, Gregory Evans and Mo McDermott to supper.

May 1. Worked on *Christopher and His Kind.* We went with John O'Shea to see *Baby Blue Marine* (Jan Michael Vincent) and then all had supper at Bellevue.

May 2. Professor Mary Gerhart[1] drove me up to Santa Barbara and back. I read Vivekananda's "Buddha's Message [to the World]" (volume 8). Worked on *Christopher and His Kind.* We saw *Inserts* (Richard Dreyfuss)[. A]te at home.

May 3. Carl Day came by with the lawyer Martin Katz. Worked on *Christopher and His Kind.* We took Bob Chetwyn and Howard Schuman to supper at the Bellevue.

May 4. Finished writing *Christopher and His Kind.* We had supper with Chris Wood and Paul Sorel at L'Ermitage.

May 5. Got *Christopher and His Kind* xeroxed. We saw Swami. We had supper with Jack Larson and Jim Bridges at their house.

May 6. Mailed *Christopher and His Kind* to Candida Donadio. We saw Martin Katz. Natalie Leavitt cooked and we had Bobby and Doris Littman and Bob Chetwyn and Howard Schuman to supper.

May 7. Jim Bridges showed us bits of his new picture *9.30.55* at Universal. We talked to Carl Day and Walter Winslow[2] (contractor) about starting building. We had Jim Gates and Warren Neal to supper at El Adobe. Natalie Leavitt cleaned house.

May 8. We drove to Laguna Beach to see Jack Fontan and Ray Unger, stopping on the way for Don to visit Beverly Baerresen.

[1] Professor of Religious Studies at Hobart and William Smith Colleges, Geneva, New York, since 1972.

[2] Not his real name.

Billy Cooper was staying with Jack and Ray. We had supper with them at Tortilla Flats and drove back late.

May 9. Don cleared out the studio with Peter Schlesinger. We saw Paul Sorel, went to the opening of David Hockney's show at Nick Wilder's. Then we had supper at Patrone's with Peter. Then went with him to Maurice Tuchman's[1] house where there was a party for David.

May 10. Heard that Larry Holt died his morning. Don drew Douglas Chapin. Don out with his parents. I ate at home.

May 11. Walter Winslow came to prepare work on studio and carport. Candida Donadio called to say manuscript of *Christopher and His Kind* had arrived. Ms. Jeanne Dobrin came to talk about Hilldale property. We went to party given by Truman Capote and John O'Shea at their new house.

May 12. Collected second unemployment check $208 (two weeks). Walter Winslow started work on studio. Don and I saw Swami. I had supper at monastery and read in temple with Chetanananda. Don saw Chris Wood and Paul Sorel.

May 13. Perry Young and David Kopay[2] came to see us. We had supper with John and Christel Boorman.

May 14. Candida Donadio called to say she had read *Christopher and His Kind.* We saw Lee Brevard. Went to gym. We saw Chris Wood and Paul Sorel. We went to party given by Frank Mc[Ca]rthy and Rupert Allen[3] for Marguerite Lamkin.

May 15. We had lunch with Marguerite Lamkin at Beverly Hills Hotel. We saw Swami. We ate at home.

[1] Curator (b. 1936, New York) of the Los Angeles County Museum. His books include *American Sculpture of the Sixties* (1967). He later organized the 1988 Hockney retrospective which travelled to the Met and the Tate.

[2] Kopay (b. 1942) was a running back for the San Francisco Forty-niners and other National Football League teams until 1972; in 1975, he was the first professional athlete in America to come out publicly. Perry Dean Young, non-fiction author and playwright, was co-writing *The David Kopay Story* (1977) with him, a best-seller.

[3] Allan (1913–1991), a journalist until 1955, handled public relations for Marilyn Monroe and other major stars.

May 16. We saw *The Missouri Breaks* (Jack Nicholson, Marlon Brando) at the Academy. Don drew. We ate at home.

May 17. 153¾. Don drew Warren Neal. I showed Jim [Gates] details of the house, which he and Warren will move into when we go. We had supper with Barry Krost, Doug Chapin, John Osborne, Jill Bennett,[1] Peter Finch and [Eletha] Finch[2] at Le Restaurant.

May 18. Did birthday shopping. We saw Chris Wood and Paul Sorel. We went to gym. Went to party given by Barry Krost and Douglas Chapin for John Osborne and Jill Bennett. Later we had supper with David Hockney at Mr. Chow's—Henry Geldzahler, Peter Schlesinger, Nick Wilder, Dwight[.]

May 19. To Dr. Kafka. Did various chores in preparation for journey. Don out with his parents. I ate at home.

May 20. To Dr. Kafka. Had supper with Jo Lathwood at Casa Mia.

May 21. We had supper with Ted Bachardy at Casa Mia. He drove us to the airport. We left at 10:30 p.m. by TWA for New York. Jim Gates and Warren Neal are to stay in the house while we are away.

May 22. We landed in New York (Kennedy) at 6:30 a.m. We stayed at the Chelsea, room 611. Slept several hours. We saw Jack Larson, ate at El Quijote. We saw *The Lady from the Sea* (Vanessa Redgrave, Kipp Osborne, Allison Argo). Then to birthday party for the nurse of Vanessa's son.

May 23. We had breakfast with Jack Larson at Virgil Thomson's apartment. I was interviewed by Carolyn Heilbrun. We saw *Pacific Overtures.* We moved from 611 to Jack Larson's room 814; he had left for Los Angeles. We had drinks with Bill Caskey at Bobby

[1] Osborne's fourth wife; see Glossary.

[2] London-born Australian actor of stage and screen (1916–1977) and his third wife; his films include *The Heart of the Matter* (1953), *A Town Like Alice* (1956), *Far from the Madding Crowd* (1967), *Sunday Bloody Sunday* (1972), and *Network* (1976), for which he posthumously won an Academy Award for best actor.

Ossorio's[1] house. We had supper with Mike Van Horn and Brian Campbell.

May 24. We had lunch with Candida Donadio at Fontana di Trevi. We had supper with Roger and Dorothea Straus[2]—Candida, Michael di Capua, Robert Giroux,[3] Susan Sontag,[4] Barbara Epstein.

May 25. Had lunch with Ed Mendelson, Joe LeSueur and Don at El Quijote, then Ed read through the parts of my book which refer to Auden. We had drinks with Paul Sorel at the Plaza. We had supper with Ron Holland, Joe LeSueur, J.J. Mitchell, Paul Schmidt,[5] Brad and Jan at the Four Seasons.

May 26. We saw Mike Van Horn and Brian Campbell. Went with them to art galleries in Soho. We had supper at La Grenouille with Terry [K]ramer, her daughter, Ton[i], and Harry Rigby. Lunch with Anita Loos at Russian Tea Room. Started a cold and cough.

May 27. We had lunch with Michael di Capua at Pete and Jerry's, after being shown around the Farrar Straus offices. We had supper with Mike Van Horn and Brian Campbell at the Empire Diner where J.J. Mitchell works.

May 28. We had lunch at Bob Regester's apartment on Sutton Place with his friend Vaughan.[6] Stephen Spender came by and I gave him *Christopher and His Kind* to read. We went to a party given by Stephen Spender at the Coffee House Club—Vera Stravinsky, Bob Craft, etc.

[1] Robert Ossorio (1923–1996), ballet dancer and patron, born in the Philippines, son of a wealthy sugar refiner; he founded the Manhattan School of Ballet and co-founded Manhattan Festival Ballet.

[2] Straus (1917–2004), co-founder and longtime chairman of Farrar, Straus and Giroux, Isherwood's new American publisher, and his wife.

[3] (1914–2008), Straus's partner since 1955, and from 1964, chairman.

[4] American philosopher, critic, novelist, playwright, screenwriter, director (1933–2004). She contributed to *The New York Review of Books*, *The Nation*, *Partisan Review*, *The New Yorker*, *The Times Literary Supplement*, *Granta* and others. Her books, including the next, *On Photography* (1977), were published by Farrar, Straus and Giroux.

[5] Poet, playwright, actor, essayist (1934–1999), born in Brooklyn. He was a professor of Slavic Languages at the University of Texas at Austin, translated French and Russian poetry, then returned to New York to write and translate for the stage and act in avant garde theater, on T.V., and in film.

[6] Vaughan Edwards (b. 1955), Welsh stage and film set designer working in the U.S.

May 29. Don had lunch with Harry Rigby. I lunched at the Algonquin Hotel with Stephen Spender and Keith Milow. Then to see *Knock Knock* with Don, produced by Harry Rigby, directed by José Quintero. Drinks with Harry afterwards. Don and I took the revised manuscript of my interview with Carolyn Heilbrun to her apartment and had supper with Vera Stravinsky—Bob Craft, Ed Allen and Barbara Epstein were there.

May 30. We went with Anita Loos and Paulette Goddard to see *The Man Who Fell to Earth.* Then we all had lunch at La Crêpe. We saw Virgil Thomson at the Roosevelt Hospital. We went out to the airport. Our BOAC plane was to be two hours late, so we switched to Pan Am, which was also very late. We took off at 11 p.m.

May 31. We arrived in London about 10:15 a.m. We went to stay in Neil Hartley's house—31 Maida Avenue. Neil left soon after we arrived to drive back to the *Joseph Andrews* location. We slept for a while, then had supper with Bob Chetwyn and Howard Schuman at Eaton's.

June 1. Cough bad. Lunch at home. We had supper with Francis Bacon, Lucian Freud and John Simmonds at Wheeler's on Old Compton Street.

June 2. Cough bad. Lunch with David Hockney at Sheekey's—Peter Schlesinger and Nikos Stangos. To British Museum. To David's studio at Pembroke Studios, to Peter's studio and to see the new studio which is being built for David. We had drinks with John Lehmann and Tommy Urqu[h]art[-]Laird. We saw *La Grande Eugène*[1] with Peter and Eric Boman.

June 3. Cable from Anandaprana to say that Swami is very ill. We had lunch with Peter Grose of Curtis Brown at Our Mutual Friend. Phoned Jim Gates in Santa Monica about Swami. We saw *Otherwise Engaged* by Simon Gray. To party given by John Lehmann—Rosamund Lehmann, John Morris,[2] Eric White.[3]

[1] The play by Frantz Salieri, at the Round House.
[2] BBC radio director; in 1952–1953, John Lehmann broadcast a literary magazine for him called *New Soundings*, on the Third Programme.
[3] Eric Walter White (1905–1985), poet and musicologist. He worked at the Arts Council, produced many volumes of verse and critical works about music and poetry, including studies of Benjamin Britten and Stravinsky which Lehmann published in the late 1940s.

We had supper with Bob Chetwyn and Howard Schuman, with Andrew Brown.

June 4. We had lunch at Canaletto. We saw Pasolini's *Canterbury Tales.* We went to Patrick Procktor's exhibition. We had drinks with Robert Medley and his friend Gregory [Brown]. We had supper with Neil Hartley at Eaton's.

June 5. Got a letter from Michael Barrie to say Chris Wood died on May 31. We saw *The Bed Before Yesterday* by Ben Travers. We talked to Joan Plowright backstage. We had supper at Spot Three with Neil Hartley, George Thomas and his friend Linda.

June 6. Brunch with Peter Schlesinger, Eric Boman and George Lawson. We went to see Amiya Sandwich and her sister Joy. We saw Pasolini's *Arabian Nights.* We saw Paul Dehn, Jimmy Bernard, Ken and José. We had supper at home and watched Jill Bennett in "Hello Lola" on T.V.

June 7. Warm. We saw *Aces High* (Peter Firth and Malcolm McDowell). We saw Richard Buckle and David Dougill. We saw *The Other Side of the Swamp* (Royce Ryton) at the King's Head Theatre, Islington. We had supper with John Schlesinger. Michael Childers, Barney W[an], Drew Okun,[1] at The Grange. Saw Lauren Bacall and Cecil Beaton at the restaurant.

June 8. Very warm. I went to see John Lehmann to talk about his book and mine. We went to Howard Schuman to see part of his "Rock Follies" on T.V. Tim Curry[2] was there. We had dinner with Marguerite Lamkin. David Hockney was there.

June 9. Very warm. Drove with David Hockney to have lunch with the Beesleys, leaving at 10:30 a.m. and arriving 12:20. We left for Aldeburgh at 3:00, arrived 5:00. We stayed at the White Lion Hotel. We had supper with Benjamin Britten and Peter Pears

[1] Gay porn star (1952–1992), born in Natick, Massachusetts, settled in Los Angeles; his professional name was Al Parker. He died of AIDS.
[2] British actor, singer, composer (b. 1946), Stevie Streeter in "Rock Follies"; he was in the original London cast of *Hair*, played Dr. Frank N. Furter in *The Rocky Horror Show* (1973) and in *The Rocky Horror Picture Show* (1975) and on Broadway, Tristan Tzara in Stoppard's *Travesties* (1975), and Mozart in *Amadeus* (1980).

at The Red House—Pauline Princess of Hess[e].[1] We saw *Paul Bunyan* at the Maltings.

June 10. We left with David at 9:30 for Stow-on-the-Wold, arriving 1:30 (via London)—Tony Richardson was filming *Joseph Andrews* there. In the afternoon we saw Blenheim, drove into Oxford and punted on river. We stayed at the Hillside Farm, Adlestrop, with Tony, Neil Hartley, David Ford,[2] Bob Newman. Peter Firth was also at supper.

June 11. We left with David at 7:45 a.m. and drove to Edinburgh, via Leicester (The Fosse Way), Durham, where we stopped to see the cathedral and have lunch, and Berwick-on-Tweed. We arrived at 5:45 p.m., stayed at Caledonian Hotel. We had supper at the Café Royal, ate haggis.

June 12. We went with David to the National Gallery in the castle, had lunch with George Lawson at the Café Royal; then saw the portrait gallery, drove up Arthur's Seat and all had tea with George's mother,[3] aunt,[4] and Evan Crui[c]kshank.[5] Supper with David, George, Evan, and George's uncle, Robin Stark,[6] at Chez Jul[es]. Then Robin took us to see the Water of Leith,[7] then to see his apartment.

June 13. Left at 7:45, driving via Glencoe, Fort William (lunched Alexandra Hotel), Urquhart Castle on Loch Ness, to Inverness. Back via Blair Atholl, Perth (supper at Station Hotel), Forth bridges, arriving back in Edinburgh 10:30 p.m.

[1] British-born Margaret "Peg" Geddes, Princess of Hesse and by Rhine (1913–1997), wife of Prince Ludwig. He was a cousin of Lord Harewood, who introduced them to Britten in the early 1950s, beginning a close friendship.

[2] American assistant on *Joseph Andrews*, then in his twenties; he went into publishing, became head of Little Brown's Books for Young Readers in New York and later a Managing Director of Walker Books U.K.

[3] Gladys Lawson.

[4] Peggy Stark, her unmarried sister.

[5] Student teacher (b. 1948, Montrose, Scotland), then at Moray House preparing to teach English as a foreign language. As an undergraduate reading history at Edinburgh University, he had lodged in Mrs. Lawson and Miss Stark's basement flat.

[6] Playwright and local theater owner; he and his wife, novelist and playwright Christine Orr, helped to start the Edinburgh Festival in 1947.

[7] I.e., the large stream flowing through the suburbs of Edinburgh to the port of Leith.

June 14. Left Edinburgh at 10 a.m. Visited Lindisfarne Castle on Holy Island, lunch at Stable (White Swan) near Warenford, walked on Roman wall near Housesteads (Cuddy's Crags)[.] 7:00 reached Keswick, stayed at Royal Oak Hotel, took motor boat on Derwent Water.

June 15. We left Keswick at 8:00 a.m., saw Wordsworth's Dove Cottage. Drove via Hawes to Haworth—bad road over the fells above Semer Water. We had lunch at Ponden Hall (original of Thrushcross Grange) with Roderick and Brenda Taylor.[1] We saw the Brontës house at Haworth. Saw David's mother at her home in Bradford. Then to Thornway, Highlane, arriving 6:15, to stay with the Bradleys and Richard. (David and Don drove back to Bradford, where they stayed the night and saw David's father in hospital.)

June 16. Rain, on and off. Walked down to the bank to cash traveller's check. Watched T.V. with Richard, Dan and Evelyn Bradley. (Meanwhile David Hockney drove Don back from Bradford to London. Don returned to stay at 31 Maida Avenue.)

June 17. Left by 12:20 train from Stockport, arrived Euston 3:00. Rejoined Don at 31 Maida Avenue. We had supper with Patrick Woodcock; Keith Vaughan, Edward Albee and his friend John.[2]

June 18. We went to see John Cullen at Methuen's about my book. We saw Franju's *The Sin of Father Mouret*. We had supper with Francis Bacon at Wheeler's in Old Compton Street, met Lucian Freud and his son.[3] We all went to Muriel's[4] then to an after-hours bar.

June 19. We had drinks with Beatrix Lehmann at her home; then had supper at Odin's with David Hockney, George Lawson and Wayne Sleep. Peter Lang[a]n was there.

June 20. David Hockney and we had lunch with Stephen and Natasha Spender. Don drew Wayne Sleep. Then Don joined David

[1] I.e., in *Wuthering Heights*; the Taylors ran it as a weaving center and a bed and breakfast.
[2] Longtime friend who was a painter, now dead.
[3] Alexander "Ali" Boyt (b. 1957).
[4] I.e., Muriel Belcher's bar, The Colony.

and me for drinks with Ron Kitaj. We and David had supper with Patrick Woodcock. Peter Coni[1] and John Brewer[2] came in later.

June 21. Don drew Stephen. We went with him and Natasha Spender to have lunch with Marguerite Lamkin. Dudley Moore,[3] Tuesday Weld (his wife)[,][4] Mary Dunlop[5] were there. Don drew Natasha. We had supper with David Hockney and Peter Schlesinger at Tandoori's, then walked in Hyde Park, then watched some films at David's studio. David Chambers measured Don for a suit.

June 22. Very hot. We had lunch with Alec Guinness at L'Étoile. We went to National Portrait Gallery and met Dr. Hayes.[6] We saw Nick Furbank. We had supper with Andrew Brown. Howard Schuman, Leo Madigan, Graham McDonald[7] and Ron Davis,[8] Neil Kennedy[9] were there.

June 23. Very hot. To Platypus Films to see the film Alan Wallis made about us. We had lunch with him at the Inigo Jones. Jane Burkett was there. David Hockney picked us up in his car at 5:30 and drove us to John Gielgud and Martin Hensler's house, The South Pavilion, Wotton Underwood, where there was a dinner party.

[1] Barrister and, later, Queen's Counsel (1936–1993), from New Zealand. He rowed for Cambridge University and became Chairman of Henley Royal Regatta and a member of Britain's Olympic Committee.

[2] A friend of Coni, known as "Mrs." Brewer; Charles Bronson's chef in Hollywood. He died of AIDS in the early 1990s.

[3] British actor, comedian, musician, composer (1935–2002), first known in *Beyond the Fringe*; he appeared often on T.V. and, later, made films, including *10* (1979) and *Arthur* (1981).

[4] New York-born actress (b. 1943), in *Play It as It Lays* (1972), *Looking for Mr. Goodbar* (1977), and others. His second of four marriages, her second of three.

[5] Wife of Scottish art dealer and author Ian Dunlop, whose books include *The Shock of the New* (1972); he worked at Sotheby's in New York in the 1970s.

[6] John Hayes (1929–2005), the director.

[7] Aka Graeme McDonald (d. 1997), British T.V. executive. He produced BBC's "The Wednesday Play" and its successor "Play for Today." From 1981, he was head of BBC drama, then, 1983–1987, controller of BBC 2.

[8] McDonald's longtime companion, perhaps a lawyer, now dead.

[9] British actor and aspiring playwright, now dead; he had small roles in Derek Jarman's films *Sebastiane* (1976), co-directed with Paul Humfress, and *Jubilee* (1977), and in one episode of "Rock Follies." His T.V. play "The Judy Queen" was never produced. Later, he performed in porn in Los Angeles.

June 24. Very hot. Press conference at National Portrait Gallery—Young Writers of the Thirties exhibition. We had lunch with Richard Buckle at La Scala. We lay in the sun in the garden. We went to the opening party of the Thirties exhibition (saw John Auden,[1] Mrs. C. Day-Lewis,[2] William Coldstream, etc.). We had supper at Wheeler's with Robert Medley, Gregory Brown, Peter Schlesinger.

June 25. Very hot. Got plane tickets made out. Don drew Mrs. Mary Montague. We had supper with Neil Hartley at Spot Three.

June 26. Very hot. We went to *Watch It Come Down* (John Osborne). We saw Julian Jebb. We went to a party at David Hockney's. Peter Schlesinger, Eric Boman, Nikos Stangos, George Lawson, Wayne Sleep, Maurice Payne, Norman Stevens,[3] David Plante, Celia Birtwell.

June 27. Very hot. Patrick Cain[e][4] (Neil's friend) arrived to stay. We had lunch with Sir Alfred and Lady Dee Ayer. 3:20 left Waterloo for Isle of Wight. To stay with Edward and Hilda Upward at 3 Hill Street, Sandown. We arrived at 6:10.

June 28. Very hot. Don drew Edward. We walked with him along the cliffs. We left Sandown at 5:46, arrived Waterloo 8:57. Had supper with Marguerite [Lamkin and her companion. He] was leaving for Ulster in the morning.

June 29. A little cooler. We had lunch with Michael Laughlin and Jim Katz at Mr. Chow's in Knightsbridge. We had tea with Verity Lambert[5] and Andrew Brown at the Ritz. With Don to National Portrait Gallery, where Stephen Spender and I spoke. Professor

[1] Cambridge-educated British geologist (1903–1991) and, from 1960, an official of the World Health Organization. He worked mostly in India and retired to London in 1970. Older brother of W.H. Auden.

[2] British actress Jill Balcon; see Day-Lewis in Glossary.

[3] British painter and printmaker (1937–1988), fromYorkshire; he studied with Hockney at the Bradford School of Art and the Royal College of Art.

[4] Los Angeles school teacher, a regular member of Tony Richardson's tennis foursome.

[5] British T.V. and, later, independent film producer (1935–2007); she launched BBC's "Dr. Who" in the early 1960s and, from 1974, was controller of drama at Thames T.V., where she oversaw production of "Rock Follies" and "Edward and Mrs. Simpson" for Brown, as well as "Rumpole of the Bailey" and others.

R.D. Smith[1] was the chairman. Then there was a dinner given by Dr. John Hayes at the gallery's branch at Carlton House.

June 30. We went to the National Gallery, had lunch with John Lehmann and Tommy Urquhart-Laird. To see *Three Sisters* directed by Jonathan Miller; left at intermission. To National Gallery. We had drinks with Waris Hussein—Ian Goodjohn was there. We had supper with Patrick Woodcock—Rosamond Lehmann and Brian Masters[2] were there.

July 1. Jim Gates phoned to say that our house had just been burgled. Don drew Freddy Ayer. Dee (Wells) Ayer drove us to the Star and Garter at Putney, where we met Dicky Buckle and went with him to Hampton Court. On the way back we saw Nancy [Ackerley] West. We had supper with Stephen and Natasha Spender. Baron Philippe de Rothschild,[3] Andy Warhol.[4]

July 2. David Chambers brought Don's suit. We had lunch with Erik Falk at the Reform Club. We talked to Frank Konigsberg about Gertrude Stein film.[5] We went to the National Gallery. We went to see Amiya Sandwich. We went to see David Hockney—Anne Upton, her son Byron and her friend, David [Graves].[6] We saw *Blithe Spirit.* Supper with David Hockney and Peter Schlesinger at Odin's.

July 3. Left London at 9:30 a.m. by British Airways, arriving Tangier 11:45 p.m. Gavin Lambert met us and took us to stay with him at the Villa Tingitane. He took us to lunch with Peggy

[1] Reginald Donald Smith (1914–1985), Professor of Liberal Studies at the New University of Ulster, ex-communist friend of Louis MacNeice. He formerly wrote and produced drama for BBC radio.
[2] British biographer and journalist (b. 1939). In the 1970s, he wrote about British royals and aristocrats; later, he focused on serial killers.
[3] The French winemaker (1902–1988), also a poet, translator, playwright, screenwriter, stage and film producer.
[4] Warhol designed the label for the 1975 vintage of Château Mouton Rothschild.
[5] Independent American T.V. producer (b. 1933), of specials for Bing Crosby in 1976 and 1977, then serials, mini-series and made-for-T.V. movies. The Gertrude Stein project did not go forward.
[6] Upton, née McKechnie, divorced British painter Michael Upton (1938–2002) in 1964, and their son Byron died at sixteen in 1982. Her second husband, David Graves, managed Hockney's London studio and did research and writing for him. Hockney painted Anne and David Graves several times.

Hubrecht. David Herbert and Adolfo de Velasco were there. Later we had a drink with Gavin's friend Mohamm[e]d Cherrat. Then we had supper at the Nautilus.

July 4. We drove with Gavin to Asilah and had lunch there, then returned to Tangier. About 5:00 p.m., Jim Gates called to say that Swami died, early today. Later we visited Paul Bowles, Mrabet[1] and Dan Bente[2] were there. We had supper at Gavin's.

July 5. Saw about plane tickets home. We had lunch at Chuck Pringle's house in the Casbah with Noël Mostert, Peggy Hubrecht, Adolfo de Velasco. Supper with Gavin at the Nautilus.

July 6. David Herbert showed us his house. We had lunch with Marguerite McBey.[3] Peggy Hubrecht was there. At 5:30 we left with Gavin to drive to Chaouen and arrived there at 7:45. We had dinner and stayed at the Parador Hotel.

July 7. We left Chaouen at 10:15 and drove to Fez, arriving there at 1:15. We had lunch at the Tour d'Argent. Stayed at the Grand Hotel. In the afternoon we visited the Medina.

July 8. We left Fez at 7:00 a.m. We lunched at Hotel Chems outside Beni Mellal. Arrived Marrakech at 4:30. Stayed at Grand Hotel Tazi. Walked through the souks and had supper at restaurant overlooking the market place.

July 9. We left Marrakech at 9:30 a.m. We had lunch at the Auberge Savoyard outside Rabat. Arrived back in Tangier at 6:00 p.m. We had dinner at the Nautilus with Gavin, Peggy Hubrecht and Noël Mostert.

July 10. We flew by Iberia to Madrid, leaving Tangier at 11:00 a.m. and arriving an hour later. Stayed at the Ritz. We spent the afternoon at the Prado. We had supper at the Ritz.

[1]Mohammed Mrabet (b. 1936), Moroccan storyteller and painter; Paul Bowles transcribed some of his stories and translated them from Moghrebi into English, including *Love with a Few Hairs* (1967), *The Lemon* (1969), *M'hashish* (1969), *Harmless Poisons, Blameless Sins* (1976), *The Big Mirror* (1977).

[2]Young American teacher at the American School in Tangier.

[3]American watercolorist (1905–1999), born Marguerite Loeb to a wealthy Philadelphia family, widow of Scottish painter and printmaker James McBey (1883–1959), with whom she settled in Tangier in 1932.

July 11. We went to the Prado again in the morning. We left Madrid by TWA at 1:50 p.m., arrived New York at 3:00 p.m. (N.Y. time). We stayed at the Chelsea Hotel, room 814. We saw *Murder by Death* (Truman Capote). We had supper at the Sheraton Pavilion coffee shop.

July 12. We had lunch with Bob Regester at the Rotisserie. We saw *The Tenant* (Roman Polanski). We had supper at El Quijote. We saw *Chicago* (Gwen Verdon and Chita Rivera).

July 13. We spent the morning in the Metropolitan Museum of Art. We had lunch at the Fontana di Trevi. We left New York by TWA at 5:00 p.m. and arrived Los Angeles at 7:00 p.m. Jim Gates and Warren Neal met us at the airport.

July 14. We talked to Carl Day about the studio building. We went to the gym. Ate at home. We had breakfast before 6:00 a.m. at Zucky's.

July 15. We went to gym. Don had supper with his parents. I had supper with Jo Lathwood at Casa Mia; Don joined us later.

July 16. Special worship and homa fire for Swami. Swamis Asheshananda (Portland), Shraddhananda (Sacramento), Sarvagatananda (Boston), Swahananda (Berkeley), Prabuddhananda (San Francisco), Adiswarananda (New York), and Bhaskarananda (Seattle) were present. Don and I both went.

July 17. Don and I drove up to the Santa Barbara temple, where the swamis and I spoke at a memorial service for Swami. We had lunch at Oxnard on the way back. We had supper at home.

July 18. To Vedanta Place for the memorial service. Don and I came later. The swamis and I spoke. I had lunch at the monastery. We had supper with Paul Sorel (who is now living in Chris's former house) at the Brown Derby. He gave us a tape recorder and cufflinks.

July 19. 154½. Went through accounts, to bank, etc. We saw *Stay Hungry* (with Jeff Bridges), went to gym, ate at home. Started watering plants. Saw Dr. Glenn Dayton about my sore left eye.

July 20. 152. We ran down to the beach, went in water. Don bought new T.V. set—a nineteen-inch Sony.[1] We had supper with Billy Al Bengston and Penny Little. Karen Carson[2] was there.

July 21. We re-registered for unemployment payments. We went to the gym. We had supper with Jo Lathwood. Anne Baxter was there. Saw Dr. Dayton.

July 22. 153½. We went to gym. Greg Harrold came and we had supper with him at Casa Mia. Don went with him to see films. Natalie Leavitt cleaned house.

July 23. We went to a dinner party for Tennessee Williams at the house of Mr. and Mrs. Jerry Hellman[3] in the Malibu Colony (68). Jack Larson came with us. Some of the other guests were John Schlesinger, Joan Didion and John Dunne, Leonard Rosenman.[4]

July 24. 152½. We had supper with Brian and Marlene Finney at their house.

July 25. 152¼. We had lunch with Marti Stevens. Joseph Cotten, Patricia Medina,[5] Jean Feldman[6] and Tony Santoro were there. We stayed home in the evening.

July 26. 150¾. We ran down to the beach, went in water. Don had supper with his parents. He and I went to see *The Sailor Who Fell from Grace with the Sea* (Kris Kristofferson, Sarah Miles).

July 27. 151½. We had supper with Marti Stevens at her house. Saw John Kniest at Pelican Records about recording. Don drew me for the jacket of *Christopher and His Kind.*

[1] To replace the one taken by the burglars, July 1; only replaceable items such as the T.V. and cameras had been stolen.

[2] American artist (b. 1943), educated at the University of Oregon, UCLA and Claremont; she showed at the Cirrus Gallery in the 1970s and, from 1979 onward, at Rosamund Felsen and taught at California State University at Northridge, 1972–1977, and, later, UCLA.

[3] Jerome Hellman (b. 1928, New York), talent agent and T.V. and film producer, including *Midnight Cowboy* (1969) and *The Day of the Locust* (1975) for John Schlesinger.

[4] American composer (1924–2008), from Brooklyn. His many T.V. and film scores include *East of Eden* (1955), *Rebel Without a Cause* (1955), *Barry Lyndon* (1975, Academy Award), *Bound for Glory* (1976, Academy Award), and *Star Trek IV: The Voyage Home* (1986).

[5] The British actress and Cotten's second wife; see Glossary under Cotten.

[6] I.e., Jean Santoro.

July 28. 152¼. Don drew me for the jacket of *Christopher and His Kind.* Saw Chetanananda at Vedanta Place. We ate at home.

July 29. 151¼. We ran down to the beach and went in the water. Brian Finney came to see me. We took Jim Gates and Warren Neal to dinner at the Bellevue.

July 30. 151¾. We ran down to the beach and went in the water. We ate at home and watched *Mr. Lucky* on T.V.

July 31. 151¾. We ran down to the beach and went in the water. We saw Ingmar Bergman's *Face to Face* (Liv Ullmann) with Billy Al Bengston and Penny Little, then we all had dinner at the Masukawa. Don drew Doug Chapin.

August 1, 1976—June 9, 1980

August 1. We have settled down, more or less, into the routine of discomfort caused by the building—or better say, work in suspension. Don's new studio's still only a skeleton but it looks awfully big from the road below; almost bigger than The Casa itself.

Terrific resentment against the architect and the contractor—[...] Carl Day and that slob Walter Winslow with his piggish crew who leave beer cans on the roof and all over the place. (One exception, cute, boyish Gary who works stripped to the waist and has softening smiles to calm our annoyance when he fills the back porch so full of tools that you can't open the closets.) We revenge ourselves by keeping the house doors locked, so none of them can use the bathrooms any more. Our excuse is a lie: that the police have told us we must do this. But we do both strongly suspect that some workman, having cased the place, tipped off friends who were the actual burglars. One's resentment against a burglary is against the intrusion far more than the loss of property. As somebody so rightly said, it's a rape.

Four days of gorgeous beach weather in a row; we've been in the water every day.

Have been timing and cutting the passages I am to record for Pelican Records—they chose the bits from *Sally Bowles*, *A Single Man* and "Anselm Oakes"; the good taste shown by this last choice surprised me.[1]

August 8. Four more days of sea and beach in a row. Jonathan Fryer has just turned up again. He is much more attractive than he used to be; he has the body of a big strong girl. I gave him the typescript of *Christopher and His Kind* to read.

[1] The recording was released in 1976 in the U.S. as "Christopher Isherwood Reads."

A bit of paper has been lying on my desk for months. On it I have written:

> For the homosexual, as long as he lives under the heterosexual dictatorship, the act of love must be, to some extent, an act of defiance, a political act. This, of course, makes him feel apologetic and slightly ridiculous. That can't be helped. The alternative is for him to feel that he is yielding to the compulsion of a vice, and that he is therefore dirty and low. That is how the dictatorship wants him to feel.

I have copied this here so I can throw the paper away. It isn't clearly expressed but it means something important to me.

August 17. Today I finished correcting the first set of proofs. I don't know what I feel about the book. Jonathan Fryer didn't seem exactly shattered by it, but then he's so enigmatic. Bill Scobie called it "a cliffhanger" approvingly, but thinks that I shall be attacked for worrying about my private life instead of the Nazis, which doesn't seem perceptive, since the two were so closely interrelated.

We have now put the Hilldale property up for sale, and so are hoping to ditch our demon tenants. The builders are still very much with us. They block the back porch every night with their heavy dirty tools.

I'm consistently failing to meditate properly. And yet I feel that Swami is very much "there," in a sense that he never was while he was alive. He knows everything now, I say to myself; there is no concealment. And this is reassuring. It puts my "sins" and "impurity" in a proper perspective. What do they matter, as long as I don't forget him?

August 27. So now I'm seventy-two. A happy peaceful birthday with my darling yesterday and a birthday supper in bed; salmon cakes and champagne. I dreamt that I was shot dead by a firing squad—I forget why—and that I then went around telling everybody that there was nothing to it, it didn't hurt and it wasn't at all unpleasant, and I didn't feel any different after it had happened. Was this whistling in the dark?

I think about death more and more—I prod myself into thinking about it; and always the thought brings me to the *almost* certainty that Swami will be present with me, when it comes.

A very attractive young endodontic dentist named Thomas

Rauth, recommended by Dr. Kurtzman, told me today that my upper left first molar won't hold together much longer. It is the anchor tooth for that side of my bridgework. Luckily, there is another molar behind it which is in better condition. Otherwise, I might have to have one of those upper plates which only remain in place by sticking to the roof of your mouth.

Don is out dancing tonight with the Nick Wilder bunch—Gregory Evans, etc. Around 10:30, a young man named Gene Hendrickson called me long distance from Albuquerque. Last March he sent me a piece called "Memories of a Queer Life," which I rather liked, and another called "The Feeling of Negentropy," which bored me. He tried to explain negentropy[1] to me, but I hadn't turned off the T.V. and simply could not concentrate. We spent nearly an hour chatting with great good humor.

September 7. Well, my upper left first molar is out and Labor Day (yesterday) is over, thank God. Now we enter the beautiful fall–winter season of work. I still can't make up my mind whether or not I want to start editing my 1939–1944 diaries, to make a sequel to *Christopher and His Kind*. Don insists that they mustn't be cut, or even much commented on—otherwise, he says, they will lose their flavor. But the trouble is, there are so many passages which could cause offence to people still alive.

It has also come to me that I should write something about Swami, without delay. Not a formal biography, beginning at the beginning. I think I should start now—or, at any rate, with his death—and then keep looking back. The difficulty is that this kind of portrait lends itself to posing; the author is tempted at every point to present himself as the disciple the guru loved, the disciple who betrayed him, the disciple who helped him most, the disciple who was the lowest, morally, of them all and therefore avoided being a hypocrite, like all the others.... Still and all, even as I wrote the above sentences, I began to feel a stirring excitement: Yes, yes, this *is* something. Why shouldn't I at least try it? What can I lose? At worst, it'll be an unusual document. I am bound to say a few interesting things on the subject.

More about this, I hope, in the days to follow.

On the evening of Sunday the 5th, we were about to sit down to dinner with Nick Wilder, Gregory Evans, Mo [McDermott],

[1] A state in which work can be carried out with least waste; characterized by positive feelings, psychological activation, spontaneous motivation, complete concentration.

Mark Lipscom[b], Carlos Sagui and John Ladner, when a terrific thunderstorm burst over the Canyon. I think the telephone pole outside the house was struck; anyhow, it started to spurt a little crackling flame and all our lights went out. They weren't put back on again until yesterday afternoon. When the lightning struck, nobody panicked—thanks to a subnormal state of awareness in some cases, due to drink, pot or nature, and a sanguine temperament in others. Indeed, this rather made our evening, which was lit by candle stumps.

September 20. Don, talking about drawing a portrait: "The whole point is setting up this thing—these two idiots looking at each other. It has all the earmarks of significance and at the same time it's absurd—and it's the absurdity which is *so marvellous*!"

A strange dark, perhaps Asian, woman puts in several sessions of chanting every day, in the bushes on the slope below number 147, Elsa's house. She uses what used to be a hideaway set up by the boys and girls from one of the houses on the other side of Adelaide Drive, to smoke pot in, presumably, and screw. The woman has a suitcase, which I've seen lying on the pathway which runs along the top of the slope, and sometimes she spreads clothes on the surrounding bushes—to air them out, I suppose. She chants, or, rather, shouts loudly enough to be tiresome and distracting when the wind is blowing her voice toward us and we are having our own morning sit. One day, it seemed to me that she was repeating "egotism ... egotism ... egotism." But I couldn't be sure, and certainly the words vary. She may belong to the Nichiren Shoshu or some similar sect which teaches you how to chant to obtain a lover or a Lincoln Continental. I feel inclined to leave a note in her nest, reminding her that Christ told us to pray in secret.

September 22. The Autumnal Equinox. (Just as I was starting to type those words—at about 3:23 p.m.—the velvet case containing the miniature of Richard as a child, which he gave me when I was at Wyberslegh last June, fell off my desk and damaged itself even more than it was already. But the prop, which I hadn't been using, is still intact. So I pulled the doors of the case off and now it looks and stands much better without them. An omen? Of what?)

I am really only writing today because I want to make a token act of work on this auspicious date. There's nothing special to report. At present, a door-installing man is in the studio. We are promised the stairs and the deck next week. Still no plumbing and no electricity. Both Don and I find Walter Winslow the contractor so repulsive that we cannot bring ourselves to look at him while we're talking to him.

I had meant to begin my memoir of Swami today, but that would be just a compulsive gesture. What I will do, until I do actually begin, is to discuss the project with myself, here.

For example: I originally thought I would start with getting the news of Swami's death by phone from Jim Gates at Gavin's house in Tangier, the day after our arrival from London. But I feel that this approach would have a certain vulgarity. Because it would necessarily hit a note of drama.... No, I should begin at the very beginning, quite undramatically. I should have to begin with Gerald Heard and, in fact, follow the line of my diary. I must be shown to have met Swami *through* Gerald—not merely in the sense that Gerald introduced me to him, but in the sense that Gerald presented him, Gerald's image of him, to me. At first, I certainly saw Swami through Gerald's eyes.

Another thing I realize is that I must read right through my diaries—all of them, down to the present day, in order to get an overview. By an overview, I mean a sense of how the relationship between these two people, Swami and me, developed and changed. In this way, I shall probably find out a great deal which I don't know, am not aware of, yet. Okay, good, that's how I'll begin.

October 14. Have just finished reading the two typewritten diary volumes, 1939 through 1944. There is a great deal of good material in them but I still feel it would be a mistake to publish them in their present form. Not only because so many of those written about are still living. Because the material itself is too *dense.* There are so many minor characters whose portraits follow each other boringly, I fear. For example, the people at the Vedanta Center, and the people at the Haverford refugee hostel. One virtue of the material used in *Christopher and His Kind* is that it composed much more easily into the form of a nonfictitious novel. Its major characters are most of them extremely active and there is a sense, all through the narrative, of outside menace—from The Others in general and the Nazis in particular. The major characters in the 1939–1944 material are both contemplatives rather than actives—Swami and Gerald Heard—and there is very little sense of outside menace, even though the war is on during nearly all of the period.

I come back, therefore, to the feeling that I should write exclusively about Swami. But, as yet, I don't know how to do that. I can't strike the right tone. I am certainly not aiming to write a biography. What I should try for is a highly subjective memoir—always stressing the idea that what I am describing is a personal impression, a strictly limited glimpse of a character very different from myself and therefore often quite mysterious to me. Without being fake humble, I should also—even for purely artistic reasons—stress the materialistic, gross, lustful, worldly side of myself—but without making Swami appear merely "better" than me. The real artistic problem is to find a way to do that.

Now that Don's studio has been more or less hammered together, the poor thing looks as crude as Frankenstein's monster; its thick graceless skeleton is held together by oversize bolts. It's quite clear that both Day and Winslow are amateurs. Day, having attached the stairway lights in an unreachable position, actually suggested that we should climb a nearby tree when we wanted to change the bulbs in them. (There was a don't-care impudence in the way he said this.) Winslow is just a slob, trying to get by with his miserable fudging and lack of foresight. What they can't take away from us—unless they build skyscrapers opposite—is our magnificent view, which is about a third better from the upper floor of the studio than it is from the balcony of the house.

October 15. I was talking to Barada on the phone this morning, because I'll be coming up to the Santa Barbara convent on Sunday to read. She told me that they would all like Vandanananda to return here and be head of the Hollywood center. I hadn't heard this before. Barada also told me that Swami, shortly before he died, told Abhaya that it would be a good thing if Vandanananda succeeded him. Previously, he had been determined that Vandanananda *shouldn't* succeed him and had even stated this in his will.

I then asked Barada if she believed that Vandanananda had really had affairs with women in the congregation. She wouldn't quite admit this, but she said that Vandanananda had been denounced to Swami by two women who were jealous of each other because of Vandanananda—adding "where there's smoke there's fire."

The woman who chants in the bushes below us had a passionate solo argument this morning, at the top of her voice, just as it was getting light. Maybe she was addressing her enemies. She kept shouting, "It's not happening, it's not happening, it's not happening!"

October 16. Yesterday afternoon, I spent nearly two hours with Dr. Elsie Giorgi—whom Gavin discovered years ago and the

Boormans more recently—telling her my medical history and being examined, as part of a checkup. I shall hear more when my blood and urine samples have been tested, next week.

I didn't go to her because of the injury caused by my falling down the studio steps, on the 5th, or because of any particular anxiety I have been feeling about my health. But because she appeals to me as a possible successor to Dr. Kolisch—someone to ease me onto my deathbed, when the time comes. (This is the only medical function about which I'm seriously concerned, nowadays. Any fool can "save" a patient; but only one doctor in thousands is capable of "losing" him—seeing him off on his journey in the proper manner. I would trust Patrick Woodcock to do this, but Patrick isn't around and soon he may be retiring.) Actually, Elsie seems to think I'm in good shape, at present, but that's neither here nor there.

She is a nonstop talker, an egomaniac, a show-biz snob and extremely sympathetic. Don is in favor of her, too.

October 17. Last night, we had the Boormans, Tony Richardson, Neil Hartley and Bob Newman to supper. Natalie wasn't free to come and cook so Don fixed a stew which turned out to be a masterpiece, *the* classic lamb stew, followed by a crème brulée which was worthy of it. Tony was a bit aggressive toward John Boorman, his usual competitive attitude toward a colleague, but John took it very well and the evening was a success. Tony talked intelligently about politics, saying that Ford should have balanced his pardon of Nixon by also pardoning all the Vietnam deserters.[1] And he said that China is the only country which is preparing for the future world oil shortage by organizing its local industries not to rely on oil. (I have no idea whether this is true or not.)

Today I drove to Santa Barbara with [one of the boys from the Hollywood center] to give a reading at the temple, and returned with Shelly Lowenkopf. More about this tomorrow.

October 19. Couldn't write anything yesterday because Bob Gordon came down from San Francisco to talk to me about his novel *Leaks*, and the friend who was to drive him back up there after our meeting arrived so late that we had to spend four hours together.

[1] When Ford pardoned Nixon (see September 11, above), he proclaimed that a fair trial was impossible, that Nixon had suffered enough, and above all that he wished "to firmly shut and seal this book" rather than allow a nationally divisive matter to drag on through legal channels. All three arguments might equally have supported a pardon for the deserters.

I like Bob and I quite like this novel—at any rate as a possible movie story. We already showed it to Tony Richardson without telling Bob and he turned it down, but not very decidedly and I think maybe we can sell him on it. Don is reading it now. If he likes it we'll try.

As for the two who drove me to the Santa Barbara temple and back, [the boy] interests me far more than Shelly, who is a Jewish teacher of literature, brimful of energy and enthusiasm.[1] [The boy] is in rather the same situation that I was in, latterly, at the Hollywood center, when I had already decided to leave and was having sex and clap on the side and feeling guilty, simply because I happened to be staying at a religious institution and knew how terribly shocked the others would be if they found out. I think I made [him] feel better—if I did, the trip was worthwhile.

On the 17th, Don made a valuable assertion of his will and set up an important historic landmark by drawing his first sitter in the reconstructed studio (Louise Fletcher[2]) although Doug Walsh was hammering away downstairs.

October 21. Last night, while he was having dinner with us, Tony Richardson announced, with a perfectly straight face, that his one real reason for preferring to live in Los Angeles was that it's the best place on earth to play tennis in.

Today, for the first time in months, I have actually been doing some writing! But it's only the notes for the sleeve of the reading record I made for Pelican.

I saw Elsie Giorgi yesterday morning. She told me that my tests turned out almost perfect for my age. Nearly every count was within the normal limits; I only have too much iodine. This may be due to excessive use of iodized salt or eating too many shrimps. As for the pains in my knee since my fall down the steps, she is trying to cure them with some medicine, Butazolidin, which they give to racehorses to make them run faster—at least, I *think* that was what she told me! She is sort of like a witch; you can imagine her prescribing "finger of birth-strangled babe." Kolisch was like that, too.

October 31. I don't know if the pains are really any better, aside

[1] He was on the faculty at USC for over thirty years, taught at UCSB, Westmont College, and Antioch University in Santa Barbara, published fiction, essays and book reviews, and worked as an editor.

[2] American film and T.V. actress (b. 1934); she won an Academy Award and a Golden Globe as Nurse Ratched in *One Flew Over the Cuckoo's Nest* (1975).

from the natural healing of the injury, but going to a doctor, especially a new one, is in itself a rite of exorcism and I feel better for that.

The last ten days have been a sloth period. Meditation zero. Much energy wasted on aggression against noisy neighbors, including the chanting woman, dogs and Walter Winslow, whose latest contribution has been to borrow our bucket and somehow make a hole in it. Little Tom Shadduck, now parted from his friend Eric [Moore], is still in our favor; he is repotting our plants and cleaning paint off the fronds of the sago palm.

Part of my sloth seems related to that doldrums period which always occurs when I am waiting for the first bound copy of one of my books to arrive. This one is due tomorrow, so maybe I'll get myself back to work again, at least on rereading my diaries.

Doug Walsh continues to work nights in the studio with one or both of his skinny boy helpers, the pretty one and the weird one who pulls compulsive grimaces.

November 21. The doldrums period has prolonged itself, with the excuse that I really can't settle down to any kind of work until I have been to San Francisco and San Diego (November 28–December 1) and to New York (December 7–11, approximately) to make propaganda for my book. There will be more propaganda to be made here in Los Angeles, between these two trips.

So far, I have read only three notices, two of them very short; the American Library Association *Booklist* and the *Publishers Weekly*. The first of these is bland and perfunctory, the second rather hostile: "[A]lways his ultimate measure of the world seems to be sexual." The third notice, written by Penelope Mesic in the *Chicago Tribune Book World*, is quite vicious. It ends: "The big, bold book of fact has become a fairy tale." But its viciousness has a tone which will probably sell a lot of copies and inspire loyal indignation in the hearts of Chicago queers.

November 28. This afternoon we're scheduled to take off for San Francisco. We're not going on to San Diego, though, because the publicity manager, Jay Allen,[1] got the word that three powerful news producers there, "Sun-Up,"[2] the *San Diego Union* and KGO

[1] A Los Angeles press agent organizing the publicity campaign for *Christopher and His Kind*.

[2] A T.V. show.

News,[1] felt that their owners or backers, while agreeing that Mr. Isherwood was indeed a distinguished writer, would veto mention of his homosexuality or discussion of the subject in general!

Well, so be it. As usual, before embarking on a campaign of this kind, I feel passive. (That's probably why I so seldom have stage-fright.) Let the situation carry me wherever it wants to. This is all *maya*. Already, the orgasm of having produced the book itself has left me weak and almost indifferent to its reception—*almost* isn't *quite*, however. I'll probably bristle up if it gets some sufficiently nasty stinging insults.

It helps immensely, having Don with me. On such occasions, I see the whole thing through his eyes, and that makes it interesting and worth experiencing.

December 23. San Francisco was drastic and New York even more so. Both were reassuring, because I found I could hold my own in the rat race. Indeed, I often surprised myself and Don because I was so quick on the uptake during interviews. I am, more and more, ascribing this to the effects of the K.H.3 capsules which I have been taking regularly since November 10. But I couldn't possibly have gotten through the New York trip without Don, who was sustaining me throughout. I have *never* known him to be more marvellous and angelic.

Perhaps the most moving experience was going down to the Oscar Wilde Memorial Bookshop in the village and signing copies of my book, with a line of people, mostly quite young, stretching all the way down Christopher Street and around the corner. I had such a feeling that this is my tribe and I loved them. ("They're beginning to believe that Christopher Street was named after you," Gore said, with his sly grin, half flattering, half mocking, at dinner the night before last. He is dieting, in order to lose twenty pounds. Howard needs to lose thirty, but isn't. I am happy to say that Howard seems ready to forget and forgive the fuss we had about the Jews.)

The night we got back from New York (December 11) our taxi driver (who was young and cute) instantly informed us that a major earthquake had been predicted for December 20. This seemed like such farcical antipublicity for Los Angeles, especially when addressed to two arriving travellers, that we were more amused than alarmed. Still, I was a tiny bit relieved when the 20th passed without the quake. (The explosion of the oil tanker,

[1] A local radio station.

which killed several people and broke windows all over the Long Beach area, *might*, perhaps, have been misread as an earthquake by a precognizing psychic.[1])

December 29. Paul Bailey came down from Oakland to spend a few hours here yesterday. We both liked him. He seemed truthful, intelligent, modest, with a nice funny face and blue eyes, rather fat. When he taped an interview with me, for the BBC—it was at the KABC[2] studios—I noticed that his hands were shaking, just before we started, which seemed touching and sympathetic. But he obviously has plenty of gumption. He has been teaching at a college in North Dakota and told everyone there he was gay. With the result that the prize superstud student came to him and played a drunken scene in order to get Paul to go to bed with him.

Paul told us a wonderful story about Gore being interviewed by some reactionary character, who said: "We're getting far too lenient. We ought to bring back the cat-o'-nine-tails for these homosexuals. (*aggressively*) Don't you agree?" Gore (poker-faced): "Certainly I do. But only for consenting adults." At such moments Gore is in the class of Wilde.

Talking of Wilde, a news announcer on the car radio this morning said, predicting rain: "Our fine-weather picture is turning into Dorian Gray."

And, talking of being too lenient to queers, I got a letter yesterday from the Reverend Paul Trulin, Chaplain of the California Senate, appealing to his fellow Californians to help repeal the recent laws "whose effect have (*sic*) been to legalize adultery, sodomy, homosexuality, and fornication (to name just a few) ... all of which according to the Holy Scriptures are an abomination to the Lord.... Those of us who are fed up with this deplorable rejection of God by our government leaders have formed a committee to replace these law makers and leaders with god-fearing individuals." So he appeals for money to support the campaign of the California Christian Campaign Committee.

How deeply depressing this sort of thing is—such a glimpse of dreary spiritual squalor. This was a form letter of course. I wonder who gave them my name? A practical joker?

[1] An 810-foot Liberian tanker, *Sansinena*, exploded at a Union Oil berth in Los Angeles Harbor on December 17; eight crew and a security guard were killed, twenty-two crew and thirty-six members of the public were injured.
[2] ABC's Los Angeles radio station.

December 30. The predicted rain has fallen heavily and there's more to come. This morning, trying to concentrate on my meditation, I picked up *The Eternal Companion* to get some encouragement, opening it at random. I certainly got some! "After leaving the body the true guru lives on in the invisible realm; sometimes he reveals himself to his disciples, but at all times he helps and guides them ..."

Looking through Evelyn Waugh's diaries (lent me by Ken Tynan, who is now living in the Canyon and said by Tony Richardson to be in a near-suicidal state) I found this note, made toward the end of Waugh's life:

> In all discourses on prayer one is told not to expect an answer or a perceptible sense of nearness to God. Only very rarely are "consolations" given and they are not to be sought or highly valued. "You cannot pray? Did you mean to do so? Did you try? Do you regret failing? Then you *have* prayed." To the sceptic this must be the essence of deception.[1]

Why does the above surprise and move me? Because it seems to prove that Waugh actually was religious in a way that I can understand. In my lofty-snobbish way I have been taking it for granted that his religion was merely a function of *his* kind of snobbery. Now I want to read the whole book or at least much more of it. I think it was a very good sign that people often disliked and were bored by him, even when he tried to be pleasant. That makes his character seem more genuine, less compromised by posing. I have always thought of him as a poseur.

Last night, we had supper with the two men who own the arrestingly angled pink house across the Canyon on Amalfi, designed by Charles Moore. Inside, there are dramatic effects of height; bookcases with inaccessible upper shelves; a staircase which might have been built by Gordon Craig for a production of *Macbeth*.[2] There are also many intimate nooks, charmingly, too charmingly furnished. It is just as much a showplace, in its miniature way, as Chatsworth or Blenheim. One couldn't live in it with any sense of privacy or snugness. The architect's real achievement is to have packed so much drama and variety into such a relatively small

[1] March 26, 1962.

[2] Edward Gordon Craig (1872–1966) first designed *Macbeth* in 1908 for Beerbohm Tree, and his last work for the stage was a New York production in 1928; his models and drawings for Shakespeare, with their stark, vertiginous backdrops, were widely exhibited and reproduced in books.

space. Such houses are far too rare. I would like to have at least a dozen of them in this canyon but I would hate to live in one.

We went on to a show of paintings by a nearly attractive redhead Mark K[e]iserman[1] whose work Don rather admires. Billy Al and Penny were there. Billy told me, "You two are the most distinguished people we know," slapping my back with drunken laughter. I often feel really fond of him. He sends me up a bit with his compliments, but in exactly the right tone.

On the 27th, I finished reading the manuscript of Jonathan Fryer's book about me. I'm afraid it is hopelessly dull. There is no impression of me as a character, although he talked to so many of my friends. Caskey alone could have provided dozens of lifelike glimpses, and Jonathan told me that Caskey was the most interesting of all his informants. You could never guess this from Jonathan's book; it just goes on and on, doggedly listing events, and missing the *point* of them almost always.

December 31. More rain last night and this morning, followed by a rainbow over the Canyon. Don said it was a good omen for 1977.

Last night we had supper at Trader Vic's, with Alan Searle, George Cukor and some friends of theirs. I was trapped in a corner with Alan, who was the host. I am fond of him, largely for old times' sake, but oh dear he is like a wealthy widow. Getting a bit maudlin, he said he hoped he had made Willie happy—also that Willie was so fond of me and of Don. Also that he would send us a copy of Willie's unexpurgated autobiography, including all the homosexual part. Don doubts that he will really do this.

Three parties in prospect tonight—at Jack and Jim's, at the pink house belonging to Lee Burns and John Smith, at Peter Bogdanovich's. Bogdanovich called this afternoon—I don't really know why, unless it was because he has just read my book; we have never met properly—only, according to him, at a showing of *Bringing Up Baby*, at which, he says, I laughed even louder than he did. I don't remember being so amused by the film as all that. Probably it was somebody else.

1977

January 1. This is merely to wish an unworthy hungover old Dubbin a soberer New Year. I woke up so shaky that it took me

[1] He had several shows in Los Angeles in the 1970s before moving to Las Vegas and giving up painting. He sat for Don Bachardy several times.

nearly three-quarters of an hour to figure out which way to attach the new refill to the calendar stand. And even now it looks odd to me, the way I've done it.

I have only dim memories of the Bogdanovich and the Burns–Smith parties. Jim Bridges, with characteristic self-indulgence, failed to cancel their party although he already felt sick in the afternoon. Instead he went to bed and left Jack to cope with us, Zizi Jeanmaire, her daughter Valentine,[1] Ustinov's daughter Pavla,[2] Nellie Carroll and Miguel. He failed to make us jell and nobody raised a finger to help him except Don and me. I can't help it, I do so dislike Frogs.

My darling, in contrast to me, made an admirable start on 1977, writing his diary, running down to the beach and, this afternoon, painting me—the first painting he has done in the studio since its rebuilding.[3]

January 2. Because of yesterday's failure, today must be my symbolic start of the work year. At least I can get on with reading through my diaries. I have three more volumes ahead of me, beginning with the trip to Austria on the "Silent Night" project, with Danny Mann, in September 1966.

Other jobs: Revision of the Avadhuta Gita;[4] continuing the reconstruction of the diaries from the beginning of 1945—thus far, I've reached May 1951 and would like to carry the narrative on until at least the end of 1952.

Last night we had supper with Tony Richardson, Bob Newman and Larry Gonzales at Au Petit Joint. Tony was in good spirits and full of enthusiasm for Jackie Onassis whom he seems to regard with an almost yin-feminine admiration as a yang power-lady, stressing her utter ruthlessness. One pictured him being raped by her. I like Bob much more now than I used to; he is a sex comedian with

[1] Renée "Zizi" Jeanmaire, French ballerina (b. 1924), left the Ballet de l'Opéra with her husband, dancer and choreographer Roland Petit, to star in his company and in Hollywood musicals—*Hans Christian Andersen* (1952) and *Anything Goes* (1956). Their daughter, Valentine Petit (b. 1955), twice sat for Bachardy.

[2] Pavla Ustinov (b. 1954) appeared in some of her father's films, including *The Thief of Baghdad* (1978), and produced, wrote and directed for movies and T.V.

[3] The main objective in rebuilding the studio was to give Bachardy more room to paint.

[4] Isherwood was evidently helping Chetanananda with his translation; see Glossary.

a sadistic act, often very funny and also sexy. Larry is a gracious, softly flattering but quite intelligent young man with certainly some Negro in him. Attended by these two, Tony appears a sultan with an ambisexual harem. (I don't know if he actually goes to bed with Larry, who is described as Neil Hartley's friend.)

Tony is obviously sincerely worried about *Joseph Andrews*; the English distributors are said to hate it. He predicts another "disaster" and laughs nervously and apologetically. Don and I are praying that we like it, when we're shown it the day after tomorrow.

A flat tire on the way home. We had to call the Auto Club to change the wheel. One really should know how to do such things.

Today we may see *Casanova*. I woke up this morning and was disturbed to find I couldn't remember the name of its director, although I knew the names of some of his other films, *La Dolce Vita, The Satyricon* etc.[1] I worry about these seeming advances of senility, and yet I know that it's also a sort of game I'm playing with myself.

January 16. On the 12th, I read at the Vivekananda breakfast puja. Looking through my diaries, not very carefully, it seems that I first read the Katha Upanishad at the breakfast puja in 1960. Or did I first read it in 1944, while I was actually living at the center? I simply can't remember. The only part of the breakfast ritual which I remember clearly from those days is Sister's pouring of the coffee and lighting of the cigarettes.... Maybe Swami didn't introduce the reading until much later. It occurs to me that Swami may have "invented" this role of puja reader for me because he felt that I was spiritually "astray" and knew that the duty of reading would form a link between me and the society for at least one day every year. That would be characteristic of his kind of slyness.

This year, Chetanananda officiated. I didn't feel any powerful spiritual emanation from him, but I did feel one from the shrine itself. If I could have just sat beside it, all by myself, for maybe half an hour, I believe I would have been at least somewhat spiritually refreshed and restored—the publication of my book and the resulting ego-jitters have dulled me.

It was maybe a couple of days after the puja that Jim Gates told me he has heard that Krishna has already had two malignant tumors removed from his face and that there is another which may cause him to go blind, either in one eye or in both. I didn't

[1] Federico Fellini.

actually get to speak to Krishna on the 12th. Jim is going to try to find out more about this.

Yesterday I heard that Anaïs Nin has just died. Spoke to Rupert [Pole] who was perfectly in control of himself and ready to talk. He said it was a great mercy, she had been in terrible pain, on and off, for two years. Now he was going to devote himself to editing and publishing the rest of her diaries, translating a much earlier diary from the French, etc. He said that Anaïs's great fear was not of dying but of being separated from Rupert forever. Sister Mary Carita[1] had come to talk to her about this and had made her believe that it wouldn't be so. I got the impression—I don't know exactly how—that there is some other woman close to Rupert who will help him with the diary editing and generally give him her support.

Jim Gates just called to say that Krishna had this third tumor removed and it was malignant but the doctor thinks his eye is saved.

January 28. Oh horror—at the end of next week, Sunday February 6th, we leave to tour the frozen cities of Rochester, Toronto, Chicago, Minneapolis, where blizzards are still raging. Someone said that there may actually be a 100 degree difference in temperature between here and there! Ordinarily, Rochester would be fun; we are going there to view Pabst's *Diary of a Lost Girl* and to meet its star, Louise Brooks, now poor, arthritic, cranky and as old as I am; also, weather permitting, to dash over for a glimpse of Niagara Falls. Then we fly across the lake to Toronto for one night, then to Chicago for one night, then to Minneapolis for one night, then home. And then—after stopping here for *one day*, we are supposed to fly down to a Pacific coast port near Oaxaca, where Billy Al Bengston and Penny Little are staying! The port is called Puerto Escondido.

I can hardly believe any of this, and I fear drastic psychosomatic sabotage to prevent me from carrying out the plans. My bad thumb is paining me already. And yet maybe old show-horse Drub will respond to the T.V. camera and the mike, if they can ever get him through the snow and ice to them. It will be a severe trial for both the Animals, that's for sure.

February 1. I have just finished reading right through my diaries,

[1] Sister Mary Carita Pendergast (1904–2002), Roman Catholic nun from New Jersey, a missionary in China, 1924–1951. Nin was Roman Catholic.

from the beginning of 1939. There is still one gap I want to fill—never mind how inadequately; from January 1, 1976, until the next entry, on August 1. A lot of this period was just slogging away on the final version of *Wanderings* (I must find out from Don when it was that he suggested and I adopted *Christopher and His Kind* as its title). Then there were the not very interesting legal proceedings and hearing before the Coastal Commission, in order to get our permits to put the upper floor on Don's studio and thereby cut about a foot out of a worm's-eye, street-level view across the Canyon. Then there were the arrangements for starting the construction work and our ill-fated agreement that Walter Winslow should be the contractor. Then there was the making of Alan Wallis's film about us (which is now just about to be shown on T.V. in England, we hear).[1] Then there was our takeoff for New York, England and later Morocco, to stay with Gavin Lambert. And then our return to Los Angeles on July 13, via Madrid and New York.

What is fatally missing from the diary as the result of this gap are any entries about our last few meetings with Swami. I shall also have to describe the two memorial services for him which I did attend and particularly my meeting with Krishna when he gave me back my rosary with Swami's bead added and I took the dust of his feet. I hope that Don will have some details in his diary about all this.

Having read through the diaries, I still don't have any clear idea how I am going to do this book about Swami. No doubt, as usual, I'll have to start writing before I begin to see the way ahead.

Meanwhile, the frozen North awaits us at the end of the week—followed, presumably, by the septic South. I long to have all this behind me and start work—work which I ought to have started months ago. Don, meanwhile, sets me an example by working away in his studio. I feel that he now rather loves it and has gotten used to all its structural defects. A very significant psychological barrier was broken when he first dropped paint on the floor and let it stay there. (He only paints downstairs; upstairs he draws.)

Just one thing I'd like to record now, because it gave me so much satisfaction. At Gore and Howard's big party on the 28th, I ran into Sue Mengers and she told me how marvellous she thinks my new book is. "And you said such nice things about the Jews,"

[1] The half-hour documentary, *Christopher Isherwood: Over There on a Visit*, was broadcast February 5, 1977 by London Weekend Television on "Aquarius," series 10.

she added. For the first time since our famous clash, I felt that everything was really and truly all right between us—and *not* because she was kidding herself that I had had a change of heart or that I hadn't really meant what I said to her then. I felt that, after reading the book, she had put my earlier remarks in perspective and had somehow understood exactly how much I meant by them.

February 18. The frozen North wasn't so frozen after all. Rochester was really the coldest—the only place where my face ached—and that was only after dark. Toronto was merely cool, Chicago was windless for the first time since I began visiting it, Minneapolis was so springlike that the Minneapolitans were positively apologetic, insisting that they were just between blizzards.

Our best experience: the meeting with weird old Louise Brooks and seeing her marvellous young self in *The Diary of a Lost Girl* and *Miss Europe* (*Prix de Beauté*); the view from the window of our room at the Drake Hotel up the Chicago lakeshore at dusk, with the traffic describing a great golden curve and the lake misty blue beyond. And then there was Don's heroic feat; his skiddy drive out from Buffalo to see Niagara under ice. (We parted and rejoined in Chicago because he had forgotten to pack his passport and we didn't want to risk his getting stuck in Canada.) My biggest emotional thrill was my reception at the Gay Academic Union meeting at the University of Toronto. This kind of thing could easily go to one's head; it's what happens to successful old actors.... Oh, and one other thing: the schoolchildren who were watching a television program in Minneapolis being sent out while I was being interviewed and brought back after I had finished.

Now I'm back home and so glad to be; we unanimously decided not to go down to Puerto Escondido and join Billy Bengston and Penny.

On February 12, the day after our return, I formally made a start on my Swami memoir. (I tried doing the opening draft in pencil on Gerald Heard's old writing board, and again it seemed to give forth some power—at least, I scribbled several pages.) The great problem at the beginning is to keep to the point—which also means that I want to reserve all the rest of the material for the published version of my 1939–1944 diaries and not leave too large holes in that book. But, all the way through, the problem will be to keep to my relation with Swami and only let other people appear in the background. I have a feeling that (a) this memoir is going to be very short, maybe not even as long as *Prater Violet*, and that (b) it will speed up terrifically through the later years. Perhaps the best thing about it will be its final passage, a description of me in old age and of what Swami means to me now that he is dead and of how I view my approaching death and of the phenomenon of happiness near the end of life.

A thought—referring back to what I wrote above about my reception at Toronto: the kind of love which young people feel for old figurehead people like me is perfectly healthy, beautiful indeed, not in the least silly and woe unto young people who are incapable of feeling it; they are emotionally lame. *But* it is so important for the old figureheads *not* to take this love personally; to understand that it is simply an effect of the interaction between age groups—to understand this makes it more beautiful, not less.

March 16. Yesterday, I got a card from the UCLA Department of Anatomy which I am to carry in my billfold. It announces that I have willed my body to the department and wish to have it sent there if I drop dead in the street. So now that's taken care of; I'd never have got accepted if Elsie Giorgi hadn't used her influence at UCLA. I think I really began to decide to take this step because of my horror of the mortuary atmosphere which was inspired by a visit to Forest Lawn when Tony Richardson was about to film *The Loved One.* I am very glad that I've done it—although there are occasional absurd qualms at the thought of Dobbin's old pickled carcass hanging up on a hook ready to be carved. But far far stronger is the satisfaction of knowing that there won't be any kind of "resting place" to which my darling might feel obligated to come, on anniversaries, with wreaths. Then, too, I like to think that I'm following Gerald Heard's example; that maybe, even, my ashes will be dumped with his in some communal hole after the cadaver has served its purpose.

And while we're on such themes, today came an announcement of a performance of Edward Albee's "most profound play yet, *All Over*, which deals with the awareness that we are forever alone as we enter our twilight years"!!

March 20. Yesterday, we went to see Tom Stoppard's *Travesties.* Don said, its humor is verbal, not dramatic at all—which meant that the director (Ed Parone) evidently felt he had to eke it out by making his actors skip around and mug and strike attitudes—with the result that the show ran nearly three hours.... Howard and Fran Warshaw and Judith Anderson were there. At intermission, Don suggested that I should ask them to have supper with us later. I hurried over to their seats and did so—the second act was just about to start. Sitting in a row just below them was a pretty elderly woman with carefully fixed white hair in a halo around her head. I thought to myself that she was a typical rich Beverly Hills matron, rather like Dee Katcher (who is again putting the Hilldale property on the market for us).[1] The woman gave me a look which might have meant that we knew each other from somewhere, so I started a tentative smile, then shut it off again, largely because I was in such a hurry. On my way back to my seat, it hit me: "That was Peggy Kiskadden." I hadn't recognized her—perhaps because she had hardly changed, wasn't looking nearly old enough. Don said yes, he was sure it was her. I felt shocked and shaken. It was like suddenly realizing that the thing which had looked like a bead necklace, and which you'd nearly bent down and picked up, had really been a coral snake.

Howard has changed a lot; skinny and sunken-faced and old-Jewish, after his cancer operation. And Judith has become quite tiny and humped over with an ailing back. Only Fran looked just the same, like a powdered silly smiling goose. But they were all very affectionate and full of chatter, and we had a charming supper.

At the end of the week we fly to England. Am dreading this.

But at least I've now made a good start on the Prabhavananda book—more than thirty pages.

April 17. We did fly to England on March 26 and got back here on April 11th. More about that by degrees, maybe. For the moment, what's chiefly important to me is that I was able to get restarted right away on the Prabhavananda book. Now I must simply drive ahead with it until I've gone so far that I *have* to finish.

A couple of days ago, Prabha called from the Santa Barbara convent, with two objectives—to get me to promise to fill in some reading dates during the summer, because Chetanananda is off to India for a three-month pilgrimage (as they call it—may this mean that he is escaping and doesn't intend to return?) and because Swahananda is also leaving for a holiday on the East Coast—and also to get me to convey a warning to the boys at Trabuco that they have got to shape up and face the facts of monastic life and open a bookshop and entertain the public and have lectures and stop being so reclusive and mystical and self-indulgent, or else the property may be sold over their heads and they may be ordered back to Hollywood to *work*. Prabha's silvery southern coo was hateful with malice as she told me this, putting it all onto the will

[1] She was a local real estate agent.

of the Lord and of the Belur Math trustees—as if she gave a blue shit for either when *her* will was involved! So I said I'd think about the reading but could promise nothing, and that I'd speak to the boys if I went down there, as I probably may.

April 18. This morning, Tom Shadduck, who had come by to do our watering, noticed that the tall cactus plant—one of the ones Jim Tyndall[1] gave us before he moved to North Carolina—has been stolen. We both suspected the kids who trespass continually through this property and felt the usual fury. I called George Caldwell, flattered him a bit, and suggested that he should speak to one of his next-door neighbors who is also his tenant and has a lot of teenagers in the house; I just want him scared a bit by the prospect of a scandal, because I blame him utterly for anything his sons and their friends do, it's all this shit-permissive mealy-modern parent stuff. Rather to my surprise, Caldwell was impressed, saying that, "This kind of thing affects all of us." He agreed to speak to his tenant.

A really interesting and horribly depressing talk, last night on T.V., about the approaching oil famine within thirty years and consequent plans for transmitting solar energy via satellites, etc. I got such a sense of a future which I don't want to, and anyway can't, live on into. At the same time, I quite realize that my aversion is merely romantic; I hate to part with the notion of space as something awesome, of the moon as a shining mysterious orb, etc., and contemplate a time when the earth will be surrounded by a sort of backyard full of skyjunk.

Nick Wilder brought down a New York dealer this morning named Robert Miller to see Don's work. Nick says he was impressed. Oh, if only—

April 25. Struggles with letter and check writing have stopped me from making daily entries, as I'd hoped. But I have kept on with the Prabhavananda book—a page a day, which would mean finishing a rough draft around the end of this year, if the book is about the length I expect it to be.

Today, I heard from the other abbess of the Vedanta Society; Anandaprana. She wants me to get started on the recording of the

[1] A generous Isherwood fan, son of a North Carolina state senator; he worked at I. Magnin in Beverly Hills, got bathrobes for Isherwood and Bachardy at cut-rates, found paying sitters for Bachardy, and even offered to be a vigil nurse at Bachardy's deathbed.

Gita. Usual complaints about being overworked. She seemed to be running the place, just as usual. No mention of either of the swamis.

I do hope I can settle down a bit, now, and regain some slight sense at least of—I won't venture to call it "contact with"—just ordinary awareness that the word Vedanta means anything at all. I have been incredibly alienated by these publicity trips and sealed up tight inside my "image."

April 30. Vistas of work are now opening up. On the 26th, we had a talk to George Weidenfeld, who was visiting here, and it seems that he is quite serious about commissioning a book from us—*Drawings and Dialogue* is a provisional title suggested by Candida Donadio; the only objection to it being that there may be a few of Don's paintings as well. I was prejudiced against Weidenfeld for becoming a lord but then prejudiced in his favor by his foreign accent and courtly foreign charm and his appreciation of Don's work.... Also my book about Swami, now at page 53. Also the diaries of the forties and early fifties, to be brought up to 1953, at least. Also this recording of the Gita.

One memory of our English trip came back vividly to me this morning. It was on the night of April 1, which I spent up at High Lane, with Richard. I went into the bedroom in his house where I was to sleep, and opened the wardrobe, and there, hanging all alone, was Frank's red army parade uniform! It was almost as startling as seeing his ghost, especially since I had assumed that Richard, who really hates his memory—not Frank himself, whom he barely knew—would have destroyed or given it away, long ago. Now I come to think of it, I was so shocked that I never even asked him why he had kept it. Must remember to do so.

May 1. Discovered for the first time this morning, from the dictionary, that "mayday," the international radiotelephonic signal for help, derives from "*m'aider*." This suggests a clue for May Day in a crossword puzzle: "Not waving but drowning" and reminds me that we saw Glenda Jackson as Stevie Smith in a play, while we were in London.[1] She did her best, but Smith didn't make much of a character to work with. We also saw three other plays with people we know in them—Alec Guinness as Swift,[2] Gielgud

[1] *Stevie* by Hugh Whitemore. "Not Waving but Drowning" is the title poem in the 1957 collection by British poet and novelist Stevie Smith (1902–1972).
[2] In *Yahoo*, which Guinness co-wrote with Alan Strachan.

as Julius Caesar, John Mills and Jill Bennett in *Separate Tables*; all disappointments.

Suddenly, we are on the brink of selling the Hilldale property. It will be marvellous to be rid of our tenants and their dishonesties and the sheer nuisance of paying bills and keeping an eye on the rents. And yet—we found ourselves sad and a bit apprehensive. There's something painful about parting with *land*.

And this morning Don and I decided to ask John Ladner to arrange Don's adoption for us. He seemed quite pleased to do so. He is very bright and confident, without being smart alecky. Maybe he'll become our "family lawyer."

When I talked to Prabha the day before yesterday, she announced approvingly that *all* the devotees were getting visions of Swami and "feeling his presence." Well, I have neither felt nor seen. Just once, he appeared in a dream, but it was confused and I forget all the details. Does this distress me? No. I keep calling on him. And my mood, a lot of the time, is wonderfully happy.

May 9. The "triple anniversary."[1] Shall I ever make it quadruple?

It seems that the Hilldale property really is going to be sold. We go into escrow today or tomorrow.

All this time, I have kept on with the Prabhavananda book. Reached page 70 today. Of course, this is just jogging; I'm not really getting down to the big narrative problems.

Items from a sex catalogue (Blueboy Products) I got in the mail yesterday: Dr. Richard's Ring (to maintain an erection), Jewelled Cock Rings, Electric Cock Rings, Leather Harness, Extasx (suck vibrator), Anal-egg Vibrator, Big Squirt (the ultimate pulsating dildo) and Big Billy—the sexy doll ("the chicken who never says no.... almost better than the real thing").

June 8. Yesterday morning we went through the adoption ceremony in Santa Monica, before Judge Mario Clinco. It was done in his chambers, not in the courtroom, quickly and politely, with no awkward questions asked. He said "good luck" to us both when it was over, which made the ceremony seem like a marriage. John Ladner probably had a good deal to do with the speed and smoothness of the proceedings. We both like him very much now and think of him as "our" lawyer. We had lunch with him afterwards

[1] Death date of Isherwood's father, Frank, in 1915, of his grandfather, John Bradshaw-Isherwood, in 1924, and of his aunt, Muriel Bradshaw-Isherwood-Bagshawe, wife of Frank Isherwood's older brother, Henry, in 1966.

and drank champagne. And much much more in the evening, to celebrate Billy Al Bengston's birthday. This would have been an auspicious day indeed, if it hadn't been for Anita Bryant's sweeping victory in Florida and her threat to carry her persecution of our tribe into California.[1] I feel dull with forebodings. Courage. My job is to get on with my book about Swami, now at page 118.

July 4. Swami's deathday, his first. This morning, some journalist called to tell me that Vladimir Nabokov has just died. I could give him no comment.

Have got to page 157—more than half, surely, of the first draft? I wonder if it is because I am writing this book that I feel so cut off from Swami's presence.

Have reached a point in the story which is very difficult to handle. It is my meeting with Don. [Vernon] and Caskey I have been able to handle, because [Vernon]'s involvement with Swami didn't last long and Caskey never had any. But Don ended by becoming his disciple. His impressions of Swami and his impressions of my relationship to Swami are of enormous importance to the book; they give me the only opportunity to get a temporary binocular view of the situation.

Aside from working with him on that, we may possibly soon be working on some project for John Frankenheimer. We met him at a lunch given by the Housemans on the 26th, and he suggested that we should meet and discuss stories. He is said to be unreliable. But, if he isn't serious, why did he make the suggestion, out of the blue, without any prompting?

Then there is the possibility of our doing the commentary together for Weidenfeld's book of Don's drawings. Weidenfeld seems fairly serious about this. At least he is starting to argue that my name ought to be first on the title page, for reasons of publicity. Don insists that this doesn't matter. I find it a bit humiliating. But I suppose it doesn't.

And then there is still Harry Rigby on the horizon. He called, not long ago, to declare that he wants to produce our play this fall.

We'll probably regret and curse all of these projects, if they materialize, and feel let down if they don't.

The escrow on the Hilldale property closed on the 29th. No

[1] On June 7—following the "Save Our Children" campaign of Southern Baptist, beauty queen, singer, and Florida orange juice promoter, Anita Bryant (b. 1940)—sixty-nine percent of the turn-out in Dade County (now Miami-Dade County) voted to repeal its ordinance prohibiting discrimination based on sexual orientation.

more tenants, no more writing checks for the Hazel Allen loan,[1] no more keeping accounts of water bills, property taxes, etc.! But also no more real estate other than the bit we live on, to keep increasing steadily in value. And the payments I'll receive from the sale will be increasingly trimmed down by inflation.

August 7. Today I reached page 203, which is almost certainly much more than two-thirds of this draft. I still haven't the least idea what is caught in the net. It is still entirely possible that the question, "Why are you telling me all this?" won't be adequately answered. But, in all my long experience, I have never been able to find anything better than this fumbling way of getting down to the nerve.

I am making an entry today merely to try to start a sequence of diary keeping. These isolated entries are almost worthless. And, I must say, the more I read the later diaries, the more I see how worthwhile diary keeping is.

Here are a few news flashes to be commented on later: Both Truman Capote and Johnny O'Shea are going to A.A. meetings. Weidenfeld seems near to making us a firm offer for our book of drawings and text. Harry Rigby says he has the money for a production of *Meeting* and they'll do it this winter with someone, but *who* (Bedford and Moriarty are seemingly out)? It seems nearly certain that I must have a contracture operation on my right hand soon. It seems most uncertain that Frankenheimer will do any film business with us. Howard Warshaw is dying—if not already dead—of cancer. Don's next show seems fairly set for the end of September at Nick Wilder's. A show in Houston ditto, sometime this winter. The prospect of a show in Chicago has dimmed.

August 11. The California Assembly voted to override Governor Brown's veto, thus restoring the death penalty.

There are serious difficulties about publishing Don's book of drawings, according to Candida Donadio. It looks like the books would be so expensive to produce that nobody would buy them—thirty dollars at least.

No word from Frankenheimer. And we have turned down an extract from a book (called, I think, *The Fools in Town Are on Our Side*[2]) about an American boy of six who is looked after by the

[1] Taken out expressly to finance the property; they did not know her personally.
[2] By Ross Thomas, published 1970.

girls of a Shanghai whorehouse in the late nineteen thirties, during the Japanese invasion. It is both cute and dirty—Disney porn, in fact.

But all is not gloom. The very sympathetic doctor, Moulton Johnson, to whom Elsie Giorgie sent me, told me today that he doesn't think it is necessary to operate on my contracture nodule at present and that maybe it won't get worse.

Also, the day before yesterday, I got my driver's license renewed. For some obscure reason, I always dread taking the test for it. Perhaps because I have a snobbish contempt for all these arbitrary regulations and therefore I feel that I shall never never be able to memorize them. As it turned out, I made one mistake—and it was the same mistake I made in my last test. But in 1973 the speed laws were different. Then, if you were towing a light trailer behind your passenger car, you were limited to fifty-five miles an hour. Now, you can go the maximum, which is fifty-five miles an hour. Last time I guessed sixty; this time, fifty.[1]

September 5. Labor Day—and all the usual vows to restart journal keeping regularly.

Anyhow, the Swami book is plodding on. I've got to page 236. Theoretically, the end is in sight, but I still have quite a lot of material to deal with.

Just changed the ribbon. We didn't go down on the beach today. But we did yesterday, and saw something perhaps historic. For years now, there have been antifag inscriptions written up in the pedestrian tunnel under the coast highway and on the wall of the tennis courts. Last time that I remember, it was "Kill All Fags" and "A Good Fag is a Dead Fag." This year was milder: "Fags Go Home." That appeared sometime last week. Then, yesterday we saw that the inscription had been changed overnight, to: "Welcome fags, you *are* home." This is on the tennis court wall only. The hostile inscriptions in the tunnel remain. We are wondering if there'll be a backlash.

Saw Elsa Laughton yesterday, down at the house next door, with several guests. She was rather drunk and seemed more aggressive than usual. As I was leaving, she came outside with me and told me that the postsurgical cancer tests she periodically has have shown cause for alarm—up from two to four, whatever that means. Poor

[1]In 1973, Congress imposed a national speed limit of 55 mph in response to the energy crisis. Before that, California required "reasonable and prudent" speed, which could be much faster than 55 on highways. Cars towing trailers were limited to 55 both before and after the change.

woman, I could feel her dread—she remembers what it was like for Charles. She hasn't even told Ray Henderson yet.

Don now has the catalogue for his show. It is the most attractive of all his catalogues, with two really good reproductions of color portraits. For the first time he has a male nude and a bottomless boy.

September 13. Elsa's examination proved negative, after all. We have just heard that Truman fell off the wagon while he was in the East and now he's in hospital, drying out.

On Sunday 11th, I drove up to read Vivekananda at the Montecito temple with Jim Gates and two friends of his; one of them, Robert Berg, is a *joli laid* flax-headed Canadian who is living at Vedanta Place intending to become a monk.[1]

Talked to Prabhaprana about Swami. She seemed very devotional and unusually emotional, altogether different. During lunch, when we got onto the subject of Swami again, she began to cry and disappeared into the kitchen for a few minutes. Later, when Jim and I were alone together, he told me that she has become notorious for her drinking!

From Jim I also heard something which moved me greatly. It seems that Krishna has now become Swahananda's attendant, looking after him as carefully as he looked after Swami. I had always thought—much as I admire Krishna—that his service to Swami had a great deal of possessiveness, and therefore egotism, in it. That he is prepared to do the same thing for Swahananda *proves* that he is a saint.

There *was* a reaction to the altering of the antifag inscription on the beach, but a wretchedly mild one. The opposition didn't even attempt to erase the "Welcome fags" sign. Instead they wrote at the side: "Fags suck." So what else is new?

The Swami book is going very slowly. It seems to lack all substance and to consist only of protestations of admiration, love etc. All the more reason to finish it quick.

September 17. I am trying to type this while listening with half an ear to a cassette of the Ram Nam led by seven guest swamis on July 18, the evening of the memorial service for Swami, in 1976. It sounds terribly straggly and jangly. You can hear that the girls are taking it very professionally, however—in fact, I suspect that

[1] He did become a monk for two or three years and remained a devotee; he works in film editing, on dialogue sound effects.

the sloppy effect is created by the disharmony of the western and Hindu approaches to the music.

Tomorrow evening, we are going to the "Star-Spangled Night for Human Rights" at the Hollywood Bowl. Have just heard that trouble is expected from right-wing groups.

And then, on the 21st, Don's show. Quite aside from all the excitements and anxieties connected with this, I always feel that I am seeing his work, on these occasions, quite newly. What usually happens, when he has just done a painting or a drawing, is that I go into his studio and he holds it up for me to look at; the amount of time I get to look at it is only a few seconds, and of course it is neither framed nor hung on the wall. Furthermore, I feel obliged to make some comment, and this in itself is an embarrassment unless one particular picture really hits me hard on first sight. The moment the pictures are on walls in a gallery, I can take my time; I can come back to them again and again. It isn't until then that I really know which ones I like best—or what impression the show as a whole makes on me. Here, I am at a disadvantage in relation to our work. Because, when I show a manuscript to Don, he can take it away by himself and read and reread it, and there is no need of any comment until he is ready, after due consideration, to make one. Maybe this difficulty could be solved if Don staged minishows for me—twenty or so pictures at a time. Must talk to him about it.

September 20. A new doctor has told Jess Bachardy that his cancer is spreading and that it can be only a matter of weeks or months. The new doctor hasn't even examined Jess properly, yet, but, as Don said, the fact that he is *Japanese* makes the verdict horribly convincing. We at once began to wonder if Ted could be induced to live with and look after Glade, if we paid his rent for him.

Jim Gates has pneumonia. He's in bed in his apartment but seems cheerful—at least when speaking on the phone—and says he feels better already.

The Bowl evening was a flop in one sense; nobody quite made it jell, and it was hideously mismanaged, with huge pauses and several fifth-rate acts. In another sense, the mere getting together of all those thousands of gays was, in itself, a triumph and a mutual assurance that we all really exist, along with our demands and our wrongs and our hopes. To me, and to several others I talked to later, the playing of "The Star-Spangled Banner" was extraordinarily moving. You felt the eagerness of all those thousands to be accepted, to belong, to have a place of their very own in this land, which *isn't* free enough to accept them.

Bette Midler was as always a success of the biggest and most symbolic kind, and she brilliantly managed to scrape her bare knees in a fall so that they bled and had to be fixed up with Band-Aids in full view of the audience. This, of course, made her seem gigantically human. But she really is very sympathetic.

As for Richard Pryor, his display of aggression puzzled more people than it offended. It seems he was probably zonked. I think maybe he was merely trying to point out how regrettable it is that the minorities don't support each other properly. Anyhow, he ended by showing us what he called "my rich black ass," but with his pants on.[1] And then the real mistake was made. The producer, an ass named Aaron Russo,[2] and the director, Ron Field,[3] both got up and apologized for Pryor, as if he had been a blackout or a failure in the sound system—assuring us that they had no responsibility for what he had said. Whereas, a few funny civilized words at this point would have pulled the whole show together and given it political meaning.

Talking of politics, the miserable bunch of Baptists waving banners at us about Sodom as we came in seemed impotent and absurd—on this particular occasion. This was, in a way, to be regretted. Because they are neither.

September 25. Last night, driving home along the Santa Monica freeway after dark, we were hit by a car wheel, maybe a spare which had fallen off another car or truck. We were in the extreme left lane, and it whizzed across the freeway from our right and hit us. Don managed to drive over it without swerving too much, though the bump was terrific. We might easily have crashed into the divider or been suddenly slowed down so that another car ran into us from behind. We might, in fact, have been badly injured or killed. As it was, we got the underpart of the Fiat somehow damaged, so that you can't turn the wheel far without scraping. And Don hurt his nose a little and I slightly grazed my shin.

Don's show opening, on the 21st, was crowded, though a lot of people didn't appear—mainly Hollywood and Movieland

[1] Black comic Richard Pryor (1940–2005) attacked the audience of about 17,000 for ignoring the plight of blacks during the 1965 Watts riots, waved his backside saying, "Kiss my rich, black ass," and was ferociously booed off the stage.

[2] Rock promoter, filmmaker, and, later, political activist (1943–2007); he was Midler's manager and planned the benefit.

[3] Dancer and Broadway choreographer (1934–1989), for instance of *Cabaret*, which won him a Tony Award.

characters. Don thinks they were scared off by the catalogue, with its nude Rick [Sandford] and cocknude Mark Valen. Anyhow, the show certainly impressed a lot of Don's colleagues, which is what really matters. There was a feeling that this show is his best to date. I felt great excitement and relief (you always feel that when you're rooting for someone, and he brings it off)[.]

No critics yet. One sale only, but an impressive one. A collector named Taubman(?),[1] who had come to the gallery to buy a de Kooning and an [Olitski], went to look Don's show over and then and there bought his painting of Paul Sorel. (Paul's reaction to this was negative, because he thinks it so unattractive. He asked Don not to divulge his name to the buyer! On such occasions, Paul's vanity seems so grotesque that one suspects it of being a camp attitude.)

Yesterday I finished the twenty-first chapter of my book about Swami. This contains the last batch of journal extracts, ending on the last day of 1975. Which means that chapter 22 will probably be the last chapter. It will have to cover very briefly the first four and a half months of 1976, when I was here and saw Swami at least a dozen times if not more—all these meetings, alas, unrecorded because I had given up journal writing altogether under the pressure of finishing *Christopher and His Kind*—then our stay in New York, England, Morocco etc.—during which period Swami died and I have to quote secondhand reports—and then our return in July, to take part in the two memorial services. What will follow all this is still vague. It may be in the same chapter, or separated, as an afterword. Some kind of glimpse of myself and how I feel about the whole thing now, and about the prospect of my own death, and Swami's relation to that.

Never have I felt as dependent on Don's judgment and taste and intuition as I do now. But I am going to read this material through and make my own notes before I show it to him. There'll also, anyhow, be quite a lot of research to be done up at Vedanta Place before I can start a rewrite.

By and large, the atmosphere at Don's opening, besides being airless because of poor ventilation, was predominantly queer, very kissy. More kisses than handshakes, yet most of them purely social. After greeting some near stranger in this manner, I became aware

[1] Possibly A. Alfred Taubman (b. 1924), the Michigan shopping-mall magnate who later acquired Sotheby's and served a prison sentence for price-fixing with Christie's. He was building an art collection and bought Rothkos, Calders, Warhols, Lichtensteins, and Betty Ashers from nearby gallery Newspace around this time.

that a lady beside me was slightly shocked. So I observed, in a bright conversational tone, "I sure wish I had a dollar for every boy I've kissed this evening!" This at least got me a laugh. Incidentally, this more-or-less socially accepted kissing between males is still apt to be more involving than kissing women. Woman kissing is mostly done on the cheek, to save their lipstick. But, with men, there is no reason not to kiss on the lips, so one often does.

Future prospects, hopeful or not so: After a long silence, George Weidenfeld says that Willie Abrahams, Caskey's old friend who is now the Senior West Coast Editor of the publishers Holt, Rinehart and Winston, is greatly interested in our book of drawings and memories. Abrahams is coming to see us on the 30th. Harry Rigby has gone to England to try to cast *Meeting by the River.* Not one word from Frankenheimer; neither he nor Rick McCallum came to Don's show. The lump on my right palm seems definitely bigger but no fingers are contracted. My left eye seems to be dimming appreciably. And of course there is the prospect of Jess Bachardy's death, with all the problems it will bring.

Jonathan Fryer's book arrived at last and I have just reread it. I was astonished to get a letter from Edward Upward praising it mildly: "It makes your life sound very exciting and extraordinary." To me, it reads like a careless rehash of my own books; in order to avoid quoting me, he simply inserts other, less suitable adjectives.

October 5. Here are some extracts from William Wilson. His "Art Walk" column in the *Los Angeles Times* of September 30 reviews four shows. Don's show is reviewed first and gets more words than any of the others—this, says Nick Wilder, is all that really counts. The others are Paul Jasmine, Allen Jones and Chris Burden.

Wilson is as weird as ever. He starts off with a paragraph saying that four galleries are showing work which "involves itself with some combination of glamor, exhibitionism, narcissism or self-conscious stylishness" and that these values are regarded in the art world as "ominous bellwethers of rigid stylization, autocannibalistic self-indulgence and incipient kinkiness." And he adds, "Please God, don't let it be a *trend.*"

Now Don gets some not unkind words:

> He's always been an absorbing, virtually literary talent, despite a problematically slick style. Recently he's opened up to allow his vision to flood with some of the shadowed, troubled side of this cultural enclave.
>
> With this exhibition, Bachardy begins to look like our top

> probers into the psychology of people whose style of life threatens to turn them into cultural robots while eating them up from the inside out. He begins to remind us of Egon Schiele, Francis Bacon, David Hockney or Richard Avedon. His subjects are as decorous as if they were posing for Ingres or Holbein, but each is haunted with contemporary malaise. . . .
>
> Finally, Bachardy has committed himself to seeing the obsessive fantasies that can animate and destroy "interesting" people. Whether his insights are "true" to his subjects is as immaterial as whether a novelist's characters are "real" people. Bachardy is finally using visual means to make sulphorous, painfully believable spectres.
>
> The artist is as mannered as ever. There is ground for aggravation that his line refuses to be simple, his eyes are so big and glassy, mouths so consistently trying to be Mick Jagger's big, sexy, crybaby mouth. (This is probably a reference to Mark Valen's kissable lips. Did they turn Wilson on?) His expressionist watercolors' apparent looseness, sometimes looks like conventional work passed under a water faucet.

Then, right at the end, Don gets the accolade: "All that said, Bachardy still makes the orchestration work in a way that demands we take him seriously."

Wilson then grudgingly says a kind word for Jasmine which his show certainly doesn't deserve, and is nasty to Jones, whose show we both liked quite a lot: "The work seems to assume that (a) we are kinky and too chicken to admit it or (b) he's going to use sex to sell us good art because we are too dumb to take it straight." Wilson is very ready to take offence on behalf of the viewer.

And finally he's quite horrid to poor little Chris Burden, "his public image is such a mess that everything he now does has its intentions and meanings blotted out by the name 'Chris Burden,' which has turned into an independent entity prowling in an omnivorous search for any limelight that will shine on it."[1]

I'd meant to write a whole lot more today about various happenings but now I must get ready to appear at an NBC interview for the "Tomorrow" program.[2]

[1] Burden (b. 1946) became famous with his first show, "Shoot" (1971), in which a friend shot him in the left arm as he stood against a blank gallery wall. The 1977 show presented Burden's 1976 income tax return and other personal financial documents with a T.V. spot giving "Full Financial Disclosure."

[2] Isherwood was interviewed on tape by Kelly Lange for Tom Snyder's

October 17. A good moment to pick up on this journal, because we've just been released from a formidable chore. One of the T.V. outfits had decided to do a really bang-up production of *Uncle Tom's Cabin*, a darkly undercover project since the material is in the public domain and liable to be thought of by many competitors, after the huge success of "Roots." Well, Keith Addis, Bobby Littman's assistant and far more enterprising than Bobby, as it seems to us personally—either Bobby is pissed off because we keep turning things down or else, as we suspect, he has lapsed into his former alcoholism—called us and suggested we take the job. So I read the book, for the first time, and was quite charmed by it; it has a real Dickens flavor both for better and for worse—all the way from the "dead, your majesty" style to the death of Little Nell—and there are social ironies worthy of Shaw: Miss Ophelia the northern abolitionist who hates to touch black people; Phineas the Quaker, pushing the slaver over a cliff with the words, "Friend, thee isn't wanted here." But what a construction job to get it all together! And what a huge script it would have had to be! So I was relieved beyond words when Addis called up yesterday and apologetically told us that somebody else had already grabbed the job—before they had even interviewed us—and that he was getting a minimum payment for doing it.

So now I'll try to finish off the rough draft of the Swami book as quickly as I can, so Don can at least get an impression of the whole thing.

Much more news, but it must wait till tomorrow.

October 23. It had to wait much longer, because of various distractions and preoccupations. Well, at least I have finished a rough account of Swami's death, ending on page 257. Now comes the great problem of writing that afterword, summing everything up. When I began working on this book, on February 12, I wrote that the best thing about it would perhaps be its final passage. I thought I knew almost exactly what I wanted to say in it. I only wish I had sketched something out at the time, for now I really am at a loss. Yesterday evening, a young man named Don Slaikjer (pronounced Slaker, so why take all that trouble writing it differently?) was asking me about Swami—what his influence and teaching have done for me, and I found myself stumbling and faking defensively. I couldn't say anything really clearly and

late-night "The Tomorrow Show" (following Johnny Carson's "The Tonight Show") on NBC.

I knew he was deciding that the whole thing was self-delusion.

What is holding me up? Is it that I feel an obligation to declare that Don and I are somehow "saved," that we have been given "the pearl beyond price" and that we'll never lose it and that death has lost its terrors in consequence? Obviously, I mustn't end on a note of vulgar uplift, like the ordinary, self-satisfied "religious" writer. But surely I can make *some* statement?

This block which I feel is actually challenging, fascinating. It must have a reason, it must be telling me something.

Am I perhaps inhibited by a sense of the mocking agnostics all around me—ranging from asses like Lehmann to intelligent bigots like Edward? Yes, of course I am. In a sense, they are my most important audience. Everything I write is written with a consciousness of the opposition and in answer to its prejudices. Because of this opposition, I am apt to belittle myself, to try to disarm criticism by treating it with a seriousness it doesn't (in my private opinion) deserve. I must not, cannot, do this here. I must state my beliefs and be quite quite intransigent about them. I must also state my doubts, but without exaggerating them. Yes, that's it. I must give the reader a glimpse of myself in a transitional stage, between Swami's death and my own.

The doubts, the fears, the backslidings, the sense of alienation from Swami's presence, all these are easily—too easily—described. One mustn't overemphasize them. What's much more important is a sense of exhilaration, remembering, "I have seen it," "I have been there." For some absurd reason, as a sort of epigraph, I keep thinking of Arne Saknussem's code message in Verne's *Journey to the Center of the Earth*: "Descend into the crater of Yocul of Sneffels, which the shade of Scartaris caresses, before the kalends of July, audacious traveller, and you will reach the center of the earth. I did it."[1]

As of this moment I feel that I must reread the whole of my manuscript before even trying to write another draft of that afterword.

The above-mentioned Don Slaikjer had come to drive me to a[n] ACLU garden party where I had agreed to speak. Oh, those liberals who think it indecent to mention and show interest in somebody's racial background—saying that, "All that matters is what people really are," and, "Why can't you take people as you find them, without digging into their past history?" Some of the Jews present didn't like it at all when I said that I had encountered

[1] Chptr. 3.

very little anti-Semitism among non-Nazi Germans during the pre-Hitler period, but one old Jew agreed with me. Two others, in particular, reprimanded me, saying that the Germans had always persecuted the Jews. So, when it came my turn to be asked about my attitude to antigays, I said firmly that I was far more interested in justice than in being loved, that I could very well understand that some people just naturally feel a distaste for being with gays, that we are just another minority, a part of nature with a right to existence—"We don't think we're the chosen people," I nastily added. I had a feeling that I left a faint stink behind me, though I was profusely thanked by many.

Heard from Joanne Carson that Truman got so drunk before leaving for New York and the Smithers clinic[1] that they wouldn't let him board the plane. He was persuaded to do so next day, but then, on arrival, he went to his apartment, not the clinic, and Jack Dunphy was unable to cope with him. He is at Smithers now, apparently, and is committed to stay there for at least twenty-five days; but Joanne feels that he'll most probably go right back on the bottle as soon as they let him out. Her reaction to this is to get weepingly, self-pityingly sick—her "Baby" (Truman) is responsible, but of course she loves him, etc.

October 24. Returning to this problem of the afterword—how has Swami's death affected me, what difference has it made to my life?—this morning I get a kind of Zen feeling: why should there be *any* difference if he isn't really dead, if he hasn't really left us? Coming from worldly self-indulgent old me, this sounds merely heartless, but it mustn't be forgotten that Swami himself said of Brahmananda, "After his passing away I felt no void."

What I'm getting at is this—isn't it much more to be expected that there will be no immediately apparent results—that life will go on much as ever, on the surface? Perhaps until I actually die, or at least become terminally ill?

Anyhow, I think I'll postpone trying to write the afterword until I've talked to Don and read his notes and myself read through the manuscript again.

November 8. This is a period of great pressure. Don's father seems to be dying, at last, and the whole problem of what to do with his mother is unsolved. She can't be left alone because she is really gaga in many respects. And Ted can't be relied on. He might flip at any time—just to duck out of any responsibility.

[1] I.e., The Smithers Alcoholism Center.

Then there are our projects:

Albert Marre who's to direct our play is a prima donna and a bore and we may have to say definitely that we don't want him to do it. But Rigby is terrified of this because he fears that then the Kramers, who are his backers, will drop it. Tomorrow his suggestions will arrive, and we shall have to make a decision, will we accept them or not. We fear that he will try to get all the homosexuality out of the play, for openers. He says it's "dreary."

The book of Don's drawings and our memories of his sitters seems to be definitely agreed on. Billy Abrahams, of Holt, Rinehart and Winston, wants to do it. This will be fun, perhaps, but an awful lot of work.

Then there's a man named Dick Clements[1] who's prepared to pay us out of his own money to do a treatment of *After Many a Summer*, followed by a teleplay if he can get the backing. One's always suspicious of these freelancers. But I guess we'll have to talk to him, at least.

Then there's my book on Swami, which I'm very very slowly beginning to revise. Ordinarily, this in itself would be quite enough to keep me at work all winter.

And, in the background, in case one should venture to relax, is the assurance that we can't much longer escape an important earthquake in the area. There's some kind of a mound or ridge or small hill which is steadily rising, right on the San Andreas Fault near Palmdale. One's nerves are probably on the alert for this, all the time, even during sleep.

Am reading—you'd never guess—Borrow's *The Bible in Spain* and loving it. It is astounding how his magic personality makes these very ordinary adventures so magic. Am also reading Flaubert's travel diaries of his visit to Egypt.[2] Which are so consciously literary by comparison, interesting but unmagical.

My darling, under tremendous strain, is wonderfully active. He draws, paints, goes to the gym, runs miles daily, and finds enough surplus energy for mice.[3] Our relationship is indescribable—though I guess I must try to describe it someday. It has moments which seem to blend the highest camp with the highest love with the highest fun and delight.

[1] Probably Dick Clemen*t* (b. 1937), British screenwriter and director; with Ian La Frenais, he wrote T.V. comedy for the BBC in the 1960s and 1970s—"The Likely Lads," "Porridge"—and films—*The Jokers* (1967) and the 1992 adaptation of Roddy Doyle's novel *The Commitments* (1987).

[2] *Flaubert in Egypt*, mentioned above, May 8, 1973.

[3] I.e., pursuing other men, in his role as Kitty.

I have various ailments, and yet I find it unbelievable that I am seventy-three. I don't feel in the slightest prepared for death—whatever that means. A lot of the time I am so marvellously, calmly happy.

December 17. Jess Bachardy died in hospital on the morning of December 7. He had been in great discomfort and some pain, vomiting, unable to eat; but I don't think there was any great agony. Don, Ted and I all happened to see him the night before—I, because Don's car stalled right in front of the hospital and I had to drive in and pick him and Glade up. (This in itself was kind of weird, because the Fiat was so very much Jess's cared-for creature, he always looked after it when it went out of order—so this stalling on the eve of his death and as near as it could possibly get to his deathbed was like the behavior of some faithful hound which knows when it is going to lose its master.) Both Don and I privately said a Ramakrishna prayer over him.

Don is being marvellous about it all, taking on the cares of a son and executor, driving into town again and again to have meals with poor old Glade and clear out the junk in the apartment with Ted. Glade's memory has nearly shrunk to nothing, a lot of the time she seems to live in the Now like a saint and is therefore cheerful and contented to be alone. But then memory returns with a cruel stab and she's scared and thinks she's been deserted by her sons and cast adrift. This is heartrending.

It would be hypocritical to say that I personally am unhappy, because the situation produces an extra closeness between Don and me, and closeness between us is my life.

Albert Marre, with whom we were both feeling extremely unclose at the time of my last diary entry, is now our hope and our leader. His notes after reading our script were astonishingly intelligent, and we responded by writing in some new scenes—a couple of them lifted from our own filmscript, written in the fall of '73. Now we are looking for an actor to play opposite Keith Baxter. Maybe John Hurt. Maybe Michael York. If we can get either of them. We are supposed to start rehearsals in the early summer.

The book of Don's drawings is seemingly agreed on, but no contract yet signed. *After Many a Summer*, ditto. And we have met two very nice young producers at Fox, Walter Hill and David Giler, who are friends of Peter Viertel and whom we may be able to work on some project with.

December 19. On the 16th, I finished revising chapter 1 of the book about Swami. It took me fifty-two days to do this, and the chapter is only sixteen pages! At this rate, revising the whole book would take between two and three years. Gee up, Dobbin!

December 25. Christmas blues. Damn it I am depressed, as I haven't been in weeks—and for no good reason. We probably both drank far too much last night, first at Billy Al Bengston's, with a lot of artists and art fanciers I mostly don't care for, then in a much seedier corner of Venice with fags, fag molls, a black or two whom I worked very hard getting along with. I only feel senile when I'm frisking about, not acting my age. Also, dancing, or rather, stomping, I made my leg worse, it's been bad for a week and must see Elsie Giorgi the day after tomorrow.

Don has gone off to see his mother and help Ted clear Jess's things out of the apartment. This awful problem of her future remains an awful problem. Don is really wonderful about it, wonderfully brave and competent.

Have just finished volume seven of Byron's letters. Wish they'd hurry and publish some more. What do I get from reading him? Reassurance.

My meditation has been a dead telephone line for months and months. But a thought (message?) came to me. Just don't bother about that and make japam; that's the basic minimum response to everything, and it can get you through anything, I believe. Only I don't do it nearly enough. And I bet I don't think of Swami more than ten times a day—unless I'm actually writing about him.

1978

January 20. Here's a belated start to the diary year. The beginning of 1978 has been so depressing, rainstorm following rainstorm, that I just haven't had the gumption to begin. Now the sun is shining. Also, I've cultivated a little bit of an appetite for diary keeping through reading Virginia Woolf's first volume, 1915–1919. She is fascinating. Somehow, the abridged diary that was published in 1953 left me with a quite different and much less stimulating impression. She seemed to be so preoccupied with people's opinions of her books. This volume makes her seem much more positive, much less self-preoccupied. But as soon as I've finished this volume I'll go back through the abridgement and see if I get the same impression as I got before.

Bother with my knee. I do hope I don't have to have an operation. And it's so frustrating not to be able to jog. Jogging is most necessary if I'm to keep in one piece.

I'm crawling along with the Swami book—still provisionally called *Another Kind of Friend*. I don't know why I am so slow and inhibited. I don't like my tone of voice in it, the tone in which it's written. The prose seems so bland. But I must get on.

At dinner with Vincent and Coral Price last night, Frank McCarthy told me he thinks that Israel's lack of real enthusiasm for Egypt's peace efforts[1] is disgusting a lot of Jews in this country. Interesting if true. But I have never trusted him or liked him. Coral and Vincent were both charming, particularly Vincent describing his Oscar Wilde play.[2] He plied us heavily with some wine called WHITE which has the extraordinary property of not giving you a hangover. Coral's charm was slightly neutralized by anti-queer overtones as she described the huge amounts of money lavished by Roddy [McDowall] on the training of Paul [Anderson] as a singer, when he has absolutely no talent. He now calls himself Adam.

Don's problems with his mother and the settling of his father's estate continue, but more about that another time. Sometimes I think he is the toughest person I ever met in my life; then suddenly he shows the strain by blowing up about some trifle. There is really nothing I can do to help him. Except love him and make japam.

January 22. Don is gradually building up to telling Ted he has to take over responsibility for visiting their mother, or else she will have to go into a home. This would be workable with an ordinary person, but the danger with Ted is that he'll checkmate Don by going mad. Oh, Jesus.

Last night we had supper with Billy Bengston. He was preparing to run a footrace early this morning with assorted males of all ages. But he feared his knee might go back on him. Well, it didn't. Today he told me on the phone that he did the course, six miles,

[1] Egypt's President Sadat addressed the Knesset in Jerusalem on November 19, 1977 while working all year with U.S. Secretary of State Cyrus Vance to prepare peace talks, but on January 19, 1978, he withdrew his Foreign Minister from negotiations in protest over Israel's tactics.

[2] *Diversions and Delights*; see Glossary.

in forty-six minutes, which was even better than he'd hoped. I do like his chirpy macho manner. The two women, Penny and her sister Barbara,[1] also present, seemed tiny and sexless beside him.

Worry about my knee. Last night they gave me a plaster containing a magnet, made in Japan. If you put the magnet exactly over a place in your body which hurts, it is supposed to draw out the pain. It didn't, with me, but I didn't put it on quite the right place. Must try again tomorrow. Just out of scientific curiosity.

Because of my knee, I have to meditate sitting upright in a chair. This morning, after many many days of dryness, I at least felt some emotion. What I do not feel is any sense of visitation. I have never once really felt Swami's presence since he died.

From the blurb on the Avon paperback edition of *A Meeting by the River*: "With stunning delicacy, Christopher Isherwood portrays the dramatic clash of two complex men. . . ." etc. Could that idiotic phrase ever be used appropriately? Yes. "With stunning delicacy, I reminded him that he owed me twenty thousand dollars."

A visit to gloomy old Harry Brown, anxious June his wife and rather sweet Jared, their son, aged thirteen. Harry has been threatening to have a breakdown and retire to St. John's Hospital. But when I was asked what I'd been doing and mentioned how we'd turned down the Huxley *After Many a Summer* project, June at once suggested that I should get the job for Harry. So I guess I'll try to.

This is neither here nor there, but I got a letter of April 10 of last year and have kept it all this time, meaning to copy it down because I think it's so funny. So—

> Dear Mr. Isherwood, I am sorry to trouble you again. I had written you in February and you were kind enough to sign a copy of one page of your "Hypothesis and Belief." When I went to get my mail in my front yard I saw a dog chewing on my mail, and when I tried to take it away the dog attacked me. I had to have thirty-nine stitches in my leg and right arm. It was one of the most frightening days of my life. I have retyped the page and hope you will be so kind as to sign it for me again. Wishing you the best of health and happiness.
>
> Sincerely, Marilyn Fisher.

January 25 [Wednesday]. Yesterday I went to see Dr. John

[1] Barbara Little, Penny's older sister; administrator at the James Corcoran Gallery for a few years.

McGonigle. Having read the X-ray pictures of my knee he says that I have a badly torn ligament and that it should be repaired. Elsie Giorgi supports this view, so I've agreed to have the operation, at St. John's Hospital. I'm to go in there on Tuesday next, January 31, be operated on next day and probably stay till the following Saturday or Sunday. January 31 is auspicious as it is the day of Swamiji's puja this year. I hope to read the Katha Upanishad as usual, if I can manage to sit in the shrine without too much discomfort.

Am sort of unquiet about this but not really scared or even unhappy. It seems to be part of my pattern, to get myself shaken up from time to time by something of this kind. I think it's partly an atonement, not only for my bad behavior but also for my success and my extraordinary good luck. I am purified by passing through the ritual death of the general anesthetic.

What I really do feel anxious about is whether I'll get a room to myself. And this isn't even self-indulgence, because Albert Marre is coming the week after and we shall never get the rewriting of Rafferty's character done—turning him into a woman—unless we can have these days while I'm at the hospital to work together in privacy.

About ten days ago, after correcting a whole lot of errors in Brian Finney's wretched book on me, I got a letter from him saying he'd just had a long letter from Bill Caskey, and as the result of it he wanted to make three insertions:

To mention Jo Lathwood's name as "a G.I. bride," to say that Lincoln Kirstein was my model for Charles in *The World in the Evening*, and to describe "Vernon" as "a typical hustler from a male brothel."

This last really amazes me, it's so viciously untrue. *Any* mention of Lincoln will offend him. And the statement about Jo is false. All this Caskey knows perfectly well. It makes me feel that he is in his worst "situation queen" mood, and I dread the thought of his coming here, as he proposes. I must make up my mind in advance, how to handle this.

January 28. Blues—largely due to getting drunk last night at Neil Hartley's. It's very unwise for me to drink when there really are things to be blue about—so often, there aren't, in my so much above average happy life. It is unworthy of Dobbin to get depressed at the prospect of an operation; this is the sort of thing I'm supposed to be good at accepting.

Speaking of happiness reminds me of a passage I liked in Virginia

Woolf's first volume of diaries. This was written on April 20, 1919:

> .... I can pardon these self-seeking impulses on the part of Desmond [MacCarthy] very easily. My tolerance in this respect is far greater than poor Mary Sheepshanks'.... But then I am happy; and Mary S. is not; every virtue should be natural to the happy, since they are the millionaires of the race.

Uneasiness about the construction of *Another Kind of Friend*, I seem to be meandering through my narrative; it's so ill-constructed, compared with well-made *Christopher and His Kind*.

February 2. On Monday 30, the day before I was scheduled to go into St. John's Hospital, Don suddenly declared that he didn't want me to have the operation, he'd been worrying about it terribly, we must cancel it somehow. So he called Elsie Giorgi, who encouraged him, saying that there was no hurry and if I didn't mind having the pain for a few more weeks I could start some therapy and see if my knee didn't get better without surgery. I was glad of course. I now realize that I simply agreed to the surgery in the first place because I believed it could be done at once. Then they postponed it and I couldn't back out.

This reprieve means that I must work much harder. Yesterday and today, we've been rewriting the character of Rafferty for our play, turning him into Consuela Rafferty, a jet-set lady newsperson. As for *Another Kind of Friend*, I keep digging at it; still very tough but I feel it will gradually ease up. Albie Marre is still expected at the weekend to discuss the play.

Have just restarted Stevenson's *In the South Seas* with surprised pleasure.

Don's name for Jim Gates, in the days when he was up at Vedanta Place—Our Man in Nirvana. (Incidentally, Jim tells me he has just fallen in love with someone; but so has Warren [Neal]. They are quite frank about it with each other.)

I now begin to suspect June Brown. She called me the other day on the phone to please call Harry at once. So I did and Harry acted mad at her for doing this, says she's disturbed and that his marriage is falling apart. So which one of them is *really* disturbed? Or is it both?

February 7 [Tuesday]. To see Dr. Blair Filler yesterday about my knee. He is a fellow runner of Billy Al and Penny's, and was therefore especially well able to pass a verdict on the ligaments

I have torn jogging. But the verdict was dismaying: he saw no alternative to an operation, saying that the knee will otherwise merely keep getting better and worse, whenever I rest it and use it too much. Came home very depressed. It's hard to screw myself up again to submit to the surgery; the first time it was relatively easy.

Now, however, on Don's urging, I'm going to see a man Jack Larson knows. He isn't a surgeon and only uses therapy, and he's said to have cured all manner of fallen stuntmen. I'm to see him next Monday, the 13th. Meanwhile, Marre is coming—if the snowstorm in New York lets him take off—to see us tomorrow and pass judgment on our substitute for Rafferty, Consuela. I do think she's a vast improvement.

Lots of rain has been falling. Don out tonight, so I'll eat and watch T.V. I am really and truly not depressed, most of the time. It's odd—something to do with Swami, or with my mantram (if there's really a difference) and yet not seemingly much connected with either. No, that isn't true, I do feel a lift almost always when I make japam.

Criminal neglect of *Another Kind of Friend*. Must force myself tomorrow morning, early. I know it'll be easier later along in the rewrite.

A weird discovery we have both made: since using the tape recorder to record our discussions of the play, we have both realized that we cannot be certain which of our voices is which!

February 11. On February 9, I got a call on the message speaker of my telephone answerer. It began in a very sinister sex-maniac kind of tone: "You cocksucking goddam sonofabitch, fuckass prick...." I hastily switched off the on button so I could talk to the caller (male), but I lost him. Pretty soon the phone rang again. This time, the caller said, in a much more natural voice: "Hey, Chris, how's it going? I still think you're a dick, okay. And I'm not going to forgive you." Don thought this message was genuine, not some kind of joke. I don't know. It was spooky, it worries me a bit.

Albert Marre has been here, twice, plane hopping between San Francisco, where he had an opening, and New York. He doesn't like our Consuela—that's to say, he wants her to give Oliver a hard time before she tells him he's "the real thing." I'm not sure what I feel about this. Isn't it just dramatically tricky? Marre's other idea, that Tom shall be much younger, eighteen, rather appeals to both of us. Marre wants him to be a more powerful figure—the son perhaps of a state senator (our idea not his) and therefore able to use his father's position to help him commit irresponsible acts, hence creating a greater threat of scandal to Patrick.

February 20. Depressed. I don't feel nice Dr. Richard Tyler's treatments are doing anything particular for me. I get the impression that he just tries what he has in his box of tricks and hopes for a wonder cure. Today we tried a Japanese kind of acupuncture, the needle gives you a shock but doesn't pierce the skin.

Today, after more snailing along, I did get to the end of chapter 4, which is page 41. This is a bad period. Patience.

The weather is glorious, and I can't run down to the beach. Maybe I never will again.

February 26. Today I am seventy-three and a half. Grey chilly weather. Don out somewhere with someone. My knee no better. I had a sort of love scene about it with sweet faded-cute Dr. Richard Tyler who pouted and said he couldn't bear it when his patients weren't instantly cured. So we had some electricity of a different kind which made the muscles of the leg jerk, not unpleasantly. I would love to produce a miracle for him. Later he showed me photos of himself as a cute boy with Ingrid Bergman in *The Bells of St. Mary's*[1] and as a wrestler in the Highland Games, a year or two ago. He has a terrific body, but his face is drawn and pale because he starves himself—daren't eat supper because he puts on weight so easily.

Morale good today because I am on a slightly easier part of the book and can see my way ahead. Good also because relations with Don are very good. Also I've taken a Dexamyl. Big resolutions to make lots of japam, during every spare moment, and to *work*. There is a curious strength to be found in the terminal condition of being old—it cuts out such a lot of shit. But I am subject to senile resentments, mostly against strangers, who bother me to sign books, etc. It is very seldom that I really enjoy being a celebrity, but I guess I'd miss it if I weren't one. The young need so much support and one should give more and more. That's the only creative way of taking one's mind off oneself and one's ailments.

All right, Dobbin, you're dismissed. Let me see you do better. Let me be proud of you.

March 13. Slightly more optimistic today. Because the sun has shone, after all these days of rain, and the alarming earthslides

[1] Tyler had been a child actor and worked in T.V.

close on the other side of the studio and even just below the house (smaller ones). Because, though I know my knee isn't really better, it is in a better phase and allowed me to walk down to the beach—down and then up all those stairs—and get it wet in the ocean and dry it in the sun—without particularly acting up afterwards. Also because I do see now, at least for the immediate future, an easier part of my book. I have passed page 50.

Meanwhile, much talk about our book; Don's drawings and our memories of the sitters. But we are standing pat on the demand that they must first find us someone who can make satisfactory reproductions of the drawings. If the artwork is to be substandard for commercial reasons, then we won't go along with the book at all. Still no actor to play Oliver in our play, but that does seem to be at least probably about to get produced. Harry Rigby has been here.

I love my darling Kitty so. Am I living off his vitality? Yes, to some extent. But he has plenty to spare. I still have some of my own. But my old left eye is getting dimmer, and I'm losing my hair, and I don't suppose my knee will ever really recover. Keep calm, Dobbin, and make japam, and prepare for landing.

March 26. I meant to write an Easter Statement, but have spent nearly all day fussing around with the end of chapter 6, to page 61; I think it's nearly right now. And it's time to go and see two friends of Paul Bowles who may or may not be a good idea. One of them has the disadvantage of a difficult Basque name, Leonardo Arriza-Balaga(?)[.][1]

March 27. While "meditating" this morning, I began to think of that unsolved problem, the postscript to my Swami book. And it came to me that I should point out that this book differs in kind from books on other subjects because there is no failure in the spiritual life. This sounds trite, but I feel there is a very valuable statement to be developed here—namely that one cannot make a statement. I can't say that my life has been a failure, as far as my attempts to follow Prabhavananda go, because every step is an absolute accomplishment. So I have neither succeeded nor failed, and I am always the Atman anyhow. More to be thought about this, I hope.

[1] Leonardo de Arrizabálaga y Prado, Spanish poet (b. 1946), from Madrid. He became friendly with Bowles in Tangier, where he lived for several years, and was dedicatee of a chapbook edition of Bowles's story "In the Red Room." He published *Poemas del Silencio* (1976) and *Nueve Prosas Ejemplares* (1983; *New Exemplary Prose*).

While getting breakfast ready, I created a poem:

> Ten kittens came in crowns
> From different towns.

Don capped this without hesitation:

> Ten dobbins came in gowns
> And looked like clowns.

April 6. It is highest time that I pulled myself together. I have let days drift by, and with so much to be done. Am now only at page 64, and really I am just delaying myself by being picky. I condemn sentences because they are too verbose and yet fail to rewrite them. Stephen is coming to see us next week and I would have liked to be able to show him some of the book; but I know now that I can't reach a suitable spot to stop reading at. Anyhow, I feel that he can't like it—it isn't at all his kind of thing.

This is really a work-impending period of my life, and I ought to be pressing on and not moaning. Because there are at least two major chores ahead, our play and our picture book. Albert Marre phoned yesterday from England to say that Simon Ward wants to play Oliver. Albie is greatly in favor of him, and he loves the play, and is a former pupil "of Upward"—this could mean either Edward or Mervyn[1]— and Albie saw him on the stage and he was very interesting. We are planned to open at the University of Tennessee in the late summer, then Kennedy Center, then New York.

Meanwhile, after so much foot dragging, everybody has accepted enthusiastically the idea that the reproduction of Don's drawings for our book could be done here—thus solving all the problems of Don's having to go to New York or elsewhere to supervise them. And, what do you know, Nick Wilder's friend Jack Woody (*only* a friend now, alas, we are told) found us two firms within twenty-four hours, and Don has interviewed them both and found both seemingly suitable—so now comes the test: both must submit specimen reproductions.[2]

Oh yes, and some people (talents and financial status unknown) are interested in getting the film rights to "Paul."

[1] Edward Upward taught at Alleyn's School, Dulwich, where Ward studied; John Mervyn Upward, Edward's younger brother, taught at Port Regis and was headmaster for many years.

[2] Woody himself eventually published a different version of the book; see Glossary.

Storm clouds: Don has more spots on his face, to be examined today for possible malignancy. We await the income tax bill, known already to be supermalignant, because of the sale of the Hilldale property. My old Mr. Right Knee is still a mess, but I begin to think I shall just let it alone until it utterly collapses under me. Left Eye is a bit worse too, but I let it ride.

At a party, two days ago, we saw the Garretts. Jean is now in a wheelchair; the cancer is spreading and she hasn't long to live. I am about to get involved in this, she wants me to go over and read to her. At the party, she made a point of having me eat off her plate—saying she wasn't hungry and didn't want to waste the food. I couldn't help remembering how Larry Holt half-seriously accused me of not wanting to sit too near him on the couch lest I should catch his cancer. Was Jean testing me? I used her fork.

Tonight we have to go to a David Bowie concert, which I look forward to about as much as a puja. Courage.

April 16. Don's spots were serious but were removed in time. The income tax was very serious—about $22,000, including federal plus state plus first federal and state payments of estimated tax. All one can do in the face of this persecution is earn and earn and earn. But what when Dub gets real old?

Bowie's silent acting, in the midst of the noise made by his assistants[,] was again impressive. Afterwards we were photoed at a party, but not one word was said about the project he proposed to us when we last saw him.[1]

Stephen has just left, after two nights (only) staying with us—how short and yet how long! It is utter bliss, having my bathroom free again. Don really hates him, and I think Don is right in saying that Stephen doesn't really like either of us, or anybody else, and that he's guiltily antifag. I am however basically sorry for him because, devoured as he plainly is by both envy and ambition, I see little prospect of his producing a masterpiece before curtain-fall; the drafted scenes for his new *Trial of a Judge* play seem to me stodgy and dull.

At the Bowie party, a palmist told Don he is good and me that I am lucky. He also said Don is coming into a lot of money somehow, and that I shall have an affair next year!

Oh, if only I could get along faster with the Swami book! I have the most brutal block, and terror whispers in my ear that that's because there is something wrong with the whole project.

[1] Probably a film set in Berlin; see Glossary.

As for my relations with Don—beautiful most of the time BUT the truest of all sayings, "Nothing burns in Hell except self-will," is still true for both of us, and our self-will seems practically fireproof.

May 12. The block has continued, all this while. What I'm also aware of, and this rather alarms me, is that my brain is clearer when I use Dexamyl—I still keep this down to every third day, however. Some days I feel really a bit nuts, I cannot concentrate, and I cannot write one line which I don't begin to pick to pieces as soon as it's written, a truly fiendish and sick sort of perfectionism. I ask myself, have I maybe had some very slight strokes. However—today I did finish chapter 7, to page 77, and now I begin the account of my life at the Vedanta Center, which means that a whole section of the book is finished, and probably more than a quarter of it.

Tongue has been sore for days, and my knee is acting up, but fuck all that if only I can function. The play will apparently go ahead unless they fail to agree with Keith Baxter on his terms. Billy Abrahams is still farting around with his silly cunt of an art director in New York, about our picture book. And now there's a nibble on the line for a T.V. version of *Lady Chatterley's Lover*!

Am reading Wordsworth, *The Prelude*. The marvellous things are nearly all bits that I know; the bald spots are very very bald.

June 2. Tongue very sore again, despite treatment. Left eye a bit dimmer. Right knee acting up, but I persist in running it down to the beach. Old age is still a sort of camp about the ailments I used to have when I was young. I chatter away chirpily about death to interviewers and am aware that it isn't all quite real to me. But, oh my God, how thankful I am for my good spirits, so immensely increased by my angel and his life going on beside me.

Have just finished two real downers, the biography of Monty Clift, by Patricia Bosworth, and Robin Maugham's *Conversations with Willie*. They would cheer one up on one's deathbed. Especially the Willie book—that awful determination to be miserable. I think Willie put everything of himself into his plays and novels; there seems to be nothing left over. Whereas Virginia Woolf, only four months before suicide, burns so brightly in her "A Sketch of the Past," in *Moments of Being*. The optimism there is in first-rate writing, no matter about what!

Keep crawling on with the Swami book. But still, nearly a third of it is now revised, and later there will be many journal extracts. It now seems we really are going to start soon on our book of Don's drawings and our combined memories. And Harry Rigby wants us to come east at the end of this month and look at the designs for the set of *Meeting*.

June 7. First anniversary of Adoption Day. Oh, I am so happy with my darling, when I don't have to watch him being tortured by the red tape of his executorship and Glade's immense passive nuisance-presence and Ted's selfishness and the worry of finding them somewhere to live.

All I fear is becoming gaga, when I need to work more than ever and have so much work in prospect. Just wrote an outline of our play for Harry Rigby to use in some publicity or other. And today I finished chapter 8, to page 89. I do feel a lack of energy. Elsie Giorgi says I may have pernicious anemia, which causes the tongue to become "denuded," i.e. unnaturally smooth. But the only treatment she prescribes for this is B12 shots. Had the first one today.

Tonight we go out to celebrate Billy Al's birthday—which is also Gauguin's, and Forster's deathday.

June 22. Day after the solstice, full of resolves and temporary energy. I have finished page 100 of the Swami book (still haven't got a good title for it). Now I *must* get ahead.

A depressing dinner with Paul Sorel, last night. His steady squandering of the last of his money is like the ebbing of a life; he's on his financial deathbed, it seems. We simply cannot imagine what will happen to him and he knows this and it pleases him. But has he *really* thought about what the rest of his life is going to be like? No. And I somehow admire him for it. This is a sort of parody of the life of a holy pauper whose whole religion is not to think about the future.

Now we have Harry Rigby in town, with Albert Marre due to arrive soon. Then the play will have its other main parts cast here, or else we shall have to go to New York for the casting. Neither of us wants to do this.

Also, we await the go-ahead on the book of Don's drawings. This could arrive at any moment—or not at all.

August 27. My birthday was celebrated in bed in the evening watching T.V.—Senator Briggs in dubious battle with Troy Perry over Proposition 6, each side quoting the Bible, such a mistake. Scobie thinks Briggs will win. The real danger are these parents, who hate the schools anyway because they, the parents, think it is *their* right to tell their children what "values" they should have.[1] Two bottles of excellent Mumm, which left no hangover whatsoever, and delicious kedgeree fixed by Don.

Major breakthrough on Pussns' civil liberties front: we switched places in bed and slept with the second window also uncurtained; which will be permanent, I guess. Actually, the light didn't bother me.

From a birthday greeting from a fan: "Our gay pride is built upon the dignity of your life and work. Our love finds honesty in your words." This should poultice my smarting vanity because of having lost some writing award by USC to—Joan Didion!

Very important decision (if kept): To make japam *with beads*, no matter how briefly, at noon as well as morning and evening.

Nearly a decision: not to write foreword for Vividishananda's book. It is quite quite unreadable.[2]

August 28. Shed tears as I read Masefield's poem "Biography" this morning while shitting. Perhaps the saki hangover helped to make me "tender" (in the Quaker sense).[3] We had supper with Nick Wilder at the Imperial Gardens, with three others; it's an unsnug Jap clip-joint. One of the many sinister features of this period is the fantastic amount of money spent by most of the people I know on restaurant meals. This period isn't—I mean, in its *feel*—a prewar period like the thirties. It's much more a predisaster period. One expects some huge single catastrophe with a very short timespan—an earthquake, a collision with an asteroid, a rocket attack which wrecks everything instantly. This is much less depressing.

Noon japam made today.

August 29. A reaction to my *At One With* interview with Keith Berwick:

> Dear Mr. Isherwood: I had nightmares all night. I was so disappointed to find out that you are a faggot. How do you think you got here, because your parents were faggots? You have the

[1] Proposition 6, initiated by State Senator John Briggs, aimed to bar gays, lesbians, or anyone advocating a homosexual lifestyle from teaching in California state schools.

[2] *A Man of God* (1957), being republished in paperback; Isherwood did write the Foreword; see Glossary under Vividishananda.

[3] Receptive to the light of God or the inspiration of the Spirit; see February 28, 1961 in *D.2*.

power to lead many, the pity is that it is in the wrong direction. God will get you for that. Sincerely, Bonnie Purchase(?) P.S. There's still time to change.[1]

Dumb Dobbin on the beach irritated Kitty because Kitty started discussing Tony Sarver, whom they'd seen the night before, and Dobbin thought they were talking about Tony Richardson. Dobbin said, "I think he's the most difficult person I know"!

Noon japam made today. Henceforth I'll only remark on it if I miss making it.

August 30. A colorblind model from Bolton, Lancashire, who has just become Miss England in the beauty competition, is named Beverley Isherwood.

Last night, Don said that he really doesn't want to do this book of his drawings and our comments on the sitters, because he feels that it is the wrong way to appear before the public as an artist. I think he's right, but we're going to discuss it some more.

August 31. I meant to mention yesterday a good example of a really "in" joke. On the evening of the 29th, we went with Rick Sandford to see *The Sound of Music,* which Rick admires greatly and we very much didn't. As we drove home, I was saying how Rick's enthusiasms are always to be respected because they are so earnestly held and supported by so many impressive arguments. Don said, "Yes, he really believes in Diana." I wonder how many people in this whole city would have recognized the allusion to [Shaw's] *Androcles and the Lion*?

Edward writes today to say that his son Christopher has multiple sclerosis, and that he himself is having attacks of vertigo which prevent him from working on his short prose pieces. In fact he is going through another period of block, just as he did when he was trying to finish his trilogy—only this time not so violent. As for Christopher, that's too cruel to bear thinking about, and I would be really upset about it, except that I scarcely know him.

[1] Berwick—Canadian-born, American-educated historian, journalist, T.V. broadcaster, teacher, and former Santa Monica neighbor—recalls that Isherwood said on the show (taped May 13 for NBC): "I believe that by being absolutely frank about oneself, one builds the most surprising bridges with other people. If you make the kinkiest confession, reveal the secret of secrets, expose something that no one else has experienced—sure as hell you're going to get a letter from a stranger saying, 'That's exactly how I feel—how did you know?'"

The seventh volume of Byron's letters and journals has just arrived and I've started to read it.

September 1. When I asked Elsie Giorgi about multiple sclerosis on the phone today, she told me that it is sometimes wrongly diagnosed because there is a tumor (benign) which sometimes forms in (?) on (?) something called the foramen magnum and produces symptoms of sclerosis until it is removed. She claims that she once detected such an error in diagnosis and that the doctor who had made it would never forgive her for it. Am writing to tell Edward this. There is just the mad millionth chance—and I keep thinking how he has so often said in the past that I've brought him luck with publishing his books—

September 2. Starting to read John Lahr's life of Joe Orton,[1] in a ready-to-become-hostile mood because, in his foreword, he says of his wife Anthea: "... her insight into the neurotic patterns of Orton and Halliwell are (*sic*) present on every page." So we are to bow to the authority of this cunt. Well, we'll see.

I have missed two writing days, disgracefully. And all because of a problem which has cropped up again and again in writing this book—the brute problem of arrangement: which bit of information comes where, how should information be fed to the reader. I suppose historians have this problem all the time, and indeed I have been a historian before, but here it seems particularly irritating and acute. And am I, also, in writing this book, continually making an apology for the subject matter to Edward? Maybe so, and maybe this is as it should be. Most writers about holy men just go ahead, taking their own devotion for granted and not giving a shit how it will impress the unbelievers.

Harry Rigby called and told us, this morning, that Keith Baxter now has his "green card" and could therefore play in New York. But nothing has yet been settled for Simon Ward.

September 3. I have been toiling all day and am exhausted, but at least my book is restarted, after an absurd block caused by difficulties over a time question—should I or should I not refer to a later event in the same sentence. Really, such nitpicking seems scarcely sane.

Tonight we have a huge dinner party which was really begun in order to prove to tiresome Tony Richardson that we are not

[1] *Prick Up Your Ears* (1978).

"neglecting" him, as he told Gavin. Well, offer it up. Don has been drawing "Divine," who is an enormous soft-voiced charming curiously dignified roly-poly pudding.

September 4. A bad hangover after a grimly dull party. Everybody seemed dreary, including the usual drears. Gavin gallantly tried to make conversation, telling how he went to a man who told him about his previous incarnations, he was a dolphin, a hater of Jews, a lover of a sea captain. Gavin was impressed because the man said he had very often been born in Africa, which would explain his love of Morocco, he thinks. Oh, I am so depressed, and Don is too but won't tell me why, yet. And this should be a wonderful day, hopeful and positive, Labor Day the start of the working season of my year.

Tony said he doesn't like Orton's plays, finds them "thin."

September 5. Another night of disaster. A Chinese restaurant downtown, the Changsha, tables all crowded together. Cukor weary and old, feeling his bad back. Two dragged-in guests, an assistant producer and his boyfriend. Kitty ready to explode with resentment, nervous irritation, and the strain of driving us all when he was at the limit of exhaustion after last night. Also, the airless wet heat had a lot to do with it. Today it's raining.

Have introduced another bit of self-discipline: to do Wittenberg's alternate isometric program every day.[1] (Will mention whenever I miss it.) The ankle press I must omit, however. I did it yesterday and it hurt my bad knee a little.

I have talked Amohananda into accepting the fact that I will not write an introduction to Vividishananda's book. He was very understanding about this, maybe because nobody at Vedanta Place really thinks it's important. Maybe they just tried it on me, not with any bad intentions but just because they cannot, will not, understand that it could be the least trouble for me, a writer, to scribble off a little something about anything, anytime, in no seconds flat.

September 6. Had an unpleasant dream—not quite a nightmare—and woke last night about 4 a.m. It is very rare nowadays for me to wake in the middle of the night or to have this kind of a dream; yet it wasn't altogether unpleasant. I was in some kind of

[1] Devised by American wrestler Henry Wittenberg (1918–2010), 1948 Olympic gold medallist.

a gothic-looking old house, English rather than American, with lots of ivy, and there I was visited by a group of somewhat sinister youngish creatures, I guess they were demons. The atmosphere was menacing but with smiles, I was aware that it might get out of hand. I remember groping what I thought would be the buttocks of one of the young creatures and finding that it hadn't any; in fact, there was a concave curve covered over with what looked like tinfoil. The dream ended with my saying somewhat coyly, "You know, I've got a *word* which would make you get out of here"—meaning Ramakrishna. They seemed to agree that it would.

On the 3rd, Don did three magnificent drawings of Divine—one of them when he was half asleep, looking like an adorable zonked seal—in another, one of his hands sticks out like a flipper. Don says they are good because Divine is his kind of person, which so few of his sitters are.

September 7. Last night we saw Visconti's *The Leopard.* Although this was a terrible print, faded to sepia, I was far more moved by the film, this second time; it seemed almost a kind of cynical *War and Peace.* This made me read about Garibaldi this morning, starting with D.H. Lawrence's *Movements in European History,* a sloppily written book.

I love this, from Joe Orton's *The Ruffian on the Stair:*

> Wilson (speaking about his dead brother): We had separate beds—he was a stickler for convention, but that's as far as it went. We spent every night in each other's company. It was the reason we never got any work done.
>
> Mike: There's no word in the Irish language for what you were doing.
>
> Wilson: In Lapland they have no word for snow.

September 8. Speaking of *Tennessee Williams' Letters to Donald Windham* (which he is just now reading), Don, on the beach today, described Windham's literary style as "exquisitely minimal." A few moments later, a huge wave shot up the beach slope and drenched our blanket. A glorious day—the kind on which one can't exactly feel old, only full of aches and stiffness. As I hobbled back, along familiar Mabery Road, where I've passed back and forth, drunk and sober, ever since I used to go swimming with the Viertel family in 1939, I met old Madge MacDonald, another survivor, and was complimented by her on my healthy looks. My survival reassures her, and yet, if I were to die, that would be another kind of reassurance—that she was going to live longer than anybody. As for my aches, they are comforting, in a way; they make me feel lazy and inclined to stretch out and doze. Gerald Heard had a dictum, of which I can't exactly remember the wording—to the effect that the chief pleasure of old age is relief from pain. He put it like that because he loved pessimistic-sounding utterances. The statement in itself isn't really pessimistic; all it means is that there is a deep pleasure in relaxation and relief from the feeling that it is one's duty to do something or other. Also, let it never be forgotten, relief from pain *is* happiness, because happiness is our nature—except, one sometimes suspects, in the case of Jews.... Such happiness isn't *necessarily* of an inferior kind.

September 9. Today I've finished, though not properly polished, chapter 12. Eleven pages long, bringing the total revised manuscript up to page 138, which means that I've dropped four pages behind the first-draft manuscript.

Waiting for Don to get back from his many chores to find if I shall have time to attend a party which is being held by Henry Jaglom[1] against the Briggs initiative. It is to convince heterosexual liberals in show biz that they ought to oppose the initiative. Jaglom, a heterosexual Jew, says he is amazed to find how scared a lot of notoriously liberal actors are of attending a "pro-fag" party. I do feel I ought to go, if possible.

September 10. I couldn't go, because there was no way of communicating with Don and finding out when he was coming home. Then we had another dreary party—Jack and Jim, Gordon Hoban and Bill Franklin—to watch the first episode of the T.V. series based on Jim's *Paper Chase.* It didn't amount to much, but Jim reports this morning that it has been very well received. The party was dreary because three of the guests didn't drink and because Bill, who did, is dreary anyhow.

Stathis Orphanos, back from Greece, reports seeing Bill Caskey in Athens, drunk and belligerent and beat-up and with legs caked with dirt, because his water in the apartment had been cut off.

Prema just sent me a book of spiritual talks by the Dutch swami Atulananda who died in India aged ninety-six. I think it's going

[1] Actor, film director, screenwriter (b. 1940, London), a protégé of Orson Welles; he edited *Easy Rider* (1969) for his friend Dennis Hopper, produced documentaries about Vietnam, and later directed films based on his life in the movie business.

to be valuable for me in considering the later part of Swami's life. He says: "When the mind is on the ordinary plane, then all, even the saints, are mere human beings and behave like ordinary people. Thus it is said that it is difficult to live with sages."[1] But, surely, the memory of the higher plane must, to some extent, affect their "ordinariness"?

September 11 [Monday]. Talked to Robert Rauschenberg last night, at a party given for him in Venice, at Chuck Arnoldi's studio. Rauschenberg reminded me so much of Tennessee, but louder, vulgarer. We laughed a lot—I out of embarrassment. Rauschenberg said, "It's wonderful how much you can do with laughter," which made him seem suddenly very shrewd.

Dobbin got the week off to an auspicious start by remembering to call Natalie Leavitt early this morning, to ask her if she had seen one of the two small squares of black rubber which Kitty uses as grips for the barbells when he goes to the gym. She said no, she hadn't. But then Dobbin found it wedged inside the rim of the cover of the washing machine. Triumph! Dobbin was kissed and hugged and made much of. But what will he lose or forget before the day's out?

September 12. While polishing chapter 12, I found that I dropped some more material and ended on page 137 instead of 138. Today, reading ahead through the first draft of the book, it seems to me that I shall have to cut a lot as I go along; far better to have the book shorter, unless the material is going to be really to the point and really about Swami and our relationship—nothing else has any place in it.

Last night we had a goodbye supper with Gavin, who leaves tomorrow for Morocco. He had chosen to entertain us in an Algerian restaurant called Moun of Tunis—tacky and overpriced, with inferior food, tables too near together, seating uncomfortable. Jack Larson and Jim Bridges were with us. Jim has let himself get involved in the No on Briggs movement and now he's a bit scared. He gave them a thousand dollars and appeared at that party and feels he's a marked man. And Jack talked pessimistically about how rough the campaign is going to be, with the mafia and big

[1] *With the Swamis in America* (1938) by Gurudas Maharaj, also known as Atulananda; edited in the 1970s by Pravrajika Brahmaprana and reissued as *With the Swamis in America and India.* Atulananda, one of the first western swamis in the Ramakrishna Order, lived in monasteries in the U.S. and India.

business involved on Briggs's side. He remarked that he's worried because their house is made of wood; some hostile activist might set it on fire. Got drunk and depressed.

September 13. (A big holdup here, because I noticed that the bottom has fallen out of the e on my typewriter, so had to switch on Don's, and now, fuck it, Don's ribbon is so faint that I'll have to changc it, but not yet.)

Last night we had supper with Billy Al Bengston and Penny. Bob Graham was the only other guest. I have come around to liking him quite a lot, largely because he seems genuinely impressed by Don's work.

Billy showed us watercolors he did while in Hawaii. They seemed even better than usual, but this was partly because I had a sudden insight into the quality of Billy's art in general. Can't describe this, except by saying that it all seemed to relate to itself as one enormously long chain of variations on his themes. And the big red painting which was near the dining table seemed so powerfully lively and exhilarating that it became like a band playing loud brassy seaside music. I was drunk, yes; but drinking doesn't ordinarily give me perceptions of this kind. It was as if I had been stoned.

Couldn't stop reading Armistead Maupin's *Tales of the City*, which I got in the mail from the author yesterday. The mood of it, the kind of campy fun, is perfect of its kind. I kept thinking how Wystan would have loved it. Somehow, though so different, it made me think of the novels of Ada Leverson.

Just changed the ribbon. Have been listening to the radio but no news so far about the threatened postal strike.

September 14. Michael di Capua called this morning from New York, wanting to know when my Swami book would be finished. So I kind of promised the end of this year, which has now thrown me into a flap. Can I really manage that? Theoretically, yes—practically, I'm not sure. My "Muse" is such a *bitch* sometimes, and it knows that it has the whip hand.

Also a call from James P. White. After all these years, his wife has suddenly become pregnant. And can he please name the child after me? He takes it for granted that it'll be a boy, apparently. I had to say yes, of course, and was then told that the second name would be—*Jules*!

September 15. Last night, we ate with Paul Sorel as his guests at La Ruche, surrounded by politely insolent Frog waiters—the only people with whom Paul seems truly at ease. He made one of his strange boasting self-deprecating speeches about how half of him is crazy or evil or something, and will spend absolutely any money it can get its hands on. This whole speech keeps bringing in Chris Wood, so that one feels his presence powerfully. Paul speaks with great affection of him, yet keeps pointing out that Chris really despised him and would only give him money in the form of tips, as it were, never in the form of, say, income property to provide for his future. Once, they decided to split up altogether and Chris gave him thirty thousand dollars. Paul spent it in a few months. Now, he seems to be getting through most of his available inheritance money, according to him only about twenty-nine thousand dollars remain; after that, he has absolutely nothing except the house, and his car. Today on the beach (which was glorious but already autumnal) I asked Don if he thought Paul is hoping to get us into the mood to help him when he is finally broke. Don said he didn't think so but maybe; anyhow we ought not to give him a cent, although we could continue to buy him meals.

Have just reread Gavin's *A Case for the Angels.* It is actually the Case for Lambert versus Clint Kimbrough. I found it curiously depressing and lowering, although it's so well done.

September 16. Was so delighted when dear old Muhammad Ali won last night and became champion for the third time.[1] What does one demand of a hero? That he shall be admired, prepared to risk his fame by doing something unpopular, that he shall be sexy, that he shall be a comedian, that he shall be vulnerable, losing sometimes as well as winning, that he shall have weaknesses, such as boasting, that he shall nevertheless not be corrupt.

Running down to the beach, I noticed, as so often nowadays, the bad state of our hillside with its dangerous slippages, the bulge of the brick wall below our deck, and all the cracks. Thought, it's like me, with my eye going dim and my head getting bald and my sore tongue and the tweaking pains in my bad knee. But the house alarms me more: it's collapse seems a much more catastrophic event than my impending death.

Reading Laurens van der Post's *Jung and the Story of Our Time.* I think this is going to be really interesting, but his sentences are like

[1] Leon Spinks (b. 1953) won the heavyweight title from Ali on February 15, 1978, and on September 15, Ali won it back from him.

long limp bits of wet string. Does he think it's somehow *insincere* to write well?

September 17. Amazing, what a block I have against getting on with this book. It's a question of tone. I don't know exactly the right tone of voice in which to narrate it. Never before have I been so aware of the truth of [Cyril] Connolly's accusation that I desire to ingratiate myself with the reader. And yet, just because of that, I know that there is a danger in the opposite direction: The lady doth confess too much. (Just as I was about to write this misquotation, a doubt struck me. I'd thought it was from *Macbeth*—no, *Hamlet*.[1]

September 18. Last night we saw *Days of Heaven*, directed by Terrence Malick. The story lacks force because it is a heavy sex-drama in the Italian tradition, treated artily. The treatment, the artiness, in itself is beautiful—a shot of a train crossing a high bridge against the sky, the clouds of a thunderstorm over the prairie, a wineglass lying underwater in a stream—the first of these made me actually gasp out loud. The effect is curiously like something by Virginia Woolf, especially in *Mrs. Dalloway*. The trouble is that, if you attempt this, if you try to show all kinds of outside sights and sounds of nature surrounding the human drama, then you must somehow relate the whole thing to its parts—that's the difference between art and real life. In real life, all the parts *are* interrelated, but not evidently—only by their existence within Brahman.... Still and all, this film was truly an experience, visually—otherwise it wouldn't have made me write all this rather confused stuff.

A glorious brilliant windy day. We ran down to the beach and went in the water.

September 19. A hangover from Clytie's birthday party last night combines with hot Santana weather to rob me of my strength. Only the ocean was bracingly cold. An interesting intimate talk with Don on the beach—he was in an unusually happy mood, saying that everything at that moment was perfect. I see now much more clearly why he needs someone like Bill Franklin, a sort of familiar; also I felt how heterosexual he is in his homosexuality. We agreed that he had had only one boyfriend with a good character, Bill Bopp. Bill Franklin, says Don, is ugly, undersexed,

[1] "The lady doth protest too much, methinks." (*Hamlet* III.ii).

Jewishly competitive, sorry for himself, always worried about his health; but Don feels he can connect with Bill (that's the word he used) and he likes Bill's ass. Don also said, "I couldn't possibly be an artist in the way that David Hockney is." He describes his being an artist at all as almost accidental; it arose out of an interest in people, not in art.

Natalie was very drunk last night while cooking. I fear she may have smashed some more glasses. She always disarms us with her joy over Thaïs's success. Thaïs is now to dance the lead in *Death and the Maiden*.[1]

September 20. Another hot brilliant day, with the ocean water even colder. Walking back from the beach, I overtook a little blonde girl, five, six, maybe seven years old, pretty but unmistakably weird. She said "Hi," I said "Hi." She said, "I'm wearing two lots of clothes, I had a heart attack." She *was* wearing some kind of a tank-top shirt with a jacket over it. She said, "It wasn't really a heart attack," and then mumbled something I couldn't understand. Then we got to the house she was evidently living in, one of the smaller ones, on the south side of the street. She said, "Do you want to come into my house and do it?" Maybe this had some quite innocent meaning, which I'd missed by not having heard all of her dialogue. But I didn't want to get involved, so gave her a smile and waved my hand and went on.

September 21. We had a birthday party for Penny Little last night—her birthday is really on August 7, but she was away then in Hawaii. Except for Billy Al, all the guests were female, which made, for us, an agreeable change. The girls got together in a bunch and entertained each other merrily, with a certain amount of lesbian flirtation. Billy Al said to me, "Don's twice the artist David is." Don says that Billy Al is very shrewd—which I know—and that he sees into the real David, whose greatest inspiration is ambition.

We scream at the neighbors' dogs. Last night, being a bit drunk, we screamed a lot, at midnight, and this resulted in a furious call on my recording machine from a Mrs. Lawrence Davidson, who lives below at 242 Maybery Road. However, when I called her up today, she seemed quite on our side. Her bedroom is right next to the neighbors with the dogs and she hates them as much as we

[1] Natalie was not drunk, but taking tranquilizers prescribed by her doctor. Thaïs Leavitt, her daughter, was a ballerina; see Glossary.

do. In fact, she contradicted herself completely. She promised to go round to the neighbors tonight and speak to them about the barking. I doubt if she has. Anyhow, the barking continues.

Another wonderful beach morning today.

September 22. A great "face the brute" drive this morning, against my accumulated mail. Not that I answered any of it, but by sorting it and tearing up quite [a] few letters, and noting that others could be answered by phone, I somehow reduced the pile from around fifty to around thirty.

Worried in the night—I was awake an hour or two, unusual for me—over my book. Have I explained to the reader enough about Ramakrishna, Vivekananda, etc.—and, if I haven't, does it matter?

September 23. Worrying about the Swami book. To date, I have one hundred forty-two pages revised, which leaves one hundred eight pages of the first-draft typescript. Of these, the last forty-eight pages consist almost entirely of direct transcription from my diaries, so that very little alteration will be necessary. The real work will be on the next sixty pages, and even these contain a fair amount of transcription. So courage, Dobbin. Whether the manuscript will add up to being a book, that's another question.

This weather is too hot for me, I rather hate it, it is getting me down. Rectal bleeding this morning; hope it's just a pile.

September 24. The "rectal bleeding" was merely the coloration caused by eating beets, it seems!

Very hot today, maybe the hottest of this hot spell. Decided not to go to the beach; a hot Sunday is intolerable there. But I often think romantically, on such weekends, that, by the law of averages, there must at least be two or three couples of boys who will meet for the first time and plunge into a real love affair, maybe *the* affair of their lives.

We saw Frederick Brisson yesterday, because he's considering becoming our producer for the play. Don thinks he's maybe queer. We both think he may turn out to be dictatorial and clash with Marre. For the first time, we heard from him of a possible alternative to New York or London for our opening—to go on a tour in this country, of Los Angeles, San Francisco, Chicago, etc. and not come into New York until the fall of 1979.

Have just finished John Lehmann's *Thrown to the Woolfs.* All you can say in its favor is that it's a necessary document for the record. But does John really have to write so boringly? I don't say this merely as a put-down. He makes me ask myself, am I perhaps being more boring than I realize, in my Swami book? The writer of any kind of autobiographical book is in deadly danger whenever he is trying to get from point A to point B in a hurry—when, that's to say, he isn't interested in what he's immediately writing. Somehow or other, one must *make* such bridge passages interesting. There are many of them in my narrative, and that is really what's worrying me.

September 25. This heat is really something! We neither of us felt up to running down to the beach today; there's quite a bit of smog in the air, even down here. The temperature *at the edge of the ocean* is said to have been 87 yesterday.

Last night, in deep heat, we saw poor old Vincent Price do his Oscar Wilde show down on the USC campus. We both felt he was wrong for it, despite his technical know-how. He certainly did not make you feel that Wilde was a very good man, really compassionate, although he had the lines to speak of this, and he didn't show Wilde's strength. Sometimes, his vanity was wrong too, it seemed the vanity of a man unsure of himself and therefore aggressive, more like Paul Sorel's.

September 26. Don has gone out to see *Marie Antoinette*, but I can't face Norma Shearer, and this foul heat which continues. Have just been to dear old Dr. Maxwell Wolff, who pulled a little growth out of my nose, not serious but something, and Don was worried about it. It was surprisingly hard and long. Now, in an odd, somehow rather reckless mood, I am drinking a bottle of Superior (Mexican [beer]) before watching a segment of Jim's T.V. "Paper Chase."

Last night, a truly dreadful party at Leslie Wallwork's—I can never imagine where he *finds* his guests, all middle-aged closet queens; maybe they're business associates. Really I do *loathe* nearly all parties. I don't think Don realizes quite how much I loathe them. It's because I am always being fed to someone or other and I am charming to him, her, them, and I *bore* myself being charming. The only exceptions are beautiful young men, and even with them I feel subtly humiliated by the energy of my performance. I am perfectly happy by myself, much as I love my darling and his furry nearness, and I nearly never want to go out on these social evenings. If only I could spend the last bit of my life alone with Don or alone.

September 27. Perhaps a shade cooler. But we have to go to Pasadena tonight. Keith Baxter has arrived, we see him and his hostess, Brenda Vacaro(?)[1] tomorrow. And now Keith Addis has come up with another of his improbable jobs—to make a T.V. spectacular out of a sci-fi bestseller called *Dune, Doone*(?).[2] And I am still being plagued by the editor of the revised *Gay Liberation Book* to somehow fix up the miserable mess of an interview produced by that lazy sloppy boy, Len Richmond.[3] I think I shall refuse to let him print it—even if I did improve it, it would still be dull and weak in comparison with the quite good one I did for the *Gay Sunshine* collection.

September 28. A really bad hangover after that long hot evening in Pasadena. First a party at Harry Montgomery's,[4] the kind in two acts—act 1 squares with an academic flavor, act 2 screaming fags in the pool; the difficulty is getting the curtain down and up again between acts. We only took in part of act 1. At a very very dull party like that, a newly arrived friend seems far larger than life, so exciting, so delightful, so utterly welcome. Then on to Cal. Tech., to see Laddie Dill's pictures, including the one they borrowed from us. I do like most of them very much and am rather proud of being able to like them so sincerely, as I have my blind spots in this kind of art. Then to have supper with Don Sorenson,[5] with Nick Wilder [and] Jack Woody and his lover Tom Long. Nick has done his first (and hopefully last) stint in jail, two days. We deserved all of that for getting so drunk and driving home. Don, like Jim Charlton, is almost totally trustworthy drunk, partly because his head is very strong; but he felt bad about it afterwards. Incidentally, this morning, Kitty was very very sweet to old Drub. So sweet that Drub is wondering—?

September 29. Another hangover—this is getting to be dreary.

[1] Vaccaro; see Glossary.

[2] *Dune* (1965), by Frank Herbert.

[3] *The Gay Liberation Book* (1973) and *The New Gay Liberation Book: Writings and Photographs about Gay (Men's) Liberation* (1979), edited by Len Richmond and Gary Noguera. Noguera evidently redid the interview; see October 17 below.

[4] D. Harry Montgomery, partner in a local commercial and fine art printing company, Typecraft, Inc., started by his father-in-law, former publisher of the *Pasadena Independent Newspaper.*

[5] American artist (1948–1985), trained at California State University, Northridge; he first showed at Nicholas Wilder's gallery in 1975 and worked prolifically in many styles until his early death from AIDS.

And there was no need for it, we only had two people to dinner, Keith Baxter and Brenda Vaccaro. And Keith I think I really like. Anyhow, as Don says, he is perfect for the part of Patrick. He has hot dark sexy eyes and looks immensely strong. He is very polite—remembering, for instance, to address both of us when talking about the script. This morning he came again and Don drew him and he was full of praise. His suggestions for script changes were woolly, however; both of us thought so. A very pretty, tall boy named Mark came to pick him up.

Don liked Vaccaro too. I wasn't so sure. A funny misunderstanding when she came in, and exclaimed, "What a *warm* house!" I thought she meant it was stuffy and said something defensive about the weather being so hot. She had meant that it had a warm feeling—i.e. hospitable, cozy, etc.

September 30. Sore tongue, tiredness from the heat, although it seems to be letting up. Last night, we had supper with Paul Sorel, who was to leave for England today. Again he brought up all this stuff about believing in his luck; somehow or other, he would get some more money. I suggested that he should call Mirandi Levy, whom he knows well and who is very practical and worldly-wise, and tell her the exact facts of his financial situation [and] ask her what she advises him to do. Don thought this a good idea too. Paul did call her, but she was out—he told me this on the phone last night. Now, presumably, he has left.

Some worry that Ted may be going mad again. He seemed very odd to me on the phone a couple of days ago, when he talked of the Pope's death and said, "Now we'll have to play pick-a-pope again!"[1] with a wild uncharacteristic laugh. Don has noticed something too. But we saw him on the beach today and he seemed fairly normal.

Awful grinding at my book. I get into moods when I simply don't feel the slightest interest in what I am writing; it all seems mere word splicing.

October 1. It has been foggy all day down here by the beach, but there is still a feeling of heat behind it. A long talk with Kurt Yount. I always feel how much I wish he lived around here, then I know I would be over at his place a lot, reading aloud to him, for one thing. It is tragic, the authors he hasn't been able to read

[1] Pope John Paul, elected August 26, 1978, died thirty-three days later on September 28, the shortest reign in papal history.

because, really, so little is published in braille or recorded. In his cheerful unsentimental way, he tells how he gets "bored" and therefore drinks himself drunk all by himself.

Don is again worried about Ted—especially as he is about to take a vacation, which is always a dangerous period. If Ted flips, how shall we manage about Glade? She keeps referring to "the person who lives with me." I said to Don, wouldn't it be funny if he looked in one day without warning and found Glade sitting up with a real old-fashioned guardian angel wearing huge white wings?

October 2. Bad news. Oliver Andrews got a heart attack while diving and died of it or drowned, we haven't heard which. Worse news—let's hope it's not true—Ingrid Bergman said to be dying of cancer. That really shocked us both.

It's cooler today, but I feel very tired and am slipping behind with my work.

Alan Bates is really one of the best actors alive; his Mayor of Casterbridge is magnificent.[1] He could play Lear in a new way, I feel.

Woke up in the middle of last night. Foggy, dead silent, rather too warm. As so often, lately, I thought there was going to be an earthquake—felt everything waiting for it.

October 3. A girl named Nancy Monk[2] came here with a girl-friend this morning. She is blonde and boyishly good-looking; probably they are lovers. The sculpture, about the size of a small book, is called *Send Me Some Sand for My Shoes.* Part of it is grey blue, most of it is white. I didn't know what to say about it and so covered my embarrassment with facile chatter about the slide situation of the cliffs. Nancy had written me a very nice letter of admiration for my work. Was it worth seeing me? No—under the circumstances—but they will never admit that, or not until I am dead and they are old.

A fun rumor from Rome—that the late Pope may have been murdered. There was apparently no autopsy; it was just announced that he died of a heart attack.

October 4. At present I'm living an exceptionally compulsionistic life. It is certainly better than my usual laziness, but it *is* compulsionistic, like square hopping. I have to do my various specimen

[1] In the 1978 BBC T.V. adaptation of Hardy's novel.

[2] Artist and teacher; she later showed photographs and mixed media work at the Craig Krull Gallery in Santa Monica.

tasks each day—my midday beads, my isometric stretchings, my bit of the Swami book, and this diary. Perhaps it is the only way I can function now. Well, at least I'm functioning. It's around eight in the evening, Darling has gone out mousing and I'm alone for the night, quite snug, sipping a glass of Superior Light Mexican beer. Later I shall watch *Network*, which we both hated when we saw it as a feature film, and probably enjoy it, because all values on T.V. are scaled down, especially when you're eating a T.V. supper—Mexican, tonight.

A 5.6 earthquake in Owens Valley. And mysterious earthslides in Laguna Beach. We're getting a bit worried about Jack Fontan and Ray Unger, having called them both at home and at their gym and gotten no answer.

Last night I dreamt that Don had painted a portrait of somebody sitting in a garden. I was rather excited, feeling that the introduction of the garden was some sort of a breakthrough for him. Actually, he is painting very well indeed just now, without showing the need of any such adjuncts.

October 5. Hated *Network* as much as ever, it's such desperately strained satire. Earlier in the evening, Ray Unger called and said he and Jack are all right; the slides took place in another canyon altogether.

Today, we await a visit from Armistead Maupin (pronounced More-pin), the author of *Tales of the City*, and his friend Ken Maley. Am curious, at least to see what they'll be like.

Kitty and his mouse smoked till the mouse fell asleep, so there was no gobbling; and then, this morning, the mouse's mousefriend appeared without warning and so Kitty had to pitpat away pronto. The mouse has proposed a return gobble this evening.

This morning I did something odd, read nearly all of Byron's "The Corsair," *for the story*. Which is how it was originally read, I suppose, and why it was so popular. I'd never read it before.

October 6. The evil breath of the Briggs Proposition 6 begins to be smelt terribly strong. The polls say it will win. And it's certain to be followed by worse. Nevertheless, feeling miserable about this and feeling rage against Briggs and the other swine is a deadly temptation. A. One has to do something about it. B. One has to get on with one's own work. Otherwise, this despair and indignation evaporates in mere tamas.

Armistead Maupin was unexpectedly attractive and youthful. Not all that much so, but I realize I was expecting something terribly closet-elegant. His friend and promoter was also young and pleasant. They might become friends, don't know yet. They've gone back to San Francisco today.

The return gobble of the mouse took place and was a success.

October 7. A somewhat reassuring talk to Bill Scobie today—I mean, reassuring about Briggs. He is still ahead but only a point or two—although there are many undecided, and, as Don says, undecided are usually conservatives.

This afternoon, out of the blue air, David Hockney called. He is right here in town despite Tony Richardson's news that he would remain in New York for a long while. And he's in the house he has rented, although he was said to have given it up. David is the Rommel of the art world, with his bold surprise moves and sudden appearances on one's flank. . . . Just the same, we're both happy that he's here.

October 8. Yesterday evening wasn't a success. David was tired and not his usual cheerful self. Also he'd brought with him dreary Maurice Payne. And we had to have dreary Bill Franklin. Mark Lipscomb, admittedly undreary but uninvited, came too. Sometime during the evening, Mark whispered a pass at me, rain-checkwise, he wanted us to do it sometime—"Before you die" was unspoken. He hinted at this once before. I was pretty drunk and now can't remember what I answered.

October 9. Oh, the sadness of going out with Glade yesterday! Don and I gave her her dinner, we just drank wine and watched her. All she could talk of was the weather, she was afraid it would be bad next day. And her poor ankle, turning over like rubber and nearly making her fall down the steps. It seems touching that she always knows who I am when I come to see her, she even knows my name.

A glorious beach day, with refreshing waves. But we both have mysteriously upset stomachs. And a party looms grim on the evening horizon. David Hockney, the Tynans, Cukor, Dagny Corcoran, Tony Richardson and a woman he asked to bring, Tamara Asseye[v],[1] Leslie Wallwork, and *The Downer.*[2]

[1] American film producer, educated at Marymount College and UCLA; she was a production assistant for Roger Corman in the 1960s and worked in T.V. Her films include *Norma Rae* (1979) and, for CBS, Richardson's "The Penalty Phase" (1986). Isherwood typed Asseyed.

[2] I.e., Bill Franklin.

October 10. I will perhaps write about the deeper significance of the party some other time, when I see it at more of a distance. Superficially it was a "success." Ken Tynan carried on about Louise Brooks, thereby oddly annoying George Cukor, who seems to disapprove of her chiefly because she left the movies before they ceased to want her, finding the life of sound pictures boring because you had to learn your lines in advance of time and therefore couldn't stay up all night at parties, as you could during the silent picture days. Ken says he had a four-hour sex talk with Louise, during which she told him that women can ejaculate just as men do; she herself had been masturbating recently and had shot a spurt of liquid right across the room. He also told me that she is anti-queer and that this is characteristic of women whose lives are centered on sex—they have no use for men who don't desire them.

October 11. I think the sun has now definitely set beyond the headland, into the sea, but can't be certain because of low-lying clouds. I creep on with the Swami book. My old head is so thick and stupid it's brutal. I fight my way on, sentence by sentence, and always a cold scornful remnant of reason waits for the next morning, when it looks through the latest page and says, idiot, can't you *see* that that sentence ought to be the other way around, and that that adjective is utterly wrong? Are you really so senile? And it's right—I do see it.

Chat with Cukor, who, like ourselves, is invited to have dinner with the Tynans to meet Princess Margaret on Friday. We've been told to call her "Ma'am," and George is working himself up into a nervous cramp, like a young actor with only one line to say; he feels sure he'll get it wrong and call her Mum or Mom or something.

An attorney put his hand into his mailbox—I think it was in the Pacific Palisades—and was bitten by a diamondback rattlesnake. It is thought that the rattler was planted there, as a revenge. I shouldn't like to have the job of putting a rattler into a mailbox. Oddly enough, I was objecting to an almost similar situation in [Agatha Christie's] *Death on the Nile*, where a large cobra is planted in Poirot's cabin. I pictured the practical difficulties of doing this. Also, this morning, *before* hearing about the mailbox rattler, I had been thinking of *The Speckled Band*, which ends with Sherlock Holmes putting the swamp adder into the safe and locking the door—and imagining how Dodie Smith would have exclaimed, "What a *horrible* story!" meaning simply that the snake gets shut in without enough air—and grinning to myself.

October 12. Gavin Lambert called unexpectedly to say he is back in town. We had supper with him yesterday evening. He says that Tangier is becoming more scandalous than ever. The American Consul, who is, admittedly, just about to retire anyhow, has started an affair with a thirteen-year-old boy, while his wife has gone off to live with a seventy-five-year-old lesbian artist (whom we met while we were there). Gavin also described how the customs examiner—who is required to search outgoing travellers—actually groped him and kissed him.

A nightmare, last night, that the manuscript of my Swami book had been destroyed—by fire, I think.

October 13. The German translation of *Christopher and His Kind* just reached me. Now I have to compose a long letter in German to Heinz, to go with the manuscript, and, although I don't really expect any trouble—at least, not from him personally—and although I don't *really* care even if he says it mustn't be published in Germany at all—I feel a bit tense about the whole thing, as I always do before a showdown of any kind.

No time for more now. The Tynan dinner impends and I still haven't done my stint on the Swami book. How characteristic of me—thinking of Princess Margaret, I remember Browning: "next moment, I dance at the King's!"[1]

October 15. Missed a day because I fell and sprained my right wrist and one of my fingers. It is better today but still much swollen. I also missed out on my isometric exercises. But not my midday beads.

Meeting Princess Margaret was hardly a dance. We only got her in very short hops, and although Ken Tynan honored me by putting me at her table, I hardly got to tell her more than that I had been at a school which produced three archbishops of Canterbury in a row—that was all the royalty talk I knew; and Neil Simon's wit,[2] as he sat opposite me, was altogether silenced. She seemed quite a common little thing, fairly good-humored but no doubt capable of rapping your knuckles.

[1] From "The Laboratory: Ancien Régime."

[2] Simon (b. 1927) was already famous for his Broadway comedies, especially *Barefoot in the Park* (1965) and *The Odd Couple* (1965), and a dozen screenplays, most recently *The Goodbye Girl* (1977) and *California Suite* (1978); his newest work, *They're Playing Our Song* (1979), was about to preview at the Ahmanson in Los Angeles before opening on Broadway in February.

Yesterday, James White came from Texas on a pilgrimage to ask The Old One if it was okay to write detective stories on the side if you want to be a "serious" novelist. The Old One told him to use another name. He is rather a sweet little randy rabbit, cruising the hedges and ditches for male mating-partners, but thrilled that he and his wife are about to have a baby after many years of marriage. He talks a poor mouth but one smells lots of money. We went to a Gays Against Proposition 6 auction at the Carriage Trade Restaurant. Prospects are better, the polls show a swing toward us, but the situation is still very dangerous. One great coup, the Catholic archbishop is on our side.[1]

October 16. The capsules Elsie Giorgi prescribed, Butazolidin Alka, certainly did their job. Hardly any traces left of my sprain except a little stiffness of one finger. Yesterday night was a double miss—we went to one party and Princess Margaret didn't arrive before we left; we went on to another, where we found that Governor Brown had already been, made a speech and left. Got fairly drunk and took [Guy] Dill for [Laddie] Dill and complimented him on one of Guy's pictures!

Missed my isometrics again because I don't want to flex my wrists until I'm certain they're both all right.

Billy Al Bengston, at the second party, warned me against the dangerous effects of Butazolidin—I forget what they are. "They give it to horses," he said. "But I *am* a horse," I told him. He thought this was brilliantly funny.

October 17. We had nine people to supper last night; God how I hate these chattering mobs. Up very late, slipped behind with all my chores. And now a young man named Gary Noguera is coming to interview me for *The Gay Liberation Book.* He will be late, and that will make us late for the movies, and Kitty will rage. And meanwhile the dogs down below keep barking as never before. Father forgive.

Missed my isometrics.

October 18. At least I accomplished one psychologically gigantic feat today, I roughed out a letter in German to Heinz about *Christopher and His Kind*, telling him he could change it, or even veto its German publication, if he cared to. Otherwise, dogs

[1] Cardinal Timothy Manning (1909–1989); Los Angeles was among the largest dioceses in the U.S. for enrollment in Catholic schools.

nearly as bad, and no isometrics, because my wrist is still sensitive. Thanks.

October 19. The letter to Heinz is typed up and I will mail it tonight. Today I fussed with a passage describing how I prayed to Ramakrishna at Trabuco to help me get on with *The World in the Evening* or else stop me trying to write it. So difficult, this semi-ironical, spiritual-psychological material. So hard to hit the right tone.

Boy, this is a hard dogged climb! So slow, and yet I don't really feel any doubts. It *must* be Swami's will that I finish this book. But no more prayers for literary aid!

A talk with Don on the beach—it's still warm enough to go in the water—about the above subject. We considered my life as a writer's deliberate hunt for material—beating the bushes, as it were, to flush out the game. Every artist's doings can be considered from this angle, however much he may produce quite different and convincing reasons for them. It's a good subject for farce. Especially when the doings are of the kind called good and noble.

Last night we saw *Autumn Sonata*, with Rick Sandford. Liv Ullman[n] and Ingrid Bergman are wonderful, but oh God I am tired of Ingmar Bergman's delight in guilt.

No isometrics, wrist still hurts.

October 20. Last night I called Mrs. Lawrence Davidson, to find out what was going on about the dog barking, which has been fierce. She says she called the next-door neighbor, a Mr. Kline, and told him about the dogs and that he wasn't at all nice but the dogs have stopped barking at night since then. I offered to go see Mr. Kline, but she said please not to until we had waited to see if the dogs will be better. Mrs. Lawrence is a truly remarkable case of a woman who switches from being a nasty screaming harridan to a tiny weak crushed put-upon mouse.

Jim Gates called to say he will very soon be breaking up with Warren, and that he'll anyhow be going to live in Morocco next summer; a friend has invited him—not a lover. The tonic effect of this on him is such that he has started to read Vivekananda again!

Gavin has rented an apartment here for a year, with an option to renew. The apartment is empty, but he has gone out and got some furniture. What always amazes me about Gavin is his gumption. All alone, with a bad leg, he has the *life-belief* to go through with enterprises of this kind, to demand to be comfortable, to create a home place without anyone else to create it for. No isometrics.

October 21. Talking of Gavin, with whom we're to have supper tonight, the latest we hear of him is that he thinks he may have Legionnaire's disease but assures us it isn't catching!

A very good day of work, I think, surmounting the first of three biggish literary "cliffs" in the narrative landscape; a description of my literary block at Trabuco, which resulted in my praying to Ramakrishna about it and then finishing the first draft of *The World in the Evening.* (The second cliff is a first description of Don, at the time of our meeting; the third is a first description of John Yale.)

My darling has been his most angelic self, today and yesterday. No isometrics.

October 22. Did isometrics today, by cheating the stress on my right hand slightly, so as to avoid the sprained parts. No pain at all.

A nice run down to the beach and plunge in ocean, in my new sporty blue and white Adidas trunks. They made me run better down Mabery, such is the power of vanity, even at my age; I felt quite light footed and springy jointed.

We saw the Olivier–Jones film of [Dreiser's novel] *Sister Carrie* last night. Decided, as so often before, that tragedy *as* tragedy (and comedy *as* comedy) bores me. Even when it is well done.

A phrase from one of the questionnaire statements in *The Gay Report,*[1] which has just been sent me for a blurb: "He takes off his shoes, perhaps, and my shirt is unbuttoned or off. Or I help him take off his shoes and I feel the shoes, enjoy them, smell the leather. *I like to set an air of genteel shamelessness.*" (Italics mine.)

October 23. Really commanding pillars of smoke from two brush fires, one up north, the other much nearer, around Mandeville Canyon and the Sepulveda Pass. Hot violent Santana wind, said to be blowing seventy miles an hour in the Angeles National Forest. I have stomach upset and the shits from a meal last night at Lucky's, with Tony Richardson, who claims that this is the only Chinese restaurant which serves food like the food you get in China. The fire makes me nervous, sort of, and I didn't run down to the beach today, as I should have, because of the high wind, which I hate—but already it's dropping. All I want is to get the hell on with my book. My darling is upset too, but that's from other causes—chiefly this feeling which he still gets from time to time that people don't take his art seriously, or him either. And of

[1] By activist English professor Karla Jay and journalist Allen Young, based on 5,400 questionnaires completed by gays and lesbians, published in 1979.

course, there are always reasons, good ones, for him to feel like this. So I don't know what to say, because I know that, if I do say anything, it'll be the wrong thing. What he really needs is a good disinterested friend of his own age, preferably an artist, to talk him out of this mood and into some self-confidence. *Late bulletin*: Peter Plagens,[1] who at least counts as a serious critic, called to offer Don a trade—one of Peter's pictures for one of Don's paintings. So morale is up a bit, especially as Peter asked for a painting, not a drawing.

October 24. Another morale-raiser today, a citation sent to Don by the *Art Direction* magazine for his drawing of me on the cover of the Avon paperback of *The Memorial*!

The fire last night got really quite scary. We watched it from our deck while eating an omelette and drinking with Bill Franklin. Don was seriously concerned; I either couldn't or wouldn't believe it would get to us, though, actually it seems that it might well have if the wind had been stronger. It was coming down Mandeville Canyon, which is an extension inland of our canyon.

Yesterday I did no work at all on the Swami book, which is sinful tamas, and today I shall only do a token stint. This is disgraceful. My Chinese gut-ache persists.

October 25. Jim Gates phoned this afternoon to say that he is moving out of their apartment and going to live on his own. He will soon have a salaried job with his real estate company and no longer have to depend on percentages from closed escrows.

Am feeling (and look) bulgingly fat after a rich dinner at Perino's, given us by David Hockney last night; Mark Lipscomb and John Ladner were the other guests. In honor of the occasion, Mark was done up in a neat suit, instead of wearing those terrifically sexy cutoffs in which he often appears on the most unsuitable occasions. I am really very fond of him, except when aiding and abetting Carlos [Sagui]. And John was charming as always, and David very lovable in his Bradford millionaire persona.

October 26. Last night, The Downer declared that Los Angeles would be a dull dreary place if it wasn't for the big Jewish population. At which I spoke out, having had a few drinks—not nearly as

[1] American painter, art critic, novelist, teacher (b. 1941); he was an associate editor at *ArtForum*, 1966–1976, and chairman of the board at the Los Angeles Institute of Contemporary Art, 1977–1979. Later, he was senior art critic at *Newsweek*, 1989–2003.

violently as I did that time to Sue Mengers, but violently enough. I'll have to watch myself when he's around, because I don't really like him and I shall show it if I'm not careful, which'll be upsetting and inconvenient for my darling. Ah, *why* couldn't dear Rick Sandford be in The Downer's place? But that's Rick's fault.

October 27. Feeling depressed. Morris Kight just admitted to me that it looks like we'll lose on Proposition 6. I certainly realized this to begin with, but had then begun to hope—and indeed things have been looking much better. Of course it's also true, and not just a phrase, that we have won a victory of sorts just by getting ourselves into so much prominence.

By today's mail I got a letter from a Craig Smith, living in Illinois, apologizing for an attack on my writing and character. I do dimly remember such a letter but I remember it as being anonymous. Should I answer him? I know it might well entangle me in correspondence; he sounds like a neurotic, maybe psychotic nuisance.

A sweet evening with my darling, snugly watching *Dead End*, which seemed almost incredibly stagey, though politically daring, I suppose, in 1937. When one contrasts those *Dead End* kids with the beautiful young black murderers John Ladner has to deal with, it becomes farcical.

Jim Gates just called to tell me he has found an apartment. He told me how, after he had been away from Vedanta Place for six months and had already met Warren, Swami had asked him to come back. So Jim said to Ramakrishna: "I feel responsible for Warren. If you want me to come back to Vedanta Place, then find someone to be with Warren instead of me[.]" Which Ramakrishna has now done—but much much too late—or so it seems.

October 28. A cable from Peter Viertel yesterday to say that Salka died in her sleep on the 26th. Gavin tells me today that Peggy Hubrecht, whom we saw a lot of in Tangier while staying with him in 1976, has cancer of the liver.

We went to see *The Razor's Edge* film last night. The religious part is insufferable, all sweetness and cardboard, and Tyrone Power has a stunned look throughout, but the worldly Maugham comedy-drama scenes really work; I liked Clifton Webb and [Anne] Baxter best. Don preferred [Gene] Tierney. When we go to the movies with Rick, the atmosphere is one of extreme aesthetic seriousness. He and Don notice every nuance of direction, lighting, cutting, acting, set design; they concentrate on the film as though it were a chess game. I can't properly participate in this, because I can't seem to get a meaningful high, as they do before we go into the theater. Still I enjoy it at secondhand.

October 29. We had supper with Leslie Caron last night, at Jeanne Weimer's(?) apartment.[1] Despite the cuisine of la belle France, the lamb was tough. Leslie seemed much more Frenchified and even had difficulty finding English words[. She] frequently hesitated. Aaron Co[pl]and was there, deaf and vague and dull; altogether another of these utterly unfruitful parties, which Don hates just as much as I do.

A truly horrendous new threat: a dog has been barking from inside (apparently) the garden of 147 next door! But, if there is going to be a showdown about this, I think I can rely on Elsa's tenants, Dick Shawn[2] and his nice-looking son, to be understanding about it.

Speaking of Elsa reminds me that I happened to notice, in my little date book, that the 28th was Elsa's birthday. So I phoned her yesterday afternoon. We had a sad little talk. She is scared of Michael Hall, who has been phoning people out here from New York, uttering hysterical expressions of despair over his friend Bill Mills's death and threatening to come out here and jump into Elsa's bed and "just hold" her. She dreads seeing him [...]. And yet she is lonely and terribly bored, and she hardly eats anything on account of what she calls her hernia (the doctors told her that that is what it is, but I can't help suspecting cancer); anyhow, it makes her vomit whenever she eats. I know I ought to go and see her sometimes but I am nearly certain that I shan't. The millionaire feels guilty about the starving poor but doesn't, in practice, visit them.

October 30. After another party at Leslie Wallwork's—not as bad as the previous one because I was interested to meet Jeremy Brett[3]

[1] Jeanne Weymers, wife of lighting expert Robert Weymers, who lit the Getty Museum and also worked in Paris; both Belgian. They were friends of filmmaker Jean Renoir and his wife and close friends of Caron; she often stayed in their Beverly Hills apartment.

[2] American comedian (1924–1987), he did stand-up routines and appeared in about twenty-five films, including *It's a Mad, Mad, Mad, Mad World* (1963), *The Producers* (1968), and *Love at First Bite* (1979).

[3] British actor (1933–1995), educated at Eton and the Central School of Speech and Drama. He played classical roles at the Old Vic and with the Royal National Theatre and appeared in *My Fair Lady* (1964). He later starred on T.V. in "The Adventures of Sherlock Holmes" (1984–1994).

again—though rather disappointed both by his present looks and personality—I came home drunk, with Don, and fell down the outside stairs from the carport. I hurt my right arm just above the elbow, and my knee slightly, and I feel lousy. And Don is cross with old Drub for his clumsiness. And The Downer will be with us this evening. Didn't do my isometrics today because of my arm.

Gavin has heard that Peggy Hubrecht died, the very day after she talked to him on the phone.

October 31. Thanks to The Downer's presence, another gloomy evening, which I cut short rather than have supper with him and Don. And this morning I wake up with strange new symptoms; the muscles of my throat are so tender they feel as if someone had tried to strangle me, and the muscles at the back of my neck hurt so much that I can hardly raise my head from a pillow if I lie down.

A letter from Heinz, saying yes, please send him the German translation of *Christopher and His Kind*. Heinz has now retired on a pension. After remarking how many of our friends are dead, he goes on: "We'll see how much longer *we'll* be tolerated on the earth."

Two parties to go to this evening, but at least I'll have Pussn all to myself on the way there and back.

No isometrics, because of all these muscle pains.

Had to look something up in the life of Aldous Huxley today. I don't know why it is—both such wonderful people—but for me they both simply reek of pessimism, doubt and death. Is it their Frenchness?[1]

November 1. Pains annoying—no worse than that—in neck and back of head. Last night we had a quite nice short visit to Vaccaro, with a highly theatrical performance by Roscoe Lee Brown[e],[2] including a recitation of an e.e. cummings poem. Then a Halloween reading by dim light at the Rosamund Felsen Gallery by a gracious-voiced monotonous female amateur. It was about a séance; I thought it must be an inferior Poe, but could not find it in his works. Don thinks it might have been by Borges, who

[1] Huxley and his first wife, a Belgian, lived in France during the 1920s and 1930s; see Glossary.

[2] Black American actor (1925–2007), a literature professor before he took his first Shakespearian role in 1956. He also appeared in films—*Black Like Me* (1964), *The Comedians* (1967), *Topaz* (1969), *The Cowboys* (1972)—and on T.V.

is much admired in the Felsen circle.... Well, I got to eat supper with my darling at Casa Mia later.

This afternoon I mailed off the German translation of *Christopher and His Kind* to Heinz. No isometrics.

November 2. I seem to do nothing but complain about parties. Very well, I complain. Last night it was William Burroughs, at the Tropicana. I like old Burroughs, he's a friendly man, not in the least demonic, or sex maniacal in a tiresome way. One could become very fond of him, I expect. But last night two spidery photographer girls skipped about taking pictures of us—for a collection called *Bad Boys*! Meanwhile poor [K]athy Tynan told me that Ken is very seriously sick, and their children want to be educated in England, where Ken's health won't allow him to live, and she doesn't know what to do.

Pains much less bad today, despite the kind of windy weather which excites them. So I did my isometrics.

November 3. Yesterday, Armistead Maupin and Ken Maley came to see us, and while Don was drawing Armistead I had to entertain Ken. He is a nice boy but a compulsive talker and very proud indeed of being the man who runs Armistead's life for him. I must say, however, that I feel this is a real justification for someone's existence. Because Armistead, on second acquaintance, impresses me greatly. He seems to be absorbing impressions constantly, which means that he is tremendously "responsive" in Kathleen's use of the word, and kind of mediumistic in the way he has psychic feelers out, testing the atmosphere.

Today we had William Burroughs over to be drawn again (brilliantly) by Don. Was not charmed when the gate buzzer buzzed and I opened to admit *five* people—Burroughs and his friend James Grauerholz,[1] Victor Bo[c]kris the skinny little English journalist,[2] a photographer he'd brought with him, and Paul Getty. Does Burroughs always go around in mobs?

We had, however, known that Paul Getty was coming;

[1]Burroughs's bibliographer and literary executor (b. 1953); a University of Kansas drop-out. He edited some of the later novels and managed Burroughs's chaotic life and career.

[2]Bockris (b. 1940, England) graduated from the University of Pennsylvania; he worked for Andy Warhol, contributed to his magazine, *Interview*, and wrote biographies of Lou Reed, Keith Richards, Debbie Harry, Patti Smith, and Warhol himself—*The Life and Death of Andy Warhol* (1989). With Burroughs, he co-wrote *A Report from the Bunker* (1981).

Grauerholz had asked if he might bring him. Indeed Don and I had had a joke. Don wanted me to ask him about his kidnapping. "Say to him, if you'll tell me about it, I'll lend an ear."[1] Paul proved to be not only fairly pretty, though spotty and looking much older than twenty-two (his claimed age), but also really charming and genuinely interested in our collection of pictures. A mop of curly reddish hair concealed his ear-lack. He emitted almost, but not definitely, flirtatious vibes which reminded me somewhat of Mark Lipscomb. He asked about the Vedanta Society and asked, "Do they need money?" as though he had a few millions to spare, though we had supposed him penniless. I said, "Doesn't everybody?"

Which reminds me that Anandaprana called to nag me about recording the Gita. Would I be able to do it before I left for New York? "If God wills," I bitchily replied. I bet she could have slapped me.

Title ideas Don and I have had for my Swami book. *Guru and Disciple. A Guru and His Disciple. My Guru and His Disciple. Guru and Friend.* I sort of feel these are on the right lines.

November 5. A day missed, chiefly because I had to go out and see a Canadian from London, Ontario who had read *Ramakrishna and His Disciples* and was only here for a couple of days. Told him to go up and see the Vedanta Center today. He was a quite nice, youngish, skinny man; a bit too intense. His name, John Kennedy.

On the evening of the 3rd, we went with Billy Bengston and Penny to Nick Wilder's first show back of the same building, in the rooms which are to be his new gallery, or shared with Jim Corcoran. Harry Brown and June were there, I can't imagine why. Harry seemed to be relatively prosperous—that is, he had just finished a job and was starting another. From there we went on to the "Change Inc. West" show at the museum—it's somehow in aid of beginner artists—to which Don has contributed a very beautiful painting (of Muff Brackett) which I fear he'll be simply gypped out of. Guy Dill was there, very drunk, and greeted me with maximum enthusiasm, insisting on kissing me and giving me a hug. My identity mistake has mysteriously paid off.

Then we ate with Billy and Penny at that nasty Frog restaurant,

[1]John Paul Getty III (1956–2011), grandson of the oil tycoon, was kidnapped in Rome in 1973 and chained in a cave in the Calabrian mountains for five months. His grandfather loaned his father the multimillion-dollar ransom only after the kidnappers cut off the boy's ear and mailed it to a Roman newspaper.

St. Michel. They both described the Brooklyn marathon,[1] which they took part in. Penny did so well she qualified for the Boston marathon, and Billy finished more leisurely. They gave us a vivid impression of this great folk migration, through all the boroughs of New York, which was really moving and only horrible when one imagined oneself taking part in it.

Yesterday morning, Scobie and Sarver came by, to photograph selected nude drawings of Don's to be shown in *The Advocate*. Don already regrets agreeing to this. Scobie told us that [Jimmy] Carter has just spoken against the Briggs proposition in public, while supporting Mervyn Dymally. I don't know if a president carries much weight when he is admittedly campaigning for his party, but at least his attitude may be regarded as impressive, being taken by a Baptist. According to Scobie, the latest poll is over sixty percent in our favor.

Last night, driving home, we were hit on the side of the car by two nurses riding to work at a hospital on a tiny motorcycle, or moped. One of them seemed fairly seriously hurt. She lay in the road, her head on the pillow from our car, and was examined by a passing doctor, some paramedics, the police, etc. Things looked dreary, though it really wasn't Darling's fault at all, but no charges were made.

Today poor Angel has a bad throat and fever, and has just drunk a whisky toddy and lain down to take a nap. But, wonder of wonders, the two nurses called a while ago—just as he was about to call *them*—to say that they weren't seriously hurt. They were quite apologetic, not a word of blame. Alas, how unlike most accident encounters!

November 6. Canadian John Kennedy got very bad marks from me today by calling up grandly disillusioned by his visit to Vedanta Place. He had found them, I forget the exact expression he used, but it meant that they were way off the beam, absorbed in details, red tape, business, whatever. He said, more or less in those words, that they wouldn't know Ramakrishna if he walked in the door. So His Holiness takes off for Canada this evening. His verdict on me? I was the only thing worth coming down here for—but this, too, sounded like a put-down.

November 8. Missed another day, because of preparing for a party yesterday evening. Divine and a lot of others, including David Hockney and The Downer, and Gregory Evans, who bids fair to

[1] Evidently the New York marathon, which passes through Brooklyn.

grow into another Downer, now that he's getting old and dull. Well I was charming, but drank a lot and got a bad hangover, and the evening was only saved for me by the fairly crushing defeat of Briggs's Proposition 6. (Today he is declaring that he'll try again next year, because by then a new law imposing equal, undiscriminating employment will put a gay into every business, and so everybody will find out what they're like and will know their true rottenness and loathe them.)

Yesterday afternoon, Dr. Wolff dragged another, bigger rootlet out of the side of my nose and also fixed three trouble spots on Darling.

Nearly had a serious row with Darling because I saw him looking into one of my old (1956–1958) diaries which was open on the desk—it's the period I'm now covering in the Swami book. Don says I spoke to him as if he were "a chambermaid" caught snooping. I carefully explained to him what is the truth, that I was afraid he might find some slighting reference to himself—there are several in all those early diaries—and not be able to forget it. And I reminded him of that travel diary of his, covering our trip to Asia in 1957, which he let me see, having probably forgotten that it contained a most wounding outburst against the misery of our relationship at that time and his longing to be free of it. I have never forgotten how much I minded, when I read it—even though I realized that he had written it in a violent black hysterical mood, and that I was quite capable of writing something similar. . . . I *think* I got him to understand all this. At the time it happened, I had said I didn't want him to read the diaries until after my death. But I amended this, saying that he could read them any time provided he would read right through, not just dip into them and thus take statements out of context.

November 9. Albie Marre called yesterday from New York. He is coming out here around Thanksgiving time. His news is slightly disturbing but maybe all for the best in the long run; Terry Kramer is probably backing out of backing us, but Brisson is ready to take her place and Marre is in favor of this. Still nothing definite about Simon Ward, but nothing disastrous; he just hasn't got his permit yet.

Am still battling to get on with this wearisome book. When I reread bits I have written, they seem good individually, but what will it all add up to? I don't feel I am establishing a real character (whatever that means) for Swami. I fear, when all is finished, he will have slipped out of my net.

November 10. Another party last night. The Downer is in San Francisco so the party was technically "mine" (Leslie Caron, Jeanne Weymers, Jack and Jim, Gavin) and Kitty could say *he* hated it—which he proceeded to do, with a will, and much bitchery, clawing Drub's muzzle. Today, Drub is in the doghouse and will hardly be let out till after the weekend. Rain tonight, and a dinner party given by Michael Laughlin—at *midnight*! This morning, Elsie told me I am in permanent danger of developing pernicious anemia, but it's all right as long as I go on with the vitamin B shots. Hurrah.

November 11. I was wrong, as I so often am about Darling. He came back yesterday afternoon and kissed me sweetly, so we made up, at least seventy-five percent. Then we went to movies followed by a late-night party given by Michael Laughlin and his new woman, they're not married yet, Su[s]ann[a] Sylbert.[1] It was a messy crowded affair with a steel band making conversation difficult and a clutter of furniture making eating difficult, it mostly had to be on your lap. Susanna made a bad first impression on both of us, but I suppose she was embarrassed to death. Those tragic and presumably dying women, Lenn[y] Dunne and Joan Didion, were around, no doubt feeling as sick as they looked[2]. How can they martyr themselves by going to these get-togethers? Is it really preferable to staying home? Are they so afraid of loneliness?

November 13. Missed another day—why? Just because of sloppy idling, but I did at least get a fairly big chunk of the Swami book done—nearly three pages. Today I have another hangover from drinking, but at least I enjoyed the party we went to; supper up at Mark [Lipscomb] and John [Ladner]'s, with several cute boys. Carlos [Sagui] made one of his biggest love scenes with me, mock, but he is so sexy that I am turned on nearly as much as I am bored. Even when he's fully dressed I am so aware of that incredible body. He made some kind of magical sign on my forehead. I joked about this to Mark as we left, saying he had tried to hex me. Today I got a call from him and Mark, much disturbed and apologetic. I realize that Mark is a fearful mischief-maker. It is such a pity about that, and a pity that Carlos behaves so badly when he comes here, guzzling drinks etc., because we will have

[1] I.e., Susanna Moore, then divorcing Richard Sylbert; see Glossary under Laughlin.

[2] Didion and her sister-in-law, Ellen Griffin Dunne (1932–1997), wife of Dominick Dunne from 1954 to 1979; both had multiple sclerosis. Lenny Dunne was later confined to a wheelchair.

to avoid having him and that will mean breaking with Mark and John, whom we both rather love.

Oh, and such a bitterly sad thing happened. Before we went up to have supper, we visited Dan Luckenbill, who lives in an old house near Echo Park, at the top of a flight of steps. Coming back down them, my darling little Hopi turquoise ring, which Don gave me, slipped off my finger which had shrunk in the cold night air, and fell into deep weeds. It was awful. Don gave a cry of despair. He searched and searched but of course couldn't find it. It was as if I'd lost a little bit of his love—I felt sick.

November 14. Lunch with Gottfried and Silvia Reinhardt, a happy reunion. He has changed, looks like Evelyn Hooker; she hasn't, except the appearance of film makeup, the leading lady gets older by powdering her hair and drawing in lines on the forehead and around the mouth. She still nags him; he's still placid. Much talk about the death of old friends, Gottfried said there should be two parties, one for the living and one for the dead, but you should give the one for the dead first; otherwise, a lot of the living will die in between and you'll have to entertain them twice. This, wouldn't you know, was a joke Kaper had made.

November 20. I'm not going to comment any more on my diary lapses. What has happened during the gap? Notably that I started recording the Gita with Mitchell and [Monday] on the 16th.[1] The Downer has been around, of course. It was particularly depressing to have him actually cooking spaghetti in the kitchen on the 17th; I had such a vision of his doing this after I'm dead. How silly can you get? Well, Dobbin can get it.

November 22 [Wednesday]. Yesterday, Don told me he is planning a ten-day trip to Yucatan, with The Downer, starting next Monday the 27th. Also, a letter from Heinz arrived, expressing horror and dismay after reading the German translation of *Christopher and His Kind*. Also, we heard from Harry Rigby, who's in town, that Equity has approved Simon Ward's application to be allowed to act in our play. And Marre is arriving tomorrow. I talked to him on the phone this morning. He wants more rewrites. It's been pouring down rain and I'm getting increasingly worried about the cracks in our ramp to the carport. No more for now.

[1] *The Bhagavad Gita: The Song of God, Selections Read by Christopher Isherwood*, recorded by Charlie Mitchell (Krishnadas) and Jon Monday (Dharmadas); see Glossary under Dharmadas.

November 26. The ominous feel of predeparture, sadness, tenderness. This'll be the first time Don went off by himself in more than four years, and why the hell shouldn't he? Am indulging in some interesting physical symptoms; I begin to feel a certain weakness in my right knee, the first twinges in many months. Also, today, I peed blood twice. But I won't say anything about this to Don unless it gets really bad before he leaves.

We have seen Marre and he is being tiresome about script changes. But I'll write more about this when it develops. Also other details. Cheer up old horse.

November 28. Darling went off yesterday—it seems like ages, already—after shedding tears and being so loving and cuddly. I didn't have to wish The Downer a pleasant trip, he came to the airport under his own steam. Last night Kitty called from Merida to say they were safely arrived, and that it was very hot, even at night.

I feel very very alone, but without the least wish to go anywhere or see anyone. I feel old and, oddly enough, a bit nervous by myself—another dream about a fire, and I jumped several times at strange sounds and was even worried because some men were talking somewhere up at the top of the ramp and seemed vaguely menacing.

Today has been lovely and I jogged a little on San Vicente, refusing to pay any attention to a few twinges. No more blood peeing.

Marre was to have come and discussed the play with me this afternoon, but he has called off because he suddenly has to go to Chicago (I think) because of some opera (for [his wife] Joan, I suppose). This will have the good effect of making him work on the play himself, during the long plane rides. So he'll probably come back with much more definite suggestions.

7:55 p.m. I suddenly heard a little cry, *exactly* like the greeting cry Kitty so often utters when he returns home. I'm just getting ready to watch Gordon Hoban's episode (I mean he wrote it) of "The Paper Chase" on T.V. (*Later.* It wasn't bad—much better than the others.)

November 29. Worried. Nothing from Don all yesterday, although he'd said he'd call. And now this evening I heard on the car radio that there have been *six* earthquakes in central and south Mexico—one of them at least had a Richter scale reading of 7.5 approximately. Of course I realize that the telephone lines will be tied up for hours and hours, maybe days.

November 30. Paul Savant, who worked on the Larson-Bridges house and came around today to arrange about painting the outside of ours, is not only sexy but nice, I think. I was worried all morning about Don, and more worried when I heard on the car radio that there has been another earthquake in Mexico. Then, at sunset here, Darling called—they have moved to Cancún, in Quintana Roo—and when I asked him about the earthquakes he didn't even know there had been any!

Crawling along on the Swami book. Half a page written with maximum effort.

December 1. I miss Darling more and more wretchedly, yet I'm not exactly unhappy and not at all lonely; I am full of my thoughts. Work went slowly and with great difficulty but will soon pick up, I think. Paul Savant and Gregg Gerren—a really beautiful big blond—and a cute small unidentified boy whom I wasn't introduced to began preparing to paint the outside of the house with an off-white paint called Navajo White. But rain at midday made them call it off. They left. The rain stopped.

Around five, Albie and Joan Marre arrived. Joan, exhausted with an utterness of which only actresses are capable, because she'd seen a very bad opera with the book by Christopher Fry about *Paradise Lost*, retired to the bedroom and slept on top of the basket, with the comforter over her. Albie gave me a lot of vague notes—it is a terrible disadvantage, not having Don here.

Then they left and I am about to eat my fifth solo supper, voluntarily, almost aggressively—I could have eaten out with someone every single night if I'd wanted, but that would have meant drinking and I'll be fucked if I'll be forced into drinking if I say I won't. Don't know how long I shall keep this up. The only people I even consider seeing are Gavin and Jim Gates. I will *not* go to any party—I do loathe them so.

I am at my best with total strangers—with Paul Savant today and an English girl who called me from London. She was so breathless with joy that she was actually talking to me that she kept weeping, or maybe just gasping. And Paul said I was "super interesting." And I am, too. I play back all the stuff I've got from Heard, Huxley, Stravinsky, Auden, everyone, and, oh, with such charm. I feel I am being of some use, even. The Old Man reassuring them all that Life doesn't always end in despair, terror, yearning for lost youth. What they don't know is, why I can do this. It's all because Kitty exists. Merci.

December 4. After six solo suppers, I had supper last night at Casa Mia with Jim Gates and his new lover Kenny Howards, rather a pretentious queen. So Dub drank, much wine. And will drink again tonight, supping with Gavin. The dinner last night was a reward for Jim's helping me by typing up the notes, many of them highly obscure, which I had taken at my last session with Marre.

The trio of painters is madly sexy. Big blond beautiful Gregg, with nearly perfect legs in becomingly short shorts. Obviously he's aware of them or he wouldn't wear the shorts in this freezing weather. The small cute boy is named Tim Powers. He boldly climbs the high ladder, while Gregg, who suffers from vertigo, holds it for him, pushing it dangerously along the wall of the house, amidst wild laughter. They are such truly sweet boys and I like Paul greatly and am even beginning to like Dori his girlfriend, though she seems a bit sulky sometimes.

A.L. Rowse, on the Dick Cavett program a few nights ago, announced that Tennessee Williams is almost certainly a descendant of Emilia Bassano Lanier, his candidate for The Dark Lady of the *Sonnets*![1]

December 6. Two grim firsts—today and also yesterday (I believe) I weighed 158 on the bathroom scale. And today snow fell, little pellets of it from a long long cloud, on Palos Verdes.

Lunch with Paul Sorel, who is now in clear sight of the end of his money, after a long extravagant trip to Europe. Again I urged him to consult Mirandi Levy, who is so sensible, about what to do with his remaining assets. I doubt if he will, though. It seems that he's embarrassed to let her know how crazy he has been in his squandering. Yet, to me, he was still half boasting of his craziness.

It is hideously cold. Today, Paul and Dori worked away on their own; it seems that Gregg and Tim are only up to scraping old paint off, not painting itself. Paul has severe standards.

The audition for the part of Tom in our play has been postponed till the day after tomorrow. So, all being well, Kitty will be able to hear Rick Sandford and Keith McDermott[2] read. He and The Downer are now in Mexico City and scheduled to arrive that

[1] Williams's full name was Thomas Lanier Williams. A.L. Rowse (1903–1997), English historian, poet and Shakespeare scholar, published his third highly controversial book about Shakespeare in 1977.

[2] American actor (b. 1953, Texas), trained at the Ohio University School of Theater; he was a replacement in the Broadway production of *Equus*, October 1974–October 1977, appeared in the film *Without a Trace* (1983), and published a novel, *Acqua Calda* (2005).

morning around noon. My Angel Fur, how delicious to snuggle up to him in this weather!

December 14. Angel got home on the 8th, feeling sick. He slept all afternoon. He hadn't slept at all in Mexico City. But the trip had been a success and he and The Downer had gotten along very well indeed, better than he'd expected they would. I have to admit that this disappointed me, but that's just dumb human nature. What is more serious is that The Downer is now off to San Francisco to try for a part in Divine's play which is opening up there soon.[1] So Kitty will feel lonely for him and I shall be made to regret this. Meanwhile, Angel's Mexican bug is manifesting itself in shits. This has had the good effect of getting Elsie Giorgi to become his doctor as well as mine.

Otherwise, things are going well. Suddenly we're to start rehearsing in New York fairly early in January. Now we're doing rewrites on the script. This has held up my Swami book. Its latest title, *My Guru and His Disciple*, has been told to Michael di Capua, who likes it.

The whole of the outside of the house is painted, as of yesterday.

Bad. I got a ticket yesterday for shooting a red light. If I could have lasted without one until January 21, it would have been seven years.

December 18. Yesterday—triumph—we finished yet another draft of our play and gave it to Harry Rigby to take with him back to New York on the red-eye flight last night. There will be much fussing around with it still, but they can hardly waste more time altering the structure, as they now have to start casting and rehearsals within two-three weeks. We are supposed to join them in early January.

Maybe Don will go ahead of me, so I can really get down to finishing my book. I can't literally finish it. But I could just possibly get through to the end of my visit to India with Swami.

Darling seems much better. He has given up drinking and says this makes him feel good.

In the play audition, Rick read poorly. The other boy was much better but he isn't quite right physically.

Terrific windstorm and rain this morning. Paul Savant had offered to paint the deck table and chairs for us in exchange for

[1] *The Neon Woman* (1978) by Tom Eyen, which had already had an off-Broadway run.

our two armchairs which we don't want. But he propped the glass top of the table against the rail of the deck and it got blown over and shattered.

December 22. We have now heard that the draft of our play isn't satisfactory to Marre—Rigby told us this. So I said well then Marre must write out exactly what he does want and send it to us, because we obviously don't understand. So that's what we're waiting for. Probably it's not so serious as it sounds.

Poor old George Cukor—we saw him on the 20th at Leslie Wallwork's and he practically collapsed. Leslie says he suffers from insomnia and takes far too many sleeping pills for it and has outbursts of temper. It was so sad. Alan Searle was with him, just as usual; *he* has made himself a personality which includes being gaga, so he isn't at all embarrassing.

A victory, also on the 20th: feeling energetic, I got to the end of the Gita recording, thus frustrating Anandaprana's nagging. Of course there will be a lot of picking up loose ends, slips, slurps, etc.; but the brute task is done.

I dream of getting to the end of chapter 17 (the visit to India with Swami) before we leave for New York. But this would be a huge feat and I mustn't force it through or the quality will suffer.

Paul Bailey was here today, with an Italian friend. Paul is writing a life of Henry Green.[1] He wants to use one of Don's portraits of Henry.

December 25. Happy Christmas. I have spent nearly all its daylight hours making a draft of what appears to be going to be one of the last really difficult passages in the book, a discussion of Ramakrishna and sex. Meanwhile, Don is in the studio with The Downer. Peace on Earth. Later we are to go to two parties—one at Billy Bengston's, the other at Tony Richardson's which is apparently going to be a replay of a party at Neil Hartley's last night, where I got very drunk and nearly fell down the outside stairs *again*. Last night, we also saw Mark and John and Carlos, all of them looking their best, which was absolutely adorable in Mark's case and very sexy in Carlos's.

Darling hastily did a portrait of Cici Huston yesterday. She asked for it and paid him for it (not enough) and her lover, Maurice Jarre, to whom she gave it for Christmas, loved it.

The last two weeks I have been missing out on my isometric

[1] Abandoned; see Glossary.

exercises. Started them again today. The one thing I have kept up, almost without a single break (am not *quite* sure about this), is my midday japam.

1979

January 7. The year began with a nasty bump, an earthquake at 3:15 in the afternoon, with a few small aftershocks. It nearly made Paul Savant, who was painting Don's studio, fall off his ladder. When we got to Billy Al's a bit later for a party, we found that there had been a near panic among the guests.

Today Don is in the studio with The Downer, who is really, I think, looking less attractive than he ever has; poor creature, I suppose I must learn to live with him, or worse may befall. I do hate it when he comes in the house and they talk in low voices in the kitchen. Kitty was very tense and aggressive for a few days but sweet last night and today. Am terribly feeling the strain of getting some rewrites on the play finished for Judith Anderson. Meanwhile, Harry Rigby sent the play to Elsa, which has caused a tiresome crisis; I, of course, had to be the one to lower the boom on him, telling him that he has just got to refuse to have her—if she wants to do it—pray God she doesn't. Oh yes, and then the ass who was organizing Don's Redlands show has printed the announcement so faint that it's a disaster, and this is probably too late to get redone. *And* we have had some very heavy rain, causing a new leak in the studio and revealing that Paul's paint job on the deck can be washed off.

That's all the bad news, which, I admit, I superstitiously interpret as a working off of bad karma—to be welcomed. At least I will try to get chapter 17 finished before we leave.

April 1. I did get chapter 17 finished, but that was NOT the end of the bad karma. Am not sure if I care to relate all the bumpy and grindy times between then and now.[1] Suffice it to say today that I am simply writing this to mark our return to California and the beginning, I hope, of a new period of more constructive work,

[1] *A Meeting by the River*, directed by Albert Marre and produced by Terry Allen Kramer and Harry Rigby, had ten previews on Broadway beginning March 20 and opened for a single performance on March 28. Simon Ward played Oliver, Keith Baxter played Patrick, Keith McDermott was Tom, Siobhán McKenna, Margaret, Sam Jaffe, the Swami, Meg Wynn-Owen, Penelope.

finishing the rest of *My Guru and His Disciple*, which is about fifty more pages with the deadline of mid-May.

May 26. Am starting diary keeping again (I hope) now that the drive to finish *My Guru and His Disciple* is over. That is to say, I have sent off the corrected pages of the manuscript to Michael di Capua, along with the ending. That was on the 22nd. There are, however, two possible difficulties ahead. Anandaprana has been given a copy to read and she may well raise all kinds of picky points to be corrected. And Vernon has a copy too. I had supper with him last night. He may object that certain passages about our relationship are too intimate, but he did tell me, last night, that he had read *Christopher and His Kind* and had liked it, which at least means he doesn't object, in principle, to being mentioned as my lover.[1]

Don is in Texas—or maybe, by now, in Louisiana. Having had a show at the Texas Gallery in Houston, he is going on to spend a night at Speed Lamkin's house in Monroe. Speed found him a commission, which sounds doomed to failure, because he has to draw some woman quick like a bunny before a wedding. Then he is going on to New Orleans, with Bill Franklin, planning to return on Monday night, the 28th.

Bill, I beg to announce, is no longer The Downer. I decided in the middle of April to see if I couldn't throw myself into another gear and take a different attitude toward him. It was a psychological experiment, really. The problem was how to do this effectively without kidding myself that I loved him, that he is one of God's creatures and therefore my dear little brother, etc. etc. That kind of goo gets you nowhere. So, instead, I concentrated on him as a sex object. After all, he does have a sexy ass and a good skin and there is a kind of overall sleek sexiness about him, and his thighs are thick and sexy, though nearly cancelled by his disproportionately slender lower legs. So I got my fantasy working and imagined having sex with him until I was sufficiently aroused and jerked off and had an orgasm. Having thus, so to speak, "had" him, I started treating him with the intimacy of a bed partner. That was on April 17, when Don and he and I had a couple of parties to go to and also the opening of a big show of portraits at the Municipal Art Gallery, where Don was showing some paintings. I drank a good deal and came on very strong—that is to say, the way I normally come on with Mark Lipscomb, pinching his ass, kissing him, putting my

[1] Though Isherwood uses the pseudonym Vernon in both books.

arm around his shoulder, whispering intimately into his ear, etc. He was pretty surprised but he responded, and since then we have kept it up in a modified way, which pleases Darling of course. I haven't explained my technique to him yet, however. I want it to become second nature, if possible, through constant practice.

Richard died on the 15th, in Dan Bradley's arms. It was a heart attack, without warning. Since then there have been calls, about money. I talked to Mrs. Bradley and to her daughter Susan and to her son Ronald, who is businesslike. I can't be bothered to write down at length what the problem is, but it is already plain that Tom Isherwood is determined to refuse to help out, and that I shall have to, one way or another. Tom's letter arrived today, simply announcing that Richard left him Wybersle[gh], and that he doesn't think he can afford to restore it.

May 27. A grey-skied holiday. Darling still away. I have been numbering the pages of the last two xerox copies of my book because I have to give one copy to Peter Grose to take back with him to England. He and his wife are coming here tonight, if they ever find the way.

I keep wondering how people will receive the book. I have almost no idea. Sometimes I think, after all, the material itself is so remarkable that some people must find it fascinating. But I think many others will be repelled by the weepy devotional tone of certain passages. Also by a very private kind of camp, which sometimes expresses itself in stock religious phrases which aren't meant to be taken literally—though how is the reader to know that? Oh, I am glad that it is written, though!

May 28. Darling is supposed to arrive home in the middle of the night from New Orleans, where he has been with Bill Franklin.

The Peter Groses were nice and we had a pleasant evening; Grose a bit loud and possessive about "his" authors, but that somehow suited his Australian persona, which he seemed to wish me to be aware of—perhaps he is used to being made to feel inferior about it by the British.

Saw Bill Roerick this afternoon. He is staying at the house of Martin Manulis[1] in Bel Air and playing in Shaw's *Pygmalion* at the Ahmanson. Don has never liked him and, this time, I saw why very

[1] American director and producer (1915–2007). He worked on Broadway before launching "Playhouse 90," a live T.V. drama series which won CBS six Emmy awards in 1956. Later, he was head of production at Twentieth Century-Fox Television, then founded his own production company.

clearly; he is such an operator and so possessive about relationships and so pushy—he actually tried to get me to say I'd undertake to edit, with comments, a book of John Swope's photographs—John has recently died and I saw so clearly how Bill would have collected the credit as Mr. Fixit from Dorothy,[1] if I had agreed to it.

Am now in the recuperation stage from my book. (Not that I'm not vividly aware that I may have to make a whole lot of alterations as the result of having shown it to Anandaprana and the others.) My eyes see bad[ly]—even my good one seems dimmer. I hope my psoriasis may slowly clear up; it does seem a little better. Meanwhile, I'm trying to restore various disciplines—to keep this diary every day, to make midday japam, to do my isometric exercises and at least a little daily jogging. (Have jogged for five days in a row, now, and at least there's no reaction from my right knee.) And then, of course, there's the piled mail, which I shall probably never entirely answer, only keep weeding out and throwing away, letter after letter.

May 30. First slip-up; missed a diary day. But japam, isometrics and jogging maintained so far.

Don Amador[2] called this morning to tell me that a city ordinance has been passed 11–2 (but by whom? I neglected to ask him exactly what body, council, board, committee passes these things and of course he took it for granted that I'd know) abolishing all restrictions of race, sex orientation, creed, etc., in employment, renting, etc., for the city of Los Angeles. This is much the biggest city in the nation to pass such an ordinance. New York hasn't passed one, to its utter but usual disgrace. Don was in touch with me a couple of days ago to ask if I'd ride with him in the next gay parade here. Don't know if I will or not. Darling seems to be against it.

Darling has been such an angel since he returned. When he is like this, I feel such terrific communication with him that even the prospect of dying seems hardly a separation. I write *seems* advisedly. Talk of this kind must always be fair-weather talk.

A very touching letter from Evelyn Bradley arrived today. I might copy bits of it into this book, but not yet because the spelling is bad in places and it's hard to know exactly what she means.

[1] Swope's wife was the actress Dorothy McGuire; see Glossary.

[2] Local activist (b. 1942), raised in Troy, New York, as Donald Grace. In 1976, he began teaching gay studies at California State University Long Beach and Northridge campuses and at L.A. Community College.

Also, I haven't much time. A New York critic-columnist, Herbert Mitgang, is coming to interview me shortly.[1]

July 4. Am resuming chiefly because it's a landmark day. Well, at least the revised pages of *My Guru and His Disciple* are finally xeroxed and have gone off to Farrar Straus. They are now busy getting it ready but it won't be published till February 1980. Anandaprana raised no shattering objections. I think she rather hated the whole tone of the book and was embarrassed to death at the prospect of having to excuse it to the congregation and the Belur Math. Told her she didn't have to. Adjemian wasn't shocked and as for Vernon Old he thought it was beautiful and wrote me a sweet but embarrassing letter saying that I am one of the few people one meets in a lifetime that one can really trust and signing himself my disciple.

Have heard nothing yet from Methuen. Publishers always take longer than one can believe possible, I remind myself.

My health seems better. Elsie Giorgi sent me to another skin doctor, a young woman named Olsen. She pulled out my keratosis and the lab tested it and said it was nonmalignant. She also gave me some other ointment for my skin condition, which she assured me wasn't psoriasis but due to poor circulation, which Elsie pooh-poohs. Anyhow, the skin condition began to clear up instantly and is now nearly invisible. Dr. Olsen says it will return.

I won't write about Bill Franklin now, because my attitude toward him is complicated again. Shall wait till it simplifies itself.

Will try to keep this record going for a while, now that the book is out of my hands.

July 6. Saw Vernon last night. He was gracious but condescending to Don about his drawing and painting. Really, he is quite without humor. He also says that he doesn't quite understand what love is. He takes himself nearly as seriously as poor John van Druten. Don went out to see a mouse and Vernon sat up with me, on and on. Still, I did feel his affection, such as it was. I gather that Vernon has some fairly young girl devoted to him.

Much nicer than this encounter was a lunch given to George Cukor to celebrate his eightieth birthday, which is tomorrow, by Leslie Wallwork. We were the only other guests. The lunch was

[1] Mitgang—magazine writer, novelist, biographer, editor, T.V. news documentary producer (b. 1920)—was on the editorial board of *The New York Times*. He published several articles in the 1980s about the Auden–Isherwood–Spender circle; he and Isherwood began corresponding in 1965.

at La Ruche and it dragged on for hours, but I have to admit, in all fairness to the Frogs, that the food was hot and even good. Old George was making an effort, no doubt; on the whole he bore up well. We laughed quite a lot, over old show-biz gossip.

The press is beginning to seethe with worldwide nervousness about the impending fall of bits of the skylab.[1]

July 7. I'm slowly beginning to think first thoughts about my next book. Chiefly, how it will be constructed. Should I have chapters dealing with specific subjects, rather than a continuous narrative advancing through time? A whole chapter (or two, or three) dealing with the periods of work at the movie studios; a chapter on the Quakers; a chapter on Gerald, including the La Verne seminar; a chapter on the Huxleys.... This sounds sensible but somehow it doesn't charm me. I'd rather wander all over the place, from topic to topic, like George Moore or Yeats. But this requires great artfulness.

July 8. Have been studying van Druten's *The Widening Circle*, to get hints for the structure of my next book. Johnnie is like a rather weepy head of the Supreme Court, delivering verdicts which pain him deeply. He's so *sorry* to have to decide that Arnold Bennett's reputation isn't what it was.... I don't say this out of bitchiness, it's just that I must know how to avoid other people's mistakes. Will try Moore next.

The radio (5 p.m.) says that Michael Wilding is dead but doesn't utter a word about Skylab. Are they trying to hush up an impending disaster?

July 17. This is in great haste because we have to go out to a new very successful Frog restaurant called Michael's, which is so grand that it demands you shall arrive there at 12:15 for lunch.

This morning, just after I'd gotten out of bed, having had a long sweet sleep with my darling, and was entering the bathroom, the Firebird swooped down and whispered into my ear that the key to the construction of my new book is this house. Just as John van Druten, in *The Widening Circle*, used the ranch as the setting for all his thoughts and memories, so I shall use this house. The house will be the image of me, the old man, in constant

[1] Abandoned by the last of three crews in 1974, *Skylab 1* was meant to orbit until the mid-1980s, but it eluded NASA's control and returned to earth July 11, 1979, scattering debris over the Indian Ocean and western Australia.

danger of collapse. This won't be played up—unless of course, the house does actually collapse while I'm working on the book, as the result of a fire or an earthquake or a flood—in which case it might provide an amusing symbolic climax (but let's not even think such a thought!). That's really all I have to say, right now, but I do see that the fact that I lived in this Canyon, in various houses, right from the beginning of my Californian life, will help me to relate this book to almost any of the memories of people I may want to put into it. Also—as Don pointed out when I told him about this at breakfast—there are objects in the house, pictures, etc., which will serve as promptings to memories which aren't directly connected with the Canyon.

July 22. Today, I've made up my mind to break the ice by putting down some thoughts about the outline of my new book. At present, it seems to me that its narrative span, pulled out thin like elastic, *could* extend right from the beginning to the present moment of my American life, just as *My Guru and His Disciple* does. But this is really a question of how much usable material I find I actually have. I must remember that I originally supposed that *Christopher and His Kind* would include my first years in the States!

Well, anyhow.... Let's suppose that I begin in the narrative present—the *now* of writing this book—with myself standing on the deck of this house and looking out over the Canyon at the various buildings I have lived in, here, which are actually (more or less) visible—Amalfi Drive, Denny's apartment on Entrada, the Viertels' former house on Mabery, 333 East Rustic Road (and 322 opposite, which was Michael Barrie's, but we stayed there too, and Gerald Heard lived and died there), and then the two houses Don and I had before this one, the rented house on Mesa and the house we bought on Sycamore.

The Amalfi Drive section would include a description of how I left with Vernon for California—after, maybe, some impressions of our life in New York, including some memories of Lincoln Kirstein, unless it seems advisable not to offend him by mentioning him. Also, our first home in Hollywood, the apartment on Franklin, where we were visited by Chris Wood and Gerald. Then, getting to Amalfi, there would be a natural place to describe the Huxleys' house, higher up that same street. And all the people who came there, Bertrand Russell, Anita Loos, Cukor, etc.

And that would lead to Mabery and a description of the household there, the Viertels and all their film friends—which would lead to my first film jobs and life in the studios.

Awkward gaps after this, because of my increasing involvement with the Vedanta Center, and my work with the Quakers in the East. Here is where big problems of narrative technique must be solved, so I somehow avoid constant references to *My Guru and His Disciple*. I have to bridge over to meeting Caskey and occupying Denny's apartment and then the Viertels' garage and then 333 East Rustic, with its hauntings, and drunken hospitality. (I suppose our stay in Laguna must be touched on too.)

Then I meet Don and we move into the Mesa house and the Sycamore house. And finally into this house. And then this house itself can be described. How far we go on from that point is anybody's guess.

Well, as usual, I come to the same old conclusion: I shall only find answers by starting a more or less slapdash rough draft of the book.

July 28. I tried to make a start on the book but suddenly lost confidence. Isn't this approach via the various houses I've lived in a bit too formalistic? And yet, even as I write this, I do see that it's at least a method of starting, and that's worth a lot. It's unlikely that I'll get any deeper insights until I pressure myself by writing something.

A mad letter from Caskey, full of aggression. Maybe he's mad at me because I haven't answered his letters. He demands the pre-Columbian figure back; I'd always regarded this [as] part of my share in the division of property when we parted.[1] He also wants some Greek photographs of his back and so far I can't find them. He also makes threats about a book he says he's writing. If I don't show myself "sympathetic" to his book, he will show me up as a freak. He also claims to have amazed John Lehmann by telling him how sexually "naive" I was when Billy met me. This disturbs me just a little bit because it is so obviously a version of the truth twisted to suit Billy, and now he believes it.

July 29. Am depressed, not exactly about Caskey but in general. It's so sad about Tony Sarver having cancer and (almost certainly) requiring [a] colostomy[.] It was also sad seeing Anita Loos so much older and deafer, at lunch with her niece Mary, the day before yesterday. However I think the climate of my depression was created by too much drinking, especially wine drinking which is so lowering and acid.

[1] The Tarascan figure, from Mexico, is mentioned again in the entry for August 7, 1980, below.

Now I'm waiting for some German journalist to arrive, Adalbert Reif (or Adelbert; I see from the German dictionary that it can be either).

They left a couple of hours ago and I've forgotten already if it was Adal or Adel.[1] Anyhow, their visit was quite enjoyable because they have an adorable blond baby which slept fatly, and quite worthwhile because they want to do something about publishing more of my books in German.

Have just finished reading Stephen's play *The Corporal*, a revised and much (too much) expanded version of his *Trial of a Judge*.[2] I can see that parts of it would play, but the tragedy seems weary and empty and formalistic—I mean, the characters are idea mouthpieces first and characters second, or not at all. However, I find plays extraordinarily hard to judge. I must try reading it again.

August 7. *The Corporal* is better on second reading, but it will have to be cut drastically. Otherwise I've accomplished nothing but a few letters answered. I know I should make some kind of a start with my new book, but I drift, excusing myself that I can't think of a proper opening. Well, at least let's try. For instance—

How would it be to open with a diary passage? Not faked (well, not very much faked) about the happenings of this particular day, now, in August 1979? The passage would begin factually, with details of some happening, and then reveal that I have this underlying feeling of guilty compulsion to get my book of memories started.... No, no, that's nothing. Or rather, it's that cliché battle cry of tired authors: I don't want to write but, by jingo, if I do—

Today I must send off a letter telling Mr. Sidebotham that my share of the money willed to Richard through the Trevor Will Trust[3] is to go to the Bradleys. That's because I just heard from Tom Isherwood that he wants to keep the whole of his share. Fuck Tom, but I can see no way out of this.

August 15. On August 8, Anita and Mary Loos came to dinner with us, accompanied by Mary's so-called beau, a decorator named

[1] Adelbert Reif (b. 1936, Berlin) wrote for German newspapers and magazines and published books based on interviews with European intellectuals—Hannah Arendt, Roland Barthes, Erich Fromm, Claude Lévi-Strauss, Golo Mann, Albert Speer—and about music and composers. His wife, Ruth Renée Reif, co-authored several.

[2] Produced by the Group Theatre in 1938 and published by Faber the same year; the new version was never performed.

[3] Created by Isherwood's godmother, Agatha Trevor, who died in 1933.

Dewey Spriegel. They arrived late, Dewey complaining that they couldn't find the house (actually, Mary lives within sight of us). Dewey said to Don—whom he already met at lunch at Mary's house: "I hear you're an artist. Why can't you repaint the number on your mailbox? It's illegible." This doesn't sound all that bad, written down. But it was, the way he said it, one of the most insulting things I have heard anyone say to anyone, unprovoked, sober and at the beginning of a party. Later he was publicly rude to Mary, at table.

On August 10, an English journalist named Ian Brodie, who works for the London *Telegraph*, called to tell me that Stephen Spender has cancer—implying that it was terminal. He then went on immediately to ask me to make a statement for his newspaper about Stephen's achievements and place in the literary world.

Shock followed by incredulity gave way to rage. I told him to tell his editor that I'd said his behavior was disgusting: "You're just telling me this for your own commercial reasons." "*Commerical* is hardly the word I would have used," he said, primly. I slammed down the phone. Then I got through to Patrick Woodcock, who had heard nothing. As far as he knew Stephen was well and on Corfu. Patrick, equally indignant, said he knew several people on the *Telegraph* and would make enquiries. Since then, I haven't heard any more from anyone. I don't believe the story, and yet I find it hard to imagine that a big newspaper wouldn't have double-checked on its facts.

An invitation: Juleen Compton's and Rod Steiger's All White Party. Une Après Midi Dans La Jardin. Quiche and champagne.[1]

August 28. Well, at least my birthday's over. The most heartless part of it were the gifts of flowers, ordered by telephone or telegram long-distance—a dress rehearsal for a funeral which—if I'm lucky and make it to UCLA dissecting room—will never take place. But quite aside from birthday blues, I do feel seriously perturbed and undermined by all this mail. I am now hopelessly in the red, or should I say read, as regards answering it. I think I

[1] Perhaps Isherwood mistyped *La* Jardin for *le* Jardin, but his French was good; more likely he replicated errors, including the capitalization, to highlight the absurdity of the invitation. Compton, second wife from 1960 until his death in 1980 of Broadway director Harold Clurman, wrote, directed and starred in *Stranded* (1965) and wrote, directed, and produced *The Plastic Dome of Norma Jean* (1966). Steiger (1923–2002) became famous acting with Marlon Brando in *On the Waterfront* (1954) and won an Academy Award for *In the Heat of the Night* (1967). He married five times.

owe more than I did when I started writing letters again after the long gap early this year.

And I'm stuck in preparations for beginning my next book. It really is clearer to me than ever that I have no "plot" for it. The best I can come up with is still the idea of a memoir based on my various dwelling places in the Canyon, which is so Proustian and sentimental.

Just heard from Stephen, very touched because he had discovered from Patrick Woodcock about the call from the *Telegraph* and my reactions to it. He is perfectly all right and full of his play.

Encouraging: both Prema and Claude Summers have written very encouragingly about *My Guru*. Prema came up with only one factual mistake, which can be easily and briefly corrected—when you beg for alms after becoming a swami, you get given the alms in a gerua napkin, not in the fold of your own robe. (Since I actually witnessed this, and wrote about it in my diary at the time, this shows how unobservant I was—and still am.)

September 17. Still this heat, which takes all the zing out of life; just when I feel so anxious to get going. Since Jack Woody produced this project of a picture book of Don's drawings accompanied by some sort of text by me, I have resolved to keep a detailed daily diary for the month of October. It's a crazy project because it will surely be almost impossible to relate the text to the drawings. Don has a sort of mystic faith that the drawings and text will do this of themselves—and who shall say he's wrong?

The evening in San Francisco, when I gave "readings and conversation" to raise funds for Gay Advocates, was an extraordinary experience. In a way it is almost more difficult to cope with an audience which has made up its mind to adore you than with one which says "you show us." Amidst the cheers and standing ovations, I felt somehow like a conductor conducting a piece of music which isn't his composition. I don't mean that I was consciously trying to say only things I thought they'd like to hear. But I *was* trying to be one of them, whether in agreement or in disagreement. I felt that I was a member of the tribe; the fact that I was addressing it was hugely exciting and joyful but still only of secondary importance.

Well, anyhow, the show was nearly sold out. And the posters announcing it[1]—the ones that were left over, signed by Don and me—were all sold.

[1] A reproduction of a Bachardy drawing of Isherwood.

San Francisco seemed utterly delightful, especially after the smog down here, such refreshing sea-clean air. And all the little old houses looked so charming. Don and I agreed that it is the most beautiful city anywhere.

September 18. In the midst of a slightly drunken argument last night, Don came out with a line which I thought was so marvellous I wrote it down immediately: "What do you think *I* am—some tawdry little *recipient*?" And now I simply cannot remember what the argument was about.

Yesterday, we were visited by two young men, one English, one Irish—Christopher Corr and Eamonn Leddy—Christopher a young artist who has a scholarship from the Royal Academy of Art to draw landscapes, etc., here in the States,[1] Eamonn his friend, probably lover, describes himself as a mathematician, twenty-four, one year older than Christopher. Both have metal spectacles and perfect pink skins and white teeth and are blondish. If the angels who visited Sodom had worn modern dress, that's exactly how I imagine they would look.

Today another near-angel, Stuart Timmons,[2] arrived to be drawn and painted by Don; dressed to be undressed—very short cutoffs which showed all of his scarred, slightly disfigured yet extraordinarily sexy legs. Later he told me that he had lied to Don, saying that he had posed for art classes in the nude. When Don suggested a nude pose he was embarrassed but did it anyway. He is a curly-haired blond with a kiss-pouting mouth, determined to reactivate his gay sex life this fall—on principle, one supposes; he is a deadly serious gay-political boy, determined to show solidarity with the tribe.

September 19. The anniversary of Caskey's and my departure for South America, thirty-two years ago. Which reminds me that I haven't yet heard from Caskey since I sent him a half-reconciliatory letter enclosing the chapter which refers to him in *My Guru*. It will be very tiresome if he takes me up on my offer to cut him right out of the book!

[1] Corr (b. 1955, London, of Irish parents) was the Leverhulme Travelling Scholar; he won the Royal College of Art Drawing Prize after his return, followed by later prizes for his drawings, designs, and book illustrations.

[2] Journalist—for *L.A. Weekly*, *The Advocate*, *Frontiers*, *Vibe*, *Spin*—social historian, and biographer; later, he wrote *The Trouble with Harry Hay* (1990) and co-wrote *Gay L.A.: A History of Sexual Outlaws, Power Politics, and Lipstick Lesbians* (2006).

Last night, Don and I went to the movies and were much annoyed by a noisy baby in a party of four. Don went over to complain to them and one of the party, a young man, told him to "get lost." So then I yelled "shut up" when next the baby cried. As we were leaving, the young man came up beside Don and glared at him, so Don glared right back. The young man told Don, "You're ugly." Don said, "That's *my* problem." The young man faded away. Actually, I think *you're ugly* was a backhanded compliment, in that the young man was unwillingly impressed by Don's readiness to be hostile, plus his muscular superiority. It was one of those situations which develop readily in the steamy hot weather we're having.

September 20. Mike Van Horn has just arrived on holiday from New York. He came to supper last night, a bit fatter and with a bit less hair, but otherwise his usual sturdy, outwardly placid, deep-feeling, Dutch-obstinate self. He is very close to us emotionally and yet there isn't much sparkle in our encounters.

Beth Ann Krier[1] (whom I do like) was also among the guests. She is studying female millionaires and how they got their money (marriage is cheating, they have to be self-made). Tried to interest her in making us into millionaires without our doing anything about it. Suggested that this would be interesting simply as an experiment. But she didn't see it that way. Later in the evening, I didn't see anything, got stupidly drunk, and I don't know why because I like everybody who was present—Penny, Billy Al, Jack Larson, and sympathetic silly-goose Sydney Cobb[2] (who is still rather in love with Mike, after all these years).

On the steps below the house we found a plastic container made to hold liquids. The container was empty but arranged on it were some bits of plant which Mike and Don both thought might be pot. So we took them for future examination, leaving the container. That happened this afternoon as we were returning from the beach. It may all be very harmless, but I don't like the idea that this might be some sort of arranged "drop." The police would probably descend on us and fuss and search the house.

September 21. Played the tape of my talk in San Francisco, which shows me that I still say "er—er—," which I thought I'd cured

[1] American journalist and longtime staff writer for the *Los Angeles Times*.
[2] Artist and graphic designer, living in Pasadena. He sat for Bachardy several times.

myself of and that I also have some nasty nasal tones in my voice when reading aloud. Also, it created a really bad impression, not knowing instantly who Dan White was.[1] But I might have been a lot worse.

When Mike (who stayed the night in the studio) ran down to the beach with Don today, they found that the plastic container was still on the steps.

Yesterday evening, some unfortunate Mexican, drunk or zonked on something, wrecked a car[,] which was also (so we are told) stolen[,] and then fled. And now the helicopters came after him, with police-car support on the ground. The big rattling airborne insects, swinging their search beams like probing stings, seemed unspeakably futile. It seems impossible that they could ever catch anyone.

November 11. During October, Don and I carried out the scheme which Don worked out with Jack Woody—Don to do a drawing or painting of someone every day, me to keep a daily diary throughout the month. Don got the whole of his share of the work done—not just thirty-one but forty-six pictures, four of which are black and white paintings. My daily diary is only a rough draft, however. Don is rereading it to make notes and suggestions.

Paul Wonner came down on November 1 for the opening of his show of paintings. On the 2nd, Don and I went to look at them—he had already seen them the day before, and had decided to buy one. When we got to the Corcoran Gallery we happened to meet Nick Wilder who took Don aside and confirmed that he is definitely giving up his own gallery. He also showed Don a letter from Jim Corcoran to Nick, telling him that he only wants to take over four of Nick's artists—Ron Davis, Sam Francis, David Hockney and Don. Which pleased us, needless to say.

On the 4th, we phoned Vera in New York, having just been told that she has had several small strokes. She seemed shaken and a bit vague, but still very much her usual self—greatly indignant because a specialist had come to examine her after her strokes—she referred to them as "illnesses"—and had shown her pictures of various animals, asking her to tell him what they were.

On the 8th, we gave a party—the first in quite some while, because of our involvement in our *October* project. We told Tony

[1] The San Francisco City Supervisor who shot and killed Mayor George Moscone and gay fellow Supervisor Harvey Milk on November 27, 1978. See Glossary.

Richardson, whom we'd already invited, to bring a guest with him. He suggested Dean Skipworth[1] and Don said no, absolutely not; he wasn't to be allowed in the house, ever again. The reason is he came to us once before and dumped a great bag of garbage, which he produced from his car, into our garbage can. It contained something really rotten and stank up the whole area so much that Don had to drag it out again and dump it somewhere else.... I think Tony was sulking about this throughout the meal.

On the 9th, we were invited by Richard Burton's wife Susan to a birthday party at Chasen's—that is to say, *I* was invited. (I suppose Susan, who has only met Don once before, and was doing the organizing herself because this was a surprise party, couldn't be accused of actual rudeness in omitting him, she probably didn't realize what our relationship is and how long ago Richard knew both of us.) Anyhow, I quite easily got Don included. But then, what do I do? Sitting right next to Richard—who is now on the wagon—I drank and drank until I passed out and sat slumped in the chair and couldn't rise to take part in the birthday toast. It's a terrible situation, especially as I know Susan hardly at all, and it's to her I must write an apology. And I can't do that until tomorrow at the earliest because I don't have an address to send the letter to until the offices reopen after the weekend and I can make enquiries!

Yesterday, I drove to Vedanta Place to see Swami Ritajananda from Gretz—he had been sick while up at the Santa Barbara convent but seemed fairly all right. He is very impressive. I felt a lot of love in him and also a great shrewdness. I think he was being careful not to criticize Prema to me, he became a bit guarded when I asked about Prema's part in their life. Several times, he remarked—maybe to reassure *me*—that you could be a monk for a while and then stop being one—that wasn't so important. As for me, I was still feeling my hangover—had, indeed, been a bit scared while driving on the freeway because I kept getting cramps in my hands and slight attacks of dizziness. When Ritajananda and I sat alone in the living room of the old bungalow and I started talking about Swami and about Sister, they both seemed intensely present, and I shed tears, quite spontaneously but yet I felt like some overemotional fakey female devotee, and that was when I became aware of the sharpness of Ritajananda's glance. Still, I am very glad indeed that I saw him again—it may very possibly have been for the last time.

[1] Costumer on several Hollywood films, including, later, Francis Ford Coppola's *One from the Heart* (1982) and *Hammett* (1982).

Afterwards, I talked to the boys, including Bob [Adjemian], who has now got several of the other monks to join him in long-distance runs. Bob, who is a real little fighting cock, is longing for the showdown which he anticipates over *My Guru and His Disciple*, when it appears. According to him, Swahananda's only fear is that my book should give the impression that Vedanta endorses homosexuality. Anandaprana anticipates big trouble with the congregation. And Bob is just praying for it!

November 14. On the 12th, backing out of our driveway, I bumped a car which was standing on that fatal spot, exactly where a backing-out car from our driveway will hit, with only a tiny bit of bad luck. This time it was a brand-new Subaru Brat, belonging to the lady who lives opposite, a Mrs. Karlow, who was peevish about it but not venomous. Her son, also present, is named Jeff and is cuteish. Later I called their number to find out the extent of the damage and found myself talking to Jeff, who had been researching for a term paper on Aaron Copland and had discovered that I was mentioned as one of his friends. So I was on the map, and the temperature rose.

That evening, after we'd seen *Cabin in the Sky* and liked it—I with more reservations than Don, because it speaks, albeit with Yiddish and black accents, the language of the Southern Baptists—Don suggested we eat at El Adobe, and I agreed, only predicting that it would be full up as usual. It *was* nearly full, only one table, but then out rushed one of the owners, Patricia Casado, who always receives us as though we were royalty. And the next thing we knew, we were being introduced to Governor Brown, who just happened to be in there. He looked just "like" himself, very thin faced, with a bright burning in the eyes and an air of being obedient to an inner voice. We were both impressed by him, he seemed extremely ambitious, perfectly serious and quite worthy of our respect. It was easy to imagine him robed as a priest. He gave us a copy of his announcement of candidacy. When I asked him to autograph it, he wrote, "Chris and Don, Peace! Jerry Brown." Somehow, we got into talking about our play and hence the temptation of Jesus in the wilderness and Swami's book, *The Sermon on the Mount According to Vedanta*. He seemed genuinely interested.[1]

December 16. Am now in the midst of revising my *October Diary*. I

[1] Brown, who once studied for the Roman Catholic priesthood, was running for president. See Glossary.

still don't know what I think of it but at least it does represent an accomplished task.

What has happened since I last wrote? [The gardener's helper] has had hepatitis, contracted erotically, but is nearly well now. I have had an attack of dizziness—that was quite recently, on the 12th of this month. It was embarrassing, because it took place while I was being visited by a terribly square-dull psychologist named Tony Joseph,[1] and I feel sure he believed that I invented the attack in order to get rid of him. Elsie prescribed a seasick medicine, so Don got Merezine, because it suggested the word "mare," and it was effective, apparently. I've had no more symptoms. It was rather unpleasant, as though my head had become so heavy that it kept making me lose my balance. We've been to a vast number of parties—about which I can at least say that I never once fell asleep or otherwise misbehaved. We went to Merle Oberon's memorial service on the 28th, chiefly because George Cukor had asked me to write a few words for him to say and so I was obligated to go and hear him say them, which he didn't do very well, poor old thing. The Downer remains a presence in the background—his absence is a lasting reproach to Dobbin's meanness. But time will heal, or make things worse in some quite new way. Meanwhile Kitty is Kitty, the one without a second.

December 21. Solstice blues. It has rained today and is gloomy and Dobbin is gloomy because of bulging gassy bowels and pains around the hips and groin. Elsie, on the phone, recommended charcoal tablets and now his tongue is black as the tongue of a devil horse. However, I have made various assertions of the will—written a blurb for Warhol's Pop memoirs,[2] which I have just finished and found so wonderfully good-humored and so wise; sent off the publisher's copy of my British contract, duly signed; told the Black Sparrow Press that I will not sign the copies of the special edition of Stephen Spender's letters to me during the thirties—I still don't know if it was Stephen who put them up to asking me to do this. It seems to me an indecent idea, a kind of built-in blurb. If Stephen is publishing all of the letters, that means several quite vicious ones are included, and, knowing Stephen, it probably also means that he has added bitchy footnotes which I am not about to endorse. There'll probably be a fuss. Let there be.

[1] Astrologer and therapist (1946–1986), born in Cleveland, settled in San Francisco; a student of Jung's work and Executive Director of the National Council for Geocosmic Research. He died of AIDS.

[2] *POPism, the Warhol 60s* (1980), by Andy Warhol and Pat Hackett.

I should mention that Susan Burton has never replied to my letter of apology for getting drunk on Richard's birthday. That proves, says Don, that she isn't a lady.

December 24. Last night I had a dream about Kathleen. I was wandering around somewhere and suddenly came face to face with her. I recognized her—that is to say, she had changed a bit, for an instant I wasn't quite certain, then I knew it was she. The *shock* of meeting her was really the chief experience of the dream. She seemed sad and rather vague. I felt that she was *astray*, didn't quite know where she was going. I asked her, knowing from the first that she was "dead," if she had come to tell me that I would soon die, myself. But no, she didn't seem to know that; she didn't seem sure of anything.

As soon as I woke, it struck me, self-reproachfully, that I hadn't asked her about herself, or about Frank, or about Richard. I think I was kind and more or less protective in my manner toward her. But she couldn't stay with me, it seemed, or didn't want to. She drifted away.

When I spoke about my dream to Don, I began to feel that this hadn't actually been Kathleen but something disguised in her appearance. She wasn't quite convincing. She was a sort of clone, I said to Don.

The feel of this dream was pessimistic, disconcerting, something was wrong.

1980

January 24. A rolling 5.5 earthquake up north at Livermore. Went to see poor miserable old Jo who has to go into hospital, suffering from diverticulitis. What really upsets her is that they want to put a tube into her with a bag to drain off the shit—at least for several months. She shed tears, exclaiming that she wouldn't be able to go in the ocean for all that time: "I couldn't possibly wear it with a bikini."

At last, the galley proofs have been corrected, after long sessions with Michael di Capua. Today for the first time in over two weeks, I restarted work on the revision of our *October Diary* project.

February 2. A dreary crisis period, with Carter "leading" the U.S. against Russia for all-too-evident election-year reasons. Suddenly, an act of atomic idiocy seems possible.[1] Meanwhile I try to plod along with our *October Diary* project and fail, and blame it on pains in my back. Well, at least I'll try to keep this diary for a while.

Largely thanks to Don's vigorous persuasions, old Jo decided at the very last moment not to have the operation, but to stay home and try the effect of a diet. Whenever you dial her number, the phone starts whining, but she seems better.

Very happy with Darling. He is working hard for his show at the Corcoran at the end of the month; it's to be all of women.

February 6. Morale much better—chiefly, I do believe, because we have stopped drinking since the day before yesterday. Our modest goal is February 14, when the Animals *have* to celebrate their 27th anniversary with champagne. Don agrees that he feels so much better without it. Yesterday evening we had a curiously boring supper with Tony Richardson and David Hockney. Tony held forth at too much length about an involvement with some Mafia-connected bird dealers, in whose defence he was conned into appearing in court as a witness: the story didn't hold together. As for David, he just blandly sat there and made corny jokes. I am fond of them both, and of Gregory Evans and of Peter Schlesinger (who has shown up here in town unexpectedly). I am even quite fond of Keith Coleman, who is one of three part-time successors to The Downer. But I would have been much fonder of them with a few drinks inside me. Keith is fat but sexy, silly and greedy but sympathetic.[2]

Now I really must pull myself together and restart revision of the *October Diary*.

February 14. Now it's Animals' Anniversary, their twenty-seventh. Rain on and off, and no particular plans for celebration, unless we go to a Bette Midler concert this evening.

I've been correcting proofs over the phone with Michael di Capua and still have a final chore, underlining the words which would normally be italicized, to make them stand out from the italics in which the diary extracts are printed. This underlining has to be done *by hand* on the plates!

[1] Responding to the Soviet invasion of Afghanistan in December 1979, Carter asserted in his State of the Union Speech on January 23, 1980, that the U.S. would protect its "vital interest" in the Persian Gulf region with force if necessary; see Glossary under Carter.

[2] American musician; he often sat for Bachardy, and they sometimes had sex. He died of AIDS in his early thirties. Coleman was an assumed name.

I *haven't* gotten on with the *October Diary*—this is the kind of chore which I can only accomplish by brute willpower; something in me is resisting it. And yet I know that some of this stuff isn't bad. I suppose what I resist is the lack of design, the irrelevance of these entries. They suggest an air of dainty caprice—the tone of a writer whose motto is "I write as I please," meaning aren't I cute, aren't I delightfully irresponsible, don't you wonder what I'll be up to next? In fact, I am evading the iron question which I try to live by as a writer: *Why* are you telling me this?

February 17. We've had a clear day since the big rainstorm, and now, at five in the evening, the second storm is coming over and things don't look good. We're always worried when rain comes, being mindful of our last big slide and the fact that there isn't anything *more* to slide without bringing down a chunk of our retaining wall just beyond Don's studio. Meanwhile I've finished marking all the places where there are italics in the diary extracts; now I have to decide how many of these are really necessary.

The rain's getting heavier. Ah, how I hate it. Don's studio has been flooding—not only over the floor but down the walls of the storage room, threatening his portfolios.

I don't know what I think of *My Guru.* I can imagine really savage attacks on it and yet in a way I think it is the most worthwhile book I have written *and* probably one of the best modern books of its kind.

February 21. This miserable fugue of rainstorms continues—there's said to be a whole procession of them stretching back nearly to the international date line, advancing steadily on southern California.

After the slide on the night of the 17th–18th which swept a tree and a lot of mud down onto the road below, from the hillside just across the steps beyond Don's studio, we found that one of the hoses used by Tom Shadduck to water the upper lot had been turned on; presumably it had been running all night or maybe even for several days and nights—in which case it could well have actually caused the slide. This was spooky and a bit scary. We wonder if it was a deliberate act of sabotage by some enemy. It is very very unlikely to have been due to carelessness on Tom's part.

February 23. A pause in the rainstorms during yesterday and the day before, but more rain is promised for tomorrow—gleefully, one feels, by the T.V. weather people.

Tom Shadduck, who showed up yesterday for the first time

since the rains began, assured me that the hose we'd found pouring water onto the slide area didn't belong to him and wasn't ever used by him. (It now seems to me that this hose looks like one which we had a long while ago and which got stolen. All of which supports the sabotage theory.)

Yesterday we saw a rough cut of *Urban Cowboy* at Paramount—Jim and Jack were there and also John Travolta. John was certainly first-rate—indeed he seemed the perfect actor for this particular part. And yet we had our misgivings. The picture seemed to lose momentum several times. The mechanical bull became a bit tedious, it just wasn't quite dangerous enough to be the number-one thrill and threat. There *was* the fire at the plant, of course, which was certainly dangerous but wasn't sufficiently led up to—or down from—to assume great importance in the story.... Well, maybe we were just oversevere critics because we were rooting for Jim and Travolta so hard. And it *was* long—two hours and forty-five minutes.

Travolta was very sympathetic this time, as before when we saw him, on the 6th. He campily performed the final scene in *Cruising*, when the undercover detective entraps the stabber by proposing a sex act.[1] He knew the dialogue more or less verbatim, as far as I could judge, and he actually dropped his pants to make it extra realistic. (He kept his jockeys on, however.)

After this we went to the county museum and saw two olden moldies: *Duel in the Sun* and *The Wild Heart*.[2] So that, when it was all over, we'd clocked nearly seven hours' viewing time!

Yesterday, I dictated to Michael di Capua's assistant Carol Kino all of the words to be underlined in the italicized sections of the book. So that, roughly speaking, my part in the work seems to be finished, with the exception of the blurb.

The day before yesterday, I walked down through the Canyon to see the flood damage. Both the Friendship and Casa Mia got a lot of mud inside them, carried down the road, not down the channel, which seems to have functioned very well. A lot of the beach has been washed away and all of it has been littered with driftwood and other debris. Some of the canyons are said to be still blocked. During my walk I was hailed by Lyle Fox, from a big car. He has white hair now but seemed as vigorous as ever. He is probably going to become the manager of the Georgian Hotel for old folks.

[1] The 1980 film about a serial killer targeting gay men starred Al Pacino, not Travolta.

[2] Re-edited U.S. version of *Gone to Earth* (1950); both films starred Jennifer Jones.

Something I saw the other day and haven't recorded: two youths are jogging on San Vicente, both of them well-made and sexy, in shorts and T-shirts. As they run, one of them, without breaking his stride, strips off his shirt and tosses it into the branches of a coral tree. The abandon with which he did this was extraordinarily erotic. (When I told Peter Schlesinger about the incident, he claimed that the shirt was still there in the tree; he had noticed it as he drove by.)

March 2. Sunday, muddy Sunday. It's raining again, hopefully not for another week-long period, or we'll lose more hillside. This is a bad dark day of depression. Yesterday was the reception for the opening of Don's show, which was, officially, the day before, February 29. Crowds of people came, including Marguerite Lamkin, just arrived in town, but only one picture was sold. Darling deeply depressed—why do I go through this, what's the use if I can't make a hit at my age, and many other unanswerable questions. I feel we're let down by Jim Corcoran, who couldn't even be bothered to express either concern or optimism. Then, driving home in a mood of extreme nihilism, Don bumped the back of another car—not hard at all. The other driver, with the worst kind of Jewishness, began by accusing Don of being "inebriated," and threatening to call the cops, then gradually, contemptibly, backed down and started to address Don as "my friend." "Don't you call me that," Don told him. "I'll call you anything I damn well like," the man blustered. So Don called him "a jerk" and he took it meekly, having lost steam. Now, this morning, the man calls to tell Don that he has taken a look at his rear bumper and can't find any mark on it at all. Don reports that he behaved with the utmost politeness, thanking the man for calling.

Jim Gates has entered a home for gay alcoholics—it seems that he has been a constant solitary drinker, which I'd never known. Talked to him on the phone. He has done thirty days of his ninety-day stint and is in a good state of mind. He is even considering rejoining the monastery. If he doesn't do this, he is anyhow determined to give up the real estate business and get some other kind of job, in which you don't have to tell so many lies.

Now, I'm slowly signing copies of a specially printed edition of *A Single Man*, and preparing to restart *October Diary*, maybe even this afternoon, God willing.

March 6. Another heavy rainstorm last night and another tree down. God, how this weather depresses me. And how burdened

I feel by letters (which I hardly ever answer anyhow) and by the dreary empty chore of writing this *October Diary*. The only thing I feel inspired to write in it is why I don't want to write it at all. It should be a deliberate study in triviality, the miseries of attempting to write a potboiler.

March 30. Well, at least Don sold two drawings and a painting and got two commissions. And the *Los Angeles Times* critic, Suzanne Muchnic, wrote: "Don Bachardy's portraits look marvellous in any season. Now that human likenesses are enjoying a revival of interest, his work looks better than ever. In an exhibition of new portraits of women, Bachardy is up to his usual impressive level as he chisels features, freezes character and eliminates odd details. He makes beautiful drawings that are endlessly interesting because they offer insight into the substance behind appearance."

However, after a brief summer of good spirits, Don is down down down again, and so am I. We both have our reasons, and God knows, reasons hang from every tree in this gloomy epoch of inflation and crisis,[1] ready to be plucked.

The day after tomorrow, I have to haul ass up to San Francisco to give a lecture at Marin College,[2] because they will give me $2,000.

Am stuck in *October Diary* again, and my gut hurts and the usual pile of letters has accumulated and I still haven't finished the long complicated census form. Well, enough of all that.

Old Jo is in hospital because they have at last persuaded her to have the operation—in fact she's had it and now is lying there groaning with tubes in her.

My Guru has arrived in page proofs but Farrar Straus still don't seem to have decided if it is worth their while to pay for us to come to New York and make propaganda for it. Methuen is far more enthusiastic.

Oh Swami, be with us and help us and cheer us up.

April 6. Happy Easter and, Swami, I do think my morale is better and Don's also, although Darling has a painful stiff neck, and Elsie

[1] Oil prices doubled in 1979, and inflation rose to nearly fifteen percent by March 1980; Iran still held the U.S. hostages; the war in Afghanistan continued; and the U.S. and eleven other countries planned to boycott the 1980 Olympic Games in Moscow.

[2] I.e., College of Marin, a two-year college in Marin County, north of San Francisco.

Giorgi diagnosed my gut pain as a hernia but didn't insist on surgery, only advising me to lose weight and not wear my belt tight.

The talk at the College of Marin was quite a success. Armistead Maupin was there with Ken Maley and they thought it better than the Gay Rights Advocates talk.[1] I got a very intense fan letter about it yesterday [...], the kind of letter Don would say warns of trouble: "Often times the words seemed secondary, it was your presence, your smile, your laugh, your love that spoke to me. In fact perhaps the most inspiring moment was when you first walked on stage, before you had spoken a word. I felt such powerful energy radiating from within me and from you...." etc. etc.

Much more important, my few hours of separation from Darling won me such a meltingly sweet welcome home from him—he had taken the trouble to park his car and come right up to the gate at the airport to meet me as I came off the plane, after midnight.

Yesterday night, the Oliviers—Larry, Joan [Plowright], Richard, Julie Kate—came to supper, and it was really cozy and nice. Joan went into the kitchen afterwards and kissed Natalie, which was stylish and much appreciated. And, when they went into Don's studio, pretty plumpish Richard went into raptures and Larry commissioned Don to paint Joan.

April 17. Deeply depressed again—partly too much wine drinking, partly because of Gavin, who was at the party last night, has now read *My Guru* and doesn't like it. Don continues to say that he is sure people in general aren't going to—are going to hate it, in fact. The other day he declared that all the stuff I have written for the *October* book is of inferior quality and obviously made-to-order. He said we should drop the whole project and tell Jack Woody at once. But now he says well, after all, it'll do, it's better than he thought. So I must struggle on. Meanwhile, I feel the hernia nearly all the time. Courage, old horse. My Darling is very sweet to me—that keeps me going.

May 18. Yesterday, I finished the rewrite of the *October Diary* for Jack Woody, all except for its ending—there has to be a paragraph which somehow rounds the whole thing off, and I can't quite decide what it should be. I don't want anything which suggests that this whole job was done to order, although, of course, that would be obvious to anyone who knows anything about writing. There is nothing spontaneous about it. No flow. Each entry is

[1] See the entry for September 17, 1979 above.

just "making conversation." However, I promised Darling that it would be ready for his birthday, and now it is, practically speaking.

On April 27, Marti Stevens gave a lunch for Gavin Lambert—one of those fatal alcoholic lunches which I ought never never to attend under any circumstances. I was bored and got drunk and fell down the steps when we got home and hurt my head, rather worse than I have in other similar falls. The lump on it is still there and I get pains over the top of the skull and my neck is chronically stiff.

Darling finally decided he had to have spectacles. He got them the day before yesterday. I think he looks attractive in them, but then I admit I have a thing about spectacles. Having to wear them depresses him—it's all part of his feeling of getting older, which the birthday reminds him of sadly. God, all of us need so much courage, even when things aren't going particularly badly!

May 19. Boswell, Gladstone and T.E. Lawrence died today—and why not? The Canyon is full of fog. I'm really not in the least depressed. The *October Diary* is practically finished. *My Guru* is out in the world on its own, and we've got to follow it and fuss around, making publicity—but not for another week. And even then, I just have to do one book-signing session, next Sunday.

Darling's birthday was at least brightened for him by seeing a really good print of *2001*, which he and Rick and I all thoroughly enjoyed. What an extraordinary masterpiece this is! This time around (the 4th-5th?) I was most strongly impressed by the sequence of the man-apes—their chattering and the curious tension which Kubrick creates around them; they seem so threatened. (What they are threatened by is an enlargement of their consciousness!) The other sequence which seemed specially remarkable this time was the conversation between Dr. Floyd and the Russian scientists, on board the space station; the banality and polite bitchery of it.... But, overall, I felt that it's preeminently a *visual* artwork—not merely because of its great scenic passages but often in details of interior decoration. Again and again, glimpses of daily life on the spacecraft compose into static pictures you might exhibit in a gallery.... One noticed that yesterday because the print was so good.

Our birthday supper was baked potatoes with caviar and champagne. A small pot of the caviar was Russian, I'd bought it at the Farmer's Market the day before, for $34. The rest was Icelandic lumpfish $2.29 at Vicente Foods. The Russian was noticeably better, which was reassuring, but the lumpfish wasn't bad, at all.

The champagne was Mumm's. Mount Saint Helens honored the occasion by exploding, but there's no lava yet.

May 20. Drizzle in the morning, foggy all day. I've spent most of today reading the typescript of Gavin's still incomplete novel, *Running Time*—343 pages of it, so far. I found it heavy going, the characters seemed to have little life, and the device of mixing them in with real people, such as L.B. Mayer, didn't help. I have a hunch that this novel is made out of material which Gavin collected for his nonfiction book about Hollywood.

We had Paul Sorel to dinner last night, at the Café Swiss. His money affairs are getting worse and worse. This time, we had the impression that he is losing faith that anything will happen to "save" him. But he still keeps saying that he won't give up the house.

Last night, and the night before, we slept out in the studio—the reason for this is that the mattress on the studio bed is newer and much firmer and springier. Don feels it helps his bad back to sleep on it. I feel it helps my neck, which has been bad ever since my fall down the steps. I do have the impression that the bump on my head is now getting smaller and perhaps the neck is better too. Right after my fall, I did have some rather alarming symptoms, which I don't think I've ever had before—a kind of amnesia during which I couldn't be sure what I had seen or heard or what I had perhaps dreamed the night before. Such images kept melting into each other in a disconcerting way.

May 21. When Tom Shadduck came today, he discovered that another hose had been attached to the same water faucet on our upper lot and the water turned on; this after some rain had fallen yesterday. In fact, this was a repetition of the apparent act of sabotage on the night of February 17–18; the end of the hose had been left hanging over the wall and it had washed down a little mud from our slide area. This really is scary and we are seriously considering calling the police in. If only we could believe that they'd be the smallest use!

We had Robin French to lunch because we wanted to ask him if he thought we could handle our money affairs better than at present. On the whole he was reassuring—that is to say, he didn't tell us to sell everything and buy gold, or send our money to Switzerland, or invest in more real estate.

Today Don finished reading my *October Diary* and was quite pleased with it. He had made a lot of notes, so now I can get to work revising the whole thing.

Weather still damp and threatening rain. Pains in the top of my skull, they worry me a bit.

June 8. The pains went on worrying me, throughout our visit to San Francisco, from the 2nd through the 4th, to promote *My Guru.* After we got back here, I went to see Elsie. She doesn't think I need have X rays at present, but she had me buy a cervical collar. The collar is quite a conversation piece—when I saw other people wearing them, I always thought, in my ignorance, that their necks were broken—but I feel embarrassed to wear it during public confrontations, like for example this evening when I have to sign books at the National Gay Archives. It's really comforting when you wear it in a movie theater though; it rests your neck and you could easily sleep in it if you wanted to.

No more reviews of *My Guru*, and I'm beginning to feel that it has flopped. Oddly enough, I have already heard the opinions of two lapsed Catholics—one an interviewer in San Francisco, the other a gay activist in Philadelphia. Both admitted that they could scarcely read the book, because of their prejudice against "religion." What depresses me is that neither of them saw that *My Guru* is actually an attack on the whole Catholic mentality, and the religion of the God of Law.

The interviewing in San Francisco was drastic. On the 3rd it ran for nearly nine hours without any breaks except for transportation between points. On the 4th, about the same—plus all the fuss of getting onto the plane and flying home. Don was amazing. He shared several of the interviews like a professional, saying all the right things but never merely echoing.

June 9. Last night I appeared at Jim Kepner's homosexual library in Hollywood, grandly called the National Gay Archives. I answered questions, read aloud from *October Diary* and signed copies of *Guru.* It all went off quite well. I was surprised by the quickness and smoothness of my answers. It was as if part of my mind is still much smarter than the rest of it, like some area of a computer which goes on functioning when a lot has been unplugged. I noticed this occasionally when we were up in San Francisco.

But I still feel that this is a losing battle. Today I hear from Bob Gordon that the *San Francisco Chronicle* has given the book a lukewarm notice; and a notice from New York today was half-assed. Don is going to be proved right—and what else was to be expected? Perhaps the book is only going to be intelligible to faggots.

One thing I was spared—Don Kilhefner didn't show up, as Gary Hundertmark feared he would, to attack me. The last thing I wanted was a battle over my carcass—which Gary was fully prepared to fight.

[*Isherwood's typed diary ends here. After a gap during which he travelled to London and Amsterdam with Bachardy to launch* My Guru and His Disciple, *he wrote the remaining entries, from July 16, 1980 to July 4, 1983, by hand on the lined folios of a black, bound ledger.*]

July 16, 1980—July 4, 1983

July 16. Shortly before 8 a.m. this morning, while Don was out at the gym, Humphrey Carpenter (Auden's biographer) called from England. He'd been asked to tell me that Bill Caskey is dead. He was found dead in his apartment, in Athens. That was all Carpenter knew.

I said I wasn't greatly surprised. He must have been in terrible shape, with all that drinking. No doubt he'd had a heart attack.

Since I'd been meaning to write Carpenter anyway about his book—which I read most of while in London—I took this opportunity to tell him that some of his statements were inaccurate.[1] I had *not* disliked Chester when I first met him, had *not* left New York in May 1939 *because* of Chester—had, in fact, done my best to help bring Wystan and Chester together. Having said this, I became aware that Carpenter was a bit shocked by my changing the subject. He thought I didn't really give a damn about Bill's death.

July 17. Well, what *do* I feel about it?

Relief, sadness, relief, sadness.

Relief that we're rid of the tiresome menace of Billy's jealousy of Don. I could never be quite sure that he wouldn't suddenly show up and subject Don to some ugly idiotic scene. Not to mention his claims to various possessions, allegedly stolen from him—nearly all of them, in fact, lost years ago. And the ugly scene would have been followed by a drunken reconciliation, equally tiresome—and false.

Sadness because Billy's life in Athens looked so sad—at least

[1] *W.H. Auden: A Biography* was published in 1981; Isherwood read a typescript.

from the little I knew about it. But maybe it wasn't, really. Maybe he kept busy with photographic projects, and still enjoyed eating and drinking, and had friends he was fond of.

Do I feel guilty about him? No, hardly at all. But I do feel grateful to him—I always have—for making the break between us when it had to be made and I perhaps wouldn't have made it myself.

What's the best thing I remember about Billy? That he created such a powerful liveliness around him. His liveliness drew people into his orbit to eat and drink and chatter and enjoy themselves. All kinds of people were fascinated by his liveliness—including Forster and Stravinsky. Stravinsky used to say, "He's my type." Also, he was very much of a Nanny. Under favorable circumstances, he might have become a legendary male nurse.

July 18. A specimen bit of contemporary art talk, announcing a show of Jane Reynolds's work at the Jancar-Kuhlenschmidt gallery.[1] "The peaceful coexistence of opposites and similarities is a continuing imperative; (in so far as their mixture is an unavoidable reality)."

Meanwhile, I try to heal Dobbin by dipping him every day in the ocean—four days in a row, so far. Beautiful weather.

Yesterday evening, the possibility of a tidal wave was announced, because of an earthquake in some islands east of Australia. But nothing happened to us. I saw Old Jo and got drunk. Don saw his mother. Her apartment is filthy and there are cockroaches in it. It looks like the dreaded moment is nearing when she will have to be provided with a visiting nurse—whom she'll almost certainly hate.

July 19. We didn't go down to the beach today—partly because Ted was down there and relations between him and Don are bad; he made a silly jealousy scene because Don didn't see him at once after returning here, but did see Rick. And then David Dambacher came to sit for Don but yakked so mercilessly—maybe he was zonked on something—that Don couldn't work.

Result, I haven't been out of doors and feel fat and drowsy and guilty. I meant to take a beach and ocean cure, not a sleep cure.

This morning I called Bill Scobie who told me that Tony Sarver's cancer has reappeared and that he must soon have another operation.

[1] Los Angeles gallery run in a basement by Tom Jancar and Richard Kuhlenschmidt.

July 20. I really do become pathologically lazy, the minute I cease to have a work project. The only constructive thing I've done today has been to run down to the ocean and dip in. The rest of the time I've also been dipping—in Maupin's *Tales of the City.* That book is a modern Dickens novel—tearjerking love affairs (I have a terrific crush on Michael Mouse and weep when he and John think he's going to die of paralysis)—corny comedy and an ingeniously disgusting horror climax which would have shocked Dickens, I suppose, but is in his best manner—oh yes and also violent political indignation, though Dickens wouldn't have approved of that either, because it's indignant support of gay lib.

Tonight, Jack and Jim are bringing John Travolta to supper. Natalie is cooking, which is a shame, because this is the evening of her daughter Thaïs's return from Europe, on holiday from her ballet career.

July 21. I guess our supper party went off all right, last night. Peter and his wife Brooke Schjeldahl brought their five-year-old daughter but she turned out to be one of the most agreeable children imaginable, neither sulky nor sly nor pushy nor ugly, with a charming trustful smile for all of us, she went off without the slightest protest and slept in our bed until it was time for everybody to leave. But I did feel that both Brooke and Peter cramped Travolta's style. Their intellectual talk gave him no openings to shine, and Brooke was downright tactless in mentioning a bad review of *Urban Cowboy* by Pauline Kael in *The New Yorker.* (Actually, this review was chiefly directed against Jim Bridges.)[1] Don and I were both aware of Travolta's extreme vulnerability; he seemed pathetically insecure. I wish now that we could have had him alone with Jack and Jim, so that he could have been encouraged to show off to his heart's content.

This morning, Michael di Capua called, with goodish-baddish news of my book's fortunes. They have sold nearly ten thousand copies. But business is bad—many, many books don't get republished as paperbacks at all, nowadays. I had a contract with Avon, but the offer they came up with was so low that Candida Donadio and Michael agreed to try another firm. However, the best they could get was from Penguin—$8,500—which is only about a

[1] Peter Schjeldahl, American poet and journalist (b. 1942), wrote about art for *The New York Times, The Village Voice,* and other publications; later, in 1998, he became chief art critic for *The New Yorker.* His wife, Brooke Alderson, had a small role in *Urban Cowboy.*

thousand better than Avon's. As for the reviews, they haven't been bad, on the whole. But it seems that there won't be any from *Time*, or *Newsweek*.

July 23. Last night we had supper with Mark Lipscomb and John Ladner at the Black Forest. Mark has shoulder-length hair and he was dressed in a sort of Three Musketeers black costume. He looked exactly like a woman playing the part of Principal Boy in a pantomime. He and John were down at Palm Springs recently and, when they went to a motel, they were told, "We're gay here, we don't have mixed couples." Mark says he doesn't at all mind being taken for a woman, indeed rather likes it. He looks older, but still coarsely sexy in a way which turns me on, rather. He and John seem to have made their home into a regular dormitory of boys; Mark woke up the other night and found someone in bed with him and wasn't sure at first who it was. Both he and John seem to have plenty of money—John says he's earning a lot, Mark presumably inherited some from his mother.

On the 21st Don drew a really brilliant portrait of William Wyler.[1] Yesterday he drew Kris Johnson, Edmund White's very attractive artist friend, a nice but too serious boy, or so he seems to us.[2]

I still haven't begun work on my next autobiographical book. There are all these diaries to read through. Admittedly I'm lazy, and glad to let the days slip by. What if I die? Oh well, the diaries are there—let someone else fuss around with them. Fie, Dobbin! Perhaps I'll start tomorrow—or even today.

July 25. A couple of days ago, I noticed that the metal cable which has been wound around the half-fallen section of supporting wall below the house—to prevent the wall from collapsing altogether—had been slipped off. The wall hadn't collapsed, but now there was nothing to stop it from doing so.... Was this another mysterious act of aggression by the same person who twice put

[1] Film director (1902–1981), born in Alsace, working in Hollywood from the early 1920s; he won Academy Awards for *Mrs. Miniver* (1942), *The Best Years of Our Lives* (1946), and *Ben-Hur* (1959). His many other films include *Wuthering Heights*(1939), *Carrie* (1952), *Roman Holiday* (1953), and *Funny Girl* (1968).

[2] Johnson (d. 1985), a Norwegian-American painter, worked in a gay bookstore on Santa Monica Boulevard, where he met the American writer White (b. 1940), visiting from New York. Johnson died in his early thirties, of AIDS.

the hose over the wall above the cliff and turned on the water to provoke a cliff slide? Tom Shadduck got the cable back around the wall without much difficulty, however.

July 27. Now, after two whole weeks at home, I must absolutely get to work. I hope to make a token beginning on the read-through of my old diaries, later today.

Needless to say, this morning, which I was planning to spend indoors working, is sunny, for the first time in more than a week. The thick white sea-fog, which has been blocking the Canyon, rolled away early. So Dobbin's away to the waves.

Juan Szczesny came by here, yesterday afternoon. He is really quite a sweet boy who bores but also rather charms me by talking obsessively about all the money he plans to make—prospecting for precious minerals in Brazil or breeding horses in Argentina. He reminds me very much of Berthold, his father, but I don't think he is nearly so wild or so imprudent and apt to get into trouble.

There is one sad photograph of Berthold himself, with his wife and a son (maybe Juan, maybe his brother; I forgot to ask). The picture is sad because of the sadness in Berthold's middle-aged face; he has lost his belief in himself as the young stud, maybe he is even already aware of his cancer.

Tracy Tynan just called this afternoon to say that Ken is dead. Kathleen wants me to speak at a memorial service and I said I would, "Quaker style," not making the customary funeral "tribute." So I asked Don, what did he think Ken's chief characteristic was. He answered without hesitation: "Appetite." A very good answer.

July 28. I had another conversation about Ken's memorial service—this time with [K]athleen Tynan. She was rather apologetic because Penelope Gilliatt is coming from London and will have to be asked to speak, too. I'm rather glad, because that means I will have to say less. Kathleen was worried because she is afraid that Penelope will get drunk and make an exhibition of herself—at least that's how I interpreted her tone of voice.

Thaïs Leavitt came round today, for the first time since her visit to Natalie began. She is noticeably changed—more self-assured, better looking, much more evidently a young woman with a career she is happy in. And this made her more likeable. Her frank ambition wasn't in the very least repulsive. She *is* a ballerina.

July 30. I'm in a curious state of anxiety about what I'm to say at Ken's memorial service. (He is being buried in England, because that was his wish. I must say, this preoccupation with one's grave place does seem strange to me, if it's anything more than a merely sentimental caprice. Sure, it's kind of nice to leave a monument—but, in the case of writers and of makers and creators in general, isn't that to be found in the work they leave behind? And why preserve, even temporarily, the remains of a mortal envelope?)

Ah—I knew it, or at least I hoped it: as I sat down to write this entry, I began to see how I could speak about Ken....

Begin by saying that I believe in survival of some kind. Speak of Swami's death and what I feel about his continuing presence.

Speak also perhaps about Wystan and how I read his poetry after hearing he was dead.

Say I believe that Ken is with us here, at this service—but that the degree to which he is present depends quite as much on us as on him. Therefore we should dwell on the aspects of Ken which each of us finds most positive and most inspiring. As far as I'm concerned, Ken's most inspiring aspect is his appetite for experience, etc. etc.

And then maybe I'll quote the William Penn extract from *Some Fruits of Solitude*: "They that love beyond the world cannot be separated by it...."[1]

August 1. Well, it didn't turn out like that. I decided to read Ken's diary extract from *Show People* about his feelings for Louise Brooks. I said a few words about this and then read the William Penn, and that was that. Not a distinguished tribute, but many of the guests liked it because I spoke briefly and clearly, which neither Shirley MacLaine nor Penelope Gilliatt did. Both of them were embarrassing. Shirley looked upward and addressed Ken directly as though he was hovering under the roof beams.[2]

Afterwards we went to the Tynans' house, where a secretary told us about a pornographic screenplay which is one of Ken's last works—very S. and M.—and Kathleen began to tell us about Ken's last days—saying that he had been terribly scared. Unfortunately she was interrupted in the midst of this.

From there, we went to the Veterans' Hospital to see Tony

[1] Isherwood quotes a different line from this passage, above, February 23, 1974.

[2] MacLaine, the movie actress and Broadway musical star (b. 1934), whose films include *Sweet Charity* (1969), *Terms of Endearment* (1983, Academy Award), *Madame Sousatzka* (1988), *Steel Magnolias* (1989), and *Postcards from the Edge* (1990), believes in reincarnation and has written about it in several autobiographical books.

Sarver, who will have his tumor removed in a few days. Bill Scobie had already told me that he thinks Tony's case is nearly hopeless—the cancer is almost certain to recur, and anyhow Tony will be an invalid for a long time after the operation. Tony was much thinner but seemed in fairly good spirits. Bill Scobie was with him. God, the sadness of it. And Tony talking brightly of how he wants to sell his business and devote his time to Art, as soon as he's better.

We cheered ourselves up (partially) by going to see *Strangers on a Train.*

August 3. *Newsweek* publishes a summary of a report on the world's prospects in the year 2000, prepared by the State Department and the Council on Environmental Quality: The world's population will have increased by fifty-five percent, nearly all in the less developed countries. Hundreds of millions will be hungry. The world's forests will have been largely cut down, farmlands will have become sand dunes. Prices will be enormous. Hundreds of thousands of species of plant and insect will have become extinct.

And that's even if we don't have any nuclear wars!

Meanwhile, fat old Dobbin eats and drinks and trots down to the beach and often dips in the ocean, and is starting preliminary reading of diaries for his next book—because, after all, what else can he do?

August 7. A letter from Alan Ansen[1] (July 24) tells me that Caskey had been dead *for about a month* when he was found in his apartment. There were no signs of foul play. But the police did find a telegram dated early June informing him of the death of his mother.

What shocks me—I don't mean that I'd necessarily mind it *for myself*—is the vision of Billy leading a life so utterly without friends. He did refer to himself in one letter as being lonely. No wonder he got drunk and let his old resentments seethe in him. Today I've been going through the carbon copy he sent me of a letter in which he furiously attacks me for not having answered his letters and then goes on to demand the Tarascan figure (which was a birthday present but which, according to me, he agreed should be included in my share of our belongings when we split up). It's so characteristic of him that he sent me a *copy* of the letter—getting

[1] American poet and scholar (1922–2006), educated at Harvard. He was Auden's secretary, 1946–1948, and lived in Athens from the early 1960s.

all ready for a lawsuit, it seems, in his Irish way. Now the copy has faded so much that parts of it are unreadable. I have "restored" them in ink where I am able, but there are some sentences I can't reconstruct. I suppose the originals will go to Billy's surviving sister, whom he hated. However, the sister must have at least two later letters showing that we made it up. And now, are we to haggle with the Caskey family over the Tarascan figure?

August 10. Yesterday was the first day of a Golden Anniversary celebration of the Vedanta Society of Southern California. So I went up to the center for the first time in months, and read snippets from my commonplace book and answered questions—many of them about the early days of the society.

During these doings I kept feeling what I've recurringly felt this year—while making publicity for *My Guru* in San Francisco, here, and in London—that I was seeing myself as others see me. Seeing the spry little old man who is "wonderful," merely because he hops around, is still quick with his answers and jokes, and looks perhaps ten years younger than his age. These external glimpses of myself aren't reassuring, indeed they make me feel all the more death-conscious, that is to say, all the more conscious that I am *surviving*, rather than just simply living. This is a deadly bad attitude, and I have to rally Swami and Holy Mother and the power of the mantram to my aid. I have to remember, as Forster advises, to behave as though I were immortal—which means get the hell on with my work.

Darling has been so wonderful lately. He has shown me his love so strongly, and that *never* seems dated or different. He has a genius for expressing love—the genius is in his utter lack of embarrassment.

August 20. Angel has had terrible pain from strained muscles in the chest and shoulder, but got almost instantly better yesterday after he'd at length persuaded Elsie Giorgi to prescribe him a muscle relaxant instead of codeine.

This morning I went to Dr. Dayton because my vision in the left eye is getting worse; I wondered if this could be because the opticians had made a mistake in preparing the lens. No, says Dr. Dayton, my left eye has now got quite a big cataract in it. And my right eye also has a small cataract. I'm to show them to him again in January.

Then I went to see a Richard Meador who wrote and phoned me some time ago—while in Detroit, he met Collins George, who told him about Denny Fouts.[1] (Collins has died since them.) This made Meador want to write a book about Denny—or rather, to write a book about the androgyne, having decided that Denny was an example of one. By this, he means a person who gets artistic insights out of being androgynous.

Meador is a slim, quite nice looking coffee-colored black boy in his early thirties, soft mannered, with a wispy, much-too-long moustache. He sculpts—representationally, but one couldn't judge how well from the tiny transparencies he showed me. He looks after his grandfather, who is blind and very deaf and nearly bent double, but does barbell exercises every day and cooks for himself. He is 102, and white.

Am depressed because Candida Donadio's office tells me that they still haven't sold *October Diary* to any magazine. If they can't do this, maybe Farrar Straus won't agree to publish it in book form, and then even Methuen may back down.

August 24. Almost the whole of my morning was consumed by two monster phone calls—one from Phil Grant,[2] the other from Haridas (I can never remember who he is except that he belongs to the Vedanta Society and is a nonstop [...] gossip). He had been in India, where, according to him, *My Guru* has shocked some, pleased others. He threatens to visit me, with Bob Baldwin,[3] whom I'm supposed to know and doubtless do. He says that Amiya wrote or (more likely) intended to write a letter protesting against a review of *Guru* in either *The Sunday* or *The Daily Telegraph*. The one in the Sunday was by John Lehmann—the worse of the two and (I thought) intolerably superior, thick-skinned and offensive. So I hope that's the one Amiya dislikes. But I fear she'll never write to say so.

Phil Grant's call—much the longer of the two—was full of deathly serious holy Jewish scruples about love and sex. But at least he's got himself a lover who's passionate and probably as humorless as he is. Anyhow, they seem to be most sacredly involved.

[1] Fouts and George, pacifists during W.W.II, met at Civilian Public Service Camp in 1942 and shared an apartment; as Isherwood tells in *D.1*, George was a professor of French before the war. He was a very light-skinned black, could easily "pass," and introduced Fouts to his black friends.

[2] Not identified.

[3] Ordained in 1978 as a minister of the Universal Christ Church, later called the Universal Church of God, an interfaith community based in Burbank. He was a member of the Vedanta Society from 1957, when he was initiated by Prabhavananda.

August 30. I'm writing this under rather odd circumstances, sitting on a bench outside Melnitz Hall, inside which Darling is watching an Alice Faye movie. When it gets dark, I'll probably go in and join him, unless there's enough light in the foyer to write or read by.

My birthday was a bit tense, as my birthdays usually are. They always seem to upset Don in a way of which he is perhaps not even aware. Anyhow, it's over and all is happiness again.

I think I'll risk making a confession here—I say "risk" because there is a risk that, by making it, I'll somehow break the spell of my successfully (so far) sustained resolve.

Immediately after I got back here from England, on July 12, I resolved to stop what I call "making faces." This is a habit which I have had for many many years—at least since my early manhood, if not earlier. My face making is of two kinds, which might be called "gloating" and "hating."

I "gloat" over possessions, situations which please me, achievements of which I am proud. For instance I might gloat over our home, or the life we lead in it, or a manuscript which I have just written.

I "hate" individuals who have done either of us harm, or whole groups of people whom I regard as hostile and hateworthy.

The face making which accompanies these emotions is different, of course. But it is expressed by tense grimacing, a flexing of the facial muscles, clenching of the teeth. I never look at myself in the glass when I am doing this. I, in fact, don't wish to admit that I am doing it. But Don has often caught me at it—especially when I'm "hating." When he does, he checks me by saying, "Dobbin's thinking something bad."

I have always been conscious that "face making" is a negative unconstructive occupation. It wastes time, and, in the case of "hating," it tires me. Yet I have very seldom managed to stop it for long. I do clearly remember that I once gave it up for just over a month in 1963, while I was in San Francisco during part of April and May, lecturing at U.C. Berkeley and working on *A Single Man*.

At that time I was so aware of my will to abstain from face making that my awareness actually tempted me to "gloat." This time, my resolve didn't require such a conscious act of willpower. It was founded on weariness, after the exhausting ordeal of book signing and interview giving in England and Holland. It was an instinctive move to conserve my remaining strength. There was in it, as it were, an acknowledgement of old age.

But, oh, it *is* worthwhile. As long as I don't relapse into face making, I can fill much of my time with japam—with the result that I feel calm, loving, nervously relaxed. My mind is far clearer than before, because it's no longer dulled by the toxins of hatred. You might say, in fact, that "face making" is a kind of antijapam.

I have very seldom described the practice of "face making" to anyone. I can only clearly remember telling one person about it and that was Gerald Heard, to whom I confided so much. To my amazement, Gerald said that he had practised it but that he had now given it up. I wish I had questioned him more—to be quite certain that we were talking about the same thing.

September 2. Well, once again I give thanks that the summer is over and the beautiful fall, the work season, begins. Only, now I *have* to work. This morning I wasted in financial arrangements which seem nearly intolerable and also quite difficult to me—just transferring money from one account to another; the sort of thing which even Don, who God knows is no financier, takes care of frequently without the least fuss or complaint.

I have kept slowly drawing up a list of contents of my 1939–1942 diary—this is a preliminary to getting started on the next volume of my autobiography. But, thus far, I haven't managed to feel any appetite for my material. Then there is the prospect of doing a screen treatment of "Paul" for Michelle Rappaport(?)—I'm still not sure how she spells it; the telephone book shows that there are at least three acceptable ways: pp, ap, op.[1]

My eyes are getting worse all the time—particularly the left one; it rather scares me, and yet I can't somehow for one moment believe in blindness as a condition I could find myself in.

September 9. Rather depressed—partly because of the deterioration of my eyes—partly, I think, because I have been reading Tom Cullen's book on Gerald Hamilton[2] and the Gold–Fizdale book on Misia Sert[3]—both of them ultimately downers. Also, we've got to go to New York at the end of this month. I wouldn't dream of not going—it's for Darling's show at the Robert Miller Gallery. But I dread New York, that hell of squalor and hostility.

[1] Rappaport; see Glossary.

[2] Cullen (1913–2001) published books about the murderers Jack the Ripper and Dr. Hawley Crippen and about the conman Maundy Gregory but nothing about Gerald Hamilton.

[3] *The Life of Misia Sert* (1980), about the Polish-French pianist and saloneuse (1872–1950), by Arthur Gold (1917–1990) and Robert Fizdale (1920–1995), piano duo and longtime companions.

A letter from the American vice-consul in Athens, saying that Caskey's body was only discovered after several weeks, when the neighbors reported his absence, that no cause of death was diagnosed by the coroner because the body was in an advanced state of decomposition, and that it was buried in the Zografou Cemetery on July 21.

The letter ends, "Please accept my sympathy for the tragic loss of your friend. Richele Keller, American Vice-Consul." Now I must try to find out if Richele is a male or female name, so I can thank him or her.[1]

We have been intimidated by a rat which eats the dry cereal when it can get into the closet. Otherwise, it shits over the beds and floors. Now we're trying to deter it by keeping all the doors and windows shut—which means sleeping in Don's studio!

September 15. I got around the problem of Richele Keller's sex by addressing him/her as "Dear Vice-Consul Keller," even though that isn't the prescribed form and sounds a bit like a character in an Ibsen play.

Sometime soon—maybe as early as next week—we'll be taking off for New York. If we fly direct from here to there we wouldn't have to leave until Sunday the 28th. But we may make a short whiz-tour with David Hockney—one which he has made already and strongly recommends—to drive from Los Angeles through the mountains to Denver and take the plane from there to New York.

Today I finally got Basil Bunting's *Collected Poems* out of the Santa Monica library and have read some of "Briggflatts," which Thom Gunn specially recommended, saying that it's "about as good as Eliot when you get down to it." Cyril Connolly, I now find, praised it too. Don't know what I think, yet. But at least and at last I've looked up "oxter" which I first came upon in early Auden, "axe under oxter,"[2] and now know—but for how long?—that it means "armpit." Bunting uses it here.

[The gardener's helper] has a sore throat, is afraid it may be clap. Poor angel, he really is a martyr to V.D.

No sign of the rat or rats lately. But Tom Long is to put a screen door on our bedroom window.

September 25 [Thursday]. Yesterday, Curt Klebaum[3] phoned and

[1] Her; she served in Athens from 1979 to 1981.

[2] In "The Decoys," beginning, "There are some birds in these valleys," *The Orators*, Book II.

[3] Matthew Curtis Klebaum; he worked at Nicholas Wilder's gallery for a while and sat for Bachardy several times.

said, "At David's party on Saturday, I kissed you." "That was a rare pleasure," I told him politely, not getting his drift. He then told me that, since the party, he has developed hepatitis [...]—so now Don and I would have to get gammaglobulin shots. Elsie Giorgi gave them to us yesterday afternoon. She said we were probably immune anyway, since we'd already had hepatitis—at the time, we were told that this would make us *more*, not less, liable to reinfection—but gammaglobulin is to be recommended also as a protection against flu.

Well, we're still due to leave on Sunday—that is, unless we switch the reservation to Saturday night. A series of foggy mornings makes us fear we might arrive late and not get to see David Bowie's performance in *The Elephant Man*, to which we've been invited for Sunday evening.

Oh, the fuss and the tension, the minute one leaves this haven of peace!

On Sept 19, Don had lunch with The Downer—after their long separation. He got the impression that The Downer would be happy to come back to him, if asked. No comment.

However, Don has now met a really attractive and bright and *rich* young man named Don Carr, whom he likes, and who likes him. They were together last night. Dobbin is keeping his hoofs crossed. Carr worked for the Mark Taper Forum people, in the casting department.

October 7. On September 28, we were driven to the airport by Jim White where we met Jack Woody, complete with two just-bound copies of our book *October*, and we took off with him for New York. A perfect flight with a tail wind so strong that we put down nearly three quarters of an hour early, with plenty of time to move into Paul Sanfaçon's apartment, get changed and washed and still arrive at the theater punctually to see David Bowie in *The Elephant Man*. He was quite good despite the special cutely distorted "elephant" voice which he used throughout his performance. But his supporting actor and actress were far inferior to the ones in the off-Broadway production, and the play itself seemed preachy and second-rate, especially toward its end.

There was a reception afterwards at which Bowie appeared and made himself charming, and I don't mean that snidely. As I told him, when he phoned a few days later, he is the politest person I have ever met, and that includes Michael York with his great line, "Why, you're just like me—no, I mean I'm just like you."

This reception was held at the top of some grim old warehouse which had had its top floor dressed for the occasion with hangings and pictures and potted trees and buffets. During this affair, both Don and I became aware that relations between David Hockney and Peter Schlesinger were still much more tense and complicated than we had imagined. [...]

On the 29th, we had lunch with Andy Warhol and Bob Colacello[1]—largely because they had just published extracts from the *October* diary in their magazine *Interview*, along with Don's drawing of me projecting hostility. Was much impressed by the dining room of their office, which is walled with eighteenth-century(?) paneling from England. However, it smelled unpleasantly because of the plates of unfresh cold meat which were to be our lunch. Andy seemed unexpectedly benign—almost what one might describe as a "good man." But that didn't mean that the usual tape recorder wasn't in action. The difference was that, today, Andy entered into the conversation in a way he never used to—at least, not in my presence. He seemed almost benign, very friendly to us.

October 8. One more day of the thick smoggy fog which has wrapped itself around the area throughout the time we were away in New York. At the moment, Darling is (I suppose) in Cleveland, drawing a wealthy and domineering old lady because one of her sons offered him the job—there are to be three drawings at a thousand dollars each. But to continue my narrative of our stay in New York—

The afternoon of September 29 was spent largely at Bob Miller's gallery, where Nick Wilder was supervising the hanging of Don's pictures. I wondered if Bob and his partner [...] John Cheim[2] minded Nick's interference. They didn't seem to and they expressed great enthusiasm for the work. However, the larger half of the gallery was occupied, as we'd known it would be, by the geometrical abstractions of Nabil Nahas—a quite sympathetic and handsome man with a foreign (Egyptian?) accent[3] who seemed to be gay himself—or maybe he acted that way around us out of politeness.

[1]Colacello (b. 1947) was editor, 1971–1983, of *Interview* and contributed a regular column, "Out," accompanied by his party snapshots. He also helped Warhol run his studio, The Factory. Later, he wrote for *Vanity Fair* and published a biography of Warhol, *Holy Terror: Andy Warhol Close Up* (1992), before undertaking a two-volume work on the Reagans.

[2]Cheim designed and edited catalogues for the gallery; in 1996, he opened a new gallery, Cheim and Read, with Howard Read, also a Robert Miller co-director and the photography curator.

[3]Born 1949 in Beirut; educated at Louisiana State University and Yale.

In the evening we had supper with Mike Van Horn and Brian Campbell. Mike, who seemed to be his usual sweet devoted friendly self—quite unchanged—had actually postponed a trip to Puerto Rico in order to be at Don's opening the next day. Brian did seem to have changed and become faggier—preoccupied with his jewelry made of gilded fishbones. I'm fond of him too and their union seemed solidly domestic but not restrictive.

Don made his opening on the 30th into a typical marathon event by taking us first to spend the whole afternoon at the Picasso show. This was its last day—we had to use Kynaston McShine's power to get admitted;[1] but actually the crowds weren't any bigger than when we saw it in late June. Perhaps my greatest impression was of the massiveness of the two young men reading the letter—that monumental hand on the brother's shoulder.[2]

Darling's own show drew a satisfactory crowd. No sales, but about fourteen of our *October* books were pre-ordered. (Later Miller spoke handsomely to Don, saying that he was proud to exhibit his work anyway. It was its own reward, sales or no sales. So now I love Miller dearly—unless he makes a wrong move later.) The new paintings made me prouder of my angel than ever.

No drinks at the opening—just as well. Drub stayed on his feet throughout, as was his duty. But afterwards, at the party at Miller's surprisingly sumptuous home, they served 100-proof vodka and Drub drubbed out.

I think it was next day, October 1, that the New York glooms closed in around me. Paul Sanfaçon's apartment has its charms—the Maurice Grosser paintings of Morocco and the South of France. They lead you away into far distances—some of them are at least five miles deep. But they can't distract you for long from the skylessness of the walled-in streets—the oppressive sense of neighbors all around.

So I began to sulk and scowl. Darling was patient at first, but he had lots to do—portraits of famous fags for the collection Bruce Voeller has commissioned[3]—and he really could not be expected to endure Old Drag without protest; especially as he doesn't much like city life himself. (I think it was our first morning that some workmen, being lowered or hauled upwards on a suspended

[1] McShine was a longtime curator at the Museum of Modern Art where the Picasso Retrospective filled the whole museum, May 16–September 30. It was curated by William Rubin.

[2] *Reading the Letter* (1921), the men are not necessarily actual brothers.

[3] The Mariposa Portraits, of twelve gay and lesbian leaders; see Glossary under Voeller.

platform, looked in and saw us lying in the same bed, and laughed and made jokes.)

Don was also repelled by the circle of Gay Elite in which he had to move—Voeller and his friends. They all seemed to be psychiatrists, either professional or amateur, and were archly knowing about the problems of gay married life. You were made immensely conscious, however, that at least one out of each pair had *money*—lots. And that this, and not psychology, was what ultimately settled all the marriage problems. That's bitchy and simplistic, of course. But what I'm trying to describe is an atmosphere.

(I forgot to mention a rude shock I got at Don's opening—the totally unexpected appearance of The Downer. He had decided to surprise Don by coming to New York for the show. So, after this, he had to be invited to come along with us to several dinners, parties or openings. I think I behaved correctly.)

On October 2, Paul Sanfaçon came by the apartment in the morning, to pick up letters, etc. I was taking a bath. He asked wouldn't I like to have my back scrubbed. I said yes—feeling pretty sure what this would lead to and reflecting well hell it's a return for hospitality and who is a seventy-six-year-old Drub to complain that Paul isn't the most beautiful of mortals? Drub gave satisfaction, I think, by the vigor of his response. Paul said, "Don's very lucky."

On October 3 we had lunch with Alan Stern and talked about writing a screenplay for "Paul." I realize that I shall have to take all the initiative in doing this, although I know Darling will become most supportive, once I have something down on paper.

On October 5, I managed to leave New York—just. Although I set off for the airport one and a half hours early, I picked a Latin taxi-driver who was pretending to understand English. I told him I wanted to go by way of the Triborough Bridge; he drove me clear up to the George Washington Bridge; then admitted that he was lost, turned around and came right back down and around, through the most crowded streets available. Finally, having reached the airport, he couldn't find the United Airlines terminal for about twenty minutes.... And then, having gotten myself and my bags unloaded, and having paid this maddening but innocent creature about thirty-five dollars, I found myself at the end of a line of some three hundred people, waiting to board a plane for England; they all had to pass through the same electronic barrier; it was the only one functioning. Without consciously deciding to cheat, I wandered up to the head of the line and joined it without getting any protest—I suppose because of my vulnerable senior citizen appearance. When I passed through the barrier, the control signal buzzed. I gave them my keys; passed through it again. It buzzed. One of the guards tested my clothing all over with a hand device. It didn't buzz. On the plane, I ordered double scotches until *I* buzzed.

(Again I forgot something important— Just before I started for the airport on the afternoon of October 5, we went to see Vera Stravinsky. Ed Allen had warned us on the phone that her condition varies from day to day. As we were leaving, Ed and Bob Craft sadly admitted that she had been much worse than usual. Since her strokes, last February, Vera has been subject to periods of confusion, loss of memory, acute anxiety. Our first glimpse of her, that afternoon, was reassuring. She no longer wears a wig but her own hair is still pretty and her face hasn't greatly changed; it still has some beauty. She smiled charmingly but it was clear at once that she wasn't sure who we were—me she dimly remembered, perhaps, but not my name. Don she seemed less sure of.

The way we were sitting painfully crisscrossed the lines of our attention—I on one side of Vera, Ed on the other, Don beyond Ed, Bob at some distance beyond Don, over near the door. Ed, who seemed more than usually female and nurselike, felt it necessary to keep Vera interested in us and to make her aware of our identity. I felt it necessary to keep kissing Vera and holding her hand while regarding her with that beaming switched-on empty smile, like the headlights of a car, which one always finds oneself using on such occasions. Bob Craft, meanwhile, was competing for my attention, just as he used to in the old days, when he talked, as it were, behind the Stravinskys' backs, although in their presence. Bob was telling me that Vera could still enjoy smoking and drinking, that she was still a good hostess, smiling charmingly at her guests without knowing who they were, and pressing them to eat. (The cook they now had was marvellous.) She also went out, went to shops, bought things, etc. etc. Bob is working on a new project—a volume of letters to (and from?) Stravinsky. As with all his projects, he had a martyred air about this—he was utterly exhausted—but now this seemed only natural, because he and a nurse are exposed to Vera's condition almost without respite. Ed can only get to join them at weekends.

As this exquisitely painful interview continued, Ed created a new kind of tension by telling Vera about Don's show at the Miller Gallery. Whereupon Vera complained that she hadn't been told about its opening. Ed evaded this by saying that the show was still open. Vera said they should go to it at once—and bring people with them, important people, who would make propaganda for it. All her acquired know-how about organizing such events when her own paintings had been shown made her seem quite lucid and sensible for a moment. But she wasn't lucid, alas—she wasn't for one moment herself—she was her own ghost—a pathetically anxious ghost, anxious because she was confused, and on the verge of being scared. I insisted on leaving as soon as we decently could. When saying goodbye to Bob and Ed, I suddenly shed tears.)

October 19. Yesterday morning, toward midday, I was working on the outline of the "Paul" screenplay, when I heard someone outside in the passage. I came out of my room and two Spanish youths came running up to me, asking, "Where is the money?" One of them held a slim-bladed knife in his hand—and yet, they weren't altogether alarming; at first they seemed more like cab-drivers or porters bothering a passenger at a foreign airport. It took me an appreciable moment to realize that this was a holdup. When I did, I foolishly began to shout, "Help! Help! Don!"—not because I was in a panic but because I consciously wanted to scare them. Don was in fact in the studio with Don Carr, hopelessly out of earshot. Now they grabbed me and one of them hit me in the face, making my upper lip bleed. They tied me up with the belt of my bathrobe. By this time I was adjusting to the situation and thinking sensibly. I decided that it would be wise to let them have the money in my billfold—it was between 100 and 200 unfortunately—because one is always warned to give burglars something and because the billfold was lying on top of the bureau in the closet, exposed to their view. (Some money still remained hidden between the pages of Stendhal's *Le Rouge et le Noir*; Darling's little hideaway.) In addition, they took my rings and my watch but mercifully didn't go off with the billfold itself and all my identification cards.

Now they had me down on the floor of my room beside one of the big bookcases—tied up, gagged with a sock. They also wrapped my light Japanese robe around my head. The gagging was unpleasant, of course, but I could actually breathe quite well. My situation seemed to me corny in the extreme—something which only belongs in a book. Instinctively I began playing up to it, fretting like a baby and groaning like an old man. I even tried to worry them by crying "*O mi corazon!*"[1] in the hopes I'd seem to be having a heart attack. I repeat, they were never really alarming.

[1] Oh my heart!

One of them told me, in accented but educated-sounding English, "If you are quiet we won't hurt you." I had the impression that they were nervous, however. It was only their nervousness which was dangerous, like the nervousness of a snake.

After a short while, they went into Don's little room, where they hunted in drawers but didn't take anything. Then there was silence; they had left the house through his open window. I managed to get my feet out of their bonds and then shuffled out to the studio and banged on the door with my elbow. As ill luck would have it, Darling and Don Carr hadn't finished fucking. Darling came to the door naked, followed by Don Carr pulling on his pants. Darling cried out in dismay when he saw Dobbin all bloody—all the rest of the day he was so sweetly concerned. Then the police came. Don Carr was impressed by Dobbin's sangfroid; not having seen the old ham perform before.

Today I have a sore rib—not much other damage.

October 22. Talked to Bill Scobie on the phone this morning. He now feels pretty certain that Tony Sarver will die soon. He can't have any more surgery, although he has a new malignant growth. He is very thin but in fairly good spirits. Oh God, these awful dreary long-drawn-out deaths, with their respites of false hope! And then, when it's over, what will Bill do? Is he really rooted here? Or will he go back to England? His life seems so horribly lonely—I may be wrong but I don't feel he has any other real friends—certainly no one to take Tony's place. Oh the poignancy of these separations! And of course this keeps bringing me back to the impending separation from Darling. My encounter with the burglars, I know, has given him a much greater shock than it gave me. He has been so loving, since that happened; we're very close now, almost painfully so.

Last night we had supper with Paul Sorel. The Internal Revenue Service has now presented its ultimatum—if he can't pay his arrears of taxes by the end of this month (and of course he can't, possibly) they will sell Chris Wood's house by auction, not caring how much it goes for as long as the taxes are paid. All Paul's friends are urging him to put the house up for sale through an agent *before* this deadline; in which case, he might not only get money for the taxes but as much as a hundred thousand for himself.

Paul declares that he won't do this. It would be against what he calls his "mystique." His mystique makes him certain that he mustn't do the prudent thing, mustn't provide for himself, because this would disappoint Chris Wood who, while he was alive, always expected Paul to behave recklessly, imprudently, outrageously. Such behavior, says Paul, was what kept Chris fascinated and amused by him. If Paul sells the house through an agent, he will cease to amuse Chris. If, however, Paul refuses to sell and lets the IRS auction it, Chris will be amused and will be compelled to help Paul in one way or another.

Paul got quite excited and aggressive as he announced his decision to us. He seemed a bit crazy but one was aware of his cunning, too. Even if he half believed that he could work on Chris's feelings in the land of the dead, he was also working on *our* feelings—trying to amuse *us*. And indeed he later admitted that one of his rich friends, who found his recklessness amusing, had offered him an arrangement—the friend would buy the house and let Paul live in it for the rest of his life. Paul said, however, that he didn't trust this friend, feeling that there might be strings attached to his offer. Don himself is rather interested in this possibility of our buying the house—he even said jokingly to Paul that his offer was to pay Paul's taxes—no more—in payment for the house. But today Don says that he would never get into any such deal with Paul because he knows how bitchy Paul would be about it, later.

October 26. A week after the event, one of my ribs is still sore but nothing else. The small wound on my upper lip shrank to a scab and then peeled off, leaving no trace.

Mary Miles Minter O'Hilderbrandt phoned—seemingly to ask me about our robbery, actually to tell me about hers; she never saw the robbers, being in bed upstairs. But they had left a jar (?) containing muriatic acid[1] which presumably they were planning to throw over her. Her narrative was confused by references to an incendiary bomb which had been used, on a different occasion, to cover up a "much larger theft." But she broke off the explanation of this because her housekeeper had arrived, and I haven't heard from her since.

Yesterday, Jim White phoned, having just returned to his house with Janice and discovered that it had been burglarized—the television set, etc. etc. And then Old Jo called to say that the residents of the Uplifters[2] have been forced to band together and subscribe to a private security guard, because robberies have been so frequent, there.

[1] Generally called hydrochloric acid.

[2] The former Prohibition-era drinking club and ranch retreat in Rustic Canyon.

October 31. My rib is still sore. The weather was so hot yesterday that I felt no pleasure in going down to the beach and returned exhausted. While we were down there, a couple of young men, quite presentable though no great beauties, began lip-kissing while gently stroking each other's backs, hair and arms. They were lying only a few yards away from us, with other beachgoers all around. Although the young men were nearly naked, in swimming trunks, this didn't appear to be merely an audience-conscious scene of exhibitionistic lust. It was romantic, and that was what made it odd. Public romance is embarrassing, public lust isn't.

I was reminded of it yesterday evening by two of the Alexanders' cats. They lay on the couch beside me, head to head, and kept touching each other on the mouths with their tongues—just like kissing.

We managed to get up Tuna Canyon, which is bumpy but just passable, and thus avoided the long detour to the Alexanders' house. They have added a great deal to it, but the place still makes me feel uncomfortable, exposed to the elements. The big windows are all naked to the darkness; I would hate to be in it during a windstorm. Peter and Clytie were very sweet as usual, but we both felt that they and their two daughters were unhappy, stuck there in the hills, far from the bright lights and the people.

November 2. On the evening of the 31st, we went to a dinner party given at the Corcoran Gallery for Laddie Dill's opening. I was given a seat at table next to a youngish man and his girlfriend—I don't remember their names but will try to find out.* The man reminded me that we had talked while I was signing copies of *My Guru* at Brentano's bookstore in Beverly Hills on June 14, and that he had told me that they had bought a house where I had once lived, 333 East Rustic Road. I had said to him, then, "Call me after you've been there a while, I'd like to talk to you about it." "And now," he told me, "we know what you meant." They keep hearing voices—two people in conversation. So I told them about the various psychic happenings which we'd experienced while living there. And I added, "As long as you love each other, you'll be quite all right." Yes, alas, I was drunk when I said that—and getting rapidly drunker. I thereby missed some fascinating confidences which Don was receiving. Guy Dill told Don that Laddie

*Joe Hacker—they are actually man and wife; Joan Quinn told me. [Joseph Hacker acted on T.V. and in commercials; later, he became a lecturer at USC School of Theater. He did not marry until 1986.]

had told him that he mustn't bring his wife to the party—[...]. In addition to this situation, Peter told Don that he wanted to come by and have a talk with us, and Don is sure that Peter wants to discuss the unsatisfactory state of his marriage with Clytie.

Old Drub got messily drunk, spilling wine over his pants and having to be helped to leave the premises by Darling, with everybody watching. "I played the scene very well," Darling told me, "but I know they were all thinking, 'How awful it must be for Don.'" So Drub swore to reform.

Old Jo reports that two black boys tried to mug a young man and a girl right on Channel Road, a few days ago, but the young man shouted and they ran away. The same pair tried to hold up The Golden Bull, but one of the employees went for them with a knife, and again they ran away.

November 4. Thick coastal fog tonight, suitable for election day—although it seems clear already that Reagan will win. Last night at supper Gavin Lambert declared that this would be better than Carter—partly because it will force us gays to make a stand and show the forces of reaction that we won't be bullied—partly because Reagan is a realist, he will always modify his political attitudes and retreat from them altogether if he sees it's necessary. I wish I could accept Gavin's optimistic view that the gays are now firmly established and can't be pushed around. A pessimistic voice whispers in my ear, "It's all very easy for him to talk like that—he doesn't live in this country."

Also at dinner were Joan Didion and John Dunne, about to leave on an Asian tour paid for by lecturing. I really quite like them both, though Joan's look of misery gets me down. Don Carr was again a social success; he had brought one of Joan's books for her to sign, and he was charming to Marsh Ahunt[1] and Marilyn Goldin. And Darling was so beautiful, and in such control of the situation. Don's eyes kept fastening on him admiringly, as well they might.

As for Drub, he was in the dry doldrums. Resolved not to drink, he nevertheless had a bellyache and was so bored he could have neighed with anguish. He simply isn't meant for parties, only intimate encounters. And, perhaps because of his general condition,

[1] Black actress; she was in *Hair* in London, had a daughter by Mick Jagger, and appeared in Lindsay Anderson's *Britannia Hospital* (1982). Her real name was Marsha Hunt, but she changed the spelling to join the Screen Actors Guild because a white actress with the same name (b. 1917) already belonged. She and her daughter both sat for Bachardy.

the meat and vegetables seemed quite tasteless. It was scary—a sort of blindness of the palate.

Today is the first day in about a week that I haven't dictated a rough draft of scenes for our "Paul" screenplay to the tape recorder. I mustn't falter.

When you are getting senile like Dobbin, any assertion of the will seems valuable and cheering. For example, I felt sure that I would never be able to follow the instructions for using a waterpik, but now I use it, though I still spray water all over the bathroom.

Then, yesterday, I realized that I'd lost the most recent bank statement, together with my cancelled checks. At the bank they agreed to send me a copy of the statement but added that the checks, couldn't, of course, be duplicated. I had a nagging feeling that I might have thrown out the statement in its envelope unopened. Today is trash-collection day, so I had to hunt all through the garbage yesterday afternoon—and I found the statement. This was a real triumph of the will. It raised my morale absurdly high.

November 10. Morale low again—Reagan's victory, sad sea-fog, something the matter with my right (better-seeing) eye, and the horrible murder of poor pretty foolish Mark Bernolak, who seems to have gotten himself in wrong with the dope-dealing underworld and was tortured before being killed; neighbors heard his screams. (Jim Bridges told me about this a couple of days ago.)[1] And yesterday was a downer—two huge parties, one given by the Michael Yorks, the other by Barney Wan for Michael Childers, up at the Schlesingers' house. During the first party, Don discovered that Pat had put his two drawings away off in an uttermost corner, where the damp is getting to them. I am utterly utterly bored by myself as a party celebrity. I don't want to hear what I am about to say—and neither do most of the people I'm about to say it to. Also, I spilled some bright red syrup on my suit—only a silly pretentious cunt like Pat would have served such an unmanageable dessert at a buffet lunch. I don't know which I hate more—getting drunk in order to endure such a party or having to endure it anyhow, sober. Yesterday I was sober. The only good thing that happened—David gave me one of his blue-faced comic watches—it has a comic-strip character* on it—also some little

*Fred Flintstone.

[1] Bernolak, a friend of Bridges and Larson, worked as a carpenter building film sets; he was murdered for dealing drugs in someone else's territory. Bridges wrote *Mike's Murder* (1984) about him.

smears of Hockney paint on its wrist strap, which make it into a relic. He is one of the people I get fonder and fonder of. There aren't many.

Darling is outside all of this. I just want to be with him always, every minute. (Though I don't fret when I'm alone). Yesterday he got us out of the second party in record time—about five minutes. We slipped out by the back way, leaving our glasses on the terrace, but having already been photographed, so we could *prove* we had been there—like an alibi in a murder mystery.

November 28. Yesterday morning, Ted Bachardy called and said, in his characteristic tone of gloomy satisfaction: "They're dropping like flies"—this referred to the recent deaths of Mae West and George Raft[1]—"Now it's Rachel Roberts." Very soon afterwards, Gavin called, bubbling with details: how Rachel had been to see Rex Harrison and had then announced to Darren Ramirez that she would like to move back into their house and kill herself there. She followed this by telling Darren that she was leaving for New York, where she'd been offered a play. Then she packed a bag (to make this seem convincing, I suppose) and went into hiding up the hill behind the house. There she took a big dose of drugs. Having done this, she got scared, it seems, and hurried back to the house hoping to get help of some sort. But she collapsed, falling through a plate-glass window, cutting herself, and died sometime later. The coroner was able to avoid saying that this was suicide; officially the cause of death has been called a cardiac arrest.

I've probably got some of these details wrong, but the sum of them is that this poor miserable woman seems, consciously or unconsciously, to have attempted to deal Rex Harrison's morale a death-blow on the eve of his opening in *My Fair Lady* tonight.[2] Good old Rex, the nicest man's man you could hope to meet, has probably always been a selfish bastard with women; but Darren, who brought Rachel so much Latin charm and devotion in her early middle age, didn't deserve this final whack.

I never saw this side of Rachel very much—my clearest memory is of a long ago dinner party at which I was forced to sit next to her

[1] American Broadway star (1895–1980) who played Hollywood gangsters during the 1930s and 1940s. Mae West made her film debut opposite him in *Night After Night* (1932); they were reportedly lovers, and they both appeared in her last film, *Sextette* (1978). He died on November 24, and she died on November 26.

[2] A Los Angeles revival.

fearing for my best suit because she seemed about to throw a plate of soup at Rex. Most of my meetings with her were lively—she could be very funny.

A beautiful milky blue calm ocean this morning, with fishing boats dotted about on it in what you can imagine as being significant relations to each other, like chessmen.

Old-age pains, a girdle around my loins on waking. But oh, the wonder of the depths of Kitty's fur!

December 6. Two days ago, Don and I had one of our sudden emotional openings-out to each other, prompted by a couple of light scotches. Don was saying how happy he feels in his work—which also means, in our life together. He spoke with intense, slightly incoherent excitement about the ways he glimpses of developing his drawing and painting: "forcing the same old tools to be versatile" was how he put it. Oh the *pride* of feeling that Kitty is airborne! He doesn't need Dobbin any longer. He can fly. (That reminds me how, back in the early 1960s, he used to say to me, "I wish I loved you more and needed you less[.]")

On December 3, we went with Tom Siporin, my ex-pupil from UCLA,[1] and Marilyn Goldin, to see Muktananda,[2] at the ashram which his followers have set up for him in Santa Monica. As far as I was concerned, I suppose the visit was foredoomed to be a flop. I was in a bad mood, I now realize, arising out of mixed ego-attitudes. I didn't like the crowd, although it was very well behaved. I didn't want to bow down to Muktananda or stand in line to do so, and yet I would have been equally unwilling to meet him in a private one-to-one interview. So I behaved rather rudely to poor Tom, who is bubbling over with a new convert's enthusiasm, and insisted on leaving before Muktananda spoke. (Admittedly, this could have been no big deal, since he only does so through an interpreter.) Was I *afraid* of Muktananda? Perhaps a little. They all stress his extraordinary power, and speak of shocks and visions transmitted by his hand-touch. I don't want any of that from anyone outside the Ramakrishna Order. It would be a sort of disloyalty to Swami, who was always inclined to be jealous of his disciples' contacts with other gurus, even Ramakrishna monks.

I suppose, to be quite candid, I also have a prejudice against

[1] In 1965; afterwards, Siporin (1942–2001) became a lawyer for the American Indian movement. He appears in *D.2*.

[2] Hindu monk (1908–1982), a disciple of Bhagawan Nityananda; he brought Siddha Yoga to the West in the 1970s, giving *shaktipat* initiation to thousands and setting up hundreds of meditation centers and many ashrams.

Muktananda merely because his ministry is so hugely successful. I equate any kind of large-scale success with falsity of some sort—at any rate, in the West.

Yesterday, Joan Quinn, I think it was, told me that a new version of my adventure with the Mexican boys on October 18 is going around town. Instead of my saying, "O mi corazon!" I am being quoted as having said, "O mi parasol!" which is supposed to mean that I was pretending to be crazy.... Don felt that this was giving the story an anti-gay twist—the silly old fag resorts to camping, even when in danger.

Yesterday night, at a party given by Kiki Kiser for the pianists Arthur Gold and Robert Fizdale, Kiki confessed that she is terrified of her house (157 Kingman Avenue).[1] So I told her that we have always felt it had a bad atmosphere and might well be haunted. Luckily, she has already managed to sell it to a family who simply love it!

I must record that I have been having a slight lapse into face making—of both kinds—these past few days. I must be on my guard against this.

December 12. Tony Sarver died on the night of December 7. His funeral was yesterday, at the chapel in the veterans' cemetery. Tony had been a very devout Catholic while he was in the navy, and a Catholic priest had visited him at the hospital—the terminal ward section known as the hospice—although, by then, Tony had lost much of his faith and didn't care to receive the last sacrament.

This priest spoke at the funeral. He had funny wavy hair which looked as if it had been curled with curling irons—but which Don, the expert, said was natural—and rather campy overemphatic gestures. We were both curiously impressed by him, because he talked about death so naturally, as a consummation of life rather than an embarrassing medical accident, a doctor defeat.

The get-together afterwards at Bill Scobie's house made Bill's predicament painfully evident, because so many of the mourners were black and one got the impression that they were Tony's friends rather than his. What will he do with himself now? Will he stay here or go far far away? Is there anyone around who might be prepared to make him want to stay?

Last time Tom Shadduck was here, he asked me what does the word karma really mean. (I find that he, too, has had a Catholic upbringing; and lately he has been seeing a psychiatrist who is

[1] The number was 757 Kingman Avenue; see Glossary under Kiser.

Catholic and, I believe, a priest.) So, to illustrate my answer, I told him—maybe not quite accurately—the story of the two men who went into the forest; one of them picks up a coin of the lowest value; the other treads on a thorn and hurts his foot. Later, the first of these men is told (by a holy man?) that his karma is very bad—otherwise, he would have discovered a gold mine; and the second man is told that *his* karma is very good—otherwise he would have been bitten by a cobra.

A few minutes after I'd told Tom this story, he picked up a penny, while watering the plants down at the back of the house. He at once came up to the carport, where I was, exclaiming: "I've got bad karma!" He said this jokingly yet with a certain note of superstitious dismay. As he spoke, I looked down and saw, lying at my feet, another penny. What made this all the odder was that *this* penny, judging from its appearance, had been lying there for a long time; it was rusted. I must have walked very close to it over and over again.

December 25. A model Californian Christmas Day—quite warm, the sunshine lighting up the external scene. But not the internal. Despite the uplift of a Dexedrine tablet, I'm down. And so is Darling—in one of his states of acute frustration: "What's in all this for *me*?" To which the answer is always "nothing" or "something," according to mood. What we're both suffering from, right now, is pressure. This "Paul" screenplay, which we don't really want to write and yet feel we should. And the abiding nuisance-presence of poor old gaga Glade—and of me too, as far as Darling's concerned. I am, I suppose, "wonderful for my age," but nevertheless still an old crabby killjoy drag, and often a bore.

Well, as usual, let's just repeat it, the only possible watchword is *courage*. Let's get on with the show.

Perhaps one cause of my downed spirits is that, despite pretty regular sits, night and morning, I feel a dryness, even a lack of faith. Am I getting anywhere with my religious life? Where's the joy in it? To this, the answer is: Don't ask, simply keep at it. And, when I say that to myself, I find that my faith *consists* in keeping at it. I have no other source of strength. So I echo Jung's, "I do not need to *believe*—I *know*."[1]

[1] In a 1959 interview on BBC T.V., John Freeman, host of "Face to Face" asked Jung, "Do you now believe in God," and Jung replied: "Now? ... Difficult to answer. I know. I needn't—I don't need to believe. I know."

1981

January 1. I woke up heavily depressed, this morning—and yet I don't quite know why—unlike so many New Year's Eves, this one had been pleasant. We were with Rick, whom I'm fonder of than I am of almost any of our friends. It was Rick's birthday, so we also invited one of Rick's close buddies, Jeff Capp. Jeff's maybe a bit dull but adequately intelligent and (for my taste) more than adequately sexy—blond, feminine, but muscular. I'm turned on by his arms, which he usually exposes by wearing only T-shirts without a jacket. The film we saw was adequately interesting though Frog-cute, *Mon Oncle d'Amérique*, and then we ate at El Adobe which I always like, and then we went to a quite small party of young men, many of them attractive and all friendly.

So why my complaints?

January 3. I *do* know why I was depressed; it was because these hernia pains are lowering. Also I guess I am worried about the cyst in my mouth. It hasn't grown larger but it hasn't grown smaller either; and, now that we're past the holidays, Elsie will want a specialist to look at it.

Anyhow, this is the time for a pulling of myself together. This morning is beautiful and my mood is up—admittedly this may be partly because I just took a Dexedrine tablet which seems to be doing its job.

A big flap right after breakfast because I got confused and gave David Hockney the wrong message—that it was all right for him to bring his two friends from Bradford to dinner tomorrow night, when Virgil Thomson (at his own suggestion) is to cook at our house. Darling, who has been in pain from one of his muscular spasms in the shoulder, exploded in fury against his/our life, with all the work load that it puts on him. This time, he did something he has never done before—picked up a bunch of papers in my room and flung them around over the floor. His recurring complaint on such occasions is that I've upstaged him, grabbed the role of dear old saintly celebrity who always emerges from every crisis "smelling like a rose," and leaving him to play the nasty little vixen. This is true. It's also true that I, deep down, feel put upon, groan and picture myself as a martyr to our entertaining—a heavily burdened social secretary. The truth is that we are both highly talented charming amusing attractive (yes, Dobbin too) individuals and selfish as such creatures nearly always are. Why should we stay together? Only if we love each other—which I sincerely believe we do. And that doesn't alter the fact that my death—I sincerely believe—would set Darling free. He wouldn't collapse. He would spread his wings and fly higher.

Well, courage, Dobbin. Pray to Swami for love and devotion and acceptance of dying. And get on with your work and don't sulk, and watch that face making. As I noted last month, it has been recurring.

January 4. The weather of the household is sunshiny again—thanks to those two sweet boys, Bob Drennon and Billy Faught. We met Bob about the middle of December through Mark Valen; he's a curly-headed, funny-faced, well-built, bespectacled *joli laid*, with a cast in one eye—intelligent and very sexy. Billy we met at the New Year's Eve party—in fact, I think both Bob and Billy were among the hosts of it. After a while, Billy got drunk and slept, which enabled us to admire his beauty—he is Texan, American-Indian Irish, with goldish-red hair. His eyes are blue. Darling wasn't charmed when Bob showed up to be drawn bringing Billy with him, unannounced. But the two of them had soon endeared themselves to both of us. They are very married. At least, Billy is very married to Bob, though we gather that the marriage isn't sexually exclusive. Billy told me, "Sex isn't so important—what matters is to love and know you're loved." Looking through our *October* book with Bob and seeing the double portrait of David Dambacher and Gene [Martin], he said, "That's us, twenty years from now." So far, they have only been together a couple of weeks; and I wouldn't be too sure of Bob's devotion. But they make a quite adorable pair. They returned later to have dinner with us at Casa Mia and then drink some more at The Friendship.

January 9. Yesterday, I went to see Dr. Alfred Katz, whom Elsie Giorgi recommended, and showed the cyst in my mouth. He said that it is almost certainly nonmalignant but should be taken out. He guaranteed that the entire operation—driving to the Cedars-Sinai hospital and having the surgery—would take less time than a trip to the airport and back! I was impressed by him, although he works very hard at keeping up a line of facetious chatter.

The year is opening grimly with confident forecasts of economic disaster—inflation, a fall in real estate values, unemployment, increase in crime. Not to mention the prospect of nuclear-war scares plus nuclear accidents.

Against all this I have whatever I can muster of my religious faith, plus Don's love. Yesterday, for example, his love seemed extraordinary, as good as new, no, far better. It's almost incredible to realize its strength. And to realize that there must be other people who enjoy such love. (They'd *better*!)

Am plugging along with an outline of the rest of the "Paul" screenplay, dictating it to the tape recorder. This is a *very* stiff push, but I must have something to show our producers.

January 13. A perfect day. Ran down to the beach, where I saw—for the first time, almost, that I can remember—someone reading a book of mine (*Guru*)!

Have now scrambled somehow through the "Paul" outline (on tape) as far as where he leaves for the forestry camp. The ending itself is still uncertain. But I do feel there's *something*—

January 24. I still have a sore mouth after the removal of the cyst on the 21st. Dr. Katz correctly said that the surgery itself would be brief and painless—it lasted about eight minutes and I felt nothing while it was going on. But the whole experience was unpleasant because of the scuffle and fuss of getting a parking space, filling out medical forms, having to undress completely and put on a robe, slippers and cap, lie on a trolley keeping up a line of bright chatter with swarms of cute Asian nurses and being strenuously "nice"—as was expected of me by those (quite a few) who knew "who I am"—and it took two and a half long hours!

Kitty has been adorable about it all. And yesterday, after so many career disappointments, he had a reward: Bob Miller phoned to say that a Swedish art dealer, who had seen Don's New York show and was greatly enthusiastic, returned and told Bob he wants to buy *twenty-five* Bachardy paintings! Admittedly, he'll get them at a cut price. But this is a first European beachhead. And the dealer will be obliged to make propaganda in the art world to secure his investment.

Not only this, but Kitty has also been invited to compete for the art prizes offered annually by the Academy-Institute. He'll be able to exhibit some of the pictures he has at the Robert Miller Gallery. I don't know who it was who suggested his inclusion. Maybe faithful old Glenway.

The other day, when we were feeling very loving, I asked: "What'll Kitty do when Dobbin has to depart?" Kitty answered, without hesitation, "Give him a great send-off."

February 12. I *must* try to write more often in this book. Because this is an extremely interesting phase in my life—one might call it the pre-terminal phase, meaning that the event of death is *more or less* recognized and accepted although health remains still pretty good, with regular bowel movements, full nights of untroubled sleep, adequate amounts of energy available.

The terminal phase would be, of course, the phase in which one's principal occupation is dying—that is to say, a phase in which death becomes a constant threat and, to some degree, a desired release.... What was I beginning to say in this paragraph? I forget—because, in the midst of writing it, I was interrupted by a call from tiresome but admirable Jim White to tell me that Harry Brown is in hospital having tests for his emphysema. This led to Jim's enthusing over *Exhumations*, which he is teaching from, at USC, and to a declaration that he was going to try to get the book republished in paperback. Bless his heart. His buzzing often drives Don and me to distraction but he is a tireless ally.

I started writing in this book this morning as part of a drive to get myself out of a bog of inaction-inertia and start work again—on the "Paul" screenplay outline and also on my next autobiography. This is first and foremost a question of "art for my sake."

Later— Just back from seeing Dr. Katz. He examined my mouth. All is well. But he made the extraordinarily discouraging admission that this could happen again at any time. We are continually biting the insides of our mouths and are therefore continually in danger of having a bite cause a lump. Once caused, such a lump should be examined by a doctor; and, if the doctor feels that there is *any* possibility of the presence of cancer, he should intervene surgically. In theory, it is therefore possible that a patient might undergo such surgery several times every year.... On the plus side of this depressing calculation, I must put the fact that Dr. Katz's fee was relatively small, fifty dollars. Of course, I'm taking it for granted that that's with medicare allowed for and deducted.*

February 24 [Tuesday]. Yesterday we returned from a weekend in San Francisco, staying with Bill Brown, so I could appear at a fundraiser for Gay Rights Advocates on Sunday evening, the 22nd. The crowd would have looked adequate if it hadn't been in a depressing old barn, the Nourse Auditorium—there was nobody at all in the gallery. Still they were very sweet, quick to pick up on laughs and appreciate points I made. (My chief "message" to them was that we must attack the opposition on religious rather than political grounds—tell our enemies that their "religion" is rotten, blasphemous, anti-Christian and philosophically fucked up.)

*Famous first words! It cost *plenty*!

Armistead Maupin introduced me, going much too heavy on the praise, exalting me, in fact, to the position of America's Old Mr. Queer. But he is basically a friend I feel—not one of those flattering foes. As for Darling, he looked absolutely dazzling and brilliantly held up his side of our partnership during the get-together at the Hayes Street Grill which followed.

I read from our *October* book, and, I must say, I do now feel that it has real merits as read-aloud material—having tried it out several times. It even scored one powerful delayed take (or whatever such an effect should be called). I was reading the passage (October 25) on my physical likes and dislikes and came to the paragraph beginning: "Beautiful boys with long hair are beautiful boys with long hair—it doesn't bother me. And sometimes it gives them the magic, ferociously disheveled aspect of young barbarians, riding out of the primeval forest to rape us," (at this point there was laughter and applause, which lasted until I said to myself, that's killed the end of the sentence for sure—nevertheless, when it was over, I had to continue), "we hope." I was quite wrong—they hadn't forgotten what I'd been saying, and now the applause was a major explosion.

All in all, this San Francisco visit was among our pleasantest. Paul wasn't there, and it was easier being with Bill alone. Don, who has never really liked Paul, complains that he's a downer. Bill is much livelier, more socially flexible. He also arranged for us to meet Robert Johnson, who is the director/curator of the department of drawings and prints at the Achenbach Foundation and a rare portrait-art enthusiast, greatly impressed by Don's work. So something might come out of that.... The two beautiful cats at Bill and Paul's house were more beautiful than ever, posing with insolent aristocratic grace on the beds, or imperiously entering our room as a shortcut to jumping out the window to attend to some business in the garden—they are constantly making the rounds, and, if anyone objects to their intrusions, they couldn't care less.

March 12. This is such a strange period in my life. In some respects one of the very happiest, because of Darling's so frequently demonstrated love—the basket is a paradise of loving snugness, and, throughout the day, we exchange dozens of looks and words and kisses and hugs. And yet, at the same time, there is a sense of constant menace—terminal old age with all its aches, and warning symptoms. I am continually being made aware of my bad knee and of something wrong with my hip, and of the cataract dimming my left eye. (Yet even this isn't constant—I am writing this sentence with only my left eye open.) Somebody* said that one is only completely sane (that may not be the right word) when one *knows* one is going to die. I am pretty well convinced of this but of course I hope for more years—maybe even a dozen—*plus* adequate health; which is asking a lot. It is also asking a lot of Darling, who has to face up to this, at best, grim experience of watching me fade away while his own most valuable years are squandered. Oh, what is the use of writing all this down?

My only salvation lies in Swami and the mantra he gave me, and in purposeful activity—getting this screenplay outline finished for the "Paul" film, and then starting my next autobiographical book (which more and more appears to me as a collection of interrelated but quite separate pieces about different aspects of my life in the U.S., from 1939 onward—including life with the L.A. German-Jewish refugee colony, life as a screenwriter, life at the Haverford Friends' refugee hostel, life with the Huxleys and their circle, memories of journeys taken, glimpses of California as a *haunted wilderness*, memories of Gerald Heard and Chris Wood).

We have just had visits from Stephen Spender (here from February 26 to March 2) and Speed Lamkin (from March 2 to March 7). I enjoyed seeing both of them, but of course it was a severe energy demand. Stephen's demand is related to his bigness—he is *there* like a huge crate which has just been delivered; its very presence makes you feel something must be done about it. Speed's only (*only*!) demand was that I should read the novel he has just finished—it took me two whole days, and made me see double. Oh yes, it's really quite good—he knows everything about everybody, and his talent is for endlessly sustained gossip. Only, I found the beginning and end of it dismayingly sentimental.

It must be added that both Speed and Stephen behaved well. Stephen was delighted by everything we did to entertain him, and we have just received thank-you letters from him. (Stephen writes to me: "Our walk on the beach was a real recovery of past time. I felt we were near the pier at Sellin.[1] I continue to think that your relation with Don is one of the Seven Wonders of the World.")

Speed seems to have grown taller. There is something rigid in his face, carved into a grin. He is maniacally self-obsessed and quite

*It was Jim Charlton's friend, Ben Weininger, the psychologist.

[1] On Ruegen Island, north of Berlin in the Baltic Sea, where they holidayed in 1931 and 1932; the third section of *Goodbye to Berlin*, "On Ruegen Island," is based on Isherwood's first summer there.

a bore, but he is generous with his money and, in his own crude way, very polite. I still feel fond of him. This morning, I had to call and tell him what I think about the sentimentality of his novel's ending. He took this very well—that's to say, very practically, accepting the criticism and at once planning how he could rewrite the chapter.[1]

April 1. I find that I need to make another pep-talk entry—that is, to pull myself together by reviewing the situation. The situation is, on the minus side, more of same—hernia pains, bad right leg, resulting sloth and failure to get the "Paul" film treatment finished; on the plus side, Don, who sustains everything by making his love continually evident, and by setting me the example of his own untiring art work—not to mention all he does to keep the household going. He seems to me to be at a peak of his art. His show at Pierce College[2]—only twenty pieces, drawings or paintings, in a small room, opened on the 23rd. I think it is one of his best. As I looked at it, my eyes kept spilling tears of joy and pride. The boldness of his paint strokes, their decision and their energy (e.g. the head of Jim Teel), the richness and magic strangeness of color in the skins of the faces (Jerry Stille,[3] Kiki Kiser); the seeming delicacy and immense strength of the drawings: As I told Jim White, who came with us, the full-length drawing of him has such vital togetherness that it might as well be of a rattlesnake, his clothes belong to him and express his vitality like a skin.

One can say with heartfelt gratitude and no shit that Reagan's escape from death two days ago was a mercy for all of us. If he'd been killed, Haig and the other extremists would have used it as an excuse for a campaign of fascistic "vengeance."[4]

The triumph of *Ordinary People*, at the Academy Awards yesterday, depresses me—not only because the film was bad but because I feel that its title—far from expressing the irony it no doubt intended—meant, to those who voted for it, just *that*. "Ordinary" equals wholesome, and it was for *wholesome* that my fellow academy members—a preponderantly reactionary bone-stupid bunch—voted.

[1] The novel has not been published.
[2] Woodland Hills, California.
[3] A frequent subject for Bachardy during this period, with whom Bachardy freely experimented because Stille never objected to any of the resulting portraits.
[4] John Hinckley Jr. shot President Reagan on March 30; see Glossary under Reagan.

April 12. The day before yesterday, I was hit by the stunning news that my federal income tax this year will be $26,548—last year it was $8,491! This it seems is because my three time deposits, totalling $225,000, have been earning such a lot of interest. If I withdraw any of these deposits prematurely I shall have to pay a penalty. If I scrape together the money which *is* available, it will hardly be enough, and then we shall have nearly nothing to live on, because Don has *his* money even more extensively tied up in time deposits. The only glimmer of hope is that the rest of Richard's legacy—which *could* be as much as $30,000—might arrive; but that is very very unlikely.*

April 17. Crucifixion Day. Very depressed—not on account of Christ, who, God knows, can take care of himself. Am full of aches, from my hernia and resultant pains up and down my right leg.

And yet there is much to rejoice about.

Don is being adorable and loving—as indeed he usually is, but nowadays it really counts.

Yesterday I at last finished a very rough outline of the "Paul" film (begun February 13). It is forty pages long, so fairly detailed and substantial, except at the very end.

On April 15, I spoke at Pomona College, twice, which was tough, because a Japanese dinner with saki made a reflex-slowing interruption between talk one and talk two. Just the same, I managed to keep the flag aloft.

Amazing how your warmest supporters are the ones who do the most—consciously or unconsciously—to make you look ridiculous; in this case by billing me as, "Among the century's most insightful observers of the human condition," "One of the most controversial figures in contemporary literature," and (thanks to Gore) "The best prose writer in English."[1] (Following this bombardment of bouquets came a last and least, so anticlimactic that it sounded like a put-down: "He also is widely regarded as an exciting and witty speaker"!) Never mind, Dobbin came home with a thousand-dollar check, which helped delay the Animals' ruin. We have now paid our federal and state taxes but have postponed paying the first installments of the estimated.[2]

*When it *did* arrive, in July, it was only $5,686.75!

[1] In his review of *Christopher and His Kind* in *The New York Review of Books*, December 4, 1976, vol. XXIII, no. 20, p. 10.

[2] I.e., advance part payments of estimated income tax for the coming year.

A nice lunch with Alan Stern and Michelle Rappaport yesterday. We feel, as of now, that we couldn't have more sympathetic and intelligent producers. But of course we haven't got down to work with them yet.

Work, work—that's the cure, even if temporary, for so many ills.

Now I have no excuse not to find time for at least some kind of stab at my next autobiographical book, more about this *soon*.

April 19. An Easter of rainy skies, hernia belly pains and slight sickness, mixed melancholy and happiness.

Darling is being adorable but he is under great pressure because old Drub is becoming increasingly senile, slipshod, self-indulgently invalid, unwilling to leave the house or see people. Our relationship is largely a nursery world of sleeping together, cuddled like children.

I feel I might die quite suddenly—the vital supports are beginning to give way. As far as Don is concerned, this would be for the best. I am *not* writing this pathetically, only practically.

A few nights ago, I dreamt about Swami. I haven't done this in a long while. It was definitely an "appearance"—that's to say, he didn't merely play a role in a dream about other people and happenings. His presence was the whole dream. What I dreamed was just being with him—in his room at Vedanta Place, I suppose—and being moved by the faith and joy with which he spoke about God. You could call it a vision or just a vivid memory, since I experienced the same thing many times in his presence when he was still alive.

Loss of hair. Loss of taste and consequent loss of weight—down to 147 and ½—no big deal, this.

Cuisine note: the oddly disgusting taste of coffee drunk from an incompletely cleaned mug in which powdered Aka Miso (red soybean paste soup) has been prepared.

Thoughts about the next autobiographical book. Yes, it seems *almost* certain that it should be made up of separate chapters—I mean chapters which might be published separately in magazines—an economic consideration. You sell the material twice over.

But, at the same time, I rather wish the book could have the riverlike flow of George Moore's memories, or Yeats's, with their artful-simple air of going nowhere.

Of course, a lot depends on how much of the available time span—1939 to the present day—is to be used; does it or doesn't it demand two volumes?

April 24. Yesterday I suddenly found myself being sent by Elsie Giorgi to a surgeon called Fred Marx, who told me that I ought to have a hernia operation right away, because otherwise the gut might burst and there'd be a mess. Don is dead against this, and I don't want it of course. Actually the hernia itself is no big deal but I get real electric shock pains in the upper legs. Last night, when I got into bed, the pain was miserable, until I'd blocked it with a couple of Anacin tablets.

Is Dobbin going to collapse? Having these pains makes me remember what cruel pain I often had when I was young. Probably the rheumatic fever pains were the worst, when I was ten.

April 25. A bad dip. I feel pretty sure that I'll end up having to have the hernia operation, in spite of everything. Don is terribly rattled, as he nearly always is when I'm sick. He wants us to drop the "Paul" project. I feel that, if I do, it'll be a serious psychological defeat. And what's the alternative? Beginning the new autobiographical book would be just as much of a sweat.

Also, we need money, and "Paul" is the only dependable way of earning any, right now.

April 26. Last night, I dreamt that a gang was trying to steal the money out of my time deposits. I was at a party at an embassy—whose, where, I don't know—and I kept seeing people who looked exactly like *me*. Whenever I saw one of them I experienced a thrill of terror, I don't know why. But, each time, the likeness faded after a few moments. Don came in and I pointed some of these people out to him.

Yesterday I started wearing a truss—of a kind which Peter Gowland recommended when he came here to supper with Alice and Old Jo on the 23rd. Well, it's another "first," like being gagged, on October 18. Maybe I have several more such experiences ahead of me!

April 28. Am still waiting around for a decision on what's to be done with old Dub's carcass. I must say, I would like to avoid an operation—not so much for itself as for the misery of being in hospital several days, woken in the midst of sleep and stabbed with needles. For the moment, I'm letting things slide—trying to take adequate walks every day, as the Gowlands and Old Jo recommend. But walks are such a bore. I realize, now I'm old, that the greatest part of my physical exercise used to be fucking, which was largely wrestling and which kept me in very good shape.

I know instinctively that one way back to health would be to get another project going. I really do want to turn out a draft of the "Paul" film. At present, the difficulties seem to be all at the beginning—establishing who Chris is, why he is in Los Angeles, etc.

May 2. Two days ago, I did indeed make a beginning on a new draft of the "Paul" film, which is not only psychologically reassuring but seems to make some sense—it follows suggestions just made by Don; and I see, once again, how valuable his suggestions nearly always are. What he says, in effect, is that we must instantly begin getting to know the Chris character—what his predicament is, as a newcomer from England to California. Thus we find what kind of a companion he needs, and thus we are prepared for the entrance of Paul—even though Chris won't be ready to recognize Paul as his companion until quite a good while later.

Don is absolutely right about this. The only problem is a technical one—that all these explanations have to be made by means of soliloquy, and a little soliloquy goes an awful long way.

Meanwhile, Elsie Giorgi has decided not to have the hernia operation right now, maybe to postpone it indefinitely.

What with this damp weather and Dobbin's generally infirm condition, last night was one of the most wearisomely uncomfortable I have spent in a long while. It was like a nightmare journey of painful slowness—I couldn't sleep, and every move triggered some muscle twinge. But Darling was there.

I keep trying and trying to get some contact with what is within the mantram—my only resource and safety. Consciously, I very seldom do get it, but I guess the effort is a form of contact anyway. And then, occasionally, these sudden tears of joy.

May 7. Today we came to a decision—not to go on working on the "Paul" treatment for Alan Stern, because Don wants to get ready a batch of new paintings to show at the West Beach Café. Don has been looking at the paintings based on photographs of movie stars which he did around 1970–1975. At the time, I thought them extraordinary. Now they seem even more so. Darling feels inspired to begin working again along those lines—only this time he may base the paintings on existing portraits.

Meanwhile, I'm seriously considering starting my next autobiographical book right away—*California* would do for a working title. I think I'd begin with movie studio experiences—my first few jobs, anyhow.

May 10. Dreary gut-pains from the hernia. Disinclination to do anything, and yet a desperate nagging of conscience toward *work*—any work.

May 9—the family death-day—brought us some good news, however: Robert Miller phoned from New York that the Swedish dealer has reappeared and that Don is to mail off a selection of thirty paintings at once.[1]

May 11. Depressed. Pains in back and hernia discomfort as bad as ever. I begin to feel that they won't stop. Am also jittery because of threats from the Writers Guild Strike Committee—we're to be fined if we don't cooperate, etc. etc.[2] (What makes me jittery isn't the threat but the prospect of having to get into a towering rage about it.) And then I'm still in a dither about starting my *California* book. Where to start? How? I suppose I'll snap out of this somehow.

May 16. A new scare. My weight has been dropping steadily and now it's lower than it has been in years, to 146 and ¼. As usual, when I get scared, I'm embarrassed to tell Elsie.

On the other side of the ledger, a real important success—on May 12, I made a start on my new autobiographical book *California*; it's opening is nothing very inspired but quite okay, I think. And I can feel how powerfully it's a pro-life force, countering my body aches. Kitty is the *real* life force and he knows instinctively how much the effort to write this book will help me.

May 21. Another very painful night. I felt so discouraged by the pain that I began to feel I'd have the hernia operation, to get some relief. Must talk to Elsie.

However, I am at least crawling on with my *California* book—have got to page 8.

May 23. Two nights ago, Don, Jim White and I drove over to the art museum at Newport Beach,[3] where there was a group show, with Don the featured artist—the reproduction on the poster was of his painting of me against the cushions. Also he appeared on a video tape, painting me in the studio. Typically, he objected

[1] The dealer bought all thirty.

[2] The Writers Guild struck April 11–July 16 against major movie and T.V. producers for a greater share of pay T.V., videos, and disks.

[3] Now the Orange County Museum of Art.

to one shot and refused to be impressed by the next—he really looked magically beautiful, both in the box and in person, that evening—and his work was worthy of him. It looked magnificent.

Last night, he had to go back there and mingle at a meaningless party—all the artists were supposed to show up but most didn't, because they aren't professionals and have no style.

Poor Angel, he gets back to a dreadful old tiresome pain-ridden Dub. My back is better but I had cramps in the early hours, leg cramps that made me cry out. What must Kitty be thinking to himself? "How much longer do I have to endure this? And what will be left for me by the time that I don't have to?" Oh, he is such an angel of love.

The light side: I have done my page a day on the *California* book so far. Also, I am at last getting on with Chetanananda's huge typescript of edited Vivekananda lectures, to which I have to write a preface.[1] Work gives courage for more work.

I keep praying to Swami to be with us both and help us.

June 7. The day before yesterday, after we'd been out for the evening with Wayne[2] and Juan[3] and Jack Woody and Tom Long and a friend of Tom's from Pittsburgh whom we didn't like, Darling, who was a bit drunk, flew into a rage because he'd heard me talking to Jim Charlton (in Ho[n]olulu) on the phone and I sounded (Darling says) so thrilled—"Oh *Jim*!" I exclaimed. Darling mimicked my gushy-delighted tone. My tone with Jim—as Darling knows well enough by now—always tends to be gushy because I'm always a bit on the defensive with Jim, who is *such* a dog person, itching to be rejected so he can whimper and growl reproachfully. Well, anyhow, Darling worked himself up into a cat snit and threw things at my desk, with the result that he slightly damaged The Totem Horse, knocking off his sacred saddle.[4] He did not damage, but did knock over, The Totem Kitty.[5]

So yesterday there was great sweetness and purring. But I was

[1] *Vedanta: Voice of Freedom*; see Glossary under Chetanananda.

[2] Shimabukuro, a photographer. He photographed one of Bachardy's last sittings with Isherwood.

[3] Castellani, his lover at the time, from Argentina.

[4] The Japanese papier-mâché horse which Bachardy gave Isherwood for his birthday in 1962; mentioned above, February 9, 1972.

[5] About seven inches long, made from fabric like a child's stuffed toy and covered with long, white fake fur; it had pale blue glass eyes and a crisscross nose and mouth sewn with pink thread. It was already dirty and bedraggled when Bachardy found it and gave it to Isherwood.

rather pissed off—I'd hurt my hernia tidying up the mess. So I said I wasn't going to Honolulu on the 29th, because Darling had created an impossible situation with regard to Jim.

Of course the truth is that now I've *got* to go—and I couldn't dread it more. This has nothing whatever to do with Jim. I simply dread spending all that time cooped up in a house with Billy Al and Penny.[1] They're all right—I'm quite fond of them—but they're *aliens*. All the more so now I'm old and shaky and pain-shot. And how can I face their gang of heterosexual swimmers? What could be more utterly ungay? Oh, if only something would force us to cancel! But don't get me wrong—I'm not praying for an accident or illness or any other unpleasant crisis.

Oh God—and this evening I have to face dinner with Mrs. Misery and Mr. Know-All—Joan and John Dunne. And dreadful old Weidenfeld will be there.

Good news: I'm keeping up my stint of pages on the new book—to the first third of page 28.

And the doses of Butazolidin really have helped my back.

June 26. Well, ha ha, the prayer was answered—that's to say, I've got to go into hospital (St. John's) on the day after tomorrow and stay there at least six ~~years~~ days (note sinister slip!) and have my hernia operated on and also be thoroughly tested, because Elsie Giorgi is concerned about my liver and my steady loss of weight—I'm down now to 139 and ½. So now I'm being, as it were, self-punished for my tiresomeness. Darling has been angelic about this. We are so utterly each other's. It is poignant and makes me cry, sometimes. I have no way of knowing if he is seriously alarmed or not. I guess not—except that he's worried about my weight loss; he sees how scraggy I look.

Ah, hospitals—how I shudder at the breath of them! I don't believe I have ever spent one whole week in a hospital in my life—which just gives you an idea how healthy and lucky my life has been. I must try to make something out of this experience—a dress rehearsal of what must so soon come.

(Incidentally, we have already switched dates and arranged to go to Hawaii and visit Billy Al and Penny sometime around the end of July. And now that's okay. I'll even enjoy it, I think. Oh crazy bad Dobbin!)

June 27. It's tantalizing, how clearly I seem to see my way ahead

[1] Bengston had rented a house in Honolulu for an extended period.

on the *California* book just when I have to stop work because of going into hospital. Yesterday, I had a really deep insight—it occurred to me that the letter I wrote to Gerald Hamilton in 1939, attacking the war propaganda made by Erika and Klaus Mann and others, was really a device, to get myself regarded as an enemy in England and therefore make it impossible for me to "repent" and return. That was why I chose Gerald Hamilton to send my letter to—I knew he would broadcast it.[1]

Another "thread" in this weave is my continuing effort to turn myself officially into a pacifist—by associating with Aldous Huxley, Gerald Heard, Allan Hunter, and then joining the Quakers. It comes back to me now that, soon after Pearl Harbor, when I was already working at the Haverford Quaker hostel, I got some sort of offer from Washington—promoted, I'm nearly certain, by Lincoln Kirstein. I think it was an intelligence job—was in an office—not amidst cloaks and daggers, but dignified by a military rank (major?). This I refused, on pacifist grounds. When I remember Wystan's ill-advised appearance in London wearing a U.S. major's uniform, I get a glimpse, not altogether pleasing, of my own foxiness. I would never have made such a miscalculation. And Wystan seems to me, at such moments, fundamentally more innocent than I could ever be.

July 17. Maybe this is a propitious date—what with my look-alike "Wrong Way" Corrigan's landing in Ireland (July 17, 1938)[2]—and the eclipse of the moon last night—anyhow, I woke up this morning with a curiously strong feeling of assurance that I am at last about to get better—that's to say, stop feeling so miserable. My back doesn't hurt as much as usual and I'm not so limp and lazy, even though this heat hasn't let up and the air's still smoggy.

Anyhow, this nonsense has been going on quite long enough. (Reading back over this account, I see that I have been ailing since March.) Now I must really push ahead with my *California* book.

July 26. Ah, this weary summer of hot weather! My old body aches

[1] Hamilton showed the letter to Tom Driberg, then writing William Hickey's gossip column in the *Daily Express*; see the entry for August 20, 1961 in *D.2.* and Glossary.

[2] Douglas Corrigan, American aviator (1907–1995), took off from New York headed for California but landed in Ireland claiming he had misread his compass. He had been refused permission for the transatlantic flight because his plane seemed inadequate. He and Isherwood look remarkably alike in photographs.

and I feel sick to my stomach so much of the time, and nauseated by the sight and smell of food. Darling is being truly wonderful. He cooks for me, waits on me and grumbles just sufficiently to remain human, not intimidatingly angelic.

There is no way through this wretched period except by working at the *California* book and other projects. One welcome helper is Lyle Fox. I was somehow inspired to call him—he's managing the old folks' hotel in Santa Monica, The Georgian. He came at once—instantly announcing that he didn't want to be paid—and gave my back a deep massage. He married a French wife while living for two years at Cap Ferrat. He seemed quite unchanged, cheerful, silly, a truly kind and good man.

August 2. Being sick, in the way I've been for the past few weeks, means that you experience the life journey as conscious effort. Instead of spinning along the road almost without effort, as if on a bicycle, you feel like a character from *The Grapes of Wrath*, coaxing your broken-down, rattling, over-heated Ford to keep going, mile after mile—hardly even expecting to reach "California" ever.

This is old age. And I must say I'm well aware that—despite all my complaints—I know I'm travelling deluxe. There is more love in this house now than there ever has been before. And, in a strange way, I feel Don's love and Swami's love as two combined forces, not always distinct from each other. I'm not expressing this properly, but my sense of it is powerful. Thinking about this love while I was walking in the park before breakfast today, my eyes streamed tears of joy. And, oh, the beauty of the breakers on the shore! My lifelong urge has been to plunge into them, but I don't really care if I never do again, because the weakness of the body is a merciful self-adjustment.

August 27. Despite good resolves, I have been dragging on with my tiresome back, etc. However, ten days ago, after trying a nice but somehow unconvincing Japanese therapist with needles and other techniques, I came upon a Dr. Fichman, right here in Santa Monica, who got me walking without a stick after a week of therapy with ultrasound (I think it's called). I still have severe twinges if I make a wrong move, but I'm in better shape. But oh, if only this exhaustingly hot weather would let up!

I have to admit that I felt curiously scared on my birthday yesterday, to realize I was seventy-seven. Why, I don't know. The number must have some occult significance for me.

I also got into an utterly ridiculous flap over the prospect of renewing my driver's license. I nearly always flunk the written test, because I have convinced myself that I just cannot understand any kind of bureaucratic language (the secretly underlying assumption is that I'm "above" all that kind of thing). Well, I flunked this time. But the young black clerk at the desk helped me out, asking me additional questions verbally, in such a tone that he indicated the proper answers. He may have done this out of kindness of heart, or compassion for old folk, or contempt for the system as such. That doesn't matter. To me it was a victory, just because I had resisted defeatist impulses to surrender before I was defeated and simply resign myself to living the rest of my life without driving a car—no big deal for me but an extra load on Don, who is already carrying most everything else.

The only way I can repay him for all he's done is to get well *quick* and get on with my book. The first is easily said but not done. The second shouldn't be too difficult. Because what I basically have to begin with is to copy out *all* the material I need from the diary, just editing and annotating. I restarted work yesterday, continued today.

October 16. Well, the moment has come when I must recognize and discuss the situation with myself, which means, as usual, writing it down and looking at it in black and white. I have got some sort of malignancy, a tumor, and that's what's behind all this pain. They will treat it, of course, and so we shall enter the cancer-recognition phase and its gradual retreat to the terminal. I shall get used to the idea, subject to fits of blind panic. The pain may actually be lessened but there will be the constant awareness of it. Before all, there will be the need to accept what is going to happen. My goodness—at my age, should that be so difficult? No, it shouldn't be. Yes, but it will be.

Don is heroic, heartbreaking in his devotion. He keeps me off the pain pills as much as possible, to prove to me that I can do without them—which I can, so far. But this isn't a problem of how to bear pain; you can at least relieve that with drugs. What I have to face is dying.

This ought to be easiest of all. I pray and pray to Swami—to show himself to me, no matter how—as we've been promised that he will, before death.

I feel that I wish I could talk to Krishna, or someone from the center who was really close to Swami. But to arrange to do this is a huge psychological effort, and it might not be a success. I get fits of being very very scared.

Don says I should work, and he is absolutely right. I remember how wonderful Aldous Huxley was, working right to the end. I have promised Don that I'll get on with my book.

The love between me and Don has never been stronger, and it is heartbreakingly intimate. Every night he goes to sleep holding the old dying creature in his arms.

1982

January 1. A big gap here, because I gave way to the dreariest, most cowardly depression and thus stopped writing in this book. I couldn't settle down sensibly and matter-of-factly to having cancer—although Elsie Giorgi kept reassuring me that, practically speaking, I didn't have it, since it could be kept at bay indefinitely by the medication that she and Dr. Brosman have prescribed.

Then, on December 23, Dr. Brosman did a biopsy—sticking some kind of an appliance up my rectum and cutting a sample of tissue from the prostate—unpleasant but not as unpleasant as it sounds. Since then, I have heard nothing from him directly but Elsie called me on the 28th to tell me he'd called her and told her that the biopsy was negative. He has never called me, however, and Elsie thinks this is odd and insensitive of him. He *is* odd—kind of take-it-or-leave-it in his medical manner—yet, at the same time, I get the impression that he regards me with respect, even awe, as a writer. Maybe he's simply shy.

As a layman, I feel temporarily relieved but still quite unclear about my condition. Did I ever have cancer at all? Can cancer come and go so casually? Or did Brosman merely mean that he *thought* I had cancer, originally? (I had got the impression that his original diagnosis was based on an earlier biopsy performed while I was having tests at St. John's.)

Enough, Dobbin. If you can't be brave about all this and rise above it, then you must deliberately distract your mind from it by concentrating on your book—which is *still* only about one-third written—and that only in rough draft.

January 4. This morning comes a charming pair of letters from Dodie and Alec Beesley, saying how pleased they were because Don and I called them long-distance on December 27. So glad we did. Dodie writes: "Again and again throughout the book" (this is the volume of her autobiography she's at present working on) "I have had a sense of happiness shared with you, followed by a sense of loss because those days were so long ago. And your unexpected call somehow restored the happiness, made me feel you are part of our lives for ever." She and Alec are eighty-five and seventy-eight respectively. (Alec tells us this—I think it is the first time he has ever spoken of their ages.)

On New Year's Eve, Don painted Rick Sandford because it was his birthday. Rick asked me, "How long was it after you met Don that you and he had sex?" I said: "We had sex and *then* we met."

February 19. Today warm and windless—the first day I've walked down to the beach this year. I look so grotesque, with my huge belly and skinny arms, that I'm a bit embarrassed to show myself there—that's just senile vanity. Darling doesn't care how I look, as long as I look healthy—and I certainly am much better; my legs quite strong again, no excessive back pain, only a little nausea when about to eat.

Meanwhile—oh so slowly but fairly steadily—I progress with the *California* book—still only at page 94. I do believe that this is interesting material, though. Actually it's material I've always intended to use, in one way or another.

April 11. Easter Day. Heavy rain all morning. We are still waiting to hear what is to be done about the Falkland Islands, toward which the British ships are speeding, with Prince Andrew on board—let the Argentine invaders beware of harming one hair of his adorable head![1]

I still feel sickened by the smell of certain kinds of food and my back still aches a lot. Sometimes I feel the death fear bothering me again. I pray hard to Swami, asking him to make me feel his presence, "Now and in the hour of death." The response I get from this is surprisingly strong. I'm moved to tears of joy and love. I pray for Darling also, seeing the two of us kneeling together in his presence. Religion is about nothing but love—I know this more and more. And who's to say exactly what I mean by "his presence"? As long as I feel love at the hour of death, then I pray it will save me from fear—and save Darling too.

May 3. A slip of the tongue made by a radio newscaster today:

[1] The Falkland Islands, a British Crown Colony 250 miles southeast of Argentina, were occupied on April 2. Prince Andrew (b. 1960), second son of Queen Elizabeth II and Prince Philip, was serving in the Royal Navy aboard HMS *Invincible*.

"The British announce that the Argentine crooner, *General Belgrano*, has been sunk."[1]

May 26. This entry—just exactly three months before my birthday—is an attempt to establish the start of a diary-keeping period which I'll try to maintain unbroken till I become seventy-eight.

I've been as bad as I've ever been in a long while. Very little work on *Scenes from an Emigration*[2] (as I'm still calling it); I've only reached page 106. But I still find the material worth the work and I still want to go through with it. I'm certainly not being hindered by my health or external circumstances. My back is better, my legs are stronger—quite up to walking, even running; and I don't feel nearly so sick to my stomach. So now, let's see some improvement. Darling sets me such an example—drawing and painting tirelessly, and even seeming far less depressed than usual.

I do understand one thing: Instead of looking ahead into an obscure future—which means dwelling on death—I should keep quite short periods in view, just a few months, and try to fill them with effort to carry out clearly defined projects. Aside from this, my strength should be devoted to constant acts of recollection of Swami, his presence and his grace. That is the very best way, also, of showing my love for Darling.

May 31. Memorial Day—on which I should certainly give thanks that I haven't to count a whole bunch of war dead among my dear ones. Indeed, my father, after all these many years, is still at the head of a very short list.

June 2. Yesterday morning, we got the news that Roddy McDowall and Paul Anderson have split up. When I called them, they confirmed this, in tones of dignified regret, not giving me the faintest hint of a reason for the split. I guess we shall get the whole story sooner or later—probably from Vincent and Coral Price. Coral is said to dislike Paul intensely and to despise him for being such an ungifted actor. I am sorry for Paul. I like him and I think he stands to suffer more from the split. Roddy is pretty tough and can take care of himself.

Twice we've been to the Spago restaurant (formerly the Club

[1] Argentina's second largest warship and only cruiser, the *General Belgrano*, was torpedoed by a British submarine on May 2, southeast of the Falklands; 800 of the 1,042 crew survived.

[2] Previously titled *California*.

Gala), which I used to regard almost as *my* club, in the early 1940s. Neil Hartley regards it as his club now, and Dagny Corcoran also. But now it's horribly overcrowded and has the atmosphere of a restaurant at which the management is no longer in complete control—the service is slow, the waiters wild-eyed, the food under- or overcooked. Why then do Neil and Dagny like it? Just *because* of the overcrowding, *because* of the uncomfortable tension, the feeling that *something is going on*. Our dear little local Casa Mia is often equally crowded in its tiny way—yet, the moment you sit down, you feel relaxed, snugly at home, taken care of.

July 5. Well, this is the last day of the hateful threefold holiday. And at least I have the satisfaction of knowing that *Scenes from an Emigration* is restarted.

Yesterday our street was peacefully, very politely, invaded by dozens of darkly clothed Iranians, on their way out to the park to view the fireworks which are discharged from Santa Monica pier. Could they leave their car in our carport for an hour while they did this? I couldn't say no, because there's plenty of room now that our Honda is away, recovering from its collision with the blacks on Sunset Boulevard, June 29. The blacks were dignified and courteous—but their legal representative later disgraced them by claiming that a pregnant lady passenger in their car had perhaps suffered effects which would compromise her childbirth—which was about as cheap as you can get.

Well, anyhow, I told the Iranians that they could go ahead and park their car. Whereupon, their spokesman told me, "We are your guests." He said this solemnly and his tone suggested that a state of mutual obligation had been declared. (When we returned, much later—after supper at Tony Richardson's—their car was gone.) Incidentally, the Iranians told me that they had come to this country after living in Manchester, England. (This struck me, for some reason, as being a good omen.)

The people in the house directly below us, facing on Mabery—that's to say, the young (federal?) police officer plus several male and female friends, spanned a tarpaulin over the tiny bit of their open back yard. I suppose this was done to get into the holiday mood—though why, I couldn't say. They then proceeded to barbecue. Alas, the wind wasn't blowing the right way. Clouds of smoke were trapped and driven back and forth and the tarpaulin became like the chimneyless roof of one of those miserable hovels one sees in the worst slums of South America—crammed with coughing, half-suffocated merrymakers.

August 16. Shameful lapses—have failed to get on with my book (*Scenes from an Emigration*). My back is definitely better. Now it's up to me to celebrate my prebirthday week by starting work again, cheering up, damping down my death fears and taking (at least *some*) daily outdoor exercise. I have *no* excuse to moon and idle, especially considering how much Darling helps me. Yes—and *stop sulking*, you horrible old horse. You sulked so much yesterday because you had to see a film that didn't appeal to you, *Love Me Tonight*. A lot of my objection to musicals is sheer snobbery, and it's unkind to Don because he loves them.

August 17. Supper last night with Marguerite [Lamkin and her companion]—just the four of us—at La Scala. There is something snug about that place, even when it's crammed. And I felt a real affection from [her companion]. (He confessed, for no particular reason, that he suffers terribly from vertigo.)

We've received a joke present from Jim White—a broken cup which is inscribed: "I got smashed in Texas."

August 19. I found a piece of paper slipped in between the pages of one of my journals: "Why *ever* do anything you don't want to? The moments of joy are life in the present—no matter with Auden or Heinz. The rest is apprehension, guilt—which helps absolutely nobody. When Christopher went off with Heinz he should have constructed an absolutely self-contained world and lived in it. One thinks one is atoning for something when one is miserable and guilty. Actually one is just a bore."

On the other side of the paper is written: "Just as people poured their superior scorn on Christopher's homosexuality so they scorned his belief in Vedanta."

My guess is that these two statements are notes for a book I was writing—most probably for *Christopher and His Kind*. Both statements seem crude and flat and obvious—the second even more so than the first. That's partly perhaps because some kind of psychological block often seems to prevent me from setting down *exactly* what the point is of the note I want to make. (Why should this be so? Because of an ingrained secretiveness, which makes me subconsciously regard such notes as "classified material" which has to be obscurely expressed because it might possibly fall into "enemy" hands?)

And yet there are insights here—not very extraordinary but quite valid, I think.

One of the dangers that threaten me as a writer is of lapsing into self-accusation. Self-accusation is often necessary and valuable as long as it confines itself to answering the question: "What did I do wrong?" A gym-instructor who has fallen off a trapeze may make his slip educational for his pupils by explaining to them exactly what the mistake was which caused his fall. And one may analyze literary slips in the same manner.

August 20. How I love our flowers on the deck! Never before have we had so many, so beautifully blooming—cineraria "Dusty Miller," yellow and orange French marigolds, chrysanthemums, "Busy Lizzie," scarlet sage, sweet alyssum, and the coleus, which are my favorites. I feel a strong communication with them—as if they were animals rather than vegetables. I feel that they are aware of my presence.

September 1. Just back from a walk in the park, which I love. The park always makes me aware of old age—probably because there are so many old people walking there. And the sea makes me think of death, but without terror—this is death in the aspect of an element which receives—takes you to its bosom soothingly. It nearly always brings Edward Upward to mind, because he, too, lives by it. I say to myself: "We're two old men looking at the sea." (He sent a cassette of himself reading some short stories—his voice still so clear and strong, with that same melodious subtly ironical Mortmere tone—a tone which seems to be always between quote marks.)

September 12. Am starting to struggle out of a sloth block—I let myself get into the usual jitters. Nothing new in that. But I do detect the possibility of a more serious senile state. It's absolutely necessary that I refuse to recognize any excuses based on old age. I must get on with my work *for the sake of getting on.* No other reason is required. My health is improved, actually.

Both Don and I were instructively impressed by an evening at David's house, the day before yesterday. He seems to be less than himself. He continues to work but it's somehow forced—a token demonstration—not up to his standard. The place seems full of spongers. Don thinks he's taking quite a lot of cocaine.

Of course I know the unreliability of such subjective impressions. David has his characteristic kind of strength—quite different from ours. He has never minded living in a crowd. And he seems to suffer spongers gladly—well, anyhow, he suffers them. None of this is really dangerous, no doubt. He is immensely strong and probably he's relaxed inside of himself in a way I can't even imagine.

October 1. I can't record that my sloth block has been struggled out of. I have messed a little with my book but I still haven't given myself the decisive order to pull myself together and get the hell on with it. Very well—that order has now been given. Really, I have no excuses. I can't pretend that I am sick or incapacitated in any way. Sure, my back hurts a bit, but not when I'm sitting down.

Death thoughts hover around continually, but I only feel fear now and then. Darling keeps me going—sometimes by his sweetness, sometimes by his scolding—he is always a massive support. I know that he is terribly rattled by Ted, who is indulging in what Ted calls a "high." A short while ago—with infuriating self-indulgence—he attacked an elderly woman and got himself locked up. His trial for this is to take place soon. One simply can't take him seriously as a madman, although I realize that he just possibly might get into a crazy state in which he would harm Darling.

I'll try to describe—just as a psychological observation—something which I've experienced several times lately, while doing my morning "sit." I pray to Swami, as usual, to "be with me and take away my fear when I am dying." At first, when I'd said this prayer, I felt reassured and soothed. Then doubts began. The doubts scoffed at the prayer. "Why do you assume that Swami is present and can answer your praying?" But then I knew how to answer the doubts: "Swami really *is* present within me because I remember him. Why shouldn't I remember him when I am dying and get strength and reassurance from that memory? In the last analysis, strength and reassurance are all I'm asking for. Looked at from this point of view, my problem doesn't have to be solved by "faith." I just have to relive one of those powerful experiences of reassurance which I quite often had in his presence while he was still alive. My memory of this reassurance is gradually getting dimmer, of course. But it is right here inside me, just as vivid as my memory of Auden or Forster, or any other inspiring human being. By dwelling on it, I think I can make it last my time.

October 23. The last few days have been astonishingly hot. Nearly no wind. I'm not in a good state. Death fears—that's to say, pangs of foreboding—recur often. They seem to be part of a quite normal physical condition; the pangs of a dying animal, thrilling with dread of the unknown.

Darling is probably under greater strain than he has ever been. I'm so dull witted and forgetful and often so sulky—and then he has to cope with his senile mother in the nursing home, and with Ted who is busy having a "breakdown"—partly self-advertisement, partly jealousy of Darling. He bashed on the carport gate so hard that he cracked its wood. This rattles our nerves of course, until we long to have him taken away and locked up some place for the rest of his life. There is very little I can do to help Don, even when I'm in a relatively unselfish mood. I just try to keep assuring him how much I love him.

1983

January 1. Yesterday I was in one of my lousiest negative moods. A boy named Dan Turner sat for Don. He carried on about the treatments he'd had for cancer and created a quite powerful death gloom, which he thickened by showing us two lesions on his legs.[1] I allowed myself to get into a kind of sulk, in which I hated him for doing this and felt that this whole vital and decisive New Year's Eve was being hexed. We went on to a couple of parties, at which 1983 was drunk to and I sulked and said to myself that we'd gotten off on the wrong foot, just because of this jerk. (Only I don't seriously believe that.)

February 7. Heavy rain, depressing and yet work inspiring. Darling has gone out and there is no reason why I shouldn't get my book restarted, having wasted one whole month already. He is so good—urging me to work and setting me such an example by working so tirelessly himself—he seems never to be idle.

April 14. This is a mere token restart.

What have I been doing? Failing to get on with my work. Old age, viewed from this angle, is nothing but an excuse for laziness. I'm not feeble or fragile. I walk in the park with vigor and enjoy my walks and thoughts. I adore my Darling and feel the keenest interest in his thinkings and doings. I'll only be seventy-nine in August—there's a whole year tucked away in there, inviting me to show the wonders I can perform before I'm eighty. Even my

[1] It later proved that Turner had AIDS; he was the first sufferer Isherwood and Bachardy knew and died after ten years of illness.

tendency to senile melancholia seems to have weakened. All right, come on, you idle Naggin, gee up!

Now a bit of senility—just for laughs. I picked up the torn-off top part of my ballot ticket for the April 12 primary elections, printed in English and Spanish, and dull-wittedly took the Spanish—*elecciones primarias*—to mean "prime ribs"!

July 4. Yes, it's that certain day. Don's with a sitter in the studio and I'm at the table on the deck, having decided to waste my time doing something which at least belongs in the category of work—writing in this book.

It's classic Canyon summer holiday weather—grey skied, warm, blah. Hardly a firework heard so far. Only one flag visible.

The joy of being with Darling remains almost constant, nowadays—expressed by huggings, kisses, loving jokes. There is a constant sense of communication between us. What we communicate to each other is that we're sharing a day-by-day experience of life; everything happens to both of us. This is, of course, far from being true, but it's true *enough.*

Have just finished a draft of a letter to Stephen, congratulating him on having been knighted. He hasn't written to me about this but he let me find out about it through his friend Br[y]an Obst.[1] I think I entirely understand Stephen's feelings. I had to be told, and yet he felt apologetic, maybe remembering how he'd smirked and sneered when Wystan was awarded the King's Gold Medal for poetry and had to go to the palace to receive it.[2]

(Edward Upward, in a letter, made us roar with laughter by quoting Banquo's line: "Thou hast it now. . . ."[3])

[*Isherwood made no more diary entries. He died of prostate cancer on January 4, 1986, after many months of illness. Bachardy gave up his other sitters during the last six months of Isherwood's life, and he drew Isherwood almost daily. He was with him, drawing him, when Isherwood died, and he drew several more drawings after he was dead.*]

[1] Zoologist, with a special interest in ornithology. Spender met him in 1976 in Gainesville, Florida, and encouraged him to give up teaching school there to pursue a graduate degree at UCLA, where he became a professor in 1987. He died of AIDS in 1991.

[2] From King George VI, November 23, 1937; it was the medal for 1936.

[3] *Macbeth*, III.i.

Glossary

Abedha. American disciple of Swami Prabhavananda, born Tony Eckstein. He spent many years at Trabuco and at the Hollywood Vedanta Society, but never took sannyas and eventually left to work for Parker Pens.

Ackerley, J. R. (Joe) (1896–1967). English author and editor; he wrote drama, poetry, fiction, and autobiography. He is well known for his intimate relationship with his Alsatian, described in *My Dog Tulip* (1956) and *We Think the World of You* (1960). Other books include *Hindoo Holiday: An Indian Journal* (1932) and *My Father and Myself* (1968). He was literary editor of *The Listener* from 1935 to 1959 and published work by some of the best and most important writers of his period; Isherwood contributed numerous reviews during the 1930s. Their friendship was sustained in later years partly by their shared intimacy with E.M. Forster. Ackerley was also close to his sister, Nancy West, who was a great beauty and a drunk, as Isherwood records. Ackerley appears in *D.1* and *D.2.* and is mentioned in *Lost Years.*

ACLU. American Civil Liberties Union, non-partisan, non-profit organization founded in 1920 to protect and preserve individual liberties guaranteed by the U.S. Constitution and its Amendments.

Addis, Keith. Hollywood agent, educated at Columbia University. When Isherwood first mentions him in 1977, he was working with Bobby Littman at The Robert Littman / Keith Addis Company. In 1980, he founded his own agency and later, in 1989, co-founded Industry Entertainment, which manages major film and T.V. stars and produces films and T.V. series.

Adjemian, Bob (b. 1947). American monk of the Ramakrishna Order, also called Nirvana after he took his brahmacharya vows. He has a B.A. in communications and oversees the Vedanta Press and the retail mail order department for the Vedanta Catalogue. Isherwood first met him in 1969 or 1970, when Adjemian came to Adelaide Drive to talk about living in a monastery. As Isherwood mentions, he is a committed runner; he has completed nine 100-mile mountain trail races.

Albee, Edward (b. 1928). American playwright, educated at Lawrenceville, Choate, and Trinity College in Connecticut. He had his first success with *The Zoo Story* off-Broadway in 1958, won Pulitzer Prizes for *A Delicate Balance* (1966),

Seascape (1975), and *Three Tall Women* (1994), and is best known for *Who's Afraid of Virginia Woolf?* (1962).

Albert, Eddie (1906–2005). American actor, on Broadway from the mid-1930s. his films include *Carrie* (1952), *Roman Holiday* (1953), *Oklahoma* (1955), and *The Heartbreak Kid* (1972), and he played Oliver in the T.V. series "Green Acres," which first aired in 1965. He appears in *D.2*. His wife of forty years, the actress and dancer Margo (1917–1985), was born in Mexico City as Maria Marguerita Guadalupe Teresa Estel Bolado Castilla y O'Donell. She had a brief first marriage. Her films include *Winterset* (1936), *Lost Horizon* (1937), and *Viva Zapata!* (1952). She sat for Bachardy in 1973.

Albert, Edward (1951–2006). Actor; son of Eddie and Margo Albert. He appeared in his first film when he was fourteen. Later he went to UCLA and Oxford. He starred in *Butterflies Are Free* (1972), as Isherwood records, and eventually became a photographer. He sat for Bachardy in 1973. He appears in *D.2*.

Aldous. See Huxley, Aldous.

Alec. See Beesley, Alec and Dodie Smith Beesley.

Alexander, Clytie (b. 1940). American artist, from Kansas, educated in Quebec and Dhaka, Bangladesh. She was an architect's apprentice and studied art and engineering at Antioch College in Ohio and at San José State College in California before focusing on fine arts at UCLA. Much of her work—abstract, minimalist, linked to colorfield—evokes the light and landscapes of southern California. Since the 1970s, she has exhibited in countless solo and group exhibitions on the West and East Coasts, Europe and Japan; her work is held in many public and corporate collections, and she has received numerous grants and awards. She was married to Peter Alexander for more than fifteen years, until the mid-1980s, and had two daughters with him: Hope, later an architect and designer, and Julia. For a number of years she was Bachardy's closest woman friend. She divides her time between New York and Los Angeles.

Alexander, Peter (b. 1939). American artist; born in Los Angeles, educated at the University of Pennsylvania, the Architectural Association in London, at Berkeley, at the University of Southern California, and at UCLA. His murals, posters, and sculptures—mostly landscapes and figurative work—are in many public venues in California, including Disney Hall, and he has exhibited all over the United States. Collections holding his work include the Corcoran Gallery of Art, Washington, D.C.; the Fogg Art Museum, Harvard University; the Getty Museum, Malibu; the Los Angeles County Museum of Art; the Norton Simon Museum of Art, Pasadena; the Metropolitan Museum, the Museum of Modern Art, and the Guggenheim, New York; the Walker Art Center, Minneapolis. His first wife was the painter Clytie Alexander, and he later married Claudia Parducci, a painter and musician. He first sat for Bachardy in 1969, and Isherwood met him around that time.

Allen, Alan Warren. Isherwood's general practitioner from April 1961 until the mid-1970s; he was about forty when they met, tall, handsome, soft-spoken, easygoing, and married. As Isherwood tells in *D.2*, his first wife committed suicide.

Allen, Edwin (Ed). American librarian, at Wesleyan College; once a salesman for Oxford University Press. He met the Stravinskys in 1962, when he was about thirty, and catalogued Igor Stravinsky's library. He also helped with errands and domestic tasks in California and New York, becoming a weekend fixture in the Stravinskys' Fifth Avenue apartment when they moved permanently to New York. He appears in *D.2*.

Altman, Dennis (b. 1943). Australian academic, author, gay activist. He did graduate work at Cornell in the mid-1960s and published one of the first accounts of the gay liberation movement in the U.S., *Homosexual: Oppression and Liberation* (1971), followed by many books on sexuality and political culture. Later, he became a politics professor at La Trobe University in Melbourne and President of the AIDS Society of Asia and the Pacific. He appears in *D.2*.

Amaya, Mario (1933–1986). Canadian curator and art critic. He founded the London magazine *Art and Artists* and is credited with bringing Yoko Ono's work to London. He was a member of Andy Warhol's circle, was with Warhol the day Warhol was shot in 1968, and was himself also shot. In 1972, he became Director of the New York Cultural Center where he staged a show of Bachardy's drawings, February 15–April 7, 1974.

Amendt, Rudolf (Rudi) (1895–1987). German actor, in Hollywood from the 1930s, at first sometimes using the name Robert O. Davis. He played small parts in thrillers, dramas, war and horror movies—often as a Nazi—before moving into T.V. in the 1950s and 1960s.

Amiya (1902–1986). English Vedanta devotee, born Ella Sully, one of ten daughters of a handsome Somerset farm laborer whose well-born wife chose scandal and poverty in order to marry him. Amiya travelled to California in the early 1930s with an older sister, Joy, who married an American artist called Palmerton. She was hired by Swami Prabhavananda and Sister Lalita as housekeeper at Ivar Avenue. By the time Isherwood met her at the end of the decade, she had received her Sanskrit name from Swami and become a nun. She became a particular friend of Isherwood's when he lived at the Vedanta Society during the 1940s. Amiya had married in the late 1920s, becoming Ella Corbin, but the marriage failed; in the early 1950s she met George Montagu, Earl of Sandwich (then in his sixties), when he visited the Vedanta Society. A few weeks later Swami gave them permission to marry, and Amiya returned to England, divorced Corbin, and became Countess of Sandwich. She grew close to Isherwood's mother and his brother Richard. She was also close throughout her life to her younger sister, Sally Hardie (1906–1990). Bachardy drew Amiya twice. She appears in *D.1* and *D.2*.

Amohananda. American monk of the Ramakrishna Order. Until he took sannyas in 1971, he was called Paul Hamilton.

Anamananda. See Arup Chaitanya.

ananda. Sanskrit for bliss or joy; an aspect of Brahman. It is used as the last part of a monk's sannyas name in the Shankara Order, for example, Vivek*ananda*, "whose bliss is in discrimination."

Anandaprana (Ananda). See Usha.

Anderson, Judith (1898–1992). Australian-born actress; she made her first appearance on the New York stage in 1918 and played major roles throughout the 1930s and 1940s, including the lead in *Mourning Becomes Electra* (1932), Gertrude to Gielgud's Hamlet in 1936, Lady Macbeth twice, and Medea twice. She also had many movie roles, including Mrs. Danvers in Hitchcock's *Rebecca* (1940) and other often chilling parts. She took the lead in the brief Broadway run of Speed Lamkin's play *Comes a Day* at the end of the 1950s and toured with Bill Roerick in 1961, performing snippets from her most famous shows. She appears in *D.1* and *D.2*.

Anderson, Lindsay (1923–1994). British director; born in India and educated at Cheltenham College and Oxford. After World War II, he founded the British film magazine *Sequence* with Karel Reisz and Gavin Lambert, a friend since school; he wrote film criticism for it, for *Sound and Sight*, and for *The New Statesman*. He made documentaries as well as features and was at the heart of the British Free Cinema Movement, which gave rise to the new realism in British films in the early 1960s. Many of his stage productions were at The Royal Court where he was an artistic director from 1969 to 1975. His films include *Thursday's Child* (1954), *This Sporting Life* (1963), *If...* (1968), *O Lucky Man!* (1973), *Britannia Hospital* (1982), and *The Whales of August* (1987).

Anderson, Paul. American would-be actor and singer; as Isherwood records, he was a boyfriend of Roddy McDowall, who supported his efforts to establish himself professionally, bringing celebrity friends to Anderson's nightclub performances. As well as singing at The Little Club and appearing in a production of Pinter's *The Homecoming*, he had a few tiny movie parts as Adam Anderson. But his career never took off.

Andrews, Oliver and Betty Harford. California sculptor, on the art faculty at UCLA; American actress, his wife until the 1970s. Andrews knew Alan Watts well and travelled with Watts to Japan. Harford was a close friend of Iris Tree and acted at Tree's High Valley Theater in the Upper Ojai Valley. She also worked for John Houseman in numerous stage productions, appeared in a few movies, including *Inside Daisy Clover* (1965), and had numerous T.V. roles from the 1950s onward, including Professor Kingsfield's secretary, Mrs. Nottingham, in "The Paper Chase" (1978–1986) and the cook, Hilda Gunnerson, in "Dynasty" (1981–1989). They had a son, Christopher, born in the 1950s and named after Isherwood. They are mentioned as a couple in *D.1* and *Lost Years*, where Isherwood called her Betty Andrews, and separately in *D.2*. After they split, she lived with Hungarian actor Alex de Naszody until he died in the early 1980s. As Isherwood tells, Andrews died suddenly of a heart attack in 1978, while still in his forties.

Anhalt, Edward (Eddie) (b. 1914). American screenwriter. He wrote war, crime, and spy thrillers but also adapted *Beckett* (1964), *The Madwoman of Chaillot* (1969), and many others. He began writing with his first wife, Edna Anhalt, during the 1950s and later worked successfully by himself. His second wife, whom he married later in the 1950s, was called Jackie George. In 1958, Selznick hired

Anhalt to work on *Mary Magdalene*, replacing Isherwood, but the film was never made. He appears in *D.1*.

the Animals. Isherwood and Bachardy. Also, homosexuals in general, as against human beings or heterosexuals. Isherwood and Bachardy called their Adelaide Drive house La Casa de los Animales (The House of the Animals). See also Dobbin for Isherwood in his identity as a horse and Kitty for Bachardy.

***Apollo 13*.** The third lunar landing attempt was launched April 11, 1970; on April 13, 200,000 miles from earth, an oxygen tank blew up, causing the other to fail; the three astronauts had to evacuate from the command module into the lunar module, and travel back to earth in near-freezing temperatures with little food and only six ounces of water a day. They splashed down safely near Samoa on April 17.

Arizu, Betty. The daughter Jo Masselink had with Ferdinand Hinchberger. She appears as a grown woman in *D.2*. and in this diary, married to Fran Arizu, a Mexican, with whom she had two children.

Arnoldi, Charles (Chuck) (b. 1946). American sculptor, painter, and graphic artist; born in Ohio, educated briefly at the Art Center in Los Angeles and at the Chouinard Art Institute. He became recognized in the early 1970s for his wall reliefs and objects made of sticks, rope, and bamboo. His works are in the Los Angeles County Museum of Art, the Norton Simon Museum, the Fine Arts Museum of San Francisco, the Metropolitan Museum and the Museum of Modern Art in New York, the Smithsonian American Art Museum, the Art Institute of Chicago, the Albright-Knox Art Gallery in Buffalo, the Guggenheim Museum in Bilbao, and the National Gallery of Australia, among others.

Arup Chaitanya. A disciple of Swami Prabhavananda, born Kenneth (Kenny) Critchfield. He arrived at the Vedanta Society towards the end of the 1940s and lived there and at Trabuco. He took his brahmacharya vows in 1954, becoming Arup Chaitanya; then in 1963, on taking sannyas, he became Swami Anamananda. He worked for many years in the Vedanta Society Hollywood bookstore and spent his last days at Trabuco, where he died in the early 1990s. He appears in *D.1* and *D.2*.

Asaktananda, Swami (1931–2009). Indian monk of the Ramakrishna Order. He was groomed by Swami Prabhavananda to take over the Hollywood Vedanta Society, until Prabhavananda unexpectedly decided that Asaktananda had the wrong personality for the role and sent him back to the Belur Math in India against Asaktananda's wishes and amid much controversy. Asaktananda later headed the Narendrapur Center, an enormous educational establishment of the Ramakrishna order outside Calcutta. He appears in *D.2*.

asanas. Yoga postures, or the mat on which they are performed.

Assembly Bill 489 (A.B. 489). California state legislation decriminalizing adultery, oral sex, and sodomy between consenting adults. It was introduced in Sacramento every year for about a decade by liberal black assemblyman Willie Brown until it was approved in March 1975 by the state assembly and on May 2 by the more conservative state senate. Mervyn Dymally, newly elected, also black, rushed back from out of state to break the tie while the state senators were kept locked in the building to ensure a quorum. Governor Jerry Brown signed the bill, and it took effect on January 1, 1976.

Atman. The divine nature within man; Brahman within the human being; the self or soul; the deepest core of man's identity.

Auden, W. H. (Wystan) (1907–1973). English poet, playwright, librettist; perhaps the greatest English poet of his century and one of the most influential. He and Isherwood met as schoolboys towards the end of Isherwood's time at St. Edmund's School, Hindhead, Surrey, where Auden, two and a half years younger, arrived in the autumn of 1915. They wrote three plays together—*The Dog Beneath the Skin* (1935), *The Ascent of F6* (1936), *On the Frontier* (1938)—and a travel book about their trip to China during the Sino-Japanese war—*Journey to a War* (1939). A fourth play—*The Enemies of a Bishop* (1929)—was published posthumously. They also wrote a film scenario, "The Life of an American," probably in 1939. As well as several stints of schoolmastering, Auden worked for John Grierson's Film Unit, funded by the General Post Office, for about six months in 1935, mostly writing poetry to be used for sound tracks. He and Isherwood went abroad separately and together during the 1930s, famously to Berlin (Auden arrived first, in 1928), and finally emigrated together to the United States in 1939. After only a few months, their lives diverged; Auden settled in New York with his companion and, later, collaborator, Chester Kallman, but he remained close friends with Isherwood. Auden's libretti include *Paul Bunyan* (1941) for Benjamin Britten, *The Rake's Progress* (1948) with Kallman for Stravinsky, and *Elegy for Young Lovers* with Kallman for Hans Werner Henze. Auden is caricatured as "Hugh Weston" in *Lions and Shadows* and figures centrally in *Christopher and His Kind*. There are many passages about him in *D.1*, *D.2*, and *Lost Years*.

Austen, Howard (Tinker) (1928–2003). Companion to Gore Vidal from 1950. He worked in advertising in New York and studied singing, then devoted most of his time to Vidal, managing Vidal's business and social life. He appears in *D.1* and *D.2*.

Avis, Annie (1869–1948). Isherwood's nanny, from near Bury St. Edmunds in Suffolk. She was hired by Isherwood's mother when he was two months old and Avis herself was about thirty; she remained with the family for the rest of her life and never married. In childhood, Isherwood and his younger brother Richard spent more time with Nanny than they did with their mother. In *Kathleen and Frank*, Isherwood writes that he had loved Nanny dearly; he bullied her in adolescence, but she never criticized him, and he shared intimate secrets with her.

Ayer, A.J. (Alfred, Freddie) (1910–1989). British philosopher, educated at Eton and Oxford. He married four times and had many affairs. His second wife was also his fourth, American journalist Dee Wells, née Chapman (b. 1925), author of the best-selling novel *Jane* (1973). They met in 1956, married in 1960, divorced in the early 1980s, and married again in 1989. Ayer's third wife was Vanessa Lawson, formerly wife of Nigel Lawson; she and Ayer married in 1982 but were involved with one another from 1968 onward; she died of cancer in 1985. Another long affair, in the early 1950s, was with Jocelyn Rickards, who remained a friend. Ayer was also a close friend of Tony Bower, whom he met in New York during World War II, and with whose half-sister, Jean Gordon-Duff, he was briefly involved around the same time. He appears in *D.2*.

Bacall, Lauren (b. 1924). American stage and screen star, born in the Bronx and trained at the American Academy of Dramatic Arts. She appeared on Broadway before making her first film *To Have and Have Not* (1944) with Humphrey Bogart, whom she married in 1945. Her other films with him are *The Big Sleep* (1945), *Dark Passage* (1947), and *Key Largo* (1948), and she also appeared in *How to Marry a Millionaire* (1953), *Sex and the Single Girl* (1965), and *The Mirror Has Two Faces* (1996), among others. She returned to Broadway in *Cactus Flower* (1965), *Applause* (1970), for which she won a Tony Award, and *Woman of the Year* (1981). Bogart died in 1957, and in 1961 Bacall married Jason Robards, Jr., but they later divorced. Isherwood first mentions her in *D.1* in 1954 when Peter Viertel was friendly with her, and they met again several times with the Viertels and at the Selznicks' during the 1950s.

Bachardy, Don (b. 1934). American painter; Isherwood's companion from 1953 onwards. Bachardy accompanied his elder brother, Ted Bachardy, to the beach in Santa Monica from the late 1940s, and Isherwood occasionally saw him there. Ted Bachardy, with other friends, first introduced them in November 1952. They met again in early February 1953 and, on February 14, began an affair which quickly became serious. Bachardy was then an eighteen-year-old college student living at home with his brother and his mother. His parents were divorced. He had studied languages for one semester at UCLA, then transferred at the start of 1953 to Los Angeles City College in Hollywood, near his mother's apartment. At first he studied French and Spanish but dropped French for German as a result of Isherwood's influence. He had worked as a grocery boy at a local market, and, like Isherwood in youth, spent most of his free time at the movies. In February 1955, Bachardy went back to UCLA to begin his junior year and almost immediately changed his major from languages to theater arts. In July 1956, he enrolled at the Chouinard Art School, supplemented his instruction there by taking classes with Vernon Old, and within a few years got work as a professional artist, drawing fashion illustrations for a local department store and then for newspapers and magazines. During this period he began to do portraits of Isherwood, close friends, and favorite film stars, and to sell his work. His first major portrait commission, from Tony Richardson, was to draw the cast of the 1960 stage production of *A Taste of Honey*. In 1961 he attended the Slade School of Fine Art in London, supported partly by his patron Russell McKinnon and partly by *Women's Wear Daily* as their London fashion illustrator. His work at the Slade led to his first individual shows, in London in 1961 and in New York in 1962. Since then, he has done countless portraits, both of the famous and the little-known, and exhibited in many cities. His work is held in numerous public and private collections, including the Smithsonian Museum of American Art in Washington, D.C. and the National Portrait Gallery in London, and he has published his drawings in several books, including *October* (1981) with Isherwood, *Last Drawings of Christopher Isherwood* (1990), and *Stars in My Eyes* (2000). Together, Isherwood and Bachardy wrote several stage and film scripts, including their award-winning screenplay for the T.V. film "Frankenstein: The True Story" (1973). He figures centrally in *D.1* and *D.2*.

Bachardy, Glade De Land (1906–198[8]). Don Bachardy's mother, from Ohio. Childhood polio left her with a limp, resulting in extreme shyness. Her father was the captain of a cargo boat on the Great Lakes, and she met her husband, Jess Bachardy, on board during a summer cruise with her sister in the 1920s. They married in 1928 in Cleveland, Ohio, and travelled to Los Angeles on their honeymoon, settling there permanently. The Bachardys divorced in 1952, but later reconciled; once Don and his brother Ted Bachardy had moved out of their mother's apartment, their father moved back in, early in 1955. An ardent movie-goer, Glade took Don and Ted to the movies from their early childhood because she could not afford babysitters, thus nurturing an obsession which developed differently in each of them. According to Don, Glade did not know what homosexuality was until her elder son Ted had his first breakdown in 1945. She appears in *D.1* and *D.2*.

Bachardy, Jess (1905–1977). Don Bachardy's father, born in New Jersey, the youngest of several brothers and sisters in an immigrant German-Hungarian family. Jess's mother, who never learned to speak English, was pregnant with him when she arrived in the U.S.; his father drowned accidentally shortly before. Jess was an automobile enthusiast and a natural mechanic and took several jobs as a uniformed chauffeur when he was young. Afterwards, he worked on board a cargo boat on the Great Lakes, where he met his future wife. They moved to California, and he turned his mechanical skills to the aviation industry, working mostly with Lockheed Aircraft for the next thirty years. His progress was limited by the fact that he never finished high school, but he worked his way up to the position of tool planner before he retired in the 1960s. He never allowed his sons to learn Hungarian, and they barely knew their Bachardy grandmother or any of her family. For fifteen years, he refused to meet Isherwood, but he finally relented and came to like him. He was a lifelong smoker and died of lung cancer. He appears in *D.1* and *D.2*.

Bachardy, Ted (1930–2007). Don Bachardy's older brother. Isherwood spotted him on the beach in Santa Monica, probably in the autumn of 1948 or spring of 1949, and invited him to a party in November 1949 (Ted's name first appears in Isherwood's diary that month). Isherwood was attracted to Ted, but did not pursue him seriously because Ted was involved with someone else, Ed Cornell. Around the same time, Ted experienced a mental breakdown—about the third or fourth he had suffered since 1945, when he was fifteen. Eventually he was diagnosed as a manic-depressive schizophrenic. He was subject to recurring periods of manic, self-destructive behavior followed by nervous breakdowns and long stays in mental hospitals. Isherwood continued to see Ted intermittently during the early weeks of his affair with Don, but a turning point came in February 1953 with Ted's fourth or fifth breakdown, when Isherwood sympathized with Don and intervened to try to prevent Ted from becoming violent and having to be hospitalized; nevertheless, Ted was committed on February 26. When well, Ted took odd jobs: as a tour guide and in the mail room at Warner Brothers, as a sales clerk in a department store, and as an office worker in insurance companies and advertising agencies. Isherwood writes about him in *D.1*, *D.2*, and *Lost Years*.

Bacon, Francis (1909–1992). English painter, born in Dublin. He worked as an interior decorator in London during the late 1920s and lived in Berlin in 1930, around the time that he taught himself to paint. He showed some of his work in London during the 1930s, but came to prominence only after the war, when his *Three Studies for Figures at the Base of a Crucifixion* made him suddenly famous in 1945. He appears in *D.1*, where Isherwood records some of his remarks on art, and in *Lost Years* and *D.2*. The story Bacon told Isherwood in 1972, about his friendship with the Kray twins, is told in Bacon's own words in Michael Peppiatt's biography, *Francis Bacon: Anatomy of an Enigma*. Bacon was especially interested in Ronnie Kray, a homosexual and the more sadistic of the two notoriously brutal, murderous brothers. Bacon had kept some paintings in his studio because he considered them poor work; these were the paintings stolen, as Isherwood mentions. The theft was discovered when the Krays tried to sell them, and Bacon eventually bought them back and destroyed them because he didn't want them shown publicly.

Baddeley, Hermione (1906–1986). British actress, dancer, comedienne; she first appeared on Broadway as a replacement for Angela Lansbury in *A Taste of Honey* (1960–1961). Among her films are *Brighton Rock* (1947), which she previously played on stage, *Room at the Top* (1958), and *Mary Poppins* (1964). She was married twice and lived for a time with actor Laurence Harvey.

Bailey, Jeffrey (Jeff). American writer, of short stories for little magazines. He interviewed Paul Bowles, James Leo Herlihy, Isherwood, Gavin Lambert, Anaïs Nin, John Rechy, Muriel Spark, Gore Vidal, and Edmund White, some for *The Paris Review*. As Isherwood records, Bailey visited in 1972 with another aspiring writer, Michael McDonagh, after he and McDonagh spent their junior year abroad at the Loyola University Center in Rome in 1971–1972. Later, he lived in Morocco, working for the Peace Corps there, and in Europe. McDonagh settled in San Francisco where he became an arts journalist, mostly for the *Bay Area Reporter*, and published some of his poems.

Bailey, Paul (b. 1937). British novelist; educated at the Central School of Speech and Drama. He worked as an actor and wrote a radio play before publishing his first novel, *At the Jerusalem*, in 1967. Isherwood was on the panel which awarded Bailey the first E.M. Forster Prize given by the American Academy of Arts and Letters in 1974. Bailey's other novels include *Peter Smart's Confessions* (1977), *Old Soldiers* (1980), *Gabriel's Lament* (1986), *Kitty and Virgil* (1998), *Uncle Rudolf* (2002), and *A Dog's Life* (2003). There are varied works of biography and autobiography. He has won numerous prizes and twice been on the Booker Prize shortlist. He was authorized to write a life of Henry Green for Chatto & Windus, as he mentioned to Isherwood in 1978, but after some preliminary research, abandoned it for other projects. Later, he supplied an introduction to a paperback reprint of Henry Green's novel *Living*.

Balanchine, George (1904–1983). Russian-born choreographer, son of a composer. He studied ballet at the Maryinsky and piano at the St. Petersburg music conservatory. In 1924, he emigrated via Berlin and spent a decade working in Europe, mostly for Diaghilev and the Ballet Russe. In 1933, Lincoln Kirstein persuaded him to emigrate again, to New York, and together they founded the American School of Ballet, struggling off and on for another decade to finance and house the company which would eventually become the New York City Ballet. Balanchine made over four hundred ballets and is known for his Modernist approach—abstract, technically demanding, and based on a committed understanding of music. He was to twentieth-century ballet what Picasso was to painting and Stravinsky to music, and he collaborated with Stravinsky a number of times. He married five times.

Bangladesh. In the Pakistani elections in 1970, the Awami League, led by Mujibur Rahman, won almost all the seats in East Pakistan (formerly East Bengal, and sharing a long border with West Bengal, in India). Despite winning no seats in West Pakistan, the league had a clear constitutional majority; nonetheless, it was prevented by the government in power from forming a national government with Rahman as Prime Minister. On March 1, 1971, the president of Pakistan, Yahya Khan, indefinitely postponed the next session of the national assembly, provoking a General Strike in East Pakistan and sympathetic protests in India. The Awami League was banned, Mujibar Rahman was arrested, and his close supporters fled to India where they formed a provisional government. In East Pakistan, the Pakistani army initiated bloody reprisals against supporters of Bengali independence. As Isherwood mentions, Indira Gandhi was elected Prime Minister of India on March 10, in a landslide victory for her Congress Party. On March 26, the Awami League declared independence for Bangladesh, and India joined with Bangladesh in the ensuing war of liberation from Pakistan.

Barada. A senior nun at the Santa Barbara convent, born Doris Ludwig; after sannyas she was called Pravrajika Baradaprana. Barada was interested in music and composed Vedantic hymns. Isherwood first met her at the Hollywood monastery in 1943; she appears in *D.1*.

Barnett, Jimmy. American monk of the Ramakrishna Order, also known as Sat and as Swami Buddhananda. He lived at Trabuco during the 1960s and later at the Hollywood Vedanta Society. Eventually, he left the order and settled in Sedona, Arizona, where he became a Native American chieftain and worked as an artist, counsellor, and medicine man. He is mentioned in *Lost Years* and *D.2*.

Barrie, Jay Michael (1912–2001). A one-time singer with financial and administrative talents; friend and secretary to Gerald Heard from the late 1940s onward. He met Heard through Swami Prabhavananda and lived at Trabuco as a monk until about 1955. He was friendly with Isherwood and Bachardy throughout the 1950s, and they rented Barrie's house, at 322 East Rustic Road, for roughly two months in 1956. Barrie nursed Heard through his five-year-long final illness until Heard's death in 1971. He appears in *D.1* and *D.2*.

Baxter, Anne (1923–1985). American actress, a granddaughter of Frank Lloyd Wright; educated in New York private schools. She studied acting with Maria Ouspenskaya, debuted on Broadway at thirteen, and had made her first movie by seventeen. Her films include *The Magnificent Ambersons* (1942), *The Razor's Edge*

(1946), for which she received an Academy Award as best supporting actress, *Yellow Sky* (1949), *All about Eve* (1950), for which she received an Academy Award nomination, *The Outcasts of Poker Flat* (1952), *The Blue Gardenia* (1953), *The Ten Commandments* (1956), *Cimarron* (1960), and *Walk on the Wild Side* (1962). From 1971, as Isherwood records, she returned to Broadway, replacing Lauren Bacall in *Applause*. She also acted on T.V., including, from 1983 to 1985, "Hotel." Her first husband was the actor John Hodiak, with whom she had a daughter; the second, from 1960 to 1968, was Randolph Galt, an outdoorsman and adventurer with whom she had two daughters; the third was David Klee, an investment banker. With Galt, Baxter went to live in the Australian outback on a cattle station; after the marriage failed, she published a book about her experience there, *Intermission: A True Story* (1976). She was a client and friend of Jo Masselink, and she appears in *D.1* and *D.2*.

Baxter, Keith (b. 1933). Welsh actor, trained at the Royal Academy of Dramatic Art (RADA). His first film role was in *The Barretts of Wimpole Street* (1957), and he later worked in T.V. His theater roles include Orson Welles's *Chimes at Midnight* (1960), and, on Broadway, *A Man for All Seasons* (1961) and *Sleuth* (1970).

Beaton, Cecil (1904–1980). English photographer, theater designer, author, and dandy. He photographed the most celebrated and fashionable people of his era, beginning in the 1920s with the Sitwells and going on to the British royal family, actors, actresses, writers, and others. From 1939 to 1945 he worked successfully as a war photographer. Isherwood and Beaton were contemporaries at Cambridge but became friendly only in the late 1940s when Beaton visited Hollywood with a production of *Lady Windermere's Fan* and was helpful to Bill Caskey, then trying to establish himself as a photographer. Returning later to Hollywood, Beaton designed costumes and productions for *Gigi* (1958) and *My Fair Lady* (1964) and both times won the Academy Award for costumes. He collected many of his photographs into books and travel albums, often with commentary, and he published five volumes of diaries. He appears in *D.1*, *D.2*, and *Lost Years*.

Bedford, Brian (b. 1935). British stage actor and, later, director; an American citizen from 1959. He trained at RADA and starred in the West End and on Broadway in Shakespeare and other classic dramas as well as new plays by Stoppard, Shaffer, and others. During 1969, he was in revivals of *The Cocktail Party* and *The Misanthrope* for Ellis Rabb's APA-Phoenix Theater repertory program on Broadway. In 1971, he won a Tony Award for his role in *The School for Wives*. He acts regularly at the Stratford Festival in Ontario, Canada, and on T.V. and in films. He appears in *D.2*.

Beesley, Alec (1903–1987) and Dodie Smith Beesley (1896–1990). She was the English playwright, novelist and former actress, Dodie Smith. He managed her career. They spent a decade in Hollywood because he was a pacifist and a conscientious objector during World War II. She wrote scripts there for Paramount and her first novel, *I Capture the Castle* (1949). Isherwood met them in 1942 through Dodie's close friend John van Druten, once a Christian Scientist like both the Beesleys. When Isherwood left the Vedanta Society in August 1945, his first home was the Beesleys' chauffeur's apartment. Dodie encouraged his writing, and he discussed *The World in the Evening* with her extensively. It was Dodie Beesley who challenged John van Druten to make a play from *Sally Bowles*, leading to *I Am a Camera*. In the summer of 1943, the Beesleys mated their dalmatians, Folly and Buzzle, and Folly produced fifteen puppies—inspiring Dodie's most famous book, *The Hundred and One Dalmatians* (1956), later filmed by Walt Disney. Her plays include *Autumn Crocus* (1931) and *Dear Octopus* (1938). The Beesleys returned to England in the early 1950s and settled in their cottage, The Barretts, at Finchingfield, Essex. They appear in *D.1*, *D.2*, and *Lost Years*.

Behrman, S.N. (1893–1973). American playwright, producer, screenwriter, short story writer, journalist. His successes on Broadway include *The Second Man* (1927), *End of Summer* (1936), *No Time for Comedy* (1939), the book (with Joshua Logan) for *Fanny* (1954), and *Lord Pengo* (1962). He also adapted work by others, including *Serena Blandish* and Maugham's short story "Jane." He worked for the Hollywood studios off and on from 1930, specializing in dialogue, and was known for his contributions to Garbo's films *Queen Cristina*, *Conquest*, and *Two-Faced Woman*. He also wrote for *The New York Times* and published a number of long profiles in *The New Yorker*, including the excerpts that Isherwood writes about from his memoir, *People in a Diary* (1972). He is mentioned in *D.1* and *D.2*.

Bell, Larry (b. 1939). American painter and sculptor, born in Chicago, raised in California and trained at the Chouinard Art Institute. In 1958 and 1959, he worked in a picture framing shop in Burbank, where he grew intrigued by the reflective and refractive qualities of glass; during the 1960s, he became known for his glass cubes, in various sizes, made from clear and treated glass. His first solo show was at the Ferus Gallery in Los Angeles in 1962, and he began showing at the Pace Gallery in New York a few years later, followed by countless solo and group exhibitions and installations in many cities. His work is held by major museums including the Albright-Knox, the Art Institute of Chicago, the Hirshhorn in Washington, the Los Angeles County Museum of Modern Art, the Guggenheim and the Museum of Modern Art in New York, the Tate Gallery and the Victoria and Albert Museum in London, and the Walker Art Institute in Minneapolis. In the 1970s, he settled with his wife in Taos, New Mexico, where he sometimes taught at the Taos Institute of Arts. Later, he also continued to work in Los Angeles.

Ben. See Masselink, Ben.

Bengston, Billy Al (b. 1934). American artist, born in Kansas, educated at the California College of Arts and Crafts in Oakland, at Los Angeles City College, and at the Los Angeles County Art Institute (now Otis Art Institute). He had his first one-man show at the Ferus Gallery in 1958, followed, from the 1960s onward, by shows and public and private commissions throughout the United States, Canada, Germany, and Japan. His work includes painting, sculpture, textiles, lithography, and architectural design. He has been a guest artist and a professor at the Chouinard Art Institute, UCLA, and elsewhere, and has held numerous fellowships and grants, including a Guggenheim. Based for years in Venice, California, he moved in 2004 to Victoria, British Columbia, with Wendy Al, his Japanese-American wife of many years, but they returned in 2007.

Isherwood met him through Bachardy who was commissioned to do Bengston's portrait, along with other prominent Los Angelinos, for *Harper's Bazaar* in 1967. He appears in *D.2*.

Bergman, Ingrid (1915–1982). Swedish star of stage and screen. David Selznick brought her to Hollywood in 1939 to remake Gustav Molander's *Intermezzo* (1936) for American audiences. She debuted on Broadway a year later in *Liliom* and went on to star in *Casablanca* (1942), *For Whom the Bell Tolls* (1943), *Gaslight* (1944; Academy Award), *Spellbound* (1945), *The Bells of St. Mary's* (1945), *Notorious* (1946), and *Joan of Arc* (1948), adapted from another Broadway role. In 1949, she fell in love with Italian director Roberto Rossellini while they were filming *Stromboli* and had his son although she was still married to Swedish dentist Petter Lindstrom. As a result, she was attacked by American women's groups, religious groups, and even in the Senate, lost custody of her daughter, Pia Lindstrom, and disappeared from Hollywood films. She married Rossellini in 1950 and had twin daughters, one of whom is the actress, Isabella Rossellini (b. 1952). In 1956, Bergman was welcomed back to Hollywood with an Academy Award for *Anastasia*. She split with Rosselini in 1958 and married a Swedish stage producer, Lars Schmidt, whom she divorced in 1975. She died of breast cancer after seven years of illness during which she continued to act. Later films include *Elena et les Hommes / Paris Does Strange Things* (1956), *The Inn of the Sixth Happiness* (1958), *Murder on the Orient Express* (1974, Academy Award), and *Autumn Sonata* (1978); she had numerous other stage roles and also appeared on T.V. As he tells in *D.1*, Isherwood first met her in 1940 on the set of *Rage in Heaven*.

Berlant, Anthony (Tony) (b. 1941). American artist, born in New York, raised and educated in Los Angeles where he got B.A., M.A., and M.F.A. degrees at UCLA. He is known for his large collages made from painted sheets of metal. His first solo show, in 1963, was at the David Stuart Gallery in Los Angeles; his work is held by the Hirshhorn, the Los Angeles County Museum of Art, the San Francisco Museum of Modern Art, and the Whitney, and he has executed public commissions for the Fox Network Center in Los Angeles, the San Francisco Airport, the Minneapolis Institute of Art, the Mayo Clinic, and the Target Corporation in Minneapolis. Much of his work reflects his preoccupation with the desert, and he is also a collector and dealer in Navajo blankets, Mimbres pottery, and early human artifacts.

Bertrand, Marcheline (1950–2007). American actress and model, part French-Canadian, part Iroquois Indian. She studied with Lee Strasberg and had bit roles in *Lookin' to Get Out* (1982) and the Blake Edwards remake of *The Man Who Loved Women* (1983). In 1971, as Isherwood records, she married Jon Voight, with whom she had two children, James Haven and Angelina Jolie, before divorcing in 1978. She was later married for five years to Tom Bessamra.

Bhadhrananda. See Worton, Len.

Bhavyananda. Indian monk of the Ramakrishna Order; head, when Isherwood visited there in 1970, of the London Ramakrishna-Vedanta Centre. As a jnani, his discipline was to identify with God rather than to worship God dualistically. Also, he was a medical doctor, trained at the Karachi Ramakrishna Mission under

Swami Ranganathananda, who emphasized lecturing and social service above devotional activities. Many jnanis perform pujas, but Bhavyananda had relatively recently arrived from India, where heads of centers rarely perform pujas.

Birtwell, Celia. See Clark, Celia.

Blanch, Lesley (1904–2007). English journalist and author. She studied painting at the Slade, designed book jackets, and from 1937 to 1944 was an editor at *Vogue*. Her books include *The Wilder Shores of Love* (1954), *The Sabres of Paradise* (1960), *The Nine-Tiger Man* (1965), *Pavilions of the Heart* (1974), biography, travel essays, cook books, and an autobiography titled *Journey Into the Mind's Eye* (1968). She married twice, the second time to the Russian-born French novelist, diplomat, and film director, Romain Gary (1914–1980), whom she met in England during World War II. She was posted with him to Bulgaria and Switzerland, and she travelled widely elsewhere before they divorced in 1962 in Los Angeles, where he was the French consul. Blanch was a close friend of Gavin Lambert who introduced her and Romain Gary to Isherwood and Bachardy during the 1950s; Isherwood also tells about their friendship in *D.1* and she appears in *D.2*. Later she settled in France.

Blum, Irving (b. 1930). American art dealer. As a salesman for Hans Knoll, purveyor of modernist furniture, he helped Knoll's Cranbrook-trained wife, Florence, carry out corporate decorating assignments, which often included paintings, and he frequented the Manhattan art scene before joining the Ferus Gallery on La Cienaga Boulevard in 1957. His efforts to create a clientele included organizing classes with co-owner Walter Hopps to educate West Coast collectors. In 1967, he opened his own gallery, where he continued to show contemporary Californian artists including Ed Moses, Billy Al Bengston, and Don Bachardy, and more widely known talents like Diebenkorn, Stella, Lichtenstein, Warhol, and Johns. Later he opened a New York gallery with Mark Helman. He appears in *D.2*.

Blum, Shirley Neilsen (b. 1932). American scholar of Renaissance painting, educated at Stockton College, the University of Chicago, and the University of California; she was a professor of art history at U.C. Riverside for eleven years, and afterwards on the East Coast. Irving Blum was her first husband; when she remarried, she took the name Resnek.

Bobo, Wallace (Bobo). Neighbor of Denny Fouts at 137 Entrada Drive, where he lived with his friend Howard Kelley. He worked as a gardener. Bobo and Kelley attended all Fouts's parties in the late 1940s and were often in his apartment. Bobo later became an alcoholic. He appears in *D.1* and in *Lost Years*.

Bogdanovich, Peter (b. 1939). American director, producer, screenwriter. He studied acting with Stella Adler, appeared in Shakespeare, directed off-Broadway, wrote features and profiles for *Esquire* and monographs and books on Fritz Lang, John Ford, Orson Welles, and others before starting to work in film himself. *The Last Picture Show* (1971) made him famous, and he had a few other successes, including *What's Up Doc?* (1972) and *Paper Moon* (1973).

Boorman, John (b. 1933). British film director; educated at Salesian College, Chertsey, Surrey. He was a film critic and a T.V. news editor before achieving recognition for his BBC documentaries. His first feature was *Catch Us If You Can / Having a Wild Weekend* (1965) about the British rock group The Dave Clark Five, and he went to Hollywood to make *Point Blank* (1967) and *Hell in the Pacific* (1968), both starring his friend Lee Marvin. He won Best Director at the Cannes Film Festival for *Leo the Last* (1970) and became famous with *Deliverance* (1972), which he directed and produced and for which he was nominated for an Academy Award. He then wrote, directed and produced *Zardoz* (1974), a science-fiction film starring Sean Connery as a barbarian warrior who worships a stone head representing the god, Zardoz, and which had no success. His later films include *Excalibur* (1981), *The Emerald Forest* (1985, nominated for Academy Awards for Best Director, Best Producer and Best Screenplay), *Hope and Glory* (1987, New York Film Critics Circle Best Director and Best Screenplay, Academy Award nomination), and *The General* (1998, Best Director award at Cannes). In the late 1950s, he married Christel Kruse, costume designer on several of his films, and they had a son and three daughters, all of whom have been involved in his films.

Bopp, Bill (b. 1932). A friend of Bachardy; he worked in administration for Burroughs Corporation, the data processing company, and lived in an apartment in Hollywood.

Bowen, Karl (d. 1992). American would-be actor and painter, a patient of Patrick Woodcock. He was a member of the Kellogg cereal family and received income from a trust fund. He lived in an impressive flat in London above the Shaftesbury Theater, where *Hair* was playing from 1968 to 1973. For a time, Wayne Sleep had a room there. Later, Bowen lived in New York. He appeared in Derek Jarman's film *In the Shadow of the Sun* (1980). He lived with AIDS for over a decade before deciding to commit suicide.

Bower, Tony (1911–1972). American editor and art dealer; educated in England at Marlborough College and Oxford. His mother became Lady Gordon-Duff through a second marriage, and his accent and manners gave the impression he was English. He is said to have made a living by playing bridge for money, and he was a spectacular gossip. He worked briefly for *Horizon*, was drafted into the U.S. Army twice during World War II, and trained on Long Island and later in San Diego. After the war he worked at New Directions, and in 1948 he became an editor at the New York magazine *Art in America*. Eventually, he became an art dealer. Isherwood met Bower in Paris in 1937 through Jean and Cyril Connolly. He appears in *D.1*, *D.2*, and *Lost Years*, and, as Isherwood tells, was the model for "Ronny" in *Down There on a Visit*.

Bowie, David (b. 1947). British rock star. He lived in Los Angeles when he was filming *The Man Who Fell to Earth* (1976), and Isherwood met him the year the film came out. Bowie then lived in Berlin, where he collaborated with Brian Eno and Tony Visconti on *Low* (1977) and *Heroes* (1977), the first two albums in his Berlin triptych, completed with *Lodgers* (1979). His next U.S. tour, starting March 1978, featured the music from the first two Berlin albums and resulted in the live album *Stage* (1978). Bachardy recalls that Bowie proposed working together on a film, possibly *Just a Gigolo* (1979), about a Prussian World War I veteran; it is set in Berlin and also stars Marlene Dietrich in her final role, a cameo. This may be the project Isherwood refers to in his entry for April 16, 1978. Bowie's first wife, Angela (Angie) Barnett (b. 1949), to whom he was married from 1970 to 1980, was an American model born in Cyprus; her father was English and her mother of Greek Cypriot descent.

Bowles, Paul (1910–1999). American composer and writer, best known for his novel *The Sheltering Sky* (1949), filmed by Bertolucci. In addition to fiction, he wrote poetry and travel books and made translations. Isherwood first met Bowles fleetingly in Berlin in 1931 and used his name for the character Sally Bowles without realizing that he would later meet Bowles again and that Bowles would become famous in his own right. Bowles and his wife, the writer Jane Bowles (1917–1973), lived in George Davis's house in Brooklyn with Auden and others during the 1940s. They later moved to Tangier, where they lived separately from one another but remained close friends. As Isherwood tells in *D.1*, he and Bachardy visited them there in 1955. Bowles also appears in *D.2*.

Brackett, Charles (1892–1969) and Muff. American screenwriter and producer and his wife. He was from a wealthy East Coast family, began as a novelist, then became a screenwriter and, later, a producer. He often worked with the Austro-Hungarian writer-director Billy Wilder. He was one of five writers who worked on the script for Garbo's *Ninotchka* (1939); he won an Academy Award as writer-producer of *The Long Weekend* (1945); and he produced *The King and I* (1956), as well as working on numerous other films. When Isherwood knew him best during the 1950s, Brackett worked for Darryl Zanuck at Twentieth Century-Fox, where he remained for about a decade until the early 1960s. His second wife, Lillian, was called Muff; she had been the spinster sister of Brackett's first wife, who died, and Muff was already in her sixties when Brackett married her. Brackett also had two grown daughters, and one, Alexandra (Xan), was married to James Larmore, Brackett's assistant. The Bracketts appear in *D.1*, *D.2*, and *Lost Years*.

Bradbury, Ray (b. 1920). American novelist, poet, playwright, and screenwriter; he finished high school in Los Angeles and never went to college. He is best known for his science-fiction classics *The Martian Chronicles* (1950) and *Fahrenheit 451* (1953). Other works include *Something Wicked This Way Comes* (1962) and, among his collections of stories, *I Sing the Body Electric!* (1969). From the mid-1980s, he adapted his short stories for his T.V. series, "The Ray Bradbury Theater." He married and had four daughters. He appears in *D.1*, *D.2*, and *Lost Years*.

Bradley, Dan and Evelyn. English laborer and his wife; they cared for Richard Isherwood. He was the younger brother of Alan Bradley, who worked at Wybersley Farm and befriended Richard there after World War II. Alan Bradley and his wife looked after Richard when Kathleen Isherwood died; they passed the role to Dan after an accident at work made Dan otherwise unemployable. Richard called these Bradleys "the Dans" and his previous carers "the Alans." After Marple Hall was torn down, the Dans lived next door to Richard in one of the new houses built on the estate. Richard left much of his property and money to the two Bradley families in his will. They appear in *D.2*.

brahmachari or **brahmacharini.** In Vedanta, a spiritual aspirant who has taken the first monastic vows. In the Ramakrishna Order, the brahmacharya vows may be taken only after five or more years as a probationer monk or nun.

Brahman. The transcendental reality of Vedanta; the impersonal absolute existence; infinite consciousness, infinite being, infinite bliss.

Brahmananda, Swami (1863–1922). Rakhal Chandra Ghosh, the son of a wealthy landowner, was a boyhood friend of Vivekananda with whom, ultimately, he was to lead the Ramakrishna Order. Later he was also called Maharaj. Married off by his father at sixteen, he became a disciple of Ramakrishna soon afterwards. Like Vivekananda, Brahmananda was an *Ishvarakoti*, an eternally free and perfect soul born into the world for mankind's benefit and possessing some characteristics of the avatar. He was an eternal companion of Sri Krishna, and his companionship took the intimate form of a parent/son relationship (thus reenacting a previously existing and eternal relationship between their two souls). After the death of Ramakrishna, Brahmananda ran the Baranagore monastery (two miles north of Calcutta), made pilgrimages to northern India, and in 1897 became President of the Belur Math and, in 1900, of the Ramakrishna Math and Mission, founding and visiting Vedanta centers in and near India.

Brainard, Joe (1942–1994). Arkansas-born artist. He worked in New York City from the early 1960s and in Vermont.

Brando, Marlon (1924–2004). American actor; raised in Nebraska. He was kicked out of military school in Minnesota, studied acting for a year in New York, debuted on Broadway in *I Remember Mama* in 1944, and became a star as Stanley Kowalski in Tennessee Williams's *A Streetcar Named Desire* in 1947. He continued to explore Method Acting as a professional and joined the Actors Studio in the late 1940s. Once he arrived in Hollywood, his stardom became phenomenal; his blend of defiance and charisma gave him the stature of Garbo and few others, and the violent eccentricities of his private life (including three failed marriages and murderous and suicidal off-spring) did not diminish his fame. His films include: *The Men* (1950; Academy Award nomination), *Viva Zapata!* (1952), *Julius Caesar* (1953, Academy Award nomination), *The Wild One* (1953), *On the Waterfront* (1954, Academy Award), *Guys and Dolls* (1955), *The Young Lions* (1958), *Mutiny on the Bounty* (1952), *Reflections in a Golden Eye* (1967), *The Godfather* (1972), *Last Tango in Paris* (1972), *Apocalypse Now* (1979). Isherwood first met Brando when Tennessee Williams came to Hollywood to polish the film script for *A Streetcar Named Desire* (1951), for which Brando received an Academy Award nomination; he tells about this in *Lost Years*, and he also mentions Brando in *D.1* and *D.2*. Sacheen Littlefeather, an Apache actress, refused on Brando's behalf his 1973 Academy Award for Best Actor for *The Godfather* in order to highlight the mistreatment of American Indians in the movie and T.V. industries and also to draw attention to the ongoing armed siege at Wounded Knee, South Dakota. Members of the American Indian Movement had occupied Wounded Knee on February 27, 1973 to protest federal government policies towards the Indians, high unemployment, police brutality, and corruption among tribal leaders. The siege lasted seventy-one days and resulted in two deaths and 1,200 arrests.

Brandt, Stuart. American painter. He lived at the Vedanta Society in Santa

Barbara for a time and remained a devotee afterwards, settling in a nearby house which he rented from the society. His Sanskrit name is Sadhak. He later ran a business specializing in faux finishes, trompe-l'oeil, and hand-painted furniture and moved away from the society though he keeps a studio in Santa Barbara.

Brevard, Lee (1949–1994). American painter and jewelry designer; born outside Pittsburgh, educated at Arizona State and Cooper Union. He worked as a watercolorist in Paris, where he met David Hockney, then relocated to Los Angeles around 1977 and began designing jewelry. Elizabeth Taylor was one of his first and best clients; his pieces were sought out by other stars, too, and appeared in a number of films. The business was called The House of Brevard; his studio was in Venice.

Bridges, James (Jimmy, Jim) (1936–1993). American actor, screenwriter and director; raised in Arkansas and educated at Arkansas Teachers College and USC. He was frequently on T.V. in the 1950s and appeared in a number of movies, including *Johnny Trouble* (1957), *Joy Ride* (1958), and *Faces* (1968). He lived with the actor Jack Larson from the mid 1950s onward, and through Larson became close friends with Isherwood and Bachardy. In the early 1960s, he was stage manager for the UCLA Professional Theater Group when John Houseman recommended him as a writer for a Hitchcock suspense series on T.V. He turned out plays constantly, some of which were shown only to Larson, and many of which were never staged. Bridges came to prominence in the 1970s when he directed and co-wrote screenplays for *The Babymaker* (1970), *The Paper Chase* (1973), *The China Syndrome* (1979), *Urban Cowboy* (1980), and, later, *Mike's Murder* (1984), *Perfect* (1985), and *Bright Lights, Big City* (1988). He directed the first production of Isherwood and Bachardy's play *A Meeting by the River* for New Theater for Now at the Mark Taper Forum in 1972, and he directed the twenty-fifth-anniversary production of *A Streetcar Named Desire* at the Ahmanson in 1973. He appears in *D.1* and *D.2*.

Brisson, Frederick (1912–1984). Danish-born Broadway producer, husband of Rosalind Russell. He produced *The Pajama Game* (1955) and *Damn Yankees* (1956), both winners of Best Musical Tony Awards, *Alfie!* (1964), *Jumpers* (1974), and others.

Britten, Benjamin (1913–1976). British composer. Auden worked with him briefly from September 1935 at John Grierson's GPO Film Unit in Soho Square and introduced him to the Group Theatre; at Auden's instigation, he composed the music for *The Ascent of F6*, and Isherwood perhaps first met him at rehearsals in February 1937. By March 1937, the two were friendly enough to spend the night together at the Jermyn Street Turkish Baths, though they never had a sexual relationship. Britten also wrote the music for the next Auden–Isherwood play, *On the Frontier*. In the summer of 1939, he went with Peter Pears to America, where he collaborated with Auden on their first opera, *Paul Bunyan* (1941); then, Britten and Pears returned to England halfway through the war, registering as conscientious objectors. Britten composed songs, song cycles, orchestral music, works for chorus and orchestra such as his *War Requiem* (1961), and nine operas including *Peter Grimes* (1945), *Albert Herring* (1948), *Billy Budd* (1951), *A Midsummer Night's*

Dream (1960), and *Death in Venice* (1973). He appears in *D.1* and in *Lost Years*. After *Paul Bunyan*, he set Auden's "Three Songs for St. Cecilia's Day," their last collaboration; Britten resented Auden's affectionate but bossy advice not to settle for the comfortable fame available in England. Once the friendship with Auden ended, Britten also gradually withdrew his friendship from Isherwood and from the Spenders because they were close to Auden. But a reunion with Isherwood was brought about by Pears at Aldeburgh in 1976. Britten, already frail, wept when he saw Isherwood.

Brookes, Jacqueline (b. 1930). American actress, trained at RADA. She was acclaimed on the New York stage in the mid-1950s, worked steadily in daytime T.V., and made a few films, including *Without a Trace* (1983).

Brown, Andrew (1938–1994). T.V. producer, from New Zealand; he worked in the U.K. in the 1960s and 1970s, for the BBC and then for Thames T.V., and afterwards co-produced from New Zealand. Among his shows are "Rock Follies" (1976) and "Selling Hitler" (1991), both by Howard Schuman, "Edward and Mrs. Simpson" (1978), "Prick Up Your Ears" (1986) by Alan Bennett, and "Anglo-Saxon Attitudes" (1992) adapted from Angus Wilson's novel. He was a member of Peter Adair's Mariposa Film Group, which created the gay and lesbian film *The Word Is Out* (1977).

Brown, Bill (1919–2012). American painter. He has used the professional names W.T. Brown, W. Theo. Brown, W. Theophilus Brown, and Theophilus Brown. He was born in Illinois and made his career on the West Coast with his longterm partner, Paul Wonner, also a painter. Brown and Wonner were companions from the late 1950s until the mid-1990s, sharing apartments and houses in Santa Monica, Malibu, New Hampshire, Santa Barbara, and finally San Francisco, where, after twenty years, they settled into separate apartments in the same building. Along with Wonner, Richard Diebenkorn, Wayne Thiebaud, David Park, Nathan Oliveira and others first perceived as a group in the early 1950s, Brown has been characterized as an American or a Californian Realist, a Bay Area Figurative Artist, a Figurative Abstractionist. Isherwood met Brown and Wonner in August 1962, when they attended Don Bachardy's first Los Angeles show at the Rex Evans Gallery with Jo and Ben Masselink. He appears in *D.2*.

Brown, Harry (1917–1986). American poet, playwright, novelist, screenwriter. He was educated at Harvard and worked for *Time Magazine* and *The New Yorker*. His first novel, *A Walk in the Sun* (1944), was filmed in 1945, and afterwards he worked on numerous Hollywood scripts, especially war movies. He won an Academy Award for co-writing *A Place in the Sun* (1951), and he also wrote *Ocean's Eleven* (1960), among others. In the early 1950s, he worked at Twentieth Century-Fox and MGM, and he was married for a few years to Marguerite Lamkin. Later he married June de Baum. His other novels are *The Stars in Their Courses* (1968), *A Quiet Place to Work* (1968), and *The Wild Hunt* (1973); he published five volumes of poetry. He appears in *D.1* and *D.2*.

Brown, Jerry (b. 1938). Governor of California from 1975 to 1983, and again from 2011, and the only son of the previous Governor Brown. When Isherwood met him in November 1979, he was running for president against Jimmy Carter,

Brown's second of three unsuccessful attempts to secure the Democratic nomination. He went to a Roman Catholic high school, St. Ignatius, run by Jesuits, and to a Jesuit college, Santa Clara University. He then joined a Jesuit seminary, Sacred Heart Novitiate, in 1958 with the intention of becoming a priest. But he left the seminary to study classics at Berkeley, and later got a degree in law from Yale.

Brown, Rick. Aspiring actor, born and raised in rural West Virginia. He left high school to join the navy, married and had a son before divorcing. In 1971, he had an affair with Truman Capote after they met in a bar on the West Side of Manhattan where Brown then worked. Capote introduced him to Isherwood.

Brown, Susan (b. 1946). English actress. She had a few stage and film roles and appeared on British T.V. in "Making Out," "Prime Suspect," "Casualty," "East Enders" and "The Ruth Rendell Mysteries."

Browne, Coral. See Price, Vincent and Coral Browne.

Buchholz, Horst (1933–2003). German stage and screen actor, son of a shoemaker. He starred in European films in the 1950s, then achieved Hollywood fame as a gunslinger in *The Magnificent Seven* (1960). His wife, Myriam Bru (b. 1932), was an actress, and, later, a talent agent in Paris.

Buckingham, Bob (1904–1975) and May (1908–1997). British policeman and his wife, a nurse. E.M. Forster met and fell in love with Bob Buckingham in 1930. In 1932 they made a radio broadcast together, for a BBC series "Conversations in the Train," overseen by J.R. Ackerley who had introduced them. When Buckingham then met and married May Hockey, it caused turmoil in his relations with Forster, but the three eventually established a lifelong intimacy. Forster even gave the Buckinghams an allowance as they grew older. In 1951, Buckingham retired from the police force, joined the probation service, and settled with May in a new post in Coventry in 1953. They had one son, Robin, who married and had children of his own, before dying in the early 1960s of Hodgkins Disease. The Buckinghams appear in *D.2*.

Buckle, Christopher Richard Sanford (Dicky) (1916–2001). British ballet critic and exhibition designer; educated at Marlborough and, for one year, Oxford. He was ballet critic for *The Observer* from 1948 to 1955 and for *The Sunday Times* from 1959 to 1975. He designed an influential exhibition about Diaghilev and the Ballets Russes in 1954, a less successful one about Shakespeare in 1963, and a Cecil Beaton exhibition at the National Portrait Gallery in 1968. He also created the 1976 exhibition "Young British Writers of the Thirties," at which Isherwood spoke at the Portrait Gallery. He published biographies of Nijinsky (1971), Diaghilev (1979), and Balanchine (1988), as well as three volumes of autobiography. He appears in *D.2*.

Buddha Chaitanya (Buddha). An American disciple of Swami Prabhavananda; born Philip Griggs. He lived as a monk both at the Hollywood Vedanta Society and at Trabuco during the 1950s and took brahmacharya vows with John Yale in August 1955, becoming Buddha Chaitanya. In 1959 he left Vedanta for a time, but eventually took sannyas and became Swami Yogeshananda. Later he led a Vedanta group in Georgia. He appears in *D.1* and *D.2*.

Burroughs, William (1914–1997). American writer and painter, born in St. Louis, educated at Harvard; he lived with Jack Kerouac in Greenwich Village and for many years abroad in Paris, Mexico City, Tangier and elsewhere. He was also close to Allen Ginsberg and Byron Gysin. His extreme and destructive bohemianism, obsessed by heroin and sadistic sex, is charted in his novels, *Junkie* (1953), *The Naked Lunch* (1959), *The Wild Boys* (1971), and *Queer* (1985, written in 1953), among others. *A Report from the Bunker* (1981) was about his life in New York in the 1970s.

Burton, Richard (1925–1984). British actor, born Richard Jenkins in a Welsh coal-mining village; he took the surname of his English master and guardian, Philip Burton. He made his professional stage debut in the early 1940s then served in the air force and briefly studied English at Oxford, where he acted in Shakespeare. He starred in *Hamlet* in London and New York in 1953 and 1954, followed by other Shakespearian roles and, later, the Broadway musical *Camelot* (1960) and *Equus* (1976). His films include *My Cousin Rachel* (1952), *The Robe* (1953), *Alexander the Great* (1956), *Look Back in Anger* (1959), *Becket* (1964), *The Night of the Iguana* (1964), *The Spy Who Came in from the Cold* (1965), *Who's Afraid of Virginia Woolf* (1966), *The Taming of the Shrew* (1967), *Where Eagles Dare* (1969), *Equus* (1977), and *California Suite* (1978). He was nominated seven times for an Academy Award. His first wife was a Welsh actress, Sybil Williams, with whom Isherwood met him in the late 1950s, and with whom he appears in *D.1*. He began a tempestuous public romance with Elizabeth Taylor when he played opposite her in *Cleopatra* (1963); they married twice, in 1963 and 1975, and divorced both times. His third wife, whom he married in 1976, was an English model, Susan Hunt, and he met his fourth wife, Sally Hay, when she worked as an administrator on his television film "Wagner"; they married in 1983. His legendary drinking hastened his death. As Isherwood records in *D.2*, he worked for Burton on a screenplay of "The Beach of Falesá" by Robert Louis Stevenson, but nothing came of the project. Around the same time, the Burtons loaned their Hampstead house to Don Bachardy when he studied at the Slade, and Isherwood lived there with Bachardy during 1961.

Butazolidin. Brand name for phenylbutazone, a nonsteroidal, anti-inflammatory and analgesic drug.

Byrne, John (b. 1945). English bibliophile and scholar, educated at Marlborough and Cambridge, where he abandoned his classics scholarship after two years. He worked on manuscripts and archives for Bertram Rota from 1967 until 1995, and occasionally contributed reviews to *The Book Collector* and the *Times Literary Supplement*. He edited *James Stern: Some Letters for His Seventieth Birthday* (1974), prepared and introduced an edition of Lord Berners's roman à clef, *The Girls of Radcliff Hall* (2000), and wrote *Aiding & Abetting: An Alphabet for AIDS*.

Cadmus, Paul (1904–1999). American painter of Basque and Dutch background, trained by his parents and at the National Academy of Design in New York; he joined the U.S. government Public Works of Art Project in 1933. Lincoln Kirstein became interested in his work in the mid-1930s and later married Cadmus's sister, Fidelma, also a painter. Jon Andersson, a model and would-be singer and Cadmus's long-time companion, was the subject of many of his works. Cadmus and Andersson eventually settled in Weston, Connecticut. As Isherwood tells in *D.1*, Cadmus drew Isherwood in February 1942 in New York, and the two became friends.

Caldwell, George and Philadelphia. Neighbors, across the street and two doors east, on Adelaide Drive. As Isherwood records, he and Bachardy met with them several times in the spring of 1976, in preparation for the March 15 Coastal Commission hearing at which they obtained the permit to enlarge Bachardy's studio. The Caldwells initially opposed the building work, which was roughly between their own house, in which she grew up, and the ocean view.

Calley, John (1930–2011). American film producer and studio executive, born in New Jersey. He began his career in T.V. in the 1950s, worked in advertising, and then became an executive at Filmways, Inc. From 1968 to 1981, he was a senior executive at Warner Brothers and president for a time. During the 1980s, he was an independent producer, working closely with Mike Nichols, and then he ran United Artists and Sony Pictures. Isherwood worked for him when Calley was co-producer with Haskell Wexler of *The Loved One* (Neil Hartley was associate producer). Calley's other films include *Ice Station Zebra* (1968), *Catch-22* (1970), *Postcards from the Edge* (1990), *The Remains of the Day* (1993), *Goldeneye* (1995), *Closer* (2004), and *The Da Vinci Code* (2006). He appears in *D.2*.

Calthrop, Gladys (1894–1980). British artist and stage designer; she designed most of Noël Coward's plays and films from the 1920s onward.

Cameron, Roderick (Rory) (1913–1985). American shipping heir; he published history and travel books about Africa, India, the South Pacific, South America, Australia, and the French Riviera where he lived. After World War II, he and his often-married Australian socialite mother—when she was Countess of Kenmare—restored La Fiorentina and its gardens at St. Jean on Cap Ferrat. As Isherwood records, the villa was near Somerset Maugham's house, Villa Mauresque. In the 1960s, Cameron and his mother sold La Fiorentina, and he moved into a smaller adjacent property, an eighteenth-century farmhouse, Le Clos Fiorentina, the oldest house on Cap Ferrat. This is the house Isherwood visited with Hockney and Schlesinger in 1970.

Camilla. See Clay, Camilla.

Campbell, Alan (1904–1963). Actor and screenwriter, second husband of Dorothy Parker. They first married in 1933, divorced after the war, and later remarried, eventually settling on Norma Place, West Hollywood, the heart of Boys Town, as it was known among the gay community (Campbell was rumored to be homosexual). They worked on more than a dozen screenplays together, including, with Robert Carson, *A Star Is Born* (1937) and Lillian Hellman's film adaptation of her play *Little Foxes* (1941). Campbell was involved with the Hollywood Anti-Nazi League and other leftist causes, and he was blacklisted by the House Un-American Activities Committee during the 1950s. He appears in *D.2*.

Campbell, Brian. American model, actor, women's clothing designer. He was Mike Van Horn's boyfriend for a time from the mid-1970s. Campbell fell ill with

AIDS, and, although they broke up soon afterwards, Van Horn looked after him during his last months. He sat for Bachardy twice.

Capote, Truman (1924–1984). American novelist, born in New Orleans; his real name was Truman Persons. In *Lost Years*, Isherwood describes meeting Capote in the Random House offices in May 1947 shortly before the publication of Capote's first novel, *Other Voices, Other Rooms*. They quickly became friends, and Capote also appears in *D.1* and *D.2*. Capote wrote for *The New Yorker*, where he worked in the early 1940s, and other magazines. His books include *The Grass Harp* (1951), *Breakfast at Tiffany's* (1958), and the non-fiction novel *In Cold Blood* (1966). He never finished his last novel, *Answered Prayers*, though a chapter, "La Côte Basque, 1965," was published in *Esquire* magazine in 1975, forever alienating rich and powerful friends like the Paleys who were portrayed in it. The rest of what he had written of the novel was published posthumously. Capote's companion for many years was Newton Arvin, a college professor; afterwards, he lived and travelled with Jack Dunphy, and then later picked up new boyfriends with increasing frequency, including John O'Shea. Drink and drugs hastened his death.

Caron, Leslie (b. 1931). French dancer and actress. Her father was a chemist, her American-born mother a dancer. Caron studied ballet from childhood, performed in Paris as a teenager, and was discovered by Gene Kelly, who made her a star in *An American in Paris* (1951). Isherwood first met her during the 1950s, when she was appearing in Hollywood musicals such as *The Glass Slipper* (1955) and *Gigi* (1958); he mentions her in *D.1*, and she appears in *D.2*. She received British Film Academy Awards and was nominated for Academy Awards for *Lili* (1953) and *The L-Shaped Room* (1962). Later, she appeared in *Damage* (1992), *Jean Renoir* (1993), *The Reef* (1997), *Chocolat* (2000), and *Le Divorce* (2003). She also worked on the stage in New York, London, and Paris. She was married briefly to George Hormel in the early 1950s, then for ten years to British director Peter Hall, with whom she had two children. Her third husband, from 1969 to 1980, was American producer Michael Laughlin.

Carroll, Nellie (d. 2005). American artist, born Jean Dobrin; she designed and drew greeting cards. She was a close friend of Jim Bridges and Jack Larson. Bachardy drew and painted her many times after they met in 1963. She married once and had a daughter, Amy, who died of cancer in the early 1990s. For the last forty or so years of her life, she lived with a Mexican man about fifteen years her junior, who also had a wife and son. They appear in *D.2*.

Carson, Joanne. Pan Am stewardess and RKO starlet. She was "Tonight Show" host Johnny Carson's second of four wives. They separated in 1970 and divorced in 1972. She launched her own talk show and later became a therapist. She was one of Truman Capote's most loyal friends, and he died in her house while she was with him.

Carter, Jimmy. In his diary entry for February 2, 1980, Isherwood mentions his anxiety that Carter's response to the Soviet invasion of Afghanistan will lead to a nuclear face-off. On February 1, *The New York Times* reported on a seventy-page Pentagon study, "Capabilities in the Persian Gulf" (completed before the invasion of Afghanistan), which said that the U.S. could not repel a Soviet thrust into northern Iran without the possible use of tactical nuclear weapons. Carter had already imposed a grain embargo on the USSR, and his new military budget was the largest proposed in fifteen years. U.S. military presence in the Gulf had been building ever since Iranian militants took more than sixty Americans hostage in the U.S. Embassy in Tehran on November 4, 1979. The longterm concern was protecting the flow of oil from the region.

Caskey, William (Bill) (1921–1981). American photographer, born and raised in Kentucky; a lapsed Catholic of Irish background, part Cherokee Indian. Isherwood met him in 1945 when Caskey arrived in Santa Monica Canyon with a friend, Hayden Lewis, and joined the circle surrounding Denny Fouts and Jay de Laval. They became lovers in June that year and by August had begun a serious affair. Caskey was briefly in the navy during World War II and was discharged neither honorably nor dishonorably (a "blue discharge") following a homosexual scandal in which Hayden Lewis was also implicated. Caskey's father bred horses, and Caskey had ridden since childhood; he had worked in photo-finish at a Kentucky racecourse, and in about 1945 he took up photography seriously. He took portraits of his and Isherwood's friends, and he took the photographs for *The Condor and the Cows*, which Isherwood dedicated to Caskey's mother, Catherine. Caskey's parents were divorced, and he was on poor terms with his father and two sisters. He and Isherwood split in 1951 after intermittent separations and domestic troubles. Later, he lived in Athens and travelled frequently to Egypt. As well as taking photographs, he made art objects out of junk, and for a time had a business beading sweaters. There are many passages about him in *D.1*, some in *D.2*, and he is a main figure in *Lost Years*.

chaddar. A length of cloth worn on the upper body, often draped on the shoulders as a shawl, by monks and nuns of the Ramakrishna Order and by many other Hindus. Some Western Vedantists meditate in it, to keep warm, and to conceal their rosary.

Chaikin, Joe (1935–2003). American actor, stage director, writer. In the early 1960s, he founded the Open Theater which became one of the most influential experimental groups in New York. In the late 1970s, he wrote plays with Sam Shepard.

Chamberlain, Richard (b. 1935). American actor and singer; born in Beverly Hills and educated at Pomona College before serving in Korea for a year and a half. He became famous in the series, "Dr. Kildare," in which he starred from 1961 to 1966, and he never shook off the role, despite ambitious appearances on the London stage (for instance as Hamlet) and in a number of films, including *The Madwoman of Chaillot* (1969), *Julius Caesar* (1970), *The Three Musketeers* (1974), and *The Four Musketeers* (1975). He returned to T.V. successfully in the miniseries "Shogun" (1980) and "The Thorn Birds" (1985). He appears in *D.2*.

Chandlee, William H., III (Will). American historian; raised in Philadelphia, son of an architect. He was Assistant Chief of the Division of Education at the Philadelphia Museum of Art, organized lecture series there, and wrote seasonal publications. He joined Woodfall Productions for a time, then returned to Philadelphia and ran an antique store in nearby Germantown.

Chapin, Douglas (Doug) (b. 1950). American actor and film and T.V. producer, younger brother of actor Miles Chapin and son of Betty Steinway, a descendant of the founder of the piano company. He appeared on T.V. in the 1960s and 1970s. Later, with his partner and companion Barry Krost, he produced *When a Stranger Calls* (1979), *American Dreamer* (1984), *What's Love Got to Do with It* (1993), *Love! Valour! Compassion!* (1997), and the T.V. series "Dave's World."

Chapman, Hester (1895–1976). Novelist and biographer of royal figures such as Anne Boleyn, Caroline Matilda of Denmark, and the Duke of Buckingham. She was a cousin of Dadie Rylands, a habitué of Bloomsbury, and a longtime friend of Rosamond Lehmann. With her first husband, she ran a boys' prep school in Devon. Her second husband, Ronnie Griffin, a banker, died in 1955. She appears in *D.2*.

Charlton, Jim (1919–1998). American architect, from Reading, Pennsylvania. He studied at Frank Lloyd Wright's Taliesin West in Arizona and also at Wright's first center, Taliesin, in Wisconsin. He joined the air force during the war and flew twenty-six missions over Germany, including a July 1943 daylight raid. Isherwood was introduced to him by Ben and Jo Masselink in August 1948 (Ben Masselink had also studied at Taliesin West), and they established a friendly–romantic attachment that lasted many years. Towards the end of the 1950s, Charlton married a wealthy Swiss woman called Hilde, a mother of three; he had a son with her in September 1958. The marriage ended in divorce. Afterwards he lived briefly in Japan and then, until the late 1980s, in Hawaii, where he wrote an autobiographical novel, *St. Mick*. Charlton was a model for Bob Wood in *The World in the Evening*. He appears in *D.1*, *D.2*, and *Lost Years*.

Chester. See Kallman, Chester.

Chetanananda. Indian monk of the Ramakrishna Order. He was sent to Hollywood at the start of the 1970s to be Swami Prabhavananda's assistant in place of Vandanananda. When Prabhavananda died, Chetanananda was too young and inexperienced to take over; Belur Math appointed Swahananda, with Chetanananda continuing as assistant. The arrangement was not successful, and Chetanananda was eventually appointed head of the St. Louis center. He often travelled back to southern California, spending several months a year at Laguna Beach. Isherwood wrote forewords for two volumes edited by Chetanananda, *Meditation and Its Methods* (1976), a selection on meditation from the works of Vivekananda, and *Vedanta: Voice of Freedom* (1986), a selection from Vivekananda's lectures with a preface by Huston Smith. He also helped Chetanananda with his translation of the Avadhuta Gita of Dattatreya, published in India by Advaita Ashrama.

Chetwyn, Robert (b. 1933). British director; educated at the Central School of Speech and Drama; he began his career as a repertory actor in the early 1950s. From 1960 onward, he directed for the stage in London and abroad, including works by Shakespeare, Shaw, and Wilde, Terence Frisby's *There's a Girl in My Soup*—which he took from the West End to Broadway in 1967—and, in 1979,

Martin Sherman's *Bent* at the Royal Court. He also directed and produced for BBC television and for ITV. In 1970, when he worked with Isherwood on the proposed London production of *A Meeting by the River*, he was also considering directing Lynn Redgrave in Michael Frayn's series of short plays, *The Two of Us*, and planning to take his West End production of Tom Stoppard's *The Real Inspector Hound* to New York, but these deals fell through.

Childers, Michael (b. 194[4]). American photographer; born in North Carolina, educated at UCLA Film School. He was the longtime companion of film director John Schlesinger, whom he met in 1968. Schlesinger asked him to assist on *Midnight Cowboy* (1969), and they afterwards collaborated on other productions. Childers was a photographer for Andy Warhol's magazines *Interview* and *After Dark* from their inception and for *Dance* magazine, and he created the multimedia presentation used in *Oh! Calcutta!* He was also the photographer for a number of productions at the National Theatre in London during the 1970s. Other work includes hundreds of record album sleeves; film posters and stills for Hollywood productions such as *Grease*, *Marathon Man*, *The Year of Living Dangerously*, *Coal Miner's Daughter*, *The Terminator*, and *Torch Song Trilogy*; and covers for *Life*, *GQ*, *Esquire*, *Vogue*, *Elle*, and *Paris Match*.

Clark, Ossie (1942–1996) and Celia (b. 1941). British dress designer and his wife, British fabric designer Celia Birtwell. He studied at the Manchester School of Art, where he met David Hockney in 1961, and later at the Royal College of Art when Hockney was also there. She trained at the Salford School of Art and later taught at the Chelsea College of Art. He used her fabrics in his designs and together they ran Quorum, a popular clothing boutique. Hockney was best man at the Clarks' wedding in 1969, and Celia posed for him so often that she is sometimes called his muse. During 1970 and 1971, Hockney made a large painting of them with their cat, *Mr. and Mrs. Clark and Percy*, now in the Tate. The Clarks had two sons before separating. Later, she ran her own fabric shop in London and launched a clothing line.

Clarke, Gerald (b. 1937). American journalist and biographer; raised in California, educated at Yale, settled in New York. He wrote magazine profiles of artists and celebrities for many years and then published *Capote: A Biography* (1988) and *Get Happy: The Life of Judy Garland* (2000), both adapted as films.

Claxton, Bill (1927–2008) and Peggy Moffitt (b. 1939). He was a photographer known for his work with musicians and actors. His wife, Peggy Moffitt, was a model and actress. She was muse to fashion designer Rudi Gernreich, modelling his topless bathing suit in the mid-1960s, and she had a small role in Antonioni's *Blow-up* (1965). In 1991, they published *The Rudi Gernreich Book*, full of Claxton's photographs of Gernreich's designs worn mostly by Moffitt in her signature white pancake makeup with heavily blacked eyes. One shot shows Bachardy drawing her portrait while Gernreich looks on. Isherwood met Bill Claxton through Jim Charlton, and he appears in *D.1*. They both appear in *D.2*.

Clay, Camilla (d. 2000). American stage director. She assisted Ellis Rabb at the APA Repertory Company in 1966 and occasionally later. In 1967, she assisted José Quintero when he directed O'Neill's *More Stately Mansions* with Ingrid Bergman and Colleen Dewhurst at the Ahmanson before bringing it to New York. From 1967 to 1972, she rented a house in Malibu with writer Linda Crawford and before that, briefly, they lived at the Château Marmont. In 1972, the pair moved back east where Clay directed *Cabaret* at a community theater on the North Fork of Long Island in 1974 and *Stuck* by Sandra Scoppettone in 1976. She lived in Los Angeles again for a few years from 1979 onward before finally settling in New York, where she died of cancer. Isherwood met her through Gavin Lambert. She appears in *D.2*.

Clement. See Scott Gilbert, Clement.

Clytie. See Alexander, Clytie.

C.O. Conscientious objector.

Cockburn, Jean. See Ross, Jean.

Cohan, Robert (Bob) (b. 1925). American dancer from New York; he was a longtime soloist with the Martha Graham Dance Company. In the late 1950s, he also became a choreographer and teacher, launching his own school and company in Boston, where he joined the faculty of Harvard's Loeb Drama Center. Then, in 1966–1967, he founded the London Contemporary Dance Theatre and School, bringing Martha Graham's technique to Europe. Isherwood generally misspelled his surname as Coh*e*n.

Cohen, Andee (b. *circa* 1946). American photographer. She began taking pictures of her friends—actors, artists, and rock-and-roll musicians in London and Los Angeles—when her boyfriend, James Fox, gave her a camera in 1966. Her work appeared on album covers for Frank Zappa, Joe Cocker, Tom Petty, and others. Later she married Rick Nathanson, a film producer. She appears in *D.2*.

Collier, John (1901–1980). British novelist and screenwriter. He is best known for *His Monkey Wife* (1930) and also wrote other fantastic and satirical tales. Isherwood admired his short stories. Collier was poetry editor of *Time and Tide* in the 1920s and early 1930s and came to Hollywood in 1935. Isherwood met him in the 1940s, perhaps at Salka Viertel's, and they became close friends while working at the same time at Warner Brothers during 1945. In 1951, Collier moved to Mexico, though he continued to write films, including the script, deplored by Isherwood in *D.1*, for the film version of *I Am a Camera*. He appears in *Lost Years*.

Connolly, Cyril (1903–1974). British journalist and critic; educated at Eton and Oxford. He was a regular and prolific contributor to English newspapers and magazines, including *The New Statesman*, *The Observer* (where he was literary editor in the early 1940s) and *The Sunday Times*. He wrote one novel, *The Rock Pool* (1936), followed by collections of criticism, autobiography, aphorisms, and essays—*Enemies of Promise* (1938), *The Unquiet Grave* (1944), *The Condemned Playground* (1945), *Previous Convictions* (1963), and *The Evening Colonnade* (1973). In 1939, he founded *Horizon* with Stephen Spender and edited it throughout its publication until 1950. He was perhaps the nearest "friend" of Isherwood and Auden who publicly criticized their decision to remain in America during World War II. He blamed them for abandoning a literary-political movement which he was convinced they had begun and were responsible for. Connolly married three times: first to Jean Bakewell, who divorced him in 1945, then to Barbara Skelton from 1950 to 1956, and finally, in 1959, to Deirdre Craig with whom he had a son, Matthew, and a daughter, Cressida. From 1940 to 1950 he lived with Lys Lubbock, who worked with him at *Horizon*; they never married, but she changed her name to Connolly by deed poll. He appears in *D.1*, *D.2*, and *Lost Years*.

Cooper, Billy. See McCarty-Cooper, Billy.

Cooper, Douglas (1911–1984). London-born heir to an Australian fortune, which he spent on Modernist art, especially early French Cubism. He was a curator at the Mayor Gallery in London, investigated Nazi art thieves after World War II, and lectured and wrote widely on Léger, Picasso, Juan Gris, and others. Among the works at his Château de Castille, and which Isherwood mentions, was a design commissioned by him from Picasso and sandblasted onto one of the walls. American decorator Billy McCarty was a longtime companion, and Cooper adopted him in 1972.

Cooper, Gladys (1888–1971). British stage and film star; she was a teenage chorus girl, World War I pin-up, and silent film actress before establishing her reputation on the London stage. As Isherwood tells in *D.1*, he first met her in Los Angeles in 1940 when she was past fifty and had made few films. She had a supporting role in *Rebecca* that year and afterwards appeared in *The Song of Bernadette* (1943), *Green Dolphin Street* (1947), *The Secret Garden* (1949), *Madame Bovary* (1949), *The Man Who Loved Redheads* (1955), *Separate Tables* (1958), and *My Fair Lady* (1964), among many others. She also appears in *D.2*.

Cooper, Wyatt (1927–1978). Actor, screenwriter, editor, from Mississippi; educated at Berkeley and UCLA. He appeared on stage and T.V., had a small role in *Sanctuary* (1961), and wrote the screenplay for *The Chapman Report* (1962). In *D.1*, Isherwood describes meeting Cooper when Cooper was involved with Tony Richardson; he also appears in *D.2*. He became the fourth husband of Gloria Vanderbilt (b. 1924), only granddaughter of Cornelius Vanderbilt and in girlhood the subject of a headline-making custody battle between her widowed, reportedly lesbian mother and her forceful, richer aunt, Gertrude Vanderbilt Whitney, who raised her in lonely splendor on Long Island. At seventeen, Gloria married Pasquale "Pat" DiCicco, a Hollywood agent; at twenty-one she inherited four million dollars. Her two other husbands were conductor Leopold Stokowski and film director Sidney Lumet. She again made her name a household word with her designer jeans in the 1980s. She had two sons with Cooper: the younger one, Carter, committed suicide in 1988; the older one is CNN anchor Anderson Cooper. Beginning in 1969, Wyatt Cooper edited a magazine, *Status*, to which, as Isherwood mentions, Bachardy contributed an interview and, later, some drawings, but the magazine ceased publication before the work appeared. Cooper also wrote *Families: A Memoir and a Celebration* (1978).

Copland, Aaron (1900–1990). American composer; born and raised in Brooklyn, trained in New York and Paris. He drew on jazz, folk, and other popular idioms to give his orchestral and choral works an American character. *Appalachian Spring* (1944), composed for Martha Graham's dance company, won a Pulitzer Prize. His film score for *The Heiress* (1949) won an Academy Award, and three of his many later film scores were also nominated. His numerous other works include *Our Town* (1940), *Rodeo* (1942), *The Tender Land* (1954), and *Lincoln Portrait* (1942).

Corcoran, James and Dagny. Los Angeles gallery owner and art dealer and his wife. He took over Nicholas Wilder's gallery in 1979 and for several years represented Bachardy. She is an art collector and founder of the West Hollywood bookstore Art Catalogues which specializes in current and out-of-print exhibition catalogues and books on modern art and photography. She is the daughter of real estate developer and art collector Edwin Janss, Jr. who helped expand Sun Valley and Snowmass ski resorts and whose father and grandfather developed parts of Los Angeles and its suburbs and donated 385 acres to UCLA for its campus.

Coricidin. A brand-name cold remedy. Some versions contain a cough suppressant (dextromethorphan) attractive to recreational drug users and dangerous in combination with the antihistamine ingredient (chlorphenamine maleate). When Isherwood used it, Coricidin contained a decongestant (pseudoephedrine) and not an antihistamine. Some versions also contain an analgesic (acetaminophen), for fever and pain.

Cotten, Joseph (1905–1994). American actor. He worked on Broadway from the early 1930s and played the lead opposite Katharine Hepburn in *The Philadelphia Story* in 1939 and 1940. He was also a member of Orson Welles's Mercury Theater from 1937 to 1939, and Welles brought him to Hollywood to appear in *Citizen Kane* (1941). He went on to star in *The Magnificent Ambersons* (1942) and *Journey Into Fear* (1943) for Welles and then, for Hitchcock, in *Shadow of a Doubt* (1943). His many other films include *Portrait of Jennie* (1948), *The Third Man* (1949), and *Hush ... Hush, Sweet Charlotte* (1964). He appears in *D.1* with his first wife, Lenore Kipp; she was wealthy in her own right and a friend to Isherwood and Bachardy until her death from leukemia in 1960. The year Lenore died, Cotten married British actress Patricia Medina (b. 1920), who was in *The Three Musketeers* (1948) and Welles's *Mr. Arkadin* (1955) and starred with Cotten on Broadway. They appear in *D.2*.

Courtenay, Tom (b. 1937). British actor, educated at the University of London and RADA; he came to prominence in Richardson's 1962 film *The Loneliness of the Long Distance Runner.* His many other stage and film roles include Pasha in David Lean's *Doctor Zhivago* (1965), *The Dresser,* which he took from stage to screen in 1983, and *Last Orders* (2001).

Craft, Robert (Bob) (b. 1923). American musician, conductor, critic, and author; colleague and surrogate son to Stravinsky during the last twenty-three years of Stravinsky's life. Isherwood first met Craft with the Stravinskys in August 1949 when Craft was about twenty-five years old and had been associated with the Stravinskys for about eighteen months. Craft was part of the Stravinsky household, and travelled everywhere with them, except when his own professional commitments prevented him. Increasingly he conducted for Stravinsky in rehearsals and supervised recording sessions, substituting entirely for the elder man as Stravinsky's health declined. In 1972, a year after Stravinsky's death, Craft married Stravinsky's Danish nurse, Alva, who had remained with Stravinsky until the end, and they had a son. Craft published excerpts from his diaries as *Stravinsky:*

Chronicle of a Friendship 1948–1971 (1972; expanded and republished 1994), edited three volumes of *Selected Correspondence* by Stravinsky, which appeared in 1981, 1984, and 1985, and produced other books arising from his relationship with the Stravinskys as well as articles, essays, and reviews on musical, literary, and artistic subjects. He appears in *D.1*, *D.2*, and *Lost Years*.

Crawford, Joan (1904–1977). American movie star, born in Texas and discovered by MGM in a Broadway chorus line. Isherwood first met her when he worked on the script for *A Woman's Face* in 1940–1941, and she appears in *D.1*. Her many films include *Possessed* (1931 and 1947), *Grand Hotel* (1932), *Rain* (1932), *Mildred Pierce* (1945, Academy Award), *Sudden Fear* (1952), and *Whatever Happened to Baby Jane?* (1962). She married four times, finally to Pepsi chairman Alfred Steele in 1956, and when he died, she joined the board of Pepsi. She is the subject of *Mommie Dearest* (1978), by her adopted daughter Christina Crawford, made into a 1981 film starring Faye Dunaway.

Crawford, Linda (b. 1938). New York writer. She shared a house in Malibu with Camilla Clay from 1967 to 1972, then returned to New York and began to publish novels in the mid-1970s; they include *In a Class by Herself* (1976), *Something to Make Us Happy* (1978), *Vanishing Acts* (1983), and *Ghost of a Chance* (1985). She appears in *D.2*.

Cribb, Don. American photographer, born in North Carolina; he got a film degree from USC in 1969. Later, he restored antiques. He founded the Santa Ana Council of Arts and Culture, an arts advocacy group behind the creation of the Grand Central Arts Center and Santa Ana Arts Village and behind the Orange County Center for Contemporary Art. He posed for Hockney during the 1970s and again in the late 1990s, and he sat often for Bachardy in the mid-1970s.

Cukor, George (1899–1983). American film director. Cukor began his career on Broadway in the 1920s and came to Hollywood as a dialogue director on *All Quiet on the Western Front* (1930). In the thirties he directed at Paramount, RKO, and then MGM, moving from studio to studio with his friend and producer David Selznick. He directed Garbo in *Camille* (1936) and Hepburn in her debut, *A Bill of Divorcement* (1932), as well as in *Philadelphia Story* (1940); other well-known work includes *Dinner at Eight* (1933), *David Copperfield* (1934), *A Star is Born* (1954), and *My Fair Lady* (1964). Isherwood tells in *D.1* that he met Cukor at a party at the Huxleys' in December 1939. Later they became friends and worked together. Cukor also appears in *Lost Years* and in *D.2*.

Cullen, John (1909–1977). British publisher, educated at Cambridge. He worked as a schoolmaster before going into the educational side of Methuen in the 1930s. He returned as General Editor after serving in World War II, developed Methuen's list of British, Irish, and European literature, and launched the paperback originals series, Methuen Modern Plays, beginning with Shelagh Delaney's *A Taste of Honey*. He was Isherwood's editor after Alan White retired, until he himself retired in 1976.

Curtis Brown. Isherwood's first literary agency in London and in New York, from the mid-1930s. In September 1935, Curtis Brown's London office oversaw the contract between Isherwood and Methuen committing Isherwood to deliver his next three full-length novels to Methuen; 1935 is also the year Isherwood first appears on the books of Curtis Brown in New York. (At the time, Isherwood was still being published by the Hogarth Press, also represented by Curtis Brown, but Methuen began publishing him with *Prater Violet* after the war.) Isherwood evidently formed a relationship with Curtis Brown's play department for *The Ascent of F6* in the mid-1930s, and Curtis Brown also represented Auden from about this time. In the New York office, Alan Collins was Isherwood's agent until 1959 when Perry Knowlton took over and continued as Isherwood's American agent until 1973. In the mid-1970s, Isherwood left Curtis Brown, New York, for Candida Donadio. But he stayed with Curtis Brown in London, where he was represented by John Barber, James McGibbon, Richard Simon, Peter Grose, and Anthea Moreton-Saner.

Cuthbertson, Tom (194[5]-2005). American writer. He studied German at U.C. Santa Cruz and got a masters in English at San Francisco State University. He was a conscientious objector during the Vietnam War. Isherwood read several of his stories as a favor to Peter Schlesinger, praised them, and later met Cuthbertson and his first wife, Pat Zylius. In 1972, Cuthbertson, a cyclist, published a best-selling book on bicycles, *Anybody's Bike Book*, which gave him financial independence.

Dambacher, David. American artist. Once a year, he created a tableau honoring Marilyn Monroe, either on her birthday or on the anniversary of her death. He became a professional trucker with his friend Gene Martin, who was about fifteen years older than he. They were sex addicts and told stories of elaborate seductions on the road. Both died of AIDS.

Dare, Diana. Secretary to Oscar Lewenstein at Woodfall Productions. Later, she was the second wife of British fashion photographer Terence Donovan.

darshan. In Hinduism, a blessing or sense of purification which is achieved by paying a ceremonial visit to a holy person or place; also, the ceremonial visit itself.

David. See Hockney, David.

Davidson, Gordon (b. 1933). Theater director, raised in Brooklyn. He worked as stage manager and director at the American Shakespeare Festival, where he met John Houseman who invited him to UCLA to work with the Theater Group in 1964. In 1967, the Theater Group moved into the new Mark Taper Forum, and Davidson became artistic director. He continued in the job for thirty-eight years, opening the new Ahmanson Theater in 1989 and the Kirk Douglas Theater in 2004. By the time he retired, he had won eighteen Tony Awards, three Pulitzer Prizes, and sent thirty-five productions to Broadway. He also won a Margo Jones Award for his contribution to the development of American regional theater.

Davis, Ronald (Ron) (b. 1937). American painter and lithographer, born in Santa Monica and raised in Wyoming, where he studied at the University of Wyoming before attending the San Francisco Art Institute in the early 1960s. He was influenced there by Abstract Expressionism. His first show was at Nicholas Wilder's gallery in 1965. In 1966, he taught at the University of California at Irvine and had a solo show at the Tibor de Nagy Gallery in New York, followed in 1968 by a show at Leo Castelli. His optical geometric works soon began to be acquired by major museums, including the Museum of Modern Art, the Tate, the Los Angeles County Museum, and the Art Institute of Chicago.

Day-Lewis, Cecil (1904–1972). Irish-born poet, novelist, translator, editor; educated at Sherborne School and Oxford, where he became friends with Auden and, through him, met Isherwood. He was a schoolmaster during the 1930s, wrote for leftist publications and joined the Communist party in 1936. His poetry from the period reflects his political involvement, but he later returned to personal themes and abandoned his radical opinions during World War II. He wrote roughly twenty detective novels under a pseudonym, Nicholas Blake, as well as three autobiographical novels, and he published verse translations of Virgil and of Paul Valéry. He was Professor of Poetry at Oxford from 1951 to 1956 (just before Auden) and became poet laureate of Britain in 1968. He married twice, the second time to actress Jill Balcon (b. 1925), with whom he had three children (one is the actor Daniel Day-Lewis); he had two children with his first wife. He also had a long affair with Rosamond Lehmann during his first marriage. He appears in *D.1* and *D.2*.

Deepti. See High, Beth.

Dehn, Paul (1912–1976). British film critic, playwright, lyricist, librettist, and screenwriter; raised in Disley, Greater Manchester, and educated at Oxford. He won an Oscar for his first film story, *Seven Days to Noon* (1951), co-authored with his live-in partner, film composer James Bernard (1925–2001), and he won a British Film Academy Award for *Orders to Kill* (1958). He later worked on *Goldfinger* (1964), *The Spy Who Came in from the Cold* (1965), *Beneath the Planet of the Apes* (1970), and the many *Planet of the Apes* sequels. His lyrics for musicals and films include the song for the film version of *I Am a Camera*.

"de Laval, Jay" (probably an assumed name). American chef; he adopted the role of the Baron de Laval. In the mid-1940s he opened a small French restaurant on the corner of Channel Road and Chautauqua in Santa Monica, Café Jay, frequented by movie stars seeking privacy. In 1949, he opened a second restaurant in the Virgin Islands, and in 1950 he was briefly in charge of the Mocambo in Los Angeles before opening a grand restaurant in Mexico City. There, he also planned interiors with Mexican designer Arturo Pani and created a menu for Mexicana Air Lines and crockery for Air France. Isherwood met de Laval through Denny Fouts. He was a lover of Bill Caskey before Isherwood and a friend of Ben and Jo Masselink. He appears in *D.1* and *Lost Years* and is mentioned in *D.2*.

de Velasco, Adolfo. Spanish-born antique dealer and socialite, friendly with the Moroccan royal family. He ran two shops, one in Tangier and one in the Hotel La Mamounia in Marrakech, and he also had two palaces, one in the Casbah in Tangier and the other in Marrakech. The palaces were lavishly decorated with oriental antiques, and he threw stupendous parties in them. He died in the 1990s.

Dexamyl. Dextroamphetamine, an antidepressant or upper, combined with amobarbitol, a barbiturate to offset its effect. Isherwood was introduced to it by Bachardy, and they both used it when tackling big swathes of work; for many years they shared Bachardy's prescription since neither of them relied on it habitually. Bachardy was first given it at eighteen by a friend, Alex Quiroga.

(Dexedrine, which Isherwood also mentions, is a brand name for a preparation of dextroamphetamine without the barbiturate.)

dharma. Vocation, duty, including religious duty, station or role in life; also morality, righteousness.

Dharmadas. Jon Monday, an American devotee of Prabhavananda. His wife, Anna Monday, was known as Urbashi, and they were both members of the Venice Group. They had a daughter, Rachel. He built a recording studio at Takoma Records from which he ran MondayMedia, and where, with Charlie Mitchell, he recorded the 1979 L.P. *The Bhagavad Gita: The Song of God, Selections Read by Christopher Isherwood*. Isherwood read from the translation he made with Swami Prabhavananda. Indian flutist Harprasad Chaurasias contributed musical interludes, accompanied by Rijram Desad and Sultan Khan. The L.P. was later remastered as a C.D.

di Capua, Michael. Isherwood's editor at Farrar, Straus Giroux, where di Capua worked from 1966 until 1991. Afterwards, he moved with Isherwood's work to HarperCollins until 1999. Di Capua also published children's books by prizewinning authors and illustrators such as Jules Feiffer, Randall Jarrell, Maurice Sendak, and William Steig. He took his imprint, Michael di Capua Books, to Hyperion in 1999 and then to Scholastic in 2005, but Isherwood remained at HarperCollins.

Didion, Joan (b. 1934). American writer, raised in the Sacramento Valley and educated at Berkeley. Her career began as a *Vogue* staff writer and film critic. She is known for her essays on American cultural decline collected in *Slouching Towards Bethlehem* (1968), *The White Album* (1979), and *After Henry* (1992), and for the journalism and critical pieces she still contributes to *The New Yorker* and *The New York Review of Books*. Her best-selling novels include *Run River* (1963), *Play It as It Lays* (1970), and *The Book of Common Prayer* (1977). She wrote film scripts with her husband, John Gregory Dunne. After his death, she published a memoir of her grief, *The Year of Magical Thinking* (2005), staged as a play starring Vanessa Redgrave. Five weeks before the memoir came out, their only child, a daughter called Quintana Roo, died after being critically ill off and on for some years; Didion incorporated Quintana Roo's death into the play.

Diebenkorn, Richard (1922–1993). American painter, born in Portland, educated at Stanford, at the California School of Fine Arts in San Francisco, and at the University of New Mexico. He travelled and studied widely, absorbing influences from Edward Hopper, Rothko and Clyfford Still (with whom he taught in the 1950s), Arshile Gorky, de Kooning and, later, Matisse, among others. His style evolved in alternating phases of figurative and non-figurative interest so that he helped to launch a West Coast movement in Abstract Expressionism and later was at the center of the Bay Area Figurative Movement. He taught at the California School of Fine Arts, the University of Illinois, Stanford, and in 1966 moved to Santa Monica to take a job at UCLA. His studio in Ocean Park became the focus of his last development as an abstract painter with the Ocean Park series, which he continued to work on into the late 1980s.

Diehl, Digby. American book critic, mostly for *The Los Angeles Times* where he was the founding editor of *The Los Angeles Times Book Review*. He has written and co-written over three dozen books, including the autobiographies *Million Dollar Mermaid* (1999), about Esther Williams, and *Angel on My Shoulder* (2000), about Natalie Cole, and the novel *Soapsuds* (2005) with actress Finola Hughes. He is also a broadcast commentator on media and entertainment.

Dill, Guy (b. 1946). American sculptor, educated at the Chouinard Art Institute. By the 1970s, he was having several one-man shows each year, and he has won major fellowships and prizes. He taught sculpture at UCLA, where he was head of the department from 1978 to 1982. His work is in the Los Angeles County Museum, the Norton Simon Museum, the Minneapolis Institute of Art, the Smithsonian, the Guggenheim, the Whitney, and the Museum of Modern Art in New York, and many smaller public and private collections. He is the younger brother of Laddie John Dill.

Dill, Laddie John (b. 1943). American painter and sculptor; older brother of Guy Dill. Born in Long Beach, California; trained at the Chouinard Art Institute. He ran a framing company with Chuck Arnoldi while he was still at Chouinard, and afterwards became an apprentice printer at Gemini in West Hollywood, where he assisted Robert Rauschenberg, Claes Oldenburg, Roy Lichtenstein, and Jasper Johns. He lived with Johns in New York for a few months during the 1970s. His father was a scientist who designed lenses, and Dill refers to that as an inspiration for his interest in technical experimentation and in unconventional materials such as neon and argon tubing, sand, cement, and plate glass. He has had countless one-man shows, and his work is held by many major museums.

Divine (1945–1988). Transvestite singer and actor; his real name was Harris Glenn Milstead. He appeared in John Waters's underground films, including *Pink Flamingos* (1972), *Female Trouble* (1974), *Polyester* (1981), and the mainstream hit *Hairspray* (1988).

DMSO. Dimethyl sulfoxide, a by-product of wood pulp manufacture, used as a solvent and, experimentally from about 1963, as a topical pain-killer.

Dobbin. A pet name for Isherwood, known in his lifetime only to himself and Bachardy. Other names included Dubbin, Dub, Drubbin, and Drub, all associated with his private identity as a reliable, stubborn old workhorse.

Dobyns, Dick. A friend of Paul Millard who lived in and helped to manage the apartment building Millard owned. He appeared in *D.1*.

Dodie. See Beesley, Alec and Dodie Smith Beesley.

Don. See Bachardy, Don.

Donadio, Candida (1929–1971). Isherwood's New York literary agent, from the mid-1970s when she opened her own agency. Later, she formed a partnership with Eric Ashworth and finally with Neil Olson. Her other clients included Thomas Pynchon, Joseph Heller, for whom she sold *Catch-22*, and Philip Roth, for whom she sold *Goodbye, Columbus*. The first book she handled for Isherwood was *Christopher and His Kind*, which she sold away from Simon & Schuster to Farrar, Straus and Giroux. After her death from cancer, Neil Olson ran the Donadio and Olson Literary Agency on his own and continued to represent Isherwood. He also became a novelist with *Icon* (2005).

Donoghue, Mary Agnes (b. 194[6]). American screenwriter and director; born in Queens. She first moved to Los Angeles to become Assistant Director of Publicity at the Los Angeles County Museum of Art and later scripted *The Buddy System* (1984), *Beaches* (1988), *Paradise* (1991), which she also directed, *White Oleander* (2002), and *Veronica Guerin* (2003), among others. She lived with artist Joe Goode for several years and then married British writer Chris Robbins in 1976.

Doone, Rupert (1903–1966). English dancer, choreographer and theatrical producer; founder of The Group Theatre, for which Isherwood and Auden wrote plays in the 1930s. His real name was Reginald Woodfield. The son of a factory worker, he ran away to London to become a dancer, and then went on to Paris where he was friendly with Cocteau, met Diaghilev, and turned down an opportunity to dance in the corps de ballet of the Ballets Russes. He was working in variety and revues in London during 1925 when he met Robert Medley, his longterm companion. He died of multiple sclerosis after years of increasing illness. He appears in *D.1*, *D.2*, and *Lost Years*.

Downer, The. See Franklin, Bill.

Druks, Renate (1921–2007). Austrian-American painter, actress, film director, scenic designer; born in Vienna, where she studied at the Vienna Art Academy for Women. Later, she studied at the Art Students League in New York. She settled in Malibu in 1950. Her paintings were mostly allegorical portraits of women friends—Anaïs Nin, Joan Houseman, Doris Dowling—in naive, magic-surrealist style. Druks had a role in Kenneth Anger's experimental film *Inauguration of the Pleasure Dome* (1954) and in other underground films. She often sat for Don Bachardy. She appears in *D.2*.

Dub, Dub-Dub, Dubbin. Isherwood; see under Dobbin.

Dunne, John Gregory (1932–2003). American writer, raised in Connecticut and educated at Princeton. Younger brother of novelist and journalist Dominick Dunne. He worked at *Time Magazine* until 1964, when he married Joan Didion and moved with her to Los Angeles to write screenplays, including *Panic in Needle Park* (1971), *Play It as It Lays* (1972), adapted from one of Didion's novels, *A Star is Born* (1976), and *True Confessions* (1981), adapted from one of his own novels. Dunne wrote two non-fiction books about Hollywood, *The Studio* (1969) and *Monster: Living Off the Big Screen* (1997), as well as an autobiography, *Vegas* (1974), and political and cultural essays.

Dunphy, Jack (1914–1992). American dancer and novelist; born and raised in Philadelphia. He danced for George Balanchine and was a cowboy in the original production of *Oklahoma!* He was married to the Broadway musical-comedy star Joan McCracken, and from 1948 he became Truman Capote's companion, although in Capote's later years they were increasingly apart. He published *John Fury* (1946) and *Nightmovers* (1967). He appears in *Lost Years* and is mentioned in *D.2*.

Dupuytren's Contracture. A disease of the hand in which the connective tissue underneath the skin of the palm and fingers develop fibrous bumps or cords. The cords gradually shorten, contracting the fingers into a bent position so they cannot be straightened. It usually affects only the third and fourth fingers. Cortisone injections can alleviate the condition and wearing a splint at night can slow its progress, but eventually, the bumps and cords have to be removed surgically, in particular to prevent the middle joint of the fingers from becoming fixed in a bent position and in severe cases, where nerve and blood supplies are cut off, to prevent the fingers from requiring amputation. The cords can grow back after surgery and are more difficult to remove the second time. The disease is more frequent in men than in women, and more common in middle age.

Duquette, Tony (191[4]–1999) and Elizabeth ("Beegle") (d. 1994). Los Angeles interior designers. Bachardy worked for them in 1955. They appear in *D.1*.

Durga. A name for the divine mother, consort of Shiva. She is shown with ten arms, riding a lion, and sometimes with her four children. Durga puja, the biggest of the Hindu festivals in Bengal, is celebrated in September or October over a ten-day period. On the first day of the puja, the goddess leaves the Himalayas to visit her parents below. Clay idols of the goddess are newly made each year and stood on pandals; on the last day of the festival, the idols are immersed in the ocean or a river to symbolize Durga's return to her celestial abode. Vedanta Society devotees in southern California worship a photograph of Durga; the photo is not immersed, but stored for the following year.

Eckstein, Tony. See Abedha.

Edward. See Upward, Edward.

Elan, Joan (1929–1981). British actress. She settled in Hollywood after making her first film, *The Girls of Pleasure Island* (1953), but subsequent appearances in film and T.V. were undistinguished, apart from a small role in the Broadway production of Jean Anouilh's *The Lark*. She was a friend of Marguerite Lamkin, and in the mid-1950s she had a love affair with Ivan Moffat. She frequently sat as a model for Don Bachardy in the 1950s and early 1960s. Later she married an advertising executive, Harry Nye, and lived with him in New York until she died young, of a heart attack. She appears in *D.1*.

Elsa. See Lanchester, Elsa.

Emily, or Emmy. See Smith, Emily Machell.

Energy Crisis. Energy prices rose two hundred percent during the autumn of 1973 after Arab oil-exporting countries imposed an embargo in retaliation for U.S. support of Israel in the Yom Kippur War. Long lines at gas stations became ubiquitous, and stringent energy-saving measures were implemented throughout the cold winter of 1973–1974. But on the very day that Isherwood mentions the fuel shortage, January 6, 1974, *The New York Times* reported that large quantities of crude oil could be seen flowing from tankers into U.S. refineries in New Jersey and that the major U.S. oil companies were refusing to tell government officials or anyone else how much refined oil they were producing and how much they already had stored. This contributed to widespread suspicion that the oil companies were contriving the shortage to keep prices high for their own benefit.

Ennis, Bob. Black actor and dancer, his professional name was Exotica. He was tall, glamorous, and feminine looking. He had a part in Jim Bridges's *The Babymaker*, and he also worked as an extra cameraman and miscellaneous crew on other films.

Epstein, Barbara (1928–2006). American editor and journalist, born in Boston and educated at Radcliffe. She was one of the five founders, during the *New York Times* strike in 1963, of *The New York Review of Books*, and she co-edited it with Robert Silvers for forty-three years. Before that she was a book editor, oversaw Anne Frank's *Diary of a Young Girl* (1952) for Doubleday, and worked at *The Partisan Review*. She was married from 1954 to 1980 and had a son and a daughter.

Evans, Gregory. Longtime companion and assistant to David Hockney; raised in Minnesota. Isherwood met him in Los Angeles, where he surfaced as a lover of Nick Wilder. He became Hockney's lover by 1974, his most frequent model and, later, managed Hockney's studios, travels, and complex array of interests and commitments. He had problems with alcohol and drugs until the mid-1980s, when he joined Alcoholics Anonymous and changed his way of life.

Exotica. See Ennis, Bob.

Fairfax, James (b. 1933). Australian art collector and philanthropist, educated at Oxford. He was the last family chairman, from 1957 to 1987, of the Fairfax newspaper empire. He appears in *D.2*.

Falk, Eric (1905–1984). English barrister, raised in London. Falk, who was Jewish, was a school friend from Repton, where he was in the same house as Isherwood, The Hall, and in the History Sixth. He helped Isherwood edit *The Reptonian* during Isherwood's last term, and they saw one another during the school holidays and often went to films together. Falk introduced Isherwood to the Mangeots, whom he had met on holiday in Brittany. He appears in *Lions and Shadows*, *D.1*, *D.2*, and *Lost Years*.

Faye, Alice (1915–1998). American actress, singer, comedienne; born and raised in New York, where she went on the stage at fourteen. She starred in Hollywood musicals from 1933 to 1945—including *Every Night at Eight* (1935), *Poor Little Rich Girl* (1936), *Alexander's Ragtime Band* (1938), *The Gang's All Here* (1943)—but quit movies over conflicts with Darryl Zanuck at Twentieth Century-Fox and focused back on radio and stage. She made only a few further films. In 1972 and 1973, she revived the musical *Good News* on Broadway, then toured in it for a year, including to Los Angeles. She was a childhood favorite of Bachardy. She appears in *D.2*.

Finney, Albert (b. 1936). English actor, trained at RADA; son of a bookie. He acted in Shakespeare from the mid-1950s for the Birmingham Repertory Theatre and came to prominence on the London stage in *Billy Liar* (1960). Afterwards, he appeared in several John Osborne plays directed by Tony Richardson, receiving great praise for *Luther* in 1961, and taking the role to Broadway in 1963. In 1965, he joined the National Theatre Company and appeared in Peter Shaffer's *Black Comedy* (1965) and Peter Nichols's *A Day in the Death of Joe Egg* (1967); then, after a hiatus, he returned to the company to star in *Hamlet*, *Tamburlaine*, *Macbeth*, and others. His film career was launched with *Saturday Night and Sunday Morning* (1960), and he became an international star in Richardson's *Tom Jones* (1963). His other films include: *The Entertainer* (1960), *Night Must Fall* (1964), *Scrooge* (1970),

Murder on the Orient Express (1974), *The Dresser* (1983), *Under the Volcano* (1984), *The Browning Version* (1994), *Erin Brokovich* (2000), and *Traffic* (2000). He appears in *D.2*.

Finney, Brian (b. 1935). British scholar and professor of literature, educated at the University of Reading and the University of London. He served in the Royal Air Force and worked in the electrical and telephone industry then, in 1964, took an administrative job in the extension section of the University of London, where he went on to teach for many years. He got his Ph.D. at Birkbeck College in 1973. In the summer of 1976, he was a visiting professor at UCLA, and was already at work on *Christopher Isherwood: A Critical Biography* (1979). In 1987, he emigrated permanently to southern California, where he has taught at U.C. Riverside, UCLA, USC, and U.C. Long Beach. His work on twentieth-century British fiction includes books and articles on Beckett, D.H. Lawrence, and Martin Amis. His first wife, Marlene (1936–1989), a figurative painter, died of cancer. Bachardy drew her in 1976. His second wife is a photographer and floral designer.

Firth, Peter (b. 1935). British child actor, born and raised in Yorkshire. He starred in *Equus* on the London stage in 1973, on Broadway, and in the film adaptation. In 1976 he appeared in John Osborne's BBC T.V. adaptation of *The Picture of Dorian Gray* and in *Aces High*, as Isherwood mentions; during the same year, he made *Joseph Andrews* (1977) for Tony Richardson. Other films include Zeffirelli's *Brother Sun, Sister Moon* (1972), Polanski's *Tess* (1979), *The Hunt for Red October* (1990), and *Shadowlands* (1993). He has appeared often on British T.V.

Flamini, Roland and Janet. Maltese journalist and author and his British wife, née Morton. He was a foreign correspondent for *Time Magazine* from 1968 to 1994 and a bureau chief in Rome, Bonn, Paris, Lebanon, Jerusalem, and for the European Community, which he covered from London. When Isherwood mentions him in 1976, he was finishing his tour as Hollywood correspondent and had recently published his first book, *Scarlett, Rhett, and a Cast of Thousands, the Filming of Gone with the Wind* (1975). He wrote eight more non-fiction books and went on to become Chief International Correspondent for UPI. He and his wife had two children.

Foch, Nina (1924–2008). American actress, born in Holland and raised in Manhattan. Her films include *The Return of the Vampire* (1944), *Johnny Allegro* (1949), *An American in Paris* (1951), *Scaramouche* (1952), *Executive Suite* (1954, Academy Award, Best Supporting Actress), *The Ten Commandments* (1956), *Spartacus* (1960), and *Mahogany* (1975). She had roles on Broadway, was a member of the American Shakespeare Festival, and appeared regularly on T.V. in John Houseman's "Playhouse 90," "The Outer Limits," and others. She also directed for stage and screen and, from the 1960s, taught acting at USC and at the American Film Institute. Her third husband, from 1967 to 1993, was stage producer Michael Dewell (b. 1931). She appears in *D.2*.

Fonda, Jane (b. 1937). American actress, born in New York, raised in Hollywood and Greenwich, Connecticut, educated at Vassar; daughter of actor Henry Fonda and his socialite second wife, Frances Seymour Brokaw, who committed suicide in 1950. She worked as a model before studying with Lee

Strasberg at the Actors Studio. Attracted to the vanguard of cultural trends, she opposed the Vietnam War during the 1960s, toured American G.I. camps with Donald Sutherland and other actors as the Anti-War Troop and, in 1972, travelled through North Vietnam followed by press and making radio broadcasts. In 1988, she apologized publicly for supporting the enemy and allowing herself to be photographed at the controls of a North Vietnamese anti-aircraft gun. Her films include *Walk on the Wild Side* (1962), *Period of Adjustment* (1962), *The Chapman Report* (1962), *Cat Ballou* (1965), *Barefoot in the Park* (1967), *They Shoot Horses, Don't They?* (1969), *Klute* (1971; Academy Award and New York Film Critics Award), *Vietnam Journey: Introduction to the Enemy* (1974), a documentary about her visit to North Vietnam, co-directed with Tom Hayden and Haskell Wexler, *Julia* (1977), *Coming Home* (1978, Academy Award), *California Suite* (1978), *9 to 5* (1980), and *On Golden Pond* (1981). She married three times: in 1965 to Roger Vadim, who directed her in *La Ronde/Circle of Love* (1964) and *Barbarella* (1968), then from 1973 to 1990 to political activist Tom Hayden, and from 1991 to 2001 to CNN tycoon Ted Turner. She had one child with Vadim and another with Hayden. She appears in *D.2*.

Fontan, Jack. American actor, artist, astrologer. As Isherwood tells in *Lost Years*, he became known to queers as "The Naked Sailor" when he appeared in *South Pacific* in nothing but cut-off blue jean shorts which displayed his magnificent physique. He sprawled center stage during the song "What ain't we got? We ain't got dames!" and attracted complaints from ladies in the front rows who could see his genitals, but the director, Joshua Logan, refused to alter his costume. He was photographed in youth by George Platt Lynes. Once a lover of Bill Harris, he spent fifty-three years with his companion, Ray Unger.

Forbes, Bryan (b. 1926). English actor, director, producer, screenwriter, novelist; born in London and educated at RADA. He worked on the stage from seventeen, had film roles during the 1950s, and appeared in *The Guns of Navarone* (1961) and *A Shot in the Dark* (1964). From the 1960s, he turned mostly to directing—including *The L-Shaped Room* (1962), *King Rat* (1965), *The Madwoman of Chaillot* (1969), and *The Stepford Wives* (1975)—and contributed some of his own screenwriting and producing. His second wife is the English actress Nanette Newman (b. 1934), with whom he has two daughters, Emma, an actress, and Sarah, a fashion journalist. He appears with his wife in *D.2*.

Ford, Glenn (1916–2006). Canadian-born actor raised in Santa Monica. He was already making movies by 1939, served in the marines during World War II, and afterwards became a star opposite Rita Hayworth in *Gilda* and opposite Bette Davis in *A Stolen Life*, both in 1946. Among his many other films are *The Big Heat* (1953), *The Blackboard Jungle* (1955), *Cimarron* (1961), *The Courtship of Eddie's Father* (1963), *The Four Horsemen of the Apocalypse* (1962), and *Midway* (1976). He also acted in T.V. films and in the series "Cade's County" (1971) and "The Family Holvak" (1975). As Isherwood tells in *D.2.*, they met in 1960, when Ford was beginning a love affair with Hope Lange. From 1943 to 1959 Ford was married to the American tap dancer, Eleanor Powell (1910–1982); they had a son, Peter Ford. Later, Ford was married to actress Kathryn Hays from 1966 to 1968, and then to actress Cynthia Hayward from 1977 to 1984, and to Jeanne Baus from 1993 to 1994.

Forster, E.M. (Morgan) (1879–1970). English novelist, essayist and biographer; best known for *Howards End* (1910) and *A Passage to India* (1924). He was an undergraduate at King's College, Cambridge, and one of the Cambridge Apostles; afterwards he became associated with Bloomsbury and later returned to King's as a Fellow until the end of his life. He was a literary hero for Isherwood, Upward, and Auden from the 1920s onward, and Isherwood regarded Forster as his master. They were introduced by William Plomer in 1932. Forster was a supporter when Isherwood was publicly criticized for remaining in America during World War II. He appears in *D.1*, *D.2*, and *Lost Years*. He left his papers and copyright to King's College with a life interest to his literary executor, the psychologist and translator of Freud, Professor W.J.H. Sprott (d. 1971). Isherwood was also named in his will, as the heir to the American rights of Forster's unpublished homosexual novel *Maurice*, written 1913–1914 and heavily revised 1959–1960; it was published posthumously in 1971 under Isherwood's supervision. Forster and Isherwood shared an understanding that any proceeds would be used to help English friends in need of funds for U.S. travel; Isherwood assigned the proceeds to the National Institute of Arts and Letters where an E.M. Forster Award was created to support English writers on extended visits to the U.S.

Fouts, Denham (Denny) (*circa* 1914–1948). Son of a Florida baker; he worked for his father as a teenager then left home to travel as companion to various wealthy people of both sexes. Among his conquests was Peter Watson, who financed *Horizon* magazine, and Fouts helped solicit some of the magazine's earliest pieces. During World War II, Watson sent Fouts to the U.S. with Jean Connolly, and she and Tony Bower introduced Fouts to Isherwood in mid-August 1940 in Hollywood. Fouts determined to begin a new life as a devotee of Swami Prabhavananda, but Swami would not accept him as a disciple, so, after a spell in the East, Fouts moved in with Isherwood in the early summer of 1941, and they led a spartan life of meditation and quiet domesticity. Isherwood describes this in *Down There on a Visit* where Fouts appears as "Paul," and there are many passages about Fouts in *D.1* and *Lost Years*. In August 1941, Fouts was drafted into Civilian Public Service camp as a Conscientious Objector; on his release in 1943, he lived with a friend from the camp while studying for his high-school diploma; afterwards he studied medicine at UCLA. In 1945 and 1946, Isherwood and Bill Caskey lived in Fouts's apartment at 147 Entrada Drive while Fouts was mostly away; eventually, when Fouts returned, Caskey quarrelled with him, ruining Isherwood's friendship. Soon afterwards, Fouts left Los Angeles for good. He became an opium addict in Paris, and Isherwood saw him there for the last time in 1948 before Fouts died in Rome.

Fox, James (Willie) (b. 1939). British actor, from childhood; his real name is William; he is a younger brother of actor Edward Fox. Tony Richardson gave him a small role in *The Loneliness of the Long Distance Runner* (1962), and he later starred in *The Servant* (1963), *The Chase* (1966), *Isadora* (1968), *Performance* (1970), *A Passage to India* (1984), *Absolute Beginners* (1986), *The Remains of the Day* (1993),

The Golden Bowl (2000), *Sherlock Holmes* (2009), and other films. He left acting for Christian evangelism for a time during the 1970s. He appears in *D.2*.

Fox, Lyle. He ran the gym in Pacific Palisades attended by Isherwood, Bachardy, and the Masselinks from the start of the 1960s. He was blond and muscular. In 1967, he married an attractive younger woman called Rez. Eventually he became personal trainer and masseur for Gregory Peck and travelled with Peck on location all over the world. He appears in *D.2*.

Frank. See Isherwood, Frank Bradshaw.

Frankenheimer, John (1930–2002). American movie director, educated at Williams. He served in an air force film squadron during the Korean War, worked in T.V. through most of the 1950s, and became famous in 1962, when he had three films released: *All Fall Down*, *Birdman of Alcatraz*, and *The Manchurian Candidate*. Other work includes *Seven Days in May* (1964), *The Train* (1964), *Grand Prix* (1966), *The Fixer* (1968), *I Walk the Line* (1970), *The Iceman Cometh* (1973), and *Reindeer Games* (2000).

Franklin. See Knight, Franklin.

Franklin, Bill. American would-be actor. He briefly worked as a movie critic. He died of AIDS around the year 2000.

French, Robin (b. 1936). Hollywood agent and, later, producer and T.V. syndicator; educated at boarding school in England and briefly at college in California. He worked with his father, Hugh, and increasingly represented Isherwood in the film business, taking over entirely in about 1970. He presided over the "incredible rights mess" of the play *I Am a Camera* and the musical and film *Cabaret*, securing Isherwood a substantial income for many years. By 1974, he left the agency business to become head of domestic production at Paramount Pictures. He later produced a few films, but worked primarily as a T.V. syndicator; eventually he operated and part-owned several T.V. stations before retiring in the late 1990s. French is mentioned in *D.1* and appears in *D.2*.

Freud, Lucian (1922–2011). British painter, born in Berlin, a grandson of Sigmund Freud. He emigrated with his family in 1933, and became an English citizen in 1939. He trained at the Central School of Art in London and at the East Anglian School of Painting and Drawing. His first solo show was in 1944 at the Alex Reid and Lefevre Gallery. He was a foremost figurative artist of his generation, especially of portraits and nudes, and, from the late 1980s, was internationally acclaimed. His second wife, from 1953 to 1957, was Caroline Blackwood, a subject of four early works. They both appear in *D.1*, but not together. He had two daughters with his first wife, Kitty Epstein (to whom he was married from 1948 to 1952), and a number of children with women he did not marry: a son, Alexander—whom Isherwood evidently met in June 1976—and four daughters with Suzy Boyt; two daughters and a son with Katherine McAdam, from whom he became estranged; two daughters, later well-known as fashion designer Bella Freud and novelist Esther Freud, with Bernardine Coverley. Most of these children were born in the late 1950s and early 1960s; more were reportedly born later.

Friebus, Florida (1909–1988). American actress and writer. She debuted on Broadway in an Ibsen revival in 1929 and appeared, among other things, in *Alice in Wonderland*, which she adapted for the stage with her longtime companion Eva Le Gallienne. She also had numerous T.V. roles, notably in "The Many Loves of Dobie Gillis" and "The Bob Newhart Show."

Fry, Basil. A bachelor cousin of Kathleen Isherwood, sixteen years younger than her; educated at Oxford. In 1928, Isherwood made his first trip to Germany to stay with Fry in Bremen, where Fry was British vice-consul. This visit was the basis for "Mr. Lancaster" in *Down There on a Visit*, and Fry was the model for Lancaster.

Furbank, P.N. (Nicholas, Nick) (b. 1920). English scholar, critic, editor, author. He was a longstanding friend of E.M. Forster and was asked by Forster to write his biography, which appeared in 1978. Furbank went on to edit, with Mary Largo, two volumes of Forster's letters. His other books include a prize-winning biography of Diderot and, with W.R. Owen, a biography of Defoe. He is a regular contributor to *The New York Review of Books* and other publications.

Gage, Margaret. A rich, elderly patroness of Gerald Heard; she loaned him her garden house on Spoleto Drive in Pacific Palisades, close to Santa Monica, from the late 1940s until the early 1960s. She also provided Will Forthman with a room in her house during the same period. She appears in *D.1* and *D.2*.

Gain, Richard (Dick). American ballet dancer. He danced in the original Broadway chorus of *Camelot* (1960) and in Martha Graham's company and later worked as a choreographer and teacher. He became friendly with Richard and Sybil Burton during *Camelot* and shared their Hampstead home with Isherwood and Bachardy in 1961, when he toured with Jerome Robbins's "Ballets: USA." Gain's friend, Richard (Dick) Kuch, also danced in *Camelot*. Eventually the pair moved to East Bend, North Carolina, to teach dance at the North Carolina School of the Arts, where Kuch became an assistant dean. They resigned in 1995. They appear in *D.2*.

Gambhirananda, Swami (1899–1988). Indian monk of the Ramakrishna Order; philosopher, scholar, translator. A powerful General Secretary of the Ramakrishna Order for many years, in charge of its practical daily operation. Then, from 1985 to 1988, he was the eleventh president of the order, the spiritual leader, with no role in temporal matters. He appears in *D.2*.

Garrett, Stephen and Jean. A British couple, a few years younger than Don Bachardy. He had an administrative job in Los Angeles. After she died of cancer, he moved back to England with their two daughters, and Isherwood and Bachardy lost touch with them.

Gates, Jim (1950–*circa* 1990). American non-conformist, violinist, monk of the Ramakrishna Order; born in Washington State and raised in Claremont, California; his father taught Latin and English. He moved out of his parents' house before the end of his sophomore year in high school and after high school went with Peter Schneider to Los Angeles, where they shared various living arrangements in Venice, San Marino and Hollywood, and where he eventually revealed to Schneider that he was gay. He was obsessed with Isherwood and hoped to run into him on the beach; Schneider looked up Isherwood's telephone number in the phone book and called him to explain this, and Isherwood invited them to Adelaide Drive. Gates attended Santa Monica College briefly and worked as a bus boy, library clerk, and live-in assistant to the husband of Marlene Dietrich. He also joined the Hollywood Vedanta monastery for a time. He died of AIDS. He appears in *D.2*.

Gavin. See Lambert, Gavin.

Gaynor, Janet (1906–1984). American film star. She appeared in her first movie in the 1920s and by 1934 was the biggest box office attraction in the U.S. Her films include *The Johnstown Flood* (1926), *Sunrise* (1927), *Seventh Heaven* (1927), *Street Angel* (1928), *State Fair* (1933), *A Star is Born* (1937), *The Young in Heart* (1938), and *Bernadine* (1957). For many years, she used the name of her second husband, fashion designer Gilbert Adrian (known as "Adrian," d. 1959), with whom she appears in *D.1*. In 1964, she married producer Paul Gregory. Gaynor was also an accomplished painter and showed her still lifes in New York in 1976. She appears with Gregory in *D.2*.

Geldzahler, Henry (1934–1994). American art historian and curator; son of a Belgian diamond dealer; educated at Yale and Harvard. He zealously promoted contemporary artists and supervised the creation of the Department of Twentieth Century Art at the Metropolitan Museum of Art, where he organized the major exhibition, "New York Painting: 1940–1970." He also served as the first director of the Visual Arts Program at the National Endowment for the Humanities, where he was responsible for grants to young artists. Later he became Commissioner of Cultural Affairs for New York City. Hockney drew and painted him a number of times.

Gerald. See Heard, Henry FitzGerald.

gerua. Hindi for ocher, the color of the cloth worn by monks and nuns who have taken their sannyas vows and symbolizing their renunciation.

Gielgud, John (1904–2000). British actor and director; born and educated in London, trained briefly at RADA. He achieved fame in the 1920s acting Shakespeare, Wilde, and Chekhov, and as a director he had his own London company from 1937. From the 1950s onward, he also worked with contemporary British playwrights, including Peter Shaffer, Alan Bennett and David Storey. He won three Tonys for his work on the New York stage. His movies, in which he often majestically played supporting and character roles, include: *Hamlet* (1939), *Julius Caesar* (1953), *Richard III* (1955), *Becket* (1964), *Hamlet* (1964), *The Loved One* (1965), *The Charge of the Light Brigade* (1968), *Oh! What a Lovely War* (1969), *Lost Horizon* (1973), *Murder on the Orient Express* (1974), *Joseph Andrews* (1977), *A Portrait of the Artist as a Young Man* (1979), *The Elephant Man* (1979), *Arthur* (1981, Academy Award), *Chariots of Fire* (1981), *Gandhi* (1982), *The Shooting Party* (1985), *Plenty* (1985), *Prospero's Books* (1991), *Hamlet* (1996), *The Portrait of a Lady* (1996), *Shine* (1996), *Elizabeth* (1998). During the 1970s and 1980s, he also worked in television, notably as Charles Ryder's father in the series "Brideshead Revisited." His companion in the 1950s was Paul Anstee, an interior decorator. In 1960, he met Martin Hensler at an exhibition at the Tate Gallery in London; Hensler, Hungarian by background, moved in with him about six years later, and they remained together for the last thirty years of Gielgud's life. Isherwood tells in *D.1* that he first met Gielgud in New York in 1947 and didn't like him; they met again in London in 1948 and became friends. Gielgud also appears in *Lost Years* and *D.2*.

Gilliatt, Penelope (1932–1993). English critic, novelist, screen writer; born Penelope Connor in London and briefly educated at Bennington College in Vermont. She was a staff writer for British *Vogue* and later for *Queen*, and by 1961 she was film critic for *The Observer*. Later she became widely known in America as film critic for *The New Yorker*. Her first husband, Roger Gilliatt, was a London neurologist and best man at the wedding of Princess Margaret to Antony Armstrong-Jones. In 1963, she married John Osborne and had a daughter with him; the marriage broke down by 1966, and Gilliatt settled in New York in 1967, where she had a relationship with the stage and film director Mike Nichols which lasted until 1969, followed by a brief affair with Edmund Wilson. She became an alcoholic, and her career at *The New Yorker* ended when she fabricated an interview with Graham Greene for the magazine. She wrote the prize-winning original screenplay for *Sunday Bloody Sunday* (1971), more than ten volumes of fiction including short stories, and several volumes of film history and criticism.

Ginsberg, Allen (1926–1997). American poet, born in New Jersey; educated at Columbia University; a member of the Beat scene in New York, San Francisco, and Paris in the 1950s and early 1960s, and a central figure in 1960s counterculture. He was a Zen Buddhist and campaigned against the Vienam War and in favor of drugs, communism, and homosexuality. He invented the phrase "flower power" and claimed he was the first to chant Hare Krishna in North America, through his friendship with Swami Prabhupada who launched the movement in the West. He was a poetic disciple of William Blake and of Walt Whitman and is best known for his early works *Howl and Other Poems* (1956) and *Kaddish and Other Poems* (1961). He published numerous volumes of poetry and also lectures, letters, and journals telling about his friendships with William Burroughs, Jack Kerouac, Neal Cassady, and with his longterm lover, American poet Peter Orlovsky (b. 1933), whom he first met in San Francisco in 1954. Orlovsky had dropped out of high school and served as a U.S. Army medic before becoming Ginsberg's secretary; he published several volumes of his own poetry. Ginsberg and Orlovsky appear in *D.2*.

Glade. See Bachardy, Glade.

Glaesner, Ole. Danish tailor and costume designer, from Copenhagen; son of a sailor. He was trained by Yves St. Laurent and rented a room near Patrick Procktor in Manchester Street, where he ran a business making trousers known for their immaculate fit. He was evidently Procktor's lover for a time.

Glenway. See Wescott, Glenway.

Goddard, Paulette (1911–1990). American film star, once a model and a Ziegfeld Girl. She was Charlie Chaplin's third wife, from 1933 to 1942, and became famous in his *Modern Times* (1936) and *The Great Dictator* (1940). She was reportedly cleverer than other Hollywood stars and was admired by intellectuals such as Aldous Huxley and H.G. Wells. Her third marriage was to Burgess

Meredith during the late 1940s, and she virtually retired after her fourth marriage in 1958 to German novelist and screenwriter Erich Maria Remarque (author of *All Quiet on the Western Front*). Isherwood met her soon after coming to Hollywood. She appears in *D.1*.

Goldin, Marilyn. American screenwriter. She co-wrote *Souvenirs d'en France* (1974) and *Barocco* (1976) with director André Téchiné in French with English subtitles, *Camille Claudel* (1988) with Bruno Nuytten in French, and *The Triumph of Love* (2001), based on Marivaux's play, with Clare Peploe and Bernardo Bertolucci. She also had a small role in *The Conformist* (1970).

Goode, Joseph (Joe) (b. 1937). American artist; born in Oklahoma and trained at the Chouinard Art Institute. He had his first solo show at a San Fransicso gallery in 1962 and was part of the early pop art exhibition at the Pasadena Art Museum, "New Painting of Common Objects," which included work by Lichtenstein, Warhol, and others. He has won a number of grants and awards, and his work is in the collections of the Museum of Modern Art and the Whitney in New York, the National Gallery and the Victoria and Albert Museum in London, the Art Institute of Chicago, the Los Angeles County Museum, the San Francisco Museum of Modern Art, and the Smithsonian, among others.

Goodstein, David (1932–1985). American lawyer, activist, publisher; educated at Cornell and at Columbia Law School. He made a fortune on Wall Street managing investment portfolios and used his money to promote gay rights. He hired James Foster to set up and run the Whitman-Radclyffe Foundation, and he organized the Committee for Sexual Law Reform, which campaigned, through Assembly Bill 489, towards the 1975 repeal of California's sodomy law. Also, in 1974, he bought *The Advocate*, which he built into the national gay news magazine.

Goodwill. Nonprofit provider of education and training for the poor, homeless, ill-educated, and physically and mentally disadvantaged in the U.S. It was founded in Boston in 1902 by a Methodist minister, Edgar J. Helms, who trained the poor to mend and resell used household goods and clothing.

Gordon, Bob. American writer, settled in San Francisco. He was a friend of Jack Larson and Jim Bridges, who introduced him to Isherwood.

Gore. See Vidal, Gore.

Gowland, Peter (b. 1916) and Alice. Photographer and camera maker and his wife, director of his photo shoots. He is known for his photographs of celebrities and his nudes, many of which have appeared as *Playboy* centerfolds. His Gowlandflex camera, designed in 1957, is still on the market and is widely used by professionals. The Gowlands were among the Masselinks' closest friends, and Isherwood met them through the Masselinks in the early 1950s. They have two daughters, Ann Gowland and Marylee Gowland. The Gowlands appear in *D.1* and *D.2*.

Granny Emmy. See Smith, Emily Machell.

Gregory, Paul (b. 1920). American film, T.V., and theater producer; born and raised in Iowa, and, briefly, in London. He acted in two films, then turned to booking and management. In 1950, he persuaded Charles Laughton to be his client and arranged tours for him, then T.V. appearances and stage and film productions for which Laughton acted, directed, and sometimes wrote material. Gregory's Broadway shows include *John Brown's Body* (1953) and *The Caine Mutiny Court-Martial* (1954); his movies, *The Night of the Hunter* (1955) and *The Naked and the Dead* (1958). He was the third husband of Janet Gaynor. He appears in *D.2*.

Griggs, Phil. See Buddha Chaitanya.

Grose, Peter. Isherwood's penultimate agent at Curtis Brown in London. He went on to found Curtis Brown Australia, then became a publisher at Secker & Warburg, and later a non-fiction author.

Grosser, Maurice (1903–1986). American painter and writer; raised in Tennessee and educated at Harvard. He wrote about art for *The Nation* and published a number of books including *The Painter's Eye* (1956), *Painting in Our Time* (1964), and *Painter's Progress* (1971). He was the longtime companion of Virgil Thomson; both were close friends of Paul and Jane Bowles, and Grosser lived partly in Tangier. He had a Manhattan apartment which he often loaned to Bachardy, on 14th Street between Seventh and Eighth Avenues. He appears in *D.2*.

Guerriero, Henry. American painter and sculptor; from Monroe, Louisiana, where he had known Marguerite and Speed Lamkin and Tom Wright, who were roughly his contemporaries in age. In Los Angeles, he moved in different circles with his companion Michael Leopold. Isherwood met him in the early 1950s, and he is mentioned in *D.1* and *D.2*. He changed his name to Roman A. Clef in 1978.

Guinness, Alec (1914–2000). English actor, born in London; his mother's name was de Cuffe; he never knew who his father was. He trained at the Fay Compton School of Dramatic Art and began his career on the London stage in the mid-1930s, appearing in Shakespeare, Shaw, and Chekhov at the Old Vic. During World War II, he served in the navy, and afterwards began making films, generally in comic or character roles. These include *Great Expectations* (1946), *Oliver Twist* (1948), *Kind Hearts and Coronets* (1949, in which he played eight parts), *The Lavender Hill Mob* (1951; Academy Award nomination), *The Bridge on the River Kwai* (1957, Academy Award), *The Horse's Mouth* (1958, for which he wrote the screenplay, nominated for an Academy Award), *Our Man in Havana* (1959), *Lawrence of Arabia* (1962), *Doctor Zhivago* (1965), *The Comedians* (1967), *Scrooge* (1970), *Brother Sun, Sister Moon* (1973), *Hitler, the Last Ten Days* (1973), *Murder by Death* (1976), *Star Wars* (1977, Academy Award nomination), *The Empire Strikes Back* (1980), *Return of the Jedi* (1983), *A Passage to India* (1984), and *Little Dorrit* (1987, Academy Award nomination). He also played the lead in the television miniseries of John Le Carré's *Tinker, Tailor, Soldier, Spy* and *Smiley's People*. He was knighted in 1959, and he was awarded an honorary, career Academy Award in 1980. His wife—Lady Guinness, once he was knighted—was Merula Salaman, an actress and later a painter and children's author. He appears in *D.2*.

guna. Any of three qualities—sattva, rajas, tamas—which together constitute Pakriti, or nature. When the gunas are perfectly balanced, there is no creation or manifestation; when they are disturbed, creation occurs. Sattva is the essence of form to be realized; tamas, the obstacle to its realization; rajas, the power by which the obstacle may be removed. In nature and in all created beings, sattva is purity, calm, wisdom; rajas is activity, restlessness, passion; and tamas is laziness, resistance, inertia, stupidity. Since the gunas exist in the material universe, the spiritual aspirant must transcend them all in order to realize oneness with Brahman.

Gunn, Thom (1929–2004). English poet. Thomson Gunn was educated at University College School, Bedales, and Cambridge. His father edited the London *Evening Standard*; his mother, also a journalist, committed suicide when he was fifteen. He contacted Isherwood in 1955 on his way from a creative writing fellowship at Stanford to a brief teaching stint in Texas; Isherwood invited him to lunch at MGM and they immediately became friends. Gunn later taught at Berkeley off and on from 1958 until 1999. His numerous collections of poetry include *Fighting Terms* (1954), *My Sad Captains* (1961), *Moly* (1971), *Jack Straw's Castle* (1976), *The Man with Night Sweats* (1992), and *Boss Cupid* (2000). He appears in *D.1* and *D.2*.

Guttchen, Otto. German refugee. Isherwood met him in Hollywood during World War II and writes about him in *D.1* and *D.2*. Guttchen's kidneys were badly damaged in a Nazi concentration camp, and he was also tortured. He left his wife and child in Switzerland. He struggled to find employment in Hollywood, was often too poor to eat, and became suicidal late in 1939. Isherwood found it difficult to help him adequately and felt intensely guilty about it. In the mid-1950s, they met again and Guttchen appeared to have regained his hold on life.

Hall, Michael. American actor and, later, antique dealer; he appeared in *The Best Years of Our Lives* (1946). In *Lost Years*, Isherwood describes how he met Hall at a party in the winter of 1945–1946 and began a friendship which lasted for twenty years and included occasional sex. Eventually, Hall left the West Coast and settled in New York. He also appears in *D.2*.

Hamilton, Gerald (1890–1970). Isherwood's Berlin friend who was the original for Mr. Norris in *Mr. Norris Changes Trains*. His mother died soon after his birth in Shanghai, and he was raised by relatives in England and educated at Rugby (though he did not finish his schooling). His father sent him back to China to work in business, and while there Hamilton took to wearing Chinese dress and converted to Roman Catholicism, for which his father, an Irish Protestant, never forgave him. He was cut off with a small allowance and eventually, because of his unsettled life, with nothing at all. So began the persistent need for money that motivated his subsequent dubious behavior. Hamilton was obsessed to the point of high camp with his family's aristocratic connections and with social etiquette, and lovingly recorded in his memoirs all his meetings with royalty, as well as those with crooks and with theatrical and literary celebrities. He was imprisoned from 1915 to 1918 for sympathizing with Germany and associating with the enemy during World War I, and he was imprisoned in France and Italy for a jewelry swindle in the 1920s. Afterwards, he took a job selling the London *Times* in Germany and became interested there in penal reform. Throughout his life he travelled on diverse private and public errands in China, Russia, Europe, and North Africa. He returned to London during World War II, where he was again imprisoned, this time for attempting to promote peace on terms favorable to the enemy; he was released after six months. After the war he posed for the body of Churchill's Guildhall Statue and later became a regular contributor to *The Spectator*. He appears in *Lost Years* and *D.1*, where Isherwood tells that at the start of the war, he sent Hamilton a letter which was quoted in William Hickey's gossip column in the *Daily Express*, November 27, 1939, without permission. In the letter, Isherwood mocked the behavior of German refugees in the U.S. His remarks, frivolously expressed for Hamilton's private amusement but fundamentally serious, seemed to Isherwood to have triggered the public criticism which continued into 1940 in the press and in Parliament, of both his own and Auden's absence from England. Hamilton also appears in *D.2*.

Hamilton, Richard (1922–2011). British painter and printmaker, one of the earliest and most influential pop artists. The subject of the seven paintings and two prints titled "Swingeing London 67"—which Isherwood mentions in his diary entry for March 17, 1970—is the Rolling Stones drug bust at Redlands. The judge pronounced jail terms for Jagger, Keith Richards, and Robert Fraser, who was Richard Hamilton's art dealer, telling them that "a swingeing sentence can act as a deterrent." Swingeing, pronounced with a soft g, means huge or daunting. (The prison sentences were overturned on appeal.)

Hansen, Erwin. German Communist, former army gymnastics instructor. Isherwood first met him at Magnus Hirschfeld's Institute, where Hansen did odd jobs until Francis Turville-Petre hired him as cook and housekeeper for the house he rented at Mohrin, outside Berlin, and where Isherwood went to live with them in 1932. Hansen hired Heinz Neddermeyer to assist him and thus introduced Heinz to Isherwood. And he travelled with them in 1933 when they fled Berlin for Turville-Petre's island in Greece. Isherwood thought Hansen died in a Nazi concentration camp during World War II.

Harford, Betty. See Andrews, Oliver and Betty Harford.

Hargrove, Marion (1919–2003) and Robin. Novelist and screenwriter, Marion Hargrove, and his second wife; born in North Carolina where he became a journalist while he was still in high school. He was drafted into the army during World War II and wrote the best-selling books *See Here, Private Hargrove!* (1942) and *What Next, Corporal Hargrove?* (1944), which became popular movies. Afterwards, he settled in Hollywood where he wrote the script for the film of *The Music Man* (1962) and episodes of "Maverick," "I Spy," and "The Waltons." The Hargroves lived on Adelaide Drive with their three children; he also had three other children from a previous marriage.

Haridas. American monk of the Ramakrishna Order, originally called Bill Stevens. During the 1990s, he went to the Chicago Vedanta Center and then to India where he took his sannyas and became Swami Harihareananda. Afterwards he lectured in Chicago and settled in Sacramento.

Harris, Bill (d. 1992). American artist, raised partly in the USSR and Austria. Harris painted in the 1940s and later made art-objects and retouched photographs.

Isherwood met him through Denny Fouts in 1943, while still living as a celibate at the Hollywood Vedanta Society; early in 1944 they began an affair which helped weaken Isherwood's determination to become a monk. Harris was a beautiful blond with a magnificent physique, and Isherwood found him erotically irresistible; the relationship soon turned to friendship, and Harris later moved to New York. Isherwood refers to Harris as "X." in his 1939–1945 diaries (see *D.1*), and he calls him "Alfred" in *My Guru and His Disciple*. Harris also appears in *Lost Years* and *D.2*.

Harris, Julie (b. 1925). American stage and film actress, born in Grosse Pointe, Michigan; educated at finishing school, Yale Drama School, and the Actors Studio. She became a star in the stage adaptation of Carson McCullers's *The Member of the Wedding* (1950), a status she confirmed when she originated the role of Sally Bowles in *I Am a Camera* (1951). She received a Tony Award for *Forty Carats* (1969) and for *The Last of Mrs. Lincoln* (1972), and toured with a one-woman show on Emily Dickinson, *The Belle of Amherst* (1976). Altogether, she has won five Tony Awards, more than any other actor, and she has been nominated ten times. She moved to the screen with early stage roles, receiving an Academy Award nomination for her film debut in *The Member of the Wedding* (1952); later Hollywood movies include *East of Eden* (1955), *Requiem for a Heavyweight* (1962), *The Haunting* (1963), *Harper* (1966), *Reflections in a Golden Eye* (1967), *The Bell Jar* (1976), and *Gorillas in the Mist* (1988). She has also been nominated for nine Emmy Awards, and won twice. During the 1980s, she appeared in the television series "Knots Landing." Isherwood first met her in 1951 after she was cast as Sally Bowles, and their close friendship is recorded in *D.1* and *D.2*. She was married to Jay Julien, a theatrical producer, and then to Manning Gurian, a stage manager and producer, with whom she had a son, Peter Gurian. She divorced Gurian in 1967 during a long affair with actor James (Jim) Murdock. In 1977 she married the writer William Carroll.

Harrison, Rex (1908–1990). English stage and film star, educated at Liverpool College. He made his stage debut in Liverpool at sixteen and was successful in the West End, on Broadway, and in films by the mid-1930s, especially in black-tie comedies. He married six times: to Marjorie Colette Thomas (1934–1942), to actresses Lilli Palmer (1943–1957), Kay Kendall (1957–1959), and Rachel Roberts, whom Isherwood mentions both as Rachel Harrison and as Rachel Roberts (1962–1971), to Elizabeth Harris, ex-wife of actor Richard Harris (1971–1975), and to Mercia Tinker (1978 until his death). His affair with another actress, Carole Landis, was presumed to have contributed to her suicide. He won a Tony Award as Henry Higgins in *My Fair Lady* (1956) on Broadway, and the 1964 film brought him an Academy Award. His other films, many of which also reprised stage roles, include *Blithe Spirit* (1945), *The Rake's Progress* (1945), *Anna and the King of Siam* (1946), *The Ghost and Mrs. Muir* (1947), *Cleopatra* (as Julius Caesar, 1963), *The Agony and the Ecstasy* (as Pope Julius, 1965), and *Doctor Dolittle* (1967). He appears in *D.2*.

Harrold, Greg (b. 1955). American organ builder, from Santa Monica. He began working in an organ building shop in 1974 and ran his own shop from 1979 to 2003. He also composed music, played several instruments, wrote poetry and painted. He introduced himself to Isherwood and Bachardy on the beach in 1975 or 1976. He and his longtime companion, Gordon Myers (1947–2001), sat for Bachardy several times in 1976 and 1977, and Harrold sat again in 2007. Myers was born in Kansas City, got a degree in architecture from Columbia University, and settled in the early 1970s in West Hollywood, where he worked in real estate and development and had his own firm for a while.

Hartley, Neil (1916–1994). American film producer, from North Carolina; Tony Richardson's collaborator in Woodfall Productions from 1965. He was production manager for Broadway impresario David Merrick, who imported several of Richardson's stage plays, and he met Richardson in 1958 at the Boston try-out for *The Entertainer*. The pair worked together for the first time on *Luther* when it opened in New York in 1963. *The Loved One* was the first film that Hartley produced for Richardson, and the partnership lasted until Richardson's penultimate film, *Hotel New Hampshire* (1984). Hartley also produced for T.V., including "The Corn Is Green" (1979) and several Agatha Christies. He was a semi-closeted homosexual and died of AIDS. His companion for a long time was Bob Regester. He appears in *D.2*.

Harvey, Anthony (Tony) (b. 1931). English actor turned film editor, then director. He directed his first film in 1967. He was mooted to direct *Cabaret* in 1969, following a popular success with *The Lion in Winter* (1968) and an Academy Award nomination. He turned down Isherwood and Bachardy's treatment for *Cabaret*, but in the end did not direct it anyway. He appears in *D.2*.

Hathaway, Henry (1898–1985). American director and producer, born Henri Leonard de Fiennes, the grandson of a Belgian aristocrat; his mother and father were actors. He acted and assisted in silent movies and began directing in 1932, mostly for Twentieth Century-Fox. His films include *The Lives of a Bengal Lancer* (1935), *The House on 92nd Street* (1945), *Kiss of Death* (1947), *The Desert Fox* (1951), *Niagara* (1953), *How the West Was Won* (1962), and *True Grit* (1969).

Hayden, Tom (b. 1939). American activist and writer; born in Detroit and educated at the University of Michigan, where he was editor of the *Michigan Daily* and a founder of Students for a Democratic Society. He participated in the Civil Rights Movement and opposed the Vietnam War. In 1968, he was arrested at the Democratic National Convention in Chicago and charged with six others with conspiracy and incitement to riot; the group attracted huge media attention as "the Chicago Seven." He travelled to Cambodia and North Vietnam, enemy territory, taking Jane Fonda with him on one trip in 1972 and again attracting the press and much controversy. Fonda became his second wife. He served in the California State Assembly and State Senate in the 1980s and 1990s. His books include *The Port Huron Statement* (1962), which set out the principles of the SDS, *Rebellion in Newark: Official Violence and Ghetto Response* (1967), and *Vietnam: The Struggle for Peace 1972–1973* (1973).

Hayworth, Rita (1918–1987). American movie star and World War II pinup. Her parents were dancing partners in the Ziegfeld Follies, and she was a professional dancer at twelve. Her career swelled and faded like her love life, but she was Hollywood's sex goddess during the 1940s. Her films included *Only Angels*

Have Wings (1939), *Blood and Sand* (1941), *You Were Never Lovelier* (1942), *Cover Girl* (1943), *Gilda* (1946), *Miss Sadie Thompson* (1953), *Pal Joey* (1957), and *Separate Tables* (1958). She had five husbands—including Orson Welles and Aly Khan, the son of the Aga Khan—and many lovers. She appears in *D.1* with her last husband, producer James Hill, and in *D.2*.

Heard, Henry FitzGerald (Gerald) (1889–1971). Irish writer, broadcaster, philosopher, religious teacher. Auden took Isherwood to meet him in London in 1932 when Heard was already well known as a science commentator for the BBC and author of several books on the evolution of human consciousness and on religion. A charismatic talker, he associated with some of the most celebrated intellectuals of the time. One of his closest friends was Aldous Huxley, whom he met in 1929 and with whom he joined the Peace Pledge Union in 1935 and then emigrated to Los Angeles in 1937, accompanied by Heard's friend Chris Wood and Huxley's wife and son. Both Heard and Huxley became disciples of Swami Prabhavananda. Isherwood followed Heard to Los Angeles and through him met Prabhavananda. Then Heard became an ascetic, rejecting association with women and criticizing Swami's insufficient austerity; he broke with Swami early in 1941, straining his friendship with Isherwood, and set up his own monastic community, Trabuco College, the same year. By 1949 Trabuco had failed, and he gave it to the Vedanta Society of Southern California to use as a monastery. In the early 1950s, Heard's asceticism relaxed, and he warmed again to his friendship with Isherwood and, later, Don Bachardy. During this period, he shared Huxley's experiments with mescaline and LSD.

He contributed to *Vedanta for the Western World* (1945) edited by Isherwood, and throughout most of his life he turned out prolix and eccentric books at an impressive pace, including *The Ascent of Humanity* (1929), *The Social Substance of Religion* (1932), *The Third Morality* (1937), *Pain, Sex, and Time* (1939), *Man the Master* (1942), *A Taste for Honey* (1942; adapted as a play by John van Druten), *The Gospel According to Gamaliel* (1944), *Is God Evident?* (1948), and *Is Another World Watching?* (1950), published in England as *The Riddle of the Flying Saucers*. For a number of years, he obsessively documented sightings of flying saucers, which he believed were either ultra-fast, experimental aircraft kept secret by the U.S. government or, more exciting to him, visitors from Mars.

Heard is the original of "Augustus Parr" in *Down There on a Visit* and of "Propter" in Huxley's *After Many a Summer* (1939). His role in Isherwood's conversion to Vedanta is described in *My Guru and His Disciple*, and he appears throughout *D.1*, *D.2*, and *Lost Years*.

Heflin, Lee. American writer and painter. He joined Isherwood's writing class at UCLA in the spring of 1964, when he was working at the university's vast new research library, later called the Charles E. Young Research Library, which was completed that year. Isherwood writes about their friendship in *D.2*.

Heilbrun, Carolyn (1926–2003). American literary scholar; born in New Jersey, educated in Manhattan and at Wellesley. She taught briefly at Brooklyn College and then, for thirty-three years, at Columbia University, where in 1972 she was the first woman to receive tenure in the English department. From the 1960s, she published detective novels under the pen name Amanda Cross. Her other books include *The Garnett Family* (1961), *Christopher Isherwood* (1970), *Towards a Recognition of Androgyny: Aspects of Male and Female in Literature* (1973), *Reinventing Womanhood* (1979), *Writing a Woman's Life* (1988), and *The Last Gift of Time: Life Beyond Sixty* (1997). She became president of the Modern Language Association in 1984 and exercised her influence in that role to secure for Isherwood the Commonwealth Award for Distinguished Service in Literature, then still selected by an MLA committee. She was married and had three children. She committed suicide.

Heinz. See Neddermeyer, Heinz.

Henderson, Ray. American musician, educated at USC. As a pianist, he accompanied Elsa Lanchester for many years, performing in her nightclub act at The Turnabout, a Los Angeles theater, and on tour, notably, in the autobiographical revue Laughton created for her, "Elsa Lanchester—Herself," for which Henderson was billed as Musical Director, and on her T.V. show. He was also her friend and lover, although he was much younger than Lanchester. He composed the music for some operettas which she recorded privately, and he scored and wrote lyrics for a musical version of *The Dog Beneath the Skin*, which was never produced. He died young, of a heart attack. He appears in *D.2*.

Hensler, Martin. See entry for Gielgud, John.

Herbert, David (1908–1995). British author, second son of the 15th Earl of Pembroke; raised at Wilton. Among his books is *Second Son: An Autobiography* (1972). He was a close friend of Cecil Beaton and of Jane Bowles. Around 1950, he settled in Tangier where he reportedly ran the expatriate social life. Isherwood saw him there in 1955 and 1976, and he appears in *D.1*.

Herbold, Mary. A member of Allan Hunter's Congregational church. Isherwood met her in the early 1940s. She was a typist and a notary public whose services Isherwood evidently used over the years. On Isherwood's recommendation she typed *Time Must Have a Stop* for Huxley in 1944. She appears in *D.1* and *D.2*.

Heyman, Barton (1937–1996). American character actor; he worked in T.V. beginning in the early 1960s and had many small roles in films, including *The Exorcist* (1973), *The Happy Hooker* (1975), *The Bonfire of the Vanities* (1990), *Raising Cain* (1992), and *Dead Man Walking* (1995).

High, Beth. A life-long friend of Jim Gates, from the time they attended high school together in Claremont, California. He introduced her to Swami Prabhavananda in May 1969. She married Gates's friend Gib Peters, with whom she became a Vedanta devotee in the early 1970s; then she became a Vedanta nun and took the name Deepti (Isherwood often typed Dipti). Eventually she left the convent and married again, to an ex-Vedanta monk, Peter Hirschfeld, also a close friend of Gates. Hirschfeld changed his name to Jan Zaremba and became a Sumi'e (ink painting) master; she became known as Deepti Zaremba.

Hill, Charles Christopher (b. 1948). American artist; born in Pennsylvania and educated at the University of California, Irvine. He taught at UCLA and elsewhere and had solo exhibitions in many cities, including Los Angeles, London,

Paris, San Francisco, and Chicago. His paintings are held by the Los Angeles County Museum, the Albright-Knox Museum in Buffalo, the Pompidou Center, the Guggenheim, and the Museum of Modern Art.

Hirschfeld, Magnus (1868–1935). German sex researcher; founder of the Institute for Sexual Science in Berlin, where he studied sexual deviancy. He wrote books on sexual-psychological themes and dispensed psychological counselling and medical treatment (primarily for sexually transmitted diseases). He was homosexual and campaigned for reform of the German criminal code in order to legalize homosexuality between men. His work was jeopardized by the Nazis and he was beaten up several times; he left Germany in 1930 and died in France around the same time that the Nazis raided his institute and publicly burned a bust of him along with his published works. Isherwood took a room next door to the institute in 1930 and first met Hirschfeld then, through Francis Turville-Petre.

Hoban, Gordon (b. 1941). American actor. He had small parts in movies and on T.V. and later wrote scripts, plays, and novels, including *Bike Cop* (1989), *The Adventures of a Highschool Hunk* (1990), *The Marine Olaf*, and *Runaway*.

Hockney, David (b. 1937). British artist, educated at Bradford Grammar School in West Yorkshire, Bradford School of Art, and the Royal College of Art. By 1961, he was identified—with his friend R.B. Kitaj and others—as a leader of a new movement in British art. Versatility and an appetite for new projects and techniques continually energized his career, in oils, acrylics, photography, photocopying, drawing, printmaking, faxing, computer images, watercolor, stage and opera design, as well as commentaries about art and the historical development of artistic technique. Hockney's early success allowed him to travel to the USA, Europe, and Egypt; in 1964 he settled in Los Angeles, and met Isherwood soon afterwards. During the 1960s, he taught at the University of Iowa, the University of California at Berkeley, and the University of Colorado, as well as at UCLA where, in 1966, he met Peter Schlesinger, a student in his class who became an important model. Hockney and Schlesinger rented an apartment on 3rd Street in Santa Monica, a few minutes' walk from Isherwood and Bachardy. Hockney returned to England towards the end of the 1960s and then worked in Paris for a time, near Gregory Evans who became his lover by 1974 and another important model. Later, he moved around among his studios in the Hollywood Hills, Malibu, London, and Bridlington, North Yorkshire. He appears in *D.2*.

Holt, Larry. Dr. Hillary Holt, a Hollywood devotee of German or Austrian background. He lived in Hollywood and was friendly with the American monk Swami Anamananda, who assisted him with his work and errands. He died of cancer in the 1970s, as Isherwood records. He appears in *D.2*.

Holy Mother. See Sarada Devi.

homa fire. Prepared in an ancient Vedic ceremony according to scriptural instructions, the fire is a visible manifestation of the deity worshipped. Offerings to the deity are placed in the consecrated fire. The homa ritual aims at inner purification; at the end of the ritual, the devotee mentally offers his words, thoughts, actions, and their fruits to the deity.

Hooker, Evelyn Caldwell (1907–1996). American psychologist and psychotherapist, trained at the University of Chicago and Johns Hopkins; professor of psychology at UCLA. She was among the first to view homosexuality as a normal psychological condition. She studied homosexuals in the Los Angeles area for many years, through questionnaires, interviews, and discussion in various social settings, accumulating many file drawers of notes which she referred to as "The Project." She first presented her research publicly at a 1956 conference in Chicago, demonstrating that as high a percentage of homosexuals were psychologically well adjusted as heterosexuals. (Her paper was titled "The Adjustment of the Male Overt Homosexual." There were many more.) Born Evelyn Gentry, she took the name Caldwell from a brief first marriage then changed to Hooker at the start of the 1950s when she married Edward Hooker, a Dryden scholar and professor of English at UCLA, who died of a heart attack in 1957. Isherwood met her in about 1949, possibly through the Benton Way Group, a circle of writers, artists, psychoanalysts, philosophers, and others, mostly homosexual, then living together or gathering to talk at a house in Benton Way, Los Angeles. In 1952, he rented the Hookers' garden house on Saltair Avenue in Brentwood, refurbished it, and lived there until tension developed over the arrival of Don Bachardy in 1953. After an uneasy period, the friendship resumed, as Isherwood tells in *D.1* and *Lost Years*. Hooker also appears in *D.2*.

Hoopes, Ned (1932–1984). Teacher, children's anthologist, and, during 1962–1963, host of "The Reading Room," a CBS T.V. series about children's books. He worked on a biography of Charles Laughton, at first with Elsa Lanchester's approval in 1968 and later without. The book, *A Public Success—a Private Failure: The Unauthorized Biography of Charles Laughton*, was never published.

Hope. See Lange, Hope.

Hopper, Brooke. Socialite daughter of film star Margaret Sullavan and agent-producer Leland Hayward; briefly an actress in her teens. She divorced her second husband, Dennis Hopper, in 1969, just before he appeared in *Easy Rider* with her step-brother Peter Fonda. Later she married band-leader Peter Duchin.

Houseman, John (1902–1988) and Joan. Romanian-born writer, director, producer, and actor; his real name was Jacques Haussmann. His mother was British and he was educated in England, then travelled to Argentina and the U.S. as an agent for his father's grain business which collapsed during the Depression. He worked as a journalist and translated plays, then in 1934, directed Virgil Thomson's opera of Gertrude Stein's *Four Saints in Three Acts*, a Broadway hit. Afterwards, he collaborated with Orson Welles with whom he founded the Mercury Theater in 1937. He produced Welles's film *Citizen Kane* (1941), but after a disagreement over who developed the story, he went on to work for David Selznick in Hollywood, then as a producer on his own was responsible for a string of widely admired films, including *The Bad and the Beautiful* (1952). He returned often to direct on Broadway, taught at Vassar, was Artistic Director of the American Shakespeare Festival in the late 1950s, and later of the UCLA Professional Theater Group, and he ran the drama division at Juilliard from 1967. He took his first of many movie roles in *Seven Days in May* (1964), and he won an Academy Award for his supporting role in *The Paper Chase* (1973), which he reprised in the T.V. series. He divorced his first wife and, in 1950, married a beautiful and stylish French woman, Joan Courtney, with whom he had two sons. He appears in *D.1*, *D.2*, and *Lost Years*.

Howard. See Austen, Howard.

Howard, Donald (Don) (1927–1987). American literary scholar and university professor; he was born in St. Louis, raised in Boston, and educated at Tufts, Rutgers, and the University of Florida, where he wrote his dissertation on fourteenth-century English literature. He published books on Langland, the Gawain poet, and most notably on Chaucer, and he also wrote essays about many other aspects of Christian Europe in the Middle Ages and Renaissance. He taught at Ohio State and Johns Hopkins before becoming an associate professor at the University of California at Riverside in 1963. Afterwards he taught at UCLA and, from 1974, at Stanford. He died of AIDS.

Hoyningen-Huene, George (1900–1968). Russian-born photographer, also known as George Huene; son of an American diplomat's daughter and a Baltic baron who had been chief equerry to Tsar Nicholas II. By the end of World War I, he was an exile in Paris, where he studied art and sold drawings to a fashion magazine. Eventually he became a regular photographer for *Vogue* and *Vanity Fair*, and, after 1936, for *Harper's Bazaar*. He published books containing his photographs of Greece, Egypt, North Africa, and Mexico. After the war, he settled in Hollywood where he taught photography and was color consultant on films for his longtime friend George Cukor. He also made several amateur documentaries. Isherwood met him in the late 1940s or early 1950s, through Gerald Heard and the Huxleys. He appears in *D.1* and *D.2*.

Hubrecht, Peggy. British or American widow of Daniel Hubrecht (1909–1972), a Dutchman born in Cambridge, England. They lived in Indonesia in the 1950s, then settled in Tangier, where they lived at the foot of the old mountain road, below Noël Mostert.

Hundertmark, Gary. Gay activist. He was one of the first directors of the National Gay Archives when they opened officially on Hudson Street in Hollywood in 1979.

Hunt. See Stromberg, Hunt, Jr.

Hunter, Allan. Congregational minister and author with whom Gerald Heard often met and conversed about spiritual matters. Hunter's church was the Christian focal point of Heard's California milieu. As he tells in *D.1*, Isherwood met Hunter at a conference organized by Heard in 1940, and, in 1941, Hunter and his wife, Elizabeth, participated in the La Verne seminar. They had a son and a daughter.

Hussein, Waris (b. 1938). Anglo-Indian film and stage director, born in Lucknow, educated at Eton and Cambridge. He studied theater design at the Slade School of Art and trained at the BBC, where he directed the first episode of "Dr. Who." He went on to make successful T.V. plays, literary adaptations, and mini-series in the U.K. and the U.S. He also directed at the National Theatre in London and made a few feature films.

Huston, John (1906–1987). American film director, screenwriter, actor. Son of actor Walter Huston and father of actress Anjelica. As a young man, he was California lightweight boxing champion, served as an officer in the Mexican cavalry, worked briefly as a reporter in New York, and lived rough in Paris and London. He wrote a number of successful scripts in the 1930s and 1940s before his directing debut with *The Maltese Falcon* (1941). During World War II, he filmed documentaries in battle conditions as a member of the army signal corps and was awarded the Legion of Merit for his bravery. Afterwards, he directed many further celebrated films, including *The Treasure of Sierra Madre* (1947, two Academy Awards: Best Director, Best Screenplay; his father won a third: Best Supporting Actor), *The Asphalt Jungle* (1950), *The Red Badge of Courage* (1951), *The African Queen* (1952), *Beat the Devil* (1954), *The Misfits* (1960), *The Man Who Would Be King* (1975), and *Prizzi's Honor* (1985). In 1952, he moved with his fourth wife, Ricki Soma, and their family to Ireland. In 1972, he moved to Mexico. His fifth wife, from 1972 to 1977, was Celeste Shane, known as Cici. Isherwood was friendly with Huston by 1950, possibly through the Huxleys or through Gottfried Reinhardt who produced *The Red Badge of Courage*. Huston appears in *D.1*, *D.2*, and *Lost Years*.

Huxley, Aldous (1894–1963). English novelist and utopian; educated at Eton and Oxford; a grandson of Thomas Huxley and brother of Julian Huxley, both prominent scientists. In youth, he published poetry, short stories, and satirical novels such as *Crome Yellow* (1921) and *Antic Hay* (1923) about London's literary bohemia and Lady Ottoline Morrell's Garsington Manor, where he lived and worked during World War I and where he met his first wife. The Huxleys lived abroad in Italy and France during the 1920s and 1930s, partly with D.H. Lawrence—who appears in Huxley's *Point Counter Point* (1928)—and Lawrence's wife, Frieda. In 1932 Huxley published *Brave New World*, for which he is most famous.

An ardent pacifist, Huxley joined the Peace Pledge Union in 1935, and his *Ends and Means* (1937) was a basic book for pacifists. In April 1937, he sailed for America with his wife and son, accompanied by Gerald Heard and Heard's friend Chris Wood. Plans to return to Europe fell through when he failed to sell a film scenario in Hollywood, became ill there, and convalesced for nearly a year. He was denied U.S. citizenship on grounds of his extreme pacifism. California benefitted his health and eyesight—he had been nearly blind since an adolescent illness. *After Many a Summer* (1939) is set in Los Angeles, and Huxley wrote many other books there, including *Grey Eminence* (1941), *Time Must Have a Stop* (1944), *The Devils of Loudun* (1952), and *The Genius and the Goddess* (1956).

Not long after he arrived in Los Angeles, Isherwood was introduced to Huxley by Gerald Heard. Huxley and Isherwood collaborated on three film projects together during the 1940s: *Jacob's Hand*, about a healer, *Below the Equator* (later called *Below the Horizon*), and a film version of *The Miracle*, Max Reinhardt's 1920s stage production. Like Heard, Huxley was a disciple of Prabhavananda, but subsequently he became close to Krishnamurti, the one-time Messiah of the Theosophical movement. Huxley's study of Vedanta was part of his larger interest in mysticism and parapsychology, and beginning in the early 1950s he experimented with mescaline, LSD, and psilocybin, experiences which he wrote about in *The Doors of Perception* (1954) and *Heaven and Hell* (1956).

Huxley died of cancer on the day John F. Kennedy was assassinated. He appears in *D.1*, *D.2*, and *Lost Years*, and Isherwood helped the novelist Sybille Bedford with her *Aldous Huxley: A Biography, Volume 1 1894–1939* (1973) and *Aldous Huxley: A Biography, Volume 2 1939–1963* (1974).

Huxley, Laura Archera (1911–2007). Italian second wife of Aldous Huxley. Isherwood met her in the spring of 1956 at the Stravinskys' after she and Huxley married secretly in March. She was the daughter of a Turin stockbroker, had been a concert violinist from adolescence, and worked briefly in film. She became a psychotherapist, sometimes using LSD therapy on her patients, and she published two popular books on her psychotherapeutic techniques. Her 1963 bestseller, *You Are Not the Target*, was an early self-help book. She also published a memoir about Huxley, *This Timeless Moment* (1968), and a children's book. She first befriended Aldous and Maria Huxley in 1948 and used her special method of therapy on Huxley to help him recapture lost parts of his childhood. He incorporated some of her psychotherapy results into his utopia novel, *Island*. Before marrying Huxley, Laura lived for many years with Virginia Pfeiffer; after the marriage, she and Huxley settled in a house adjacent to Virginia's. After Huxley's death, she eventually became a children's rights campaigner. She appears in *D.1* and *D.2*.

Huxley, Maria Nys (1898–1955). First wife of Aldous Huxley; eldest daughter of a prosperous Belgian textile merchant ruined in World War I. Her mother's family included artists and intellectuals, and her childhood was pampered, multilingual, and devoutly Catholic. She met Huxley at Garsington Manor where she lived as a refugee during World War I; they married in Belgium in 1919 and their only child was born in 1920. Before her marriage, Maria showed promise as a dancer and trained briefly with Nijinsky, but her health was too frail for a professional career. She had little formal education and devoted herself to Huxley. Her premature death resulted from cancer. Isherwood met her in the summer of 1939 soon after he arrived in Los Angeles. She appears in *D.1* and *Lost Years*, and she is mentioned in *D.2*.

Igor. See Stravinsky, Igor.

Inge, William (Bill) (1913–1973). American playwright, born and educated in Kansas and at the University of Kansas; he earned a teaching degree in Tennessee and taught high-school and college English, then became the music and drama critic for the St. Louis *Star-Times*. In 1944, he interviewed Tennessee Williams, who befriended him and took him to Chicago to see *The Glass Menagerie*; afterwards, Inge accepted another university teaching job and wrote his first play, *Farther Off from Heaven*, which was produced in 1947 by Margo Jones at the Dallas Civic Theater. His next play, *Come Back, Little Sheba*, opened on Broadway in 1950 to great praise, and he won a Pulitzer Prize and two Drama Critics Awards for *Picnic* (1953). *Bus Stop* (1955) and *The Dark at the Top of the Stairs* (1957) were equally acclaimed, but in the 1960s his stage work failed repeatedly. Although his plays were adapted for film mostly by others, he received an Academy Award for *Splendor in the Grass* (1961), which he co-produced; his other screenplays are *All Fall Down* (1962) and *Bus Riley's Back in Town* (1965, under the pseudonym Walter Gage). In 1963, he moved from New York to Los Angeles, and in the late

1960s, he briefly returned to teaching, at the University of California at Irvine. He wrote two novels, *Good Luck Miss Wyckoff* (1970) and *My Son Is a Splendid Driver* (1971). He was depressive and had problems with alcohol. Isherwood and Bachardy first met Inge in New York in 1953 during the original run of *Picnic*. He is mentioned in *D.1* and *D.2*.

Isherwood, Frank Bradshaw (1869–1915). Isherwood's father; second son of John Bradshaw Isherwood, squire of Marple Hall, Cheshire. He was educated at Sandhurst and commissioned in his father's old regiment, the York and Lancasters, in 1892 at the age of twenty-three. He left for the Boer War in December 1899, caught typhoid, recovered, and served a second tour. In 1902 he left his regiment and became adjutant to the Fourth Volunteer Battalion of the Cheshire Regiment, based locally, in order to be able to offer his wife a home despite his meager income. He married Kathleen Machell Smith in 1903, and they settled for a time in a fifteenth-century manor house, Wyberslegh Hall, on the Bradshaw Isherwood family estate. In 1908, Frank rejoined his regiment and the family, now including Christopher, followed the regiment to Strensall, Aldershot, and Frimley; in 1911 a second son, Richard, was born and the family moved again to Limerick, Ireland, early the following year. Frank was sent from Limerick via England to the Front Line almost as soon as war was declared in the summer of 1914, and he was killed probably the night of May 8, 1915 in the second battle of Ypres in Flanders, although the exact circumstances of his death are unknown. Isherwood felt that Frank was temperamentally unsuited to the life of a professional soldier, though he was dutiful and efficient. He was a gifted watercolorist, an excellent pianist, and he liked to sing and take part in amateur theatricals. He was also a reader and a story-teller. He was shy and sensitive, mildly good-looking, and a keen and agile sportsman. He was conservative in taste, in values, and in politics, but, unlike Kathleen, he was agnostic in religion and was attracted to theosophy and Buddhism. Isherwood wrote about his father in *Kathleen and Frank*.

Isherwood, Kathleen Bradshaw (1868–1960). Isherwood's mother, often referred to as "M." in the diaries. Only child of Frederick Machell Smith, a wine merchant, and Emily Greene. She was born and lived until sixteen in Bury St. Edmunds, then moved with her parents to London. She travelled abroad and helped her mother to write a guidebook for walkers, *Our Rambles in Old London* (1895). In 1903, aged thirty-five, she married Frank Isherwood, a British army officer. They had two sons, Isherwood, and his much younger brother, Richard. When Frank Isherwood was lost in World War I, it was many months before his death was officially confirmed. Isherwood's portrait of his mother in *Kathleen and Frank* is partly based on her own letters and diaries. She was also the original for the fictional character Lily in *The Memorial*. Like many mothers of her class and era, Kathleen consigned her sons to the care of a nanny from infancy and later sent Isherwood to boarding school. Her husband's death affected her profoundly, which Isherwood sensed and resented. Their relationship was intensely fraught yet formal, intimate by emotional intuition rather than by shared confidence. Like her husband, Kathleen was a talented amateur painter. She was intelligent, forceful, handsome, dignified, and capable of great charm. Isherwood felt she was obsessed by class distinctions and propriety. As the surviving figure of authority in his family, she epitomized everything against which, in youth, he wished to rebel. He deemed her intellectual aspirations narrow and traditional, despite her intelligence, and she seemed to him increasingly backward looking. Nonetheless, she was utterly loyal to both of her notably unconventional sons and, as Isherwood himself recognized, she shared many qualities with him. There are many passages about her in *D.1*, *D.2*, and *Lost Years*.

Isherwood, Richard Graham Bradshaw (1911–1979). Christopher Isherwood's brother and only sibling. Younger by seven years, Richard was also backward in life. He was reluctant to be educated and never held a job in adulthood, although he did National Service during World War II as a farmworker at Wyberslegh and at another farm nearby, Dan Bank. In childhood, he saw little of his elder brother, who was sent to boarding school by the time Richard was three. Both boys spent more time with their nanny, Annie Avis, than with their mother. Richard later felt that Nanny had preferred Christopher; she made Richard nervous and perhaps was cruel to him. When Richard started school as a day boy at Berkhamsted in 1919, he lodged in the town with Nanny, and his mother visited at weekends. Isherwood by then was at Repton. The two brothers became closer during Richard's adolescence, when Isherwood was sometimes at home in London and took his brother's side against their mother's efforts to advance Richard's education and settle him in a career. During this period Richard met Isherwood's friends and helped Isherwood with his work by taking dictation. Richard was homosexual, but he seems to have had little opportunity to develop any long-term relationships, hampered as he was by his mother's scrutiny and his own shyness.

In 1941, he returned permanently with his mother and Nanny to Wyberslegh—signed over to him by Isherwood with the Marple Estate—where he eventually lived as a semi-recluse. Nanny died in 1948, and after Kathleen Isherwood's death twelve years later, Richard was looked after first by a married couple, the Vinces, and then by a local family, the Bradleys. He became a heavy drinker, Marple Hall fell into ruin and became dangerous, and he was forced to hand it over to the local council which demolished it in 1959, building several houses and a school on the grounds. Eventually, Richard moved out of Wyberslegh into a new house on the Marple Estate; the Dan Bradleys lived in a similar new house next door. When he died, he left most of the contents of his house to the Dan Bradleys and the house itself to their daughter and son-in-law. Richard's will also gave money bequests to the Dan Bradleys, Alan Bradley, and other local friends. Family property and other money were left to Isherwood and to a cousin, Thomas Isherwood, but Isherwood himself refused the property and passed some of his share of money to the Dan Bradleys. Richard appears in *D.1*, *D.2*, and *Lost Years*.

Isherwood, Thomas Bradshaw (1921–2002). Christopher Isherwood's first cousin, only son of Frank Isherwood's younger brother, Jack Bradshaw Isherwood. Thomas was heir to what remained of the family estate after the deaths of Richard and Christopher, and he was the last of the Bradshaw Isherwood line.

Ishvara. God with attributes; Brahman and maya united; the personal god.

Ivan. See Moffat, Ivan.

Ivory, James (b. 1928). American film director; born in Berkeley, raised and educated in Oregon. He studied film at UCLA, served in the army, then began making documentaries, including one about India, which resulted in his meeting Ismail Merchant there in the early 1960s and forming a lifelong personal and business partnership with him. Ivory co-wrote his early screenplays with novelist Ruth Prawer Jhabvala and had a number of subsequent successes with her. His work includes *Shakespeare Walla* (1965), *Savages* (1972), *The Europeans* (1979), *Heat and Dust* (1983), *The Bostonians* (1984), *A Room with a View* (1985, Academy Awards for Screenplay, Art Direction, Costumes, nominations for Best Picture and Best Director), *Maurice* (1987), *Howards End* (1992, Academy Awards for Best Actress, Screenplay Adaptation, Art Direction, nominations for Best Picture), *The Remains of the Day* (1993, Academy Award nominations for Best Picture and Best Director), *The Golden Bowl* (2001), *Le Divorce* (2003), and *The White Countess* (2005).

James, Edward (Eddie) (1907–1984). Youngest child and only son of Willie James and Evelyn Forbes; heir to an American railroad fortune. His parents frequently entertained King Edward VII at West Dean Park. Edward James was a godson and rumored to be son of the monarch, although he himself evidently thought he was a grandson, and his mother a beloved illegitimate daughter. He was an early patron of the Surrealists (especially Dali), and was married briefly in the 1930s to Tilly Losch, the Austrian ballerina, launching a ballet company to further her career. His poetry appeared in vanity editions, paid for by himself. Isherwood knew James by the start of the 1950s, probably through the Stravinskys who, with the Baroness D'Erlanger, were among his close friends in Hollywood, and James appears in *D.1.* He spent much of his time in Mexico where he was friendly with the English painter Leonora Carrington and a young Indian man called Plutarcho Gastelum, and where he had a coffee finca in Xilitla (pronounced he-heet-la), near Tampico. There, he was often accompanied by the mysterious German writer "B. Traven" (Albert Otto Max Feige). In the last decade of his life, James built an uninhabited concrete city in the jungle, a surrealist art work.

Jan. See Niem, Jan.

japa or japam. A method for achieving spiritual focus in Vedanta by repeating one of the names for God, usually the name that is one's own mantra; sometimes the repetitions are counted on a rosary. The rosary of the Ramakrishna Order has 108 beads plus an extra bead, representing the guru, which hangs down with a tassel on it; at the tassel bead, the devotee reverses the rosary and begins counting again. For each rosary, the devotee counts one hundred repetitions towards his own spiritual progress and eight for mankind. Isherwood always used a rosary when making japa. Japam is a Tamil form which came into use among Bengali swamis of the Ramakrishna Order—including Prabhavananda, Ashokananda, Akhilananda—because they spent varying periods of time in the Madras Math.

Jarre, Maurice (b. 1924). French composer, educated at the Paris Conservatory; he was musical director for the Théâtre National Populaire and began writing music for French movies in the 1950s. He became famous with his scores for

Lawrence of Arabia (1962) and *Doctor Zhivago* (1965), both of which won Academy Awards for best music, and went on to write the music for *Barbarella* (1968), *The Damned* (1969), *Ryan's Daughter* (1970), *The Man Who Would Be King* (1975), *The Year of Living Dangerously* (1982), *A Passage to India* (1984; Academy Award), *Witness* (1985), *Fatal Attraction* (1987), *Dead Poets Society* (1989), and others. Although he told Isherwood in 1973 that he would not compose the music for "Frankenstein" unless it was made as a feature film, he did compose for T.V. occasionally. He is the father of composer Jean-Michel Jarre.

Jebb, Julian (1934–1984). British journalist, filmmaker, BBC producer; educated at Cambridge. He was an alcoholic and died from an overdose of a sedative prescribed for alcoholics. The December 1973 film he made of Isherwood was used in "Hooray for Hollywood," an "Arena: Cinema" program which also featured interviews with Neil Simon and David Puttnam. "Hooray for Hollywood" was presented and produced by Gavin Millar and went out October 10, 1978 on BBC 2. Bachardy recalls that Jebb's film may also have been aired by itself on another, earlier occasion.

Jennifer. Also Jennifer Selznick and, later, Jennifer Simon; see Jones, Jennifer.

Jess. See Bachardy, Jess.

Jo. Also Jo Lathwood. See Masselink, Jo.

Johnson, Bart (not his real name). American school teacher. An amateur writer with whom Isherwood had a brief sexual relationship at the end of the 1950s. Isherwood tried to help Johnson with his writing, but without success.

Johnson, Clifford (Cliff). Professor of English. He joined the Vedanta Society in 1960 and lived at Trabuco until the late 1970s. He became a brahmachari and was managing editor of *Vedanta and the West* for some years. Swami Prabhavananda gave him the Sanskrit name Bhuma.

Johnson, Lamont (Monty) (1922–2010). American actor and, especially, director; educated at UCLA and the Neighborhood Playhouse. He worked in radio as a teenager to pay his way through college, then moved to New York where he continued in radio soaps and directed an off-Broadway production of Gertrude Stein's *Yes Is for a Very Young Man*. In 1959, he was a founder of the UCLA Theater Group, and during the 1960s and 1970s he won numerous Emmys and Screen Directors Guild awards for his T.V. miniseries and made-for-T.V. movies. He also directed episodes of popular shows like "Have Gun Will Travel," "The Rifleman," and "The Twilight Zone," and a few feature films, including *The Last American Hero* (1973) and *Lipstick* (1976). He acted on T.V. and in a number of movies. He appears in *D.2*.

Jones, Jack. American painter; a disciple and, later, close friend of Gerald Heard. Some of his best work was of Margaret Gage's garden on Spoleto Drive, where Heard lived until 1962, and he shared Heard's interest in clothing and costume. Jones was about the same age as Don Bachardy and lived nearby in Santa Monica Canyon, so they sometimes sat for each other. He appears in *D.2*.

Jones, Jennifer (Phylis Isley) (1919–2009). American actress, born in Tulsa, Oklahoma; in childhood she travelled with her parents' stock stage company and spent hours in her father's movie theaters. After a brief stint at Northwestern University, she attended the Academy of Dramatic Arts in New York, left for a radio job in Tulsa, and began her Hollywood career in B-movies in 1939. She was discovered in 1941 by David Selznick, who changed her name and took control of her career with spectacular results. She won an Academy Award for *The Song of Bernadette* in 1943, followed by *Since You Went Away* (1944, Academy Award nomination), *Love Letters* (1945, Academy Award nomination), *Duel in the Sun* (1946, Academy Award nomination), *Portrait of Jennie* (1948), *Madame Bovary* (1949), *Carrie* (1951), *Love Is a Many-Splendored Thing* (1955, Academy Award nomination), *A Farewell to Arms* (1957), *Tender is the Night* (1962), *The Towering Inferno* (1974), and others. Her 1939 marriage to the actor Robert Walker, with whom she had two sons, ended in divorce in 1945, and Selznick left his wife, Irene Mayer, for Jones; they married in 1949. His obsession with Jones combined with her own emotional instability (including suicide attempts) made a melodrama of their careers and their private lives. In 1965, Selznick died, leaving huge debts. In 1971, Jones married a third time, to Hunt Foods billionaire and art collector Norton Simon. On May 11, 1976, her only child with Selznick, Mary Jennifer, committed suicide; partly as a result, Jones created the Jennifer Jones Foundation for Mental Health and Education and trained as a lay therapist and volunteer counsellor. Isherwood first met her when he worked with Selznick on *Mary Magdalene* in 1958, and he took her to meet Swami Prabhavananda in June that year. He writes about the friendship in *D.1* and *D.2*.

Julie. See Harris, Julie.

Kali. Hindu goddess; the Divine Mother and the Destroyer, usually depicted dancing or standing on the breast of a prostrate Shiva, her spouse, and wearing a girdle of severed arms and a necklace of skulls. Kali has four arms: the bleeding head of a demon is in her lower left hand, the upper left hand holds a sword; the upper right hand gestures "be without fear," the lower right confers blessings and boons on her devotees. Kali symbolizes the dynamic aspect of the godhead, the power of Brahman: she creates and destroys, gives life and death, well-being and adversity. She has other names: Shakti, Parvati, Durga. Kali was Ramakrishna's Chosen Ideal, and for a number of years, he devoted himself to worshipping her image in her temple at Dakshineswar. Kali puja is usually celebrated in November, and in southern California the festivities all take place in Hollywood because the ritual is complex and demanding; each year a large statue of Kali is made for worship; it takes several months to make and is about three and a half feet high. About a week after the puja, the statue is taken out to sea on a hired boat and immersed, with devotees singing and celebrating on the journey. The immersion symbolizes Kali's return to her celestial abode, where her husband Shiva resides.

Kallman, Chester (1921–1975). American poet and librettist; Auden's companion and collaborator. They met in New York in May 1939 and lived together intermittently in New York, Ischia, and Kirschstetten for the rest of Auden's life, though Kallman spent time with other friends, often in Athens as he grew older. He published three volumes of poetry and with Auden wrote and translated opera libretti, notably *The Rake's Progress* (for Stravinsky), *Elegy for Young Lovers*, and *The Bassarids* (both for Hans Werner Henze). He appears in *D.1*, *D.2*, and *Lost Years*.

Kaper, Bronislau (1902–1983). Polish-born composer, trained at Warsaw Conservatory; he wrote music for German films in Berlin, then, fleeing Hitler, emigrated to Paris and Hollywood where he continued his career at MGM. His film scores include the Marx Brothers' *A Night at the Opera* (1935); *Green Dolphin Street* (1947), from which the theme became a jazz favorite; *The Great Sinner* (1949), for which Isherwood wrote the script; *Invitation* (1952) directed by Gottfried Reinhardt; and *Lili* (1953), for which Kaper won an Academy Award. He appears in *D.2*.

Kaplan, Joseph (1902–1991). Hungarian-born physicist, educated at Johns Hopkins and Princeton. He was a professor at UCLA and, for a time, chairman of the Physics department and director of the Institute of Geophysics. As a long-time chairman of the U.S. Committee of the International Geophysical Year, he encouraged the exploration of outer space with rockets and satellites in the 1950s. He also helped to found the Global Atmospheric Research Program. In 1957, he advised the press of the greenhouse effect, predicting ocean levels would rise by at least forty feet and advocating scientific control of the earth's weather and temperature. He married twice, the second time in 1977.

Kasmin, John (b. 1934). British art dealer, known by his surname, Kasmin; educated at Magdalen College, Oxford. He opened a gallery in New Bond Street in 1963, backed by the 5th Marquess of Dufferin and Ava, a Guinness heir, to show new British and American art. Hockney had his first show with Kasmin that year, followed by many others, and Kasmin quickly became a pre-eminent London dealer. The gallery closed in 1972, but he had three other galleries in the neighborhood over the next twenty years. His son, Paul Kasmin, later opened a gallery in New York and showed Hockney there.

Kathleen. See Isherwood, Kathleen Bradshaw.

Katz, James C. (Jim). Film producer; he worked with Michael Laughlin on an idea for a film called *The Nymph and the Lamp*. Later, he got into film preservation, restoring 70 mm films, including *Spartacus*, *My Fair Lady*, *Vertigo*, and *Rear Window*.

Kelley, Howard. Upstairs neighbor to Denny Fouts at 137 Entrada Drive, where Isherwood and Bill Caskey first met him with his friend, Wallace Bobo, during the late 1940s. He appears in *D.1* and *Lost Years*.

Kerr, Deborah (1921–2007). Scottish-born actress, trained as a ballet dancer at Sadler's Wells, where she joined the corps de ballet as a teenager; she soon turned to the stage and then made her film debut in *Major Barbara* (1941). She became a Hollywood star in *The Hucksters* (1947) and made a sensation playing an adulteress opposite Burt Lancaster in *From Here to Eternity* (1953). She was nominated for an Academy Award for best actress for this and for five other roles, in *Edward My Son* (1949), *The King and I* (1956), *Heaven Knows, Mr. Allison* (1957), *Separate Tables* (1958), and *The Sundowners* (1960). Her other films include *The Avengers* (1942), *Black Narcissus* (1947), *King Solomon's Mines* (1950), *The End of the Affair* (1955), *Bonjour Tristesse* (1958), *The Chalk Garden* (1964), and *The Night of the Iguana* (1964). She retired from making movies in 1969, but continued to work on stage, where she appeared in *Tea and Sympathy* (1953) and *Seascape* (1975), among others. Her T.V. roles include "Witness for the Prosecution" (1983) and Emma Harte in "A Woman of Substance" (1982). Her first husband, with whom she had two daughters, was fighter pilot Anthony Bartley, a hero of the Battle of Britain and later a T.V. writer and producer in Hollywood. In 1960, she married Peter Viertel. She appears in *D.1*.

K.H.3. An anti-aging preparation developed by a Romanian doctor, Ana Aslan, and promoted in the 1960s and 1970s; the active ingredient is procaine, the local anaesthetic widely used by dentists and typically known by its brand name, Novocaine, K.H.3 possibly has a mild antidepressant effect, but it never persuaded the medical establishment and is not approved by the Food and Drug Administration in the U.S.

Kight, Morris (1919–2003). Pioneer gay rights activist, born in Texas. He founded the Gay Liberation Front in 1969, the Los Angeles Gay and Lesbian Center, the Stonewall Democratic Club, and Christopher Street West, which organizes the West Hollywood Gay Pride Parade.

Kilhefner, Don (b. *circa* 1940). Jungian therapist, based in West Hollywood. He co-founded with Morris Kight the Los Angeles Gay Community Services Center in 1971 and Van Ness Recovery House. In 1979, he founded the pagan-inspired Radical Faeries with Harry Hay and, in 1999, the shamanistic Gay Men's Medicine Circle.

Kimbrough, Clinton (1933–1996). American actor and director, sometimes credited as Kimbro. He appeared at the American National Theater and Academy at Washington Square in New York during the first half of the 1960s; he also had small roles in a few films and narrowly missed being cast as one of the killers in the film of Capote's *In Cold Blood*. He had affairs with both sexes, including, as Isherwood records in *D.2*, Gavin Lambert. Bachardy did many drawings and paintings of him and of his wife, Frances Doel. She was a script girl on low-budget films, then began to work on her own screenplays and original stories, including *Big Bad Mama* (1974) and *Crazy Mama* (1975), in which her husband had a small part. Later she became a producer and a script development consultant.

King, Perry (b. 1948). American actor, born in Ohio, educated at Yale and Juilliard. He appeared regularly on T.V. in "Riptide," "Melrose Place," "Titans," "The Day After Tomorrow," and in numerous made-for-T.V. movies. His feature films include *Slaughterhouse-Five* (1972), *The Lords of Flatbush* (1974), *The Wild Party* (1975), which Isherwood mentions, *Mandingo* (1975), and *Lipstick* (1976). His first wife, from 1970 to 1980, was Karen Hryharrow.

Kirstein, Lincoln (1907–1996). American dance impresario, author, editor, and philanthropist, raised in Boston, son of a wealthy self-made businessman. He was educated at Berkshire, Exeter, and Harvard, where he was founding editor of *Hound and Horn*, the quarterly magazine on dance, art, and literature. He also painted, and he helped found the Harvard Society for Contemporary Art. In 1933, Kirstein persuaded the Russian choreographer George Balanchine to come to New York, and together they founded the School of American Ballet and the New York City Ballet. Kirstein was also involved in founding the Museum of Modern Art. His taste and critical judgement combined with his entrée into wealthy society enabled him to recognize and promote some of the greatest artistic talent of the twentieth century. In 1941, he married Fidelma (Fido) Cadmus, sister of the painter Paul Cadmus. He served in the army from 1943 to 1946. Isherwood's first meeting with him in New York in 1939 was suggested by Stephen Spender, who had met Kirstein in London, and Kirstein appears in *D.1*, *D.2*, and *Lost Years*. He is a model for Charles in *The World in the Evening*. His poetry was admired by Auden, and Isherwood mentions reviewing his *Rhymes of a PFC* (1964) when Kirstein reissued it as *Rhymes and More Rhymes of a PFC* (1966). But their friendship ended that year, when Kirstein commissioned Bachardy to do portraits of the New York City Ballet stars without consulting Balanchine. Bachardy was disappointed when the project was scrapped, and he later blamed himself for the fact that Kirstein thereafter refused to see Isherwood even though Auden tried to reconcile them.

Kiser, Kirsten (Kiki). With her husband, Tony Kiser, she bought and restored 757 Kingman Avenue in Santa Monica, but, as Isherwood records, felt threatened there. The house, modern and minimalist in style and boasting a secret staircase, was built in 1930 by MGM production designer Cedrick Gibbons for his wife, the Mexican actress Dolores del Rio. They divorced in 1941. Gibbons's brother, screenwriter Eliot Gibbons, married Irene, the MGM costume designer, who killed herself in 1962. Reportedly, Frances Farmer also lived in the house for a time, and Van Johnson owned it before Kiser. Kiser later divorced; eventually she became a licensed architect, launched an architecture website, and mounted a Frank Gehry retrospective in Copenhagen.

Kiskadden, Peggy. Thrice-married American socialite from Ardmore, Pennsylvania; born Margaret Adams Plummer, she was exceptionally pretty and had an attractive singing voice. From 1924 until 1933, she was married to a lawyer and (later) judge, Curtis Bok, the eldest son of one of Philadelphia's most prominent families. In the early 1930s, she accompanied Bok, a Quaker, to Dartington, England, where she met Gerald Heard and Aldous and Maria Huxley. Her second marriage, to Henwar Rodakiewicz, a documentary filmmaker, ended in 1942, and she married Bill Kiskadden in July 1943. She had four children, Margaret Bok (called Tis), Benjamin Bok, Derek Bok (later president of Harvard University), and William Kiskadden, Jr. (nicknamed "Bull"). Isherwood was introduced to her by Gerald Heard soon after arriving in Los Angeles; they became intimate friends but drew apart at the end of the 1940s and finally split irrevocably in the 1950s over Isherwood's relationship with Don Bachardy. She died in the 1990s. There are numerous passages about her and her family in *D.1* and *Lost Years* and some in *D.2*.

Kitaj, R.B. (Ronald) (1932–2007). American painter and graphic artist, born in Cleveland, Ohio; he was a merchant seaman and served in the army before studying at Cooper Union in New York, the Academy of Fine Art in Vienna, the Ruskin School of Drawing in Oxford, and the Royal College of Art in London, where he became friendly with his fellow student David Hockney. He had his first one-man show at the Marlborough New London Gallery in 1963. Isherwood describes his April–May 1970 show at Marlborough Fine Art, and Kitaj continued to exhibit there while achieving international recognition. In the early 1960s, he taught at the Camberwell School of Art, and later at the University of California at Berkeley and at Dartmouth College. He was elected to the Royal Academy in 1991. His 1994 retrospective at the Tate was harshly criticized by the British press and soon afterwards his second wife, American painter Sandra Fisher, died of a brain hemorrhage at forty-six, driving Kitaj to leave London in 1997 and settle in Los Angeles.

Kitty. Bachardy's pet name, once known only to himself and Isherwood, and denoting his identity as an exotic, temperamental feline creature in the private myth world they shared.

Kitty's birthday pre-celebration. Isherwood notes four violent events on the day before Bachardy's fortieth birthday in 1974. Los Angeles police and the FBI had a gun battle, shown live on Los Angeles T.V., with presumed members of the Symbionese Liberation Army in a house near Watts; the siege culminated in a fire which destroyed the house and left five unidentifiable bodies. Heiress Patricia Hearst, kidnapped by the SLA on February 4, was not in the house and remained a hostage and an apparent convert to the SLA's radical anti-Establishment cause. In Dublin, three car bombs exploded during rush hour, killing at least twenty-three and injuring eighty, the worst casualties since fighting over Northern Ireland began in 1968. The bombs were thought to be set by extremist Protestants. In Eastern Lebanon, Israeli planes bombed villages and Palestinian refugee camps in the second day of reprisals for the killing of twenty Israeli students by Palestinian guerrillas in a village near the Lebanese border; more than fifty refugees were killed at Nabatieh camp alone and half the houses reduced to rubble in the first day's raid; Syrian planes challenged the Israeli flights, and continuing reprisals and counterreprisals were promised on both sides. In Peru, there was a small earthquake, felt in Lima and elsewhere; it proved to be one of a series of tremors building up to a powerful and destructive earthquake October 3.

Kleiser, Randal (Randy) (b. 1946). American film director, writer, and producer. He made a prize-winning film, *Peege*, as his master's thesis at USC. Then he directed episodes of popular T.V. series, followed by T.V. films such as "The Boy in the Plastic Bubble" (1976) and "The Gathering" (1977). He became famous with *Grease* (1978). Other films include *The Blue Lagoon* (1980) and *Honey, I Blew Up the Kids* (1992). He teaches at USC and in Europe.

Knight, Franklin (Frank) (*circa* 1924–2005). American monk of the Ramakrishna Order; he was given the name Asima Chaitanya when he took his brahmacharya vows, probably in 1965 or 1966, and came to be known as Asim. He first joined the Trabuco monastery in about 1955 and settled there permanently, but Swami Prabhavananda never allowed him to take sannyas because of an episode—referred to by Isherwood in his diary entry for December 16, 1963—in which he behaved inappropriately towards a woman outside the congregation. Knight was a cousin of Webster Milam, and is mentioned in *D.1* and appears in *D.2*.

Knott, Harley. A son of Francis Bacon's sister Ianthe; she married a South African farmer, settled with him in southernmost Rhodesia, then moved to the Transvaal after his death.

Knox, Ronnie (1935–1992). American football player and aspiring writer, born in Illinois. His real name was Raoul Landry. He was a football All-American for Santa Monica High School, a star freshman quarterback for the University of California, and widely considered to be the most talented college football player in the country when in 1953 he suddenly transferred to UCLA, sacrificing a year of eligibility and generating a scandal. He was also written up as a glamor boy because of his physical beauty. As a professional, he joined the Canadian Football League and played for Montreal. Later he wrote fiction, and published at least one of his stories. He lived with Renate Druks from 1960 to 1964, and afterwards had a French girlfriend called Véronique. All three sat for Bachardy several times. Knox died unemployed and virtually homeless in San Francisco. He is the chief model for Kenny in *A Single Man*. He appears in *D.2*.

Korda, Michael (b. 1933). Editor in chief at Isherwood's American publisher, Simon & Schuster, from 1968 until he retired in 2005. Born in London and educated at Oxford, he settled in New York in his twenties and became Henry Simon's editorial assistant in 1958. He wrote several books of his own. He is a nephew of film impresario Alexander Korda.

Kramer, Terry Allen. Broadway producer. She was an heiress to her father's Manhattan investment bank, Allen and Co., which owned, among other things, a controlling interest in Syntex Chemicals Inc., developer of the oral contraceptive pill. Her second husband, Irwin Kramer, worked in his family's hotel and then became a partner in her father's firm. Irwin Kramer adopted her daughter, Toni Phillips (b. 1956), from her first marriage. Among the Broadway shows she produced, many with Harry Rigby, are *Knock Knock* (1976), *I Love My Wife* (1979), *A Meeting by the River* (1979), *Sugar Babies* (1982), *Me and My Girl* (1989), *Shadowlands* (1991), and *The Goat, or Who Is Sylvia?* (1992).

Krishna. One of the most widely worshipped Hindu gods, a hero of the Mahabharata and the Bhagavatam. Krishna was also the Sanskrit name given to George Fitts, an American monk of the Ramakrishna Order, from New England. He joined the Vedanta Society in Hollywood in 1940 and was living there as a probationer monk in 1943 when Isherwood moved in. He was then about forty years old, had some private wealth, and spent his time obsessively tape recording and transcribing Swami Prabhavananda's lectures and classes. He took his brahmacharya vows in 1947, and early in 1958, he took sannyas and became Swami Krishnananda. He lived in Hollywood, but usually accompanied Swami on trips to Santa Barbara, Trabuco, and elsewhere. He appears in *D.1* and *D.2*.

Krishnamurti (1895–1986). Hindu spiritual teacher. As an impoverished boy in India, he was taken up by the leaders of the Theosophical movement as the "vehicle" in which their Master Maitreya would reincarnate himself. He was adopted and educated in England, then in 1919 sent to an orange ranch in Ojai, California, for his health. In 1929, he renounced his messianic role and rejected the guru–disciple relationship along with the devotional and ritual aspects of Hinduism. Although he broke with the Theosophists, he went on speaking to devotees, sometimes in huge numbers, for the rest of his life all around the world.

He was extremely handsome and charismatic and had many secret sexual liaisons which introduced tension among his followers and led to a series of lawsuits with a colleague and rival, Desikacharya Rajagopalacharya, whom he cuckolded. Isherwood first met Krishnamurti in 1939 through Aldous and Maria Huxley and later went to hear him speak in Ojai. He appears in *D.1* and *D.2*

Krost, Barry. British agent and film and T.V. producer, formerly a child actor. He appeared in a few films in the 1950s. He was John Osborne's manager for about ten years, then settled in Los Angeles in the mid-1970s. He founded the Los Angeles talent agency Barry Krost Management and promotes gay and lesbian rights in the film industry. As a producer, he works with his long-time companion, Douglas Chapin.

Ladner, John (b. 194[6]). American lawyer, educated at Berkeley and Loyola School of Law, admitted to the California Bar in 1973. He worked in Washington, D.C. and New York for the Department of Health, Education and Welfare, was a Public Defender in Los Angeles from 1977 to 1979, and until 1983 ran his own practice specializing in criminal and juvenile defense. He later spent twenty years as a Municipal and Superior Court Commissioner and presided over California's criminal child support enforcement court. His longterm companion is Mark Lipscomb.

Lambert, Gavin (1924–2005). British novelist, biographer and screenwriter; educated at Cheltenham College and for one year at Magdalen College, Oxford. He edited the British film magazine *Sight and Sound*, before going to Hollywood in 1956. He was working for Jerry Wald at Twentieth Century-Fox on *Sons and Lovers* (1960) when Ivan Moffat introduced him to Isherwood; he appears often in *D.1* and *D.2*. His novel *The Slide Area: Scenes of Hollywood Life* (1959), which Isherwood read in manuscript in 1957, was influenced by Isherwood's Berlin stories. He and Isherwood worked on a television comedy project "Emily Ermingarde" for Hermione Gingold and later for Elsa Lanchester, but the series was never produced. Lambert also helped Isherwood revise the film script of *The Vacant Room*. During the 1950s and early 1960s, he planned a musical version, never produced, of Thackeray's novel *Vanity Fair*. He wrote and directed an independent film, *Another Sky* (1956), wrote the screenplay for his own 1963 novel *Inside Daisy Clover* (1965), and scripted *Bitter Victory* (1957), *The Roman Spring of Mrs. Stone* (1961), *I Never Promised You a Rose Garden* (1977), and others. His books include *On Cukor* (1972); *The Dangerous Edge* (1975), a study of nine thrillers; *The Goodbye People* (1977); *Running Time* (1983); *Norma Shearer: A Life* (1990); *Nazimova: A Biography* (1997); *Mainly About Lindsay Anderson* (2000); and *Natalie Wood: A Life* (2004). During the 1970s, as Isherwood tells, he settled in Tangier for a time, returning to Los Angeles in the early 1980s.

Lamkin, Hillyer Speed (b. 1928). American novelist; born and raised in Monroe, Lousiana. Isherwood met him in April 1950 when Speed was twenty-two and about to publish his first novel, *Tiger in the Garden*. He had studied at Harvard and lived in London and New York before going to Los Angeles to research his second novel, *The Easter Egg Hunt* (1954), about movie stars, in particular Marion Davies and William Randolph Hearst; he dedicated the novel to Isherwood who appears in it as "Sebastian Saunders." Lamkin was on the board, with Isherwood, at the Huntington Hartford Foundation. With a screenwriter, Gus Field, he tried to adapt *Sally Bowles* for the stage in 1950–1951, but Dodie Beesley criticized the project and encouraged John van Druten to try instead. In the mid-1950s, Lamkin wrote a play, *Out by the Country Club*, which was never produced, and in 1956, he scripted a T.V. film about Perle Mesta, the political hostess who was Truman's ambassador to Luxembourg. During 1957, he wrote another play, *Comes a Day*, which had a short run on Broadway, starring Judith Anderson and introducing George C. Scott. Eventually, when the second play failed, Lamkin returned home to Louisiana. He appears in *D.1*, *D.2*, and *Lost Years*.

Lamkin, Marguerite (b. *circa* 1929). A southern beauty, born and raised in Monroe, Louisiana, like her brother Speed and briefly educated at a Manhattan finishing school. She followed Speed to Hollywood, and married the screenwriter Harry Brown in 1952, but the marriage broke up melodramatically in 1955 as Isherwood records in *D.1*, where Marguerite is frequently mentioned. Bachardy had a room in the Browns' apartment during the early months of his involvement with Isherwood, and Marguerite was an especially close friend to him. In later years she also became close to Isherwood. She assisted Tennessee Williams as a dialogue coach during the original production of *Cat on a Hot Tin Roof*, and afterwards she worked on other films and theatrical productions on the East and West coasts and in England when southern accents were required. She was married to Rory Harrity from 1959 to 1963, and later settled in London, where she had a successful third marriage, became a society hostess and raised large sums of money for AIDS and HIV research and care. She also appears in *D.2*.

Lancaster, Mark (b. 1938). British artist, raised in Yorkshire, educated at Newcastle. He travelled to New York and Los Angeles while he was still an undergraduate writing a thesis on Alfred Steiglitz, appeared in Andy Warhol's films *Kiss* (1963), *Batman Dracula* (1964), and *Couch* (1964), helped Warhol with his silkscreens, and hung around The Factory taking photographs. Back in England, he taught at Newcastle and at the Bath Academy of Art in Corsham, and from 1968 to 1970 he was the first Artist in Residence at King's College, Cambridge. In 1972, he moved to New York, where he supported his painting by working as secretary and manager to Jasper Johns. He helped Johns to design productions for Merce Cunningham's dance troupe, took over the lighting, and became Artistic Advisor to the company, designing over twenty dances during the 1970s and 1980s. Isherwood first met him with David Hockney on the beach in Santa Monica in 1966. In 1985, Lancaster moved back to England briefly, then lived in Scotland, Miami, and Rhode Island.

Lanchester, Elsa (1902–1986). British actress; she danced with Isadora Duncan's troop as a child then began acting in a children's theater in London at sixteen. In 1929, she married Charles Laughton and went with him to Hollywood in 1934, settling there for good in 1940 and becoming an American citizen. Lanchester began making films before Laughton did and they acted in several together—for instance *The Private Life of Henry VIII* (1933) and *Witness for the Prosecution* (1957), for which she received an Academy Award nomination. Her most famous film was *The Bride of Frankenstein* (1935), but she was in many more, including *The*

Constant Nymph (1928), *David Copperfield* (1935), *Lassie Come Home* (1943), *The Razor's Edge* (1946), *The Secret Garden* (1949), *Come to the Stable* (1949, Academy Award nomination), *Les Misérables* (1952), *Bell, Book and Candle* (1958), *Mary Poppins* (1964), and *Murder by Death* (1976). She also worked in television and for many years she sang at a Los Angeles theater, The Turnabout, on La Cienega Boulevard. She toured with her own stage show, *Elsa Lanchester—Herself*, during 1960 and opened at the 41st Street Theater in New York on February 4, 1961 for seventy-five performances. She met Isherwood socially in the late 1950s, was greatly attracted to him and introduced him to Laughton, afterwards vying with Laughton and Bachardy for Isherwood's attention. In the summer of 1960 Laughton bought 147 Adelaide Drive, next door to Isherwood, so that he could spend time with male friends away from his wife in their house on Curson Avenue; after Laughton died, Lanchester often spent weekends at 147. She appears in *D.1* and *D.2*.

Lane, Homer (1875–1925). American psychologist, healer, and juvenile reformer. Lane established a rural community in England called The Little Commonwealth, where he nurtured young delinquents with love, farm work, and the responsibility of self-government. For Lane, the fundamental instinct of mankind "is the titanic craving for spiritual perfection," and he conceived of individual growth as a process of spiritual evolution in which the full satisfaction of the instinctive desires of one stage bring an end to that stage and lay the ground for the next, higher stage; he believed that instinctive desires must be satisfied rather than repressed if the individual is to achieve psychological health and fulfillment. In practice, Lane identified himself with the patient's neurosis in order to allow it to emerge from the unconscious; personally loving the sinner and the sin, he freed the patient from his sense of guilt. Auden discovered the teachings of Homer Lane through his Berlin friend, John Layard, a former patient and disciple of Lane's, and in late 1928 and early 1929, became obsessed with Lane, preaching his theories to his friends and in his poems.

Langan, Peter (1914–1988). Irish restaurateur; son of a Shell Oil Company director. He ran Odin's Restaurant in Devonshire Street with Kirsten Andersen, a Dane, who started the restaurant with her first husband, James Benson. (Benson died in a car crash in 1966.) Regular customers at Odin's included Hockney, Peter Schlesinger, Ossie Clark, Wayne Sleep, George Lawson and Patrick Procktor. Procktor became Kirsten Andersen's second husband in 1973. Hockney and Procktor sometimes paid for their meals with art works; Lawson paid in secondhand books; the pictures and books were used as décor in the restaurant. Langan later bought the Coq d'Or in Mayfair and reopened it as Langan's Brasserie, in partnership with Richard Shepherd and actor Michael Caine. He was an alcoholic and died of burns when he set fire to his house following a drunken argument with his wife, Susan, who escaped.

Lange, Hope (1931–2003). American actress, born and raised in Connecticut; she was twelve years old when she debuted on Broadway in *The Patriots* (1943). As a teenager, she waitressed in her mother's Greenwich Village restaurant, modelled, and continued as a stage actress and in live T.V. drama until she was brought to Hollywood with her first husband Don Murray to appear in *Bus Stop* (1956).

Afterwards she appeared in numerous other films including *Peyton Place* (1957), for which she was nominated for an Academy Award, *The Young Lions* (1958), *The Best of Everything* (1959), *Deathwish* (1974), *Blue Velvet* (1986), and *Clear and Present Danger* (1994). During her love affair with Glenn Ford, she co-starred with him in *Pocketful of Miracles* (1961) and *Love Is a Ball* (1963). She made a number of T.V. films and won two Emmy Awards for her role in the television comedy series "The Ghost and Mrs. Muir" (1968–1970); she also appeared on "The New Dick Van Dyke Show" (1971–1974). In 1977, she returned to Broadway in *Same Time Next Year* opposite Don Murray. Lange had two children with Murray, but the marriage ended in 1960. In 1963, she married the director and producer Alan Pakula; they divorced in 1969. In 1986, she married Charles Hollerith, a theatrical producer. Isherwood first met Lange with Murray in the late 1950s; she appears in *D.1* and *D.2*.

Lansbury, Angela (b. 1925). British star of stage and film; granddaughter of pacifist Labour politician George Lansbury and daughter of actress Moyna Macgill, who brought her with her twin brothers Edgar and Bruce to Hollywood to escape the Blitz. Lansbury was making feature films before the end of the war and went on to appear in *National Velvet* (1944), *The Picture of Dorian Gray* (1945), *The Three Musketeers* (1948), *The Dark at the Top of the Stairs* (1960), *The Manchurian Candidate* (1962), *Something for Everyone* (1970), *Bedknobs and Broomsticks* (1971), *Death on the Nile* (1978), *Nanny McPhee* (2006), and *Mr. Popper's Penguins* (2011) among others. Isherwood first mentions her at the time she made a hit in Tony Richardson's *A Taste of Honey* on Broadway in 1960; other Broadway successes include *Mame* (1966), *Sweeney Todd* (1979), and *Deuce* (2007). She has also had a T.V. career, especially in "Murder, She Wrote" (1984–1996). She has won eleven Tony Awards, six Golden Globes, and been nominated repeatedly for Academy Awards. She married twice, the second time, in 1949, to Peter Shaw with whom she had two children.

Lansbury, Edgar (b. 1930). Stage and film producer and, earlier, scenic designer; brother of Angela Lansbury and twin brother of stage and film producer Bruce Lansbury. Among his successful shows are *The Subject Was Roses* (1964) and *Godspell* (1976)—of which he also produced the films—and *American Buffalo* (1977). Bachardy drew Edgar, Bruce, and Angela Lansbury.

Larson, Jack (b. 1933). American actor, playwright and librettist; born in Los Angeles and raised in Pasadena. His father drove a milk truck, and his mother was a clerk for Western Union; they divorced. At fourteen, Larson was California bowling champion for his age group. He attended Pasadena Junior College, where he was discovered acting in a college play and offered his first film role by Warner Brothers in *Fighter Squadron* (1948) (also Rock Hudson's first film). He is best known for playing Jimmy Olsen in "The Adventures of Superman," the original T.V. series aired during the 1950s. He lived for over thirty-five years with the director James Bridges and co-produced some of Bridges's most successful films, *The Paper Chase* (1973), *Urban Cowboy* (1980), and *Bright Lights, Big City* (1988). As Isherwood mentions, he wrote the libretto for Virgil Thomson's opera *Lord Byron*. Larson and Bridges were close friends of Isherwood and Bachardy from the 1950s onward and appear in *D.1* and *D.2*.

Lathwood, Jo. See Masselink, Jo.

Laughlin, Leslie. See Caron, Leslie.

Laughlin, Michael. American film producer, and later, director and screenwriter; educated at Principia College in Illinois and at UCLA, where he studied law. He produced *Two Lane Blacktop* (1971) among others. He was the third husband of French actress Leslie Caron from 1969 to 1980. The thriller he was producing in 1973, *Nicole*—starring Caron with Catherine Bach, Ramon Bieri and others, and directed by István Ventilla—never reached the movie theaters, but it was released on video in 2005 as an erotic thriller, *Crazed*. (It also had another title, *Widow's Revenge*.) Isherwood and Bachardy saw it in a screening room at Goldwyn Studios. Laughlin lived with the costume and production designer Susanna Moore (b. 1945) from around the time in 1978 when she divorced her husband, production designer Richard Sylbert, but they never married. She became a successful novelist with *My Old Sweetheart* (1982), *The Whiteness of Bones* (1989), *In the Cut* (1995), and others.

Laura. See Huxley, Laura Archera.

Lawrence, Jerome (Jerry) (1915–2004). American playwright; born in Ohio, educated at Ohio State University and UCLA. He was a reporter and editor for small daily newspapers in Ohio then a continuity editor for a Beverly Hills radio station. By the time he joined the U.S. Army during World War II, he was a senior staff writer for CBS radio. In the army, he worked as a consultant to the Secretary of War then as a correspondent from North Africa and Italy, and he co-founded Armed Forces Radio with Robert Lee. They continued their partnership as playwrights after the war. Among their best-known plays are *Look, Ma, I'm Dancin'!* (1948), the prize-winning *Inherit the Wind* (1955) about the Scopes monkey trial, the stage adaptation of *Auntie Mame* (1956), and *The Night Thoreau Spent in Jail* (1971). Lawrence and Lee also wrote the book and the lyrics for the musical *Mame* (1966), adapted James Hilton's novel *Lost Horizon* as the book and lyrics for *Shangri-La* (1956), and were involved in adapting much of their work for film. Lawrence taught playwriting at several universities and was an adjunct professor at USC. Isherwood often went to parties at his house, especially to meet good-looking young men, mostly actors, whom Lawrence knew through his theater connections. Lawrence often claimed that he had introduced Isherwood and Bachardy to each other because Bachardy and his brother Ted attended a party at Lawrence's house on February 14, the date Isherwood and Bachardy marked as the start of their romance, but, in fact, Isherwood and Bachardy met earlier. Lawrence appears in *D.1* and *D.2*.

Lawson, George. Scottish antiquarian book dealer, a longtime director of Bertram Rota. He is a friend of David Hockney and a subject of several Hockney paintings. For many years he was Wayne Sleep's companion and remains, in his own phrase, Sleep's accomplice.

Layard, John (1891–1975). English anthropologist and Jungian psychoanalyst. He read Medieval and Modern Languages at Cambridge and did field work in the New Hebrides with the anthropologist and psychologist W.H.R. Rivers. In the early 1920s, he had a nervous breakdown and was partially cured by the

American psychologist Homer Lane. Lane died during the treatment, leaving Layard depressed and seeking further treatment, first unsuccessfully with Wilhelm Stekel and eventually more productively with Jung. Auden met Layard in Berlin late in 1928 and introduced him to Isherwood the following spring; for a time all three were obsessed with Lane's theories recounted by Layard. During this period, Layard had a brief and tortured triangular affair with Auden and a German sailor, Gerhart Meyer, whereupon he tried to kill himself. Isherwood used the suicide attempt in *The Memorial*, and Layard appears as "Barnard" in *Lions and Shadows*. Layard eventually recovered his psychological health so that he was able to work and write again, and he married and had a son. Like Auden, he also returned to the Anglican faith of his childhood.

Lazar, Irving (Swifty) (1907–1993). Agent and deal-maker for movie stars and authors such as Lauren Bacall, Humphrey Bogart, Truman Capote, Noël Coward, Ernest Hemingway, Vladimir Nabokov, Cole Porter, Diana Ross, Irwin Shaw, and Tennessee Williams. He practiced bankruptcy law in New York during the Depression, then relocated to Hollywood in 1936. His wife was called Mary.

Leavitt, Natalie. Isherwood and Bachardy's Romanian-born cleaning lady and cook from about 1976; she was a widow after nursing her husband, an American, through a long illness. She was first employed at Adelaide Drive around 1960, but irritated Isherwood by asking Stravinsky for an autograph while he was at the dinner table. She eventually returned as twice-monthly cleaner, then server and dishwasher, and finally chef and shopper for dinner parties about twice a week until her retirement around 2000. Her daughter, Thaïs Leavitt, trained with the Royal Ballet in Covent Garden and was a principal dancer with the Düsseldorf Ballet until she retired with a knee injury. Thaïs later taught at Ballet Petit Performing Arts Center, the Los Angeles County Council for the Arts, and elsewhere, settled with her mother in the family home near 145 Adelaide Drive, and occasionally helped with the work there.

Lehmann, Beatrix (Peggy) (1903–1979). English actress; youngest of John Lehmann's three elder sisters. She met Isherwood in Berlin in 1932, and they remained close friends. She had a London triumph in O'Neill's *Mourning Becomes Electra* in 1938 when Isherwood was in China, and he returned in time to see her in the Group Theatre's performance of Cocteau's *La Voix Humaine* in July. During 1938 she had an affair with Berthold Viertel. She appears in *D.2*.

Lehmann, John (1907–1987). English author, publisher, editor, autobiographer; educated at Cambridge. Youngest child and only son of a close family; his mother was an American from New England; his father trained as a barrister and wrote for *Punch*. Isherwood met him in 1932 at the Hogarth Press where Lehmann was assistant (later partner) to Leonard and Virginia Woolf. Lehmann persuaded the Woolfs to publish *The Memorial* after it had been rejected by Jonathan Cape, publisher of Isherwood's first novel *All the Conspirators*. Isherwood helped Lehmann with his plans to found the magazine *New Writing* and obtained early contributions from friends like Auden. He writes about this in *Christopher and His Kind* and also about Lehmann in *D.1*, *D.2*, and *Lost Years*. When Lehmann left the Hogarth Press, he founded his own publishing firm and later edited *The London Magazine*.

He wrote three volumes of autobiography, *The Whispering Galley* (1955), *I Am My Brother* (1960), and *The Ample Proposition* (1966). For many years he shared his house with the dancer Alexis Rassine.

Lehmann, Rosamond (1901–1990). English novelist, educated at Cambridge, second-eldest sister of John Lehmann. She made a reputation with the sexual and emotional frankness of her first novel, *Dusty Answer* (1927), and her later works—including *Invitation to the Waltz* (1932), *The Weather in the Streets* (1936), *The Echoing Grove* (1953)—also shocked. Her first marriage, in 1923, was to Leslie Runciman, son of a Liberal Member of Parliament, and from 1928 to 1944, she was the first wife of the painter Wogan Philipps, with whom she had a son and a daughter. Afterwards, she had a nine-year affair with Cecil Day-Lewis. Her daughter with Philipps, Sally, died suddenly of polio in 1958 when she was twenty-four; Rosamond described her continuing spiritual relationship with Sally in *The Swan in the Evening: Fragments of an Inner Life* (1967). She appears in *D.1*, *D.2*, and *Lost Years*.

Leighton, Margaret (Maggie) (1922–1976). English actress. She made her London debut as a teenager and established her reputation in the Old Vic Company in the late 1940s. From the mid-1950s until the late 1960s, she also played on Broadway, where she won Tony Awards for *Separate Tables* (1956) and *The Night of the Iguana* (1962). Her films include *The Winslow Boy* (1948), *The Sound and the Fury* (1959), *The Loved One* (1965), and *The Go-Between* (1971). Her second marriage was to Laurence Harvey, from 1957 to 1961, and her third, in 1964, to actor Michael Wilding. She had a small role in Bachardy and Isherwood's "Frankenstein: The True Story."

Len. See Worton, Len.

Leopold, Michael. Aspiring writer, from Texas; he was about eighteen when Isherwood met him at the apartment of a friend, Doug Ebersole, in December 1949. They began a minor affair soon afterwards. Leopold was interested in literature, admired Isherwood's work, and later wrote some stories of his own. During the 1960s, he lived with Henry Guerriero in Venice, California. He appears in *D.1*, *D.2*, and *Lost Years*.

Le Page, Richard W.F. (b. 1940). British medical microbiologist, Life Fellow, Caius College, Cambridge. Author and co-author of numerous research papers in many areas, including vaccine delivery.

LeSueur, Joe (192[4]–2001). Aspiring playwright, screenwriter, and critic, originally from California. He was the companion, from 1955, of American poet and art critic Frank O'Hara (1926–1966), who died in a sand buggy accident on Fire Island.

Levant, Oscar (1906–1972). American composer, pianist, and actor. He was a close friend of George Gershwin and became famous as an interpreter of his music. His film appearances include *Kiss the Boys Goodbye* (1941), *Rhapsody in Blue* (1945), *Humoresque* (1946), *You Were Meant for Me* (1948), and *An American in Paris* (1951). Levant wrote the music for several popular musicals and had a live talk show in Hollywood, "The Oscar Levant Show," broadcast out of a shed on a minor network. His show was shut down by the sponsors in the early 1960s despite its popularity, because he insulted their products for laughs and encouraged his guests to do the same. Isherwood appeared on the show in the mid-1950s, sometimes reading poetry; this led to his occasionally being recognized in the street. In 1958, he argued with Levant about Churchill and then refused to return to the show for a time because Levant attacked him for remaining in Hollywood during the war. He appears in *D.1* and *D.2*.

Levy, Miranda Speranza (Mirandi). Sister-in-law of the Italian-American jewelry designer, Frank Patania (d. 1964) whose Native American-influenced work in silver, turquoise, and coral was bought by Mabel Dodge Luhan and Georgia O'Keeffe and can be seen in museums. She ran Patania's Thunderbird shop in Santa Fe and married Ralph Levy, a Hollywood film director. She appears in *D.1* under her maiden name, Mirandi Masocco, and in *D.2*.

Lipscomb, Mark. American painter working in Los Angeles. He had a show at the Long Beach Museum of Art in the 1970s, but afterwards exhibited his work only at home in the various houses he shared with his longterm companion John Ladner. In addition to sitting for Bachardy, he was also drawn and painted by David Hockney.

List, Herbert. German photographer. Probably introduced to Isherwood by Stephen Spender who, in 1929, became friends with List in Hamburg, where List was working as a coffee merchant in his family's firm. List appears as "Joachim" in Spender's *World Within World* and as "Joachim Lenz" in Spender's *The Temple*. Isherwood writes about him in *D.1* and *D.2*.

Little, Penny. Sales representative for a fabric house and later for a large interior design firm. She was the girlfriend of painter Billy Al Bengston for about twenty years. Afterwards she was briefly married to Kevin Hearst, and then to Gregg Hawks, son of film director Howard Hawks. She sat for Bachardy several times.

Littman, Robert (Bobby) (1938–2001). Hollywood agent, from England. He began representing Isherwood and Bachardy in 1976. He was previously a head of MGM in the U.K. His wife's name was Doris.

Locke, Charles O. (1896–1977). American journalist, novelist, screenwriter. He wrote for his family's newspaper in Toledo, Ohio, while still in college then had a career in New York working for several major papers and in advertising and publicity. He wrote poetry and song lyrics as well as five novels. In 1957, he wrote the screenplay from his most famous novel *The Hell-Bent Kid* in the office next door to Isherwood's at Twentieth Century-Fox. Isherwood describes their friendship in *D.1*, where Locke's wife and his daughter, Mary Schmidt, are also mentioned. He also appears in *D.2*.

loka. In Hindu traditions, world or plane of existence.

Loos, Anita (1888–1981). American playwright, screenwriter, and novelist. Isherwood met her through Aldous and Maria Huxley soon after arriving in Hollywood and sometimes attended the Sunday lunches at which she entertained her circle of emigré friends. Loos created the art of silent film captions, and later she wrote over two hundred screenplays for sound movies. Her novel *Gentlemen Prefer Blondes* (1925)—written to amuse H.L. Mencken—became a play, then a movie, then a musical comedy for stage, and finally a film of the musical comedy. She launched Huxley in studio writing and also introduced him socially. She wrote several volumes of autobiography in which Isherwood is occasionally mentioned. She appears in *D.1*.

Loud, Lance (1951–2001). Flash celebrity. He became famous in 1973 in a twelve-part public T.V. documentary "An American Family," a precursor of reality T.V. It showed seven months in the life of his upper-middle-class Santa Barbara family and included Loud's revelation, in a shabby New York hotel, that he was gay. He later started and played in a rock band called The Mumps and then worked as a journalist. He became addicted to crystal meth and died of AIDS.

Loy, Myrna (1905–1993). Hollywood star; born in Montana, raised there and in Los Angeles, where she became a chorus girl at eighteen. She was an exotic screen vamp in over sixty films until the mid-1930s, when she began to appear as Nora Charles in movies based on Dashiell Hammett's *The Thin Man*, and became Hollywood's number-one woman star. Her later films include *The Best Years of Our Lives* (1946), *The Red Pony* (1949), and *Midnight Lace* (1960). During World War II, she worked for the Red Cross and, later, for UNESCO. She was also active in the Democratic party. She appears in *D.2*.

Luce, Henry R. (1898–1967) and Clare Boothe (1903–1987). He was born in China and educated at Hotchkiss, Yale, and Oxford. He worked as a journalist before he co-founded *Time Magazine* in 1923. In the 1930s, he launched *Fortune* and *Life*, and then *House and Home* and *Sports Illustrated* in the 1950s. She was the illegitimate, peripatetically educated daughter of a dancer mother and a violinist father who deserted them. She worked as an actress briefly, was editor of *Vogue* and *Vanity Fair*, a successful Broadway playwright—*Abide with Me* (1935), *The Women* (1936), *Kiss the Boys Goodbye* (1938)—a Republican congresswoman and U.S. Ambassador to Italy (1953–1957). Each was married once before. She converted to Roman Catholicism in 1946 after her daughter, an only child by her first husband, was killed in a car accident. The Luces were both vehement anticommunists.

Luckenbill, Dan (b. 1945). American writer and librarian. From 1970, he worked at the UCLA library in the manuscripts division. Later he curated exhibitions and wrote catalogues on gay and lesbian studies. He published short stories from the 1970s onwards.

Luckinbill, Laurence (b. 1934). American actor, writer and director. His film appearances include *The Boys in the Band* (1970) and *Star Trek V: The Final Frontier* (1989); on T.V., he had roles in "Law and Order" and "Murder, She Wrote" among others. He also wrote and performed one-man stage shows about American "heroes," such as Hemingway, Teddy Roosevelt, Clarence Darrow and LBJ. He was married to Robin Strasser until 1976 and had two children with her; later he married Lucie Arnaz, with whom he had three more children.

Ludington, Wright (1900–1992). Art collector and philanthropist; raised in Pennsylvania, educated at the Thacher School in Ojai, at Yale, at the Pennsylvania Academy of Fine Arts, and at the Art Students League in New York. In 1927, he inherited a fortune from his father, a lawyer and investment banker who worked with the Curtis Publishing Company. He also inherited an estate in Montecito—Val Verde—and spent decades improving it with the help of a school friend, the landscape architect Lockwood de Forest. Val Verde included an art gallery for Ludington's collection of modern paintings and outdoor settings for his ancient sculpture. In 1955, he sold Val Verde and built a new house off Bella Vista Drive—Hesperides—which was designed by Lutah Maria Riggs especially to display his art collection. De Forest's widow landscaped Hesperides. Ludington was a founder and board member of the Santa Barbara Museum of Art and gave the museum many pieces from his collection. He appears in *D.2*.

M. Isherwood's mother. He called her "Mummy" and began letters to her with "My Darling Mummy," and later, "Dearest Mummy," but he invariably wrote "M." in his diaries. See Isherwood, Kathleen.

Isherwood sometimes uses "M." for Mahendranath Gupta, the schoolmaster who became Ramakrishna's disciple and recorded Ramakrishna's conversations and sayings in his diaries, later compiling them in *Sri Ramakrishna Kathamrita or The Gospel of Ramakrishna*.

MacDonald, Madge. Nurse. She worked at UCLA hospital and lived in Isherwood's neighborhood. She appears in *D.1* and *D.2*.

Macht, Stephen (b. 1942). American actor. He appeared in suspense, disaster and horror films, many for T.V.; he also appeared on "Knots Landing" in the early 1980s.

Macy, Gertrude (190[4]–1983). New York stage manager and producer; secretary and biographer of actress Katherine Cornell. She co-produced the 1951 stage version of *I Am a Camera* with Walter Starcke and thereby had a substantial financial stake in *Cabaret*, which reduced Isherwood's earnings and which he several times refers to in his diaries. When the film of *Cabaret* came out in 1972, she made a new claim based on her original involvement as co-producer of van Druten's play. She is mentioned in *D.1* and *D.2*.

Madigan, Leo (Larry). Roman Catholic writer and teacher, born and educated in New Zealand. As Isherwood tells in his entry for April 6, 1970, he was a merchant seaman and first shipped out as a teenager. He also tried acting and psychiatric work. His early writing appeared in a magazine called *The Seafarer*, published by the Marine Society in London, and which he edited for a time; he also contributed to *Blackwood's Magazine*. His first novel, *Jackarandy* (1973), is about the professional gay sex scene in London. Later, he settled in Fatima, Portugal, and published religious books, *The Devil Is a Jackass* (1995), *The Catholic Quiz Book* (1995), *Fatima: Highway of Hope* (2000), *What Happened at Fatima* (2000), and an English-language guide to the shrine. He has also written Catholic novels for children.

Maharaj. See Brahmananda, Swami.

mahasamadhi. Great samadhi, usually referring to the moment of death, when the illuminated soul leaves the body and is absorbed into the divine. See samadhi.

Mailer, Norman (1923–2007). American writer, raised in Brooklyn and educated at Harvard. He fought in the Pacific during World War II and became famous with the publication of his first novel, *The Naked and the Dead* (1948), about an American infantry platoon invading a Japanese-held island. He twice won the Pulitzer Prize: for *The Armies of the Night* (1968) and *The Executioner's Song* (1979). Other novels and works blending fiction with non-fiction and personal commentary include *The Deer Park* (1955), *An American Dream* (1965), *Why Are We in Vietnam?* (1967), *Of a Fire on the Moon* (1970), *The Prisoner of Sex* (1971), *Ancient Evenings* (1983), *Tough Guys Don't Dance* (1984), *Harlot's Ghost* (1991), *Oswald's Tale: An American Mystery* (1995), *The Gospel According to the Son* (1997), and *The Castle in the Forest* (2007). He also wrote screenplays and directed films. He co-founded *The Village Voice* in 1955, and in 1969 he ran for mayor of New York. He married six times. In *Lost Years*, Isherwood tells of his first meeting with Mailer in 1950, and Mailer also appears in *D.2*.

Mallory, Margaret (b. 1911). Art collector and philanthropist. She lived with Alice Story, known as Ala, and they travelled and collected together. Mallory was a benefactor of UCSB, endowing fellowships in music and art history and donating parts of her collection. She appears in *D.2*.

Mangeot, André (1883–1970) and Olive (1885–1969). Belgian violinist and his English wife; parents of Sylvain and Fowke Mangeot. Isherwood met them in 1925 and worked for a year as part-time secretary to André Mangeot's string quartet which was organized from the family home in Chelsea. He brought friends to meet Olive when he was in London. She is the original for "Madame Cheuret" in *Lions and Shadows*, and Isherwood drew on aspects of her personality for "Margaret Lanwin" and "Mary Scriven" in *The Memorial*. Olive had an affair with Edward Upward and through his influence became a communist. Later, she separated from her husband and for a time shared a house with Jean Ross and Jean's daughter in Cheltenham. Hilda Hauser, the Mangeots' housekeeper and cook, also moved with Olive to Cheltenham, where, together, they raised Hilda's granddaughter, Amber. As Isherwood tells in *D.1* and in *Lost Years*, Hilda's daughter Phyllis was raped by a black G.I. during World War II and Amber resulted. The Mangeots also appear in *D.2*.

Mangeot, Sylvain (1913–1978). Younger son of Olive and André Mangeot. Isherwood's friend Eric Falk initially introduced Isherwood to the Mangeot family because Sylvain, at age eleven, had a bicycle accident which confined him to a wheelchair for a time, and Isherwood had a car in which he could take Sylvain for outings. They grew to know each other well during the time that Isherwood worked for Sylvain's father, and together they made a little book, *People One Ought to Know*, for which Isherwood wrote nonsense verses to accompany Sylvain's animal paintings (it was eventually published in 1982, but one pair of verses appeared earlier as "The Common Cormorant" in Auden's 1938 anthology *The Poet's Tongue*). Sylvain is portrayed as "Edouard" in *Lions and Shadows*. Later he joined the Foreign Office and then became a journalist, working as a diplomatic correspondent, an editor, and an overseas radio commentator for the BBC. He appears in *D.1* and *D.2*.

Manionis, Anthony (Tony). Aspiring actor. He understudied and was an assistant stage manager for the Phoenix Theater Broadway production of *School for Wives* in 1971 and had a part in the first New York production *A Meeting by the River.*

Mann, Erika (1905–1969). German actress and author, eldest daughter of Thomas Mann. Isherwood met her in the spring of 1935 in Amsterdam through her brother Klaus; she had fled Germany in March 1933. Her touring satirical revue, "The Peppermill" (for which she wrote most of the anti-Nazi material), earned her the status of official enemy of the Reich and she asked Isherwood to marry her and provide her with a British passport. He felt he could not, but contacted Auden who instantly agreed. The two met and married in England on June 15, 1935, the very day Goebbels revoked Mann's German citizenship. In September 1936, Erika emigrated to America with Klaus and unsuccessfully tried to reopen "The Peppermill" in New York. As the war approached, she lectured widely in the U.S. and wrote anti-Nazi books, two with Klaus, trying to revive sympathy for the non-Nazi Germany silenced by Hitler. She worked as a journalist in London during the war, for the BBC German Service and as a correspondent for the New York *Nation.* Later, she grew closer to her father, travelling with her parents and helping Mann with his work. She appears in *D.1* and *Lost Years.*

Mann, Heinrich Klaus (1906–1949). German novelist and editor, eldest son of Thomas Mann. Isherwood became friendly with him in Berlin in the summer of 1931. By then Klaus had written and acted with his sister, Erika, in the plays which launched her acting career, and he had published several novels in German (a few appeared in English translations) and worked as a drama critic. He travelled extensively and lived in various European cities before he left Germany for good in 1933; he emigrated to America in 1936 when his family settled in Princeton. He lived in New York, continued to travel to Europe as a journalist, and eventually settled for a time in Santa Monica. He founded two magazines, *Die Sammlung* (The Collection) in Amsterdam in 1933, and *Decision*, which appeared in New York in December 1940 but lasted only a year because of the war. He became a U.S. citizen and served in the U.S. army during the war. He wrote his second volume of autobiography, *The Turning Point* (1942), in English. He attempted suicide several times, finally succeeding in Cannes. Isherwood wrote a reminiscence for a memorial volume published in Amsterdam in 1950, *Klaus Mann—zum Gedaechtnis* (reprinted in *Exhumations*), and he describes their friendship in *D.1* and *Lost Years.*

mantram or mantra. A Sanskrit word or words which the guru tells his disciple when initiating him into the spiritual life and which is the essence of the guru's teaching for that particular disciple. The mantram is a name for God and includes the word *Om*; the disciple must keep the mantram secret and meditate for the rest of his life on the aspect of God which it represents. Repeating the mantram (making japam) purifies the mind and leads to the realization of God. With the mantram, the guru often gives a rosary—as Swami Prabhavananda gave Isherwood—on which the disciple may count the number of times he repeats his mantram.

Marguerite, previously Marguerite Brown, also Marguerite Harrity. See Lamkin, Marguerite.

Markovich, John (Mark) (193[2]–2008). American painter and monk of the Ramakrishna Order, born in Detroit. He was known as Mark, and later as Brahamachari Nirmal and then Swami Tadatmananda. He became the official abbot of Trabuco. He appears in *D.2*.

Marple Hall. The Bradshaw Isherwood family seat; see entries for Frank Bradshaw Isherwood and Richard Bradshaw Isherwood.

Marre, Albert (b. 1925). American theater director, writer and producer; born in New York. He acted on Broadway, moved to directing classical theater, including Van Brugh and Shaw, and had a hit in 1953 with the musical *Kismet*. He cast singer and actress Joan Diener (1930–2006) in one of the leads and she became his wife in 1956. He was nominated for a Tony Award for *The Chalk Garden* in 1956 and had another major hit with *The Man of La Mancha* in 1965 for which he won a Tony and which he frequently revived. Joan Diener again played the lead, Aldonza / Dulcinea, her most successful role.

Mason, James (1909–1984). British actor, educated at Marlborough and Cambridge. He was a conscientious objector during World War II. He joined the Old Vic Theatre Company in the 1930s, soon began making films, and became a star in *The Seventh Veil* (1945), before moving on to Hollywood where he often played villains. Later films include *Julius Caesar* (1953), *The Desert Rats* (1953), *20,000 Leagues Under the Sea* (1954), *North by Northwest* (1959), *Lolita* (1962), *Lord Jim* (1965), *Georgy Girl* (1966), and *The Verdict* (1982). On T.V. he played Franz Gruber in Isherwood and Danny Mann's "The Legend of Silent Night" and Polidori in "Frankenstein: The True Story." He was married to actress Pamela Kellino from 1941 to 1965 and, from 1971, to Clarissa Kaye. He appears in *D.2*.

Masselink, Ben (1919–2000). American writer. Probably Isherwood and Bill Caskey met Ben Masselink with his longtime companion Jo Lathwood in the Friendship Bar in Santa Monica around 1949; they appear often in *D.1* and *Lost Years*. During the war, Masselink was in the marines; one night on leave, he got drunk in the Friendship and Jo Lathwood took him home and looked after him. When the war was over he returned and stayed for twenty years. Isherwood alludes to this meeting in his description of The Starboard Side in *A Single Man*. Masselink had studied architecture, and Isherwood helped him with his writing career during the 1950s. His first book of stories, *Partly Submerged*, was published in 1957. He then published several novels: two about his war experience—*The Crackerjack Marines* (1959) and *The Deadliest Weapon* (1965), the second of which Isherwood greatly admired—and *The Danger Islands* (1964), for teenage boys. He also wrote for television throughout the 1950s and in 1960 worked at Warner Brothers on the script for a film of *The Crackerjack Marines*. As Isherwood tells in *D.2*, Masselink left Jo in 1967 for a younger woman, Dee Hawes, the wife of their friend, Bill Hawes. Dee had an adopted daughter, Heather, from her first marriage.

Masselink, Jo (*circa* 1900–1988). Women's sportswear and bathing suit designer, from Northville, South Dakota; among her clientele were movie stars such as Janet Gaynor and Anne Baxter. In youth, she worked as a dancer and was briefly married to a man called Jack Lathwood whose name she kept professionally. Also, she had a daughter, Betty (see Arizu), and a son with a North Dakotan, Ferdinand Hinchberger. From 1938 onwards, Jo lived in an apartment on West Channel Road, a few doors from the Friendship, and by the late 1940s she knew many of Isherwood's friends who frequented the bar—including Bill Caskey, Jay de Laval, and Jim Charlton. She never married Ben Masselink, though she used his surname while they lived together. She and Masselink figure through much of *D.1*, *D.2*, and *Lost Years* as Isherwood's closest heterosexual friends.

Maugham, William Somerset (Willie) (1874–1965). British playwright and novelist. He had a daughter, Liza Wellcome (1915–1998) with Syrie Wellcome, whom he married in 1917 but never lived with. In the 1960s, he tried to disinherit Liza, saying she was the daughter of Henry Wellcome, Syrie's husband at the time of Liza's birth, but the court decided Maugham was the biological father. During Maugham's marriage, his companion was Gerald Haxton, eighteen years younger; he met Haxton in 1914 when they both worked in an ambulance unit in Flanders. They travelled and entertained on Cap Ferrat at the Villa Mauresque which Maugham bought in 1926. Haxton died in 1944, and Maugham's subsequent companion and chosen heir was Alan Searle. Isherwood met Maugham in London in the late 1930s and saw him whenever Maugham visited Hollywood, where many of Maugham's works were filmed; with Bachardy, Isherwood later made several visits to the Villa Mauresque. In 1945, Isherwood worked for Wolfgang Reinhardt on a screenplay for Maugham's 1941 novel *Up at the Villa* (never made), and he enlisted Swami Prabhavananda to advise Maugham on the screenplay for *The Razor's Edge* (1944). Although Maugham did not follow their advice, Isherwood and Swami again helped him in 1956 with an essay, "The Saint," about Ramana Maharshi (1879–1950), the Indian holy man Maugham had met in 1936 and on whom he had modelled Shri Ganesha, the fictional holy man in *The Razor's Edge*. Maugham appears in *D.1*, *D.2*, and *Lost Years*.

Maupin, Armistead (b. 1944). American writer; born in Washington, D.C., raised in North Carolina, and educated at the University of North Carolina at Chapel Hill. He served in the navy and worked as a newspaper journalist in Charleston and in San Francisco, where he settled in 1971. His best-selling *Tales of the City* (1978), set in San Francisco, and the many sequels, were begun as a newspaper serial and led to a T.V. miniseries based on the first three of the novels. Other books are *Maybe the Moon* (1992) and *The Night Listener* (2001), which was also made into a film.

maya. In Vedanta, maya is the cosmic illusion, the manifold universe which the individual perceives instead of perceiving the one reality of Brahman; in this sense, maya veils Brahman. But maya is inseparable from Brahman and can also be understood as the manifestation of Brahman's power, god with attributes. Maya has a double aspect encompassing opposite tendencies, toward ignorance (avidya) and toward knowledge (vidya). Avidya-maya involves the individual in worldly passion; vidya-maya leads to spiritual illumination.

McCallum, David (b. 1933). Scottish actor. He co-starred in *The Great Escape* (1963) and became known for his role as Illya Kuryakin in the T.V. series "The

Man from U.N.C.L.E." (1964–1968) before appearing as Clerval in "Frankenstein: The True Story."

McCallum, Rick (b. 1952). T.V. and movie producer; son of Pat York; educated at Columbia University. He was a crew member on two Merchant-Ivory productions and assisted John Frankenheimer on three films in the 1970s. Later, he worked with British T.V. and film writer Dennis Potter on "Pennies from Heaven" and "The Singing Detective" among others, and at the BBC. In the 1990s, he became widely known for producing the T.V. series "The Young Indiana Jones Chronicles" and for the *Star Wars* prequels, all for George Lucas.

McCarthy, Frank (1912–1986). American film producer. He rose to brigadier general in the army during W.W.II and worked as a reporter before producing *Decision Before Dawn* (1951), *Patton* (1970; three Academy Awards), and *MacArthur* (1977). He also headed public relations for Twentieth-Century Fox.

McCarty-Cooper, Billy (193[7]–1991). American interior designer, raised in Florida; originally called Billy McCarty. He studied architecture at the University of Pennsylvania, worked for David Hicks in London and then opened his own firm. In 1972, he was adopted by Douglas Cooper, and changed his name accordingly. When Cooper died in 1984, McCarty-Cooper inherited his fortune and Château de Castille with its Picassos and other art works, many of which he sold. With the proceeds, he became a collector in his own right and entertained lavishly. He died of AIDS.

McCoy, Ann (b. 1946). American painter, born in Boulder, Colorado, educated there and at UCLA. When Isherwood first mentions her in 1976, she was teaching aesthetics at Claremont Graduate School. She later taught at the University of Pennsylvania, Barnard, and elsewhere. She has written numerous articles and contributions to catalogues, exhibited widely, and won many awards. Her work is in the Los Angeles County Art Museum; the Metropolitan Museum, the Museum of Modern Art, and the Whitney in New York; the Hirshhorn; the Art Institute of Chicago; the Dallas Art Museum; and other major collections.

McDermott, Mo (d. 1988). British fabric designer and artist, educated at Salford College of Art. He began modelling for David Hockney during Hockney's last year at the Royal College of Art and became a close friend. He went on to work as Hockney's studio assistant and continued to pose frequently for him. Later, he married and moved to California, where he settled in Echo Park.

McDowall, Roddy (1928–1998). British actor and photographer; he made his first film aged eight. During the Blitz, he was evacuated to the U.S. and became a teenage Hollywood star as the crippled boy in *How Green Was My Valley* (1941) and with Elizabeth Taylor in *Lassie Come Home* (1943). In the 1950s, he took stage and television roles in New York and won a Tony Award for his supporting role in *The Fighting Cock* (1960), then returned to Hollywood in the 1960s and starred in *Planet of the Apes* (1968), most of the sequels, and the T.V. series. His other films include *My Friend Flicka* (1943), Orson Welles's *Macbeth* (1948), *The Subterraneans* (1960), *The Longest Day* (1962), *Cleopatra* (1963), *Bedknobs and Broomsticks* (1971), *The Poseidon Adventure* (1972), *Funny Lady* (1975), and *Fright*

Night (1985). He published several books of his photographs, mostly of celebrities. He appears in *Lost Years*.

McGuire, Dorothy (1916–2001). American stage and movie actress, born in Nebraska. She appeared on Broadway from 1938 and was brought to Hollywood by Selznick to reprise a stage role in *Claudia* (1943). Other films include *A Tree Grows in Brooklyn* (1945), *The Spiral Staircase* (1946), *Gentleman's Agreement* (1947), *Old Yeller* (1957), and *Swiss Family Robinson* (1960). In 1976, she returned to Broadway in *The Night of the Iguana*, and she often appeared on T.V. She married photographer John Swope (1908–1973) in 1943. Swope came to Hollywood in 1936 as a production assistant and publicity photographer for Leland Hayward then worked as a U.S. Navy photographer during World War II, and afterwards for *Life* magazine. He published his first book of photographs, *Camera over Hollywood*, in 1939.

McWhinnie, Donald (1920–1987). British T.V. producer and director, from Yorkshire. He was a BBC script editor, then a producer, and continued to work in T.V. throughout his career. During the 1960s, he directed several plays on Broadway. He was nominated for a Tony Award for his 1962 production of Harold Pinter's *The Caretaker*, and he directed Gladys Cooper in *A Passage to India* the same year.

Medley, Robert (1905–1994). English painter. He attended Gresham's School, Holt, with Auden, and they remained close friends after Medley left for art school at the Slade. In London, he became the longtime companion of the dancer Rupert Doone and was involved with him in 1932 in founding The Group Theatre, which produced *The Dog Beneath the Skin*, *The Ascent of F6*, and *On the Frontier*. He also worked as a theater designer and teacher and founded the Theatre Design section at the Slade in the 1950s before becoming Head of Painting and Sculpture at the Camberwell School of Arts and Crafts in 1958. He appears in *D.1*, *D.2*, and *Lost Years*.

Mendelson, Edward (b. 1946). American scholar and writer, educated at the University of Rochester and Johns Hopkins; he has been a professor of English literature at Yale, Harvard, and Columbia Universities. Auden appointed him literary executor in 1972. When Mendelson visited Isherwood in 1974, he was preparing *The English Auden*, first of many posthumous editions by Auden, and he was looking for poems which Auden enclosed in his letters to Isherwood; no volume of Auden's letters has been published, although a few letters have appeared in scholarly publications. Also in about 1974, Mendelson and Stephen Spender proposed a jointly written biography to Random House, and Spender asked Isherwood for his help, as Isherwood records early in 1975. But they abandoned the idea after a year because they were concerned about the privacy of those still living and they each had too many other projects. Mendelson eventually wrote two literary-critical biographies, *Early Auden* (1981) and *Later Auden* (1999).

Merchant, Ismail (1936–2005). Indian film producer and director. He got an MBA at New York University and provided the financial expertise in his long-running partnership with American director James Ivory, who was also his companion. They became virtually a brand in the 1980s and 1990s with their adaptations of Henry James and E.M. Forster, generally working closely with German-born screenwriter Ruth Prawer Jhabvala, Indian by marriage. Merchant directed two shorts, *The Creation of a Woman* (1960, Academy Award nomination) and *Mahatma and the Mad Boy* (1973), a T.V. documentary called "The Courtesans of Bombay" (1973), and, later, feature films including *In Custody* (1973), *The Proprietor* (1996), *Cotton Mary* (1999), and *The Mystic Masseur* (2002).

Meredith, Burgess (1907–1997). American actor and director. He distinguished himself in the theater in the early 1930s, moved to film in 1936 with his stage role in *Winterset*, and appeared in many subsequent films, including *Of Mice and Men* (1939), *The Story of G.I. Joe* (1945), *Advise and Consent* (1961), and *Rocky* (1976). He also played the Penguin in the Batman T.V. series. His third of four wives was actress Paulette Goddard, from 1944 to 1949. Meredith was blacklisted in 1949, and disappeared from movies for nearly a decade. In *D.1*, Isherwood tells how his plan to play Ransom in *The Ascent of F6* was interrupted by the war; in *D.2*, Isherwood records Meredith's interest in directing *The Dog Beneath the Skin*.

Merlo, Frank (1921–1963). Italian-American companion of Tennessee Williams; raised in New Jersey by his Sicilian immigrant parents. He met Williams in 1947. He had served in the navy and for a time continued to work as a truck driver. He was handsome and capable, kept house, cooked, and made travel, social and business arrangements for Williams. Isherwood first met him in Los Angeles when Merlo accompanied Williams on a visit there in 1949, and Merlo appears in *D.1*, *D.2*, and *Lost Years*. The relationship grew less stable during the late 1950s, and the pair were often apart during Merlo's fatal illness with lung cancer; but it was the most lasting romance of Williams's life, and Williams was at his bedside when Merlo died.

Methuen. Isherwood's English publisher from the mid-1940s. Isherwood's cousin, Graham Greene, recommended *Mr. Norris Changes Trains* to E.V. Rieu, a managing director at the firm, but Isherwood's first book published by Methuen was *Prater Violet* in the spring of 1946 (well after the U.S. publication because of the war). In September 1935, Isherwood had signed a contract for his "Next three full-length available novels" and accepted half of a £300 advance on the first (styled in the contract *Prata Violet*); he had already promised his next novel to his current publisher, The Hogarth Press, and at Leonard Woolf's insistence, *Sally Bowles*, *Goodbye to Berlin* and *Lions and Shadows* were published by Hogarth. After the war, the contract with Methuen was honored, the second novel being *The World in the Evening* and the third, *Down There on a Visit*, which he delivered to Alan White in 1961. White joined Methuen in 1924, became a director in 1933, and retired as Chairman in 1966. When White retired, John Cullen became Isherwood's editor, and after Cullen, Geoffrey Strachan. Methuen remained Isherwood's U.K. publisher for the rest of his life and posthumously until 1997, when Random House took over the imprint which by then belonged to a larger group, Reed Books. Methuen achieved independence through a management buy-out, but agreed in the negotiations to let Isherwood go to Chatto & Windus at Random House. Random House was then already publishing Isherwood's *Diaries* in a Vintage paperback edition, and, by chance, Chatto had in any case been the home since 1946 of Isherwood's much earlier publisher, The Hogarth Press.

Meyer, David and Tony. Twin brothers, both actors. When Isherwood met them in London in 1970, David, the younger twin, was Mark Lancaster's boyfriend. The twins appeared together that year in Peter Brook's production of *A Midsummer Night's Dream* at The Roundhouse. Later they were in movies, *The Draughtsman's Contract* (1982) and *Octopussy* (1985), and David appeared in Derek Jarman's films, including *The Tempest* (1979). He also worked on the stage with Lindsay Kemp and later taught and acted at the Globe. Tony, who was straight, married and became a painter.

Michael. See Barrie, Michael.

Millard, Paul. American actor. He lived with Speed Lamkin in West Hollywood for a few years during the 1950s. He briefly called himself Paul Marlin, then later changed to Millard; his real name was Fink. He was good-looking and relatively successful on the New York stage and on T.V., but eventually joined his mother's real estate business and invested in property. During 1959 and 1960, he loaned Bachardy a little house in West Hollywood to use as a studio, and around this time, the two had an affair which Isherwood apparently did not know about. Millard appears in *D.1* and *D.2*; he died in the 1970s.

Miller, Dorothy (d. 1974). Cook and cleaner to Isherwood and Bachardy from 1958 until the early 1970s. On their recommendation she later kept house for the Laughtons as well, both in Hollywood and in Charles Laughton's house next door to Isherwood and Bachardy in Adelaide Drive.

Milow, Keith (b. 1945). British painter, printmaker, and sculptor. He studied at the Camberwell School of Art and the Royal College of Art, and later worked in New York and Amsterdam.

Mishima, Yukio (1925–1970). Japanese author, of novels, short stories, poems, plays and essays; educated at Tokyo University. His real name was Kimitake Hiraoka. He was already famous in Japan when Alfred Knopf published *The Sound of Waves* in English translation in 1956 and invited Mishima to the U.S. the following year. Isherwood first met him during that visit, which he tells about in *D.1*, and they met again in November 1960, as he tells in *D.2*, when Mishima returned to the U.S. for the staging in New York of three of his Noh plays. By then Mishima had married Yoko Sugiyama, a student of English literature and daughter of the painter Yagushi Sugiyama, and the couple had had the first of their two children. More of Mishima's work had been translated into English: *Five Modern Noh Plays* appeared in English in 1957, then *Twilight Sunflower* (also a volume of plays) and *Confessions of a Mask* (1949) in 1958, followed by *The Temple of the Golden Pavilion* (1959) and many others. *Confessions of a Mask* addressed Mishima's discovery of his homosexuality. His masterwork, *Sea of Fertility*, conceived in 1962 and completed in 1970, is a tetralogy about Japan in the twentieth century. Mishima was obsessed by the warrior traditions of Imperial Japan and was expert in martial arts. In 1968, he founded a military group, the Shield Society, to revive the Samurai code of honor. Disillusioned when the young did not answer his call for a return to nationalist ideals, he committed Seppuku; in his diary entry for November 25, 1970, Isherwood transcribed an account of the ritual suicide from the *Los Angeles Times*. Mishima was twice nominated for the Nobel Prize.

Mitchell, Charlie. Vedanta devotee; educated at UCLA, where he studied English and, later, law. Around 1971, he attracted followers to what he called the First Liberty Church (his own offshoot of the Universal Life Church), founded on meditation and, evidently, pot smoking, but he soon advised his devotees to follow Swami Prabhavananda; they were the core of what Isherwood refers to as the Venice Group. Mitchell ran a small recording company and oversaw its sale to Chrysalis. Afterwards, he practised law, became the lawyer for the Vedanta Society of Southern California and also served as the society's vice-president. His Sanskrit name is Krishnadas. His wife was called Mary, later Apple, and finally Sita. Their son, Christopher, known as Lal Chand because of his red hair, got a Ph.D. in physics. Christopher became a monk at Trabuco, Asesha Chaitanya.

Mitchell, J.J. A friend and lover of Joe LeSueur and afterwards of Frank O'Hara; his relationship with O'Hara drove LeSueur out of the loft LeSueur and O'Hara had shared for many years. In May 1971, Isherwood mentions a new companion of Mitchell's, Ron Holland. Holland was rich from personal success in publicity and advertising. Mitchell died of AIDS in his mid-forties.

Moffat, Ivan (1918–2002). British-American screenwriter, educated at Dartmouth; son of Iris Tree and her American husband Curtis Moffat. He worked on government-sponsored documentaries for Strand Films in London, served in the U.S. Army Signal Corps Special Coverage Unit under the American film director George Stevens during World War II, and returned to Los Angeles in 1946 as Stevens's assistant. He worked with Stevens on *A Place in the Sun* (1951), was Stevens's associate producer for *Shane* (1953), and co-wrote *Giant* (1956), before going on to work for David Selznick on *Tender Is the Night* (1962). Other screenplays include *Bhowani Junction* (1956), *Boy on a Dolphin* (1957), and *Justine* (1969). Moffat's first wife was Natasha Sorokin, a Russian, once part of a ménage à trois with Simone de Beauvoir and Jean-Paul Sartre. Their marriage broke up at the start of the 1950s, leaving a daughter, Lorna. Moffat then had a number of beautiful and talented girlfriends, including the writer Caroline Blackwood with whom he fathered a daughter, Ivana, born in 1966 during Caroline's second marriage to the composer and music critic Israel Citkowitz; Ivana's paternity was kept secret until Caroline Blackwood's death in 1996. In 1961, Moffat married Katharine Smith, known as Kate, a well-connected English heiress. As Isherwood records, the marriage ended in 1972. Moffat appears often in *D.1*, *D.2*, and *Lost Years*. Although Moffat was heterosexual, Isherwood identified with him, as an expatriate and as a romantic adventurer, and based both the main character in the first draft of *Down There on a Visit* and "Patrick" in *A Meeting by the River* partly on him.

Moffat, Kate. See Smith, Katharine.

Monday, Jon. See Dharmadas.

Monkhouse, Patrick (Paddy), John (Johnny), Rachel. Childhood friends of Isherwood, raised near Marple in Disley, where they lived in a house called Meadow Bank with another, younger, sister, Elizabeth (Mitty), whom Isherwood got to know in 1947 when he visited Wyberslegh. Their father, Allan (1858–1956) was a journalist, theater critic, and playwright. The oldest brother, Patrick, also a journalist, was Isherwood's close friend in adolescence. Patrick attended Oxford a year or two ahead of Auden and edited *The Oxford Outlook*. Later, he achieved a senior position at the *Manchester Guardian* and married. The Monkhouses appear in *D.1* and *Lost Years*.

Montel, Michael. American stage director, once an actor. He was managing director of the New Phoenix Repertory Company which produced eight revivals on Broadway between 1973 and 1975. T. Edward Hambleton, a co-founder of the first Phoenix Theater, was also a managing director; the artistic directors were Hal Prince and Stephen Porter. John Houseman was a producing director of the first Phoenix Theater, and continued to be associated with the New Phoenix. Montel directed a few other Broadway plays and musicals, and also operas, including Bernstein's *Trouble in Tahiti* and Copland's *The Tender Land*.

Moody, Robert L. (1910–1973). British psychotherapist; raised in Surbiton and educated at Bromsgrove School. He began the Bachelor of Medicine course alongside Isherwood at King's College Medical School, London, in October 1928 but left in 1930 without taking any exams. Later, he became one of the first directors of the Jungian organization, the Society of Analytical Psychology, when it was formally registered under that name in 1945. He was also an editor of the *Journal of Analytical Psychology* and wrote various articles. He married three times, last to Louise Diamond. He appears as "Platt" in *Lions and Shadows* and is mentioned in *D.2*.

Moreau, Jeanne (b. 1928). French stage and screen star and singer; daughter of an English chorus girl. She was educated at the Paris Conservatory of Dramatic Art and became a leading actress for the Comédie Française and the Théâtre National Populaire before coming to international prominence in Louis Malle's *Frantic, (Ascenseur pour l'Echafaud*, 1957, in France; released as *Lift to the Scaffold* in the U.K. and later retitled *Elevator to the Gallows* in the U.S.) and *The Lovers* (1958). Her many film roles after that, for a range of celebrated directors, include *Les Liaisons Dangereuses* (1959), *Moderato Cantabile* (1960), *La Notte* (1961), *Jules and Jim* (1961), *Eva* (1962), *The Trial* (1963), *Diary of a Chambermaid* (1964), *Chimes at Midnight* (1966), *The Bride Wore Black* (1968), and *Going Places* (1974). She was married twice, briefly both times, and had many love affairs, including a complicated one with Tony Richardson while she was working with him on *Mademoiselle* (1966) and *The Sailor from Gibraltar* (1967).

Morgan. See Forster, E.M.

Moriarty, Michael (b. 1941). American actor and jazz pianist; raised in Detroit, educated at Dartmouth and at the London Academy of Music and Dramatic Arts. He won a Tony Award for his performance in *Find Your Way Home*, which Isherwood saw in 1974, and in the same year he won his first Emmy for his supporting role in a T.V. version of *The Glass Menagerie* with Katharine Hepburn. His films include *Bang the Drum Slowly* (1973), *Pale Rider* (1985), and *The Hanoi Hilton* (1987). In the early 1990s, he became known for his regular T.V. appearances on "Law & Order." He had four wives; the second was Françoise Martinet, from 1966 to 1978; he divorced her to marry his third, from 1978 to 1997, Anne Hamilton Martin.

Mortimer, John (1923–2009). British barrister, novelist and playwright, educated at Harrow and Oxford. He wrote *A Voyage Round My Father* (1971) and *Rumpole of the Bailey* (1975) and its sequels, later a T.V. series. He is credited with adapting Waugh's *Brideshead Revisited* for T.V. (1981), and he devised the script for *Tea with Mussolini* (1999), among many other projects. Tony Richardson directed his play of Robert Graves's *I, Claudius* in the West End, opening July 11, 1972, after first proposing that Mortimer adapt it along with *Claudius the God* as a film. The production was not successful.

Mortimer, Raymond (1895–1980). English literary and art critic; he was writer and editor for numerous magazines and newspapers and wrote books on painting and the decorative arts as well as a novel. He was at Balliol College, Oxford, with Aldous Huxley and later became a close friend of Gerald Heard, introducing Heard to Huxley in 1929; he was also intimate with various Bloomsbury figures and an advocate of their work. From 1948 onward, he worked for *The Sunday Times*, spending the last nearly thirty years of his life as their chief reviewer. He appears in *D.1* and *D.2*.

Mortmere. An imaginary English village invented by Isherwood and Edward Upward when they were at Cambridge together in the 1920s; the inhabitants were satires of generic English social types and were all slightly mad. As part of their rebellion against public school and university, Upward and Isherwood shared an elaborate fantasy life which was described by Isherwood in *Lions and Shadows*. The fragmentary stories the two wrote for each other about Mortmere were eventually published as a collection in 1994; Upward's *The Railway Accident* appeared on its own in 1949.

Moses, Ed (b. 1926). American artist, born near Long Beach, California, and educated at UCLA. His early work reflects his interest in abstract expressionism, but he has explored many styles. His drawings, paintings and graphic designs are held by major museums, including the Los Angeles County Museum of Art, the Fine Arts Museum in San Francisco, the Norton Simon Museum, and the National Gallery of Art and the Smithsonian Museum of American Art in Washington, D.C. His wife, Avilda, was a Southerner from a wealthy family and was for a time a follower of a Tibetan Buddhist mystic, Dezhung Rinpoche (1906–1987).

Mostert, Noël. Canadian journalist and short story writer, born in Cape Town, settled in Tangier. He was a military correspondent for Canadian forces in Europe and a U.N. correspondent in New York for the *Montreal Star*. His books include *Supership* (1974), about oil tankers, and the major work, *Frontiers: The Epic of South Africa's Creation and the Tragedy of the Xhosa People* (1992).

Murdock, James (1931–1981). American actor; born David Baker. He worked mostly in T.V. Westerns in the late 1950s and 1960s: "Gunsmoke," "Rawhide" (as Harkness "Mushy" Mushgrove, the cook's assistant), "Have Gun, Will Travel," and "Cheyenne." Isherwood tended to misspell his name as Murdo*ch*. He appears in *D.2*.

namaskar. Salutation; see pranam.

Nanny. See Avis, Annie.

Narendra, also Naren. See Vivekananda, Swami.

Natalie. See Leavitt, Natalie.

Natasha. See Spender, Natasha Litvin.

National Institute of Arts and Letters. The National Institute of Arts and Letters had 250 members chosen in recognition of their individual achievements in art, literature, and music. It amalgamated in 1976 with the American Academy of Arts and Letters and was called the American Academy and Institute of Arts and Letters, then later simplified its name to the American Academy of Arts and Letters. Members are chosen for life, and they confer several awards of their own, including the E.M. Forster Award. The organization maintains a library and museum in Manhattan. Isherwood was elected in 1949.

National Portrait Gallery, London. After acquiring Bachardy's 1967 portrait of Auden in 1969, Roy Strong left the NPG to become director of the Victoria and Albert Museum in 1974 and went on to a career as a writer and broadcaster. He was succeeded at the NPG by the art historian John Hayes, who was director from 1974 to 1993. Hayes was an expert in the paintings of Thomas Gainsborough. It was not until 1996, when Charles Saumarez Smith was director, that the NPG acquired Bachardy's portraits of Ackerley (1961), Forster (1961), John Osborne (1968), and Thom Gunn (1996) and commissioned portraits of James Ivory, Ismail Merchant, and Ruth Prawer Jhabvala. In 1998, the NPG purchased a ninth portrait, of Dodie Smith (1961).

Neal, Warren. Black psychiatric nurse. He was Jim Gates's first real love after Gates left the monastery. Neal commuted to work at an institution northeast of Los Angeles, past Pomona, possibly in Bakersfield. He died of AIDS some time before Gates did, and a number of years after they broke up.

Neddermeyer, Heinz. German boyfriend of Isherwood; Heinz was probably seventeen when he met Isherwood in Berlin, March 13, 1932. Their love affair, the most serious of Isherwood's life until then, lasted about five years. Hitler's rise forced them to leave Berlin in May 1933, and they lived and travelled in Europe and North Africa. In a traumatic confrontation with immigration officials at Harwich, Heinz was refused entry on his second visit to England in January 1934, so Isherwood went abroad more and more to be with him. In 1936, Heinz was summoned for conscription in Germany, and Isherwood scrambled to obtain or extend permits for Heinz to remain in the ever-diminishing number of European countries that would receive him. An expensive but shady lawyer failed to obtain a new nationality for him. Heinz was expelled from Luxembourg on May 12, 1937 and returned to Germany, where he was arrested by the Gestapo and sentenced—for "reciprocal onanism" and draft evasion—to a three-and-a-half-year term combining imprisonment, forced labor, and military service. He survived, married in 1938, and with his wife, Gerda, had a son, Christian, in 1940. Isherwood did not see Heinz again until 1952 in Berlin, though he corresponded with him. It was Heinz's conscription that first turned Isherwood towards pacifism. Their shared wanderings are described in *Christopher and His Kind*, and their friendship serves as one basis for the "Waldemar" section of *Down There on a Visit*. Heinz is also mentioned in *D.1*, *D.2*, and *Lost Years*.

Neil. See Hartley, Neil.

Nelson, Allyn L. A girlfriend of Jim Gates and Peter Schneider. She lived in Oxnard, California, and attended Claremont College, where she met Gates and Schneider. Schneider later recalled she may have been studying nursing. She appears in *D.2*.

Newman, Bob. Friend, assistant, and travelling companion to Tony Richardson during the mid-1970s; he settled in Richardson's Los Angeles pool house, looked after the gardens at Richardson's various properties, and eventually started his own gardening business. Bachardy hired him to garden at 145 Adelaide Drive from the mid-1980s onward.

Newman, Paul (1925–2008) and Joanne Woodward (b. 1930). American actor, director, producer, born in Ohio, educated at Kenyon College, Yale School of Drama and Actors' Studio and his second wife, actress, film and T.V. producer, born in Georgia and educated at Louisiana State University and New York's Neighborhood Playhouse. He debuted on Broadway in *Picnic* (1953) and received many awards and nominations for his Hollywood films, which include *Cat on a Hot Tin Roof* (1958), *The Hustler* (1961), *Hud* (1963), *Cool Hand Luke* (1967), *Butch Cassidy and the Sundance Kid* (1969), *The Sting* (1973), *The Verdict* (1982), *The Color of Money* (1986, Academy Award), and *Road to Perdition* (2002). They met when Woodward was an understudy for *Picnic* and married in 1958. By then, she had won an Academy Award for *The Three Faces of Eve* (1957). Her other films include *Rachel, Rachel* (1968) and *The Effect of Gamma Rays on Man-in-the-Moon Marigolds* (1972) both directed by Newman, *The Sound and the Fury* (1959), *Mr. and Mrs. Bridge* (1991), and two Emmy Award-winning T.V. films in 1978. They were liberal Democratic activists, and his Newman's Own food products generated hundreds of millions of dollars for charity. Gore Vidal introduced Isherwood and Bachardy to the Newmans in the mid-1950s, and they appear in *D.1* and *D.2*. Both stars sat for Bachardy in the early 1960s.

Nichols, Mike (b. 1931). American actor, director, producer; born in Berlin. He emigrated to New York at seven and was educated at the University of Chicago. His real name was Michael Igor Peschkowsky. He studied acting with Lee Strasberg and became famous with Elaine May in a comedy duo which they took to Broadway as *An Evening with Mike Nichols and Elaine May* (1960). He directed many Broadway hits, including *Barefoot in the Park* (1964), *The Odd Couple* (1965), *The Little Foxes* (1967), *Plaza Suite* (1968), *The Prisoner of Second Avenue* (1971), *The Real Thing* (1984), *Death and the Maiden* (1992), and *Spamalot* (2005). His Hollywood successes were just as numerous, among them: *Who's Afraid of Virginia Woolf?* (1966), *The Graduate* (1967), *Carnal Knowledge* (1971), *The Day of the Dolphin* (1973), *Silkwood* (1983), *Working Girl* (1988), *Postcards from the Edge* (1990), *Primary Colors* (1998), *Closer* (2004), and *Charlie Wilson's War* (2007). When Isherwood knew him, his wife was Pat Scot. His fourth wife, since 1988, is newscaster Diane Sawyer. He appears in *D.2*.

Niem, Jan (d. 1973). Chauffeur to Tony Richardson, for twenty years. He was born in Poland and sent to a camp in Siberia by the Russians during World War II. He came to the U.K. after the war, on a training scheme Churchill offered

Stalin, and was made a British citizen so that he did not have to return to the USSR. He married an English woman with whom he ran a car service for the film industry. According to rumor, Tony Richardson "won" him in a poker game with Cubby Broccoli. He died on top of a prostitute while on location in the South of France.

Nikhilananda, Swami. Indian monk of the Ramakrishna Order; a longtime head of the Ramakrishna-Vedanta Center in New York, on the Upper East Side; author of numerous books on Vedanta and translator of *The Gospel of Sri Ramakrishna* from Bengali into English with the help of Joseph Campbell, Margaret Woodrow Wilson, and John Moffitt who put Nikhilananda's translations of the songs into poetic form. He appears in *D.1* and *D.2*.

Nin, Anaïs (1903–1977) and Rupert Pole (1919–2006). American writer and her second husband, an actor, forest ranger, and teacher. She was born in Paris, raised in New York after the outbreak of World War I, and spent the 1920s and 1930s back in Paris seeking out the company of writers, intellectuals, and bohemians. She became a psychoanalyst as well as writing novels, short stories, and literary criticism. Her six-volume *Diary* began to appear in 1966, and tells, among other things, about her friendship with Henry Miller. Her other books include *Children of the Albatross* (1947)—which Isherwood read and admired before he met her—*The Four-Chambered Heart* (1950), and *A Spy in the House of Love* (1954). Some of her work, like Miller's, was published in Paris years before it appeared in the U.S. She had many love affairs, and her 1955 marriage to Pole was bigamous, since she never divorced her first husband, New York banker Hugh Guiler, and kept her continuing relationship with Guiler a secret from Pole until 1966. Pole, much younger than she, was a stepgrandson of Frank Lloyd Wright. He had a Harvard music degree, played the viola and the guitar, and acted professionally in New York in the late 1940s before going to California to study forestry at UCLA and Berkeley. As Isherwood tells in *Lost Years*, Nin lived with Pole at his forest station in the San Gabriel mountains among all the other rangers in defiance of the rules. By the 1970s, the pair settled in the Silver Lake District of Los Angeles in a house designed by Wright's grandson, Eric Lloyd Wright, and Pole became a science teacher. They appear in *D.2*.

Nixon and the students. Student opposition to the Vietnam War grew in response to the introduction of the draft lottery in December 1969 and erupted after Nixon, on April 30, 1970, announced the Cambodian invasion. On May 4, four student demonstrators were shot and killed and nine others wounded by the Ohio National Guard at Kent State in Ohio. On May 8, 100,000 students marched on Washington. Nixon left the White House before dawn on May 9 to speak with a few informally at the Lincoln Memorial. On May 15, two more students were shot dead and twelve wounded at Jackson State in Mississippi. Over 400 campuses across the U.S. were closed by protests.

Oberon, Merle (1911–1979). British film star, raised in India and discovered in London by her first husband, Alexander Korda, who made her internationally famous during the 1930s. Her films include *The Private Life of Henry VIII* (1933), *The Scarlet Pimpernel* (1935), *Wuthering Heights* (1939), *Forever and a Day* (1943),

Stage Door Canteen (1943), and *A Song to Remember* (1945). She divorced Korda in 1945 and married cinematographer Lucien Ballard, whom she divorced in 1949. In 1957 she married again, to Bruno Pagliai, an Italian industrial tycoon with vast holdings in Latin America and especially Mexico, where they went to live until she divorced him in 1973. Her fourth marriage was to actor Robert Wolders, her co-star in her last film, *Interval* (1973), which she produced herself. She appears in *D.2*.

O'Hilderbrandt, Mrs. Isherwood's neighbor. As Mary Miles Minter, she was a teenage star of silent films. Her career ended in scandal before she was twenty when her director, and, by rumor, her lover, William Desmond Taylor, was murdered, possibly by her mother. The crime was never solved, and she once offered to tell Isherwood what had happened if he would write a book about it on her behalf. As he mentions in his diary on August 18, 1972, she sued Rod Serling for portraying her as a murderer on his CBS show "Wonderful World of Crime," broadcast February 15, 1970. She appears in *D.2*.

"Old, Vernon" (not his real name). American painter. During Isherwood's first visit to New York in 1938, George Davis introduced him to Vernon Old at an establishment called Matty's Cell House. Blond, beautiful, and intelligent, Vernon matched the description Isherwood had given Davis of the American boy he'd like to meet, and Vernon featured in Isherwood's decision to return to New York in 1939. They lived together in New York and Los Angeles until February 17, 1941, when they split by mutual agreement. Vernon then lived unsteadily on his own, painting, drinking, and being sexually promiscuous, until a suicide attempt later that year. During World War II, he tried to become a monk, first in a Catholic monastery in the Hudson Valley and later at the Hollywood Vedanta Society and at Ananda Bhavan in Montecito. Eventually, he turned to heterosexuality, married "Patty O'Neill" (nor her real name) in November 1948, and had a son before divorcing. His painting career was increasingly successful, and in the late 1950s he tutored Don Bachardy. He appears in *Christopher and His Kind* and in *My Guru and His Disciple* (as "Vernon," without a surname) and throughout *D.1*, *D.2*, and *Lost Years*.

Olivier, Laurence (1907–1989). British actor, director, producer; celebrated as the greatest Shakespearian actor of his time. He became a Hollywood star by the start of World War II in *Wuthering Heights* (1939), *Rebecca* (1940), *Pride and Prejudice* (1940), and *That Hamilton Woman* (1941), and was appointed co-director of the Old Vic with Ralph Richardson near the end of the war. In 1963, he became director of the National Theatre in Britain. He directed and produced himself in a number of movies, beginning with *Henry V* (1944) and *Hamlet* (1948), which together won him several Academy Awards for acting and directing, and he appeared in more than fifty other films and over a hundred stage roles in London, New York, and elsewhere. He was married three times, to actress Jill Esmond from 1930 to 1940, to Vivien Leigh until 1960, and then to Joan Plowright until his death. Isherwood became friendly with him during 1959 when Olivier was in Los Angeles filming *Spartacus* (1960). He appears in *D.1* and *D.2*.

One, Incorporated. Homosexual advocacy and support group founded in 1952;

publisher of *One* magazine. *One* was the subject of a legal struggle with the U.S. Post Office during the 1950s; in 1958 the Supreme Court ruled that gay publications were not *a priori* obscene and could be sent and sold by mail. The business manager at One, Inc., and its driving force for over a decade, was Bill Legg. He had several names: Dorr Legg, William Lambert, and Marvin Culter.

O'Reilly, Anna. Personal assistant to Tony Richardson from 1964 until 1971, and to Vanessa Redgrave during the Richardson-Redgrave marriage. She arranged airplane tickets, bank drafts, presents, bought and rented real estate, and acted as hostess for dinner parties when Richardson needed an atmosphere of conventional respectability. She also assisted Richardson on his films and worked her way up to Associate Producer before she left to marry Graham Cottle. In 1976, she returned to help Richardson with *Joseph Andrews* and with his first T.V. production. Then she became an inhouse production assistant at Warner Brothers T.V. and went on to produce T.V. shows as Anna Cottle. She settled in Los Angeles and later worked in the film and literary management business, selling book rights for movies and T.V. Among her book-to-film projects are *Capote* and *Get Happy*.

Orphanos, Stathis (b. 1940). American photographer, born in North Carolina; his parents were Greek. His photographs of writers, artists, actors, and male nudes have appeared on book jackets and in *Vogue*, *Harper's Bazaar*, and *Antaeus*. With his longterm companion Ralph Sylvester (b. 1934), he publishes fine limited-edition books under the imprint Sylvester and Orphanos.

Osborne, John (1929–1994). English playwright, born in Fulham, West London. He worked as a journalist briefly and then acted in provincial repertory until his third play, *Look Back in Anger* (1956), established him as the center of a generation of working-class realist playwrights called "the angry young men." During the 1950s, his work was mostly produced by George Devine and Tony Richardson's English Stage Company at the Royal Court. Other plays include *The Entertainer* (1957), starring Laurence Olivier, *Luther* (1961), *Inadmissible Evidence* (1964), *A Patriot for Me* (1965), *West of Suez* (1971), *A Sense of Detachment* (1972), *Watch It Come Down* (1976), and *Déjàvu* (1991), a later sequel to *Look Back in Anger*. Several of his plays were filmed. Osborne also wrote the screenplays for Richardson's *Tom Jones* and *The Charge of the Light Brigade*. Their collaborations ended with the latter because Osborne was sued for plagiarizing Mrs. Cecil Woodham-Smith's *The Reason Why*. The rights to her novel belonged to Laurence Harvey, who agreed to sell them and abandon the suit if he could be in the film. So Richardson gave Harvey a small role previously promised to Osborne. Osborne and Richardson quarrelled and never worked together again. In *D.1*, Isherwood records that he met Osborne in Hollywood in 1960, when Osborne came to join Mary Ure (his second wife) and Richardson, both working there. In September 1961, as Isherwood tells in *D.2*, he and Bachardy were guests at La Beaumette, the house which Osborne rented from Lord Glenconner during August and September in Valbonne in the South of France. Osborne married five times; first to Pamela Lane, an actress, then to Ure, then to Penelope Gilliatt. His fourth wife, from 1969 to 1978, was Jill Bennett (1931–1990), the British actress, who starred in several of his plays and died a suicide. Finally, he married Helen Dawson (d.

2004), drama critic and arts editor at *The Observer* during the 1960s, and remained with her until his death. He wrote three volumes of autobiography, *A Better Class of Person* (1981), *Almost a Gentleman* (1991), and *Damn You, England* (1994).

O'Shea, John. Truman Capote's lover from July 1973. When they met, O'Shea was a bank vice-president living in a suburb outside Manhattan, married, Roman Catholic, and a father of three. He gave up his job to become Capote's business manager.

Owens, Rodney (Rod). California-born ceramics manufacturer and, later, fashion retailer. In 1946, he began a long relationship with Hayden Lewis, Bill Caskey's navy friend; he and Lewis appear often in *D.1* and in *Lost Years*. Together they built up a business making dinnerware and ashtrays. Eventually they split, and Owens moved to New York where he sold clothing for the designer Helen Rose.

Page, Anthony (Tony) (b. 1935). Oxford-educated British actor and director, born in India. He was Artistic Director at the Royal Court from 1964 to 1973 and directed five plays there by John Osborne, three during 1968, when Bachardy contributed drawings to the programs. In 1966, he planned to adapt Wedekind's Lulu plays, *Erdgeist* (*Earthsprite*, 1895) and *Die Büchse der Pandora* (*Pandora's Box*, 1904), but the project wasn't completed. He also directed productions of Ibsen, Tennessee Williams, Albee and others in the West End and at the National Theatre and in New York. He made a few movies, including *I Never Promised You a Rose Garden* (1977) and later became known for his television documentaries, biographies, and miniseries. He appears in *D.2*.

Paget's Disease. A chronic condition which weakens bones and can enlarge or deform them, causing bone pain and sometimes leading to arthritis or fractures. It is typically localized, rather than affecting all the bones in the body. Sufferers can be asymptomatic for years or may confuse the condition with arthritis.

Pagett, Nicola (b. 1945). British actress, trained at RADA. She played Princess Mary in *Anne of the Thousand Days* (1969) and became well known as Elizabeth Bellamy in the T.V. series "Upstairs Downstairs" in the early 1970s. Her many later T.V. roles included leads in the miniseries "Anna Karenina" (1977), "Scoop" (1987), and "A Bit of a Do" (1989). She played Elizabeth Fanshawe in "Frankenstein: The True Story."

Paley, William (Bill) (1901–1990). American media mogul; son of a Ukrainian cigar manufacturer; he was educated at the University of Chicago and the University of Pennsylvania. He bought CBS in 1928 when it was a small radio network and developed it into the radio and T.V. giant which he ran for over fifty years. During World War II, he was deputy chief of psychological warfare for the Allies. He was a figure in American cultural and intellectual life, devoting time to the Museum of Modern Art in New York, to hospitals, universities, and think tanks. His second wife, Barbara (Babe) Cushing Mortimer (1915–1978), a Boston-born society beauty, was one of Truman Capote's closest friends from 1955 until 1975. She epitomized the glamor of the rich women Capote called his Swans, but she ended their friendship when she discovered that Capote was using her private life as material for his fiction.

Parker, Dorothy (1893–1967). American poet, short-story writer, journalist, and literary critic; born in New Jersey. Celebrated for her wit and associated with the Algonquin Hotel in New York where for years she lunched with writer friends. Her first brief marriage was to a New York stockbroker, Edwin Parker. She contributed to *The New Yorker* from its debut and to many other American magazines. Her 1929 short story "The Big Blonde" won the O. Henry Prize. She wrote plays—*Close Harmony* (1929) with Elmer Rice and *Ladies of the Corridor* (1953) with Arnaud d'Usseau—and screenplays—notably *A Star is Born* (1937) and Hitchcock's *Saboteur* (1942) with her second husband, Alan Campbell. She protested the execution of Sacco and Vanzetti in 1927, covered the Spanish Civil War for *The New Masses*, and was involved in the founding of the Hollywood Anti-Nazi League and other Hollywood committees opposing fascism; she also supported the Civil Rights movement and willed her estate to the National Association for the Advancement of Colored People. She was blacklisted at the end of the 1940s and later testified before the House Un-American Activities Committee where, in contrast to many of her colleagues, she cited the First Amendment (freedom of speech) instead of the Fifth (the right not to serve witness against oneself). She took over Isherwood's teaching position at L.A. State College when he left, during the 1960s. Bachardy drew her portrait a number of times during the same period. She appears in *D.2*.

Parone, Ed. American stage director. He assisted Gordon Davidson with the professional Theater Group at UCLA, where he directed *Oh! What a Lovely War*. In 1967, he moved with Davidson to the Mark Taper Forum and ran New Theater for Now to develop new plays, including *A Meeting by the River* in 1972, directed by Jim Bridges. He stayed at the Mark Taper Forum for about twelve years, and was a director in residence and eventually associate artistic director, turning his hand to producing, writing, and editing. He was also assistant to the producer on *The Misfits* and directed for T.V. He appears in *D.2*.

Pavitrananda, Swami. Indian monk of the Ramakrishna Order; head of the Vedanta Society in New York, on the Upper West Side, and a trustee of the Ramakrishna Math and Mission. He spent many years in the order's editorial center, Advaita Ashrama, at Mayavati in the Himalayas. He often paid a month-long visit to Swami Prabhavananda during the summers. Other than Prabhavananda, he was Isherwood's favorite swami. He appears in *D.2*.

Payne, Maurice. English master printer; he met Hockney in London in 1965 and worked with him on the books Hockney did for the Petersburg Press, including the *Six Fairy Tales from the Brothers Grimm* (1969). For a while, he worked full time as Hockney's assistant before setting up a studio in New York where he collaborated with other artists.

Pears, Peter (1910–1986). English tenor; longtime companion and musical partner to Benjamin Britten. He was sent down from Oxford, after failing his first-year music exams, became a prep school master, studied briefly at the Royal College of Music and then joined the BBC Singers in 1934. He shared a flat with Britten from early 1938, and they began performing together in 1939. The same year, they went to America and lived outside New York at Elizabeth Mayer's house in Amityville and then in Brooklyn at George Davis's house in Middagh Street. They also made a trip to California. Pears studied singing further in New York, where his voice developed and gained in strength. He returned to England with Britten in March 1942, and thereafter their lives became increasingly fused, with Britten writing a great deal of music for Pears, and Pears singing it expressly for Britten. He appears in *D.1*.

Peggy. See Kiskadden, Peggy.

Pentagon Papers. The 7,000-page top secret document about U.S. involvement in Vietnam which was leaked to the press in 1971. Daniel Ellsberg (b. 1931), Harvard-educated doctor of economics and former marine, was the Defense Department analyst who leaked the document to *The New York Times* and then *The Washington Post*. Anthony Russo, a RAND fellow, was a friend who helped Ellsberg copy the documents. They were charged with espionage, larceny and conspiracy and tried in Los Angeles, beginning in January 1973. On May 12, the judge dismissed their case on grounds of government misconduct. The jury had no opportunity to make a decision. Seven of the ten jurors told the Associated Press they had been inclined to acquit.

Perry, Troy D. (b. 1940). American gay activist, from Florida. A Baptist minister at fifteen, he married and had two sons before moving to California where he came out, divorced, and founded the Metropolitan Community Church of Los Angeles in October 1968. It was the first of three hundred or more predominantly gay and lesbian churches in his Universal Fellowship of Christian Churches.

Peter. See Schlesinger, Peter.

Peters, Gilbert (Gib). High-school friend of Jim Gates in Claremont in the late 1960s; they were expelled with another friend, Doug Rauch, for wearing their hair too long, and graduated from Continuation School instead. They rented a house together, worked at menial jobs, and began to meditate under the guidance of an ex-monk from the Self Realization Fellowship (brought to the U.S. in 1920 by Paramahansa Yogananda), eventually moving with the ex-monk and several other friends to the San Gabriel mountains. Gib left for Oregon with another Claremont friend, Beth High; they married, and the ex-monk followed them and moved in with them. They continued to meditate together and began to study Esperanto, but Gates introduced Gib to Swami in July 1970, as Isherwood tells, when Gib was visiting Los Angeles on his way to an Esperanto conference in Austria, and Gib and his wife, who had already met Swami, soon moved south again to be near him, living at first in a tent in the backyard of the house Jim Gates and Peter Schneider rented in Venice. The marriage gradually dissolved as the monastic life took over.

Pfeiffer, Virginia (1902–1973). A sister-in-law and friend of Ernest Hemingway; born in St. Louis to a wealthy pharmaceutical manufacturer and raised on her family's 60,000-acre farm in Piggott, Arkansas. During the 1920s, she travelled with her older sister Pauline Pfeiffer in Europe; they befriended Hemingway and his first wife, Hadley, in Paris, and Pauline married Hemingway in 1927. Virginia was often with the Hemingways in Paris, Key West, Cuba, and Bimini during the 1930s, and she renovated a barn at the family property in Arkansas for Hemingway to work in. When it burned down, she renovated it again. She sometimes took charge of their two sons and of Hemingway's son from his first marriage. Hemingway left Pauline for Martha Gellhorn in 1940, and Pauline died at Virginia's house in Hollywood in 1951. Laura Archera lived with Virginia for many years before marrying Aldous Huxley. Virginia adopted two children, Juan and Paula, whom Laura helped her raise. After they married, the Huxleys settled in a house in Deronda Drive in the Hollywood Hills a few hundred yards from Virginia's house, and Virginia and Laura continued to be frequent companions. In May, 1961, both houses burned down in a bush fire; Virginia moved, in the autumn, to a nearby house in Mulholland Drive and invited the Huxleys to join her there. Aldous Huxley died in that house a few months later. Virginia appears in *D.2*.

Plante, David (b. 1940). American writer; raised in Rhode Island and educated at Boston College. He lives in London and in New York, where he teaches writing at Columbia University. Of his more than twelve novels, the most admired are his Francoeur trilogy, *The Family* (1978), *The Country* (1981), and *The Woods* (1982). His non-fiction work includes *Difficult Women: A Memoir of Three* (1983) and essays for *The New Yorker* and *The Paris Review*. His companion for many years was Nicos Stangos.

Plomer, William (1903–1973). British poet and novelist born and raised in South Africa. He met Isherwood in 1932 through Stephen Spender after Spender showed Isherwood Plomer's poems and stories about South Africa and Japan. Plomer was a friend of E.M. Forster and soon introduced Isherwood. In South Africa, Plomer and Roy Campbell had founded *Voorslag* (Whiplash), a literary magazine for which they wrote most of the satirical material (Laurens van der Post was also an editor). Plomer taught for two years in Japan then, in 1929, settled in Bloomsbury where he was befriended by the Woolfs who had published his first novel, *Turbott Wolff* (1926). In 1937, he became principal reader for Jonathan Cape where, among other things, he brought out Ian Fleming's James Bond novels. During the war he worked in naval intelligence. He also wrote libretti for Benjamin Britten, notably *Gloriana* (1953). He appears in *D.1* and *D.2*.

Plowright, Joan (b. 1929). British actress; trained at the Old Vic Theatre School. She first appeared on the London stage in 1954 and joined Tony Richardson and George Devine's London Stage Company in 1956. Isherwood met her when she was in the New York production of *A Taste of Honey* in 1960. During the same year, she starred with Laurence Olivier in *The Entertainer* at the Royal Court; she then appeared with Olivier in the film, and later in the stage and film versions of Chekhov's *Three Sisters*. She became Olivier's third wife in 1961, and they had two children, Richard (b. 1961) and Julie Kate (b. 1966). In the early 1960s, she and Olivier joined the National Theatre Company when it was founded at the Old Vic in London, and she played leads in Chekhov, Shaw, and Ibsen. Later, she played Shakespeare and had a long West End run in De Filippo's *Saturday, Sunday, Monday* (1973), followed eventually by his *Filumena*. She also appeared in the stage (1973) and film (1977) versions of *Equus*. Among her other movies are *Moby Dick* (1956), *Uncle Vanya* (1963), *101 Dalmatians* (1996), and *Tea with Mussolini* (1999). She appears in *D.1* and *D.2*.

Poe, Barbara. American painter and art collector, née Reese. She was once married to screenwriter James Poe, who adapted *Sanctuary* for Tony Richardson and whose other work includes the screenplays for *Around the World in Eighty Days* (1956), *Cat on a Hot Tin Roof* (1958), and *They Shoot Horses, Don't They?* (1969). She later remarried and became Barbara Poe Levee.

Pope-Hennessey, James (1916–1974). British writer, educated at Oxford. He was literary editor of *The Spectator* from 1947 to 1949. His books include biographies of Queen Mary, Robert Louis Stevenson and Anthony Trollope, as well as a work on the Atlantic slave trade, *Sins of the Fathers* (1968). As Isherwood records, he was murdered by his one-time boyfriend, Sean O'Brien. O'Brien let two fellow criminals into Pope-Hennessey's flat to rob him. During the robbery, Pope-Hennessey choked to death on his own blood after being battered in the head and throat by O'Brien.

Prabhaprana (Prabha) (d. 1998). Originally Phoebe Nixon, she was the daughter of Alice Nixon ("Tarini") and after sannyas became Pravrajika Prabhaprana. The Nixons were wealthy Southerners. Isherwood first met Prabha in the early 1940s at the Hollywood Vedanta Society, where she handled much of the administrative and secretarial work, and he grew to love her genuinely. By the mid-1950s, Prabha was manager of the Sarada Convent in Santa Barbara. She appears in *D.1* and *D.2*.

Prabhavananda, Swami (1893–1976). Hindu monk of the Ramakrishna Order, founder of the Vedanta Society of Southern California based in Hollywood. Gerald Heard introduced Isherwood to Swami Prabhavananda in July 1939. On their second meeting, August 4, Prabhavananda began to instruct Isherwood in meditation; on November 8, 1940 he initiated Isherwood, giving him a mantram and a rosary. From February 1943 until August 1945 Isherwood lived monastically at the Vedanta Society, but he decided he could not become a monk as Swami wished. He continued to be closely involved with the Vedanta Society, travelled several times to its headquarters in India, and remained Prabhavananda's disciple and close friend for life. Their relationship is described in *My Guru and His Disciple*, and Prabhavananda appears in *D.1*, *D.2*, and *Lost Years*, as well as providing inspiration for *A Meeting by the River*.

Prabhavananda was born Abanindra Nath Ghosh, in a Bengali village northwest of Calcutta. As a teenager he read about Ramakrishna and his disciples Vivekananda and Brahmananda and felt mysteriously attracted to their names. By chance he experienced an affecting meeting with Ramakrishna's widow, Sarada Devi. At eighteen he visited the Belur Math, the chief monastery of the Ramakrishna Order beside the Ganges outside Calcutta. There he had another important encounter, this time with Brahmananda, and abandoned his studies for a month to follow him. When he returned to Calcutta, he became involved in militant opposition to British rule, and joined a revolutionary organization for which he wrote and distributed propaganda. At one time, he took charge of some stolen weapons, and some of his friends who engaged in terrorist activities met with violent ends. Because he was studying philosophy, Abanindra attended Belur Math regularly for instruction in the teachings of Shankara, but he regarded the monastic life as escapist and put his political duties first, until he had another compelling experience with Brahmananda and suddenly decided to give up his political activities and become a monk. He took his final vows in 1921, when his name was changed to Prabhavananda.

In 1923 he was sent to the United States to assist the swami at the Vedanta Society in San Francisco; later he opened a new center in Portland, Oregon. He was joined there by Sister Lalita and later, in 1929, founded the Vedanta Society of Southern California in her house in Hollywood, 1946 Ivar Avenue. Several other women joined them. By the mid-1930s the society began to expand and money was donated for a temple which was built in the garden and dedicated in July 1938. Prabhavananda remained the head of the Hollywood society until he died; he frequently visited the Ramakrishna monastery in Trabuco and the Sarada Convent in Santa Barbara and also stayed in the home of a devotee in Laguna Beach.

Isherwood and Prabhavananda worked on a number of books together, notably translations of the Bhagavad Gita (1944) and of the yoga aphorisms of Patanjali (1953). Prabhavananda contributed to two collections on Vedanta edited by Isherwood, and Isherwood also worked on Prabhavananda's translation of Shankara's *Crest Jewel of Discrimination* (1947). Prabhavananda persuaded Isherwood to write a biography of Ramakrishna, *Ramakrishna and His Disciples* (1964); this became an official project of the Ramakrishna Order.

pranam. Greeting of respect made by folding the palms, by taking the dust of the feet (i.e., touching the greeted one's foot and then touching one's own forehead), or by prostrating. Namaskar, from the same Sanskrit root, "nam" for a salutation expressing love and respect, can also be made in a variety of ways depending upon local tradition and social situation: verbally, by nodding, by folding the palms, by bending at the knees to touch the ground with the forehead, or by lying flat on the ground.

prasad. Food or any gift consecrated by being offered to God or a saintly person in a Hindu ceremony of worship; the food is usually eaten as part of the meal following the ritual, or the gift is given to the devotees.

Prema Chaitanya (Prema) (1913–2000). American monk of the Ramakrishna Order, originally named John Yale and later known as Swami Vidyatmananda; born in Lansing, Michigan, educated at Olivet College in Illinois, Michigan State College and later at the University of Southern California where he obtained a doctorate in education. He taught high school before moving to Chicago in 1938 and working as a schoolbook editor. In 1941, he tried unsuccessfully to join the navy, and then in 1942 he joined Science Research Associates (SRA), a publishing house which specialized in teaching tools and psychological tests, and which was later bought by IBM. He ran SRA during the war and eventually became a director. While he was there, the entire male staff was interviewed by Alfred Kinsey for his *Sexual Behavior in the Human Male*, so Yale recorded his sexual history with one of Kinsey's assistants and was made ill by recounting it. He decided to give up sex, evidently because he was homosexual.

In 1948, after reading Isherwood and Prabhavananda's translation of the Bhagavad Gita and other works on Vedanta, he moved to Los Angeles, where he began instruction with Prabhavananda that November and moved into the

Vedanta Society in Hollywood in April 1950. Isherwood met him at the Vedanta Society in the spring of 1949. In August 1955, Yale took his brahmacharya vows at Trabuco and was renamed Prema Chaitanya. He continued to live at the Hollywood society and briefly at Santa Barbara but never at Trabuco. He developed the Vedanta Society's bookshop, building a mail-order business. He also edited the Vedanta Society magazine, *Vedanta and the West*, collaborating with Isherwood on the magazine's chapter-by-chapter publication of Isherwood's biography of Ramakrishna. Prema's own 1961 book, *A Yankee and the Swamis*, describing his journey in 1952–1953 to the Ramakrishna monastery and various holy places in India, was also published serially in *Vedanta and the West* and caused a scandal with its remarks about the residents of the various places Prema visited; offending passages were deleted by Swami Prabhavananda from the final text. Prema also edited *What Religion Is: In the Words of Swami Vivekananda*, for which Isherwood wrote the introduction, and *What Vedanta Means to Me* (1960), to which Isherwood also contributed. He appears in *D.1*, *D.2*, and *Lost Years*, and Isherwood drew on his sannyas experience in India in 1963–1964 for *A Meeting by the River*, originally intended to be dedicated to him.

In his unpublished memoir "The Making of a Devotee," Vidyatmananda tells that he became estranged from Prabhavananda while living at the Belur Math in India, where he discovered that the family model of religious life adopted by Swami Prabhavananda for the Hollywood Vedanta Society was not approved by the Ramakrishna Order, which advocated a strict monastic model with the sexes separated—officially only men could join the order. Swami had initiated many women and allowed them to live alongside the men because there were not enough devotees in southern California to populate two separate orders; Vidyatmananda began lobbying to have the nuns turned out of the Hollywood Vedanta Society. He also expressed surprise over his discovery in India that despite being a Westerner, he might still hope to rise through the hierarchy of the order, and even be permitted to transfer away from the Hollywood center. He remained in India for nearly a year, hoping to live there permanently, but he failed to find useful work, and returned to Hollywood after falling severely ill with paratyphoid. Swami Prabhavananda felt Prema's discussions with the elders of the Belur Math were disloyal, and when in 1966 Prema was invited to transfer to the Centre Védantique Ramakrichna, east of Paris in Gretz, France, Swami wrote saying he wanted nothing more to do with him. They met again, however, in 1973, and Prabhavananda gave Vidyatmananda his blessing.

Vidyatmananda remained in Gretz for the rest of his life. When he arrived, the center was in decline; the first generation of devotees had died or left following the death of the founding Swami, Siddheswarananda, and only a few new devotees had appeared. Vidyatmananda, as manager, saw to reorganizing, rebuilding and modernizing the property; he learned how to speak French and how to farm. The center was run as an ashram, and eventually thrived on the physical labor of young spiritual trainees who generally returned to secular life after a period of retreat there.

Premananda, Swami (1861–1918). A direct monastic disciple of Ramakrishna, from a pious Bengali family; born Baburam Ghosh. As a young man, he was taken to meet Ramakrishna at Dakshineswar by his Calcutta classmate, Rakhal Chandra Ghosh (Brahmananda). Premananda's sister was married to one of Ramakrishna's most prominent devotees, Balaram Bose, and his mother also came to Dakshineswar as a devotee. Ramakrishna regarded him as especially pure and sweet-natured and recognized him as an Ishvarakoti (a perfect, free soul born for mankind's benefit) like Brahmananda and Vivekananda. During the last decades of his life, Premananda was manager of the Belur Math.

Price, Kenneth (Ken) (b. 1935). American printmaker and ceramic artist, born in Los Angeles, trained at Chouinard, USC, the Art Institute of California, and the New York State College of Ceramics at Alfred University. His first solo show was at the Ferus Gallery in 1960; later he showed at the Kasmin Gallery in London, the Whitney in New York, Nicholas Wilder, James Corcoran, Gemini GEL, and the Los Angeles County Museum, among others. Many major museums hold his work. He moved to Taos in 1970, then to Massachusetts in 1982, and later back to Los Angeles where he joined the faculty at USC. His wife is called Happy.

Price, Vincent (1911–1993) and Coral Browne (1913–1991). American actor, raised in St. Louis and Australian-born actress, his third wife. Price grew up in privileged circumstances and toured Europe on his high-school graduation. He studied art history at Yale and again at the Courtauld in London, where he made his stage debut in 1935. By 1938, he had a contract with Universal Studios and during the 1950s began to develop his often comic flair for villains and horror movies. His many films include *The Private Lives of Elizabeth and Essex* (1939), *The Song of Bernadette* (1943), *Laura* (1944), *Leave Her to Heaven* (1945), *The Three Musketeers* (1948), *House of Wax* (1953), *The Fly* (1958), *The House of Usher* (1960), *The Pit and the Pendulum* (1961), *The Masque of the Red Death* (1964), and *Edward Scissorhands* (1990). He appeared on T.V. in the "Batman" series, on "Hollywood Squares" in the 1960s and 1970s, and hosted the PBS series "Mystery" during the 1980s. From July 1977, he toured world-wide as Oscar Wilde in a one-man stage play, *Diversions and Delights*, written by John Gay. He was an art collector and a gourmet, and he published a number of art history books and cookbooks. He married Coral Browne in 1974 after they worked together in *Theater of Blood* (1973). She began her stage career in Australia, emigrated to England during World War II and played classic stage roles, including Shakespeare, as well as appearing in Joe Orton's *What the Butler Saw* (1969). Her films include *Auntie Mame* (1958), *The Killing of Sister George* (1968), *The Ruling Class* (1972), and *Dreamchild* (1985).

Prince, Hal (b. 1928). American theatrical producer and director; born in New York, educated at the University of Pennsylvania; winner of more than twenty Tony Awards for his many Broadway hits. He expressed an interest in 1959 in an Auden-Isherwood-Kallman musical based on *Goodbye to Berlin*, but the project didn't progress. In 1966, he produced and directed the spectacularly successful version which he commissioned from Joe Masteroff (book), Fred Ebb (lyrics), and John Kander (music). In his diary entry for December 27, 1972, Isherwood mentions his "offended feelings about *Cabaret*"; he was offended that Prince never consulted him or made contact with him about it. Isherwood did not approve of the stage musical, but he was grateful for the income it brought him.

Procktor, Patrick (1936–2003). Dublin-born painter, especially of watercolors. He studied at the Slade from 1958 to 1962, had his first exhibition at the Redfern Gallery in 1963, travelled widely, taught at the Camberwell School of Art and the Royal College of Art, and designed for the stage, including two productions at The Royal Court in the late 1960s. He was a close friend of David Hockney, with whom he visited Los Angeles for the first time in 1965, and Hockney, early in his career, drew and painted him. Bachardy met Procktor in 1961 while studying at the Slade, and Procktor also sat for Bachardy once. They slept together when Procktor arrived in Los Angeles, but Bachardy never genuinely returned Procktor's interest. Procktor married Kirsten Andersen Benson (1939–1984), Danish-born widow of James Benson, in 1973; she had a son and daughter from her first marriage, and they had another son, Nicholas (b. 1974), and then divorced. Procktor's friend, Ole Glaesner, whom Isherwood mentions, was evidently a sexual partner but not a longterm companion. Procktor lived and painted in a flat in Manchester Street which had previously belonged to one of his teachers, William Coldstream. He became an alcoholic, and, late in his career, the flat burned down, destroying much of his work and many of his possessions.

Prosser, Lee (1944–2011). American author, painter, musician, Vedanta devotee, student of ancient religions, shamanism, witchcraft, and Wicca. He composed "The Ramakrishna Waltz" and wrote a memoir, *Isherwood, Bowles, Vedanta, Wicca, and Me* (2001). He married three times: first to Mary, from whom he was divorced in 1970, then to Grace, with whom he had two daughters, and, much later, to Debra. He appears in *D.2*.

Prouting, Norman (1924–1983). Isherwood's London landlord during the spring of 1970. Prouting lived in Moore Street, north of the King's Road in Chelsea, and let Isherwood a small flat in his house. Bill Harris was a friend of Prouting and also lodged in the house when he was in London. Prouting was a real estate expert, methodically researching neighborhoods he bought into. He was also a train buff and worked on several commercial film shorts about British transport.

puja. Hindu ceremony of worship, a watch or vigil; usually offerings—flowers, incense, food—are made to the object of devotion, and other ritual, symbolic acts are carried out depending upon the occasion.

Quinn, Joan. Arts journalist and curator. She was West Coast Correspondent for *Interview* magazine from 1978 to 1989, contributed to many other publications, and later had an arts interview show on private access cable T.V.

Rabb, Ellis (1930–1998). American actor, director, and producer, from Memphis, Tennessee. He appeared in and directed Shakespeare, Chekhov, Shaw, Pirandello, and contemporary drama, including Tennessee Williams. In 1959, he founded a group called the Association of Producing Artists (APA) of which he was Artistic Director. The APA worked in affiliation with the Phoenix Repertory Company, and many of their productions had Broadway runs in the 1960s, usually as a group of plays in repertory, including *You Can't Take It With You*, *Right You Are If You Think You Are*, *The Wild Duck*, *War and Peace*, *The Show Off*, *The Cherry Orchard*, *Pantagleize*, *The Cocktail Party*, *The Misanthrope*, and *Private Lives*. Rabb starred in some of the productions. His wife from 1960 to 1967, actress Rosemary Harris, was also in the company. The APA-Phoenix lasted until 1969. Rabb continued to act and direct, and he won a Tony Award for his 1975 production of *The Royal Family*. He appears in *D.2*.

rajas. See guna.

Ramakrishna (1836–1886). The Hindu holy man whose life and teachings were central to the modern renaissance of Vedanta. He was widely regarded as an incarnation of God. Ramakrishna, originally named Gadadhar Chattopadhyaya, was born in a Bengali village sixty miles from Calcutta. He was a devout Hindu from boyhood, practised spiritual disciplines such as meditation, and served as a priest. He was a mystic and teacher, and in 1861 he was declared an avatar: a divine incarnation sent to reestablish the truths of religion and to show by his example how to ascend towards Brahman. Ramakrishna was also initiated into Islam, and he had a vision of Christ. His behavior was sometimes highly unconventional, in keeping with his beliefs and with the extreme spiritual practices which he undertook. For instance, in youth, he put his tongue to the flesh of a rotting corpse as part of his Tantric discipline, and in order to emulate the gopis, he undertook *madhura bhava*, identifying himself as a female devotee of Krishna, assuming a feminine attitude, and actually dressing in women's clothes. He several times danced with drunkards because their reeling reminded him of his own when he was in religious ecstasy. His followers gathered around him at Dakshineswar and later at Cossipore. His closest disciples, trained by him, later formed the nucleus of the Ramakrishna Math and Mission, now the largest monastic order in India. Ramakrishna was worshipped as God in his lifetime; he was conscious of his mission, and he was able to transmit divine knowledge by a touch, look, or wish. Isherwood's biography, *Ramakrishna and His Disciples* (1964), was written with the help and encouragement of Swamis Prabhavananda and Madhavananda.

Ram Nam. A sung service of ancient Hindu prayers which invoke the divinities Rama, his wife Sita, and the leader of Rama's army, the monkey god, Hanuman. In Ramakrishna practice, Ram Nam is sung on Ekadashi, the eleventh day after the new or full moon, generally observed with worship, meditation, and fasting.

Rampling, Charlotte (b. 1945). British actress, educated in Paris and at English boarding school. Her father, an Olympic swimming gold medallist, was an army officer serving with NATO; her mother was a painter. Her only sister committed suicide as a young woman. Her films include *Georgy Girl* (1966), *The Damned* (1969), *Zardoz* (1974), *Farewell, My Lovely* (1975), *Foxtrot / The Other Side of Paradise* (1976), *Stardust Memories* (1980), *The Verdict* (1982), *Max, Mon Amour* (1986), *The Wings of the Dove* (1997), *The Swimming Pool* (2003), and *Never Let Me Go* (2010). When Isherwood met her in 1976, she was married to Bryan Southcombe, an aspiring actor and writer from New Zealand with whom she had one son, Barnaby (b. 1972). She left Southcombe that year for the French composer Jean-Michel Jarre whom she married two years later, and with whom she had another son, David. She divorced Jarre in 1997.

Rappaport, Michelle. American movie producer. Her films include *Old Boyfriends* (1979) with Talia Shire, *White Men Can't Jump* (1992) with Woody

Harrelson, and, for T.V., "Something about Amelia" (1984) and "Paper Dolls" (1984).

Rassine, Alexis (1919–1992). Ballet dancer; his real name was Alec Raysman. He was born in Lithuania of Russian parents and from about ten years old was brought up in South Africa. He studied ballet there and in Paris, joined the Ballet Rambert in 1938, and danced with several other companies before, in 1942, joining the Sadler's Wells Ballet, where he became a principal and a star. He shared John Lehmann's house for many years, living in his own self-contained flat. He appears in *D.1* and in *Lost Years*.

Rauch, Doug. A high-school friend of Jim Gates; he graduated from Continuation School after being expelled with him. Later, he became president of Trader Joe's, the specialty grocery chain. He married an ex-Catholic nun, Mikele, and they had three children; Mikele became a psychoanalyst.

Rauschenberg, Robert (1925–2008). American artist; born in Texas, trained at the Kansas City Art Institute, the Académie Julian in Paris, Black Mountain College in North Carolina, and the Art Students League in New York. Some of his early work was conceptual, and he became famous in the 1950s for his Combines, in which he attached found objects and collage to his canvases. In the 1960s, he turned to silkscreen and photography, and, in 1964, he was the first American to win the Grand Prize at the Venice Biennale. Later, he experimented with performance art, and he collaborated on set design, choreography, and projects bridging between art and engineering. He worked in Florida from 1970 onward.

Reagan assassination attempt. As Reagan was leaving the Hilton Hotel in Washington, D.C. on March 30, 1981, John Hinckley, Jr., fired six shots from a .22-caliber revolver loaded with exploding bullets. One entered Reagan's left lung, failed to explode, and was removed during surgery. Vice President George Bush flew back from Austin, Texas, and while he was in the air, Secretary of State Alexander Haig—a West Point graduate, heavily decorated veteran of Korea and Vietnam, and four-star general—asserted on national T.V. that he was in charge at the White House, incorrectly citing the Constitution on succession of power and presidential incapacity.

recession. The 1969–1970 recession was associated with closing the budget deficit created by the Vietnam War. Isherwood refers several times to the much more severe recession which began in 1971 and lasted until 1975, with stagflation (in which the economy did not grow but prices rose rapidly) as well as high unemployment. The dollar weakened, as he seemed to be aware, and the Energy Crisis dramatically worsened the situation from late 1973. See also Energy Crisis.

Rechy, John (b. 1934). American writer; born in El Paso, educated at the University of Texas and then at the New School for Social Research in New York. He served in the U.S. Army in Germany and for many years taught creative writing at the University of Southern California. His novels about the homosexual communities of New York and Los Angeles generally explore marginal themes of violence, drugs, and crime. *City of Night* (1963), tells about his experiences as a hustler. Later novels are *Numbers* (1967), with thinly disguised portraits of

Isherwood, Bachardy, and Gavin Lambert, *This Day's Death* (1970), *The Vampires* (1971), *The Fourth Angel* (1973), *Rushes* (1979), *Bodies and Souls* (1983), *Marilyn's Daughter* (1988), and *The Miraculous Day of Amalia Gómez* (1991). Rechy also wrote *The Sexual Outlaw* (1977), a documentary study of urban homosexual sexual practices. He appears in *D.2*. He teaches writing from his home in Los Angeles.

Redgrave, Corin (1939–2010). British actor, brother of Vanessa. Educated at Cambridge, where he appeared in amateur theatricals before his professional stage debut in 1963. In 1998 he won an Olivier award for his role in *Not About Nightingales*. His films include *A Man for All Seasons* (1966), *The Charge of the Light Brigade* (1968), *Oh! What a Lovely War* (1969), *David Copperfield* (1969), *Four Weddings and a Funeral* (1993), *Enigma* (2001), and *Enduring Love* (2004). He married twice, the second time to actress Kika Markham. One daughter, Jemma (b. 1965), is also an actress.

Redgrave, Vanessa (b. 1937). English star of stage and screen; from the celebrated acting family, which includes her father Michael, her mother Rachel Kempson, her brother Corin, and sister Lynn Redgrave. She trained at London's Central School of Speech and Drama, made her stage debut in 1957, and established her reputation with the Royal Shakespeare Company in the early 1960s. Her films include *Morgan* (1966, Academy Award nomination), *Blow-Up* (1966), *A Man for All Seasons* (1966), *The Sailor from Gibraltar* (1967), *Camelot* (1967), *The Charge of the Light Brigade* (1968), *Isadora* (1968; Academy Award nomination), *Oh! What a Lovely War* (1969), *Mary, Queen of Scots* (1971; Academy Award nomination), *Murder on the Orient Express* (1974), *Julia* (1977, Academy Award), *The Bostonians* (1984; Academy Award nomination), *Prick Up Your Ears* (1987), *The Ballad of the Sad Café* (1991), *Howards End* (1992), *Mission Impossible* (1996), *Mrs. Dalloway* (1998), *Girl, Interrupted* (1999), *The Cradle Will Rock* (1999), *Running with Scissors* (2006), and *Atonement* (2007). Her stage roles are too numerous to name, and she has often appeared on T.V. Much of her work during the 1960s was for Tony Richardson, whom she married in 1962 and with whom she had two daughters, actresses Natasha Richardson (1963–2009) and Joely Richardson (b. 1965), before divorcing in 1967. In 1969, she had a son with actor Franco Nero, whom she married in 2007. She is well known for her leftist political activism and has unsuccessfully run for Parliament as a member of the Workers' Revolutionary Party. She appears in *D.2*.

Regester, James Robert (Bob) (19[32]–1987). American theatrical producer and advertising executive, from Bloomington, Indiana. He met Tony Richardson in Los Angeles in the 1960s and worked for him in Europe as a member of the production team for *Mademoiselle* (1966) and *The Sailor from Gibraltar* (1967). He became a longtime companion of Neil Hartley, and they shared a house in Maida Avenue. With financial backing from a friend, Louis Miano, he co-produced *Design for Living* with Vanessa Redgrave, Jeremy Brett, and John Stride at the Phoenix Theatre in 1973; *The Seagull*, in 1985, starring Vanessa Redgrave and Natasha Richardson; Gerald Moon's *Corpse*, with Keith Baxter and Milo O'Shea in 1984; and *Legends*, starring Mary Martin and Carol Channing, which toured in the U.S. in the mid-1980s. He died of AIDS. He appears in *D.2*.

Reinhardt, Gottfried (1911–1994). Austrian-born film producer. He emigrated to the U.S. with his father, Max Reinhardt, and became assistant to Walter Wanger. Afterwards he worked as a producer for MGM from 1940 to 1954 and later directed his own films in the United States and Europe; his name is attached to many well-known films, including Garbo's *Two-Faced Woman* which he produced in 1941 and *The Red Badge of Courage* which he produced in 1951. He was Salka Viertel's lover for nearly a decade before his marriage to his wife, Silvia, in 1944. Through Salka and Berthold Viertel, Reinhardt gave Isherwood his second Hollywood film job in 1940, and he remained Isherwood's favorite Hollywood boss. During the war, he enlisted and wrote scenarios for films on building latrines, preventing venereal disease, cleaning rifles, etc. Reinhardt and his wife eventually returned to Germany and settled near Salzburg. He appears in *D.1*, *D.2*, and *Lost Years*.

Reinhardt, Max (1873–1943). Austrian theatrical producer, born Max Goldman. He became world-famous as the director of the Deutsches Theater in Berlin with his 1905 production of *A Midsummer Night's Dream*. He is remembered for his extravagant showmanship, though his work included serious classical theater from the Greeks to Shakespeare, Molière, Ibsen, and Shaw. He directed a few films in Germany and one later in Hollywood. Reinhardt's European empire ended when Hitler annexed Austria. He eventually opened an acting and theater school on Sunset Boulevard in Hollywood—the Workshop for Stage, Screen, and Radio—with his second wife, German actress Helene Thimig. He appears in *D.1*.

Reinhardt, Wolfgang (1908–1979). Film producer and writer; son of Max Reinhardt, brother of Gottfried Reinhardt. He produced *My Love Come Back* (1940), *The Male Animal* (1942), *Three Strangers* (1946), *Caught* (1948), and *Freud* (1962), for which he was nominated for an Academy Award as co-writer. Isherwood probably met Wolfgang Reinhardt through Gottfried soon after arriving in Hollywood, and as he records in *D.1* and in *Lost Years*, he and Wolfgang tried to work together several times during the 1940s when Wolfgang was a producer at Warner Brothers. With Aldous Huxley in 1944, they discussed making *The Miracle*, a film version of the play produced by Max Reinhardt in the 1920s, and in 1945 Wolfgang hired Isherwood to work on Maugham's 1941 novel *Up at the Villa*, but neither film was made by him. Much later, in June 1960, Wolfgang approached Isherwood to write a screenplay based on Felix Dahn's four-volume 1876 novel, *Ein Kampf um Rom* (*A Struggle for Rome*), about the decline and fall of the Ostrogoth empire in Italy in the sixth century, but Isherwood turned the project down. Wolfgang's wife was called Lally. He also appears in *D.2*.

Renate. See Druks, Renate.

Richard. See Isherwood, Richard Graham Bradshaw.

Richardson, Tony (1928–1991). British stage and film director; educated at Oxford where he was president of the Oxford University Dramatic Society. During the 1950s, he was a T.V. producer for the BBC, wrote about film for *Sight and Sound*, and was a founder of the Free Cinema movement, collaborating with Karel Reisz on a short, *Momma Don't Allow* (1955). He co-founded The English Stage Company with British actor and director George Devine (1910–1966) and under its auspices directed John Osborne's *Look Back in Anger* at the Royal Court in 1956. Then he and Osborne formed a film company, Woodfall, and Richardson went on to make movies, many adapted from his stage productions. In 1960, when Isherwood first mentions him in *D.1* (he also appears in *Lost Years* and *D.2*), Richardson was involved with Wyatt Cooper, then a young actor, and he was directing for screen and stage virtually simultaneously. He was filming *Sanctuary* (1961)—amalgamated from Faulkner's *Sanctuary* (1931) and its sequel, *Requiem for a Nun* (1951), which he had already staged separately at the Royal Court in London in 1957—and he was also directing Shelagh Delaney's *A Taste of Honey* in New York with a mostly English cast brought over from London. As Isherwood tells in *D.2*, he worked for Richardson on film scripts of Evelyn Waugh's 1948 novel *The Loved One* (1965), Carson McCuller's *Reflections in a Golden Eye* (later directed by John Huston with a different script), *The Sailor from Gibraltar* (1967) based on Marguerite Duras' novel, and, with Don Bachardy, adaptations of Robert Graves's *I, Claudius* and *Claudius, the God,* though much of the work was never used. Richardson's other films include *The Entertainer* (1960), *A Taste of Honey* (1961), *The Loneliness of the Long Distance Runner* (1962), *Tom Jones* (1963, Academy Award), *The Charge of the Light Brigade* (1968), *Hamlet* (1969), *Ned Kelly* (1970), *Joseph Andrews* (1977), *The Hotel New Hampshire* (1984), and *Blue Sky* (released posthumously, 1994). He was married to Vanessa Redgrave from 1962 to 1967 and had two daughters with her, and he had a long affair with Grizelda Grimond, producing a third daughter in 1973. In 1974, Isherwood mentions that Richardson resigned as director of *Mahogany* (1975) starring Diana Ross; in fact, Richardson made Berry Gordon fire him over a disagreement about casting a minor character so he could collect his full salary; Gordon, founder of Motown, directed the film himself. Richardson died of AIDS.

Rick. See Sandford, Rick.

Rigby, Harry (1925–1985). Broadway producer, sometimes of plays, including *The Ballad of the Sad Café* (1963), but mostly of adapted and revived musicals. In 1968, his *Hallelujah, Baby!* won Tony Awards for Best Musical and Best Producer of a Musical. His revival of *Irene* opened in March 1973, directed by Gower Champion (not by John Gielgud as once planned). Other successes, many with Terry Allen Kramer, include *I Love My Wife* (1977) and *Sugar Babies* (1979). His Broadway production of *A Meeting by the River* in 1979 lasted only one night after ten previews.

Ritajananda, Swami. Indian monk of the Ramakrishna Order; chief assistant to Swami Prabhavananda at the Hollywood Vedanta Society from 1958 to 1961. He then went to France to run the Vedanta Center at Gretz, near Paris, until his death in 1994. As his assistant at Gretz, he later took on Prema, by then called Swami Vidyatmananda. He appears in *D.2*.

Roberts, John. British theater producer; he produced two plays on Broadway in the late 1950s and later worked in South Africa. He was to assist Clifford Williams with the proposed London production of *A Meeting by the River* in 1970.

Roberts, Rachel (1927–1980). Welsh-born actress, educated at the University of Wales and RADA. She had many stage roles, beginning in 1951. Her films included *Saturday Night and Sunday Morning* (1960), *This Sporting Life* (1963, Academy Award nomination), *O Lucky Man!* (1973), *Murder on the Orient Express* (1974), and *When a Stranger Calls* (1979). She also appeared regularly on American T.V. in "The Tony Randall Show" from 1976 to 1978. Her second marriage, in 1962, was to Rex Harrison; they divorced in 1971. She appears in *D.2*.

Roerick, Bill (1912–1995). American actor. Isherwood met him in 1943 when John van Druten brought Roerick to a lecture at the Vedanta Society. He was in England as a G.I. during World War II and became friends there with E.M. Forster, J.R. Ackerley, and others. His companion for many years was Tom Coley. In 1944, Roerick contributed a short piece to *Horizon* defending Isherwood's new way of life in America after Tony Bower had made fun of it in a previous number. He appears in *D.1*, *D.2*, and *Lost Years*.

Ronnie. See Knox, Ronnie.

Ross, Jean (1911–1973). The original of Isherwood's character Sally Bowles in *Goodbye to Berlin*. He met her in Berlin, possibly in October 1930, but certainly by the start of 1931. She was then occasionally singing in a nightclub, and they shared lodgings for a time in Fräulein Thurau's flat. Ross's father was a Scottish cotton merchant, and she had been raised in Egypt in lavish circumstances. After Berlin, she returned to England where she became friendly with Olive Mangeot, lodging in her house for a time. She joined the Communist party and had a daughter, Sarah (later a crime novelist under the name Sarah Caudwell), with the Communist journalist and author Claud Cockburn (1904–1981), though Ross and Cockburn never married. She was close to her two sisters and shared houses and flats with them over the years, especially with Margaret, known as Peggy (b. 1913), a sculptor and painter trained at the Liverpool School of Art. Ross appears in *D.1*, *D.2*, and *Lost Years*.

Ruscha, Ed (b. 1937). American pop and conceptual artist; born in Omaha, raised in Oklahoma where he met Joe Goode; trained at the Chouinard Art Institute; working in Los Angeles. In 1958, he was a printer's apprentice, and in 1960 he worked briefly in advertising as a layout artist. His first of many solo shows was at the Ferus Gallery in 1963, and he began showing with Leo Castelli in New York in 1973. As well as oils, he has painted in Vaseline, axle grease, food—caviar, cherries, carrot juice, chocolate—and he built "Chocolate Room" for the thirty-fifth Venice Biennale in 1970. Other work includes lithography, silkscreen, books and booklike objects, films, and treated photographs. Since the 1980s, he has been the subject of retrospectives at major museums, including the San Francisco Museum of Modern Art, the Whitney, the Centre Georges Pompidou, the Hirshhorn, and the Hayward Gallery in London. He has received grants from the Guggenheim Foundation and the National Endowment for the Arts, and in 2005, he represented the U.S. at the fifty-first Venice Biennale.

Russell, Bertrand, 3rd Earl Russell (1872–1970). English philosopher, mathematician, social critic, writer; educated at Trinity College, Cambridge, where he was a Cambridge Apostle; afterwards he worked as a diplomat and academic. He published countless books and is one of the most widely read philosophers of the twentieth century. Chief among his awards and honors was the Nobel Prize for Literature in 1950. Throughout his life, Russell expressed his convictions in social and political activism. When he opposed British entry into World War I and joined the No-Conscription Fellowship, he lost his first job at Trinity, and he was fined and imprisoned more than once for his role in public demonstrations as a pacifist. Partly as a result, he became a visiting professor and lecturer in America and returned to Trinity as a Fellow only in 1944. Isherwood first met him through Aldous and Maria Huxley in late 1939 in Hollywood, as he tells in *D.1*. By December 1939, Russell had renounced pacifism because of the evils of fascism. In 1949, he began to champion nuclear disarmament. In 1958, he helped to found and was elected president of the Campaign for Nuclear Disarmament, but he resigned in 1960 to launch his more militant Committee of 100 for Civil Disobedience Against Nuclear Warfare, which sponsored several public protests in 1961 and which Isherwood refers to in *D.2*.

Sachs, David (1921–1992). American philosopher and poet, born in Chicago, educated at UCLA and Princeton where he obtained his doctorate in 1953. He lectured widely and taught philosophy at a number of American and European universities, longest at Johns Hopkins. His essays on ethics, ancient philosophy, philosophy of the mind, literature, and psychoanalysis were published in many journals, as were his poems, and he edited *The Philosophical Review*. He appears in *Lost Years* as a participant in the mainly homosexual intellectual and domestic ménage known as the Benton Way Group and in *D.2*.

Sagal, Boris (1917–1981). Ukrainian-born American director, educated at Yale School for Drama. He mostly directed for T.V., including episodes of "The Twilight Zone," "Colombo," and "The Man from U.N.C.L.E.," and he made the cult science-fiction film, *The Omega Man* (1971), starring Charlton Heston.

Sagui, Carlos. Mexican friend of Mark Lipscomb and John Ladner; he lived in their house for a time in the mid-1970s and sat for Don Bachardy during the same period.

Sahl, Mort (b. 1927). Canadian-American comedian and political satirist, raised mostly in Los Angeles and educated at UCLA. As he tells in *D.1*, Isherwood first saw him perform in Los Angeles in July 1960; that year, Sahl was pictured on the cover of *Time Magazine* as the father of a new kind of comedy. He appeared in films and on T.V. and made many recordings. He is also mentioned in *D.2*.

samadhi. The state of superconsciousness, in which an individual can know the highest spiritual experience; absolute oneness with the ultimate reality; transcendental consciousness.

Sandford, Rick (1951–1995). American actor and writer raised in Tahoe, California. He became a born-again Christian in 1971–1972 and settled in Los Angeles in 1973. He appeared in porn films and as a T.V. stand-in, and he published stories in *Men 2* and *His 3*. His only novel, *The Boys Across the Street*, appeared posthumously in 2000. He is the subject of a number of portraits by Don Bachardy.

Sanfaçon, Paul (1940–1989). Lecturer in Middle Eastern and Comparative Religions. He worked at the American Museum of Natural History in New York and had an apartment on the Upper West Side nearby. He mostly lived on West

14th Street with his much older lover, the painter and art critic Maurice Grosser, and was able to loan his own apartment to Isherwood. He died of AIDS.

sannyas. The second and final vows of renunciation taken in the Ramakrishna Order, at least four or five years after the brahmacharya vows. The sannyasin undergoes a spiritual rebirth and, as part of the preparation for this, renounces all caste distinctions. In *D.1*, Isherwood's entry for March 13, 1958 refers to the way in which Krishna (George Fitts) had first to join the Brahmin caste in order to have a caste to renounce; then Krishna had to imagine himself as dead, and to become a ghost in preparation for being reborn. At sannyas, the spiritual aspirant becomes a swami and takes a new Sanskrit name, ending with "ananda," bliss. Thus, the new name implies "he who has the bliss of" whatever the first element in the name specifies, as in Vivekananda, "he who has the bliss of discrimination." A woman sannyasin becomes a pravrajika (woman ascetic), and her new name ends in "prana," meaning "whose life is in" whatever is designated by the first element of the name.

Santoro, Jean and Tony. American actress Jean Howard (1910–2000) and the Italian musician she married in 1973. She was born and raised in Texas, got a contract with MGM in 1930 and became a Ziegfeld girl around the same time. Her first marriage was to Charles Feldman, an agent and producer—including of the film adaptation of *A Streetcar Named Desire* (1951). After they divorced in 1948, they continued to live together, and Isherwood also refers to her as Jean Feldman. She met Santoro in 1964 in Capri, where he was playing in a band called The Shakers; they lived there and in Rome for nearly a decade before marrying and returning to Hollywood. She photographed Hollywood friends informally and, later, took pictures for *Life* and *Vogue* magazines. She published *Jean Howard's Hollywood: A Photo Memoir* (1989) and *Travels with Cole Porter* (1991), illustrating her many trips with him.

Sarada (d. 2009). American nun of the Ramakrishna Order; of Norwegian descent, born Ruth Folling. She studied music and dance and, while at the Vedanta Society, learned Sanskrit. Her father lived in New Mexico. Isherwood met her when he arrived in Hollywood in 1939; in the mid-1940s, she moved to the convent at Santa Barbara where he occasionally saw her. He writes about her in *D.1*, *D.2*, and *Lost Years*. She was a favorite of Prabhavananda, but she suddenly left the Vedanta Society in October 1965 when she was forty-three years old. She married a few years later and became a painter. Prabhavananda was distressed by her abrupt departure and for a long time afterwards forbade her to be mentioned by her Sanskrit name, insisting she be called Ruth.

Sarada Convent, Montecito. In 1944, Spencer Kellogg gave his house at Montecito, near Santa Barbara, to the Vedanta Society of Southern California. The house was called "Ananda Bhavan," Sanskrit for Home of Peace. Kellogg, a devotee, died the same year, and the house became a Vedanta center and eventually a convent housing about a dozen nuns. During the early 1950s, a temple was built adjacent to the grounds.

Sarada Devi (1855–1920). Bengali wife of Ramakrishna; they married by arrangement when she was five years old. After the marriage, she returned to her family and he to his temple, and their relationship was always chaste although she later spent long periods of time living intimately with him. She became known as a saint in her own right and was worshipped as Holy Mother, the living embodiment of Mahamaya, of the Divine Mother, of the Goddess Sarasvati, and of Kali herself. Isherwood was initiated on Holy Mother's birthday, November 8, 1940.

Sarrazin, Michael (b. 1940). Canadian actor; he had a success in *They Shoot Horses, Don't They?* (1969) and afterwards appeared in made-for-T.V. movies, including "Frankenstein: The True Story" as the Creature.

Sarver, Tony (1931–1980). American actor, stage manager, singer; born in Indianapolis of Afro-Cherokee-Irish parents. He served in the navy before settling in Los Angeles where he appeared in fringe productions and directed Paulene Myers in her one-woman black history production, *World of My America*, which toured colleges nationwide. He was the longtime partner of Bill Scobie from 1969 until his death, and he met Isherwood and Bachardy through Scobie. Bachardy drew him and Scobie; the portrait of Sarver appeared in *October*. When Sarver died of colon cancer, Isherwood read Auden's poem "Since" at his funeral.

Sat. See Barnett, Jimmy.

Schlesinger, John (1926–2003). British film director, educated at Oxford. He was in the army during World War II, worked as an actor during the 1950s, and began making documentaries for BBC T.V. in 1957. Throughout his career, he received critical praise and prizes, beginning at the BBC and continuing with his first two feature films, *A Kind of Loving* (1962) and *Billy Liar* (1963), followed by an Academy Award nomination for *Darling* (1965). His first Hollywood film, *Midnight Cowboy* (1969), won Academy Awards for Best Picture and Best Director and made him internationally famous. Other films include *Far from the Madding Crowd* (1967), *Sunday Bloody Sunday* (1971), *The Day of the Locust* (1975), *Marathon Man* (1976), *Madame Sousatzka* (1988), *Pacific Heights* (1990), and *Cold Comfort Farm* (1996). He also directed for the stage, including the Royal Shakespeare Company, and for T.V. throughout his career. He lived with the photographer Michael Childers for more than thirty-six years. Isherwood was introduced to them by Gavin Lambert.

Schlesinger, Peter (b. 1948). American painter, photographer, and, later, sculptor; born and raised in Los Angeles. In the summer of 1966, Schlesinger studied drawing with David Hockney at UCLA and became Hockney's lover and model. In 1968, they travelled together to England, where Schlesinger studied at the Slade. He remained in England for the better part of a decade, also travelling in Europe. Some of the photographs which he took during this period later appeared in his book, *A Checkered Past: A Visual Diary of the 60s and 70s* (2003), including many of Eric Boman, the photographer who became his lover and companion in 1971 and with whom he moved to New York in 1978. He appears in *D.2*.

Schneider, Peter (1950–2007). American writer, editor, and teacher, raised in Claremont, California, where he became friends with Jim Gates who first introduced him to Vedanta. He and Gates moved to Los Angeles together when they were about eighteen, at first living briefly with Schneider's father, Dr. Leonard Schneider, a Gestalt therapist and psychology professor at L.A. State, and afterwards in a cabin on the canals in Venice. Around this time, Schneider attended Santa Monica College; he later got a degree in English at UCLA. He introduced himself and Gates to Isherwood over the telephone; Isherwood began driving them to Wednesday night meetings at the Vedanta Society and soon introduced them to Swami Prabhavananda, whereupon Schneider joined the society, toying with the idea of becoming a monk. Instead, he married Sumishta Brahm, a Vedanta devotee; the marriage was short, but he stayed away from the Vedanta Society for some years. Afterwards, he lived with Anya Cronin, also known as Anya Liffey, with whom he had two children in the 1980s. With Cronin, he wrote and produced musicals and other theater events for the Vedanta Society and elsewhere. He contributed to Vedanta publications and served as the society librarian in Hollywood. His Sanskrit name, given to him by Vivekaprana, was Hiranyagarbha. As P. Schneidre and later P. Shneidre, he published poems in *The Paris Review*, *Rolling Stone*, *Antioch Review*, and others. He also founded and ran a literary press, Illuminati, publishing work by James Merrill, Charles Bukowski, Viggo Mortensen, and a book of Don Bachardy's drawings of artists, *70 x 1*. He appears in *D.2*.

Schubach, Scott. A wealthy doctor who lived with Michael Hall for some years in West Hollywood.

Schuman, Howard (b. 1942). American television writer and presenter; born in Brooklyn, educated at Brandeis University and the University of California at Berkeley. He settled permanently in London in 1969 to live with Robert Chetwyn. His television work includes "Rock Follies"—which he devised and wrote, including the lyrics, and which won the BAFTA Award for best serial in 1976—and "Selling Hitler" (1991). He has been a BBC T.V. presenter and writes articles and reviews. Isherwood conjectures in this diary that Schuman is partly black, but as far back as he can trace his ancestry, Schuman knows of no black forebears.

Schwed, Peter (1911–2003). Isherwood's editor at Simon & Schuster; raised on Long Island and educated at Princeton. Eventually he became editorial chairman of the firm. He wrote several books himself, mostly about golf and tennis, and published a volume of his editorial correspondence with P.G. Wodehouse. Isherwood never genuinely felt that Schwed liked or understood his writing although they worked together for about fifteen years. He appears in *D.1* and *D.2*.

Scobie, W.I. (William, Bill) (b. 1932). British journalist, born in Scotland, educated at Perth Academy and the University of London. He reported freelance for *Time Magazine* and *The Observer* in the Middle East, Cyprus, and Italy, then in 1969 moved to Los Angeles, where he continued as a stringer for *The Observer*. He met Tony Sarver that October and spent the next ten years with him—until Sarver's death—in two wooden shacks which backed onto each other on Venice Beach. In 1974, Scobie proposed to interview Isherwood for *The Paris Review*, launching a friendship with him and with Bachardy; they saw movies together every couple of weeks and ate supper at El Coyote. The interview appeared in 1975. Scobie was a regular contributor to *The Advocate*, and he published poems in

Encounter, *The London Magazine*, and *The Paris Review*. Later, he settled in France.

Scott Gilbert, Clement. British would-be theatrical producer, he was wealthy and eventually became the owner of the Pembroke Theatre in Croydon Surrey. With Ernest Vadja, he created the characters for "Presenting Charles Boyer," an NBC radio show which ran a handful of times in 1950. He backed two 1961 productions staged in Croydon: *Mother*, with David McCallum in the cast, and *Compulsion*. And he backed the proposed London production of *A Meeting by the River* in London in 1970. He appears in *D.2*.

Searle, Alan (1905–1985). Secretary and companion to Somerset Maugham from 1938; he was the son of a Bermondsey tailor and had a cockney accent. Lytton Strachey was a former lover. When he first met Maugham in London in 1928, Searle was working with convicts—visiting them in prison and helping them to resettle in the community on release—but he told Maugham he wanted to travel. Maugham invited him on the spot to do so, but for a decade they met again only when Maugham was in London. Eventually, Searle devoted his life to Maugham and became his heir. He appears in *D.1* and *D.2*.

Selznick, David O. (1902–1965). American movie producer, most famous for *Gone with the Wind* (1939). He also brought Alfred Hitchcock to Hollywood to direct *Rebecca* (1940). Among Selznick's many other movies are *King Kong* (1933), *David Copperfield* (1934), *Reckless* (1935), *Anna Karenina* (1935), *A Tale of Two Cities* (1935), *A Star Is Born* (1937), *The Prisoner of Zenda* (1937), *Intermezzo* (1939), *Spellbound* (1945), *Duel in the Sun* (1946), *Portrait of Jennie* (1948), and *The Third Man* (1949). He worked for his father's movie company until Lewis Selznick went bankrupt in 1923; in 1926, his father's former partner, Louis B. Mayer, hired him as an assistant story editor in MGM. Selznick soon moved to Paramount, then RKO, then back to MGM until 1935 when he formed Selznick International Pictures with John Hay Whitney. Selznick's aspirations were monumental, and he tried to control every detail of his pictures. Despite his box-office success, he went into debt, and by the end of the 1940s he had to close his companies. He married Louis B. Mayer's daughter Irene in 1931, and they had two sons, Jeffrey and Daniel, before separating in 1945. When their divorce was finalized in 1949, he married Jennifer Jones. During the 1950s, he took Jones to Europe to work, and her career absorbed him at the end of his life. He traded his rights in *A Star is Born* to get Jones the lead in *A Farewell to Arms* (1957); the film failed, and it proved to be his last. Isherwood worked for Selznick in 1958, developing a script for a proposed film, *Mary Magdalene*, and they became friends, as Isherwood records in *D.1* and *D.2*.

Selznick, Jennifer. See Jones, Jennifer.

Seymour, Jane (b. 1951). British actress, educated in Hertfordshire. She played Winston Churchill's mistress in *Young Winston* (1972) directed by Richard Attenborough, then appeared as Prima in "Frankenstein: The True Story" and as the Bond Girl Solitaire in *Live and Let Die* (1973). Later she had numerous T.V. roles. The first of her four husbands was Richard Attenborough's son, Michael Attenborough, from 1971 to 1973.

Shadduck, Tom. Isherwood and Bachardy's gardener for several years in the mid-1970s. He was the younger brother of Jim Shadduck, whom they knew from the gym.

Shaw, Irwin (1913–1984). American novelist, playwright, screenwriter; raised in Brooklyn and educated at Brooklyn College, where he was a football star. He began his career writing for radio during the Depression, served with George Stevens's filmmaking unit during World War II, was blacklisted during the McCarthy era, and lived largely in Europe after 1951. His plays include *Bury the Dead* (1937) and *Gentle People* (1939). His best-selling novels, many of which were adapted for film and T.V., include *The Young Lions* (1948), *Lucy Crown* (1956), *Rich Man, Poor Man* (1970), *Evening in Byzantium* (1973), and *Acceptable Losses* (1982). He appears in *D.1*. His wife, Marian Edwards, an actress and chorus girl from Los Angeles, was a friend of Peter Viertel's first wife, Jigee, and Shaw remained close to Peter Viertel all his life. The Shaws married in 1939, divorced in 1969, and remarried in 1982. Their only child, Adam (b. 1950), was a journalist and aspiring novelist in youth and later became a commercial pilot. According to Larry Collins, a friend skiing with the group on the afternoon of the fatal avalanche Isherwood tells about in his diary account for February 12, 1973, Peter Viertel himself proposed the dangerous route, whereupon Irwin Shaw and Collins dropped out of the party; Collins's version appears in Michael Shnayerson's *Irwin Shaw: A Biography*.

Sheinberg, Sidney (Sid) (b. 1935). Studio executive; born in Texas, educated at Columbia College, the University of Texas Law School and Columbia Law School. He joined a T.V. division of Universal Studios in 1959 as a lawyer and soon moved into production as an executive. When he was just thirty-eight years old, in 1973, he became President and Chief Operating Officer of MCA, the parent company of Universal. While he was head, Universal Pictures released three of the top-grossing films of all time, *Jaws* (1975), *E.T.* (1982), and *Jurassic Park* (1993), all directed by Steven Spielberg, whom Sheinberg originally hired to work in T.V. Sheinberg later started his own production company, The Bubble Factory.

Shiva. The final aspect of the Hindu triad: Brahma, the creator, Vishnu, the Preserver, and Shiva, the Dissolver. Shiva is the father aspect, balancing the creative mother aspect of the godhead personified as Shakti Parvati, Kali, Durga, and other female gods. When worshipped as the chosen ideal, Shiva represents Brahman, the transcendent Absolute. Shiva also has other names and is represented in many different ways. Shiva is the lord of renunciation and of compassion. Shiva Ratri, or Shiva Night, in the early spring about four days before Ramakrishna's birthday, is observed by worship, meditation and fasting, all day and through four pujas, from 6 p.m. until the early hours, when there is a meal.

Shone, Richard (b. 1949). British art historian, curator, author; editor of *The Burlington Magazine* since 2003. His books and monographs include *Toulouse-Lautrec* (1974), *The Century of Change: British Painting Since 1900* (1977), *Manet* (1978), *The Charleston Artists* (1984), *The Art of Rodrigo Moynihan* (1988), *Walter Sickert* (1988), *Bloomsbury Portraits* (1994), *Sisley* (1999), and *The Art of Bloomsbury: Roger Fry, Vanessa Bell and Duncan Grant* (1999).

Shroyer, Frederick B. (Fred) (191[7]–1983). Professor in the English department at Los Angeles State College, where he was responsible for Isherwood being hired to teach in 1959. He wrote novels—*Wall Against the Night* (1957), *Wayland 33* (1962), *There None Embrace* (1966)—and he produced a number of college English books and anthologies—*College Treasury: Prose Fiction, Drama* (1956) edited with Paul Jorgensen, *Informal Essay* (1961), *Art of Prose* (1965) both with Paul Jorgensen, *Short Story: A Thematic Anthology* (1965) edited with Dorothy Parker, *Types of Drama* (1970) with Louis Gardemal, and *Muse of Fire: Approaches to Poetry* (1971) compiled with H. Edward Richardson. He appears in *D.1* and *D.2*.

Simon & Schuster. Isherwood's U.S. publisher from the late 1950s until the mid-1970s; founded in 1924 by M. Lincoln (Max) Schuster (1897–1970) and Richard L. Simon (1899–1960). Isherwood moved from Random House to Simon & Schuster in 1957 in order to work with John Goodman. When Goodman unexpectedly died, Isherwood's new editor was Peter Schwed. After several changes of heart, Isherwood remained at Simon & Schuster anyway, and the firm published *Down There on a Visit, A Single Man, A Meeting by the River, Kathleen and Frank*, and *Exhumations*. He then moved to Farrar, Straus and Giroux for the publication of *Christopher and His Kind*.

Simon, Norton (1907–1993). American industrialist, art collector, philanthropist; born in Portland, Oregon. He dropped out of Berkeley to go into the sheet metal business, and at twenty-two, at the start of the Depression, bought an orange juice bottling plant which he developed with enormous success and merged with Hunt Foods in the early 1940s. Norton Simon Inc., eventually diversified to control other household brands, including McCall's Publishing, Canada Dry, Max Factor Cosmetics, and Avis car rentals. In 1969, Simon's son Robert committed suicide. He withdrew from his corporate roles, and the following year, he divorced his first wife. In 1971, he married Jennifer Jones. He began collecting Impressionist paintings in the 1950s, and over the following thirty years amassed one of the finest ever private art collections, including Old Masters and modern work, Indian and South Asian. During the 1960s, he loaned widely from his collection, especially to the Los Angeles County Museum of Art, which he helped to found; then in 1974, he agreed to give his name and financial resources to the former Pasadena Museum of Modern Art, transforming it into the Norton Simon Museum. In 1970 he ran for the Senate as a Republican.

Simon, Richard (b. 1932). One of Isherwood's literary agents, at Curtis Brown in London. In 1971–1972, he set up his own firm, Richard Scott Simon Ltd., and Isherwood briefly considered leaving Curtis Brown to go with him. Like Isherwood, Simon suffers from Depuytren's Contracture.

Sister Lalita (Sister) (d. 1949). Carrie Mead Wyckoff was an American widow who met Vivekananda on one of his trips to America and became a disciple of Swami Turiyananda (a direct disciple of Ramakrishna). Turiyananda gave her the name Sister Lalita. She met Swami Prabhavananda when he opened the Vedanta center in Portland, Oregon, and in 1929 invited him to live in her house in Hollywood. By 1938, they had gathered a congregation around them, and they built the Hollywood temple in her garden. She appears in *D.1*.

Sleep, Wayne (b. 1948). British ballet dancer; educated at the Royal Ballet School; in 1966, he joined the Royal Ballet Company, where he became a

Principal with numerous roles choreographed on him. He also appeared as a guest dancer with other ballet companies and starred in West End musicals, including *Cats* (1981). He is a choreographer and teacher and created his own review of dance, *Dash,* in which he toured world wide. He appears in *D.2.*

Smight, Jack (1925–2003). American director, mostly for T.V., beginning in the 1940s; he directed episodes of many regular shows including "The Twilight Zone," "Route 66," and "The Alfred Hitchcock Hour." He also made films—for instance, *Harper* (1966), *No Way to Treat a Lady* (1968), *Rabbit, Run* (1970), and *Midway* (1976)—and a number of made-for-T.V. movies.

Smith, David. A young admirer of Isherwood's work; he occasionally paid court at the house in Adelaide Drive and once sat for Bachardy.

Smith, Dodie. See Beesley, Alec and Dodie Smith Beesley.

Smith, Emily Machell (Granny Emmy) (1840–1924). Isherwood's grandmother on his mother's side. Emily's husband, Isherwood's grandfather Frederick Machell Smith, was a wine merchant in Bury St. Edmunds; they married in 1864 and in 1885 moved to London with Kathleen, their only child, due to Emily's unpredictable health (Isherwood describes Emily in *Kathleen and Frank* as "a great psychosomatic virtuoso"). She was beautiful, passionate about the theater, and liked to travel with Kathleen, who helped her to prepare a book of guided walks, *Our Rambles in Old London* (1895). Emily's maiden name was Greene; her brother Walter Greene was a prosperous brewer in Bury St. Edmunds, went into politics, and became a baronet. Walter Greene entertained lavishly at his country house, Nether Hall, and Kathleen enthusiastically attended house parties and dances there as a young woman. Through Emily's family, Isherwood was related to the novelist Graham Greene.

Smith, Katharine (Kate) (1933–2000). English second wife of Ivan Moffat, from 1961 until 1972. She was a daughter of the 3rd Viscount Hambleden whose family fortune derived from the book and stationery chain W.H. Smith and Lady Patricia Herbert, daughter of the 15th Earl of Pembroke, elder sister of Isherwood's Tangier friend, David Herbert, and lady-in-waiting to Queen Elizabeth the Queen Mother. Kate Smith was a bridesmaid to Princess Alexandra and a close friend of Princess Margaret. She had two sons with Moffat, Jonathan (b. 1963) and Patrick (b. 1968), a godson of Princess Margaret. In 1973, she married thriller-writer Peter Townend, author of *Out of Focus* (1971), *Zoom!* (1972), and *Fisheye* (1974). She appears in *D.2.*

Smith, Maggie (b. 1934). British stage and screen star raised in Oxford; she trained at the Oxford Playhouse School and began her stage career at the Oxford Playhouse. She has won many awards for stage and film appearances since then, including a Tony Award for *Lettice and Lovage* in 1990. Her movies include *Othello* (1965) opposite Olivier, adapted from their stage production at the Royal National Theatre, *The Prime of Miss Jean Brodie* (1969, Academy Award), *California Suite* (1978, Academy Award), *A Room with a View* (1985), *Tea with Mussolini* (1999), *Gosford Park* (2001), *Ladies in Lavender* (2004), and the *Harry Potter* movies. She was married to British actor Robert Stephens from 1967 to 1975 and had two sons with him, both actors. In 1975, she married British playwright, screenwriter, and children's author Beverley Cross (1931–1998). His plays include *Strip the Willow* (1960), in which he cast Smith when she was still unknown, *One More River* (1959), *Half a Sixpence* (1963), and the English translation of Mark Camoletti's farce *Boeing-Boeing* (1962). For the movies, he wrote *Jason and the Argonauts* (1963) and *Clash of the Titans* (1981), in which Smith also appeared.

Sorel, Paul (1918–*circa* 2008). American painter, of Midwestern background; born Karl Dibble. He was a close friend of Chris Wood and lived with him in Laguna in the early 1940s, but moved out in 1943 after disagreements over money, living intermittently in New York. Wood continued to support him, though they never lived together again. Sorel painted portraits of Isherwood and Bill Caskey in 1950, and he appears in *D.1*, *D.2*, and *Lost Years*.

Spender, Elizabeth (Lizzie) (b. 1950). British actress and writer; educated at North London Collegiate School; daughter of Stephen and Natasha Spender. She had small parts in Isherwood and Bachardy's "Frankenstein" and in Terry Gilliam's *Brazil* (1985), and she worked in publishing. In 1990, she married Australian actor and satirist Barry Humphries—best known for his character "Dame Edna Everage." She appears in *D.2*.

Spender, Natasha Litvin (1919–2010). British concert pianist; her mother was a Russian émigré. She married Stephen Spender in 1941 and had two children with him, Matthew and Lizzie. She appears in *D.1*, *D.2*, and *Lost Years*.

Spender, Stephen (1909–1995). English poet, critic, autobiographer, editor. Auden introduced him to Isherwood in 1928; Spender was then an undergraduate at University College, Oxford, and Isherwood became a mentor. Afterwards Spender lived in Hamburg and near Isherwood in Berlin, and the two briefly shared a house in Sintra, Portugal, with Heinz Neddermeyer and Tony Hyndman. Spender was the youngest of the writers who came to prominence with Auden and Isherwood in the 1930s; after Auden and Isherwood emigrated, he cultivated the public roles they abjured in England. He worked as a propagandist for the Republicans during the Spanish Civil War and was a member of the National Fire Service during the Blitz. He moved away from his early enthusiasm for communism but remained liberal in politics. His 1936 marriage to Inez Pearn was over by 1939; in 1941, he married Natasha Litvin, and they had two children, Matthew and Lizzie. He appears as "Stephen Savage" in *Lions and Shadows* and is further described in *Christopher and His Kind*, *D.1*, *D.2*, and *Lost Years*. He published an autobiography, *World Within World*, in 1951, and his *Journals 1939–1983* appeared in 1985. Spender was co-editor with Cyril Connolly of *Horizon* and later of *Encounter*, and in 1968, he helped to found *Index on Censorship* to report on the circumstances of persecuted writers and artists around the world.

Spigelgass, Leonard (Lenny) (1908–1985). American screenwriter and Broadway playwright, born in Brooklyn. The many films he worked on—some based on his stage plays—include *I Was a Male War Bride* (1949), *Silk Stockings* (1957), and *Gypsy* (1962), and he adapted his Broadway hit *A Majority of One* (1959) for film. He was also a producer and later wrote for T.V., turning out scripts for eleven annual Academy Awards ceremonies. Isherwood met him at MGM when they both worked there during the 1950s, and he appears in *D.1*.

Stangos, Nicolas (Nikos) (1936–2004). Greek poet and translator, educated at Harvard. He was an editor at Penguin and, from 1974, of art history books at Thames and Hudson. He was the long-time companion of David Plante.

Steen, Mike (1928–1983). American stuntman, actor, author, from Louisiana, where he was friendly with Speed Lamkin, Tom Wright, and Henry Guerriero. Lamkin introduced him to Isherwood in the early 1950s. Gavin Lambert became romantically involved with Steen during 1958, and Steen also had relationships, perhaps sexual, with Nicholas Ray, William Inge, and Tennessee Williams. He worked as a stuntman in Ray's *Party Girl* and did stunts or played bit parts in other movies in the late 1950s and 1960s, including a tiny part in the 1962 film of Williams's *Sweet Bird of Youth.* He published two books: *A Look at Tennessee Williams* (1969) and *Hollywood Speaks: An Oral History* (1974). He appears in *D.1* and *D.2.*

Stephen. See Spender, Stephen.

Stern, Alan (1947–1986). Film producer. He was Michelle Rappaport's associate producer on *Old Boyfriends* (1979) and began developing with her, Isherwood, and Bachardy plans for a film of "Paul" from *Down There on a Visit.*

Stevens, Marti (b. 1931). American singer and actress; she appeared in a few films, including *All Night Long* (1961), several times on Broadway, and often on T.V. Isherwood also writes about her in *D.2.*

Strasser, Robin (b. 1945). American actress, trained at the Yale School of Drama. She was a daytime soap regular on "Another World" (1967–1972) and went on to an award-winning role in "One Life to Live" from 1979 onward. She also worked on Broadway and in repertory. She was the first wife of Larry Luckinbill, with whom she had two sons.

Stravinsky, Igor (1882–1971). Russian-born composer; he went to Paris with Diaghilev's Ballets Russes in 1910 and brought about a rhythmic revolution in Western music with *The Rite of Spring* (1911–1913), the most sensational of his many works commissioned for the company. He composed in a wide range of musical forms and styles; many of his early works evoke Russian folk music, and he was influenced by jazz. Around 1923, he began a long neo-classical period responding to the compositions of his great European predecessors. During the 1950s, with the encouragement of Robert Craft, he took up the twelve-note serial methods invented by Schoenberg and extended by Webern—he was already past seventy. After the Russian revolution, Stravinsky remained in Europe, making his home first in Switzerland and then in Paris, and he turned to performing and conducting to support his family. In 1926, he rejoined the Russian Orthodox Church, and religious music became an increasing preoccupation during the later part of his career. At the outbreak of World War II, he emigrated to America, settled in Los Angeles, and eventually became a citizen in 1945. Although he was asked to, he never composed for films. His first and most important work for English words was his opera, *The Rake's Progress* (1951), for which Auden and Kallman wrote the libretto. Isherwood first met Stravinsky in August 1949 at lunch in the Farmer's Market in Hollywood with Aldous and Maria Huxley. He was soon invited to the Stravinskys' house for supper where he fell asleep listening to a Stravinsky recording; Stravinsky later told Robert Craft that this was the start of his great affection for Isherwood. He appears throughout *D.1*, *D.2*, and *Lost Years*.

Stravinsky, Vera (1888–1982). Russian-born actress and painter; second wife of Igor Stravinsky; she was previously married three times, the third time to the painter and Ballets Russes stage designer Sergei Sudeikin. In 1917, she fled St. Petersburg and the bohemian artistic milieu in which she was both patroness and muse, travelling in the south of Russia with Sudeikin before going on to Paris where she met Stravinsky in the early 1920s; they fell in love but did not marry until 1940 after the death of Stravinsky's first wife. Isherwood met her with Stravinsky in August 1949. She painted in an abstract-primitive style influenced by Paul Klee, childlike and decorative. She appears throughout *D.1*, *D.2*, and in *Lost Years*.

Stromberg, Hunt, Jr. (1923–1985). American T.V. executive; son of Hunt Stromberg (1894–1968), who was one of MGM's most profitable and powerful film producers from the mid-1920s until he retired in 1951. Stromberg Jr. began his career as a theater producer and moved to T.V. early in the 1950s. He worked with James Aubrey at CBS on the original idea for the series "Have Gun, Will Travel" (1956) and during the 1960s, he supervised "The Beverly Hillbillies," "Green Acres," and "Lost in Space." He was fired from CBS at the time of Aubrey's downfall in 1965 and for a time ran a production partnership with Aubrey until Aubrey took over MGM Studios in 1969. Later, Stromberg worked at Universal Studios. He produced "Frankenstein: The True Story" (1973), written by Isherwood and Bachardy and "The Curse of King Tut's Tomb" (1980) adapted by Herb Meadow from Barry Wynne's book. He appears in *D.2*.

Sudhira. A nurse of Irish descent, born Helen Kennedy; she was a probationer nun at the Hollywood Vedanta Society when Isherwood arrived to live there in 1943. In youth, she had been widowed on the third day of her marriage. Afterwards, she worked in hospitals and for Dr. Kolisch and first came to the Vedanta Society professionally to nurse a devotee. She enlisted in the navy in January 1945 and later married for a second time and returned to nursing. She appears in *D.1* and *Lost Years*, and she is mentioned in *D.2*.

Summers, Claude. American academic with a doctorate from the University of Chicago. As Isherwood records, he lectured on *A Single Man* in 1976; he went on to write a book-length critical study, *Christopher Isherwood* (1980). He also published books and articles on seventeenth- and twentieth-century literature and textual studies and also on gay and lesbian themes, including a critique of E.M. Forster, *Gay Fictions: Wilde to Stonewall* (1990), and gay and lesbian encyclopedias, surveys, and collections. He is William E. Stirton Professor Emeritus of Humanities and Professor Emeritus of English at the University of Michigan-Dearborn. His companion and frequent co-author is Ted-Larry Pebworth, also a Professor Emeritus of English Literature at the University of Michigan-Dearborn. Pebworth got his Ph.D. at Louisiana State University, is a scholar of English Renaissance literature, on which he has published widely, and is an editor and a member of the advisory board for *The Variorum Edition of the Poetry of John Donne*.

Swahananda, Swami. Monk of the Ramakrishna Order, from India; born near

Habiganj, now in Bangladesh, and educated at Murari Chand College, Sylhet, and at the University of Calcutta where he got an M.A. in English. He was initiated in 1937 by Swami Vijnananda, a direct disciple of Ramakrishna, joined the order in 1947 and took his final vows in 1956. He was head of the Delhi Ramakrishna Mission, and then in 1968 was sent to the San Francisco center as assistant to Swami Ashokananda. Later, he became head of the Berkeley center, and in 1976, he was moved to Hollywood to replace the late Swami Prabhavananda. As a young monk, he edited *Vedanta Kesari*, a scholarly publication of the Ramakrishna Order, and he went on to publish numerous articles and books of his own. He appears in *D.2*.

Swami. Used as a title to mean "Lord" or "Master." A Hindu monk who has taken sannyas, the final vows of renunciation in the Hindu tradition. Isherwood used it in particular to refer to his guru, Swami Prabhavananda, and he pronounced it *Shwami*; see Prabhavananda.

Swamiji. An especially respectful form of "Swami," but also a particular name for Vivekananda towards the end of his life.

Sweet, Paula. American painter, from Carmel, New York, trained at Syracuse and at the Sir John Cass College of Art in London. She is also a sculptor and a jewelry and clothing designer.

Szczesny, Berthhold. Isherwood met Szczesny in The Cosy Corner on his first brief visit to Berlin in March 1929 and returned to Germany hoping to spend the summer with him in the Harz Mountains. But Szczesny, then called "Bubi," was in trouble with the police, fled to Amsterdam, and from there shipped out to South America. Subsequently he came to London working on board a freighter and smuggling refugees into England. As Isherwood tells in *The Condor and the Cows*, Szczesny eventually returned to Argentina, became part-owner of a factory and married an Argentine woman of privileged background, with whom he had two sons. He is mentioned in *D.1* and *Lost Years*. One of his sons, Juan Szczesny, is mentioned in this diary; he sat for Bachardy, May 10, 1980.

Tadatmananda. See Markovich, John.

tamas. See guna.

Ted. See Bachardy, Ted.

Tennant, Stephen (1906–1987). British poet, aesthete, eccentric; youngest son of Lord Glenconner, a Scottish peer, and Pamela Wyndham, a cousin of Oscar Wilde's paramour, Lord Alfred Douglas. He studied painting at the Slade and worked for decades on a novel which he never completed. He was known for his extravagantly camp dress and manners, his interior decorating, and spending the last seventeen years of his life mostly in bed. He was an intimate friend of Cecil Beaton and was photographed by him several times, and he had a long affair with Siegfried Sassoon. He is said to be a model for Sebastian Flyte in Evelyn Waugh's *Brideshead Revisited* and for Cedric Hampton in Nancy Mitford's *Love in a Cold Climate*. He is mentioned in *D.2*.

Tennessee. See Williams, Tennessee.

Thakur. Literally master or lord; a familiar name for Ramakrishna among his devotees. See also Ramakrishna.

Thompson, Nicholas. American theatrical agent, based in London. He was introduced to Isherwood and Bachardy by Robin French and became the London agent for *A Meeting by the River*. He appears in *D.2*.

Thomson, Virgil (1896–1989). American composer, music critic, author; born in Kansas and educated at Harvard. During the 1920s, he lived off and on in Paris, where he studied with Nadia Boulanger and met the young French composers, including Poulenc and Auric, who were known as "Les Six" and who were, like him, influenced by Satie. He became friendly with Gertrude Stein, who supplied libretti for his operas *Four Saints in Three Acts* (1934) and *The Mother of Us All* (1947). In 1940, he returned permanently to New York. He was the music critic for the *Herald Tribune* for a decade and a half, published eight books on music, including *American Music Since 1910* (1971), and composed prolifically. His third and last opera, *Lord Byron*, on a libretto by Jack Larson, was planned for the Metropolitan Opera in New York, but instead premiered at the Juilliard Opera Center in April 1972. His film scores include *The Plough That Broke the Plains* (1936) and *The River* (1938), both directed by Pare Lorentz, and *Louisiana Story* (1948), directed by Robert Flaherty. His companion for many years was Maurice Grosser. He appears in *D.2*.

Tony. See Richardson, Tony.

Trabuco. Monastic community sixty miles south of Los Angeles and about twenty miles inland; founded by Gerald Heard in 1942. Heard anonymously provided $100,000 for the project (his life savings), and he consulted at length with various friends and colleagues as well as with members of the Quaker Society of Friends about how to organize the community. In 1940, he planned only a small retreat called "Focus," then he renamed the community after buying the ranch at Trabuco. Isherwood's cousin on his mother's side, Felix Greene, oversaw the purchase of the property and the construction of the building, which could house fifty. By 1949 Heard found leading and administering the group too much of a strain, and he persuaded the Trustees to give Trabuco to the Vedanta Society. It opened as a Vedanta monastery in September 1949.

Turner, Charles. African-American actor and director, trained at Yale School of Drama, which is among the several places he has also taught. He has performed on and off Broadway and in T.V. and films.

Turville-Petre, Francis (1904–1941). English archaeologist, from an aristocratic Catholic family. Isherwood met the eccentric Turville-Petre through Auden in Berlin in 1929, and it was at Turville-Petre's house outside Berlin that Isherwood met Heinz Neddermeyer in 1932. In 1933, when Isherwood and Heinz fled Germany, they spent nearly four months on Turville-Petre's tiny island, St. Nicholas, in Greece. Turville-Petre, known among the boys in the Berlin bars as "Der Franni," inspired the character of the lost heir in Auden's play *The Fronny* and in Auden and Isherwood's *The Dog Beneath the Skin* (1935); he is also the model for "Ambrose" in *Down There on a Visit*.

Tynan, Kathleen. See entry for Tynan, Kenneth.

Tynan, Tracy. See entry for Tynan, Kenneth.

Tynan, Kenneth (Ken) (1927–1980). English theater critic, educated at Oxford. During the 1950s and 1960s, he wrote regularly for the London *Evening Standard*, then for *The Observer, The New Yorker*, and other publications. He was literary adviser to the National Theatre from its inception in 1963, but his anti-establishment views brought about his departure before the end of the decade. His support for realistic working-class drama—by new playwrights such as Osborne, Delaney and Wesker—as well as for the works of Brecht and Beckett, was widely influential. Many of his essays and reviews are collected as books. At the end of stage censorship in 1968, he devised and produced the sex revue *Oh! Calcutta!*, which opened in New York in 1969 and in London in 1970. His first wife, from 1951, was the actress and writer Elaine Dundy, with whom he had a daughter, Tracy Tynan, later a costume designer for films. In 1963, he began an affair with the newly married Kathleen Halton Gates (1937–1995), a Canadian journalist and, later, novelist and screenwriter, raised in England; they married in 1967 and settled for a time in the U.S. with their children, Roxana and Matthew. Isherwood first met Ken and Elaine Tynan in London in 1956, and they are mentioned in *D.1* and *D.2*.

UCLA. University of California at Los Angeles.

Unger, Ray (d. 2006). American artist and astrologer. He was the companion of Jack Fontan for fifty-three years, and, like Fontan, was known for his beauty. For some years, they managed a gym together in Laguna Beach, the Laguna Health Club. Unger was well read, well informed, talented and intelligent, without ambition for public recognition. He was an especially close friend to Bachardy.

Upward, Edward (1903–2009). English novelist and schoolmaster. Isherwood first met him in 1921 at their public school, Repton, and followed him (Upward was a year older) to Corpus Christi College, Cambridge. Their close friendship was inspired by their shared attitude of rebellion toward family and school authority and by their literary obsessions. In the 1920s they created the fantasy world Mortmere, about which they wrote surreal, macabre, and pornographic stories and poems for each other; their excited schoolboy humor is described in *Lions and Shadows* where Upward appears as "Allen Chalmers." Upward made his reputation in the 1930s with his short fiction, especially *Journey to the Border* (1938), the intense, almost mystical, and largely autobiographical account of a young upper-middle-class tutor's conversion to communism. Then he published nothing for a long time while he devoted himself to schoolmastering (he needed the money) and to Communist party work. From 1931 to 1961 he taught at Alleyn's School, Dulwich, where he became head of English and a housemaster; he lived nearby with his wife, Hilda, and their two children, Kathy and Christopher. After World War II, Upward and his wife became disillusioned by the British Communist party and left it, but Upward never abandoned his Marxist-Leninist convictions. Towards the end of the 1950s, he overcame writer's block and a nervous breakdown to produce a massive autobiographical trilogy, *The Spiral Ascent* (1977)—comprised of *In the Thirties* (1962), *The Rotten Elements* (1969), and *No Home But the Struggle*. The last two volumes were written in Sandown, on the Isle of Wight, where Upward retired in 1962. They were followed by four collections of short stories. Upward remained a challenging and trusted critic of

Isherwood's work throughout Isherwood's life, and a loyal friend. He appears in *D.1*, *D.2*, and *Lost Years*.

Urquhart-Laird, Tommy (b. 1943). Secretary and lover to John Lehmann during the 1970s. He arrived in London from Scotland in 1961 hoping to become an actor. Instead, he did odd jobs, including working as a builder, until he met Lehmann in 1972 and moved into his flat in Cornwall Gardens. He remained there until 1978 when the arrangement collapsed because of Urquhart-Laird's drinking.

Usha. A nun at the Vedanta Society and later at the convent in Santa Barbara; originally called Ursula Bond and, after sannyas, Pravrajika Anandaprana, which was sometimes shortened to Ananda. She was a German Jew, educated in England, and came to the U.S. as a young refugee during the war. Until the war ended, she worked for the U.S. government as a censor. She had been married before taking up Vedanta, and she had a daughter, Caroline Bond; she left three-year-old Caroline with the child's father, but Caroline later joined the Hollywood convent where she was known as Sumitra, took brahmacharya vows, and remained for about ten years before leaving to join an ashram that emphasized Sanskrit and scriptural study. Later, Sumitra did graduate work in Sanskrit and became a freelance editor. Anandaprana appears, as Usha, in *D.1* and *D.2*.

Vaccaro, Brenda (b. 1939). American stage, screen, and T.V. actress; born in Brooklyn, raised in Dallas, trained at New York's Neighborhood Playhouse. She began her career on Broadway, where she appeared in *Cactus Flower* (1965), *How Now, Dow Jones* (1967), and *The Goodbye People* (1968). Her films include *Midnight Cowboy* (1969), *Once is Not Enough* (1975; Academy Award nomination), *Ten Little Indians* (1990), and *The Mirror Has Two Faces* (1996).

Valen, Mark. A local friend introduced to Isherwood and Bachardy by Rick Sandford; he grew up in Los Angeles and worked for the Landmark Theater chain, beginning as a ticket taker and eventually selecting and programming their films all over the country. He sat for Bachardy many times.

Vandanananda, Swami (1915–2007). Hindu monk of the Ramakrishna Order. He arrived at the Hollywood Vedanta Society from India in the summer of 1955 and eventually became chief assistant, replacing Swami Asheshananda. In 1969 he returned to India, was head of the Delhi Center, and later, at Belur Math, became Assistant Secretary of the order, and then General Secretary. He appears in *D.1* and *D.2*.

van Druten, John (1901–1957). English playwright and novelist. Isherwood met him in New York in 1939, and they became friends because they were both pacifists. Van Druten's family was Dutch, but he was born and educated in London and took a degree in law at the University of London. He achieved his first success as a playwright in New York during the 1920s, then emigrated in 1938 and became a U.S. citizen in 1944. His strength was light comedy; among his numerous plays and adaptations, many of which were later filmed, were *Voice of the Turtle* (1943), *I Remember Mama* (1944), *Bell, Book, and Candle* (1950), and *I Am a Camera* (1951), based on Isherwood's *Goodbye to Berlin*. Van Druten acquired a sixty percent share in material that he used in *I Am a Camera* from *Goodbye to Berlin*; Isherwood retained a forty percent share. Anyone acquiring rights after van Druten needed his agreement or, after his death, the agreement of his estate. In 1951, van Druten directed *The King and I* on Broadway. He also wrote a few novels and two volumes of autobiography, including *The Widening Circle* (1957). He usually spent half the year in New York and half near Los Angeles on the AJC Ranch, which he owned with Carter Lodge. He also owned a mountain cabin above Idyllwild which Isherwood sometimes used. A fall from a horse in Mexico in 1936 left him with a permanently crippled arm; partly as a result of this, he became attracted to Vedanta and other religions (he was a renegade Christian Scientist), and in his second autobiography he describes a minor mystical experience which he had in a drug store in Beverly Hills. He was a contributor to Isherwood's *Vedanta for the Western World*, and there are numerous accounts of him in *D.1*, *D.2*, and *Lost Years*.

Vanessa. See Redgrave, Vanessa.

Van Horn, Mike. American artist and fashion illustrator, especially for Lord & Taylor. A close friend of Bachardy. In the early 1970s, he left Los Angeles for New York State where he renovates country properties. He appears in *D.2*.

Van Petten, William (Bill) (d. mid-1980s). Local reporter, for *The Los Angeles Times* and elsewhere. He grew up in La Jolla and in 1965 introduced Tom Wolfe to the beach scene there, which Wolfe wrote about in *The Pump House Gang*. He was also an early source of stories, later corroborated, that Richard Nixon was a belligerent drunk and beat his wife. He lived at the bottom of Santa Monica Canyon in an apartment once inhabited by Jim Charlton. He had a small independent income and often travelled abroad, especially to the Middle East. He had radiation treatment for cancer during the 1960s, and the treatment damaged his face and eyes. He appears in *D.2*.

Van Sant, Tom (b. 1931). Californian sculptor, painter, conceptual artist, and kite maker; he produced the first satellite map of earth, the Geosphere Image, and founded the Geosphere Project to model issues of earth resource management. He appears in *D.2*.

Vaughan, Keith (1912–1977). English painter, illustrator, and diarist. He worked in advertising during the 1930s and was a conscientious objector in the war; later he taught at the Camberwell School of Art, the Central School of Arts and Crafts, and the Slade, as well as briefly in America. Isherwood met him in 1947 at John Lehmann's and bought one of his pictures, *Two Bathers*, a small oil painting, still in his collection. Vaughan's diaries, with his own illustrations, were published in 1966. He appears in *D.1*, *D.2*, and *Lost Years*.

Vedanta Place. The Vedanta Society of Southern California was able to adopt the address Vedanta Place when, in 1952, the Hollywood Freeway cut off the tail end of Ivar Avenue where the society was located, leaving it in a cul-de-sac. The property, previously 1946 Ivar Avenue, had been the home of Carrie Mead Wyckoff, later known as Sister Lalita. As Isherwood tells in *D.1*, he lived at the Hollywood society as a novice monk during World War II.

Ventura, Clyde (1936–1990). American actor and stage director, born in New Orleans. He was in the 1963 Broadway cast of Tennessee Williams's *The Milk Train Doesn't Stop Here Anymore*, and he was Artistic Director of Theater West in Los Angeles, where he directed a number of Williams plays. He was also an acting coach at Actors Studio on the West Coast, and he had a few small movie roles. He appears in *D.2*.

Vera. See Stravinsky, Vera.

Vernon. See Old, Vernon.

Vidal, Gore (b. 1925). American writer. He introduced himself to Isherwood in a café in Paris in early 1948, having previously sent him the manuscript of his third novel, *The City and the Pillar* (1948). Vidal's father taught at West Point, and Vidal was in the army as a young man. Afterwards, he wrote essays on politics and culture, short stories, and many novels, including *Williwaw* (1946), *Myra Breckinridge* (1968, dedicated to Isherwood), its sequel, *Myron* (1975), *Two Sisters: A Memoir in the Form of a Novel* (1970), *Kalki* (1978), *Duluth* (1983), *The Judgement of Paris* (1984), *Live from Golgotha* (1992), *The Smithsonian Institute* (1998), ancient and medieval historical fiction such as *A Search for the King* (1950), *Messiah* (1955), *Julian* (1964), *Creation* (1981), and the multi-volume American chronicle comprised of *Burr* (1974), *Lincoln* (1984), *1876*, (1976), *Empire* (1987), *Hollywood* (1989), and *Washington, D.C.* (1967). He also published detective novels under a pseudonym, Edgar Box. During the 1950s, Vidal wrote a series of television plays for CBS, then screenplays at Twentieth Century-Fox and MGM (including part of *Ben Hur*), and two Broadway plays, *Visit to a Small Planet* (1957) and *The Best Man* (1960). His adaptation of Friedrich Duerrenmatt's *Romulus the Great*, about Romulus Augustulus, ran on Broadway from January to March 1962. In 1960 Vidal ran for Congress, and in 1982 for the Senate, both times unsuccessfully. He bought Edgewater, a Greek Revival mansion on the Hudson River north of New York, and lived there off and on with Howard Austen, from 1950 until he sold it in 1968; later he settled in Rapallo, Italy, and finally in Los Angeles. Over the years, Vidal campaigned to become a member of the National Institute of Arts and Letters; with pushing from Isherwood, he was finally elected in 1976 but turned it down. He describes his friendship with Isherwood in his memoir, *Palimpsest* (1995) and in *Point to Point Navigation* (2006). There are many passages about him in *D.1*, *D.2*, and *Lost Years*.

Vidor, King (1894–1982). American film director, screenwriter, producer; born and raised in Texas. He began directing in the silent era with *The Turn in the Road* for Universal in 1919, then formed his own studio, and afterwards worked for MGM. His movies include *The Big Parade* (1925), *The Crowd* (1928), *Show People* (1928), *Billy the Kid* (1930), *The Champ* (1931), *Northwest Passage* (1940), *Duel in the Sun* (1947), and *The Fountainhead* (1949). In *D.1*, Isherwood describes meeting him in Italy in 1955 on the set of *War and Peace* (1956), which Vidor was directing with the assistance of Mario Soldati. Vidor retired when his next film, *Solomon and Sheba* (1959), failed. During the 1960s he taught at UCLA film school, and in 1979 he received a special Academy Award as a creator and innovator in film. His first wife was silent screen star, Florence Vidor, who came to Hollywood with him in 1915 as a newlywed; they divorced in 1924. He was married to Eleanor Boardman, also a film star, for a few years in the mid-1920s. His third wife was a writer, Elizabeth Hill. He appears in *D.2*.

Vidya, also Vidyatmananda, Swami, previously John Yale. See Prema Chaitanya.

Viertel, Peter (1920–2007). German-born American screenwriter and novelist; second son of Berthold and Salka Viertel. He attended UCLA and Dartmouth and became a freelance writer, then served in the U.S. Marines during World War II and was decorated four times. He wrote the award-winning screenplay for Hemingway's *The Old Man and the Sea* as well as other Hemingway adaptations. His own novels are in the Hemingway vein, with subjects such as soldiering (*Line of Departure*, 1947), big game hunting (*White Hunter, Black Heart*, 1954), and bull-fighting (*Love Lies Bleeding*, 1964). His first novel, *The Canyon* (published in 1941, but completed when he was just nineteen), gives a compelling adolescent view of Santa Monica as it was around the time when Isherwood first arrived there. His first marriage was to Virginia Schulberg, and in 1960 he married the actress Deborah Kerr. Like his mother and father he eventually resettled in Europe. He appears in *D.1* and *Lost Years* and is mentioned in *D.2*.

Viertel, Salka (1889–1978). Polish actress and screenplay writer; first wife of Berthold Viertel with whom she had three sons, Hans, Peter, and Thomas. Sara Salomé Steuermann Viertel had a successful stage career in Vienna (including acting for Max Reinhardt's Deutsches Theater) before moving to Hollywood where she became the friend and confidante of Greta Garbo; they appeared together in the German-language version of *Anna Christie* and afterwards Salka collaborated on Garbo's screenplays for MGM in the 1930s and 1940s (*Queen Christina*, *Anna Karenina*, *Conquest*, and others). Isherwood met her soon after arriving in Los Angeles and was often at her house socially or to work with Berthold Viertel. In the 1930s and 1940s, the house was frequented by European refugees, and Salka was able to help many of them find work—some as domestic servants, others with the studios. Her guests included the most celebrated writers and movie stars of the time. In 1946, Isherwood moved into her garage apartment, at 165 Mabery Road, with Bill Caskey. By then Salka was living alone and had little money. Her husband had left; her lover Gottfried Reinhardt had married; Garbo's career was over; and later, in the 1950s, Salka was persecuted by the McCarthyites and blacklisted by MGM for her presumed communism. In January 1947, she moved into the garage apartment herself and let out her house; in the early 1950s, she sold the property and moved to an apartment off Wilshire Boulevard. Eventually, she returned modestly to writing for the movies, but finally moved back to Europe, although she had been a U.S. citizen since 1939. She published a memoir, *The Kindness of Strangers*, in 1969. She appears in *D.1*, *D.2*, and *Lost Years*.

Virgil. See Thomson, Virgil.

Vishwananda, Swami. Indian monk of the Ramakrishna Order. Isherwood met him in 1943 when Vishwananda visited the Hollywood Vedanta Society and other centers on the West Coast. Vishwananda was head of the Vedanta Center in Chicago. He appears in *D.1* and *D.2*.

Vivekananda, Swami (1863–1902). Narendranath Datta (known as Naren or Narendra and later as Swamiji) took the monastic name Vivekananda in 1893. He was Ramakrishna's chief disciple. He came from a wealthy and cultured background and was attending university in Calcutta when Ramakrishna recognized him as an incarnation of one of his "eternal companions," a free, perfect soul born into maya with the avatar and possessing some of the avatar's characteristics. Vivekananda was trained by Ramakrishna to carry his message, and he led the disciples after Ramakrishna's death, though he left them for long periods, first to wander through India as a monastic, practising spiritual disciplines, then to travel twice to America and Europe, where his lectures and classes spawned the first western Vedanta centers. In India he devoted much time to founding and administering the Ramakrishna Math and Mission. His followers hired a British stenographer, J.J. Goodwin, to transcribe his lectures; these transcriptions, which Goodwin probably edited into complete sentences and paragraphs, along with Vivekananda's letters to friends and to his own and Ramakrishna's disciples, were published as *The Complete Works of Vivekananda*. Isherwood wrote the introduction to one volume, *What Religion Is: In the Words of Swami Vivekananda* (1960), selected by Prema Chaitanya.

Vividishananda, Swami (d. 1980). Indian monk of the Ramakrishna Order. Head of the Vedanta Center in Seattle. In *D.1*, Isherwood tells about meeting him in Portland in October 1943 and afterwards spending a few days at his center in Seattle. In 1974, Vividishananda had the first of a series of strokes which, by 1977, left him in a semi-coma; when he was not in the hospital, the monks cared for him at the Seattle center until his death. He was the author of *A Man of God*, about Swami Shivananda (also known as Mahapurush Maharak, or Tarak), a direct disciple of Ramakrishna. Isherwood contributed a Foreword to the paperback edition, which appeared with a new title, *Saga of a Great Soul: Swami Shivananda* (1986).

Voeller, Bruce (1934–1994). American biologist. He left his associate professorship at Rockefeller University for gay activism in the early 1970s and helped found the National Gay Task Force (later the National Gay and Lesbian Task Force). In 1980, he created the Mariposa Foundation, based in Topanga, California, to study human sexuality and sexually transmitted diseases. He published a book and numerous studies about AIDS, and he originated the name Acquired Immune Deficiency Syndrome as an improvement over the stigmatizing Gay Related Immune Deficiency Syndrome. At Mariposa, he also began collecting materials related to the history of the gay and lesbian movement, an archive he donated in 1988 to create the Human Sexuality Collection at Cornell University. Voeller commissioned Bachardy to do portraits of twelve gay and lesbian leaders: Charles Bryden, James Foster, Barbara Gittings, David Goodstein, Franklin Kameny, Morris Kight, Phyllis Lyon, Jean O'Leary, Del Martin, Elaine Noble, Troy Perry, and Voeller himself. These went on a national tour in 1981 and were donated to Cornell in 1995 by Voeller's partner Richard Lucik. Voeller died of AIDS.

Voight, Jon (b. 1938). American stage and movie actor, born in Yonkers, educated at Catholic University and the Neighborhood Playhouse in New York. His father was a golf pro of Czech background. Voight was in the Broadway production of *The Sound of Music* and worked in television Westerns during the 1960s before becoming a star as the hustler in *Midnight Cowboy* (1969), a role for which he received an Academy Award nomination and the New York Film Critics

Best Actor Award. He opposed the Vietnam war and worked with Jane Fonda's Entertainment People for Peace and Justice, then appeared with her in *Coming Home* (1978), for which he won an Academy Award as the paraplegic Vietnam veteran. Among his many other films are *Catch-22* (1970), *Deliverance* (1972), *The Odessa File* (1974), *The Champ* (1979), *Runaway Train* (1985, Academy Award nomination), *Ali* (2001, Academy Award nomination), and *National Treasure* (2004). He married actress Lauri Peters in 1962 and Marcheline Bertrand in 1971 and divorced both times.

Wallwork, Leslie. Wealthy American queen; he lived alone, hired hustlers for sex, and liked to give dinner parties, especially for John Ashbery when Ashbery was in Los Angeles. He died in middle age, evidently of cancer. He was gossipy, funny, and ambitious of social power.

Wan, Barney. Chinese graphic designer and illustrator, born in Hong Kong; he studied in San Francisco during the 1950s. In 1959, he went to Paris and freelanced for *Elle*, French *Vogue*, and *Marie-Claire*. He joined British *Vogue* in 1967 and was Art Director there until 1979, living between London and Paris for many years. From 1970, he also began to design books, for Cecil Beaton, Lord Snowden, and later for various other photographers from London and from Kenya; among his titles is the prizewinning *African Ceremonies* (1990). He also designed film titles for *If*... (1968), *Isadora* (1968), and *Sebastiane* (1976).

Ward, Simon (b. 1941). British actor, educated at Alleyn's School, Dulwich, where he joined the National Youth Theatre before training at RADA. He became known on the London stage for his role in Joe Orton's *Loot* in 1967. His films include *Young Winston* (1972), *The Three Musketeers* (1973), *Hitler: The Last Ten Days* (1973), *All Creatures Great and Small* (1974), and *The Four Feathers* (1977).

Warshaw, Howard (1920–1977). American artist, born in New York City, educated at Pratt Institute, the National Academy of Design Art School, and the Art Students League. He moved to California in 1942 and worked in the animation studios of Walt Disney and later Warner Brothers, then taught briefly at the Jepson Art Institute where he was influenced by a colleague, Rico Lebrun. In 1951, he began teaching at the University of California at Santa Barbara, where he remained for over twenty years. He painted murals in the Dining Commons at UCSB and at the Santa Barbara Public Library. At UCLA he painted the neurological mural, and he also did murals for U.C. San Diego and U.C. Riverside. He was blind in one eye. His wife Frances had money of her own. She had been married previously to Mel Ferrer with whom she had a daughter called Pepa and a son. Isherwood and Warshaw became friendly when Isherwood began teaching at UCSB, and Isherwood usually slept at their house when he stayed overnight in Santa Barbara. In 1954, Isherwood encouraged Bachardy to attend drawing classes with Warshaw in Los Angeles, but Bachardy found the class too theoretical and left. Years later, as Isherwood records, Warshaw sat for Bachardy. He appears in *D.2*.

Wasserman, Lew (1913–2002). Cleveland-born studio executive; he joined MCA as an agent in 1936, was president by 1946, merged it with Universal in

1962, and, as chairman and CEO, built the company into a global entertainment giant.

Watergate. In his entry for May 10, 1973, Isherwood mentions the indictments of John Mitchell and Maurice Stans, both close to President Nixon. Mitchell (1906–1988) was Attorney General until 1972 when he resigned to run the Committee to Re-elect the President; Stans (1908–1998) was Secretary of Commerce during Nixon's first administration and then became Finance Chairman for the Committee to Re-elect the President. They were indicted with financier Robert Vesco for concealing a $200,000 cash contribution received from Vesco in April 1972 in exchange for influencing a Securities and Exchange Commission lawsuit against him. Vesco was accused of looting over $200 million from foreign mutual funds. After Mitchell helped obtain Vesco's release from a Swiss jail, Vesco fled to the Bahamas and then disappeared. Mitchell and Stans were acquitted in 1974, but Mitchell was convicted in 1975 on other charges and served eighteen months in prison before being paroled. On May 24, Isherwood also mentions James McCord (b. 1924), the former CIA and FBI officer who was security advisor for the Committee to Re-elect the President and who led the Watergate burglary. McCord had already been convicted of conspiracy, burglary, and wiretapping, on January 20; on March 19, he wrote to Judge John Sirica saying that he and his cohorts had pleaded guilty under pressure and that lies had been told. This led to further investigations.

Waterston, Sam (b. 1940). American actor, educated at Groton and Yale; his many films include *The Great Gatsby* (1974), *Heaven's Gate* (1980), and *The Killing Fields* (1984, Academy Award nomination). He won a Drama Desk Award as Benedick in Joseph Papp's 1973 production, which Isherwood mentions, of *Much Ado About Nothing*. He is a longtime regular on the T.V. series "Law & Order."

Weidenfeld, George (b. 1919). Viennese-born British publisher; trained as a lawyer and diplomat. He emigrated in 1938 and worked for the BBC during World War II, mostly commentating on European Affairs; he also wrote a foreign affairs column for *The News Chronicle*. After the war, he began publishing a magazine, *Contact*, and then founded Weidenfeld & Nicolson with Nigel Nicolson. Their first books appeared in 1949, focusing initially on history and biography. In the 1950s, they turned to fiction as well; Nabokov's *Lolita* (1959) was their first best-seller. Weidenfeld & Nicolson also became known for its books by world political leaders and for diaries, letters, and memoirs of public figures. In 1949, Weidenfeld spent a year in Israel as a political adviser and Chief of Cabinet to President Chaim Weizmann, and afterwards he maintained close ties there. He wrote *The Goebbels Experiment* (1943) and an autobiography; for a time he had a column in *Die Welt*. He married four times and had one daughter with his first wife; his second wife, Barbara Skelton, had previously been married to Cyril Connolly. He appears in *D.2*.

Weissberger, L. Arnold (1907–1981). Top show-business and arts lawyer, for instance to Stravinsky, to ballerinas Alexandra Danilova and Alicia Markova, to Orson Welles, and many others. He was also a devoted amateur photographer and published *Famous Faces: A Photographic Album of Personal Reminiscences* (1973) containing nearly 1,500 pictures of his celebrity friends and acquaintances. He lived in New York with theatrical agent Milton Goldman.

Wescott, Glenway (1901–1987). American writer, born in Wisconsin. He attended the University of Chicago, lived in France in the 1920s, partly in Paris, and travelled in Europe and England. Afterwards he lived in New York. Early in his career he wrote poetry and reviews, later turning to fiction. His best-known works are *The Pilgrim Hawk* (1940) and *Apartment in Athens* (1945). Wescott was president of the American Academy of Arts and Letters from 1957 to 1961. His long-term companion was Monroe Wheeler, although each had other lovers. In 1949, Wescott went to Los Angeles expressly to read Isherwood's 1939–1945 diaries. While he was there, Isherwood introduced Wescott to Jim Charlton with whom Wescott had an affair. Wescott appears in *D.1*, *D.2*, and *Lost Years*.

Wheeler, Hugh (1912–1987). English writer. He published mystery thrillers under the name Patrick Quentin. Later he wrote plays—*Big Fish, Little Fish* (1961); *Look, We've Come Through* (1961)— the books for the Sondheim musicals *A Little Night Music* (1973) and *Sweeney Todd* (1979), and screenplays (*Something for Everyone*, 1970). According to rumor, he wrote much of the screenplay for *Cabaret* though he was credited only as a technical advisor because of Writers Guild rules. He lived on a farm in Massachusetts with his black lover, John Grubbs. Wheeler was a good friend of Chris Wood, and possibly it was Wood who introduced Wheeler and Isherwood, probably in the 1940s. He appears in *D.1*.

Wheeler, Paul. British musician. When Isherwood met him in 1970, Wheeler was an undergraduate reading English at Cambridge. He played in London clubs and in a Cambridge band called Wild Oats. Later, he lived with his wife on the estate of John Lennon and Yoko Ono, worked as a civil servant and as a management trainer, and continued to sing and write songs on the side. He also performed with another band, Ghosts.

Whitcomb, Ian (b. 1941). British pianist, singer, ukulele player, record producer; educated at Bryanston and at Trinity College, Dublin. He moved to Los Angeles in 1965 when his song "You Turn Me On" reached number eight in the American Top Ten, and he performed there with the Beach Boys, the Rolling Stones, and The Kinks, among others. Later, he hosted a Los Angeles radio show and published books about the history of popular music, including *After the Ball: Pop Music from Rag to Rock* (1972), *Tin Pan Alley: A Pictorial History (1919–1939)*, and *Rock Odyssey: A Chronicle of the Sixties* (1983). He appears in *D.2*.

White, Dan (1954–1985). Vietnam veteran, police officer, fireman, and San Francisco City Supervisor. In 1978, he resigned as City Supervisor over low pay and clashes with liberal colleagues, but, encouraged by supporters, especially among the police force, asked Mayor Moscone for his job back. Moscone refused, whereupon White shot Moscone and Harvey Milk. White's defense argued that he was depressed, as evidenced by his uncharacteristic consumption of junk food—the "Twinkie defence." On May 21, 1989, White was convicted of manslaughter, not murder, on grounds of reduced capacity, and sentenced to only five years in prison. Angry gays marched on City Hall, smashing windows and torching police cars, and the police unleashed a counter riot in the Castro District where the march had begun. The White Night Riot, as it came to be known, was followed by peaceful celebrations of Harvey Milk's birthday, May 22. White committed suicide by gassing himself in his car. Milk became the subject of the eponymous 2008 film for which Sean Penn won an Academy Award.

White, James P. (b. 1940). American writer, educated at the University of Texas, Vanderbilt and Brown. He taught writing at the University of Texas, UCLA, USC and the University of South Alabama. When his first novel, *Birdsong* (1977), was published, he sent Isherwood a copy and met him soon afterwards during a visit to Los Angeles. White has also published poems, short stories, essays and reviews. In March 1979, his wife, Janice, gave birth to a boy, whom they named Christopher Jules White and called Jules. With Don Bachardy, White edited *Where Joy Resides: A Christopher Isherwood Reader* (1989) and co-founded The Christopher Isherwood Foundation, of which he was Executive Director until 2011.

Whiting, Leonard (b. 1950). British actor and singer. He was the Artful Dodger in *Oliver!* in London's West End and became famous as Romeo in Franco Zeffirelli's *Romeo and Juliet* (1968) before appearing in "Frankenstein: The True Story."

Wilcox, Collin (1935–2009). American actress, also known as Collin Horne and sometimes credited as Wilcox-Horne or Wilcox-Paxton. She was educated at the University of Tennessee, the Goodman School of Drama in Chicago and Actors Studio. She attracted praise on Broadway in *The Day the Money Stopped* (1958) and appeared in a few films, including *To Kill a Mockingbird* (1962), *Catch-22* (1970), and *Midnight in the Garden of Good and Evil* (1997). She also worked in T.V. She was married for a time to actor Geoffrey Horne. Later, she taught acting on the East Coast. She appears in *D.2*.

Wilder, Nicholas (1937–1989). American art dealer, son of the scientist who invented Kodachrome. He was educated at Amherst and Stanford, where he studied law before turning to art history. As a graduate student, he worked in a San Francisco gallery then abandoned his studies and became a dealer by the time he was twenty-four. He opened his first gallery in Los Angeles on La Cienaga Boulevard in the 1960s, moved to Santa Monica Boulevard in 1970, and quickly became the most influential contemporary art dealer in Los Angeles. He was a close friend of Don Bachardy, whom he persuaded to leave Irving Blum's gallery and join his own. He also discovered West Coast artists Bruce Nauman, Ron Davis, Robert Graham, and Tom Holland, among others, and he showed David Hockney, Helen Frankenthaler, Barnett Newman, and Cy Twombly. He travelled constantly to see the work of new artists and to advance the reputations of the artists he represented, but his finances were often in disarray. In 1979, he closed his gallery; James Corcoran opened in the space and took on some of Wilder's best artists. Wilder left for New York, where he began to paint abstract oils inspired by the work of James McLaughlin, an artist he had brought to prominence. His longtime companion and business associate was Craig Cook. Wilder died of AIDS, as did Cook a few years later.

Williams, Brook (1938–2005). British character actor; younger son of Emlyn Williams. He worked on the London stage and had small parts in over a hundred films. He was Richard Burton's close friend, assistant, and sometimes collaborator.

Williams, Clifford (1926–2005). British actor and director, educated at Highbury County Grammar School, in London. He worked in mining, served in the army, and acted and directed in South Africa during the 1950s. In 1961, he became a staff producer at the Royal Shakespeare Company, and from 1963 until 1991, a prolific associate director. His many independent productions, some of which went on to successful Broadway runs, include an all-male *As You Like It*, Anthony Shaffer's *Sleuth*, both in 1967, and Kenneth Tynan's *Oh! Calcutta!* in 1970. He was married twice and had two daughters with his second wife. He appears in *D.2*.

Williams, Emlyn (1905–1987). Welsh playwright, screenwriter, and actor. He wrote psychological thrillers for the London stage, including *Night Must Fall* (1935), and is best known for *The Corn Is Green* (1935), based on his own background in a coal-mining community in Wales and in which he played the lead; both of these plays were later filmed. His many other stage roles include Shakespeare and contemporary theater, and from the 1950s, he toured with one-man shows of Charles Dickens and Dylan Thomas. Among his films are *The Man Who Knew Too Much* (1934), *The Dictator* (1935), *Dead Men Tell No Tales* (1938), *Jamaica Inn* (1939), *The Stars Look Down* (1939), *Major Barbara* (1941), *Ivanhoe* (1952), *The Deep Blue Sea* (1955), *I Accuse!* (1958), *The L-Shaped Room* (1962), and *David Copperfield* (1970). His wife, Mary Carus-Wilson, known as Molly (d. 1970), was an actress under her maiden name, Molly O'Shann. Isherwood first met Williams and his wife in Hollywood in 1950; they appear in *D.1* and *D.2*.

Williams, Ralph. Aspiring young actor. He appeared in the brief Los Angeles run of *Remote Asylum*, Mart Crowley's first play after *The Boys in the Band*, and he sat for Don Bachardy several times.

Williams, Tennessee (1911–1983). American playwright; Thomas Lanier Williams was born in Mississippi and raised in St. Louis. His father was a travelling salesman, his mother felt herself to be a glamorous southern belle in reduced circumstances. *The Glass Menagerie* made him famous in 1945, followed by *A Streetcar Named Desire* (1947). Many of his subsequent plays are equally well known—*The Rose Tattoo* (1950), *Cat on a Hot Tin Roof* (1955), *Sweet Bird of Youth* (1959), *The Night of the Iguana* (1961)—and were made into films. He also wrote a novella, *The Roman Spring of Mrs. Stone* (1950). When he first came to Hollywood in 1943 to work for MGM, he bore a letter of introduction to Isherwood from Lincoln Kirstein; this began a long and close friendship which Isherwood tells about in *D.1*, *D.2*, and *Lost Years*. Isherwood later told Edmund White of a 1940s love affair, but Williams once reminded Isherwood, in Bachardy's presence, that Isherwood had rebuffed Williams's one direct pass. Williams's *Out Cry*, a revised version of *The Two-Character Play*, was staged in Chicago in 1971, and Isherwood mentions a reading he attended in Los Angeles in 1972. It previewed up and down the East Coast, opened on Broadway in March 1973 starring Michael York, then closed after twelve performances.

Wilson, Angus (1913–1991). British novelist and literary critic, educated at Westminster and Oxford. He worked as a decoder for the Foreign Office during World War II and otherwise made his career in the library of the British Museum before turning to writing full time in 1955. His novels include *Hemlock and After* (1952), *Anglo-Saxon Attitudes* (1956), *The Middle Age of Mrs. Eliot* (1958), *The Old Men at the Zoo* (1961), *No Laughing Matter* (1967), and *Setting the World on Fire* (1980). He also published short stories and literary-critical books about Zola, Dickens, and Kipling. From 1966, he was a professor of English literature at the University of East Anglia, where, with Malcolm Bradbury, he oversaw the first respected university course in creative writing in Britain. In *D.1*, Isherwood tells that he met Wilson at a party in London in 1956. Wilson also appears in *D.2*. Wilson's companion from the late 1940s was Anthony Garrett.

Wilson, Colin (b. 1931). English novelist and critic. Author of psychological thrillers, studies of literature and philosophy, criminology, the imagination, the occult and the supernatural. He became well known with his 1956 book *The Outsider*—which Isherwood mentions twice in *D.1*—about the figure of the alienated solitary in modern literature. Isherwood met him in London in 1959, and Wilson appears in *D.2*.

Wilson, William (b. 193[4]). Art critic, born and raised in Los Angeles, where he graduated from UCLA. He began freelancing for *The Los Angeles Times* in 1965, joined the staff in 1968, and in 1978 became chief art critic after his predecessor, Henry Seldis, died. He taught at Cal. State Fullerton, Santa Monica City College, and UCLA Extension and contributed to various art magazines, including *Art in America*, *Art Forum*, and *Art News*.

Wingreen, Jason (b. 1919). American actor. He was a regular and a guest on popular T.V. series, including "The Untouchables" (1960–1961) and "All in the Family" (1977–1983), and he had a few small film roles.

Winters, Shelley (1922–2006). American actress, from St. Louis. She worked on the New York stage and moved to Hollywood in 1943. She was nominated for an Academy Award for *A Place in the Sun* (1951), won Academy Awards for her supporting roles in *The Diary of Anne Frank* (1959) and *A Patch of Blue* (1965), and was nominated again for *The Poseidon Adventure* (1972). She played Natalia Landauer in the film of *I Am a Camera* in 1955. From the early 1990s, she played Roseanne's grandmother in the T.V. comedy series "Roseanne." Isherwood met her at the start of the 1950s, and she appears in *D.1* and *D.2*. She was married to actor Tony Franciosa from 1957 to 1960.

Wonner, Paul (1920–2008). American painter, originally from Arizona. He settled on the West Coast, first in Los Angeles, and from the early 1980s in San Francisco, with his longtime partner, the painter Bill Brown. Wonner's work has been called American Realist or Californian Realist, and he has been grouped with other painters who came to maturity in the 1950s—Richard Diebenkorn, Wayne Thiebaud, David Park, Nathan Oliveira—as a Bay Area Figurative Artist or Figurative Abstractionist. He taught painting at UCLA and elsewhere, and, with Brown, had close friendships with a number of writers, for instance, May Sarton, Diane Johnson, and Janet Frame. He appears in *D.2*

Wood, Christopher (Chris) (d. 1976). Isherwood met Chris Wood in September 1932 when Auden took him to meet Gerald Heard, then sharing Wood's luxurious London flat. Wood was about ten years younger than Heard, handsome and friendly but shy about his maverick talents. He played the piano well, but never professionally, wrote short stories, but not for publication, had a pilot's license, and rode a bicycle for transport. He was extremely rich (the family business made jams and other canned and bottled goods), sometimes extravagant, and always generous; he secretly funded many of Heard's projects and loaned or gave money to many other friends (including Isherwood). In 1937, he emigrated with Heard to Los Angeles and, in 1941, moved with him to Laguna. Their domestic commitment persisted for a time despite Heard's increasing asceticism and religious activities, but eventually they lived separately though they remained friends. From 1939, Wood was involved with Paul Sorel, also a member of their shared household for about five years. Wood appears throughout *D.1*, *D.2*, and *Lost Years*.

Wood, Natalie (1938–1981). American actress, of French and Russian background; her mother was a ballet dancer. She appeared in her first movie when she was five years old and was a star by the time she was nine. Her films include: *Happy Land* (1943), *Tomorrow Is Forever* (1946), *Miracle on 34th Street* (1947), *Rebel Without a Cause* (1955), *The Searchers* (1956), *Splendor in the Grass* (1961), *West Side Story* (1961), *Gypsy* (1962), *Inside Daisy Clover* (1966), *Penelope* (1966), and *Bob and Carol and Ted and Alice* (1969). She was married twice to the same man, actor Robert Wagner, and in between married Richard Gregson, father of her daughter Natasha Gregson Wagner, also an actress. She drowned after falling overboard from a yacht.

Woodcock, Patrick (1920–2002). British doctor. His patients included many London theater stars—including John Gielgud and Noël Coward—and also actors based in New York and Hollywood. Isherwood and Bachardy first met him with Gielgud and Hugh Wheeler in London in 1956, and a few days later Woodcock visited Bachardy at the Cavendish Hotel to treat him for stomach cramps. Woodcock also prescribed vitamins for Isherwood at the same time and became a friend. He appears in *D.1* and *D.2*.

Woody, Jack. Grandson of actor and stuntman Frank Woody (1896–1959) and actress Helen Twelvetrees (1907–1958); his father was also called Jack Woody (b. 1932). After completing high school in Albuquerque, New Mexico, Woody worked in bookstores in Los Angeles and then at Nicholas Wilder's gallery, where he appointed himself head of photography. He was also Wilder's lover. He founded a fine art press, which he called The Twelvetrees Press, and published *October* (1980) by Isherwood and Bachardy, Bachardy's *Don Bachardy: One Hundred Drawings* (1983), which he designed with Scott Smith, and Bachardy's *Drawings of the Male Nude* (1985). Under another imprint, Twin Palms, Woody published *Lost Hollywood*, a book of early photographs he himself collected of film stars. He designed and produced other books with Tom Long, who was his companion after Wilder. Woody's many books include work by George Platt Lynes, Robert Mapplethorpe, Duane Michals, William Eggleston, Horst P. Horst, and Allen Ginsberg, as well as *Killing Fields: Photographs from the S-21 Death Camp* and *Without Sanctuary: Lynching Photography in America*.

Worton, Len (d. late 1990s). English-born monk of the Ramakrishna Order; he settled in Trabuco and became Swami Bhadrananda. He was a British army medic during World War II, stationed in Hong Kong, and was captured there by the Japanese when he remained behind tending the wounded. He was spared execution because of his medical skills and because he was wearing Red Cross insignia. He spent four years as a prisoner of war in Japan. He appears in *D.2*.

Wright, Thomas E. (Tom). American writer, from New Orleans. Isherwood met him while both were fellows-in-residence at the Huntington Hartford Foundation in 1951, and they had a casual affair which lasted about eight months. Wright was then about twenty-four years old and had taught at New York University. He was a childhood friend of Speed Lamkin and of Speed's sister Marguerite and he knew Tennessee Williams and Howard Griffin. He also became close to Edward James and lived at James's Hollywood house as a caretaker for a number of years. Bachardy drew the first of several portraits of Wright in 1960. Wright published novels, travel books, and *Growing Up with Legends: A Literary Memoir*. He eventually settled in Guatemala. He appears in *D.1* and *D.2*.

Writers Guild Strike. The 1973 strike demands included a nearly fourfold pay increase, full pay for T.V. reruns, control over health and welfare plans with a five percent contribution from producers, a percentage of profits from cable and pay T.V., and a definition of allowable script changes without writers' express consent. The strike began March 6 and ran until June 28, when the guild accepted with much dissent substantial salary increases for T.V. writers, improved health and welfare packages, protection for hyphenates (writer-producers, writer-directors), and a guaranteed share of future supplemental markets—pay T.V., cable, cassettes, and any hardware yet to be invented.

Wudl, Tom (b. 1948). Bolivian-American painter, raised in Los Angeles from age ten. He studied at Chouinard from 1966 to 1970 and later taught at UCLA, U.C. Santa Barbara, Claremont College, Otis College of Art, and elsewhere. He has exhibited at the Los Angeles County Museum of Art, the Whitney Museum of American Art in New York, and other major galleries in the U.S., Europe, and Asia.

Wyberslegh Hall. The fifteenth-century manor house where Isherwood was born and where his mother returned to live with his brother after WWII; it was part of the Bradshaw Isherwood estate. During most of the nineteenth century, Wyberslegh was leased to a family called Cooper who farmed the surrounding fields, lived in the rear wing of the house, and rented out the front rooms. When Isherwood's parents married in 1903, his grandfather, John Isherwood, divided Wyberslegh Hall into two separate houses. Frank and Kathleen Isherwood moved into the front of the house, and, for roughly half the twentieth century, the Coopers continued to live in the back overlooking the farmyard.

Wystan. See Auden, W.H.

yoga. Union with the Godhead or one of the methods by which the individual soul may achieve such union. Isherwood and Prabhavananda translated the yoga sutras or aphorisms of the Hindu sage Patanjali (4th c. B.C. to 4th c. A.D.) as *How to Know God* (1953); they describe the aphorisms as a compilation and reformulation of yoga philosophy and practices—spiritual disciplines and techniques of meditation—referred to even earlier in four of the Upanishads and in fact handed down from prehistoric times.

Yogeshananda. See Buddha Chaitanya.

York, Michael (b. 1942) and Pat McCallum York. British film star and his wife, a photographer and writer. He was educated at Oxford, where he acted for the Oxford University Dramatic Society. His real name is Michael York-Johnson. In 1965, he had a small role in Franco Zeffirelli's stage production of *The Taming of the Shrew* for Olivier's National Theatre Company and afterwards appeared in the movie with Richard Burton and Elizabeth Taylor. Subsequent films include *Romeo and Juliet* (1968), *Justine* (1969), *Something for Everyone* (1970), *Cabaret* (1972), *England Made Me* (1973), *The Three Musketeers* (1974), *Murder on the Orient Express* (1974), *The Weather in the Streets* (1983), *The Wide Sargasso Sea* (1993), *Austin Powers: International Man of Mystery* (1997), *Austin Powers: The Spy Who Shagged Me* (1999), *Borstal Boy* (2000), *Megiddo* (2001), and *Austin Powers: Goldmember* (2002). He has often appeared on television, notably in "The Forsyte Saga" and "Curb Your Enthusiasm," and on Broadway, and he has a prize-winning career recording audio books. She was born in Jamaica to an English diplomat and his American wife, educated in England and France, married for the first time as a teenager, and in 1952 had a son, Rick McCallum, later a Hollywood producer. From the early 1960s, she worked as a fashion journalist at *Vogue*, then as a photographer and travel editor at *Glamor* before becoming a freelance photographer with a few lessons from David Bailey. Her portraits of celebrities have appeared in *Life*, *People*, *Town and Country*, *Playboy*, and *Newsweek*; she has published several books and had numerous gallery exhibitions, in particular of her later avant-garde work featuring dissected cadavers and nude portraits of professional people at work. She met Michael York when photographing him for a magazine profile, and they married a year later, in 1968. Isherwood was introduced to them in 1969, probably at the Vadims' lunch party described in the diary entry for January 15, 1969 in *D.2*.

Yount, Kurt. A blind man Isherwood met at Fullerton State in 1976, and with whom he became friendly, partly because Yount is an obsessive reader and partly because Isherwood felt curious and sympathetic about his predicament. Isherwood invited him to dinner once to introduce him to Don Bachardy.

Yow, Jensen (Jen, Jens). Artist and curator. He was a friend of Lincoln Kirstein and had a room in Kirstein's house. He appears in *D.1*.

Index

Works by Isherwood appear directly under title; works by others appear under authors' names.

Page numbers in *italic* indicate entries in the Glossary (pages 689–824).

BOOKS BY CHRISTOPHER ISHERWOOD

THE SIXTIES
Diaries 1960–1969

Available in Paperback and Ebook

The second volume of acclaimed author Christopher Isherwood's diaries takes readers to the heart of the 1960s, the decade in which Isherwood's semiautobiographical novel *Goodbye to Berlin* would be adapted into the Tony Award-winning musical *Cabaret*. Against a background of cultural paradigm shifts including the advent of space travel, pop art, and mod fashion, and seminal events like the Kennedy/Nixon election, the Marianne Faithfull/Mick Jagger romance, the rise of the Hippie movement, and the riotous explosion of America's inner cities, *The Sixties* follows Isherwood's friendships with creative powerhouses such as Francis Bacon, David Hockney, Richard Burton, and Gore Vidal, and continues the saga of his great romance with portraitist Don Bachardy.

LIBERATION
Diaries 1970–1983

Available in Paperback and Ebook

"A slip of a wild boy: with quick silver eyes," as Virginia Woolf saw him in the 1930s, Christopher Isherwood journeyed and changed with his century, until, by the 1980s, he was celebrated as the finest prose writer in English and the grand old man of gay liberation. In this final volume of his diaries, the capstone of a million-word masterwork, Isherwood greets advancing age with poignant humor and an unquenchable appetite for the new; even aches, illnesses, and diminishing powers are clues to a predicament still unfathomed.

"Compulsively readable. . . . A testament to his connections to the literati and Hollywood glitterati. . . . A fitting finale to a fascinating life."
—*Publishers Weekly* (starred review)